Stochastic Mechanics

Random Media

Signal Processing and Image Synthesis

Mathematical Economics and Finance

Stochastic Optimization

Stochastic Control

Applications of Mathematics

Stochastic Modelling and Applied Probability

31

Edited by   I. Karatzas
M. Yor

Advisory Board   P. Brémaud
E. Carlen
W. Fleming
D. Geman
G. Grimmett
G. Papanicolaou
J. Scheinkman

Springer

New York
Berlin
Heidelberg
Barcelona
Hong Kong
London
Milan
Paris
Singapore
Tokyo

Applications of Mathematics

(continued after index)

Luc Devroye   László Györfi
Gábor Lugosi

A Probabilistic Theory
of Pattern Recognition

With 99 Figures

 Springer

Luc Devroye
School of Computer Science
McGill University
Montreal, Quebec, H3A 2A7
Canada

László Györfi
Gábor Lugosi
Department of Mathematics and
  Computer Science
Technical University of Budapest
Budapest
Hungary

Managing Editors

I. Karatzas
Department of Statistics
Columbia University
New York, NY 10027, USA

M. Yor
CNRS, Laboratoire de Probabilités
Université Pierre et Marie Curie
4, Place Jussieu, Tour 56
F-75252 Paris Cedex 05, France

Mathematics Subject Classification (1991): 68T10, 68T05, 62G07, 62H30

Library of Congress Cataloging-in-Publication Data
Devroye, Luc.
   A probabilistic theory of pattern recognition/Luc Devroye,
László Györfi, Gábor Lugosi.
      p.   cm.
   Includes bibliographical references and index.
   ISBN 0-387-94618-7 (hardcover)
   1. Pattern perception. 2. Probabilities.   I. Györfi, László.
II. Lugosi, Gábor. III. Title.
Q327.D5   1996
003′.52′015192 − dc20   95-44633

Printed on acid-free paper.

Production managed by Francine McNeill; manufacturing supervised by Jeffrey Taub.
Photocomposed copy prepared using Springer's svsing.sty macro.
Printed and bound by Sheridan Books, Inc., Ann Arbor, MI.
Printed in the United States of America.

9 8 7 6 5 4 3

ISBN 0-387-94618-7   SPIN 10830936

Springer-Verlag   New York Berlin Heidelberg
A member of BertelsmannSpringer Science+Business Media GmbH

Preface

Life is just a long random walk. Things are created because the circumstances happen to be right. More often than not, creations, such as this book, are accidental. Nonparametric estimation came to life in the fifties and sixties and started developing at a frenzied pace in the late sixties, engulfing pattern recognition in its growth. In the mid-sixties, two young men, Tom Cover and Peter Hart, showed the world that the nearest neighbor rule in all its simplicity was guaranteed to err at most twice as often as the best possible discrimination method. Tom's results had a profound influence on Terry Wagner, who became a Professor at the University of Texas at Austin and brought probabilistic rigor to the young field of nonparametric estimation. Around 1971, Vapnik and Chervonenkis started publishing a revolutionary series of papers with deep implications in pattern recognition, but their work was not well known at the time. However, Tom and Terry had noticed the potential of the work, and Terry asked Luc Devroye to read that work in preparation for his Ph.D. dissertation at the University of Texas. The year was 1974. Luc ended up in Texas quite by accident thanks to a tip by his friend and fellow Belgian Willy Wouters, who matched him up with Terry. By the time Luc's dissertation was published in 1976, pattern recognition had taken off in earnest. On the theoretical side, important properties were still being discovered. In 1977, Stone stunned the nonparametric community by showing that there are nonparametric rules that are convergent for all distributions of the data. This is called distribution-free or universal consistency, and it is what makes nonparametric methods so attractive. Yet, very few researchers were concerned with universal consistency—one notable exception was Laci Györfi, who at that time worked in Budapest amid an energetic group of nonparametric specialists that included Sándor Csibi, József Fritz, and Pál Révész.

So, linked by a common vision, Luc and Laci decided to join forces in the early eighties. In 1982, they wrote six chapters of a book on nonparametric regression function estimation, but these were never published. In fact, the notes are still in drawers in their offices today. They felt that the subject had not matured yet. A book on nonparametric density estimation saw the light in 1985. Unfortunately, as true baby-boomers, neither Luc nor Laci had the time after 1985 to write a text on nonparametric pattern recognition. Enter Gábor Lugosi, who obtained his doctoral degree under Laci's supervision in 1991. Gábor had prepared a set of rough course notes on the subject around 1992 and proposed to coordinate the project—this book—in 1993. With renewed energy, we set out to write the book that we should have written at least ten years ago. Discussions and work sessions were held in Budapest, Montreal, Leuven, and Louvain-La-Neuve. In Leuven, our gracious hosts were Ed van der Meulen and Jan Beirlant, and in Louvain-La-Neuve, we were gastronomically and spiritually supported by Léopold Simar and Irène Gijbels. We thank all of them. New results accumulated, and we had to resist the temptation to publish these in journals. Finally, in May 1995, the manuscript had bloated to such extent that it had to be sent to the publisher, for otherwise it would have become an encyclopedia. Some important unanswered questions were quickly turned into masochistic exercises or wild conjectures. We will explain subject selection, classroom use, chapter dependence, and personal viewpoints in the Introduction. We do apologize, of course, for all remaining errors.

We were touched, influenced, guided, and taught by many people. Terry Wagner's rigor and taste for beautiful nonparametric problems have infected us for life. We thank our past and present coauthors on nonparametric papers, Alain Berlinet, Michel Broniatowski, Ricardo Cao, Paul Deheuvels, András Faragó, Adam Krzyżak, Tamás Linder, Andrew Nobel, Mirek Pawlak, Igor Vajda, Harro Walk, and Ken Zeger. Tamás Linder read most of the book and provided invaluable feedback. His help is especially appreciated. Several chapters were critically read by students in Budapest. We thank all of them, especially András Antos, Miklós Csűrös, Balázs Kégl, István Páli, and Márti Pintér. Finally, here is an alphabetically ordered list of friends who directly or indirectly contributed to our knowledge and love of nonparametrics: Andrew and Roger Barron, Denis Bosq, Prabhir Burman, Tom Cover, Antonio Cuevas, Pierre Devijver, Ricardo Fraiman, Ned Glick, Wenceslao Gonzalez-Manteiga, Peter Hall, Eiichi Isogai, Ed Mack, Arthur Nádas, Georg Pflug, George Roussas, Winfried Stute, Tamás Szabados, Godfried Toussaint, Sid Yakowitz, and Yannis Yatracos.

Gábor diligently typed the entire manuscript and coordinated all contributions. He became quite a TEXpert in the process. Several figures were made by idraw and xfig by Gábor and Luc. Most of the drawings were directly programmed in PostScript by Luc and an undergraduate student at McGill University, Hisham Petry, to whom we are grateful. For Gábor, this book comes at the beginning of his career. Unfortunately, the other two authors are not so lucky. As both Luc and Laci felt that they would probably not write another book on nonparametric pattern recognition—the random walk must go on—they decided to put their general view of the subject area on paper while trying to separate the important from the irrelevant. Surely, this has contributed to the length of the text.

So far, our random excursions have been happy ones. Coincidentally, Luc is married to Bea, the most understanding woman in the world, and happens to have two great daughters, Natasha and Birgit, who do not stray off their random courses. Similarly, Laci has an equally wonderful wife, Kati, and two children with steady compasses, Kati and János. During the preparations of this book, Gábor met a wonderful girl, Arrate. They have recently decided to tie their lives together.

On the less amorous and glamorous side, we gratefully acknowledge the research support of NSERC CANADA, FCAR QUEBEC, OTKA HUNGARY, and the exchange program between the Hungarian Academy of Sciences and the Royal Belgian Academy of Sciences. Early versions of this text were tried out in some classes at the Technical University of Budapest, Katholieke Universiteit Leuven, Universität Stuttgart, and Université Montpellier II. We would like to thank those students for their help in making this a better book.

Montreal, Quebec, Canada   Luc Devroye
Budapest, Hungary   Laci Györfi
Budapest, Hungary   Gábor Lugosi

Contents

2-18

2.13

2.18

1

Introduction

Pattern recognition or *discrimination* is about guessing or predicting the unknown nature of an observation, a discrete quantity such as black or white, one or zero, sick or healthy, real or fake. An observation is a collection of numerical measurements such as an image (which is a sequence of bits, one per pixel), a vector of weather data, an electrocardiogram, or a signature on a check suitably digitized. More formally, an *observation* is a d-dimensional vector x. The unknown nature of the observation is called a *class*. It is denoted by y and takes values in a finite set $\{1, 2, \ldots, M\}$. In pattern recognition, one creates a function $g(x) : \mathcal{R}^d \to \{1, \ldots, M\}$ which represents one's guess of y given x. The mapping g is called a *classifier*. Our classifier errs on x if $g(x) \neq y$.

How one creates a rule g depends upon the problem at hand. Experts can be called for medical diagnoses or earthquake predictions—they try to mold g to their own knowledge and experience, often by trial and error. Theoretically, each expert operates with a built-in classifier g, but describing this g explicitly in mathematical form is not a sinecure. The sheer magnitude and richness of the space of x may defeat even the best expert—it is simply impossible to specify g for all possible x's one is likely to see in the future. We have to be prepared to live with imperfect classifiers. In fact, how should we measure the quality of a classifier? We can't just dismiss a classifier just because it misclassifies a particular x. For one thing, if the observation does not fully describe the underlying process (that is, if y is not a deterministic function of x), it is possible that the same x may give rise to two different y's on different occasions. For example, if we just measure water content of a person's body, and we find that the person is dehydrated, then the cause (the class) may range from a low water intake in hot weather to severe diarrhea. Thus, we introduce a probabilistic setting, and let (X, Y) be an $\mathcal{R}^d \times \{1, \ldots, M\}$-valued random pair. The distribution of (X, Y) describes the frequency of encountering particular pairs in practice. An *error* occurs if $g(X) \neq Y$, and the *probability of error* for a classifier g is

$$L(g) = \mathbf{P}\{g(X) \neq Y\} \ .$$

There is a best possible classifier, g^*, which is defined by

$$g^* = \underset{g:\mathcal{R}^d \to \{1,\dots,M\}}{\arg\min} \ \mathbf{P}\{g(X) \neq Y\} \ .$$

Note that g^* depends upon the distribution of (X, Y). If this distribution is known, g^* may be computed. The problem of finding g^* is *Bayes' problem*, and the classifier g^* is called the *Bayes classifier* (or the *Bayes rule*). The minimal probability of error is called the *Bayes error* and is denoted by $L^* = L(g^*)$. Mostly, the distribution of (X, Y) is unknown, so that g^* is unknown too.

We do not consult an expert to try to reconstruct g^*, but have access to a good database of pairs (X_i, Y_i), $1 \le i \le n$, observed in the past. This database may be the result of experimental observation (as for meteorological data, fingerprint data, ECG data, or handwritten characters). It could also be obtained through an expert or a teacher who filled in the Y_i's after having seen the X_i's. To find a classifier g with a small probability of error is hopeless unless there is some assurance that the (X_i, Y_i)'s jointly are somehow representative of the unknown distribution. We shall assume in this book that $(X_1, Y_1), \dots, (X_n, Y_n)$, the *data*, is a sequence of independent identically distributed (*i.i.d.*) random pairs with the same distribution as that of (X, Y). This is a very strong assumption indeed. However, some theoretical results are emerging that show that classifiers based on slightly dependent data pairs and on i.i.d. data pairs behave roughly the same. Also, simple models are easier to understand are more amenable to interpretation.

A classifier is constructed on the basis of $X_1, Y_1, \dots, X_n, Y_n$ and is denoted by g_n; Y is guessed by $g_n(X; X_1, Y_1, \dots, X_n, Y_n)$. The process of constructing g_n is called *learning*, *supervised learning*, or *learning with a teacher*. The performance of g_n is measured by the conditional *probability of error*

$$L_n = L(g_n) = \mathbf{P}\{g_n(X; X_1, Y_1, \dots, X_n, Y_n) \neq Y | X_1, Y_1, \dots, X_n, Y_n)\} \ .$$

This is a random variable because it depends upon the data. So, L_n averages over the distribution of (X, Y), but the data is held fixed. Averaging over the data as well would be unnatural, because in a given application, one has to live with the data at hand. It would be marginally useful to know the number $\mathbf{E}L_n$ as this number would indicate the quality of an average data sequence, not *your* data sequence. This text is thus about L_n, the conditional probability of error.

An individual mapping $g_n : \mathcal{R}^d \times \{\mathcal{R}^d \times \{1, \dots, M\}\}^n \longrightarrow \{1, \dots, M\}$ is still called a *classifier*. A sequence $\{g_n, n \ge 1\}$ is called a *(discrimination) rule*. Thus, classifiers are functions, and rules are sequences of functions.

A novice might ask simple questions like this: How does one construct a good classifier? How good can a classifier be? Is classifier A better than classifier B? Can we estimate how good a classifier is? What is the best classifier? This book partially answers such simple questions. A good deal of energy is spent on the mathematical formulations of the novice's questions. For us, a rule—not a classifier—is good if it is *consistent*, that is, if

$$\lim_{n \to \infty} \mathbf{E}L_n = L^*$$

or equivalently, if $L_n \to L^*$ in probability as $n \to \infty$. We assume that the reader has a good grasp of the basic elements of probability, including notions such as convergence in probability, strong laws of large numbers for averages, and conditional probability. A selection of results and definitions that may be useful for this text is given in the Appendix. A *consistent rule* guarantees us that taking more samples essentially suffices to roughly reconstruct the unknown distribution of (X, Y) because L_n can be pushed as close as desired to L^*. In other words, infinite amounts of information can be gleaned from finite samples. Without this guarantee, we would not be motivated to take more samples. We should be careful and not impose conditions on (X, Y) for the consistency of a rule, because such conditions may not be verifiable. If a rule is consistent for all distributions of (X, Y), it is said to be *universally consistent*.

Interestingly, until 1977, it was not known if a universally consistent rule existed. All pre-1977 consistency results came with restrictions on (X, Y). In 1977, Stone showed that one could just take any k-nearest neighbor rule with $k = k(n) \to \infty$ and $k/n \to 0$. The k-nearest neighbor classifier $g_n(x)$ takes a majority vote over the Y_i's in the subset of k pairs (X_i, Y_i) from $(X_1, Y_1), \ldots, (X_n, Y_n)$ that have the smallest values for $\| X_i - x \|$ (i.e., for which X_i is closest to x). Since Stone's proof of the universal consistency of the k-nearest neighbor rule, several other rules have been shown to be universally consistent as well. This book stresses universality and hopefully gives a reasonable account of the developments in this direction.

Probabilists may wonder why we did not use convergence with probability one in our definition of consistency. Indeed, strong consistency—convergence of L_n to L^* with probability one—implies convergence for almost every sample as it grows. Fortunately, for most well-behaved rules, consistency and strong consistency are equivalent. For example, for the k-nearest neighbor rule, $k \to \infty$ and $k/n \to 0$ together imply $L_n \to L^*$ with probability one. The equivalence will be dealt with, but it will not be a major focus of attention. Most, if not all, equivalence results are based upon some powerful concentration inequalities such as McDiarmid's. For example, we will be able to show that for the k-nearest neighbor rule, there exists a number $c > 0$, such that for all $\epsilon > 0$, there exists $N(\epsilon) > 0$ depending upon the distribution of (X, Y), such that

$$\mathbf{P}\{L_n - L^* > \epsilon\} \leq e^{-cn\epsilon^2} , \quad n \geq N(\epsilon) .$$

This illustrates yet another focus of the book—inequalities. Whenever possible, we make a case or conclude a proof via explicit inequalities. Various parameters can be substituted in these inequalities to allow the user to draw conclusions regarding sample size or to permit identification of the most important parameters.

The material in the book is often technical and dry. So, to stay focused on the main issues, we keep the problem simple:

A. We only deal with binary classification ($M = 2$). The class Y takes values in $\{0, 1\}$, and a classifier g_n is a mapping: $\mathcal{R}^d \times \{\mathcal{R}^d \times \{0, 1\}\}^n \longrightarrow \{0, 1\}$.

B. We only consider i.i.d. data sequences. We also disallow active learning, a set-up in which the user can select the X_i's deterministically.

C. We do not consider infinite spaces. For example, X cannot be a random function such as a cardiogram. X must be a $\mathcal{R}^d$-valued random vector. The reader should be aware that many results given here may be painlessly extended to certain metric spaces of infinite dimension.

Let us return to our novice's questions. We know that there are good rules, but just how good can a classifier be? Obviously, $L_n \geq L^*$ in all cases. It is thus important to know L^* or to estimate it, for if L^* is large, any classifier, including yours, will perform poorly. But even if L^* were zero, L_n could still be large. Thus, it would be nice to have explicit inequalities for probabilities such as

$$\mathbf{P}\{L_n \geq L^* + \epsilon\} \, .$$

However, such inequalities must necessarily depend upon the distribution of (X, Y). That is, for any rule,

$$\liminf_{n \to \infty} \quad \sup_{\text{all distributions of } (X,Y) \text{ with } L^*+\epsilon < 1/2} \quad \mathbf{P}\{L_n \geq L^* + \epsilon\} > 0 \, .$$

Universal rate of convergence guarantees do not exist. Rate of convergence studies must involve certain subclasses of distributions of (X, Y). For this reason, with few exceptions, we will steer clear of the rate of convergence quicksand.

Even if there are no universal performance guarantees, we might still be able to satisfy our novice's curiosity if we could satisfactorily estimate L_n for the rule at hand by a function $\widehat{L}_n$ of the data. Such functions are called *error estimates*. For example, for the k-nearest neighbor classifier, we could use the *deleted estimate*

$$\widehat{L}_n = \frac{1}{n} \sum_{i=1}^{n} I_{\{g_{ni}(X_i) \neq Y_i\}} \, ,$$

where $g_{ni}(X_i)$ classifies X_i by the k-nearest neighbor method based upon the data $(X_1, Y_1), \ldots, (X_n, Y_n)$ with (X_i, Y_i) deleted. If this is done, we have a distribution-free inequality

$$\mathbf{P}\{|\widehat{L}_n - L_n| > \epsilon\} \leq \frac{6k + 1}{n\epsilon^2}$$

(the Rogers-Wagner inequality), provided that distance ties are broken in an appropriate manner. In other words, without knowing the distribution of (X, Y), we can state with a certain confidence that L_n is contained in $[\widehat{L}_n - \epsilon, \widehat{L}_n + \epsilon]$. Thus, for many classifiers, it is indeed possible to estimate L_n from the data at hand. However, it is impossible to estimate L^* universally well: for any n, and any estimate of L^* based upon the data sequence, there always exists a distribution of (X, Y) for which the estimate is arbitrarily poor.

Can we compare rules $\{g_n\}$ and $\{g'_n\}$? Again, the answer is negative: there exists no "best" classifier (or superclassifier), as for any rule $\{g_n\}$, there exists a distribution of (X, Y) and another rule $\{g'_n\}$ such that for all n, $\mathbf{E}\{L(g'_n)\} < \mathbf{E}\{L(g_n)\}$. If there had been a universally best classifier, this book would have been unnecessary: we would all have to use it all the time. This nonexistence implies that the debate between practicing pattern recognizers will never end and that simulations on particular examples should never be used to compare classifiers. As an example, consider the 1-nearest neighbor rule, a simple but not universally consistent rule. Yet, among all k-nearest neighbor classifiers, the 1-nearest neighbor classifier is admissible—there are distributions for which its expected probability of error is better than for any k-nearest neighbor classifier with $k > 1$. So, it can never be totally dismissed. Thus, we must study all simple rules, and we will reserve many pages for the nearest neighbor rule and its derivatives. We will for example prove the Cover-Hart inequality (Cover and Hart, 1967) which states that for all distributions of (X, Y),

$$\limsup_{n \to \infty} \mathbf{E} L_n \leq 2L^*$$

where L_n is the probability of error with the 1-nearest neighbor rule. As L^* is usually small (for otherwise, you would not want to do discrimination), $2L^*$ is small too, and the 1-nearest neighbor rule will do just fine.

The nonexistence of a best classifier may disappoint our novice. However, we may change the setting somewhat and limit the classifiers to a certain class $\mathcal{C}$, such as all k-nearest neighbor classifiers with all possible values for k. Is it possible to select the best classifier from this class? Phrased in this manner, we cannot possibly do better than

$$L \stackrel{\text{def}}{=} \inf_{g_n \in \mathcal{C}} \mathbf{P}\{g_n(X) \neq Y\} .$$

Typically, $L > L^*$. Interestingly, there is a general paradigm for picking classifiers from $\mathcal{C}$ and to obtain universal performance guarantees. It uses *empirical risk minimization*, a method studied in great detail in the work of Vapnik and Chervonenkis (1971). For example, if we select g_n from $\mathcal{C}$ by minimizing

$$\frac{1}{n} \sum_{i=1}^{n} I_{\{g_n(X_i) \neq Y_i\}} ,$$

then the corresponding probability of error L_n satisfies the following inequality for all $\epsilon > 0$:

$$\mathbf{P}\{L_n > L + \epsilon\} \leq 8 \left(n^V + 1\right) e^{-n\epsilon^2/128} .$$

Here $V > 0$ is an integer depending upon the massiveness of $\mathcal{C}$ only. V is called the VC dimension of $\mathcal{C}$ and may be infinite for large classes $\mathcal{C}$. For sufficiently restricted classes $\mathcal{C}$, V is finite and the explicit universal bound given above can be used to obtain performance guarantees for the selected g_n (relative to L, not L^*). The bound above is only valid if $\mathcal{C}$ is independent of the data pairs $(X_1, Y_1), \ldots, (X_n, Y_n)$. Fixed classes such as all classifiers that decide 1 on a halfspace and 0 on its complement are fine. We may also sample m more pairs (in addition to the n pairs already present), and use the n pairs as above to select the best k for use in the k-nearest neighbor classifier based on the m pairs. As we will see, the selected rule is universally consistent if both m and n diverge and $n/\log m \to \infty$. And we have automatically solved the problem of picking k. Recall that Stone's universal consistency theorem only told us to pick $k = o(m)$ and to let $k \to \infty$, but it does not tell us whether $k \approx m^{0.01}$ is preferable over $k \approx m^{0.99}$. Empirical risk minimization produces a random data-dependent k that is not even guaranteed to tend to infinity or to be $o(m)$, yet the selected rule is universally consistent.

We offer virtually no help with algorithms as in standard texts, with two notable exceptions. Ease of computation, storage, and interpretation has spurred the development of certain rules. For example, *tree classifiers* construct a tree for storing the data, and partition $\mathcal{R}^d$ by certain cuts that are typically perpendicular to a coordinate axis. We say that a coordinate axis is "cut." Such classifiers have obvious computational advantages, and are amenable to interpretation—the components of the vector X that are cut at the early stages of the tree are most crucial in reaching a decision. Expert systems, automated medical diagnosis, and a host of other recognition rules use tree classification. For example, in automated medical diagnosis, one may first check a patient's pulse (component #1). If this is zero, the patient is dead. If it is below 40, the patient is weak. The first component is cut twice. In each case, we may then consider another component, and continue the breakdown into more and more specific cases. Several interesting new universally consistent tree classifiers are described in Chapter 20.

The second group of classifiers whose development was partially based upon easy implementations is the class of *neural network classifiers*, descendants of Rosenblatt's *perceptron* (Rosenblatt, 1956). These classifiers have unknown parameters that must be trained or selected by the data, in the way we let the data pick k in the k-nearest neighbor classifier. Most research papers on neural networks deal with the training aspect, but we will not. When we say "pick the parameters by empirical risk minimization," we will leave the important algorithmic complexity questions unanswered. Perceptrons divide the space by one hyperplane and attach decisions 1 and 0 to the two halfspaces. Such simple classifiers are not consistent except for a few distributions. This is the case, for example, when X takes values on $\{0, 1\}^d$ (the hypercube) and the components of X are independent. Neural networks with one hidden layer are universally consistent if the parameters are well-chosen. We will see that there is also some gain in considering two hidden layers, but that it is not really necessary to go beyond two.

Complexity of the training algorithm—the phase in which a classifier g_n is selected from C—is of course important. Sometimes, one would like to obtain classifiers that are invariant under certain transformations. For example, the k-nearest neighbor classifier is not invariant under nonlinear transformations of the coordinate axes. This is a drawback as components are often measurements in an arbitrary scale. Switching to a logarithmic scale or stretching a scale out by using Fahrenheit instead of Celsius should not affect good discrimination rules. There exist variants of the k-nearest neighbor rule that have the given invariance. In character recognition, sometimes all components of a vector X that represents a character are true measurements involving only vector differences between selected points such as the leftmost and rightmost points, the geometric center, the weighted center of all black pixels, the topmost and bottommost points. In this case, the scale has essential information, and invariance with respect to changes of scale would be detrimental. Here however, some invariance with respect to orthonormal rotations is healthy.

We follow the standard notation from textbooks on probability. Thus, random variables are uppercase characters such as X, Y, and Z. Probability measures are denoted by greek letters such as μ and ν. Numbers and vectors are denoted by lowercase letters such as a, b, c, x, and y. Sets are also denoted by roman capitals, but there are obvious mnemonics: S denotes a sphere, B denotes a Borel set, and so forth. If we need many kinds of sets, we will typically use the beginning of the alphabet (A, B, C). Most functions are denoted by f, g, ϕ, and ψ. Calligraphic letters such as $\mathcal{A}$, $\mathcal{C}$, and $\mathcal{F}$ are used to denote classes of functions or sets. A short list of frequently used symbols is found at the end of the book.

At the end of this chapter, you will find a directed acyclic graph that describes the dependence between chapters. Clearly, prospective teachers will have to select small subsets of chapters. All chapters, without exception, are unashamedly theoretical. We did not scar the pages with backbreaking simulations or quick-and-dirty engineering solutions. The methods gleaned from this text must be supplemented with a healthy dose of engineering savvy. Ideally, students should have a companion text filled with beautiful applications such as automated virus recognition, telephone eavesdropping language recognition, voice recognition in security systems, fingerprint recognition, or handwritten character recognition. To run a real pattern recognition project from scratch, several classical texts on statistical pattern recognition could and should be consulted, as our work is limited to general probability-theoretical aspects of pattern recognition. We have over 430 exercises to help the scholars. These include skill honing exercises, brainteasers, cute puzzles, open problems, and serious mathematical challenges. There is no solution manual. This book is only a start. Use it as a toy—read some proofs, enjoy some inequalities, learn new tricks, and study the art of camouflaging one problem to look like another. Learn for the sake of learning.

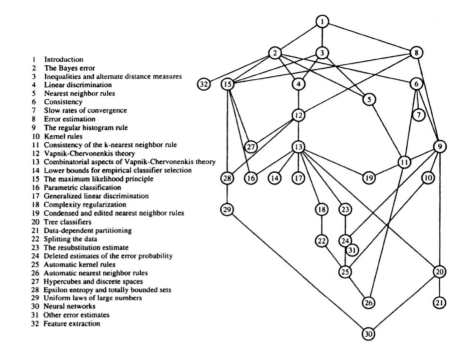

FIGURE 1.1.

2

The Bayes Error

2.1 The Bayes Problem

In this section, we define the mathematical model and introduce the notation we will use for the entire book. Let (X, Y) be a pair of random variables taking their respective values from $\mathcal{R}^d$ and $\{0, 1\}$. The random pair (X, Y) may be described in a variety of ways: for example, it is defined by the pair (μ, η), where μ is the probability measure for X and η is the regression of Y on X. More precisely, for a Borel-measurable set $A \subseteq \mathcal{R}^d$,

$$\mu(A) = \mathbf{P}\{X \in A\},$$

and for any $x \in \mathcal{R}^d$,

$$\eta(x) = \mathbf{P}\{Y = 1 | X = x\} = \mathbf{E}\{Y | X = x\}.$$

concave function ?

Thus, $\eta(x)$ is the conditional probability that Y is 1 given $X = x$. To see that this suffices to describe the distribution of (X, Y), observe that for any $C \subseteq \mathcal{R}^d \times \{0, 1\}$, we have

posteriori probability

$$C = \left(C \cap \left(\mathcal{R}^d \times \{0\}\right)\right) \cup \left(C \cap \left(\mathcal{R}^d \times \{1\}\right)\right) \overset{\text{def}}{=} C_0 \times \{0\} \cup C_1 \times \{1\},$$

and

$$\begin{aligned}
\mathbf{P}\{(X, Y) \in C\} &= \mathbf{P}\{X \in C_0, Y = 0\} + \mathbf{P}\{X \in C_1, Y = 1\} \\
&= \int_{C_0} (1 - \eta(x))\mu(dx) + \int_{C_1} \eta(x)\mu(dx).
\end{aligned}$$

As this is valid for any Borel-measurable set C, the distribution of (X, Y) is determined by (μ, η). The function η is sometimes called the *a posteriori probability.*

Any function $g : \mathcal{R}^d \to \{0, 1\}$ defines a *classifier* or a *decision function*. The error probability of g is $L(g) = \mathbf{P}\{g(X) \neq Y\}$. Of particular interest is the Bayes decision function

$$g^*(x) = \begin{cases} 1 & \text{if } \eta(x) > 1/2 \\ 0 & \text{otherwise.} \end{cases}$$

This decision function minimizes the error probability.

Theorem 2.1. *For any decision function* $g : \mathcal{R}^d \to \{0, 1\}$,

$$\mathbf{P}\{g^*(X) \neq Y\} \leq \mathbf{P}\{g(X) \neq Y\},$$

that is, g^* *is the optimal decision.*

PROOF. Given $X = x$, the conditional error probability of any decision g may be expressed as

$$
\begin{aligned}
&\mathbf{P}\{g(X) \neq Y | X = x\} \\
&= 1 - \mathbf{P}\{Y = g(X) | X = x\} \\
&= 1 - (\mathbf{P}\{Y = 1, g(X) = 1 | X = x\} + \mathbf{P}\{Y = 0, g(X) = 0 | X = x\}) \\
&= 1 - \left(I_{\{g(x)=1\}} \mathbf{P}\{Y = 1 | X = x\} + I_{\{g(x)=0\}} \mathbf{P}\{Y = 0 | X = x\}\right) \\
&= 1 - \left(I_{\{g(x)=1\}} \eta(x) + I_{\{g(x)=0\}} (1 - \eta(x))\right),
\end{aligned}
$$

where I_A denotes the indicator of the set A. Thus, for every $x \in \mathcal{R}^d$,

$$
\begin{aligned}
&\mathbf{P}\{g(X) \neq Y | X = x\} - \mathbf{P}\{g^*(X) \neq Y | X = x\} \qquad \text{negative} \\
&= \eta(x) \left(I_{\{g^*(x)=1\}} - I_{\{g(x)=1\}}\right) + (1 - \eta(x)) \left(I_{\{g^*(x)=0\}} - I_{\{g(x)=0\}}\right) \\
&= (2\eta(x) - 1) \left(I_{\{g^*(x)=1\}} - I_{\{g(x)=1\}}\right) \\
&\geq 0
\end{aligned}
$$

by the definition of g^*. The statement now follows by integrating both sides with respect to $\mu(dx)$. $\square$

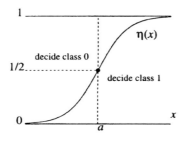

FIGURE 2.1. *The Bayes decision in the example on the left is* 1 *if* $x > a$, *and* 0 *otherwise.*

REMARK. g^* is called the Bayes decision and $L^* = \mathbf{P}\{g^*(X) \neq Y\}$ is referred to as the Bayes probability of error, Bayes error, or Bayes risk. The proof given above reveals that

$$L(g) = 1 - \mathbf{E}\left\{I_{\{g(X)=1\}}\eta(X) + I_{\{g(X)=0\}}(1 - \eta(X))\right\},$$

and in particular,

$$L^* = 1 - \mathbf{E}\left\{I_{\{\eta(X)>1/2\}}\eta(X) + I_{\{\eta(X)\leq1/2\}}(1 - \eta(X))\right\}. \quad \square$$

We observe that the a posteriori probability

$$\eta(x) = \mathbf{P}\{Y = 1|X = x\} = \mathbf{E}\{Y|X = x\}$$

minimizes the squared error when Y is to be predicted by $f(X)$ for some function $f : \mathcal{R}^d \to \mathcal{R}$:

$$\mathbf{E}\left\{(\eta(X) - Y)^2\right\} \leq \mathbf{E}\left\{(f(X) - Y)^2\right\}.$$

To see why the above inequality is true, observe that for each $x \in \mathcal{R}^d$,

$$
\begin{aligned}
&\mathbf{E}\left\{(f(X) - Y)^2|X = x\right\} \\
&= \mathbf{E}\left\{(f(x) - \eta(x) + \eta(x) - Y)^2|X = x\right\} \\
&= (f(x) - \eta(x))^2 + 2(f(x) - \eta(x))\underbrace{\mathbf{E}\{\eta(x) - Y|X = x\}}_{= \mathbf{E}(\eta(x)|x) - \mathbf{E}(Y|X) = 0} \\
&\quad + \mathbf{E}\left\{(\eta(X) - Y)^2|X = x\right\} \\
&= (f(x) - \eta(x))^2 + \mathbf{E}\left\{(\eta(X) - Y)^2|X = x\right\}.
\end{aligned}
$$

The conditional median, i.e., the function minimizing the absolute error $\mathbf{E}\{|f(X) - Y|\}$ is even more closely related to the Bayes rule (see Problem 2.12).

2.2   A Simple Example

Let us consider the prediction of a student's performance in a course (pass/fail) when given a number of important factors. First, let $Y = 1$ denote a pass and let $Y = 0$ stand for failure. The sole observation X is the number of hours of study per week. This, in itself, is not a foolproof predictor of a student's performance, because for that we would need more information about the student's quickness of mind, health, and social habits. The regression function $\eta(x) = \mathbf{P}\{Y = 1|X = x\}$ is probably monotonically increasing in x. If it were known to be $\eta(x) = x/(c + x)$, $c > 0$, say, our problem would be solved because the Bayes decision is

$$g^*(x) = \begin{cases} 1 & \text{if } \eta(x) > 1/2 \text{ (i.e., } x > c) \\ 0 & \text{otherwise.} \end{cases}$$

The corresponding Bayes error is

$$L^* = L(g^*) = \mathbf{E}\{\min(\eta(X), 1 - \eta(X))\} = \mathbf{E}\left\{\frac{\min(c, X)}{c + X}\right\}.$$

While we could deduce the Bayes decision from η alone, the same cannot be said for the Bayes error L^*—it requires knowledge of the distribution of X. If $X = c$ with probability one (as in an army school, where all students are forced to study c hours per week), then $L^* = 1/2$. If we have a population that is nicely spread out, say, X is uniform on $[0, 4c]$, then the situation improves:

$$L^* = \frac{1}{4c} \int_0^{4c} \frac{\min(c, x)}{c + x} dx = \frac{1}{4} \log \frac{5e}{4} \approx 0.305785.$$

Far away from $x = c$, discrimination is really simple. In general, discrimination is much easier than estimation because of this phenomenon.

2.3   Another Simple Example

Let us work out a second simple example in which $Y = 0$ or $Y = 1$ according to whether a student fails or passes a course. X represents one or more observations regarding the student. The components of X in our example will be denoted by T, B, and E respectively, where T is the average number of hours the students watches TV, B is the average number of beers downed each day, and E is an intangible quantity measuring extra negative factors such as laziness and learning difficulties. In our cooked-up example, we have

$$Y = \begin{cases} 1 & \text{if } T + B + E < 7 \\ 0 & \text{otherwise.} \end{cases}$$

Thus, if T, B, and E are known, Y is known as well. The Bayes classifier decides 1 if $T + B + E < 7$ and 0 otherwise. The corresponding Bayes probability of error is zero. Unfortunately, E is intangible, and not available to the observer. We only have access to T and B. Given T and B, when should we guess that $Y = 1$? To answer this question, one must know the joint distribution of (T, B, E), or, equivalently, the joint distribution of (T, B, Y). So, let us assume that T, B, and E are i.i.d. exponential random variables (thus, they have density e^{-u} on $[0, \infty)$). The Bayes rule compares $\mathbf{P}\{Y = 1|T, B\}$ with $\mathbf{P}\{Y = 0|T, B\}$ and makes a decision consistent with the maximum of these two values. A simple calculation shows that

$$\begin{aligned} \mathbf{P}\{Y = 1|T, B\} &= \mathbf{P}\{T + B + E < 7|T, B\} \\ &= \mathbf{P}\{E < 7 - T - B|T, B\} \\ &= \max\left(0, 1 - e^{-(7-T-B)}\right). \end{aligned}$$

The crossover between two decisions occurs when this value equals $1/2$. Thus, the Bayes classifier is as follows:

$$g^*(T, B) = \begin{cases} 1 & \text{if } T + B < 7 - \log 2 = 6.306852819\ldots \\ 0 & \text{otherwise.} \end{cases}$$

Of course, this classifier is not perfect. The probability of error is

$$P\{g^*(T, B) \neq Y\}$$

$$= P\{T + B < 7 - \log 2, T + B + E \geq 7\}$$
$$+ P\{T + B \geq 7 - \log 2, T + B + E < 7\}$$

$$= E\left\{e^{-(7-T-B)}I_{\{T+B<7-\log 2\}}\right\}$$
$$+ P\left\{\left(1 - e^{-(7-T-B)}\right)I_{\{7>T+B\geq 7-\log 2\}}\right\}$$

$$= \int_0^{7-\log 2} xe^{-x}e^{-(7-x)}\,dx + \int_{7-\log 2}^7 xe^{-x}\left(1 - e^{-(7-x)}\right)dx$$

(since the density of $T + B$ is ue^{-u} on $[0, \infty)$)

$$= e^{-7}\left(\frac{(7-\log 2)^2}{2} + 2(8 - \log 2) - 8 - \frac{7^2}{2} + \frac{(7-\log 2)^2}{2}\right)$$

(as $\int_x^\infty ue^{-u}\,du = (1+x)e^{-x}$)

$$= 0.0199611\ldots .$$

If we have only access to T, then the Bayes classifier is allowed to use T only. First, we find

$$P\{Y = 1|T\} = P\{E + B < 7 - T|T\}$$
$$= \max\left(0, 1 - (1 + 7 - T)e^{-(7-T)}\right).$$

The crossover at $1/2$ occurs at $T = c \stackrel{\text{def}}{=} 5.321653009\ldots$, so that the Bayes classifier is given by

$$g^*(T) = \begin{cases} 1 & \text{if } T < c \\ 0 & \text{otherwise.} \end{cases}$$

The probability of error is

$$P\{g^*(T) \neq Y\}$$

$$= P\{T < c, T + B + E \geq 7\} + P\{T \geq c, T + B + E < 7\}$$

$$= E\left\{(1 + 7 - T)e^{-(7-T)}I_{\{T<c\}}\right\}$$
$$+ P\left\{\left(1 - (1 + 7 - T)e^{-(7-T)}\right)I_{\{7>T\geq c\}}\right\}$$

$$= \int_0^c e^{-x}(1 + 7 - x)e^{-(7-x)}\,dx + \int_c^7 e^{-x}\left(1 - (1 + 7 - x)e^{-(7-x)}\right)dx$$

$$= e^{-7}\left(\frac{8^2}{2} - \frac{(8-c)^2}{2} + e^{-(c-7)} - 1 - \frac{(8-c)^2}{2} + \frac{1}{2}\right)$$

$$= 0.02235309002\ldots .$$

The Bayes error has increased slightly, but not by much. Finally, if we do not have access to any of the three variables, T, B, and E, the best we can do is see which class is most likely. To this end, we compute

$$P\{Y = 0\} = P\{T + B + E \geq 7\} = (1 + 7 + 7^2/2)e^{-7} = .02963616388\ldots.$$

If we set $g \equiv 1$ all the time, we make an error with probability $0.02963616388\ldots$.

In practice, Bayes classifiers are unknown simply because the distribution of (X, Y) is unknown. Consider a classifier based upon (T, B). Rosenblatt's perceptron (see Chapter 4) looks for the best linear classifier based upon the data. That is, the decision is of the form

$$g(T, B) = \begin{cases} 1 & \text{if } aT + bB < c \\ 0 & \text{otherwise} \end{cases}$$

for some data-based choices for a, b and c. If we have lots of data at our disposal, then it is possible to pick out a linear classifier that is nearly optimal. As we have seen above, the Bayes classifier happens to be linear. That is a sheer coincidence, of course. If the Bayes classifier had not been linear—for example, if we had $Y = I_{\{T+B^2+E<7\}}$—then even the best perceptron would be suboptimal, regardless of how many data pairs one would have. If we use the 3-nearest neighbor rule (Chapter 5), the asymptotic probability of error is not more than 1.3155 times the Bayes error, which in our example is about 0.02625882705. The example above also shows the need to look at individual components, and to evaluate how many and which components would be most useful for discrimination. This subject is covered in the chapter on feature extraction (Chapter 32).

2.4   Other Formulas for the Bayes Risk

The following forms of the Bayes error are often convenient:

$$\begin{aligned} L^* &= \inf_{g:\mathcal{R}^d \to \{0,1\}} P\{g(X) \neq Y\} \\ &= E\{\min\{\eta(X), 1 - \eta(X)\}\} \\ &= \frac{1}{2} - \frac{1}{2}E\{|2\eta(X) - 1|\}. \end{aligned}$$

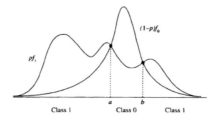

FIGURE 2.2. *The Bayes decision when class-conditional densities exist. In the figure on the left, the decision is 0 on $[a, b]$ and 1 elsewhere.*

In special cases, we may obtain other helpful forms. For example, if X has a density f, then

$$L^* = \int \min(\eta(x), 1 - \eta(x)) f(x) dx$$

$$= \int \min((1 - p) f_0(x), p f_1(x)) dx,$$

where $p = \mathbf{P}\{Y = 1\}$, and $f_i(x)$ is the density of X given that $Y = i$. p and $1 - p$ are called the *class probabilities*, and f_0, f_1 are the class-conditional densities. If f_0 and f_1 are nonoverl...ping, that is, $\int f_0 f_1 = 0$, then obviously $L^* = 0$. Assume moreover that $p = 1/2$. Then

$$L^* = \frac{1}{2} \int \min(f_0(x), f_1(x)) dx$$

$$= \frac{1}{2} \int f_1(x) - (f_1(x) - f_0(x))_+ dx$$

$$= \frac{1}{2} - \frac{1}{4} \int |f_1(x) - f_0(x)| dx.$$

Here g_+ denotes the positive part of a function g. Thus, the Bayes error is directly related to the L_1 distance between the class densities.

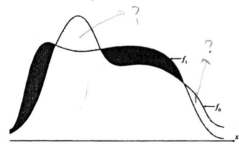

FIGURE 2.3. *The shaded area is the L_1 distance between the class-conditional densities.*

2.5   Plug-In Decisions

The best guess of Y from the observation X is the Bayes decision

$$g^*(x) = \begin{cases} 0 & \text{if } \eta(x) \le 1/2 \\ 1 & \text{otherwise} \end{cases} = \begin{cases} 0 & \text{if } \eta(x) \le 1 - \eta(x) \\ 1 & \text{otherwise.} \end{cases}$$

The function η is typically unknown. Assume that we have access to nonnegative functions $\bar{\eta}(x)$, $1 - \bar{\eta}(x)$ that approximate $\eta(x)$ and $1 - \eta(x)$ respectively. In this case it seems natural to use the plug-in decision function

$$g(x) = \begin{cases} 0 & \text{if } \bar{\eta}(x) \le 1/2 \\ 1 & \text{otherwise,} \end{cases}$$

to approximate the Bayes decision. The next well-known theorem (see, e.g., Van Ryzin (1966), Wolverton and Wagner (1969a), Glick (1973), Csibi (1971), Györfi (1975), (1978), Devroye and Wagner (1976b), Devroye (1982b), and Devroye and Györfi (1985)) states that if $\tilde{\eta}(x)$ is close to the real a posteriori probability in L_1-sense, then the error probability of decision g is near the optimal decision g^*.

Theorem 2.2. *For the error probability of the plug-in decision g defined above, we have*

$$\mathbf{P}\{g(X) \neq Y\} - L^* = 2 \int_{\mathcal{R}^d} |\eta(x) - 1/2| I_{\{g(x) \neq g^*(x)\}} \mu(dx),$$

and

$$\mathbf{P}\{g(X) \neq Y\} - L^* \leq 2 \int_{\mathcal{R}^d} |\eta(x) - \tilde{\eta}(x)| \mu(dx) = 2 \mathbf{E} |\eta(X) - \tilde{\eta}(X)|.$$

PROOF. If for some $x \in \mathcal{R}^d$, $g(x) = g^*(x)$, then clearly the difference between the conditional error probabilities of g and g^* is zero:

$$\mathbf{P}\{g(X) \neq Y | X = x\} - \mathbf{P}\{g^*(X) \neq Y | X = x\} = 0.$$

Otherwise, if $g(x) \neq g^*(x)$, then as seen in the proof of Theorem 2.1, the difference may be written as

$$
\begin{aligned}
\mathbf{P}\{g(X) \neq Y | X = x\} \quad &- \quad \mathbf{P}\{g^*(X) \neq Y | X = x\} \\
&= \quad (2\eta(x) - 1)\left(I_{\{g^*(x)=1\}} - I_{\{g(x)=1\}}\right) \\
&= \quad |2\eta(x) - 1| I_{\{g(x) \neq g^*(x)\}}.
\end{aligned}
$$

Thus,

$$
\begin{aligned}
\mathbf{P}\{g(X) \neq Y\} - L^* \quad &= \quad \int_{\mathcal{R}^d} 2|\eta(x) - 1/2| I_{\{g(x) \neq g^*(x)\}} \mu(dx) \\
&\leq \quad \int_{\mathcal{R}^d} 2|\eta(x) - \tilde{\eta}(x)| \mu(dx),
\end{aligned}
$$

since $g(x) \neq g^*(x)$ implies $|\eta(x) - \tilde{\eta}(x)| \geq |\eta(x) - 1/2|$. $\square$

When the classifier $g(x)$ can be put in the form

$$g(x) = \begin{cases} 0 & \text{if } \tilde{\eta}_1(x) \leq \tilde{\eta}_0(x) \\ 1 & \text{otherwise,} \end{cases}$$

where $\tilde{\eta}_1(x), \tilde{\eta}_0(x)$ are some approximations of $\eta(x)$ and $1 - \eta(x)$, respectively, the situation differs from that discussed in Theorem 2.2 if $\tilde{\eta}_0(x) + \tilde{\eta}_1(x)$ does not necessarily equal to one. However, an inequality analogous to that of Theorem 2.2 remains true:

Theorem 2.3. *The error probability of the decision defined above is bounded from above by*

$$P\{g(X) \neq Y\} - L^* \leq \int_{\mathcal{R}^d} |(1 - \eta(x)) - \tilde{\eta}_0(x)|\mu(dx) + \int_{\mathcal{R}^d} |\eta(x) - \tilde{\eta}_1(x)|\mu(dx).$$

The proof is left to the reader (Problem 2.9).✓

REMARK. Assume that the class-conditional densities f_0, f_1 exist and are approximated by the densities $\tilde{f}_0(x)$, $\tilde{f}_1(x)$. Assume furthermore that the class probabilities $p = P\{Y = 1\}$ and $1 - p = P\{Y = 0\}$ are approximated by $\tilde{p}_1$ and $\tilde{p}_0$, respectively. Then for the error probability of the plug-in decision function

$$g(x) = \begin{cases} 0 & \text{if } \tilde{p}_1 \tilde{f}_1(x) \leq \tilde{p}_0 \tilde{f}_0(x) \\ 1 & \text{otherwise,} \end{cases}$$

$$P\{g(X) \neq Y\} - L^*$$
$$\leq \int_{\mathcal{R}^d} |(1 - p)f_0(x) - \tilde{p}_0 \tilde{f}_0(x)|dx + \int_{\mathcal{R}^d} |pf_1(x) - \tilde{p}_1 \tilde{f}_1(x)|dx.$$

See Problem 2.10. □ ✓

2.6   Bayes Error Versus Dimension

The components of X that matter in the Bayes classifier are those that explicitly appear in $\eta(X)$. In fact, then, all discrimination problems are one-dimensional, as we could equally well replace X by $\eta(X)$ or by any strictly monotone function of $\eta(X)$, such as $\eta^7(X) + 5\eta^3(X) + \eta(X)$. Unfortunately, η is unknown in general. In the example in Section 2.3, we had in one case

$$\eta(T, B) = \max\left(0, 1 - e^{-(7-T-B)}\right)$$

and in another case

$$\eta(T) = \max\left(0, 1 - (1 + 7 - T)e^{-(7-T)}\right).$$

The former format suggests that we could base all decisions on $T + B$. This means that if we had no access to T and B individually, but to $T + B$ jointly, we would be able to achieve the same results! Since η is unknown, all of this is really irrelevant.

In general, the Bayes risk increases if we replace X by $T(X)$ for any transformation T (see Problem 2.1), as this destroys information. On the other hand, there exist transformations (such as $\eta(X)$) that leave the Bayes error untouched. For more on the relationship between the Bayes error and the dimension, refer to Chapter 32.

Problems and Exercises

PROBLEM 2.1. Let $T : \mathcal{X} \to \mathcal{X}'$ be an arbitrary measurable function. If L_X^* and $L_{T(X)}^*$ denote the Bayes error probabilities for (X, Y) and $(T(X), Y)$, respectively, then prove that

$$L_{T(X)}^* \geq L_X^*.$$

(This shows that transformations of X destroy information, because the Bayes risk increases.)

PROBLEM 2.2. Let X' be independent of (X, Y). Prove that

$$L_{(X,X')}^* = L_X^*.$$

PROBLEM 2.3. Show that $L^* \leq \min(p, 1 - p)$, where $p, 1 - p$ are the class probabilities. Show that equality holds if X and Y are independent. Exhibit a distribution where X is not independent of Y, but $L^* = \min(p, 1 - p)$.

PROBLEM 2.4. NEYMAN-PEARSON LEMMA. Consider again the decision problem, but with a decision g, we now assign two error probabilities,

$$L^{(0)}(g) = P\{g(X) = 1 | Y = 0\} \quad \text{and} \quad L^{(1)}(g) = P\{g(X) = 0 | Y = 1\}.$$

Assume that the class-conditional densities f_0, f_1 exist. For $c > 0$, define the decision

$$g_c(x) = \begin{cases} 1 & \text{if } cf_1(x) > f_0(x) \\ 0 & \text{otherwise.} \end{cases}$$

Prove that for any decision g, if $L^{(0)}(g) \leq L^{(0)}(g_c)$, then $L^{(1)}(g) \geq L^{(1)}(g_c)$. In other words, if $L^{(0)}$ is required to be kept under a certain level, then the decision minimizing $L^{(1)}$ has the form of g_c for some c. Note that g^* is like that.

PROBLEM 2.5. DECISIONS WITH REJECTION. Sometimes in decision problems, one is allowed to say "I don't know," if this does not happen frequently. These decisions are called decisions with a reject option (see, e.g., Forney (1968), Chow (1970)). Formally, a decision $g(x)$ can have three values: 0, 1, and "reject." There are two performance measures: the probability of rejection $P\{g(X) = \text{"reject"}\}$, and the error probability $P\{g(X) \neq Y | g(X) \neq \text{"reject"}\}$. For a $0 < c < 1/2$, define the decision

$$g_c(x) = \begin{cases} 1 & \text{if } \eta(x) > 1/2 + c \\ 0 & \text{if } \eta(x) \leq 1/2 - \bullet \\ \text{"reject"} & \text{otherwise.} \end{cases}$$

Show that for any decision g, if

$$P\{g(X) = \text{"reject"}\} \leq P\{g_c(X) = \text{"reject"}\},$$

then

$$P\{g(X) \neq Y | g(X) \neq \text{"reject"}\} \geq P\{g_c(X) \neq Y | g_c(X) \neq \text{"reject"}\}.$$

Thus, to keep the probability of rejection under a certain level, decisions of the form of g_c are optimal (Györfi, Györfi, and Vajda (1978)).

PROBLEM 2.6. Consider the prediction of a student's failure based upon variables T and B, where $Y = I_{\{T+B+E<7\}}$ and E is an inaccessible variable (see Section 2.3).

(1) Let T, B, and E be independent. Merely by changing the distribution of E, show that the Bayes error for classification based upon (T, B) can be made as close as desired to $1/2$.

(2) Let T and B be independent and exponentially distributed. Find a joint distribution of (T, B, E) such that the Bayes classifier is not a linear classifier.

(3) Let T and B be independent and exponentially distributed. Find a joint distribution of (T, B, E) such that the Bayes classifier is given by

$$g^*(T, B) = \begin{cases} 1 & \text{if } T^2 + B^2 < 10, \\ 0 & \text{otherwise} . \end{cases}$$

(4) Find the Bayes classifier and Bayes error for classification based on (T, B) (with Y as above) if (T, B, E) is uniformly distributed on $[0, 4]^3$.

PROBLEM 2.7. Assume that T, B, and E are independent uniform $[0, 4]$ random variables with interpretations as in Section 2.3. Let $Y = 1$ (0) denote whether a student passes (fails) a course. Assume that $Y = 1$ if and only if $TBE \leq 8$.

(1) Find the Bayes decision if no variable is available, if only T is available, and if only T and B are available.

(2) Determine in all three cases the Bayes error.

(3) Determine the best linear classifier based upon T and B only.

PROBLEM 2.8. Let η', $\eta'' : \mathcal{R}^d \to [0, 1]$ be arbitrary measurable functions, and define the corresponding decisions by $g'(x) = I_{\{\eta'(x)>1/2\}}$ and $g''(x) = I_{\{\eta''(x)>1/2\}}$. Prove that

$$|L(g') - L(g'')| \leq P\{g'(X) \neq g''(X)\}$$

and

$$|L(g') - L(g'')| \leq E\left\{|2\eta(X) - 1|I_{\{g'(X)\neq g''(X)\}}\right\}.$$

PROBLEM 2.9. Prove Theorem 2.3.

PROBLEM 2.10. Assume that the class-conditional densities f_0 and f_1 exist and are approximated by the densities $\tilde{f}_0$ and $\tilde{f}_1$, respectively. Assume furthermore that the class probabilities $p = P\{Y = 1\}$ and $1 - p = P\{Y = 0\}$ are approximated by $\tilde{p}_1$ and $\tilde{p}_0$. Prove that for the error probability of the plug-in decision function

$$g(x) = \begin{cases} 0 & \text{if } \tilde{p}_1\tilde{f}_1(x) \leq \tilde{p}_0\tilde{f}_0(x) \\ 1 & \text{otherwise,} \end{cases}$$

we have

$$P\{g(X) \neq Y\} - L^* \leq \int_{\mathcal{R}^d} |pf_1(x) - \tilde{p}_1\tilde{f}_1(x)|dx + \int_{\mathcal{R}^d} |(1 - p)f_0(x) - \tilde{p}_0\tilde{f}_0(x)|dx.$$

PROBLEM 2.11. Using the notation of Problem 2.10, show that if for a sequence of $\tilde{f}_{m,n}(x)$ and $\tilde{p}_{m,n}$ ($m = 0, 1$),

$$\lim_{n\to\infty} \int_{\mathcal{R}^d} |pf_1(x) - \tilde{p}_{1,n}\tilde{f}_{1,n}(x)|^2 dx + \int_{\mathcal{R}^d} |(1 - p)f_0(x) - \tilde{p}_{0,n}\tilde{f}_{0,n}(x)|^2 dx = 0,$$

then for the corresponding sequence of plug-in decisions $\lim_{n\to\infty} P\{g_n(X) \neq Y\} = L^*$ (Wolverton and Wagner (1969a)). HINT: According to Problem 2.10, it suffices to show that if we are given a deterministic sequence of density functions $f, f_1, f_2, f_3, \ldots$, then

$$\lim_{n\to\infty} \int (f_n(x) - f(x))^2 \, dx = 0$$

implies

$$\lim_{n\to\infty} \int |f_n(x) - f(x)| dx = 0.$$

(A function f is called a density function if it is nonnegative and $\int f(x)dx = 1$.) To see this, observe that

$$\int |f_n(x) - f(x)| dx = 2 \int (f_n(x) - f(x))_+ \, dx = 2 \sum_i \int_{A_i} (f_n(x) - f(x))_+ \, dx,$$

where $A_1, A_2, \ldots$ is a partition of $\mathcal{R}^d$ into unit cubes, and f_+ denotes the positive part of a function f. The key observation is that convergence to zero of each term of the infinite sum implies convergence of the whole integral by the dominated convergence theorem, since $\int (f_n(x) - f(x))_+ \, dx \leq \int f_n(x)dx = 1$. Handle the right-hand side by the Cauchy-Schwarz inequality.

PROBLEM 2.12. Define the L_1 error of a function $f : \mathcal{R}^d \to \mathcal{R}$ by $J(f) = \mathbf{E}\{|f(X) - Y|\}$. Show that a function minimizing $J(f)$ is the Bayes rule g^*, that is, $J^* = \inf_f J(f) = J(g^*)$. Thus, $J^* = L^*$. Define a decision by

$$g(x) = \begin{cases} 0 & \text{if } f(x) \leq 1/2 \\ 1 & \text{otherwise,} \end{cases}$$

Prove that its error probability $L(g) = P\{g(X) \neq Y\}$ satisfies the inequality

$$L(g) - L^* \leq J(f) - J^*.$$

3
Inequalities and Alternate Distance Measures

3.1 Measuring Discriminatory Information

In our two-class discrimination problem, the best rule has (Bayes) probability of error

$$L^* = \mathbf{E}\left\{\min(\eta(X), 1 - \eta(X))\right\}.$$

This quantity measures how difficult the discrimination problem is. It also serves as a gauge of the quality of the distribution of (X, Y) for pattern recognition. Put differently, if ψ_1 and ψ_2 are certain many-to-one mappings, L^* may be used to compare discrimination based on $(\psi_1(X), Y)$ with that based on $(\psi_2(X), Y)$. When ψ_1 projects $\mathcal{R}^d$ to $\mathcal{R}^{d_1}$ by taking the first d_1 coordinates, and ψ_2 takes the last d_2 coordinates, the corresponding Bayes errors will help us decide which projection is better. In this sense, L^* is the fundamental quantity in feature extraction.

Other quantities have been suggested over the years that measure the discriminatory power hidden in the distribution of (X, Y). These may be helpful in some settings. For example, in theoretical studies or in certain proofs, the relationship between L^* and the distribution of (X, Y) may become clearer via certain inequalities that link L^* with other functionals of the distribution. We all understand moments and variances, but how do these simple functionals relate to L^*? Perhaps we may even learn a thing or two about what it is that makes L^* small. In feature selection, some explicit inequalities involving L^* may provide just the kind of numerical information that will allow one to make certain judgements on what kind of feature is preferable in practice. In short, we will obtain more information about L^* with a variety of uses in pattern recognition.

In the next few sections, we avoid putting any conditions on the distribution of (X, Y).

3.2   The Kolmogorov Variational Distance

Inspired by the total variation distance between distributions, the Kolmogorov variational distance

$$
\begin{aligned}
\delta_{KO} &= \frac{1}{2}\mathbf{E}\{|\mathbf{P}\{Y = 1|X\} - \mathbf{P}\{Y = 0|X\}|\} \\
&= \frac{1}{2}\mathbf{E}\{|2\eta(X) - 1|\}
\end{aligned}
$$

captures the distance between the two classes. We will not need anything special to deal with δ_{KO} as

$$
\begin{aligned}
L^* &= \mathbf{E}\left\{\frac{1}{2} - \frac{1}{2}|2\eta(X) - 1|\right\} \\
&= \frac{1}{2} - \frac{1}{2}\mathbf{E}\{|2\eta(X) - 1|\} \\
&= \frac{1}{2} - \delta_{KO}.
\end{aligned}
$$

3.3   The Nearest Neighbor Error

The asymptotic error of the nearest neighbor rule is

$$
L_{NN} = \mathbf{E}\{2\eta(X)(1 - \eta(X))\}
$$

(see Chapter 5). Clearly, $2\eta(1 - \eta) \geq \min(\eta, 1 - \eta)$ as $2\max(\eta, 1 - \eta) \geq 1$. Also, using the notation $A = \min(\eta(X), 1 - \eta(X))$, we have

$$
\begin{aligned}
L^* \leq L_{NN} &= 2\mathbf{E}\{A(1 - A)\} \qquad \overset{one}{\xrightarrow{monotone\ increasing}} \\
&\leq 2\mathbf{E}\{A\} \cdot \mathbf{E}\{1 - A\} \qquad \xrightarrow{the\ other\ decreasing}
\end{aligned}
$$

    (by the second association inequality of Theorem A.19)

$$
= 2L^*(1 - L^*) \leq 2L^*, \tag{3.1}
$$

which are well-known inequalities of Cover and Hart (1967). L_{NN} provides us with quite a bit of information about L^*.

The measure L_{NN} has been rediscovered under other guises: Devijver and Kittler (1982, p.263) and Vajda (1968) call it the quadratic entropy, and Mathai and Rathie (1975) refer to it as the harmonic mean coefficient.

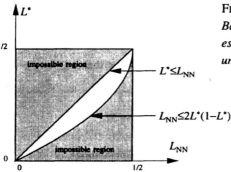

FIGURE 3.1. *Relationship between the Bayes error and the asymptotic nearest neighbor error. Every point in the unshaded region is possible.*

Special case of Chernoff Bound

3.4 The Bhattacharyya Affinity

The Bhattacharyya measure of affinity (Bhattacharyya, (1946)) is $-\log \rho$, where

$$\rho = \mathbf{E}\left\{\sqrt{\eta(X)(1 - \eta(X))}\right\}$$

will be referred to as the Matushita error. It does not occur naturally as the limit of any standard discrimination rule (see, however, Problem 6.11). ρ was suggested as a distance measure for pattern recognition by Matushita (1956). It also occurs under other guises in mathematical statistics—see, for example, the Hellinger distance literature (Le Cam (1970), Beran (1977)).

Clearly, $\rho = 0$ if and only if $\eta(X) \in \{0, 1\}$ with probability one, that is, if $L^* = 0$. Furthermore, ρ takes its maximal value $1/2$ if and only if $\eta(X) = 1/2$ with probability one. The relationship between ρ and L^* is not linear though. We will show that for all distributions, L_{NN} is more useful than ρ if it is to be used as an approximation of L^*.

Theorem 3.1. *For all distributions, we have*

$$\frac{1}{2} - \frac{1}{2}\sqrt{1 - 4\rho^2} \leq \frac{1}{2} - \frac{1}{2}\sqrt{1 - 2L_{NN}}$$
$$\leq L^*$$
$$\leq L_{NN}$$
$$\leq \rho.$$

PROOF. First of all,

$$\rho^2 = \mathbf{E}^2\left\{\sqrt{\eta(X)(1 - \eta(X))}\right\}$$
$$\leq \mathbf{E}\{\eta(X)(1 - \eta(X))\} \quad \text{(by Jensen's inequality)}$$

$$= \frac{L_{NN}}{2}$$

$$\leq L^*(1 - L^*) \quad \text{(by the Cover-Hart inequality (3.1))}.$$

Second, as $\sqrt{\eta(1 - \eta)} \geq 2\eta(1 - \eta)$ for all $\eta \in [0, 1]$, we see that $\rho \geq L_{NN} \geq L^*$. Finally, by the Cover-Hart inequality,

$$\sqrt{1 - 2L_{NN}} \geq \sqrt{1 - 4L^*(1 - L^*)} = 1 - 2L^*.$$

Putting all these things together establishes the chain of inequalities. □

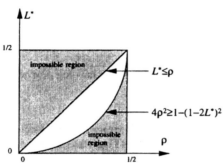

FIGURE 3.2. *The inequalities linking ρ to L^* are illustrated. Note that the region is larger than that cut out in the (L^*, L_{NN}) plane in Figure 3.1.*

The inequality $L_{NN} \leq \rho$ is due to Ito (1972). The inequality $L_{NN} \geq 2\rho^2$ is due to Horibe (1970). The inequality $1/2 - \sqrt{1 - 4\rho^2}/2 \leq L^* \leq \rho$ can be found in Kailath (1967). The left-hand side of the last inequality was shown by Hudimoto (1957). All these inequalities are tight (see Problem 3.2). The appeal of quantities like L_{NN} and ρ is that they involve polynomials of η, whereas $L^* = E\{\min(\eta(X), 1 - \eta(X))\}$ is nonpolynomial. For certain discrimination problems in which X has a distribution that is known up to certain parameters, one may be able to compute L_{NN} and ρ explicitly as a function of these parameters. Via inequalities, this may then be used to obtain performance guarantees for parametric discrimination rules of the plug-in type (see Chapter 16).

For completeness, we mention a generalization of Bhattacharyya's measure of affinity, first suggested by Chernoff (1952):

$$\delta_C = - \log \left(E \left\{ \eta^\alpha(X)(1 - \eta(X))^{1-\alpha} \right\} \right),$$

where $\alpha \in (0, 1)$ is fixed. For $\alpha = 1/2$, $\delta_C = - \log \rho$. The asymmetry introduced by taking $\alpha \neq 1/2$ has no practical interpretation, however.

3.5 Entropy

The *entropy* of a discrete probability distribution $(p_1, p_2, \ldots)$ is defined by

$$\mathcal{H} = \mathcal{H}(p_1, p_2, \ldots) = -\sum_{i=1}^{\infty} p_i \log p_i,$$

where, by definition, $0 \log 0 = 0$ (Shannon (1948)). The key quantity in information theory (see Cover and Thomas (1991)), it has countless applications in many branches of computer science, mathematical statistics, and physics. The entropy's main properties may be summarized as follows.

A. $\mathcal{H} \geq 0$ with equality if and only if $p_i = 1$ for some i. Proof: $\log p_i \leq 0$ for all i with equality if and only if $p_i = 1$ for some i. Thus, entropy is minimal for a degenerate distribution, i.e., a distribution with the least amount of "spread."

B. $\mathcal{H}(p_1, \ldots, p_k) \leq \log k$ with equality if and only if $p_1 = p_2 = \cdots = p_k = 1/k$. In other words, the entropy is maximal when the distribution is maximally smeared out. Proof:

$$\mathcal{H}(p_1, \ldots, p_k) - \log k = \sum_{i=1}^{k} p_i \log\left(\frac{1}{k p_i}\right) \leq 0$$

by the inequality $\log x \leq x - 1, x > 0$.

C. For a Bernoulli distribution $(p, 1 - p)$, the binary entropy $\mathcal{H}(p, 1 - p) = -p \log p - (1 - p) \log(1 - p)$ is concave in p.

Assume that X is a discrete random variable that must be guessed by asking questions of the type "is $X \in A$?," for some sets A. Let N be the minimum expected number of questions required to determine X with certainty. It is well known that

$$\frac{\mathcal{H}}{\log 2} \leq N < \frac{\mathcal{H}}{\log 2} + 1$$

(e.g., Cover and Thomas (1991)). Thus, $\mathcal{H}$ not only measures how spread out the mass of X is, but also provides us with concrete computational bounds for certain algorithms. In the simple example above, $\mathcal{H}$ is in fact proportional to the expected computational time of the best algorithm.

We are not interested in information theory per se, but rather in its usefulness in pattern recognition. For our discussion, if we fix $X = x$, then Y is Bernoulli $(\eta(x))$. Hence, the *conditional entropy* of Y given $X = x$ is

$$\mathcal{H}(\eta(x), 1 - \eta(x)) = -\eta(x) \log \eta(x) - (1 - \eta(x)) \log(1 - \eta(x)).$$

It measures the amount of uncertainty or chaos in Y given $X = x$. As we know, it takes values between 0 (when $\eta(x) \in \{0, 1\}$) and $\log 2$ (when $\eta(x) = 1/2$), and is concave in $\eta(x)$. We define the *expected conditional entropy* by

$$\begin{aligned}
\mathcal{E} &= \mathbf{E}\{\mathcal{H}(\eta(X), 1 - \eta(X))\} \\
&= -\mathbf{E}\{\eta(X)\log\eta(X) + (1 - \eta(X))\log(1 - \eta(X))\}.
\end{aligned}$$

For brevity, we will refer to $\mathcal{E}$ as the entropy. As pointed out above, $\mathcal{E} = 0$ if and only if $\eta(X) \in \{0, 1\}$ with probability one. Thus, $\mathcal{E}$ and L^* are related to each other.

Theorem 3.2.

A. $\mathcal{E} \leq \mathcal{H}(L^*, 1 - L^*) = -L^*\log L^* - (1 - L^*)\log(1 - L^*)$ *(Fano's inequality; Fano (1952), see Cover and Thomas (1991, p. 39)).*
B. $\mathcal{E} \geq -\log(1 - L_{\mathrm{NN}}) \geq -\log(1 - L^*)$.
C. $\mathcal{E} \leq \log 2 - \frac{1}{2}(1 - 2L_{\mathrm{NN}}) \leq \log 2 - \frac{1}{2}(1 - 2L^*)^2$.

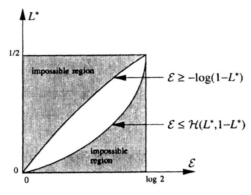

FIGURE 3.3. *Inequalities (A) and (B) of Theorem 3.2 are illustrated here.*

PROOF.
PART A.   Define $A = \min(\eta(X), 1 - \eta(X))$. Then

$$\begin{aligned}
\mathcal{E} &= \mathbf{E}\{\mathcal{H}(A, 1 - A)\} \\
&\leq \mathcal{H}(\mathbf{E}A, 1 - \mathbf{E}A) \\
&\qquad \text{(because } \mathcal{H} \text{ is concave, by Jensen's inequality)} \\
&= \mathcal{H}(L^*, 1 - L^*).
\end{aligned}$$

PART B.

$$\begin{aligned}
\mathcal{E} &= -\mathbf{E}\{A\log A + (1 - A)\log(1 - A)\} \\
&\geq -\mathbf{E}\{\log(A^2 + (1 - A)^2)\} \quad \text{(by Jensen's inequality)} \\
&= -\mathbf{E}\{\log(1 - 2A(1 - A))\} \\
&\geq -\log(1 - \mathbf{E}\{2A(1 - A)\}) \quad \text{(by Jensen's inequality)}
\end{aligned}$$

$$= \quad -\log(1 - L_{NN})$$

$$\geq \quad -\log(1 - L^*).$$

PART C. By the concavity of $\mathcal{H}(A, 1 - A)$ as a function of A, and Taylor series expansion,

$$\mathcal{H}(A, 1 - A) \leq \log 2 - \frac{1}{2}(2A - 1)^2.$$

Therefore, by Part A,

$$\mathcal{E} \quad \leq \quad \log 2 - \mathbf{E}\left\{\frac{1}{2}(2A - 1)^2\right\}$$

$$= \quad \log 2 - \frac{1}{2}(1 - 2L_{NN})$$

$$\leq \quad \log 2 - \frac{1}{2} + 2L^*(1 - L^*)$$

(by the Cover-Hart inequality (3.1))

$$= \quad \log 2 - \frac{1}{2}(1 - 2L^*)^2. \quad \square$$

REMARK. The nearly monotone relationship between $\mathcal{E}$ and L^* will see lots of uses. We warn the reader that near the origin, L^* may decrease linearly in $\mathcal{E}$, but it may also decrease much faster than $\mathcal{E}^{209}$. Such wide variation was not observed in the relationship between L^* and L_{NN} (where it is linear) or L^* and ρ (where it is between linear and quadratic). $\square$

3.6 Jeffreys' Divergence

Jeffreys' divergence (1948) is a symmetric form of the Kullback-Leibler (1951) divergence

$$\delta_{KL} = \mathbf{E}\left\{\eta(X) \log \frac{\eta(X)}{1 - \eta(X)}\right\}.$$

It will be denoted by

$$\mathcal{J} = \mathbf{E}\left\{(2\eta(X) - 1) \log \frac{\eta(X)}{1 - \eta(X)}\right\}.$$

To understand $\mathcal{J}$, note that the function $(2\eta - 1) \log \frac{\eta}{1 - \eta}$ is symmetric about $1/2$, convex, and has minimum (0) at $\eta = 1/2$. As $\eta \downarrow 0$, $\eta \uparrow 1$, the function becomes unbounded. Therefore, $\mathcal{J} = \infty$ if $\mathbf{P}\{\eta(X) \in \{0, 1\}\} > 0$. For this reason, its use in discrimination is necessarily limited. For generalizations of $\mathcal{J}$, see Rényi (1961), Burbea and Rao (1982), Taneja (1983; 1987), and Burbea (1984). It is thus impossible to bound $\mathcal{J}$ from above by a function of L_{NN} and/or L^*. However, lower bounds are easy to obtain. As $x \log((1 + x)/(1 - x))$ is convex in x and

$$(2\eta - 1) \log \frac{\eta}{1 - \eta} = |2\eta - 1| \log \left(\frac{1 + |2\eta - 1|}{1 - |2\eta - 1|} \right),$$

we note that by Jensen's inequality,

$$
\begin{aligned}
\mathcal{J} &\geq \mathbf{E}\{|2\eta(X) - 1|\} \log \left(\frac{1 + \mathbf{E}\{|2\eta(X) - 1|\}}{1 - \mathbf{E}\{|2\eta(X) - 1|\}} \right) \\
&= (1 - 2L^*) \log \left(\frac{1 + (1 - 2L^*)}{1 - (1 - 2L^*)} \right) \\
&= (1 - 2L^*) \log \left(\frac{1 - L^*}{L^*} \right) \\
&\geq 2(1 - 2L^*)^2.
\end{aligned}
$$

The first bound cannot be universally bettered (it is achieved when $\eta(x)$ is constant over the space). Also, for fixed L^*, any value of $\mathcal{J}$ above the lower bound is possible for some distribution of (X, Y). From the definition of $\mathcal{J}$, we see that $\mathcal{J} = 0$ if and only if $\eta \equiv 1/2$ with probability one, or $L^* = 1/2$.

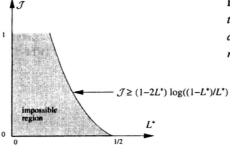

FIGURE 3.4. *This figure illustrates the above lower bound on Jeffreys' divergence in terms of the Bayes error.*

Related bounds were obtained by Toussaint (1974b):

$$\mathcal{J} \geq \sqrt{1 - 2L_{NN}} \, \log \left(\frac{1 + \sqrt{1 - 2L_{NN}}}{1 - \sqrt{1 - 2L_{NN}}} \right) \geq 2(1 - 2L_{NN}).$$

The last bound is strictly better than our L^* bound given above. See Problem 3.7.

3.7 F-Errors

The error measures discussed so far are all related to expected values of concave functions of $\eta(X) = \mathbf{P}\{Y = 1|X\}$. In general, if F is a concave function on $[0, 1]$, we define the *F-error* corresponding to (X, Y) by

$$d_F(X, Y) = \mathbf{E}\{F(\eta(X))\}.$$

Examples of *F*-errors are

(a) the Bayes error L^*: $F(x) = \min(x, 1 - x)$,

(b) the asymptotic nearest neighbor error L_{NN}: $F(x) = 2x(1 - x)$,

(c) the Matushita error ρ: $F(x) = \sqrt{x(1 - x)}$,

(d) the expected conditional entropy $\mathcal{E}$: $F(x) = -x \log x - (1 - x) \log(1 - x)$,

(e) the negated Jeffreys' divergence $-\mathcal{J}$: $F(x) = -(2x - 1) \log \frac{x}{1-x}$.

Hashlamoun, Varshney, and Samarasooriya (1994) point out that if $F(x) \geq \min(x, 1 - x)$ for each $x \in [0, 1]$, then the corresponding F-error is an upper bound on the Bayes error. The closer $F(x)$ is to $\min(x, 1 - x)$, the tighter the upper bound is. For example, $F(x) = (1/2) \sin(\pi x) \leq 2x(1 - x)$ yields an upper bound tighter than L_{NN}. All these errors share the property that the error increases if X is transformed by an arbitrary function.

Theorem 3.3. *Let* $t : \mathcal{R}^d \to \mathcal{R}^k$ *be an arbitrary measurable function. Then for any distribution of* (X, Y),

$$d_F(X, Y) \leq d_F(t(X), Y).$$

PROOF. Define $\eta_t : \mathcal{R}^k \to [0, 1]$ by $\eta_t(x) = \mathbf{P}\{Y = 1 | t(X) = x\}$, and observe that

$$\eta_t(t(X)) = \mathbf{E}\{\eta(X) | t(X)\}.$$

Thus,

$$
\begin{aligned}
d_F(t(X), Y) &= \mathbf{E}\{F(\eta_t(t(X)))\} \\
&= \mathbf{E}\{F(\mathbf{E}\{\eta(X) | t(X)\})\} \\
&\geq \mathbf{E}\{\mathbf{E}\{F(\eta(X)) | t(X)\}\} \quad \text{(by Jensen's inequality)} \\
&= \mathbf{E}\{F(\eta(X))\} = d_F(X, Y). \quad \square
\end{aligned}
$$

REMARK. We also see from the proof that the F-error remains unchanged if the transformation t is invertible. Theorem 3.3 states that F-errors are a bit like Bayes errors—when information is lost (by replacing X by $t(X)$), F-errors increase. $\square$

3.8   The Mahalanobis Distance

Two conditional distributions with about the same covariance matrices and means that are far away from each other are probably so well separated that L^* is small. An interesting measure of the visual distance between two random variables X_0 and X_1 is the so-called Mahalanobis distance (Mahalanobis, (1936)) given by

$$\Delta = \sqrt{(m_1 - m_0)^T \Sigma^{-1}(m_1 - m_0)},$$

where $m_1 = EX_1$, $m_0 = EX_0$, are the means, $\Sigma_1 = \mathbf{E}\left\{(X_1 - m_1)(X_1 - m_1)^T\right\}$ and $\Sigma_0 = \mathbf{E}\left\{(X_0 - m_0)(X_0 - m_0)^T\right\}$ are the covariance matrices, $\Sigma = p\Sigma_1 + (1 - p)\Sigma_0$, $(\cdot)^T$ is the transpose of a vector, and $p = 1 - p$ is a mixture parameter. If $\Sigma_1 = \Sigma_0 = \sigma^2 I$, where I is the identity matrix, then

$$\Delta = \frac{\|m_1 - m_0\|}{\sigma}$$

is a scaled version of the distance between the means. If $\Sigma_1 = \sigma_1^2 I$, $\Sigma_0 = \sigma_0^2 I$, then

$$\Delta = \frac{\|m_1 - m_0\|}{\sqrt{p\sigma_1^2 + (1 - p)\sigma_0^2}}$$

varies between $\|m_1 - m_0\|/\sigma_1$ and $\|m_1 - m_0\|/\sigma_0$ as p changes from 1 to 0. Assume that we have a discrimination problem in which given $Y = 1$, X is distributed as X_1, given $Y = 0$, X is distributed as X_0, and $p = \mathbf{P}\{Y = 1\}$, $1 - p$ are the class probabilities. Then, interestingly, Δ is related to the Bayes error in a general sense. If the Mahalanobis distance between the class-conditional distributions is large, then L^* is small.

Theorem 3.4. (DEVIJVER AND KITTLER (1982, p. 166)). *For all distributions of* (X, Y) *for which* $\mathbf{E}\left\{\|X\|^2\right\} < \infty$, *we have*

$$L^* \leq L_{\text{NN}} \leq \frac{2p(1 - p)}{1 + p(1 - p)\Delta^2}.$$

REMARK. For a distribution with mean m and covariance matrix Σ, the Mahalanobis distance from a point $x \in \mathcal{R}^d$ to m is

$$\sqrt{(x - m)^T \Sigma^{-1}(x - m)}.$$

In one dimension, this is simply interpreted as distance from the mean as measured in units of standard deviation. The use of Mahalanobis distance in discrimination is based upon the intuitive notion that we should classify according to the class for which we are within the least units of standard deviations. At least, for distributions that look like nice globular clouds, such a recommendation may make sense. □

PROOF. First assume that $d = 1$, that is, X is real valued. Let u and c be real numbers, and consider the quantity $\mathbf{E}\left\{(u(X - c) - (2\eta(X) - 1))^2\right\}$. We will show that if u and c are chosen to minimize this number, then it satisfies

$$0 \le \mathbf{E}\left\{(u(X - c) - (2\eta(X) - 1))^2\right\} = 2\left(\frac{2p(1 - p)}{1 + p(1 - p)\Delta^2} - L_{\text{NN}}\right), \quad (3.2)$$

which proves the theorem for $d = 1$. To see this, note that the expression $\mathbf{E}\left\{(u(X - c) - (2\eta(X) - 1))^2\right\}$ is minimized for $c = \mathbf{E}X - \mathbf{E}\{2\eta(X) - 1\}/u$. Then

$$\mathbf{E}\left\{(u(X - c) - (2\eta(X) - 1))^2\right\}$$
$$= \mathbf{Var}\{2\eta(X) - 1\} + u^2\,\mathbf{Var}\{X\} - 2u\,\mathbf{Cov}\{X, 2\eta(X) - 1\},$$

—where $\mathbf{Cov}\{X, Z\} = \mathbf{E}\{(X - \mathbf{E}X)(Z - \mathbf{E}Z)\}$—which is, in turn, minimized for $u = \mathbf{Cov}\{X, 2\eta(X) - 1\}/\mathbf{Var}\{X\}$. Straightforward calculation shows that (3.2) indeed holds.

To extend the inequality (3.2) to multidimensional problems, apply it to the one-dimensional decision problem (Z, Y), where $Z = X^T\Sigma^{-1}(m_1 - m_0)$. Then the theorem follows by noting that by Theorem 3.3,

$$L_{\text{NN}}(X, Y) \le L_{\text{NN}}(Z, Y),$$

where $L_{\text{NN}}(X, Y)$ denotes the nearest-neighbor error corresponding to (X, Y). $\square$

In case X_1 and X_0 are both normal with the same covariance matrices, we have

Theorem 3.5. (MATUSHITA (1973); SEE PROBLEM 3.11). *When X_1 and X_0 are multivariate normal random variables with $\Sigma_1 = \Sigma_0 = \Sigma$, then*

$$\rho = \mathbf{E}\left\{\sqrt{\eta(X)(1 - \eta(X))}\right\} = \sqrt{p(1 - p)}\,e^{-\Delta^2/8}.$$

If the class-conditional densities f_1 and f_0 may be written as functions of $(x - m_1)^T\Sigma_1^{-1}(x - m_1)$ and $(x - m_0)^T\Sigma_0^{-1}(x - m_0)$ respectively, then Δ remains relatively tightly linked with L^* (Mitchell and Krzanowski (1985)), but such distributions are the exception rather than the rule. In general, when Δ is small, it is impossible to deduce whether L^* is small or not (see Problem 3.12).

3.9   *f*-Divergences

We have defined error measures as the expected value of a concave function of $\eta(X)$. This makes it easier to relate these measures to the Bayes error L^* and other error probabilities. In this section we briefly make the connection to the more classical statistical theory of distances between probability measures. A general concept of these distance measures, called *f*-divergences, was introduced by Csiszár (1967).

The corresponding theory is summarized in Vajda (1989). F-errors defined earlier may be calculated if one knows the class probabilities p, $1 - p$, and the conditional distributions μ_0, μ_1 of X, given $\{Y = 0\}$ and $\{Y = 1\}$, that is,

$$\mu_i(A) = \mathbf{P}\{X \in A | Y = i\} \quad i = 0, 1.$$

For fixed class probabilities, an F-error is small if the two conditional distributions are "far away" from each other. A metric quantifying this distance may be defined as follows. Let $f : [0, \infty) \rightarrow \mathcal{R} \cup \{-\infty, \infty\}$ be a *convex* function with $f(1) = 0$. The f-*divergence* between two probability measures μ and ν on $\mathcal{R}^d$ is defined by

$$D_f(\mu, \nu) = \sup_{A=\{A_j\}} \sum_j \nu(A_j) f\left(\frac{\mu(A_j)}{\nu(A_j)}\right),$$

where the supremum is taken over all finite measurable partitions $\mathcal{A}$ of $\mathcal{R}^d$. If λ is a measure dominating μ and ν—that is, both μ and ν are absolutely continuous with respect to λ—and $p = d\mu/d\lambda$ and $q = d\nu/d\lambda$ are the corresponding densities, then the f-divergence may be put in the form

$$D_f(\mu, \nu) = \int q(x) f\left(\frac{p(x)}{q(x)}\right) \lambda(dx).$$

Clearly, this quantity is independent of the choice of λ. For example, we may take $\lambda = \mu + \nu$. If μ and ν are absolutely continuous with respect to the Lebesgue measure, then λ may be chosen to be the Lebesgue measure. By Jensen's inequality, $D_f(\mu, \nu) \geq 0$, and $D_f(\mu, \mu) = 0$.

An important example of f-divergences is the *total variation*, or variational distance obtained by choosing $f(x) = |x - 1|$, yielding

$$V(\mu, \nu) = \sup_{A=\{A_j\}} \sum_j |\mu(A_j) - \nu(A_j)|.$$

For this divergence, the equivalence of the two definitions is stated by Scheffé's theorem (see Problem 12.13).

Theorem 3.6. (Scheffé (1947)).

$$V(\mu, \nu) = 2 \sup_A |\mu(A) - \nu(A)| = \int |p(x) - q(x)| \lambda(dx),$$

where the supremum is taken over all Borel subsets of $\mathcal{R}^d$.

Another important example is the *Hellinger distance*, given by $f(x) = \left(1 - \sqrt{x}\right)^2$:

$$\begin{aligned}
H_2(\mu, \nu) &= \sup_{A=\{A_j\}} 2\left(1 - \sum_j \sqrt{\mu(A_j)\nu(A_j)}\right) \\
&= 2\left(1 - \int \sqrt{p(x)q(x)} \lambda(dx)\right).
\end{aligned}$$

The quantity $I_2(\mu, \nu) = \int \sqrt{p(x)q(x)}\lambda(dx)$ is often called the *Hellinger integral*. We mention two useful inequalities in this respect. For the sake of simplicity, we state their discrete form. (The integral forms are analogous, see Problem 3.21.)

Lemma 3.1. (LeCam (1973)). *For positive sequences a_i and b_i, both summing to one,*

$$\sum_i \min(a_i, b_i) \geq \frac{1}{2} \left(\sum_i \sqrt{a_i b_i} \right)^2.$$

PROOF. By the Cauchy-Schwarz inequality,

$$\sum_{i:a_i < b_i} \sqrt{a_i b_i} \leq \sqrt{\sum_{i:a_i < b_i} a_i} \sqrt{\sum_{i:a_i < b_i} b_i} \leq \sqrt{\sum_{i:a_i < b_i} a_i}.$$

This, together with the inequality $(x + y)^2 \leq 2x^2 + 2y^2$, and symmetry, implies

$$\left(\sum_i \sqrt{a_i b_i} \right)^2 = \left(\sum_{i:a_i < b_i} \sqrt{a_i b_i} + \sum_{i:a_i \geq b_i} \sqrt{a_i b_i} \right)^2$$

$$\leq 2 \left(\sum_{i:a_i < b_i} \sqrt{a_i b_i} \right)^2 + 2 \left(\sum_{i:a_i \geq b_i} \sqrt{a_i b_i} \right)^2$$

$$\leq 2 \left(\sum_{i:a_i < b_i} a_i + \sum_{i:a_i \geq b_i} b_i \right)$$

$$= 2 \sum_i \min(a_i, b_i). \quad \square$$

Lemma 3.2. (Devroye and Györfi (1985, p. 225)). *Let $a_1, \ldots, a_k, b_1, \ldots b_k$ be nonnegative numbers such that $\sum_{i=1}^k a_i = \sum_{i=1}^k b_i = 1$. Then*

$$\sum_{i=1}^k \sqrt{a_i b_i} \leq 1 - \frac{\left(\sum_{i=1}^k |a_i - b_i| \right)^2}{8}.$$

PROOF.

$$\left(\sum_{i=1}^k |a_i - b_i| \right)^2 \leq \left(\sum_{i=1}^k \left(\sqrt{a_i} - \sqrt{b_i} \right)^2 \right) \left(\sum_{i=1}^k \left(\sqrt{a_i} + \sqrt{b_i} \right)^2 \right)$$

(by the Cauchy-Schwarz inequality)

$$\leq 4 \sum_{i=1}^{k} \left(\sqrt{a_i} - \sqrt{b_i} \right)^2$$

$$= 8 \left(1 - \sum_{i=1}^{k} \sqrt{a_i b_i} \right),$$

which proves the lemma. $\square$

Information divergence is obtained by taking $f(x) = x \log x$:

$$I(\mu, \nu) = \sup_{\mathcal{A}=\{A_j\}} \sum_j \mu(A_j) \log \left(\frac{\mu(A_j)}{\nu(A_j)} \right).$$

$I(\mu, \nu)$ is also called the *Kullback-Leibler* number.

Our last example is the χ^2-divergence, defined by $f(x) = (x - 1)^2$:

$$\chi^2(\mu, \nu) = \sup_{\mathcal{A}=\{A_j\}} \sum_j \frac{(\mu(A_j) - \nu(A_j))^2}{\nu(A_j)}$$

$$= \int \frac{p^2(x)}{q(x)} \lambda(dx) - 1.$$

Next, we highlight the connection between F-errors and f-divergences. Let μ_0 and μ_1 denote the conditional distributions of X given $\{Y = 0\}$ and $\{Y = 1\}$. Assume that the class probabilities are equal: $p = 1/2$. If F is a concave function, then the F-error $d_F(X, Y)$ may be written as

$$d_F(X, Y) = F \left(\frac{1}{2} \right) - D_f(\mu_0, \mu_1),$$

where

$$f(x) = -\frac{1}{2} F \left(\frac{x}{1 + x} \right) (1 + x) + F \left(\frac{1}{2} \right),$$

and D_f is the corresponding f-divergence. It is easy to see that f is convex, whenever F is concave. A special case of this correspondence is

$$L^* = \frac{1}{2} \left(1 - \frac{1}{2} V(\mu_0, \mu_1) \right),$$

if $p = 1/2$. Also, it is easy to verify, that

$$\rho = \sqrt{p(1 - p)} \, I_2(\mu_0, \mu_1),$$

where ρ is the Matushita error. For further connections, we refer the reader to the exercises.

Problems and Exercises

PROBLEM 3.1. Show that for every (l_1, l^*) with $0 \leq l^* \leq l_1 \leq 2l^*(1 - l^*) \leq 1/2$, there exists a distribution of (X, Y) with $L_{NN} = l_1$ and $L^* = l^*$. Therefore, the Cover-Hart inequalities are not universally improvable.

PROBLEM 3.2. TIGHTNESS OF THE BOUNDS. Theorem 3.1 cannot be improved.
  (1) Show that for all $\alpha \in [0, 1/2]$, there exists a distribution of (X, Y) such that $L_{NN} = L^* = \alpha$.
  (2) Show that for all $\alpha \in [0, 1/2]$, there exists a distribution of (X, Y) such that $L_{NN} = \alpha$, $L^* = \frac{1}{3} - \frac{1}{2}\sqrt{1 - 2\alpha}$.
  (3) Show that for all $\alpha \in [0, 1/2]$, there exists a distribution of (X, Y) such that $L^* = \rho = \alpha$.
  (4) Show that for all $\alpha \in [0, 1/2]$, there exists a distribution of (X, Y) such that $L_{NN} = \alpha$, $L^* = \frac{1}{2} - \frac{1}{2}\sqrt{1 - 4\rho^2}$.

PROBLEM 3.3. Show that $\mathcal{E} \geq L^*$.

PROBLEM 3.4. For any $\alpha \leq 1$, find a sequence of distributions of (X_n, Y_n) having expected conditional entropies $\mathcal{E}_n$ and Bayes errors L_n^* such that $L_n^* \to 0$ as $n \to \infty$, and $\mathcal{E}_n$ decreases to zero at the same rate as $(L_n^*)^\alpha$.

PROBLEM 3.5. CONCAVITY OF ERROR MEASURES. Let Y denote the mixture random variable taking the value Y_1 with probability p and the value Y_2 with probability $1 - p$. Let X be a fixed $\mathcal{R}^d$-valued random variable, and define $\eta_1(x) = P\{Y_1 = 1 | X = x\}$, $\eta_2(x) = P\{Y_2 = 1 | X = x\}$, where Y_1, Y_2 are Bernoulli random variables. Clearly, $\eta(x) = p\eta_1(x) + (1 - p)\eta_2(x)$. Which of the error measures $L^*, \rho, L_{NN}, \mathcal{E}$ are concave in p for fixed joint distribution of X, Y_1, Y_2? Can every discrimination problem (X, Y) be decomposed this way for some Y_1, p, Y_2, where $\eta_1(x), \eta_2(x) \in \{0, 1\}$ for all x? If not, will the condition $\eta_1(x), \eta_2(x) \in \{0, 1/2, 1\}$ for all x do?

PROBLEM 3.6. Show that for every $l^* \in [0, 1/2]$, there exists a distribution of (X, Y) with $L^* = l^*$ and $\mathcal{E} = \mathcal{H}(L^*, 1 - L^*)$. Thus, Fano's inequality is tight.

PROBLEM 3.7. TOUSSAINT'S INEQUALITIES (1974B). Mimic a proof in the text to show that

$$\mathcal{J} \geq \sqrt{1 - 2L_{NN}} \log\left(\frac{1 + \sqrt{1 - 2L_{NN}}}{1 - \sqrt{1 - 2L_{NN}}}\right) \geq 2(1 - 2L_{NN}).$$

PROBLEM 3.8. Show that $L^* \leq e^{-\delta_C}$, where δ_C is Chernoff's measure of affinity with parameter $\alpha \in (0, 1)$.

PROBLEM 3.9. Prove that $L^* = p - E\{(2\eta(X) - 1)_+\}$, where $p = P\{Y = 1\}$, and $(x)_+ = \max(x, 0)$.

PROBLEM 3.10. Show that $\mathcal{J} \geq -2\log \rho - 2\mathcal{H}(p, 1 - p)$, where $p = P\{Y = 1\}$ (Toussaint (1974b)).

PROBLEM 3.11. Let f_1 and f_0 be two multivariate normal densities with means m_0, m_1 and common covariance matrix Σ. If $P\{Y = 1\} = p$, and f_1, f_0 are the conditional densities of

X given $Y = 1$ and $Y = 0$ respectively, then show that

$$\rho = \sqrt{p(1-p)}\, e^{-\Delta^2/8},$$

where ρ is the Matushita error and Δ is the Mahalanobis distance.

PROBLEM 3.12. For every $\delta \in [0, \infty)$ and $l^* \in [0, 1/2]$ with $l^* \le 2/(4 + \delta^2)$, find distributions μ_0 and μ_1 for X given $Y = 1$, $Y = 0$, such that the Mahalanobis distance $\Delta = \delta$, yet $L^* = l^*$. Therefore, the Mahalanobis distance is not universally related to the Bayes risk.

PROBLEM 3.13. Show that the Mahalanobis distance Δ is invariant under linear invertible transformations of X.

PROBLEM 3.14. Lissack and Fu (1976) have suggested the measures

$$\delta_{\mathrm{LF}} = \mathbf{E}\left\{|2\eta(X) - 1|^\alpha\right\}, \quad 0 < \alpha < \infty.$$

For $\alpha = 1$, this is twice the Kolmogorov distance δ_{KO}. Show the following:
  (1)   If $0 < \alpha \le 1$, then $\frac{1}{2}(1 - \delta_{\mathrm{LF}}) \le L^* \le (1 - \delta_{\mathrm{LF}}^{1/\alpha})$.
  (2)   If $1 \le \alpha < \infty$, then $\frac{1}{2}(1 - \delta_{\mathrm{LF}}^{1/\alpha}) \le L^* \le (1 - \delta_{\mathrm{LF}})$.

PROBLEM 3.15. Hashlamoun, Varshney, and Samarasooriya (1994) suggest using the F-error with the function

$$F(x) = \frac{1}{2}\sin(\pi x)e^{-1.8063\left(x-\frac{1}{2}\right)^2}$$

to obtain tight upper bounds on L^*. Show that $F(x) \ge \min(x, 1 - x)$, so that the corresponding F-error is indeed an upper bound on the Bayes risk.

PROBLEM 3.16. Prove that $L^* \le \max(p(1-p))\left(1 - \frac{1}{2}V(\mu_0, \mu_1)\right)$.

PROBLEM 3.17. Prove that $L^* \le \sqrt{p(1-p)}I_2(\mu_0\mu_1)$. HINT: $\min(a, b) \le \sqrt{ab}$.

PROBLEM 3.18. Assume that the components of $X = (X^{(1)}, \dots, X^{(d)})$ are conditionally independent (given Y), and identically distributed, that is, $\mathbf{P}\{X^{(i)} \in A|Y = j\} = \nu_j(A)$ for $i = 1, \dots, d$ and $j = 0, 1$. Use the previous exercise to show that

$$L^* \le \sqrt{p(1-p)}\,(I_2(\nu_0, \nu_1))^d.$$

PROBLEM 3.19. Show that $\chi^2(\mu_1, \mu_2) \ge I(\mu_1, \mu_2)$. HINT: $x - 1 \ge \log x$.

PROBLEM 3.20. Show the following analog of Theorem 3.3. Let $t : \mathcal{R}^d \to \mathcal{R}^k$ be a measurable function, and μ, ν probability measures on $\mathcal{R}^d$. Define the measures μ_t and ν_t on $\mathcal{R}^k$ by $\mu_t(A) = \mu(t^{-1}(A))$ and $\nu_t(A) = \nu(t^{-1}(A))$. Show that for any convex function f, $D_f(\mu, \nu) \ge D_f(\mu_t, \nu_t)$.

PROBLEM 3.21. Prove the following connections between the Hellinger integral and the total variation:

$$\left(1 - \frac{1}{2}V(\mu, \nu)\right) \ge \frac{1}{2}(I_2(\mu, \nu))^2,$$

and

$$(V(\mu, \nu))^2 \le 8(1 - I_2(\mu, \nu)).$$

HINT: Proceed analogously to Lemmas 3.1 and 3.2.

PROBLEM 3.22. PINSKER'S INEQUALITY. Show that

$$(V(\mu, \nu))^2 \leq 2I(\mu, \nu)$$

(Csiszár (1967), Kullback (1967), and Kemperman (1969)). HINT: First prove the inequality if μ and ν are concentrated on the same two atoms. Then define $A = \{x : p(x) \geq q(x)\}$, and the measures μ^*, ν^* on the set $\{0, 1\}$ by $\mu^*(0) = 1 - \mu^*(1) = \mu(A)$ and $\nu^*(0) = 1 - \nu^*(1) = \nu(A)$, and apply the previous result. Conclude by pointing out that Scheffé's theorem states $V(\mu^*, \nu^*) = V(\mu, \nu)$, and that $I(\mu^*, \nu^*) \leq I(\mu, \nu)$.

4

Linear Discrimination

In this chapter, we split the space by a hyperplane and assign a different class to each halfspace. Such rules offer tremendous advantages—they are easy to interpret as each decision is based upon the sign of $\sum_{i=1}^{d} a_i x^{(i)} + a_0$, where $x = (x^{(1)}, \ldots, x^{(d)})$ and the a_i's are weights. The weight vector determines the relative importance of the components. The decision is also easily implemented—in a standard software solution, the time of a decision is proportional to d—and the prospect that a small chip can be built to make a virtually instantaneous decision is particularly exciting.

Rosenblatt (1962) realized the tremendous potential of such linear rules and called them *perceptrons*. Changing one or more weights as new data arrive allows us to quickly and easily adapt the weights to new situations. Training or learning patterned after the human brain thus became a reality. This chapter merely looks at some theoretical properties of perceptrons. We begin with the simple one-dimensional situation, and deal with the choice of weights in $\mathcal{R}^d$ further on. Unless one is terribly lucky, linear discrimination rules cannot provide error probabilities close to the Bayes risk, but that should not diminish the value of this chapter. Linear discrimination is at the heart of nearly every successful pattern recognition method, including tree classifiers (Chapters 20 and 21), generalized linear classifiers (Chapter 17), and neural networks (Chapter 30). We also encounter for the first time rules in which the parameters (weights) are dependent upon the data.

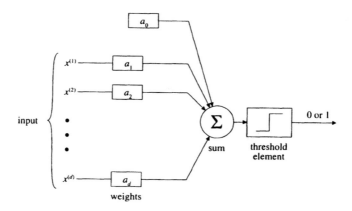

FIGURE 4.1. *Rosenblatt's perceptron. The decision is based upon a linear combination of the components of the input vector.*

4.1   Univariate Discrimination and Stoller Splits

As an introductory example, let X be univariate. The crudest and simplest possible rule is the linear discrimination rule

$$g(x) = \begin{cases} y' & \text{if } x \leq x' \\ 1 - y' & \text{otherwise,} \end{cases}$$

where x' is a split point and $y' \in \{0, 1\}$ is a class. In general, x' and y' are measurable functions of the data D_n. Within this class of simple rules, there is of course a best possible rule that can be determined if we know the distribution. Assume for example that (X, Y) is described in the standard manner: let $\mathbf{P}\{Y = 1\} = p$. Given $Y = 1$, X has a distribution function $F_1(x) = \mathbf{P}\{X \leq x | Y = 1\}$, and given $Y = 0$, X has a distribution function $F_0(x) = \mathbf{P}\{X \leq x | Y = 0\}$, where F_0 and F_1 are the *class-conditional distribution functions.* Then a theoretically optimal rule is determined by the split point x^* and class y^* given by

$$(x^*, y^*) = \arg\min_{(x',y')} \mathbf{P}\{g(X) \neq Y\}$$

(the minimum is always reached if we allow the values $x' = \infty$ and $x' = -\infty$). We call the corresponding minimal probability of error L and note that

$$L = \inf_{(x',y')} \left\{ I_{\{y'=0\}} \left(pF_1(x') + (1 - p)(1 - F_0(x')) \right) \right.$$

$$\left. + I_{\{y'=1\}} \left(p(1 - F_1(x')) + (1 - p)F_0(x') \right) \right\}.$$

A split defined by (x^*, y^*) will be called a theoretical Stoller split (Stoller (1954)).

Lemma 4.1. $L \le 1/2$ *with equality if and only if* $L^* = 1/2$.

PROOF. Take $(x', y') = (-\infty, 0)$. Then the probability of error is $1 - p = \mathbf{P}\{Y = 0\}$. Take $(x', y') = (-\infty, 1)$. Then the probability of error is p. Clearly,

$$L^* \le L \le \min(p, 1 - p).$$

This proves the first part of the lemma. For the second part, if $L = 1/2$, then $p = 1/2$, and for every x, $pF_1(x) + (1 - p)(1 - F_0(x)) \ge 1/2$ and $p(1 - F_1(x)) + (1 - p)F_0(x) \ge 1/2$. The first inequality implies $pF_1(x) - (1 - p)F_0(x) \ge p - 1/2$, while the second implies $pF_1(x) - (1 - p)F_0(x) \le p - 1/2$. Therefore, $L = 1/2$ means that for every x, $pF_1(x) - (1 - p)F_0(x) = p - 1/2$. Thus, for all x, $F_1(x) = F_0(x)$, and therefore $L^* = 1/2$. $\square$

Lemma 4.2.

$$L = \frac{1}{2} - \sup_x \left| pF_1(x) - (1 - p)F_0(x) - p + \frac{1}{2} \right|.$$

In particular, if $p = 1/2$, then

$$L = \frac{1}{2} - \frac{1}{2} \sup_x |F_1(x) - F_0(x)|.$$

PROOF. Set $\rho(x) = pF_1(x) - (1 - p)F_0(x)$. Then, by definition,

$$
\begin{aligned}
L &= \inf_x \min \{\rho(x) + 1 - p, p - \rho(x)\} \\
&= \frac{1}{2} - \sup_x \left| \rho(x) - p + \frac{1}{2} \right|
\end{aligned}
$$

(since $\min\{a, b\} = (a + b - |a - b|)/2$). $\square$

The last property relates the quality of theoretical Stoller splits to the Kolmogorov-Smirnov distance $\sup_x |F_1(x) - F_0(x)|$ between the class-conditional distribution functions. As a fun exercise, consider two classes with means $m_0 = \mathbf{E}\{X|Y = 0\}$, $m_1 = \mathbf{E}\{X|Y = 1\}$, and variances $\sigma_0^2 = \mathbf{Var}\{X|Y = 0\}$ and $\sigma_1^2 = \mathbf{Var}\{X|Y = 1\}$. Then the following inequality holds.

Theorem 4.1.

$$L^* \le L \le \frac{1}{1 + \frac{(m_0 - m_1)^2}{(\sigma_0 + \sigma_1)^2}}.$$

REMARK. When $p = 1/2$, Chernoff (1971) proved

$$L \le \frac{1}{2 + 2\frac{(m_0 - m_1)^2}{(\sigma_0 + \sigma_1)^2}}.$$

Moreover, Becker (1968) pointed out that this is the best possible bound (see Problem 4.2). □

PROOF. Assume without loss of generality that $m_0 < m_1$. Clearly, L is smaller than the probability of error for the rule that decides 0 when $x \leq m_0 + \Delta_0$, and 1 otherwise, where $m_1 - m_0 = \Delta_0 + \Delta_1$, $\Delta_0, \Delta_1 > 0$.

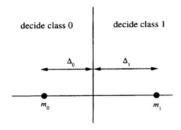

decide class 0   decide class 1

FIGURE 4.2. *The split providing the bound of Theorem 4.1.*

The probability of error of the latter rule is

$$p F_1(m_0 + \Delta_0) + (1 - p)(1 - F_0(m_0 + \Delta_0))$$

$$= \quad p\mathbf{P}\{X \leq m_1 - \Delta_1 | Y = 1\} + (1 - p)\mathbf{P}\{X > m_0 + \Delta_0 | Y = 0\}$$

$$\leq \quad p\frac{\sigma_1^2}{\sigma_1^2 + \Delta_1^2} + (1 - p)\frac{\sigma_0^2}{\sigma_0^2 + \Delta_0^2}$$

(by the Chebyshev-Cantelli inequality; see Appendix, Theorem A.17)

$$= \quad \frac{p}{1 + \frac{\Delta_1^2}{\sigma_1^2}} + \frac{1 - p}{1 + \frac{\Delta_0^2}{\sigma_0^2}}$$

(take $\Delta_1 = (\sigma_1/\sigma_0)\Delta_0$, and $\Delta_0 = |m_1 - m_0|\sigma_0/(\sigma_0 + \sigma_1)$) .

$$= \quad \frac{1}{1 + \frac{(m_0 - m_1)^2}{(\sigma_0 + \sigma_1)^2}} . \quad □$$

We have yet another example of the principle that well-separated classes yield small values for L and thus L^*. Separation is now measured in terms of the largeness of $|m_1 - m_0|$ with respect to $\sigma_0 + \sigma_1$. Another inequality in the same spirit is given in Problem 4.1.

The limitations of theoretical Stoller splits are best shown in a simple example. Consider a uniform $[0, 1]$ random variable X, and define

$$Y = \begin{cases} 1 & \text{if } 0 \leq X \leq \frac{1}{3} + \epsilon \\ 0 & \text{if } \frac{1}{3} + \epsilon < X \leq \frac{2}{3} - \epsilon \\ 1 & \text{if } \frac{2}{3} - \epsilon \leq X \leq 1 \end{cases}$$

for some small $\epsilon > 0$. As Y is a function of X, we have $L^* = 0$. If we are forced to make a trivial X-independent decision, then the best we can do is to set $g(x) \equiv 1$.

The probability of error is $P\{1/3 + \epsilon < X < 2/3 - \epsilon\} = 1/3 - 2\epsilon$. Consider next a theoretical Stoller split. One sees quickly that the best split occurs at $x' = 0$ or $x' = 1$, and thus that $L = 1/3 - 2\epsilon$. In other words, even the best theoretical split is superfluous. Note also that in the above example, $m_0 = m_1 = 1/2$ so that the inequality of Theorem 4.1 says $L \leq 1$—it degenerates.

We now consider what to do when a split must be data-based. Stoller (1954) suggests taking (x', y') such that the empirical error is minimal. He finds (x', y') such that

$$(x', y') = \underset{(x,y)\in\mathcal{R}\times\{0,1\}}{\arg\min} \frac{1}{n} \sum_{i=1}^{n} \left(I_{\{X_i \leq x, Y_i \neq y\}} + I_{\{X_i > x, Y_i \neq 1-y\}} \right).$$

(x' and y' are now random variables, but in spite of our convention, we keep the lowercase notation for now.) We will call this Stoller's rule. The split is referred to as an *empirical Stoller split*. Denote the set $\{((-\infty, x] \times \{y\}\} \cup \{(x, \infty) \times \{1 - y\}\}$ by $C(x, y)$. Then

$$(x', y') = \underset{(x,y)}{\arg\min}\, \nu_n(C(x, y)),$$

where ν_n is the empirical measure for the data $D_n = (X_1, Y_1), \ldots, (X_n, Y_n)$, that is, for every measurable set $A \in \mathcal{R} \times \{0, 1\}$, $\nu_n(A) = (1/n)\sum_{i=1}^{n} I_{\{(X_i, Y_i)\in A\}}$. Denoting the measure of (X, Y) in $\mathcal{R} \times \{0, 1\}$ by ν, it is clear that $E\{\nu_n(C)\} = \nu(C) = P\{X \leq x, Y \neq y\} + P\{X > x, Y \neq 1 - y\}$. Let $L_n = P\{g_n(X) \neq Y | D_n\}$ be the error probability of the splitting rule g_n with the data-dependent choice (x', y') given above, conditioned on the data. Then

$$
\begin{aligned}
L_n &= \nu(C(x', y')) \\
&= \nu(C(x', y')) - \nu_n(C(x', y')) + \nu_n(C(x', y')) \\
&\leq \sup_{(x,y)} (\nu(C(x, y)) - \nu_n(C(x, y))) + \nu_n(C(x^*, y^*)) \\
&\quad \text{(where } (x^*, y^*) \text{ minimizes } \nu(C(x, y)) \text{ over all } (x, y)) \\
&\leq 2 \sup_{(x,y)} |\nu(C(x, y)) - \nu_n(C(x, y))| + \nu(C(x^*, y^*)) \\
&= 2 \sup_{(x,y)} |\nu(C(x, y)) - \nu_n(C(x, y))| + L.
\end{aligned}
$$

From the next theorem we see that the supremum above is small even for moderately large n, and therefore, Stoller's rule performs closely to the best split *regardless of the distribution of (X, Y)*.

Theorem 4.2. *For Stoller's rule, and $\epsilon > 0$,*

$$P\{L_n - L \geq \epsilon\} \leq 4e^{-n\epsilon^2/2},$$

and

$$E\{L_n - L\} \leq \sqrt{\frac{2\log(4e)}{n}}.$$

PROOF. By the inequality given just above the theorem,

$$
\begin{aligned}
\mathbf{P}\{L_n - L \geq \epsilon\} & \leq & \mathbf{P}\left\{\sup_{(x,y)} |v(C(x,y)) - v_n(C(x,y))| \geq \frac{\epsilon}{2}\right\} \\
& \leq & \mathbf{P}\left\{\sup_x |v(C(x,0)) - v_n(C(x,0))| \geq \frac{\epsilon}{2}\right\} \\
& & + \mathbf{P}\left\{\sup_x |v(C(x,1)) - v_n(C(x,1))| \geq \frac{\epsilon}{2}\right\} \\
& \leq & 4e^{-2n(\epsilon/2)^2}
\end{aligned}
$$

by a double application of Massart's (1990) tightened version of the Dvoretzky-Kiefer-Wolfowitz inequality (1956) (Theorem 12.9). See Problem 4.5. We do not prove this inequality here, but we will thoroughly discuss several such inequalities in Chapter 12 in a greater generality. The second inequality follows from the first via Problem 12.1. □

The probability of error of Stoller's rule is uniformly close to L over all possible distributions. This is just a preview of things to come, as we may be able to obtain good performance guarantees within a limited class of rules.

4.2   Linear Discriminants

Rosenblatt's perceptron (Rosenblatt (1962); see Nilsson (1965) for a good discussion) is based upon a dichotomy of $\mathcal{R}^d$ into two parts by a hyperplane. The linear discrimination rule with weights $a_0, a_1, \ldots, a_d$ is given by

$$
g(x) = \begin{cases} 1 & \text{if } \sum_{i=1}^{d} a_i x^{(i)} + a_0 > 0 \\ 0 & \text{otherwise,} \end{cases}
$$

where $x = (x^{(1)}, \ldots, x^{(d)})$.

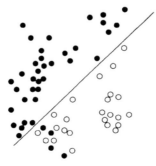

FIGURE 4.3. *A linear discriminant in $\mathcal{R}^2$ that correctly classifies all but four data points.*

Its probability of error is for now denoted by $L(a, a_0)$, where $a = (a_1, \ldots, a_d)$. Again, we set

$$L = \inf_{a \in \mathcal{R}^d, a_0 \in \mathcal{R}} L(a, a_0)$$

for the best possible probability of error within this class. Let the class-conditional distribution functions of $a_1 X^{(1)} + \cdots + a_d X^{(d)}$ be denoted by $F_{0,a}$ and $F_{1,a}$, depending upon whether $Y = 0$ or $Y = 1$. For $L(a, a_0)$, we may use the bounds of Lemma 4.2, and apply them to $F_{0,a}$ and $F_{1,a}$. Thus,

$$L = \frac{1}{2} - \sup_a \sup_x \left| p F_{1,a}(x) - (1 - p) F_{0,a}(x) - p + \frac{1}{2} \right|,$$

which, for $p = 1/2$, reduces to

$$L = \frac{1}{2} - \frac{1}{2} \sup_a \sup_x \left| F_{1,a}(x) - F_{0,a}(x) \right|.$$

Therefore, $L = 1/2$ if and only if $p = 1/2$ and for all a, $F_{1,a} \equiv F_{0,a}$. Then apply the following simple lemma.

Lemma 4.3. (CRAMÉR AND WOLD (1936)). X_1 and X_2, random variables taking values in $\mathcal{R}^d$, are identically distributed if and only if $a^T X_1$ and $a^T X_2$ have the same distribution for all vectors $a \in \mathcal{R}^d$.

PROOF. Two random variables have identical distributions if and only if they have the same characteristic function—see, for example, Lukács and Laha (1964). Now, the characteristic function of $X_1 = (X_1^{(1)}, \ldots, X_1^{(d)})$ is

$$
\begin{aligned}
\psi_1(a) &= \psi_1(a_1, \ldots, a_d) \\
&= \mathbf{E} \left\{ e^{i(a_1 X_1^{(1)} + \cdots + a_d X_1^{(d)})} \right\} \\
&= \mathbf{E} \left\{ e^{i(a_1 X_2^{(1)} + \cdots + a_d X_2^{(d)})} \right\} \quad \text{(by assumption)} \\
&= \psi_2(a_1, \ldots, a_d),
\end{aligned}
$$

the characteristic function of X_2. $\square$

Thus, we have proved the following:

Theorem 4.3. $L \leq 1/2$ with equality if and only if $L^* = 1/2$.

Thus, as in the one-dimensional case, whenever $L^* < 1/2$, a meaningful ($L < 1/2$) cut by a hyperplane is possible. There are also examples in which no cut improves over a rule in which $g(x) \equiv y$ for some y and all x, yet $L^* = 0$ and $L > 1/4$ (say). To generalize Theorem 4.1, we offer the following result. A related inequality is shown in Problem 4.7. The idea of using Chebyshev's inequality to obtain such bounds is due to Yau and Lin (1968) (see also Devijver and Kittler (1982, p.162)).

Theorem 4.4. *Let X_0 and X_1 be random variables distributed as X given $Y = 0$, and $Y = 1$ respectively. Set $m_0 = \mathbf{E}\{X_0\}$, $m_1 = \mathbf{E}\{X_1\}$. Define also the covariance matrices $\Sigma_1 = \mathbf{E}\left\{(X_1 - m_1)(X_1 - m_1)^T\right\}$ and $\Sigma_0 = \mathbf{E}\left\{(X_0 - m_0)(X_0 - m_0)^T\right\}$. Then*

$$L^* \le L \le \inf_{a \in \mathcal{R}^d} \frac{1}{1 + \dfrac{\left(a^T(m_1 - m_0)\right)^2}{\left((a^T \Sigma_0 a)^{1/2} + (a^T \Sigma_1 a)^{1/2}\right)^2}} \;.$$

Proof. For any $a \in \mathcal{R}^d$ we may apply Theorem 4.1 to $a^T X_0$ and $a^T X_1$. Theorem 4.4 follows by noting that

$$\mathbf{E}\left\{a^T X_0\right\} = a^T \mathbf{E}\{X_0\} = a^T m_0,$$
$$\mathbf{E}\left\{a^T X_1\right\} = a^T m_1,$$

and that

$$\mathbf{Var}\left\{a^T X_0\right\} = \mathbf{E}\left\{a^T(X_0 - m_0)(X_0 - m_0)^T a\right\} = a^T \Sigma_0 a,$$
$$\mathbf{Var}\left\{a^T X_1\right\} = a^T \Sigma_1 a. \quad \square$$

We may obtain explicit inequalities by different choices of a. $a = m_1 - m_0$ yields a convenient formula. We see from the next section that $a = \Sigma(m_1 - m_0)$ with $\Sigma = p\Sigma_1 + (1 - p)\Sigma_0$ is also a meaningful choice (see also Problem 4.7).

4.3   The Fisher Linear Discriminant

Data-based values for a may be found by various criteria. One of the first methods was suggested by Fisher (1936). Let $\widehat{m}_1$ and $\widehat{m}_0$ be the sample means for the two classes (e.g., $\widehat{m}_1 = \sum_{i:Y_i=1} X_i / |\{i : Y_i = 1\}|$.) Picture projecting $X_1, \ldots, X_n$ to a line in the direction of a. Note that this is perpendicular to the hyperplane given by $a^T x + a_0 = 0$. The projected values are $a^T X_1, \ldots, a^T X_n$. These are all equal to 0 for those X_i on the hyperplane $a^T x = 0$ through the origin, and grow in absolute value as we flee that hyperplane. Let $\widehat{\sigma}_1^2$ and $\widehat{\sigma}_0^2$ be the sample scatters for classes 1 and 0, respectively, that is,

$$\widehat{\sigma}_1^2 = \sum_{i:Y_i=1} \left(a^T X_i - a^T \widehat{m}_1\right)^2 = a^T S_1 a$$

and similarly for $\widehat{\sigma}_0^2$, where

$$S_1 = \sum_{i:Y_i=1} (X_i - \widehat{m}_1)(X_i - \widehat{m}_1)^T$$

is the scatter matrix for class 1.

The *Fisher linear discriminant* is that linear function $a^T x$ for which the criterion

$$J(a) = \frac{(a^T \widehat{m}_1 - a^T \widehat{m}_0)^2}{\widehat{\sigma}_1^2 + \widehat{\sigma}_0^2} = \frac{(a^T(\widehat{m}_1 - \widehat{m}_0))^2}{a^T(S_1 + S_0)a}$$

is maximum. This corresponds to finding a direction a that best separates $a^T \widehat{m}_1$ from $a^T \widehat{m}_0$ relative to the sample scatter. Luckily, to find that a, we need not resort to numerical iteration—the solution is given by

$$a = (S_1 + S_0)^{-1}(\widehat{m}_1 - \widehat{m}_0).$$

Fisher's suggestion is to replace $(X_1, Y_1), \dots, (X_n, Y_n)$ by $(a^T X_1, Y_1), \dots,$ $(a^T X_n, Y_n)$ and to perform one-dimensional discrimination. Usually, the rule uses a simple split

$$g_{a_0}(x) = \begin{cases} 1 & \text{if } a^T x + a_0 > 0 \\ 0 & \text{otherwise} \end{cases} \tag{4.1}$$

for some constant a_0. Unfortunately, Fisher discriminants can be arbitrarily bad: there are distributions such that even though the two classes are linearly separable (i.e., $L = 0$), the Fisher linear discriminant has an error probability close to 1 (see Problem 4.9).

4.4   The Normal Distribution

There are a few situations in which, by sheer accident, the Bayes rule is a linear discriminant. While this is not a major issue, it is interesting to identify the most important case, i.e., that of the multivariate normal distribution. The general multivariate normal density is written as

$$f(x) = \frac{1}{\sqrt{(2\pi)^d \det(\Sigma)}} e^{-\frac{1}{2}(x-m)^T \Sigma^{-1}(x-m)},$$

where m is the mean (both x and m are d-component column vectors), Σ is the $d \times d$ covariance matrix, Σ^{-1} is the inverse of Σ, and $\det(\Sigma)$ is its determinant. We write $f \sim N(m, \Sigma)$. Clearly, if X has density f, then $m = EX$ and $\Sigma = E\{(X - m)(X - m)^T\}$.

The multivariate normal density is completely specified by $d + \binom{d}{2}$ formal parameters (m and Σ). A sample from the density is clustered in an elliptical cloud. The loci of points of constant density are ellipsoids described by

$$(x - m)^T \Sigma^{-1}(x - m) = r^2$$

for some constant $r \geq 0$. The number r is the Mahalanobis distance from x to m, and is in fact useful even when the underlying distribution is not normal. It takes into account the directional stretch of the space determined by Σ.

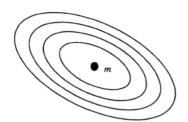

FIGURE 4.4. *Points at equal Mahalanobis distance from m.*

Given a two-class problem in which X has a density $(1 - p)f_0(x) + pf_1(x)$ and f_0 and f_1 are both multivariate normal with parameters m_i, Σ_i, $i = 0, 1$, the Bayes rule is easily described by

$$g^*(x) = \begin{cases} 1 & \text{if } pf_1(x) > (1 - p)f_0(x) \\ 0 & \text{otherwise.} \end{cases}$$

Take logarithms and note that $g^*(x) = 1$ if and only if

$$(x - m_1)^T \Sigma_1^{-1}(x - m_1) - 2\log p + \log(\det(\Sigma_1))$$
$$< \ (x - m_0)^T \Sigma_0^{-1}(x - m_0) - 2\log(1 - p) + \log(\det(\Sigma_0)).$$

In practice, one might wish to estimate $m_1, m_0, \Sigma_1, \Sigma_0$ and p from the data and use these estimates in the formula for g^*. Interestingly, as $(x - m_i)^T \Sigma_i^{-1}(x - m_i)$ is the squared Mahalanobis distance from x to m_i in class i (called r_i^2), the Bayes rule is simply

$$g^*(x) = \begin{cases} 1 & \text{if } r_1^2 < r_0^2 - 2\log((1 - p)/p) + \log(\det(\Sigma_0)/\det(\Sigma_1)) \\ 0 & \text{otherwise.} \end{cases}$$

In particular, when $p = 1/2$, $\Sigma_0 = \Sigma_1 = \Sigma$, we have

$$g^*(x) = \begin{cases} 1 & \text{if } r_1^2 < r_0^2 \\ 0 & \text{otherwise;} \end{cases}$$

just classify according to the class whose mean is at the nearest Mahalanobis distance from x. When $\Sigma_0 = \Sigma_1 = \Sigma$, the Bayes rule becomes linear:

$$g^*(x) = \begin{cases} 1 & \text{if } a^T x + a_0 > 0 \\ 0 & \text{otherwise,} \end{cases}$$

where $a = (m_1 - m_0)\Sigma^{-1}$, and $a_0 = 2\log(p/(1 - p)) + m_0^T \Sigma^{-1} m_0 - m_1^T \Sigma^{-1} m_1$. Thus, linear discrimination rules occur as special cases of Bayes rules for multivariate normal distributions.

Our intuition that a should be in the direction $m_1 - m_0$ to best separate the classes is almost right. Note nevertheless that a is *not* perpendicular in general to the hyperplane of loci at equal distance from m_0 and m_1. When Σ is replaced by the standard data-based estimate, we obtain in fact the Fisher linear discriminant. Furthermore, when $\Sigma_1 \neq \Sigma_0$, the decision boundary is usually not linear, and Fisher's linear discriminant must therefore be suboptimal.

HISTORICAL REMARKS.   In early statistical work on discrimination, the normal distribution plays a central role (Anderson (1958)). For a simple introduction, we refer to Duda and Hart (1973). McLachlan (1992) has more details, and Raudys (1972; 1976) relates the error, dimensionality, and sample size for normal and nearly normal models. See also Raudys and Pikelis (1980; 1982). □

4.5   Empirical Risk Minimization

In this section we present an algorithm that yields a classifier whose error probability is very close to the minimal error probability L achievable by linear classifiers, provided that X has a density. The algorithm selects a classifier by minimizing the empirical error over finitely many—$2\binom{n}{d}$—linear classifiers. For a rule

$$\phi(x) = \begin{cases} 1 & \text{if } a^T x + a_0 > 0 \\ 0 & \text{otherwise,} \end{cases}$$

the probability of error is

$$L(\phi) = \mathbf{P}\{\phi(X) \neq Y\}.$$

$L(\phi)$ may be estimated by the empirical risk

$$\widehat{L}_n(\phi) = \frac{1}{n} \sum_{i=1}^{n} I_{\{\phi(X_i) \neq Y_i\}},$$

that is, the number of errors made by the classifier ϕ is counted and normalized.

Assume that X has a density, and consider d arbitrary data points $X_{i_1}, X_{i_2}, \dots, X_{i_d}$ among $\{X_1, \dots, X_n\}$, and let $a^T x + a_0 = 0$ be a hyperplane containing these points. Because of the density assumption, the d points are in general position with probability one, and this hyperplane is unique. This hyperplane determines two classifiers:

$$\phi_1(x) = \begin{cases} 1 & \text{if } a^T x + a_0 > 0 \\ 0 & \text{otherwise,} \end{cases}$$

and

$$\phi_2(x) = \begin{cases} 1 & \text{if } a^T x + a_0 < 0 \\ 0 & \text{otherwise,} \end{cases}$$

whose empirical errors $\widehat{L}_n(\phi_1)$ and $\widehat{L}_n(\phi_2)$ may be calculated. To each d-tuple $X_{i_1}, X_{i_2}, \dots, X_{i_d}$ of data points, we may assign two classifiers in this manner, yielding altogether $2\binom{n}{d}$ classifiers. Denote these classifiers by $\phi_1, \dots, \phi_{2\binom{n}{d}}$. Let $\widehat{\phi}$ be a linear classifier that minimizes $\widehat{L}_n(\phi_i)$ over all $i = 1, \dots, 2\binom{n}{d}$.

We denote the best possible error probability by

$$L = \inf_{\phi} L(\phi)$$

over the class of all linear rules, and define $\phi^* = \arg\min_{\phi} L(\phi)$ as the best linear rule. If there are several classifiers with $L(\phi) = L$, then we choose ϕ^* among these in an arbitrary fixed manner. Next we show that the classifier corresponding to $\widehat{\phi}$ is really very good.

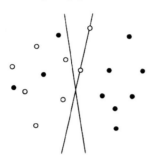

FIGURE 4.5. *If the data points are in general position, then for each linear rule there exists a linear split defined by a hyperplane crossing d points such that the difference between the empirical errors is at most d/n.*

First note that there is no linear classifier ϕ whose empirical error $\widehat{L}_n(\phi)$ is smaller than $\widehat{L}(\widehat{\phi}) - d/n$. This follows from the fact that since the data points are in general position (recall the density assumption), then for each linear classifier we may find one whose defining hyperplane contains exactly d data points such that the two decisions agree on all data points except possibly for these d points— see Figure 4.5. Thus, we may view minimization of the empirical error over the finite set $\{\phi_1, \ldots, \phi_{2\binom{n}{d}}\}$ as approximate minimization over the infinite set of linear classifiers. In Chapters 12 and 13 we will develop the full theory for rules that are found by empirical risk minimization. Theorem 4.5 just gives you a taste of things to come. Other—more involved, but also more general—proofs go back to Vapnik and Chervonenkis (1971; 1974c).

Theorem 4.5. *Assume that X has a density. If $\widehat{\phi}$ is found by empirical error minimization as described above, then, for all possible distributions of (X, Y), if $n \geq d$ and $2d/n \leq \epsilon \leq 1$, we have*

$$\mathbf{P}\left\{L(\widehat{\phi}) > L + \epsilon\right\} \leq e^{2d\epsilon}\left(2\binom{n}{d} + 1\right)e^{-n\epsilon^2/2}.$$

Moreover, if $n \geq d$, then

$$\mathbf{E}\left\{L(\widehat{\phi}) - L\right\} \leq \sqrt{\frac{2}{n}((d+1)\log n + (2d+2))}.$$

REMARK. With some care Theorem 4.5 and Theorem 4.6 below can be extended so that the density assumption may be dropped. One needs to ensure that the selected linear rule has empirical error close to that of the best possible linear rule. With the classifier suggested above this property may fail to hold if the data points are not necessarily of general position. The ideas presented here are generalized in Chapter 12 (see Theorem 12.2). $\square$

PROOF. We begin with the following simple inequality:

$$
\begin{aligned}
L(\widehat{\phi}) - L &= L(\widehat{\phi}) - \widehat{L}_n(\widehat{\phi}) + \widehat{L}_n(\widehat{\phi}) - L(\phi^*) \\
&\leq L(\widehat{\phi}) - \widehat{L}_n(\widehat{\phi}) + \widehat{L}_n(\phi^*) - L(\phi^*) + \frac{d}{n} \\
&\quad (\text{since } \widehat{L}_n(\widehat{\phi}) \leq \widehat{L}_n(\phi) + d/n \text{ for any } \phi) \\
&\leq \max_{i=1,\dots,2\binom{n}{d}} \left(L(\phi_i) - \widehat{L}_n(\phi_i) \right) + \widehat{L}_n(\phi^*) - L(\phi^*) + \frac{d}{n}.
\end{aligned}
$$

Therefore, by the union-of-events bound, we have

$$
\begin{aligned}
&\mathbf{P}\left\{ L(\widehat{\phi}) - L > \epsilon \right\} \\
&\leq \sum_{i=1}^{2\binom{n}{d}} \mathbf{P}\left\{ L(\phi_i) - \widehat{L}_n(\phi_i) > \frac{\epsilon}{2} \right\} + \mathbf{P}\left\{ \widehat{L}_n(\phi^*) - L(\phi^*) + \frac{d}{n} > \frac{\epsilon}{2} \right\}.
\end{aligned}
$$

To bound the second term on the right-hand side, observe that $n\widehat{L}_n(\phi^*)$ is binomially distributed with parameters n and $L(\phi^*)$. By an inequality due to Chernoff (1952) and Okamoto (1958) for the tail of the binomial distribution,

$$
\mathbf{P}\left\{ \widehat{L}_n(\phi^*) - L(\phi^*) > \frac{\epsilon}{2} - \frac{d}{n} \right\} \leq e^{-2n\left(\frac{\epsilon}{2} - \frac{d}{n}\right)^2} \leq e^{2d\epsilon} \cdot e^{-n\epsilon^2/2}.
$$

We prove this inequality later (see Theorem 8.1). Next we bound one term of the sum on the right-hand side. Note that by symmetry all $2\binom{n}{d}$ terms are equal. Assume that the classifier ϕ_1 is determined by the d-tuple of the first d data points $X_1, \dots, X_d$. We write

$$
\mathbf{P}\left\{ L(\phi_1) - \widehat{L}_n(\phi_1) > \frac{\epsilon}{2} \right\} = \mathbf{E}\left\{ \mathbf{P}\left\{ L(\phi_1) - \widehat{L}_n(\phi_1) > \frac{\epsilon}{2} \,\Big|\, X_1, \dots, X_d \right\} \right\},
$$

and bound the conditional probability inside. Let $(X_1'', Y_1''), \dots, (X_d'', Y_d'')$ be independent of the data and be distributed as the data $(X_1, Y_1), \dots, (X_d, Y_d)$. Define

$$
(X_i', Y_i') = \begin{cases} (X_i'', Y_i'') & \text{if } i \leq d \\ (X_i, Y_i) & \text{if } i > d. \end{cases}
$$

Then

$$
\begin{aligned}
&\mathbf{P}\left\{ L(\phi_1) - \widehat{L}_n(\phi_1) > \frac{\epsilon}{2} \,\Big|\, X_1, \dots, X_d \right\} \\
&\leq \mathbf{P}\left\{ L(\phi_1) - \frac{1}{n}\sum_{i=d+1}^{n} I_{\{\phi_1(X_i) \neq Y_i\}} > \frac{\epsilon}{2} \,\Big|\, X_1, \dots, X_d \right\}
\end{aligned}
$$

$$\leq \; \mathbf{P}\left\{ L(\phi_1) - \frac{1}{n}\sum_{i=1}^{n} I_{\{\phi_1(X_i')\neq Y_i'\}} + \frac{d}{n} > \frac{\epsilon}{2}\;\bigg|\; X_1, \ldots, X_d\right\}$$

$$= \; \mathbf{P}\left\{ L(\phi_1) - \frac{1}{n}\text{Binomial}(n, L(\phi_1)) > \frac{\epsilon}{2} - \frac{d}{n}\;\bigg|\; X_1, \ldots, X_d\right\}$$

(as $L(\phi_1)$ depends upon $X_1, \ldots, X_d$ only and

$(X_1', Y_1')\ldots, (X_d', Y_d')$ are independent of $X_1, \ldots, X_d$)

$$\leq \; e^{-2n\left(\frac{\epsilon}{2} - \frac{d}{n}\right)^2}$$

(by Theorem 8.1; use the fact that $\epsilon \geq 2d/n$)

$$\leq \; e^{-n\epsilon^2/2} e^{2d\epsilon}.$$

The inequality for the expected value follows from the probability inequality by the following simple argument: by the Cauchy-Schwarz inequality,

$$\left(\mathbf{E}\left\{L(\widehat{\phi}) - L\right\}\right)^2 \leq \mathbf{E}\left\{\left(L(\widehat{\phi}) - L\right)^2\right\}.$$

Denoting $Z = \left(L(\widehat{\phi}) - L\right)^2$, for any $u > 0$,

$$\begin{aligned}
\mathbf{E}\{Z\} \;&=\; \mathbf{E}\{Z|Z > u\}\mathbf{P}\{Z > u\} + \mathbf{E}\{Z|Z \leq u\}\mathbf{P}\{Z \leq u\} \\
&\leq\; \mathbf{P}\{Z > u\} + u \\
&\leq\; e^{2d}\left(2n^d + 1\right)e^{-nu/2} + u \quad \text{if } u \geq \left(\frac{2d}{n}\right)^2
\end{aligned}$$

by the probability inequality, and since $\binom{n}{d} \leq n^d$. Choosing u to minimize the obtained expression yields the desired inequality: first verify that the minimum occurs for

$$u = \frac{2}{n}\log\frac{nc}{2},$$

where $c = e^{2d}(2n^d + 1)$. Check that if $n \geq d$, then $u \geq (2d/n)^2$. Then note that the bound $ce^{-nu/2} + u$ equals

$$\frac{2}{n}\log\frac{nec}{2} \leq \frac{2}{n}\log\left(e^{2d+2}n^{d+1}\right) = \frac{2}{n}((d + 1)\log n + (2d + 2)). \quad \square$$

Observe for now that the bound on $\mathbf{P}\left\{L(\widehat{\phi}) > L + \epsilon\right\}$ decreases rapidly with n. To have an impact, it must become less than δ for small δ. This happens, roughly speaking, when

$$n \geq c \cdot \frac{d}{\epsilon^2}\left(\log\frac{d}{\epsilon^2} + \log\frac{1}{\delta}\right)$$

for some constant c. Doubling d, the dimension, causes this minimal sample size to roughly double as well.

An important special case is when the distribution is *linearly separable*, that is, $L = 0$. In such cases the empirical risk minimization above performs even better as the size of the error improves to $O(d \log n/n)$ from $O\left(\sqrt{d \log n/n}\right)$. Clearly, the data points are linearly separable as well, that is, $\widehat{L}_n(\phi^*) = 0$ with probability one, and therefore $\widehat{L}_n(\widehat{\phi}) \leq d/n$ with probability one.

Theorem 4.6. *Assume that X has a density, and that the best linear classifier has zero probability of error ($L = 0$). Then for the empirical risk minimization algorithm of Theorem 4.5, for all $n > d$ and $\epsilon \leq 1$,*

$$\mathbf{P}\left\{L(\widehat{\phi}) > \epsilon\right\} \leq 2\binom{n}{d} e^{-(n-d)\epsilon},$$

and

$$\mathbf{E}\left\{L(\widehat{\phi})\right\} \leq \frac{d \log n + 2}{n - d}.$$

PROOF. By the union bound,

$$
\begin{aligned}
\mathbf{P}\{L(\widehat{\phi}) > \epsilon\} &\leq \mathbf{P}\left\{\max_{i=1,2,\dots,2\binom{n}{d}:\widehat{L}_n(\phi_i)\leq\frac{d}{n}} L(\phi_i) > \epsilon\right\} \\
&\leq \sum_{i=1}^{2\binom{n}{d}} \mathbf{P}\left\{\widehat{L}_n(\phi_i) \leq \frac{d}{n}, L(\phi_i) > \epsilon\right\}.
\end{aligned}
$$

By symmetry, this sum equals

$$
\begin{aligned}
2\binom{n}{d}\mathbf{P}&\left\{\widehat{L}_n(\phi_1) \leq \frac{d}{n}, L(\phi_1) > \epsilon\right\} \\
&= 2\binom{n}{d}\mathbf{E}\left\{\mathbf{P}\left\{\widehat{L}_n(\phi_1) \leq \frac{d}{n}, L(\phi_1) > \epsilon \,\middle|\, X_1, \dots, X_d\right\}\right\},
\end{aligned}
$$

where, as in Theorem 4.5, ϕ_1 is determined by the data points $X_1, \dots, X_d$. However,

$$
\begin{aligned}
\mathbf{P}&\left\{\widehat{L}_n(\phi_1) \leq \frac{d}{n}, L(\phi_1) > \epsilon \,\middle|\, X_1, \dots, X_d\right\} \\
&\leq \mathbf{P}\{\phi_1(X_{d+1}) = Y_{d+1}, \dots, \phi_1(X_n) = Y_n, L(\phi_1) > \epsilon \,|\, X_1, \dots, X_d\} \\
&\qquad \text{(since all of the (at most } d\text{) errors committed by } \phi_1 \\
&\qquad \text{occur for } (X_1, Y_1), \dots, (X_d, Y_d)) \\
&\leq (1 - \epsilon)^{n-d},
\end{aligned}
$$

since the probability that no (X_i, Y_i), pair $i = d + 1, \dots, n$ falls in the set $\{(x, y) : \phi_1(x) \neq y\}$ is less than $(1 - \epsilon)^{n-d}$ if the probability of the set is larger than ϵ. The proof of the probability inequality may be completed by noting that $1 - x \leq e^{-x}$.

For the expected error probability, note that for any $u > 0$,

$$
\begin{aligned}
\mathbf{E}\{L(\widehat{\phi})\} &= \int_0^\infty \mathbf{P}\{L(\widehat{\phi}) > t\}dt \\
&\leq u + \int_u^\infty \mathbf{P}\{L(\widehat{\phi}) > t\}dt \\
&\leq u + 2n^d \int_u^\infty e^{-(n-d)t}dt
\end{aligned}
$$

(by the probability inequality and $\binom{n}{d} \leq n^d$)

$$
= u + \frac{2n^d}{n}e^{-(n-d)u}.
$$

We choose u to minimize the obtained bound, which yields the desired inequality.
□

4.6   Minimizing Other Criteria

Empirical risk minimization uses extensive computations, because $\widehat{L}_n(\phi)$ is not a unimodal function in general (see Problems 4.10 and 4.11). Also, gradient optimization is difficult because the gradients are zero almost everywhere. In fact, given n labeled points in $\mathcal{R}^d$, finding the best linear dichotomy is NP hard (see Johnson and Preparata (1978)). To aid in the optimization, some have suggested minimizing a modified empirical error, such as

$$
\widehat{L}_n(\phi) = \frac{1}{n}\sum_{i=1}^n \Psi\left((2Y_i - 1) - a^T X_i - a_0\right) I_{\{Y_i \neq g_a(X_i)\}}
$$

or

$$
\widehat{L}_n(\phi) = \frac{1}{n}\sum_{i=1}^n \Psi\left((2Y_i - 1) - a^T X_i - a_0\right),
$$

where Ψ is a positive convex function. Of particular importance here is the mean square error criterion $\Psi(u) = u^2$ (see, e.g., Widrow and Hoff (1960)). One can easily verify that $\widehat{L}_n(\phi)$ has a gradient (with respect to (a, a_0)) that may aid in locating a local minimum. Let $\widehat{\phi}$ denote the linear discrimination rule minimizing

$$
\mathbf{E}\left\{\left((2Y - 1) - a^T X - a_0\right)^2\right\}
$$

over all a and a_0. A description of the solution is given in Problem 4.14.

Even in a one-dimensional situation, the mean square error criterion muddles the issue and does not give any performance guarantees:

Theorem 4.7. *If* $\sup_{(X,Y)}$ *denotes the supremum with respect to all distributions on* $\mathcal{R} \times \{0, 1\}$, *then*

$$\sup_{(X,Y)} \left(L(\widehat{\phi}) - L \right) = 1,$$

where $\widehat{\phi}$ *is a linear discriminant obtained by minimizing*

$$\mathbf{E}\left\{ ((2Y - 1) - a_1 X - a_0)^2 \right\}$$

over all a_1 *and* a_0.

REMARK. This theorem establishes the existence of distributions of (X, Y) for which $L(\widehat{\phi}) > 1 - \epsilon$ and $L < \epsilon$ simultaneously for arbitrarily small $\epsilon > 0$. Therefore, minimizing the mean square error criterion is not recommended unless one has additional information regarding the distribution of (X, Y). $\square$

PROOF. Let $\epsilon > 0$ and $\theta > 0$. Consider a triatomic distribution of (X, Y):

$$\mathbf{P}\{(X, Y) = (-\theta, 1)\} = \mathbf{P}\{(X, Y) = (1, 1)\} = \epsilon/2,$$

$$\mathbf{P}\{(X, Y) = (0, 0)\} = 1 - \epsilon.$$

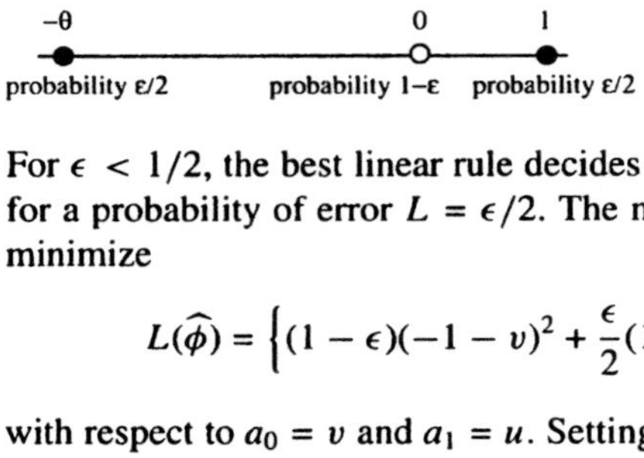

FIGURE 4.6. *A distribution for which squared error minimization fails.*

For $\epsilon < 1/2$, the best linear rule decides class 0 on $[-\theta/2, \infty)$ and 1 elsewhere, for a probability of error $L = \epsilon/2$. The mean square error criterion asks that we minimize

$$L(\widehat{\phi}) = \left\{ (1 - \epsilon)(-1 - v)^2 + \frac{\epsilon}{2}(1 - u - v)^2 + \frac{\epsilon}{2}(1 + u\theta - v)^2 \right\}$$

with respect to $a_0 = v$ and $a_1 = u$. Setting the derivatives with respect to u and v equal to zero yields

$$u = \frac{(v - 1)\theta - v}{1 + \theta^2}, \quad \text{and} \quad v = 2\epsilon - 1 + \frac{\epsilon}{2}u(\theta - 1),$$

for

$$v = \frac{(2\epsilon - 1)(1 + \theta^2) - \frac{\epsilon}{2}\theta(\theta - 1)}{1 + \theta^2 - \frac{\epsilon}{2}(1 - \theta)^2}.$$

If we let $\epsilon \downarrow 0$ and let $\theta \uparrow \infty$, then $v \sim 3\epsilon/2$. Thus, for ϵ small enough and θ large enough, considering the decision at 0 only, $L(\widehat{\phi}) \geq 1 - \epsilon$, because at $x = 0$, $ux + v = v > 0$. Thus, $L(\widehat{\phi}) - L \geq 1 - 3\epsilon/2$ for ϵ small enough and θ large enough. $\square$

Others have suggested minimizing

$$\sum_{i=1}^{n} \left(\sigma(a^T X_i + a_0) - Y_i \right)^2,$$

where $\sigma(u)$ is a *sigmoid*, that is, an increasing function from 0 to 1 such as $1/(1 + e^{-u})$, see, for example, Wassel and Sklansky (1972), Do Tu and Installe (1975), Fritz and Györfi (1976), and Sklansky and Wassel (1979). Clearly, $\sigma(u) = I_{\{u \geq 0\}}$ provides the empirical error probability. However, the point here is to use smooth sigmoids so that gradient algorithms may be used to find the optimum. This may be viewed as a compromise between the mean squared error criteria and empirical error minimization. Here, too, anomalies can occur, and the error space is not well behaved, displaying many local minima (Hertz, Krogh, and Palmer (1991, p.108)). See, however, Problems 4.16 and 4.17.

Problems and Exercises

PROBLEM 4.1. With the notation of Theorem 4.1, show that the error probability L of a one-dimensional theoretical Stoller split satisfies

$$L \leq \frac{4p(1 - p)}{1 + p(1 - p)\frac{(m_0 - m_1)^2}{(1-p)\sigma_0^2 + p\sigma_1^2}}$$

(Györfi and Vajda (1980)). Is this bound better than that of Theorem 4.1? HINT: For any threshold rule $g_c(x) = I_{\{x \geq c\}}$ and $u > 0$, write

$$
\begin{aligned}
L(g_c) &= \mathbf{P}\{X - c \geq 0, 2Y - 1 = -1\} + \mathbf{P}\{X - c < 0, 2Y - 1 = 1\} \\
&\leq \mathbf{P}\{|u(X - c) - (2Y - 1)| \geq 1\} \\
&\leq \mathbf{E}\left\{(u(X - c) - (2Y - 1))^2\right\}
\end{aligned}
$$

by Chebyshev's inequality. Choose u and c to minimize the upper bound.

PROBLEM 4.2. Let $p = 1/2$. If L is the error probability of the one-dimensional theoretical Stoller split, show that

$$L \leq \frac{1}{2 + 2\frac{(m_0 - m_1)^2}{(\sigma_0 + \sigma_1)^2}}.$$

Show that the bound is achieved for some distribution when the class-conditional distributions of X (that is, given $Y = 0$ and $Y = 1$) are concentrated on two points each, one of which is shared by both classes (Chernoff (1971), Becker (1968)).

PROBLEM 4.3. Let X be a univariate random variable. The distribution functions for X given $Y = 1$ and $Y = 0$ are F_1 and F_0 respectively. Assume that the moment generating functions for X exist, that is, $\mathbf{E}\left\{e^{tX}|Y = 1\right\} = \psi_1(t)$, $\mathbf{E}\left\{e^{tX}|Y = 0\right\} = \psi_0(t)$, $t \in \mathcal{R}$, where ψ_1, ψ_0 are finite for all t. In the spirit of Theorem 4.1, derive an upper bound for L in function of ψ_1, ψ_0. Apply your bound to the case that F_1 and F_0 are both normal with possibly different means and variances.

PROBLEM 4.4. SIGNALS IN ADDITIVE GAUSSIAN NOISE. Let $s_0, s_1 \in \mathcal{R}^d$ be fixed, and let N be a multivariate gaussian random variable with zero mean and covariance matrix Σ. Let $P\{Y = 0\} = P\{Y = 1\} = 1/2$, and define

$$X = \begin{cases} s_0 + N & \text{if } Y = 0 \\ s_1 + N & \text{if } Y = 1. \end{cases}$$

Construct the Bayes decision and calculate L^*. Prove that if Σ is the identity matrix, and s_0 and s_1 have constant components, then $L^* \to 0$ exponentially rapidly as $d \to \infty$.

PROBLEM 4.5. In the last step of the proof of Theorem 4.2, we used the Dvoretzky-Kiefer-Wolfowitz-Massart inequality (Theorem 12.9). This result states that if $Z_1, \ldots, Z_n$ are i.i.d. random variables on the real line with distribution function $F(z) = P\{Z_i \leq z\}$ and empirical distribution function $F_n(z) = (1/n) \sum_{i=1}^n I_{\{Z_i \leq z\}}$, then

$$P\left\{ \sup_{z \in \mathcal{R}} |F(z) - F_n(z)| \geq \epsilon \right\} \leq 2e^{-2n\epsilon^2}.$$

Use this inequality to conclude that

$$P\left\{ \sup_x |\nu(C(x, 1)) - \nu_n(C(x, 1))| \geq \frac{\epsilon}{2} \right\} \leq 2e^{-2n(\epsilon/2)^2}.$$

HINT: Map (X, Y) on the real line by a one-to-one function $\psi : (\mathcal{R} \times \{0, 1\}) \to \mathcal{R}$ such that $Z = \psi((X, Y)) < 0$ if and only if $Y = 0$. Use the Dvoretzky-Kiefer-Wolfowitz-Massart inequality for Z.

PROBLEM 4.6. Let L be the probability of error for the best sphere rule, that is, for the rule that associates a class with the inside of a sphere $S_{x,r}$, and the other class with the outside. Here the center x, and radius r are both variable. Show that $L = 1/2$ if and only if $L^* = 1/2$, and that $L \leq 1/2$.

PROBLEM 4.7. With the notation of Theorem 4.4, show that the probability of error L of the best linear discriminant satisfies

$$L \leq \frac{4p(1 - p)}{1 + p(1 - p)\Delta^2},$$

where

$$\Delta = \sqrt{(m_1 - m_0)^T \Sigma^{-1}(m_1 - m_0)},$$

is the Mahalanobis distance (Chapter 3) with $\Sigma = p\Sigma_1 + (1-p)\Sigma_0$ (Györfi and Vajda (1980)). Interestingly, the upper bound is just twice the bound of Theorem 3.4 for the asymptotic nearest neighbor error. Thus, a large Mahalanobis distance does not only imply that the Bayes error is small, but also, small error probabilities may be achieved by simple linear classifiers. HINT: Apply the inequality of Problem 4.1 for the univariate random variable $X' = a^T X a = \Sigma^{-1}(m_1 - m_0)$.

PROBLEM 4.8. If m_i and σ_i^2 are the mean and variance of $a^T X$, given that $Y = i$, $i = 0, 1$, where a is a column vector of weights, then show that the criterion

$$J_1(a) = \frac{(m_1 - m_0)^2}{\sigma_1^2 + \sigma_0^2}$$

is minimized for $a = (\Sigma_1 + \Sigma_0)^{-1}(M_1 - M_0)$, where M_i and Σ_i are the mean vector and covariance matrix of X, given $Y = i$. Also, show that

$$J_2(a) = \frac{(m_1 - m_0)^2}{p\sigma_1^2 + (1 - p)\sigma_0^2}$$

is minimized for $a = (p\Sigma_1 + (1 - p)\Sigma_0)^{-1}(M_1 - M_0)$, where $p, 1 - p$ are the class probabilities. This exercise shows that if discrimination is attempted in one dimension, we might consider projections $a^T X$ where a maximizes the weighted distance between the projected means.

PROBLEM 4.9. In the Fisher linear discriminant rule (4.1) with free parameter a_0, show that for any $\epsilon > 0$, there exists a distribution for (X, Y), $X \in \mathcal{R}^2$, with $L = 0$ and $\mathbf{E}\{\|X\|^2\} < \infty$ such that $\inf_{a_0} \mathbf{E}\{L(g_{a_0})\} > 1/2 - \epsilon$. Moreover, if a_0 is chosen to minimize the squared error

$$\mathbf{E}\left\{\left((2Y - 1) - a^T X - a_0\right)^2\right\},$$

then $\mathbf{E}\{L(g_{a_0})\} > 1 - \epsilon$.

PROBLEM 4.10. Find a distribution of (X, Y) with $X \in \mathcal{R}^2$ such that with probability at least one half, $\widehat{L}_n(\phi)$ is not unimodal with respect to the weight vector (a, a_0).

PROBLEM 4.11. The following observation may help in developing a fast algorithm to find the best linear classifier in certain cases. Assume that the Bayes rule is a linear split cutting through the origin, that is, $L^* = L(a^*)$ for some coefficient vector $a^* \in \mathcal{R}^d$, where $L(a)$ denotes the error probability of the classifier

$$g_a(x) = \begin{cases} 1 & \text{if } \sum_{i=1}^d a_i x^{(i)} \geq 0 \\ 0 & \text{otherwise,} \end{cases}$$

and $a = (a_1, \ldots, a_d)$. Show that $L(a)$ is unimodal as a function of $a \in \mathcal{R}^d$, and $L(a)$ is monotone increasing along rays pointing from a^*, that is, for any $\lambda \in (0, 1)$ and $a \in \mathcal{R}^d$, $L(a) - L(\lambda a + (1 - \lambda)a^*) \geq 0$ (Fritz and Györfi (1976)). HINT: Use the expression

$$L(a) = 1/2 - \int (\eta(x) - 1/2) \operatorname{sign} \left(\sum_{i=1}^d a_i x^{(i)} \right) \mu(dx)$$

to show that $L(a) - L(\lambda a + (1 - \lambda)a^*) = \int_A |\eta(x) - 1/2| \mu(dx)$ for some set $A \subset \mathcal{R}^d$.

PROBLEM 4.12. Let $a = (a_0, a_1)$ and

$$\widehat{a} = \arg\min_a \mathbf{E}\left\{\left((2Y - 1) - a_1 X - a_0\right)^2 I_{\{Y_i \neq g_a(X_i)\}}\right\},$$

and $g_a(x) = I_{\{a_1 x + a_0 > 0\}}$. Show that for every $\epsilon > 0$, there exists a distribution of (X, Y) on $\mathcal{R} \times \{0, 1\}$ such that $L(\widehat{a}) - L \geq 1 - \epsilon$, where $L(\widehat{a})$ is the error probability for $g_{\widehat{a}}$. HINT: Argue as in the proof of Theorem 4.7. A distribution with four atoms suffices.

PROBLEM 4.13. Repeat the previous exercise for

$$\widehat{a} = \arg\min_a \mathbf{E}\left\{|(2Y - 1) - a_1 X - a_0|\right\}.$$

PROBLEM 4.14. Let ϕ^* denote the linear discrimination rule that minimizes the mean square error $\mathbf{E}\left\{(2Y - 1 - a^T X - a_0)^2\right\}$ over all a and a_0. As this criterion is quadratic in (a, a_0), it is unimodal. One usually approximates ϕ^* by $\widehat{\phi}$ by minimizing $\sum_i(2Y_i - 1 - a^T X_i - a_0)^2$ over all a and a_0. Show that the minimal column vector (a, a_0) is given by

$$\left(\sum_i X_i' X_i'^T\right)^{-1} \left(\sum_i (2Y_i - 1)X_i'\right),$$

where $X_i' = (X_i, 1)$ is a $(d + 1)$-dimensional column vector.

PROBLEM 4.15. The perceptron criterion is

$$J = \sum_{i:2Y_i-1\neq\text{sign}(a^T X_i+a_0)} |a^T X_i + a_0|.$$

Find a distribution for which $L^* = 0$, $L \leq 1/4$, yet $\liminf_{n\to\infty} \mathbf{E}\{L_n(\phi)\} \geq 1/2$, where ϕ is the linear discrimination rule obtained by using the a and a_0 that minimize J.

PROBLEM 4.16. Let σ be a monotone nondecreasing function on $\mathcal{R}$ satisfying $\lim_{u\to-\infty} \sigma(u) = 0$ and $\lim_{u\to\infty} \sigma(u) = 1$. For $h > 0$, define $\sigma_h(u) = \sigma(hu)$. Consider the linear discrimination rule $\widehat{\phi}$ with a and a_0 chosen to minimize

$$\sum_{i=1}^{n} \left(\sigma_h(a^T X_i + a_0) - Y_i\right)^2.$$

For every fixed $h > 0$ and $0 < \epsilon < 1$, exhibit a distribution with $L < \epsilon$ and

$$\liminf_{n\to\infty} \mathbf{E}\{L_n(\widehat{\phi})\} > 1 - \epsilon.$$

On the other hand, show that if h depends on the sample size n such that $h \to \infty$ as $n \to \infty$, then for all distributions, $\mathbf{E}\{L_n(\widehat{\phi})\} \to L$.

PROBLEM 4.17. Given $Y = i$, let X be normal with mean m_i and covariance matrix Σ_i, $i = 0, 1$. Consider discrimination based upon the minimization of the criterion

$$\mathbf{E}\left\{\left(Y - \sigma(X^T A X + w^T X + c)\right)^2\right\}$$

with respect to A, w, and c, a $d \times d$ matrix, $d \times 1$ vector and constant respectively, where $\sigma(u) = 1/(1 + e^{-u})$ is the standard sigmoid function. Show that this is minimized for the same A, w, and c that minimize the probability of error

$$\mathbf{P}\left\{2Y - 1 \neq \text{sign}(X^T A X + w^T X + c)\right\},$$

and conclude that in this particular case, the squared error criterion may be used to obtain a Bayes-optimal classifier (Horne and Hush (1990)).

5
Nearest Neighbor Rules

5.1 Introduction

Simple rules survive. The k-nearest neighbor rule, since its conception in 1951 and 1952 (Fix and Hodges (1951; 1952; 1991a; 1991b)), has thus attracted many followers and continues to be studied by many researchers. Formally, we define the k-NN rule by

$$g_n(x) = \begin{cases} 1 & \text{if } \sum_{i=1}^n w_{ni} I_{\{Y_i=1\}} > \sum_{i=1}^n w_{ni} I_{\{Y_i=0\}} \\ 0 & \text{otherwise,} \end{cases}$$

where $w_{ni} = 1/k$ if X_i is among the k nearest neighbors of x, and $w_{ni} = 0$ elsewhere. X_i is said to be the k-th nearest neighbor of x if the distance $\|x - X_i\|$ is the k-th smallest among $\|x - X_1\|, \dots, \|x - X_n\|$. In case of a distance tie, the candidate with the smaller index is said to be closer to x. The decision is based upon a majority vote. It is convenient to let k be odd, to avoid voting ties. Several issues are worth considering:

(A) Universal consistency. Establish convergence to the Bayes rule if $k \to \infty$ and $k/n \to 0$ as $n \to \infty$. This is dealt with in Chapter 11.

(B) Finite k performance. What happens if we hold k fixed and let n tend to infinity?

(C) The choice of the weight vector $(w_{n1}, \dots, w_{nn})$. Are equal weights for the k nearest neighbors better than unequal weights in some sense?

(D) The choice of a distance metric. Achieve invariance with respect to a certain family of transformations.

(E) The reduction of the data size. Can we obtain good performance when the data set is edited and/or reduced in size to lessen the storage load?

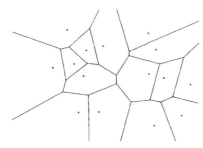

FIGURE 5.1. *At every point the decision is the label of the closest data point. The set of points whose nearest neighbor is X_i is called the Voronoi cell of X_i. The partition induced by the Voronoi cells is a Voronoi partition. A Voronoi partition of 15 random points is shown here.*

In the first couple of sections, we will be concerned with convergence issues for k nearest neighbor rules when k does not change with n. In particular, we will see that for all distributions, the expected error probability $E\{L_n\}$ tends to a limit L_{kNN} that is in general close to but larger than L^*. The methodology for obtaining this result is interesting in its own right. The expression for L_{kNN} is then studied, and several key inequalities such as $L_{NN} \leq 2L^*$ (Cover and Hart (1967)) and $L_{kNN} \leq L^* \left(1 + \sqrt{2/k}\right)$ are proved and applied. The other issues mentioned above are dealt with in the remaining sections. For surveys of various aspects of the nearest neighbor or related methods, see Dasarathy (1991), Devijver (1980), or Devroye and Wagner (1982).

REMARK. COMPUTATIONAL CONCERNS.   Storing the n data pairs in an array and searching for the k nearest neighbors may take time proportional to nkd if done in a naive manner—the "d" accounts for the cost of one distance computation. This complexity may be reduced in terms of one or more of the three factors involved. Typically, with k and d fixed, $O(n^{1/d})$ worst-case time (Papadimitriou and Bentley (1980)) and $O(\log n)$ expected time (Friedman, Bentley, and Finkel (1977)) may be achieved. Multidimensional search trees that partition the space and guide the search are invaluable—for this approach, see Fukunaga and Narendra (1975), Friedman, Bentley, and Finkel (1977), Niemann and Goppert (1988), Kim and Park (1986), and Broder (1990). We refer to a survey in Dasarathy (1991) for more references. Other approaches are described by Yunck (1976), Friedman, Baskett, and Shustek (1975), Vidal (1986), Sethi (1981), and Faragó, Linder, and Lugosi (1993). Generally, with preprocessing, one may considerably reduce the overall complexity in terms of n and d. $\square$

5.2 Notation and Simple Asymptotics

We fix $x \in \mathcal{R}^d$, and reorder the data $(X_1, Y_1), \ldots, (X_n, Y_n)$ according to increasing values of $\|X_i - x\|$. The reordered data sequence is denoted by

$$(X_{(1)}(x), Y_{(1)}(x)), \ldots, (X_{(n)}(x), Y_{(n)}(x)) \quad \text{or by} \quad (X_{(1)}, Y_{(1)}), \ldots, (X_{(n)}, Y_{(n)})$$

if no confusion is possible. $X_{(k)}(x)$ is the k-th nearest neighbor of x.

REMARK. We note here that, rather arbitrarily, we defined neighbors in terms of the Euclidean distance $\|x - y\|$. Surprisingly, the asymptotic properties derived in this chapter remain valid to a wide variety of metrics—the asymptotic probability of error is independent of the distance measure (see Problem 5.1). $\square$

Denote the probability measure for X by μ, and let $S_{x,\epsilon}$ be the closed ball centered at x of radius $\epsilon > 0$. The collection of all x with $\mu(S_{x,\epsilon}) > 0$ for all $\epsilon > 0$ is called the support of X or μ. This set plays a key role because of the following property.

Lemma 5.1. *If $x \in support(\mu)$ and $\lim_{n \to \infty} k/n = 0$, then $\|X_{(k)}(x) - x\| \to 0$ with probability one. If X is independent of the data and has probability measure μ, then $\|X_{(k)}(X) - X\| \to 0$ with probability one whenever $k/n \to 0$.*

PROOF. Take $\epsilon > 0$. By definition, $x \in support(\mu)$ implies that $\mu(S_{x,\epsilon}) > 0$. Observe that $\|X_{(k)}(x) - x\| > \epsilon$ if and only if

$$\frac{1}{n} \sum_{i=1}^{n} I_{\{X_i \in S_{x,\epsilon}\}} < \frac{k}{n}.$$

By the strong law of large numbers, the left-hand side converges to $\mu(S_{x,\epsilon}) > 0$ with probability one, while, by assumption, the right-hand side tends to zero. Therefore, $\|X_{(k)}(x) - x\| \to 0$ with probability one.

The second statement follows from the previous argument as well. First note that by Lemma A.1 in the Appendix, $\mathbf{P}\{X \in support(\mu)\} = 1$, therefore for every $\epsilon > 0$,

$$\mathbf{P}\left\{\|X_{(k)}(X) - X\| > \epsilon\right\}$$
$$= \mathbf{E}\left\{I_{\{X \in support(\mu)\}} \mathbf{P}\left\{\|X_{(k)}(X) - X\| > \epsilon | X \in support(\mu)\right\}\right\},$$

which converges to zero by the dominated convergence theorem, proving convergence in probability. If k does not change with n, then $\|X_{(k)}(X) - X\|$ is monotone decreasing for $n \geq k$; therefore, it converges with probability one as well. If $k = k_n$ is allowed to grow with n such that $k/n \to 0$, then using the notation $X_{(k_n,n)}(X) = X_{(k)}(X)$, we see by a similar argument that the sequence of monotone decreasing random variables

$$\sup_{m \geq n} \|X_{(k_m,m)}(X) - X\| \geq \|X_{(k_n,n)}(X) - X\|$$

converges to zero in probability, and therefore, with probability one as well. This completes the proof. □

Because η is measurable (and thus well-behaved in a general sense) and $\|X_{(k)}(x) - x\|$ is small, the values $\eta(X_{(i)}(x))$ should be close to $\eta(x)$ for all i small enough. We now introduce a proof method that exploits this fact, and will make subsequent analyses very simple—it suffices to look at data samples in a new way via embedding. The basic idea is to define an auxiliary rule $g_n'(x)$ in which the $Y_{(i)}(x)$'s are replaced by k i.i.d. Bernoulli random variables with parameter $\eta(x)$—locally, the $Y_{(i)}(x)$'s behave in such a way. It is easy to show that the error probabilities of the two rules are close, and analyzing the behavior of the auxiliary rule is much more convenient.

To make things more precise, we assume that we are given i.i.d. data pairs $(X_1, U_1), \ldots, (X_n, U_n)$, all distributed as (X, U), where X is as before (and has probability measure μ on the Borel sets of $\mathcal{R}^d$), and U is uniformly distributed on $[0, 1]$ and independent of X. If we set $Y_i = I_{\{U_i \leq \eta(X_i)\}}$, then $(X_1, Y_1), \ldots, (X_n, Y_n)$ are i.i.d. and distributed as the prototype pair (X, Y). So why bother with the U_i's? In embedding arguments, we will use the same U_i's to construct a second data sequence that is heavily correlated (coupled) with the original data sequence, and is more convenient to analyze. For example, for fixed $x \in \mathcal{R}^d$, we may define

$$Y_i'(x) = I_{\{U_i \leq \eta(x)\}}.$$

We now have an i.i.d. sequence with i-th vector given by $X_i, Y_i, Y_i'(x), U_i$. Reordering the data sequence according to increasing values of $\|X_i - x\|$ yields a new sequence with the i-th vector denoted by $X_{(i)}(x), Y_{(i)}(x), Y_{(i)}'(x), U_{(i)}(x)$. If no confusion is possible, the argument x will be dropped. A rule is called k-*local* if for $n \geq k$, g_n is of the form

$$g_n(x) = \begin{cases} 1 & \text{if } \psi(x, Y_{(1)}(x), \ldots, Y_{(k)}(x)) > 0, \\ 0 & \text{otherwise,} \end{cases} \qquad (5.1)$$

for some function ψ. For the k-NN rule, we have, for example,

$$\psi(x, Y_{(1)}, \ldots, Y_{(k)}) = \sum_{i=1}^{k} Y_{(i)}(x) - \frac{k}{2}.$$

In other words, g_n takes a majority vote over the k nearest neighbors of x and breaks ties in favor of class 0.

To study g_n turns out to be almost equivalent to studying the approximate rule g_n':

$$g_n'(x) = \begin{cases} 1 & \text{if } \psi(x, Y_{(1)}'(x), \ldots, Y_{(k)}'(x)) > 0 \\ 0 & \text{otherwise.} \end{cases}$$

The latter rule is of no practical value because it requires the knowledge of $\eta(x)$. Interestingly however, it is easier to study, as $Y_{(1)}'(x), \ldots, Y_{(k)}'(x)$ are i.i.d., whereas $Y_{(1)}(x), \ldots, Y_{(k)}(x)$ are not. Note, in particular, the following:

Lemma 5.2. *For all* x, $n \geq k$,

$$\mathbf{P}\left\{\psi(x, Y_{(1)}(x), \ldots, Y_{(k)}(x)) \neq \psi(x, Y'_{(1)}(x), \ldots, Y'_{(k)}(x))\right\}$$

$$\leq \sum_{i=1}^{k} \mathbf{E}\left\{|\eta(x) - \eta(X_{(i)}(x))|\right\}$$

and

$$\mathbf{P}\{g_n(x) \neq g'_n(x)\} \leq \sum_{i=1}^{k} \mathbf{E}\left\{|\eta(x) - \eta(X_{(i)}(x))|\right\}.$$

PROOF. Both statements follow directly from the observation that

$$\left\{\psi(x, Y_{(1)}(x), \ldots, Y_{(k)}(x)) \neq \psi(x, Y'_{(1)}(x), \ldots, Y'_{(k)}(x))\right\}$$

$$\subseteq \left\{(Y_{(1)}(x), \ldots, Y_{(k)}(x)) \neq (Y'_{(1)}(x), \ldots, Y'_{(k)}(x))\right\}$$

$$\subseteq \bigcup_{i=1}^{k}\left\{\eta(X_{(i)}(x)) \leq U_{(i)}(x) \leq \eta(x)\right\} \cup \bigcup_{i=1}^{k}\left\{\eta(x) \leq U_{(i)}(x) \leq \eta(X_{(i)}(x))\right\},$$

and using the union bound and the fact that the $U_{(i)}(x)$'s are uniform [0, 1]. $\square$

We need the following result, in which X is distributed as X_1, but independent of the data sequence:

Lemma 5.3. (STONE (1977)). *For any integrable function* f, *any* n, *and any* $k \leq n$,

$$\sum_{i=1}^{k} \mathbf{E}\left\{|f(X_{(i)}(X))|\right\} \leq k\gamma_d \mathbf{E}\{|f(X)|\},$$

where $\gamma_d \leq \left(1 + 2/\sqrt{2 - \sqrt{3}}\right)^d - 1$ *depends upon the dimension only.*

The proof of this lemma is beautiful but a bit technical—it is given in a separate section. Here is how it is applied, and why, for fixed k, we may think of $f(X_{(k)}(X))$ as $f(X)$ for all practical purposes.

Lemma 5.4. *For any integrable function* f,

$$\frac{1}{k}\sum_{i=1}^{k} \mathbf{E}\left\{|f(X) - f(X_{(i)}(X))|\right\} \to 0$$

as $n \to \infty$ *whenever* $k/n \to 0$.

PROOF. Given $\epsilon > 0$, find a uniformly continuous function g vanishing off a bounded set A, such that $\mathbf{E}\{|g(X) - f(X)|\} < \epsilon$ (see Theorem A.8 in the Appendix).

Then for each $\epsilon > 0$, there is a $\delta > 0$ such that $\|x - z\| < \delta$ implies $|g(x) - g(z)| < \epsilon$. Thus,

$$\frac{1}{k} \sum_{i=1}^{k} \mathbf{E}\left\{|f(X) - f(X_{(i)}(X))|\right\}$$

$$\leq \quad \mathbf{E}\{|f(X) - g(X)|\} + \frac{1}{k} \sum_{i=1}^{k} \mathbf{E}\left\{|g(X) - g(X_{(i)}(X))|\right\}$$

$$+ \frac{1}{k} \sum_{i=1}^{k} \mathbf{E}\left\{|g(X_{(i)}(X)) - f(X_{(i)}(X))|\right\}$$

$$\leq \quad (1 + \gamma_d)\mathbf{E}\{|f(X) - g(X)|\} + \epsilon + \|g\|_\infty \mathbf{P}\left\{\|X - X_{(k)}(X)\| > \delta\right\}$$

(by Lemma 5.3, where δ depends on ϵ only)

$$\leq \quad (2 + \gamma_d)\epsilon + o(1) \quad \text{(by Lemma 5.1).} \quad \square$$

5.3   Proof of Stone's Lemma

In this section we prove Lemma 5.3. For $\theta \in (0, \pi/2)$, a cone $C(x, \theta)$ is the collection of all $y \in \mathcal{R}^d$ for which angle$(x, y) \leq \theta$. Equivalently, in vector notation, $x^T y / \|x\| \|y\| \geq \cos \theta$. The set $z + C(x, \theta)$ is the translation of $C(x, \theta)$ by z.

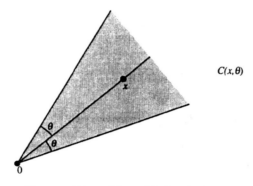

FIGURE 5.2. *A cone of angle θ.*

$C(x, \theta)$

If $y, z \in C(x, \pi/6)$, and $\|y\| < \|z\|$, then $\|y - z\| < \|z\|$, as we will now show. Indeed,

$$\|y - z\|^2 \quad = \quad \|y\|^2 + \|z\|^2 - 2\|y\|\|z\| \frac{y^T z}{\|y\|\|z\|}$$

$$\leq \quad \|y\|^2 + \|z\|^2 - 2\|y\|\|z\| \cos(\pi/3)$$

$$= \quad \|z\|^2 \left(1 + \frac{\|y\|^2}{\|z\|^2} - \frac{\|y\|}{\|z\|}\right)$$

$$< \quad \|z\|^2 \quad \text{(see Figure 5.3).}$$

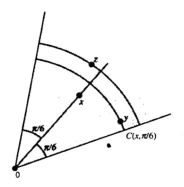

FIGURE 5.3. *The key geometrical property of cones of angle $< \pi/2$.*

The following covering lemma is needed in what follows:

Lemma 5.5. *Let $\theta \in (0, \pi/2)$ be fixed. Then there exists a set $\{x_1, \ldots, x_{\gamma_d}\} \subset \mathcal{R}^d$ such that*

$$\mathcal{R}^d = \bigcup_{i=1}^{\gamma_d} C(x_i, \theta).$$

Furthermore, it is always possible to take

$$\gamma_d \leq \left(1 + \frac{1}{\sin(\theta/2)}\right)^d - 1.$$

For $\theta = \pi/6$, we have

$$\gamma_d \leq \left(1 + \frac{2}{\sqrt{2 - \sqrt{3}}}\right)^d - 1.$$

PROOF. We assume without loss of generality that $\|x_i\| = 1$ for all i. Each x_i is the center of a sphere S_i of radius $r = 2\sin(\theta/2)$. S_i has the property that

$$\{x : \|x\| = 1\} \cap S_i = \{x : \|x\| = 1\} \cap C(x_i, \theta).$$

Let us only look at x_i's such that $\|x_i - x_j\| \geq r$ for all $j \neq i$. In that case, $\bigcup C(x_i, \theta)$ covers $\mathcal{R}^d$ if and only if $\bigcup S_i$ covers $\{x : \|x\| = 1\}$. Then the spheres S_i' of radius $r/2$ centered at the x_i's are disjoint and $\bigcup S_i' \subseteq S_{0,1+r/2} - S_{0,r/2}$ (see Figure 5.4).

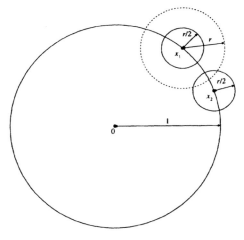

FIGURE 5.4. *Bounding γ_d.*

Thus, if $v_d = \text{volume}(S_{0,1})$,

$$\gamma_d v_d \left(\frac{r}{2}\right)^d \le v_d \left(1 + \frac{r}{2}\right)^d - v_d \left(\frac{r}{2}\right)^d$$

or

$$\gamma_d \le \left(1 + \frac{2}{r}\right)^d - 1 = \left(1 + \frac{1}{\sin(\theta/2)}\right)^d - 1.$$

The last inequality follows from the fact that

$$\sin \frac{\pi}{12} = \sqrt{\frac{1 - \cos(\pi/6)}{2}} = \sqrt{\frac{2 - \sqrt{3}}{4}}. \quad \square$$

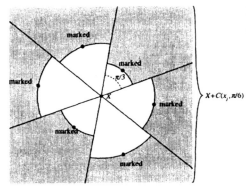

FIGURE 5.5. *Covering the space by cones.*

With the preliminary results out of the way, we cover $\mathcal{R}^d$ by γ_d cones $X + C(x_j, \pi/6)$, $1 \le j \le \gamma_d$, and mark in each cone the X_i that is nearest to

X, if such an X_i exists. If X_i belongs to $X + C(x_j, \pi/6)$ and is not marked, then X cannot be the nearest neighbor of X_i in $\{X_1, \ldots, X_{i-1}, X, X_{i+1}, \ldots, X_n\}$. Similarly, we might mark all k nearest neighbors of X in each cone (if there are less than k points in a cone, mark all of them). By a similar argument, if $X_i \in X + C(x_j, \pi/6)$ is not marked, then X cannot be among the k nearest neighbors of X_i in $\{X_1, \ldots, X_{i-1}, X, X_{i+1}, \ldots, X_n\}$. (The order of this set of points is important if distance ties occur with positive probability, and they are broken by comparing indices.) Therefore, if f is a nonnegative function,

$$\sum_{i=1}^{k} \mathbf{E}\left\{ f(X_{(i)}(X)) \right\}$$

$$= \mathbf{E}\left\{ \sum_{i=1}^{n} I_{\{X_i \text{ is among the k nearest neighbors of } X \text{ in } \{X_1, \ldots, X_n\}\}} f(X_i) \right\}$$

$$= \mathbf{E}\left\{ f(X) \sum_{i=1}^{n} \right.$$

$$\left. I_{\{X \text{ is among the k nearest neighbors of } X_i \text{ in } \{X_1, \ldots, X_{i-1}, X, X_{i+1}, \ldots, X_n\}\}} \right\}$$

(by exchanging X and X_i)

$$\leq \mathbf{E}\left\{ f(X) \sum_{i=1}^{n} I_{\{X_i \text{ is marked}\}} \right\}$$

$$\leq k\gamma_d \mathbf{E}\{f(X)\},$$

as we can mark at most k nodes in each cone, and the number of cones is at most γ_d—see Lemma 5.5. This concludes the proof of Stone's lemma. $\square$

5.4 The Asymptotic Probability of Error

We return to k-local rules (and in particular, to k-nearest neighbor rules). Let $D'_n = ((X_1, Y_1, U_1), \ldots, (X_n, Y_n, U_n))$ be the i.i.d. data augmented by the uniform random variables $U_1, \ldots, U_n$ as described earlier. For a decision g_n based on D_n, we have the probability of error

$$L_n = \mathbf{P}\{g_n(X) \neq Y | D'_n\}$$

$$= \mathbf{P}\left\{ \text{sign}\left(\psi(X, Y_{(1)}(X), \ldots, Y_{(k)}(X)) \right) \neq \text{sign}(2Y - 1) | D'_n \right\},$$

where ψ is the function whose sign determines g_n; see (5.1). Define the random variables $Y'_{(1)}(X), \ldots, Y'_{(k)}(X)$ as we did earlier, and set

$$L'_n = \mathbf{P}\left\{ \text{sign}\left(\psi(X, Y'_{(1)}(X), \ldots, Y'_{(k)}(X)) \right) \neq \text{sign}(2Y - 1) | D'_n \right\}.$$

By Lemmas 5.2 and 5.4,

$$
\begin{aligned}
\mathbf{E}\{|L_n - L_n'|\} \\
\leq\quad & \mathbf{P}\left\{\psi(X, Y_{(1)}(X), \ldots, Y_{(k)}(X)) \neq \psi(X, Y_{(1)}'(X), \ldots, Y_{(k)}'(X))\right\} \\
\leq\quad & \sum_{i=1}^{k} \mathbf{E}\left\{|\eta(X) - \eta(X_{(i)}(X))|\right\} \\
=\quad & o(1).
\end{aligned}
$$

Because $\lim_{n\to\infty}(\mathbf{E}L_n' - \mathbf{E}L_n) = 0$, we need *only* study the rule g_n'

$$
g_n'(x) = \begin{cases} 1 & \text{if } \psi(x, Z_1, \ldots, Z_k) > 0 \\ 0 & \text{otherwise} \end{cases}
$$

$(Z_1, \ldots, Z_k$ are i.i.d. Bernoulli $(\eta(x)))$ unless we are concerned with the closeness of L_n to $\mathbf{E}L_n$ as well.

We now illustrate this important time-saving device on the 1-nearest neighbor rule. Clearly, $\psi(x, Z_1) = 2Z_1 - 1$, and therefore

$$
\mathbf{E}\{L_n'\} = \mathbf{P}\{Z_1 \neq Y\} = \mathbf{E}\{2\eta(X)(1 - \eta(X))\}.
$$

We have, without further work:

Theorem 5.1. *For the nearest neighbor rule, for any distribution of (X, Y),*

$$
\lim_{n\to\infty} \mathbf{E}\{L_n\} = \mathbf{E}\{2\eta(X)(1 - \eta(X))\} = L_{\text{NN}}.
$$

Under various continuity conditions (X has a density f and both f and η are almost everywhere continuous); this result is due to Cover and Hart (1967). In the present generality, it essentially appears in Stone (1977). See also Devroye (1981c). Elsewhere (Chapter 3), we show that

$$
L^* \leq L_{\text{NN}} \leq 2L^*(1 - L^*) \leq 2L^*.
$$

Hence, the previous result says that the nearest neighbor rule is asymptotically at most twice as bad as the Bayes rule—especially for small L^*, this property should be useful.

We formally define the quantity, when k is odd,

$$
\begin{aligned}
L_{k\text{NN}} \\
=\quad & \mathbf{E}\left\{\sum_{j=0}^{k} \binom{k}{j}\eta^j(X)(1 - \eta(X))^{k-j}\left(\eta(X)I_{\{j<k/2\}} + (1 - \eta(X))I_{\{j>k/2\}}\right)\right\}.
\end{aligned}
$$

We have the following result:

Theorem 5.2. *Let k be odd and fixed. Then, for the k-NN rule,*

$$
\lim_{n\to\infty} \mathbf{E}\{L_n\} = L_{k\text{NN}}.
$$

PROOF. We note that it suffices to show that $\lim_{n \to \infty} \mathbf{E}\{L'_n\} = L_{k\text{NN}}$ (in the previously introduced notation). But for *every* n,

$$\mathbf{E}\{L'_n\}$$

$$= \mathbf{P}\left\{ Z_1 + \cdots + Z_k > \frac{k}{2}, Y = 0 \right\} + \mathbf{P}\left\{ Z_1 + \cdots + Z_k < \frac{k}{2}, Y = 1 \right\}$$

$$= \mathbf{P}\left\{ Z_1 + \cdots + Z_k > \frac{k}{2}, Z_0 = 0 \right\} + \mathbf{P}\left\{ Z_1 + \cdots + Z_k < \frac{k}{2}, Z_0 = 1 \right\}$$

(where $Z_0, \ldots, Z_k$ are i.i.d. Bernoulli ($\eta(X)$) random variables),

which leads directly to the sought result. $\square$

Several representations of $L_{k\text{NN}}$ will be useful for later analysis. For example, we have

$$L_{k\text{NN}}$$

$$= \mathbf{E}\left\{ \eta(X)\mathbf{P}\left\{ \text{Binomial}(k, \eta(X)) < \frac{k}{2} \,\middle|\, X \right\} \right\}$$

$$+ \mathbf{E}\left\{ (1 - \eta(X))\mathbf{P}\left\{ \text{Binomial}(k, \eta(X)) > \frac{k}{2} \,\middle|\, X \right\} \right\}$$

$$= \mathbf{E}\left\{ \min(\eta(X), 1 - \eta(X)) \right\}$$

$$+ \mathbf{E}\left\{ \left(1 - 2\min(\eta(X), 1 - \eta(X)) \right)\mathbf{P}\left\{ \text{Binomial}(k, \eta(X)) > \frac{k}{2} \,\middle|\, X \right\} \right\}.$$

It should be stressed that the limit result in Theorem 5.2 is distribution-free. The limit $L_{k\text{NN}}$ depends upon $\eta(X)$ (or $\min(\eta(X), 1 - \eta(X))$) only. The continuity or lack of smoothness of η is immaterial—it only matters for the speed with which $\mathbf{E}\{L_n\}$ approaches the limit $L_{k\text{NN}}$.

5.5 The Asymptotic Error Probability of Weighted Nearest Neighbor Rules

Following Royall (1966), a weighted nearest neighbor rule with weights $w_1, \ldots, w_k$ makes a decision according to

$$g_n(x) = \begin{cases} 1 & \text{if } \sum_{i:Y_{(i)}(x)=1}^{k} w_i > \sum_{i:Y_{(i)}(x)=0}^{k} w_i \\ 0 & \text{otherwise.} \end{cases}$$

In case of a voting tie, this rule is not symmetric. We may modify it so that $g_n(x) \overset{\text{def}}{=} -1$ if we have a voting tie. The "-1" should be considered as an indecision. By previous arguments the asymptotic probability of error is a function of $w_1, \ldots, w_k$ given by $L(w_1, \ldots, w_k) = \mathbf{E}\{\alpha(\eta(X))\}$, where

$$
\alpha(p) \;=\; \mathbf{P}\left\{\sum_{i=1}^{k} w_i Y_i' > \sum_{i=1}^{k} w_i(1 - Y_i')\right\}(1 - p)
$$

$$
+ \mathbf{P}\left\{\sum_{i=1}^{k} w_i Y_i' \le \sum_{i=1}^{k} w_i(1 - Y_i')\right\} p,
$$

where now $Y_1', \ldots, Y_k'$ are i.i.d. Bernoulli (p). Equivalently, with $Z_i = 2Y_i' - 1 \in \{-1, 1\}$,

$$
\alpha(p) = (1 - p)\mathbf{P}\left\{\sum_{i=1}^{k} w_i Z_i > 0\right\} + p\mathbf{P}\left\{\sum_{i=1}^{k} w_i Z_i \le 0\right\}.
$$

Assume that $\mathbf{P}\left\{\sum_{i=1}^{k} w_i Z_i = 0\right\} = 0$ for now. Then, if $p < 1/2$,

$$
\alpha(p) = p + (1 - 2p)\mathbf{P}\left\{\sum_{i=1}^{k} w_i Z_i > 0\right\},
$$

and an antisymmetric expression is valid when $p > 1/2$. Note next the following. If we let N_l be the number of vectors $z = (z_1, \ldots, z_k) \in \{-1, 1\}^k$ with $\sum I_{\{z_i=1\}} = l$ and $\sum w_i z_i > 0$, then $N_l + N_{k-l} = \binom{k}{l}$. Thus,

$$
\mathbf{P}\left\{\sum_{i=1}^{k} w_i Z_i > 0\right\}
$$

$$
= \sum_{l=0}^{k} N_l p^l (1 - p)^{k-l}
$$

$$
= \sum_{l<k/2} \binom{k}{l} p^{k-l}(1 - p)^l + \sum_{l<k/2} N_l \left(p^l(1 - p)^{k-l} - p^{k-l}(1 - p)^l\right)
$$

$$
+ \frac{1}{2}\binom{k}{k/2} p^{k/2}(1 - p)^{k/2} I_{\{k \text{ even}\}}
$$

$$
= I + II + III.
$$

Note that $I + III$ does not depend on the vector of weights, and represents

$$
\mathbf{P}\{\text{Binomial}(k, 1 - p) \le k/2\} = \mathbf{P}\{\text{Binomial}(k, p) \ge k/2\}.
$$

Finally, since $p \le 1/2$,

$$
II \;=\; \sum_{l<k/2} N_l \left(p^l(1 - p)^{k-l} - p^{k-l}(1 - p)^l\right)
$$

$$
= \sum_{l<k/2} N_l p^l(1 - p)^l \left((1 - p)^{k-2l} - p^{k-2l}\right)
$$

$$
\ge \; 0.
$$

This term is zero if and only if $N_l = 0$ for all $l < k/2$. In other words, it vanishes if and only if no numerical minority of w_i's can sum to a majority (as in the case $(0.7, 0.2, 0.1)$, where 0.7 alone, a numerical minority, outweighs the others). But such cases are equivalent to ordinary k-nearest neighbor rules if k is odd. When k is *even*, and we add a tiny weight to one w_i, as in

$$\left(\frac{1+\epsilon}{k}, \frac{1-\epsilon/(k-1)}{k}, \ldots, \frac{1-\epsilon/(k-1)}{k} \right),$$

for small $\epsilon > 0$, then no numerical minority can win either, and we have an optimal rule ($II = 0$). We have thus shown the following:

Theorem 5.3. (BAILEY AND JAIN (1978)). *Let $L(w_1, \ldots, w_k)$ be the asymptotic probability of error of the weighted k-NN rule with weights $w_1, \ldots, w_k$. Let the k-NN rule be defined by $(1/k, 1/k, \ldots, 1/k)$ if k is odd, and by*

$$(1/k, 1/k, \ldots, 1/k) + \epsilon(1, -1/(k-1), -1/(k-1), \ldots, -1/(k-1))$$

for $0 < \epsilon < 1/k$ when k is even. Denoting the asymptotic probability of error by $L_{k\text{NN}}$ for the latter rule, we have

$$L(w_1, \ldots, w_k) \geq L_{k\text{NN}}.$$

If $\mathbf{P}\{\eta(X) = 1/2\} < 1$, then equality occurs if and only if every numerical minority of the w_i's carries less than half of the total weight.

The result states that standard k-nearest neighbor rules are to be preferred in an asymptotic sense. This does not mean that for a particular sample size, one should steer clear of nonuniform weights. In fact, if k is allowed to vary with n, then nonuniform weights are advantageous (Royall (1966)).

Consider the space $\mathcal{W}$ of all weight vectors $(w_1, \ldots, w_k)$ with $w_i \geq 0$, $\sum_{i=1}^{k} w_i = 1$. Is it totally ordered with respect to $L(w_1, \ldots, w_k)$ or not? To answer this question, we must return to $\alpha(p)$ once again. The weight vector only influences the term II given there. Consider, for example, the weight vectors

$$(0.3, 0.22, 0.13, 0.12, 0.071, 0.071, 0.071, 0.017)$$

and $(0.26, 0.26, 0.13, 0.12, 0.071, 0.071, 0.071, 0.017).$

Numerical minorities are made up of one, two, or three components. For both weight vectors, $N_1 = 0$, $N_2 = 1$. However, $N_3 = 6 + 4$ in the former case, and $N_3 = 6 + 2$ in the latter. Thus, the "II" term is uniformly smaller over all $p < 1/2$ in the latter case, and we see that for *all distributions*, the second weight vector is better. When the N_l's are not strictly nested, such a universal comparison becomes impossible, as in the example of Problem 5.8. Hence, $\mathcal{W}$ is only partially ordered.

Unwittingly, we have also shown the following theorem:

Theorem 5.4. *For all distributions,*

$$L^* \leq \cdots \leq L_{(2k+1)\text{NN}} \leq L_{(2k-1)\text{NN}} \leq \cdots \leq L_{3\text{NN}} \leq L_{\text{NN}} \leq 2L^*.$$

PROOF. It suffices once again to look at $\alpha(p)$. Consider the weight vector $w_1 = \cdots = w_{2k+1} = 1$ (ignoring normalization) as for the $(2k + 1)$-NN rule. The term II is zero, as $N_0 = N_1 = \cdots = N_k = 0$. However, the $(2k - 1)$-NN rule with vector $w_1 = \cdots = w_{2k-1} = 1$, $w_{2k} = w_{2k+1} = 0$, has a nonzero term II, because $N_0 = \cdots = N_{k-1} = 0$, yet $N_k = \binom{2k-1}{k} > 0$. Hence, $L_{(2k+1)\text{NN}} \leq L_{(2k-1)\text{NN}}$. $\square$

REMARK. We have strict inequality $L_{(2k+1)\text{NN}} < L_{(2k-1)\text{NN}}$ whenever $\mathbf{P}\{\eta(X) \notin \{0, 1, 1/2\}\} > 0$. When $L^* = 0$, we have $L_{\text{NN}} = L_{3\text{NN}} = L_{5\text{NN}} = \cdots = 0$ as well. $\square$

5.6 k-Nearest Neighbor Rules: Even k

Until now we assumed throughout that k was odd, so that voting ties were avoided. The tie-breaking procedure we follow for the $2k$-nearest neighbor rule is as follows:

$$g_n(x) = \begin{cases} 1 & \text{if } \sum_{i=1}^{2k} Y_{(i)}(x) > k \\ 0 & \text{if } \sum_{i=1}^{2k} Y_{(i)}(x) < k \\ Y_{(1)}(x) & \text{if } \sum_{i=1}^{2k} Y_{(i)}(x) = k. \end{cases}$$

Formally, this is equivalent to a weighted $2k$-nearest neighbor rule with weight vector $(3, 2, 2, 2, \ldots, 2, 2)$. It is easy to check from Theorem 5.3 that this is the asymptotically best weight vector. Even values do not decrease the probability of error. In particular, we have the following:

Theorem 5.5. (DEVIJVER (1978)). *For all distributions, and all integers k,*

$$L_{(2k-1)\text{NN}} = L_{(2k)\text{NN}}.$$

PROOF. Recall that $L_{k\text{NN}}$ may be written in the form $L_{k\text{NN}} = \mathbf{E}\{\alpha(\eta(X))\}$, where

$$\alpha(\eta(x)) = \lim_{n \to \infty} \mathbf{P}\left\{g_n^{(k)}(X) \neq Y | X = x\right\}$$

is the pointwise asymptotic error probability of the k-NN rule $g_n^{(k)}$. It is convenient to consider $Z_1, \ldots, Z_{2k}$ i.i.d. $\{-1, 1\}$-valued random variables with $\mathbf{P}\{Z_i = 1\} = p = \eta(x)$, and to base the decision upon the sign of $\sum_{i=1}^{2k} Z_i$. From the general formula for weighted nearest neighbor rules, the pointwise asymptotic error probability of the $(2k)$-NN rule is

$$\lim_{n \to \infty} \mathbf{P}\left\{g_n^{(2k)}(X) \neq Y | X = x\right\}$$

$$= pP\left\{\sum_{i=1}^{2k} Z_i < 0\right\} + pP\left\{\sum_{i=1}^{2k} Z_i = 0, Z_1 < 0\right\}$$

$$+ (1-p)P\left\{\sum_{i=1}^{2k} Z_i > 0\right\} + (1-p)P\left\{\sum_{i=1}^{2k} Z_i = 0, Z_1 > 0\right\}$$

$$= pP\left\{\sum_{i=2}^{2k} Z_i < 0\right\} + (1-p)P\left\{\sum_{i=2}^{2k} Z_i > 0\right\}$$

$$= \lim_{n\to\infty} P\left\{g_n^{(2k-1)}(X) \neq Y | X = x\right\}.$$

Therefore, $L_{(2k)\text{NN}} = L_{(2k-1)\text{NN}}$. □

5.7 Inequalities for the Probability of Error

We return to the case when k is odd. Recall that

$$L_{k\text{NN}} = \mathbf{E}\left\{\alpha_k(\eta(X))\right\},$$

where

$$\alpha_k(p) = \min(p, 1-p) + |2p-1|\mathbf{P}\left\{\text{Binomial}(k, \min(p, 1-p)) > \frac{k}{2}\right\}.$$

Since $L^* = \mathbf{E}\{\min(\eta(X), 1 - \eta(X))\}$, we may exploit this representation to obtain a variety of inequalities on $L_{k\text{NN}} - L^*$. We begin with one that is very easy to prove but perhaps not the strongest.

Theorem 5.6. *For all odd k and all distributions,*

$$L_{k\text{NN}} \leq L^* + \frac{1}{\sqrt{ke}}.$$

PROOF. By the above representation,

$$L_{k\text{NN}} - L^* \leq \sup_{0 \leq p \leq 1/2} (1 - 2p)\mathbf{P}\left\{B > \frac{k}{2}\right\}$$

$$(B \text{ is Binomial } (k, p))$$

$$= \sup_{0 \leq p \leq 1/2} (1 - 2p)\mathbf{P}\left\{\frac{B - kp}{k} > \frac{1}{2} - p\right\}$$

$$\leq \sup_{0 \leq p \leq 1/2} (1 - 2p)e^{-2k(1/2 - p)^2}$$

$$(\text{by the Okamoto-Hoeffding inequality—Theorem 8.1})$$

$$= \sup_{0 \leq u \leq 1} u e^{-ku^2/2}$$

$$= \frac{1}{\sqrt{ke}}. \quad \square$$

Theorem 5.7. (GYÖRFI AND GYÖRFI (1978)). *For all distributions and all odd k,*

$$L_{k\text{NN}} \leq L^* + \sqrt{\frac{2L_{\text{NN}}}{k}}.$$

PROOF. We note that for $p \leq 1/2$, with B binomial (k, p),

$$
\begin{aligned}
\mathbf{P}\left\{B > \frac{k}{2}\right\} &= \mathbf{P}\left\{B - kp > k\left(\frac{1}{2} - p\right)\right\} \\
&\leq \frac{\mathbf{E}\{|B - kp|\}}{k(1/2 - p)} \quad \text{(Markov's inequality)} \\
&\leq \frac{\sqrt{\text{Var}\{B\}}}{k(1/2 - p)} \quad \text{(Cauchy-Schwarz inequality)} \\
&= \frac{2\sqrt{p(1 - p)}}{\sqrt{k}(1 - 2p)}.
\end{aligned}
$$

Hence,

$$
\begin{aligned}
L_{k\text{NN}} - L^* &\leq \mathbf{E}\left\{\frac{2}{\sqrt{k}}\sqrt{\eta(X)(1 - \eta(X))}\right\} \\
&\leq \frac{2}{\sqrt{k}}\sqrt{\mathbf{E}\{\eta(X)(1 - \eta(X))\}} \quad \text{(Jensen's inequality)} \\
&= \frac{2}{\sqrt{k}}\sqrt{\frac{L_{\text{NN}}}{2}} \\
&= \sqrt{\frac{2L_{\text{NN}}}{k}}. \quad \square
\end{aligned}
$$

REMARK. For large k, B is approximately normal $(k, p(1 - p))$, and thus $\mathbf{E}\{|B - kp|\} \approx \sqrt{kp(1 - p)}\sqrt{2/\pi}$, as the first absolute moment of a normal random variable is $\sqrt{2/\pi}$ (see Problem 5.11). Working this through yields an approximate bound of $\sqrt{L_{\text{NN}}/(\pi k)}$. The bound is proportional to $\sqrt{L_{\text{NN}}}$. This can be improved to L^* if, instead of bounding it from above by Markov's inequality, we directly approximate $\mathbf{P}\{B - kp > k(1/2 - p)\}$ as shown below. $\square$

Theorem 5.8. (DEVROYE (1981B)). *For all distributions and $k \geq 3$ odd,*

$$L_{k\text{NN}} \leq L^*\left(1 + \frac{\gamma}{\sqrt{k}}\left(1 + O(k^{-1/6})\right)\right),$$

where $\gamma = \sup_{r>0} 2r\mathbf{P}\{N > r\} = 0.33994241\ldots$, N is normal $(0, 1)$, and $O(\cdot)$ refers to $k \to \infty$. (Explicit constants are given in the proof.)

The constant γ in the proof cannot be improved. A slightly weaker bound was obtained by Devijver (1979):

$$L_{k\text{NN}} \leq L^* + \frac{1}{2^{2k'}}\binom{2k'}{k'}L_{\text{NN}} \quad \text{(where } k' = \lceil k/2 \rceil\text{)}$$

$$= L^* + L_{\text{NN}}\sqrt{\frac{2}{\pi k'}}(1 + o(1)) \quad \text{(as } k \to \infty, \text{ see Lemma A.3).}$$

See also Devijver and Kittler (1982, p.102).

Lemma 5.6. (DEVROYE (1981B)). *For $p \leq 1/2$ and with $k > 3$ odd,*

$$\mathbf{P}\left\{\text{Binomial}(k, p) > \frac{k}{2}\right\} = \frac{k!}{\left(\frac{k-1}{2}\right)!\left(\frac{k-1}{2}\right)!}\int_0^p (x(1-x))^{(k-1)/2}\,dx$$

$$\leq A\int_{(1-2p)\sqrt{k-1}}^{\sqrt{k-1}} e^{-z^2/2}dz,$$

where $A \leq \frac{1}{\sqrt{2\pi}}\left(1 + \frac{2}{k} + \frac{3}{4k^2}\right)$.

PROOF. Consider k i.i.d. uniform random variables on $[0, 1]$. The number of values in $[0, p]$ is binomial (k, p). The number exceeds $k/2$ if and only if the $(k + 1)/2$-th order statistic of the uniform cloud is at most p. The latter is beta $((k+1)/2, (k+1)/2)$ distributed, explaining the first equality (Problem 5.32). Note that we have written a discrete sum as an integral—in some cases, such tricks pay off handsome rewards. To show the inequality, replace x by $\frac{1}{2}\left(1 - \frac{z}{\sqrt{k-1}}\right)$ and use the inequality $1 - u \leq e^{-u}$ to obtain a bound as shown with

$$A = \frac{1}{2^k\sqrt{k-1}} \times \frac{k!}{\left(\frac{k-1}{2}\right)!\left(\frac{k-1}{2}\right)!}.$$

Finally,

$$A = \mathbf{P}\left\{B = \frac{k+1}{2}\right\}\frac{k+1}{2\sqrt{k-3}} \quad (B \text{ is binomial } (k, 1/2))$$

$$\leq \sqrt{\frac{k}{2\pi\frac{k+1}{2}\frac{k-1}{2}}}\frac{k+1}{2\sqrt{k-1}} \quad \text{(Problem 5.17)}$$

$$= \frac{1}{\sqrt{2\pi}}\frac{\sqrt{k(k+1)}}{k-1}$$

$$\leq \frac{1}{\sqrt{2\pi}}\left(1 + \frac{2}{k} + \frac{3}{4k^2}\right) \quad \text{(Problem 5.18).} \quad \square$$

PROOF OF THEOREM 5.8. From earlier remarks,

$$
\begin{aligned}
L_{k\text{NN}} - L^* &= \mathbf{E}\left\{\alpha_k(\eta(X)) - \min(\eta(X), 1 - \eta(X))\right\} \\
&= \mathbf{E}\left\{\left(\frac{\alpha_k(\eta(X))}{\min(\eta(X), 1 - \eta(X))} - 1\right)\min(\eta(X), 1 - \eta(X))\right\} \\
&\le \left(\sup_{0<p<1/2}\frac{1-2p}{p}\mathbf{P}\left\{B > \frac{k}{2}\right\}\right)L^* \quad (B \text{ is binomial } (k, p)).
\end{aligned}
$$

We merely bound the factor in brackets. Clearly, by Lemma 5.6,

$$
L_{k\text{NN}} - L^* \le L^*\left(\sup_{0<p<1/2}\frac{1-2p}{p}A\int_{(1-2p)\sqrt{k-1}}^{\sqrt{k-1}}e^{-z^2/2}dz\right).
$$

Take $a < 1$ as the solution of $\left(3/(ea^2)\right)^{3/2}\frac{1}{\sqrt{2\pi(k-1)}} = \gamma$, which is possible if $k - 1 > \frac{1}{2\pi}\frac{1}{\gamma^2}\left(\frac{3}{e}\right)^6 = 2.4886858\ldots$. Setting $v = (1 - 2p)\sqrt{k - 1}$, we have

$$
\begin{aligned}
&\sup_{0<p<1/2}\frac{1-2p}{p}\int_{(1-2p)\sqrt{k-1}}^{\sqrt{k-1}}\frac{e^{-z^2/2}}{\sqrt{2\pi}}dz \\
&\le \max\left(\sup_{0<v\le a\sqrt{k-1}}\frac{2v/\sqrt{k-1}}{1 - v/\sqrt{k-1}}\int_v^{\infty}\frac{e^{-z^2/2}}{\sqrt{2\pi}}dz,\right. \\
&\qquad\qquad \left.\sup_{a\sqrt{k-1}\le v<\sqrt{k-1}}\frac{2p\sqrt{k-1}}{1 - v/\sqrt{k-1}}\frac{e^{-v^2/2}}{\sqrt{2\pi}}\right) \\
&\le \max\left(\frac{\gamma}{(1-a)\sqrt{k-1}}, \frac{\sqrt{k-1}}{\sqrt{2\pi}}e^{-a^2(k-1)/2}\right) \\
&\le \max\left(\frac{\gamma}{(1-a)\sqrt{k-1}}, \left(\frac{3}{ea^2}\right)^{3/2}\frac{1}{(k-1)\sqrt{2\pi}}\right) \\
&\qquad\text{(use } u^{3/2}e^{-cu} \le (3/(2ce))^{3/2} \text{ for all } u > 0) \\
&= \frac{\gamma}{(1-a)\sqrt{k-1}}.
\end{aligned}
$$

Collect all bounds and note that $a = O\left(k^{-1/6}\right)$. $\square$

5.8 Behavior When L^* Is Small

In this section, we look more closely at $L_{k\text{NN}}$ when L^* is small. Recalling that $L_{k\text{NN}} = \mathbf{E}\{\alpha_k(\eta(X))\}$ with

$$
\alpha_k(p) = \min(p, 1-p) + |1 - 2\min(p, 1-p)|\mathbf{P}\left\{\text{Binomial}(k, \min(p, 1 - p)) > \frac{k}{2}\right\}
$$

for odd k, it is easily seen that $L_{k\text{NN}} = \mathbf{E}\{\xi_k(\min(\eta(X), 1 - \eta(X)))\}$ for some function ξ_k. Because

$$\min(p, 1 - p) = \frac{1 - \sqrt{1 - 4p(1 - p)}}{2},$$

we also have $L_{k\text{NN}} = \mathbf{E}\{\psi_k(\eta(X)(1 - \eta(X)))\}$ for some other function ψ_k. Worked-out forms of $L_{k\text{NN}}$ include

$$
\begin{aligned}
L_{k\text{NN}} &= \mathbf{E}\left\{ \sum_{j < k/2} \binom{k}{j} \eta(X)^{j+1}(1 - \eta(X))^{k-j} \right. \\
&\qquad \left. + \sum_{j > k/2} \binom{k}{j} \eta(X)^{j}(1 - \eta(X))^{k-j+1} \right\} \\
&= \sum_{j < k/2} \binom{k}{j} \mathbf{E}\left\{ (\eta(X)(1 - \eta(X)))^{j+1} \left((1 - \eta(X))^{k-2j-1} + \eta(X)^{k-2j-1} \right) \right\}.
\end{aligned}
$$

As $p^a + (1 - p)^a$ is a function of $p(1 - p)$ for integer a, this may be further reduced to simplified forms such as

$$
\begin{aligned}
L_{\text{NN}} &= \mathbf{E}\{2\eta(X)(1 - \eta(X))\}, \\
L_{3\text{NN}} &= \mathbf{E}\{\eta(X)(1 - \eta(X))\} + 4\mathbf{E}\left\{ (\eta(X)(1 - \eta(X)))^2 \right\}, \\
L_{5\text{NN}} &= \mathbf{E}\{\eta(X)(1 - \eta(X))\} + \mathbf{E}\left\{ (\eta(X)(1 - \eta(X)))^2 \right\} \\
&\qquad + 12\mathbf{E}\left\{ (\eta(X)(1 - \eta(X)))^3 \right\}.
\end{aligned}
$$

The behavior of α_k near zero is very informative. As $p \downarrow 0$, we have

$$
\begin{aligned}
\alpha_1(p) &= 2p(1 - p) \sim 2p, \\
\alpha_3(p) &= p(1 - p)(1 + 4p) \sim p + 3p^2, \\
\alpha_5(p) &\sim p + 10p^3,
\end{aligned}
$$

while for the Bayes error, $L^* = \mathbf{E}\{\min(\eta(X), 1 - \eta(X))\} = \mathbf{E}\{\alpha_\infty(\eta(X))\}$, where $\alpha_\infty = \min(p, 1 - p) \sim p$ as $p \downarrow 0$. Assume that $\eta(x) = p$ at all x. Then, as $p \downarrow 0$,

$$L_{\text{NN}} \sim 2L^* \quad \text{and} \quad L_{3\text{NN}} \sim L^*.$$

Moreover, $L_{\text{NN}} - L^* \sim L^*$, $L_{3\text{NN}} - L^* \sim 3L^{*2}$. Assume that $L^* = p = 0.01$. Then $L_1 \approx 0.02$, whereas $L_{3\text{NN}} - L^* \approx 0.0003$. For all practical purposes, the 3-NN rule is virtually perfect. For this reason, the 3-NN rule is highly recommended. Little is gained by considering the 5-NN rule when p is small, as $L_{5\text{NN}} - L^* \approx 0.00001$.

Let a_k be the smallest number such that $\alpha_k(p) \leq a_k \min(p, 1 - p)$ for all p (the tangents in Figure 5.6). Then

$$
\begin{aligned}
L_{k\text{NN}} = \mathbf{E}\{\alpha_k(\eta(X))\} &\leq a_k \mathbf{E}\{\min(\eta(X), (1 - \eta(X)))\} \\
&= a_k L^*.
\end{aligned}
$$

This is precisely at the basis of the inequalities of Theorems 5.6 through 5.8, where it was shown that $a_k = 1 + O(1/\sqrt{k})$.

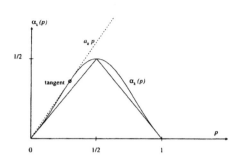

FIGURE 5.6. $\alpha_k(p)$ as a function of p.

5.9   Nearest Neighbor Rules When $L^* = 0$

From Theorem 5.4 we retain that if $L^* = 0$, then $L_{k\text{NN}} = 0$ for all k. In fact, then, for every fixed k, the k-nearest neighbor rule is consistent. Cover has a beautiful example to illustrate this remarkable fact. $L^* = 0$ implies that $\eta(x) \in \{0, 1\}$ for all x, and thus, the classes are separated. This does not imply that the support of X given $Y = 0$ is different from the support of X given $Y = 1$. Take for example a random rational number from $[0, 1]$ (e.g., generate I, J independently and at random from the geometric distribution on $\{1, 2, 3, \ldots\}$, and set $X = \min(I, J)/\max(I, J)$). Every rational number on $[0, 1]$ has positive probability. Given $Y = 1$, X is as above, and given $Y = 0$, X is uniform on $[0, 1]$. Let $\mathbf{P}\{Y = 1\} = \mathbf{P}\{Y = 0\} = 1/2$. The support of X is identical in both cases. As

$$\eta(x) = \begin{cases} 1 & \text{if } x \text{ is rational} \\ 0 & \text{if } x \text{ is irrational,} \end{cases}$$

we see that $L^* = 0$ and that the nearest neighbor rule is consistent. If someone shows us a number X drawn from the same distribution ⹁ ⸱he data, then we may decide the rationality of X merely by looking at the rationality of the nearest neighbor of X. Although we did not show this, the same is true if we are given any $x \in [0, 1]$:

$$\lim_{n \to \infty} \mathbf{P}\{x \text{ is rational } Y_{(1)}(x) = 0 \ (X_{(1)}(x) \text{ is not rational})\}$$

$$= \lim_{n \to \infty} \mathbf{P}\{x \text{ is not rational } Y_{(1)}(x) = 1 \ (X_{(1)}(x) \text{ is rational})\}$$

$$= 0 \quad \text{(see Problem 5.38).}$$

5.10   Admissibility of the Nearest Neighbor Rule

The consistency theorems of Chapter 11 show us that we should take $k = k_n \to \infty$ in the k-NN rule. The decreasing nature of $L_{k\text{NN}}$ corroborates this. Yet, there exist distributions for which for all n, the 1-NN rule is better than the k-NN rule for any $k \geq 3$. This observation, due to Cover and Hart (1967), rests on the following class of examples. Let S_0 and S_1 be two spheres of radius 1 centered at a and b, where $\|a - b\| > 2$. Given $Y = 1$, X is uniform on S_1, while given $Y = 0$, X is uniform on S_0, whereas $\mathbf{P}\{Y = 1\} = \mathbf{P}\{Y = 0\} = 1/2$. We note that given n observations, with the 1-NN rule,

$$\mathbf{E}\{L_n\} = \mathbf{P}\{Y = 0, Y_1 = \cdots = Y_n = 1\} + \mathbf{P}\{Y = 1, Y_1 = \cdots = Y_n = 0\} = \frac{1}{2^n}.$$

For the k-NN rule, k being odd, we have

$$
\begin{aligned}
\mathbf{E}\{L_n\} &= \mathbf{P}\left\{Y = 0, \sum_{i=1}^{n} I_{\{Y_i=0\}} \leq \lfloor k/2 \rfloor\right\} \\
&\quad + \mathbf{P}\left\{Y = 1, \sum_{i=1}^{n} I_{\{Y_i=1\}} \leq \lfloor k/2 \rfloor\right\} \\
&= \mathbf{P}\{\text{Binomial}(n, 1/2) \leq \lfloor k/2 \rfloor\} \\
&= \frac{1}{2^n} \sum_{j=0}^{\lfloor k/2 \rfloor} \binom{n}{j} > \frac{1}{2^n} \quad \text{when } k \geq 3.
\end{aligned}
$$

Hence, the k-NN rule is worse than the 1-NN rule for every n when the distribution is given above. We refer to the exercises regarding some interesting admissibility questions for k-NN rules.

5.11   The (k, l)-Nearest Neighbor Rule

In 1970, Hellman (1970) proposed the (k, l)-nearest neighbor rule, which is identical to the k-nearest neighbor rule, but refuses to make a decision unless at least $l > k/2$ observations are from the same class. Formally, we set

$$
g_n(x) = \begin{cases} 1 & \text{if } \sum_{i=1}^{k} Y_{(i)}(x) \geq l \\ 0 & \text{if } \sum_{i=1}^{k} Y_{(i)}(x) \leq k - l \\ -1 & \text{otherwise (no decision).} \end{cases}
$$

Define the pseudoprobability of error by

$$L_n = \mathbf{P}\{g_n(X) \notin \{-1, Y\}|D_n\},$$

that is, the probability that we reach a decision and correctly classify X. Clearly, $L_n \leq \mathbf{P}\{g_n(X) \neq Y|D_n\}$, our standard probability of error. The latter inequality is only superficially interesting, as the probability of not reaching a decision is not taken into account in L_n. We may extend Theorem 5.2 to show the following (Problem 5.35):

Theorem 5.9. *For the (k, l)-nearest neighbor rule, the pseudoprobability of error L_n satisfies*

$$
\begin{aligned}
\lim_{n \to \infty} \mathbf{E}\{L_n\} \quad = \quad & \mathbf{E}\{\eta(X)\mathbf{P}\{\text{Binomial}(k, \eta(X)) \leq k - l | X\} \\
& + (1 - \eta(X))\mathbf{P}\{\text{Binomial}(k, \eta(X)) \geq l | X\}\} \\
\overset{\text{def}}{=} \quad & L_{k,l}.
\end{aligned}
$$

The above result is distribution-free. Note that the k-nearest neighbor rule for odd k corresponds to $L_{k,(k+1)/2}$. The limit $L_{k,l}$ by itself is not interesting, but it was shown by Devijver (1979) that $L_{k,l}$ holds information regarding the Bayes error L^*.

Theorem 5.10. (DEVIJVER (1979)). *For all distributions and with k odd,*

$$L_{k,k} \leq L_{k,k-1} \leq \cdots \leq L_{k,\lceil k/2 \rceil + 1} \leq L^* \leq L_{k,\lceil k/2 \rceil} = L_{k\text{NN}}.$$

Also,

$$\frac{L_{k\text{NN}} + L_{k,\lceil k/2 \rceil + 1}}{2} \leq L^* \leq L_{k\text{NN}}.$$

This theorem (for which we refer to Problem 5.34) shows that L^* is tightly sandwiched between $L_{k\text{NN}}$, the asymptotic probability of error of the k-nearest neighbor rule, and the "tennis" rule which requires that the difference of votes between the two classes among the k nearest neighbors be at least two. If L_n is close to its limit, and if we can estimate L_n (see the chapters on error estimation), then we may be able to use Devijver's inequalities to obtain estimates of the Bayes error L^*. For additional results, see Loizou and Maybank (1987).

As a corollary of Devijver's inequalities, we note that

$$L_{k\text{NN}} - L^* \leq \frac{L_{k\text{NN}} - L_{k,\lceil k/2 \rceil + 1}}{2}.$$

We have

$$
\begin{aligned}
L_{k,l} \quad = \quad & L^* + \mathbf{E}\{|1 - 2\min(\eta(X), 1 - \eta(X))| \\
& \times \mathbf{P}\{\text{Binomial}(k, \min(\eta(X), 1 - \eta(X))) \geq l | X\}\},
\end{aligned}
$$

and therefore

$$
\begin{aligned}
L_{k,l} &- L_{k,l+1} \\
= \quad & \mathbf{E}\{|1 - 2\min(\eta(X), 1 - \eta(X))| \\
& \times \mathbf{P}\{\text{Binomial}(k, \min(\eta(X), 1 - \eta(X))) = l | X\}\}
\end{aligned}
$$

$$= \mathbf{E}\Big\{|1 - 2\min(\eta(X), 1 - \eta(X))|$$

$$\times \binom{k}{l} \min(\eta(X), 1 - \eta(X))^l (1 - \min(\eta(X), 1 - \eta(X)))^{k-l}\Big\}$$

$$\le \binom{k}{l} \frac{l^l (k-l)^{k-l}}{k^k}$$

(because $u^l(1-u)^{k-l}$ reaches its maximum on $[0, 1]$ at $u = l/k$)

$$\le \sqrt{\frac{k}{12\pi l(k-l)}}$$

(use $\binom{k}{l} \le \frac{k^k}{l^l(k-l)^{k-l}} \frac{1}{\sqrt{2\pi}} \sqrt{\frac{k}{l(k-l)}}$, by Stirling's formula).

With $l = \lceil k/2 \rceil$, we thus obtain

$$L_{k\text{NN}} - L^* \le \frac{1}{2\sqrt{2\pi}} \sqrt{\frac{k}{\lceil k/2 \rceil \lfloor k/2 \rfloor}}$$

$$= \sqrt{\frac{k}{2\pi(k^2 - 1)}} \approx \frac{0.398942}{\sqrt{k}},$$

improving on Theorem 5.6. Various other inequalities may be derived in this manner as well.

Problems and Exercises

PROBLEM 5.1. Let $\| \cdot \|$ be an arbitrary norm on $\mathcal{R}^d$, and define the k-nearest neighbor rule in terms of the distance $\rho(x, z) = \|x - z\|$. Show that Theorems 5.1 and 5.2 remain valid. HINT: Only Stone's lemma needs adjusting. The role of cones $C(x, \pi/6)$ used in the proof are now played by sets with the following property: x and z belong to the same set if and only if

$$\left\| \frac{x}{\|x\|} - \frac{z}{\|z\|} \right\| < 1.$$

PROBLEM 5.2. Does there exist a distribution for which $\sup_{n \ge 1} \mathbf{E}\{L_n\} > 1/2$ for the nearest neighbor rule?

PROBLEM 5.3. Show that $L_{3\text{NN}} \le 1.32L^*$ and that $L_{5\text{NN}} \le 1.22L^*$.

PROBLEM 5.4. Show that if C^* is a compact subset of $\mathcal{R}^d$ and C is the support set for the probability measure μ,

$$\sup_{x \in C \cap C^*} \|X_{(1)} - x\| \to 0$$

with probability one, where $X_{(1)}$ is the nearest neighbor of x among $X_1, \ldots, X_n$ (Wagner (1971)).

PROBLEM 5.5. Let μ be the probability measure of X given $Y = 0$, and let ν be the probability measure of X given $Y = 1$. Assume that X is real-valued and that $P\{Y = 0\} = P\{Y = 1\} = 1/2$. Find a pair (ν, μ) such that

    (1)   support(μ) = support(ν);

    (2) $L^* = 0$.

Conclude that $L^* = 0$ does not tell us a lot about the support sets of μ and ν.

PROBLEM 5.6. Consider the $(2k + 1)$-nearest neighbor rule for distributions (X, Y) with $\eta(x) \equiv p$ constant, and Y independent of X. This exercise explores the behavior of $L_{(2k+1)NN}$ as $p \downarrow 0$.

    (1)   For fixed integer $l > 0$, as $p \downarrow 0$, show that

$$P\{\text{Binomial}(2k, p) \geq l\} \sim \binom{2k}{l} p^l.$$

    (2)   Use a convenient representation of $L_{(2k+1)NN}$ to conclude that as $p \downarrow 0$,

$$L_{(2k+1)NN} = p + \left(\binom{2k}{k} + \binom{2k}{k+1}\right) p^{k+1} + o(p^{k+1}).$$

PROBLEM 5.7. Das Gupta and Lin (1980) proposed the following rule for data with $X \in \mathcal{R}$. Assume X is nonatomic. First, reorder $X_1, \ldots, X_n$, X according to increasing values, and denote the ordered set by $X_{(1)} < X_{(2)} < \cdots < X_{(i)} < X < X_{(i+1)} < \cdots < X_{(n)}$. The Y_i's are permuted so that $Y_{(j)}$ is the label of $X_{(j)}$. Take votes among $\{Y_{(i)}, Y_{(i+1)}\}$, $\{Y_{(i-1)}, Y_{(i+2)}\}, \ldots$ until for the first time there is agreement $(Y_{(i-j)} = Y_{(i+j+1)})$, at which time we decide that class, that is, $g_n(X) = Y_{(i-j)} = Y_{(i+j+1)}$. This rule is invariant under monotone transformations of the x-axis.

    (1)   If L denotes the asymptotic expected probability of error, show that for all non-atomic X,

$$L = \mathbf{E}\left\{\frac{\eta(X)(1 - \eta(X))}{1 - 2\eta(X)(1 - \eta(X))}\right\}.$$

    (2)   Show that L is the same as for the rule in which $X \in \mathcal{R}^d$ and we consider 2-NN, 4-NN, 6-NN, etc. rules in turn, stopping at the first $2k$-NN rule for which there is no voting tie. Assume for simplicity that X has a density (with a good distance-tie breaking rule, this may be dropped).

    (3)   Show that $L - L_{NN} \geq (1/2)(L_{3NN} - L_{NN})$, and thus that $L \geq (L_{NN} + L_{3NN})/2$.

    (4)   Show that $L \leq L_{NN}$. Hence, the rule performs somewhere in between the 1-NN and 3-NN rules.

PROBLEM 5.8. Let Y be independent of X, and $\eta(x) \equiv p$ constant. Consider a weighted $(2k + 1)$-nearest neighbor rule with weights $(2m + 1, 1, 1, \ldots, 1)$ (there are $2k$ "ones"), where $k - 1 \geq m \geq 0$. For $m = 0$, we obtain the $(2k + 1)$-NN rule. Let $L(k, m)$ be the asymptotic probability of error.

    (1)   Using results from Problem 5.6, show that

$$L(k, m) = p + \binom{2k}{k - m} p^{k-m+1} + \binom{2k}{k + m + 1} p^{k+m+1} + o(p^{k-m+1})$$

as $p \downarrow 0$. Conclude that within this class of rules, for small p, the goodness of a rule is measured by $k - m$.

(2) Let $\delta > 0$ be small, and set $p = 1/2 - \delta$. Show that if X is binomial $(2k, 1/2 - \delta)$ and Z is binomial $(2k, 1/2)$, then for fixed l, as $\delta \downarrow 0$,

$$P\{X \geq l\} = P\{Z \geq l\} - 2k\delta P\{Z = l\} + o(\delta^2),$$

and

$$P\{X \leq l\} = P\{Z \leq l\} + 2k\delta P\{Z = l + 1\} + o(\delta^2).$$

(3) Conclude that for fixed k, m as $\delta \downarrow 0$,

$$L(k, m) = \frac{1}{2} - 2\delta^2(kP\{Z = k + m\} + kP\{Z = k + m + 1\}$$
$$+ P\{Z \leq k + m\} - P\{Z \geq k + m + 1\}) + o(\delta^2).$$

(4) Take weight vector w with k fixed and $m = \lfloor 10\sqrt{k} \rfloor$, and compare it with weight vector w' with $k/2$ components and $m = \lfloor \sqrt{k/2} \rfloor$ as $p \downarrow 0$ and $p \uparrow 1/2$. Assume that k is very large but fixed. In particular, show that w is better as $p \downarrow 0$, and w' is better as $p \uparrow 1/2$. For the last example, note that for fixed $c > 0$,

$$kP\{Z = k + m\} + kP\{Z = k + m + 1\} \sim 8\sqrt{k}\frac{1}{\sqrt{2\pi}}e^{-2c^2} \quad \text{as} \quad k \to \infty$$

by the central limit theorem.

(5) Conclude that there exist different weight vectors w, w' for which there exists a pair of distributions of (X, Y) such that their asymptotic error probabilities are differently ordered. Thus, $\mathcal{W}$ is not totally ordered with respect to the probability of error.

PROBLEM 5.9. Patrick and Fisher (1970) find the k-th nearest neighbor in each of the two classes and classify according to which is nearest. Show that their rule is equivalent to a $(2k - 1)$-nearest neighbor rule.

PROBLEM 5.10. Rabiner et al. (1979) generalize the rule of Problem 5.9 so as to classify according to the average distance to the k-th nearest neighbor within each class. Assume that X has a density. For fixed k, find the asymptotic probability of error.

PROBLEM 5.11. If N is normal $(0, 1)$, then $E\{|N|\} = \sqrt{2/\pi}$. Prove this.

PROBLEM 5.12. Show that if $L_{9NN} = L_{(11)NN}$, then $L_{(99)NN} = L_{(111)NN}$.

PROBLEM 5.13. Show that

$$L_{3NN} \leq \left(\frac{7\sqrt{7} + 17}{27\sqrt{3}} + 1\right) L^* \approx 1.3155\ldots L^*$$

for all distributions (Devroye (1981b)). HINT: Find the smallest constant a such that $L_{3NN} \leq L^*(1 + a)$ using the representation of L_{3NN} in terms of the binomial tail.

PROBLEM 5.14. Show that if X has a density f, then for all $u > 0$,

$$\lim_{n \to \infty} P\left\{n^{1/d}\|X_{(1)}(X) - X\| > u|X\right\} = e^{-f(X)vu^d}$$

with probability one, where $v = \int_{S_{0,1}} dx$ is the volume of the unit ball in $\mathcal{R}^d$ (Györfi (1978)).

PROBLEM 5.15. Consider a rule that takes a majority vote over all Y_i's for which $\| X_i - x \| \le (c/vn)^{1/d}$, where $v = \int_{S_{0.1}} dx$ is the volume of the unit ball, and $c > 0$ is fixed. In case of a tie decide $g_n(x) = 0$.

(1) If X has a density f, show that $\liminf_{n \to \infty} \mathbf{E}\{L_n\} \ge \mathbf{E}\left\{\eta(X)e^{-cf(X)}\right\}$. HINT: Use the obvious inequality $\mathbf{E}\{L_n\} \ge \mathbf{P}\{Y = 1, \mu_n(S_{X,c/vn}) = 0\}$.

(2) If Y is independent of X and $\eta \equiv p > 1/2$, then

$$\frac{\mathbf{E}\left\{\eta(X)e^{-cf(X)}\right\}}{L^*} = \mathbf{E}\left\{e^{-cf(X)}\right\} \frac{p}{1-p} \uparrow \infty$$

as $p \uparrow 1$. Show this.

(3) Conclude that

$$\sup_{(X,Y):L^* > 0} \frac{\liminf_{n \to \infty} \mathbf{E}\{L_n\}}{L^*} = \infty,$$

and thus that distribution-free bounds of the form $\lim_{n \to \infty} \mathbf{E}\{L_n\} \le c'L^*$ obtained for k-nearest neighbor estimates do not exist for these simple rules (Devroye (1981a)).

PROBLEM 5.16. Take an example with $\eta(X) \equiv 1/2 - 1/(2\sqrt{k})$, and show that the bound $L_{k\text{NN}} - L^* \le 1/\sqrt{ke}$ cannot be essentially bettered for large values of k, that is, there exists a sequence of distributions (indexed by k) for which

$$L_{k\text{NN}} - L^* \ge \frac{1 - o(1)}{\sqrt{k}} \mathbf{P}\{N \ge 1\}$$

as $k \to \infty$, where N is a normal $(0, 1)$ random variable.

PROBLEM 5.17. If B is binomial (n, p), then

$$\sup_{p} \mathbf{P}\{B = i\} \le \sqrt{\frac{n}{2\pi i(n - i)}}, \quad 0 < i < n.$$

PROBLEM 5.18. Show that for $k \ge 3$,

$$\frac{\sqrt{k(k + 1)}}{k - 1} \le \left(1 + \frac{1}{2k}\right)\left(1 + \frac{3}{2k}\right) = 1 + \frac{2}{k} + \frac{3}{4k^2}.$$

PROBLEM 5.19. Show that there exists a sequence of distributions of (X, Y) (indexed by k) in which Y is independent of X and $\eta(x) \equiv p$ (with p depending on k only) such that

$$\liminf_{n \to \infty} \left(\frac{L_{k\text{NN}} - L^*}{L^*}\right) \sqrt{k} \ge \gamma = 0.339942\ldots,$$

where γ is the constant of Theorem 5.8 (Devroye (1981b)). HINT: Verify the proof of Theorem 5.8 but bound things from below. Slud's inequality (see Lemma A.6 in the Appendix) may be of use here.

PROBLEM 5.20. Consider a weighted nearest neighbor rule with weights $1, \rho, \rho^2, \rho^3, \ldots$ for $\rho < 1$. Show that the expected probability of error tends for all distributions to a limit $L(\rho)$. HINT: Truncate at k fixed but large, and argue that the tail has asymptotically negligible weight.

PROBLEM 5.21. CONTINUED. With $L(\rho)$ as in the previous exercise, show that $L(\rho) = L_{NN}$ whenever $\rho < 1/2$.

PROBLEM 5.22. CONTINUED. Prove or disprove: as ρ increases from $1/2$ to 1, $L(\rho)$ decreases monotonically from L_{NN} to L^*. (This question is difficult.)

PROBLEM 5.23. Show that in the weighted NN rule with weights $(1, \rho, \rho^2)$, $0 < \rho < 1$, the asymptotic probability of error is L_{NN} if $\rho < (\sqrt{5} - 1)/2$ and is L_{3NN} if $\rho > (\sqrt{5} - 1)/2$.

PROBLEM 5.24. Is there any k, other than one, for which the k-NN rule is admissible, that is, for which there exists a distribution of (X, Y) such that $E\{L_n\}$ for the k-NN rule is smaller than $E\{L_n\}$ for any k'-NN rule with $k' \neq k$, for all n? HINT: This is difficult. Note that if this is to hold for all n, then it must hold for the limits. From this, deduce that with probability one, $\eta(x) \in \{0, 1/2, 1\}$ for any such distribution.

PROBLEM 5.25. For every fixed n and odd k with $n > 1000k$, find a distribution of (X, Y) such that $E\{L_n\}$ for the k-NN rule is smaller than $E\{L_n\}$ for any k'-NN rule with $k' \neq k$, k' odd. Thus, for a given n, no k can be a priori discarded from consideration.

PROBLEM 5.26. Let X be uniform on $[0, 1]$, $\eta(x) \equiv x$, and $P\{Y = 0\} = P\{Y = 1\} = 1/2$. Show that for the nearest neighbor rule,

$$E\{L_n\} = \frac{1}{3} + \frac{3n + 5}{2(n + 1)(n + 2)(n + 3)}$$

(Cover and Hart (1967); Peterson (1970)).

PROBLEM 5.27. For the nearest neighbor rule, if X has a density, then

$$|E\{L_n\} - E\{L_{n+1}\}| \leq \frac{1}{n + 1}$$

(Cover (1968a)).

PROBLEM 5.28. Let X have a density $f \geq c > 0$ on $[0, 1]$, and assume that f_0''' and f_1''' exist and are uniformly bounded. Show that for the nearest neighbor rule, $E\{L_n\} = L_{NN} + O(1/n^2)$ (Cover (1968a)). For d-dimensional problems this result was generalized by Psaltis, Snapp, and Venkatesh (1994).

PROBLEM 5.29. Show that $L_{kNN} \leq (1 + \sqrt{2/k})L^*$ is the best possible bound of the form $L_{kNN} \leq (1 + a/\sqrt{k})L^*$ valid simultaneously for all $k \geq 1$ (Devroye (1981b)).

PROBLEM 5.30. Show that $L_{kNN} \leq (1 + \sqrt{1/k})L^*$ for all $k \geq 3$ (Devroye (1981b)).

PROBLEM 5.31. Let $x = (x(1), x(2)) \in \mathcal{R}^2$. Consider the nearest neighbor rule based upon vectors with components $(x^3(1), x^7(2), x(1)x(2))$. Show that this is asymptotically not better than if we had used $(x(1), x(2))$. Show by example that $(x^2(1), x^3(2), x^6(1)x(2))$ may yield a worse asymptotic error probability than $(x(1), x(2))$.

PROBLEM 5.32. UNIFORM ORDER STATISTICS. Let $U_{(1)} < \cdots < U_{(n)}$ be order statistics of n i.i.d. uniform $[0, 1]$ random variables. Show the following:
   (1) $U_{(k)}$ is beta $(k, n + 1 - k)$, that is, $U_{(k)}$ has density

$$f(x) = \frac{n!}{(k - 1)!(n - k)!} x^{k-1}(1 - x)^{n-k}, \quad 0 \leq x \leq 1.$$

(2)

$$\mathbf{E}\left\{U_{(k)}^a\right\} = \frac{\Gamma(k+a)\Gamma(n+1)}{\Gamma(k)\Gamma(n+1+a)}, \quad \text{for any } a > 0.$$

(3)

$$1 - \frac{a}{n} \le \frac{\mathbf{E}\left\{U_{(k)}^a\right\}}{(k/n)^a} \le 1 + \frac{\psi(a)}{k}$$

for $a \ge 1$, where $\psi(a)$ is a function of a only (Royall (1966)).

PROBLEM 5.33. DUDANI'S RULE. Dudani (1976) proposes a weighted k-NN rule where $Y_{(i)}(x)$ receives weight

$$\|X_{(k)}(x) - x\| - \|X_{(i)}(x) - x\|, \quad 1 \le i \le k.$$

Why is this roughly speaking equivalent to attaching weight $1 - (i/k)^{1/d}$ to the i-th nearest neighbor if X has a density? HINT: If μ is the probability measure of X, then

$$\mu\left(S_{x, \|X_{(1)}(x)-x\|}\right), \ldots, \mu\left(S_{x, \|X_{(k)}(x)-x\|}\right)$$

are distributed like $U_{(1)}, \ldots, U_{(k)}$ where $U_{(1)} < \cdots < U_{(n)}$ are the order statistics of n i.i.d. uniform $[0, 1]$ random variables. Replace μ by a good local approximation, and use results from the previous exercise.

PROBLEM 5.34. Show Devijver's theorem (Theorem 5.10) in two parts: first establish the inequality $L^* \ge L_{k,\lceil k/2\rceil - 1}$ for the tennis rule, and then establish the monotonicity.

PROBLEM 5.35. Show Theorem 5.9 for the (k, l) nearest neighbor rule.

PROBLEM 5.36. Let R be the asymptotic error probability of the $(2, 2)$-nearest neighbor rule. Prove that $R = \mathbf{E}\{2\eta(X)(1 - \eta(X))\} = L_{NN}$.

PROBLEM 5.37. For the nearest neighbor rule, show that for all distributions,

$$\lim_{n \to \infty} \mathbf{P}\{g_n(X) = 0, Y = 1\} = \lim_{n \to \infty} \mathbf{P}\{g_n(X) = 1, Y = 0\}$$

$$= \mathbf{E}\{\eta(X)(1 - \eta(X))\}$$

(Devijver and Kittler (1982)). Thus, errors of both kinds are equally likely.

PROBLEM 5.38. Let $\mathbf{P}\{Y = 1\} = \mathbf{P}\{Y = 0\} = 1/2$ and let X be a random rational if $Y = 1$ (as defined in Section 5.9) such that every rational number has positive probability, and let X be uniform $[0, 1]$ if $Y = 0$. Show that for every $x \in [0, 1]$ not rational, $\mathbf{P}\{Y_{(1)}(x) = 1\} \to 0$ as $n \to \infty$, while for every $x \in [0, 1]$ rational, $\mathbf{P}\{Y_{(1)}(x) = 0\} \to 0$ as $n \to \infty$.

PROBLEM 5.39. Let $X_1, \ldots, X_n$ be i.i.d. and have a common density. Show that for fixed $k > 0$,

$$n\mathbf{P}\{X_3 \text{ is among the } k \text{ nearest neighbors of } X_1 \text{ and } X_2 \text{ in } \{X_3, \ldots, X_n\}\} \to 0.$$

Show that the same result remains valid whenever k varies with n such that $k/\sqrt{n} \to 0$.

PROBLEM 5.40. IMPERFECT TRAINING. Let (X, Y, Z), $(X_1, Y_1, Z_1), \ldots, (X_n, Y_n, Z_n)$ be a sequence of i.i.d. triples in $\mathcal{R}^d \times \{0, 1\} \times \{0, 1\}$ with $\mathbf{P}\{Y = 1 | X = x\} = \eta(x)$ and $\mathbf{P}\{Z = 1 | X = x\} = \eta'(x)$. Let $Z_{(1)}(X)$ be Z_j if X_j is the nearest neighbor of X among $X_1, \ldots, X_n$. Show that

$$\lim_{n \to \infty} \mathbf{P}\{Z_{(1)}(X) \neq Y\} = \mathbf{E}\left\{\eta(X) + \eta'(X) - 2\eta(X)\eta'(X)\right\}$$

(Lugosi (1992)).

PROBLEM 5.41. Improve the bound in Lemma 5.3 to $\gamma_d \leq 3^d - 1$.

PROBLEM 5.42. Show that if $\{C(x_1, \pi/6), \ldots, C(x_{\gamma_d}, \pi/6)\}$ is a collection of cones covering $\mathcal{R}^d$, then $\gamma_d \geq 2^d$.

PROBLEM 5.43. Recalling that $L_{k\text{NN}} = \mathbf{E}\{\alpha_k(\eta(X))\}$, where

$$\alpha_k(p) = \min(p, 1 - p) + |1 - 2\min(p, 1 - p)|\mathbf{P}\left\{\text{Binomial}(k, \min(p, 1 - p)) > \frac{k}{2}\right\},$$

show that for every fixed p, $\mathbf{P}\left\{\text{Binomial}(k, \min(p, 1 - p)) > \frac{k}{2}\right\} \downarrow 0$ as $k \to \infty$ (it is the monotonicity that is harder to show). How would you then prove that $\lim_{k \to \infty} L_{k\text{NN}} = L^*$?

PROBLEM 5.44. Show that the asymptotic error probability of the rule that decides $g_n(x) = Y_{(8)}(x)$ is identical to that of the rule in which $g_n(x) = Y_{(3)}(x)$.

PROBLEM 5.45. Show that for all distributions $L_{5\text{NN}} = \mathbf{E}\{\psi_5(\eta(X)(1 - \eta(X)))\}$, where $\psi_5(u) = u + u^2 + 12u^3$.

PROBLEM 5.46. Show that for all distributions,

$$L_{5\text{NN}} \geq \frac{L_{\text{NN}}}{2} + \frac{L_{\text{NN}}^2}{4} + \frac{3L_{\text{NN}}^3}{2}$$

and that

$$L_{3\text{NN}} \geq \frac{L_{\text{NN}}}{2} + L_{\text{NN}}^2.$$

PROBLEM 5.47. Let $X_{(1)}$ be the nearest neighbor of x among $X_1, \ldots, X_n$. Construct an example for which $\mathbf{E}\{\|X_{(1)} - x\|\} = \infty$ for all $x \in \mathcal{R}^d$. (Therefore, we have to steer clear of convergence in the mean in Lemma 5.1.) Let $X_{(1)}$ be the nearest neighbor of X_1 among $X_2, \ldots, X_n$. Construct a distribution such that $\mathbf{E}\{\|X_{(1)} - X_1\|\} = \infty$ for all n.

PROBLEM 5.48. Consider the weighted nearest neighbor rule with weights $(w_1, \ldots, w_k)$. Define a new weight vector $(w_1, w_2, \ldots, w_{k-1}, v_1, \ldots, v_l)$, where $\sum_{i=1}^{l} v_i = w_k$. Thus, the weight vectors are partially ordered by the operation "partition." Assume that all weights are nonnegative. Let the asymptotic expected probability of errors be L and L', respectively. True or false: for all distributions of (X, Y), $L' \leq L$.

PROBLEM 5.49. GABRIEL NEIGHBORS. Given $X_1, \ldots, X_n \in \mathcal{R}^d$, we say that X_i and X_j are Gabriel neighbors if the ball centered at $(X_i + X_j)/2$ of radius $\|X_i - X_j\|/2$ contains no X_k, $k \neq i, j$ (Gabriel and Sokal (1969); Matula and Sokal (1980)). Clearly, if X_j is the nearest neighbor of X_i, then X_i and X_j are Gabriel neighbors. Show that if X has a density and $X_1, \ldots, X_n$ are i.i.d. and drawn from the distribution of X, then the expected number of Gabriel neighbors of X_1 tends to 2^d as $n \to \infty$ (Devroye (1988c)).

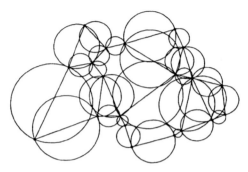

FIGURE 5.7. *The Gabriel graph of 20 points on the plane is shown: Gabriel neighbors are connected by an edge. Note that all circles straddling these edges have no data point in their interior.*

PROBLEM 5.50. GABRIEL NEIGHBOR RULE. Define the Gabriel neighbor rule simply as the rule that takes a majority vote over all Y_i's for the Gabriel neighbors of X among $X_1, \ldots, X_n$. Ties are broken by flipping a coin. Let L_n be the conditional probability of error for the Gabriel rule. Using the result of the previous exercise, show that if L^* is the Bayes error, then

(1) $\lim_{n \to \infty} \mathbf{E}\{L_n\} = 0$ if $L^* = 0$;

(2) $\limsup_{n \to \infty} \mathbf{E}\{L_n\} < L_{NN}$ if $L^* > 0, d > 1$;

(3) $\limsup_{n \to \infty} \mathbf{E}\{L_n\} \le cL^*$ for some $c < 2$, if $d > 1$.

For (3), determine the best possible value of c. HINT: Use Theorem 5.8 and try obtaining, for $d = 2$, a lower bound for $\mathbf{P}\{N_X \ge 3\}$, where N_X is the number of Gabriel neighbors of X among $X_1, \ldots, X_n$.

6

Consistency

Classification Rule : a sequence of classification functions $\{g_n\}$ based on data $\{D_n\}$

The error $L_n \to L^*$ consistency

6.1   Universal Consistency

If we are given a sequence $D_n = ((X_1, Y_1), \ldots, (X_n, Y_n))$ of training data, the best we can expect from a classification function is to achieve the Bayes error probability L^*. Generally, we cannot hope to obtain a function that exactly achieves the Bayes error probability, but it is possible to construct a sequence of classification functions $\{g_n\}$, that is, *a classification rule*, such that the error probability

$$L_n = L(g_n) = \mathbf{P}\{g_n(X, D_n) \neq Y | D_n\}$$

gets arbitrarily close to L^* with large probability (that is, for "most" D_n). This idea is formulated in the definitions of *consistency*.

DEFINITION 6.1. (WEAK AND STRONG CONSISTENCY). *A classification rule is consistent (or asymptotically Bayes-risk efficient) for a certain distribution of (X, Y) if*

$$\mathbf{E}L_n = \mathbf{P}\{g_n(X, D_n) \neq Y\} \to L^* \quad as \ n \to \infty,$$

and strongly consistent if

$$\lim_{n \to \infty} L_n = L^* \quad with \ probability \ 1.$$

REMARK. Consistency is defined as the convergence of the expected value of L_n to L^*. Since L_n is a random variable bounded between L^* and 1, this convergence

L_n. R.V. $[L^*, 1]$

is equivalent to the convergence of L_n to L^* in probability, which means that for every $\epsilon > 0$

$$\lim_{n\to\infty} \mathbf{P}\{L_n - L^* > \epsilon\} = 0.$$

Obviously, since almost sure convergence always implies convergence in probability, strong consistency implies consistency. □

$\lim_{n \to \infty} P\{ L_n - L^* < \epsilon \} = 1$

A consistent rule guarantees that by increasing the amount of data the probability that the error probability is within a very small distance of the optimal achievable gets arbitrarily close to one. Intuitively, the rule can eventually learn the optimal decision from a large amount of training data with high probability. Strong consistency means that by using more data the error probability gets arbitrarily close to the optimum for *every* training sequence except for a set of sequences that has zero probability altogether.

A decision rule can be consistent for a certain class of distributions of (X, Y), but may not be consistent for others. It is clearly desirable to have a rule that is consistent for a large class of distributions. Since in many situations we do not have any prior information about the distribution, it is essential to have a rule that gives good performance for *all* distributions. This very strong requirement of universal goodness is formulated as follows:

DEFINITION 6.2. (UNIVERSAL CONSISTENCY). *A sequence of decision rules is called universally (strongly) consistent if it is (strongly) consistent for any distribution of the pair (X, Y).*

In this chapter we show that such universally consistent classification rules exist. At first, this may seem very surprising, for some distributions are very "strange," and seem hard to learn. For example, let X be uniformly distributed on $[0, 1]$ with probability $1/2$, and let X be atomic on the rationals with probability $1/2$. For example, if the rationals are enumerated $r_1, r_2, r_3, \ldots$, then $\mathbf{P}\{X = r_i\} = 1/2^{i+1}$. Let $Y = 1$ if X is rational and $Y = 0$ if X is irrational. Obviously, $L^* = 0$. If a classification rule g_n is consistent, then the probability of incorrectly guessing the rationality of X tends to zero. Note here that we cannot "check" whether X is rational or not, but we should base our decision solely on the data D_n given to us. One consistent rule is the following: $g_n(x, D_n) = Y_k$ if X_k is the closest point to x among $X_1, \ldots, X_n$. The fact that the rationals are dense in $[0, 1]$ makes the statement even more surprising. See Problem 6.3.

6.2   Classification and Regression Estimation

based on estimated parameter

In this section we show how consistency of classification rules can be deduced from consistent regression estimation. In many cases the a posteriori probability $\eta(x)$ is estimated from the training data D_n by some function $\eta_n(x) = \eta_n(x, D_n)$.

parameter estimated from data

In this case, the error probability $L(g_n) = \mathbf{P}\{g_n(X) \neq Y | D_n\}$ of the plug-in rule

$$g_n(x) = \begin{cases} 0 & \text{if } \eta_n(x) \leq 1/2 \\ 1 & \text{otherwise} \end{cases}$$

is a random variable. Then a simple corollary of Theorem 2.2 is as follows:

COROLLARY 6.1. *The error probability of the classifier $g_n(x)$ defined above satisfies the inequality*

$$L(g_n) - L^* \leq 2 \int_{\mathcal{R}^d} |\eta(x) - \eta_n(x)| \mu(dx) = 2\,\mathbf{E}\{|\eta(X) - \eta_n(X)| \,|\, D_n\}.$$

The next corollary follows from the Cauchy-Schwarz inequality.

COROLLARY 6.2. *If*

$$g_n(x) = \begin{cases} 0 & \text{if } \eta_n(x) \leq 1/2 \\ 1 & \text{otherwise,} \end{cases}$$

then its error probability satisfies

$$\mathbf{P}\{g_n(X) \neq Y | D_n\} - L^* \leq 2\sqrt{\int_{\mathcal{R}^d} |\eta(x) - \eta_n(x)|^2 \mu(dx)}.$$

Clearly, $\eta(x) = \mathbf{P}\{Y = 1 | X = x\} = \mathbf{E}\{Y | X = x\}$ is just the regression function of Y on X. Therefore, the most interesting consequence of Theorem 2.2 is that the mere existence of a regression function estimate $\eta_n(x)$ for which

$$\int |\eta(x) - \eta_n(x)|^2 \mu(dx) \to 0$$

in probability or with probability one implies that the plug-in decision rule g_n is consistent or strongly consistent, respectively.

Clearly, from Theorem 2.3, one can arrive at a conclusion analogous to Corollary 6.1 when the probabilities $\eta_0(x) = \mathbf{P}\{Y = 0 | X = x\}$ and $\eta_1(x) = \mathbf{P}\{Y = 1 | X = x\}$ are estimated from data separately by some $\eta_{0,n}$ and $\eta_{1,n}$, respectively. Usually, a key part of proving consistency of classification rules is writing the rules in one of the plug-in forms, and showing L_1-convergence of the approximating functions to the a posteriori probabilities. Here we have some freedom, as for any positive function $\tau_n(x)$, we have, for example,

$$g_n(x) = \begin{cases} 0 & \text{if } \eta_{1,n}(x) \leq \eta_{0,n}(x) \\ 1 & \text{otherwise,} \end{cases} = \begin{cases} 0 & \text{if } \frac{\eta_{1,n}(x)}{\tau_n(x)} \leq \frac{\eta_{0,n}(x)}{\tau_n(x)} \\ 1 & \text{otherwise.} \end{cases}$$

6.3 Partitioning Rules

Many important classification rules partition $\mathcal{R}^d$ into disjoint cells $A_1, A_2, \ldots$ and classify in each cell according to the majority vote among the labels of the X_i's falling in the same cell. More precisely,

$$g_n(x) = \begin{cases} 0 & \text{if } \sum_{i=1}^{n} I_{\{Y_i=1\}} I_{\{X_i \in A(x)\}} \leq \sum_{i=1}^{n} I_{\{Y_i=0\}} I_{\{X_i \in A(x)\}} \\ 1 & \text{otherwise,} \end{cases}$$

where $A(x)$ denotes the cell containing x. The decision is zero if the number of ones does not exceed the number of zeros in the cell where X falls, and vice versa. The partitions we consider in this section may change with n, and they may also depend on the points $X_1, \ldots, X_n$, but we assume that the labels do not play a role in constructing the partition. The next theorem is a general consistency result for such partitioning rules. It requires two properties of the partition: first, cells should be small enough so that local changes of the distribution can be detected. On the other hand, cells should be large enough to contain a large number of points so that averaging among the labels is effective. diam(A) denotes the diameter of a set A, that is,

$$\text{diam}(A) = \sup_{x, y \in A} \|x - y\|.$$

Let

$$N(x) = n\mu_n(A(x)) = \sum_{i=1}^{n} I_{\{X_i \in A(x)\}}$$

denote the number of X_i's falling in the same cell as x. The conditions of the theorem below require that a random cell—selected according to the distribution of X—has a small diameter, and contains many points with large probability.

Theorem 6.1. *Consider a partitioning classification rule as defined above. Then* $\mathbf{E}\{L_n\} \to L^*$ *if*
   (1) diam($A(X)$) $\to 0$ *in probability,*
   (2) $N(X) \to \infty$ *in probability.*

PROOF. Define $\eta(x) = \mathbf{P}\{Y = 1 | X = x\}$. From Corollary 6.1 we recall that we need only show $\mathbf{E}\{|\widehat{\eta}_n(X) - \eta(X)|\} \to 0$, where

$$\widehat{\eta}_n(x) = \frac{1}{N(x)} \sum_{i: X_i \in A(x)} Y_i.$$

Introduce $\bar{\eta}(x) = \mathbf{E}\{\eta(X) | X \in A(x)\}$. By the triangle inequality,

$$\mathbf{E}\{|\widehat{\eta}_n(X) - \eta(X)|\} \leq \mathbf{E}\{|\widehat{\eta}_n(X) - \bar{\eta}(X)|\} + \mathbf{E}\{|\bar{\eta}(X) - \eta(X)|\}.$$

By conditioning on the random variable $N(x)$, it is easy to see that $N(x)\widehat{\eta}_n(x)$ is distributed as $B(N(x), \bar{\eta}(x))$, a binomial random variable with parameters $N(x)$

and $\bar{\eta}(x)$. Thus,

$$\mathbf{E}\{|\widehat{\eta}_n(X) - \bar{\eta}(X)|\,|X, I_{\{X_1 \in A(X)\}}, \dots, I_{\{X_n \in A(X)\}}\}$$

$$\leq \mathbf{E}\left\{\left|\frac{B(N(X), \bar{\eta}(X))}{N(X)} - \bar{\eta}(X)\right| I_{\{N(X)>0\}}\,\middle|\, X, I_{\{X_1 \in A(X)\}}, \dots, I_{\{X_n \in A(X)\}}\right\}$$

$$+ I_{\{N(X)=0\}}$$

$$\leq \mathbf{E}\left\{\sqrt{\frac{\bar{\eta}(X)(1 - \bar{\eta}(X))}{N(X)}}\, I_{\{N(X)>0\}}\,\middle|\, X, I_{\{X_1 \in A(X)\}}, \dots, I_{\{X_n \in A(X)\}}\right\}$$

$$+ I_{\{N(X)=0\}}$$

by the Cauchy-Schwarz inequality. Taking expectations, we see that

$$\mathbf{E}\{|\widehat{\eta}_n(X) - \bar{\eta}(X)|\} \leq \mathbf{E}\left\{\frac{1}{2\sqrt{N(X)}} I_{\{N(X)>0\}}\right\} + \mathbf{P}\{N(X) = 0\}$$

$$\leq \frac{1}{2}\mathbf{P}\{N(X) \leq k\} + \frac{1}{2\sqrt{k}} + \mathbf{P}\{N(X) = 0\}$$

for any k, and this can be made small, first by choosing k large enough and then by using condition (2).

For $\epsilon > 0$, find a uniformly continuous $[0, 1]$-valued function η_ϵ on a bounded set C and vanishing off C so that $\mathbf{E}\{|\eta_\epsilon(X) - \eta(X)|\} < \epsilon$. Next, we employ the triangle inequality:

$$\mathbf{E}\{|\bar{\eta}(X) - \eta(X)|\} \leq \mathbf{E}\{|\bar{\eta}(X) - \bar{\eta}_\epsilon(X)|\}$$

$$+ \mathbf{E}\{|\bar{\eta}_\epsilon(X) - \eta_\epsilon(X)|\}$$

$$+ \mathbf{E}\{|\eta_\epsilon(X) - \eta(X)|\}$$

$$= I + II + III,$$

where $\bar{\eta}_\epsilon(x) = \mathbf{E}\{\eta_\epsilon(X)|X \in A(x)\}$. Clearly, $III < \epsilon$ by choice of η_ϵ. Since η_ϵ is uniformly continuous, we can find a $\theta = \theta(\epsilon) > 0$ such that

$$II \leq \epsilon + \mathbf{P}\{\text{diam}(A(X)) > \theta\}.$$

Therefore, $II < 2\epsilon$ for n large enough, by condition (1). Finally, $I \leq III < \epsilon$. Taken together these steps prove the theorem. $\square$

6.4   The Histogram Rule

In this section we describe the cubic histogram rule and show its universal consistency by checking the conditions of Theorem 6.1. The rule partitions $\mathcal{R}^d$ into cubes of the same size, and makes a decision according to the majority vote among the Y_i's such that the corresponding X_i falls in the same cube as X. Formally, let $\mathcal{P}_n = \{A_{n1}, A_{n2}, \ldots\}$ be a partition of $\mathcal{R}^d$ into cubes of size $h_n > 0$, that is, into sets of the type $\prod_{i=1}^d [k_i h_n, (k_i + 1)h_n)$, where the k_i's are integers. For every $x \in \mathcal{R}^d$ let $A_n(x) = A_{ni}$ if $x \in A_{ni}$. The histogram rule is defined by

$$g_n(x) = \begin{cases} 0 & \text{if } \sum_{i=1}^n I_{\{Y_i=1\}} I_{\{X_i \in A_n(x)\}} \leq \sum_{i=1}^n I_{\{Y_i=0\}} I_{\{X_i \in A_n(x)\}} \\ 1 & \text{otherwise.} \end{cases}$$

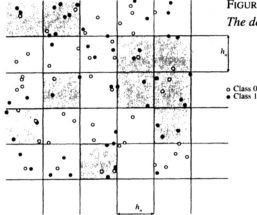

FIGURE 6.1. *A cubic histogram rule: The decision is 1 in the shaded area.*

o Class 0
• Class 1

Consistency of the histogram rule was established under some additional conditions by Glick (1973). Universal consistency follows from the results of Gordon and Olshen (1978), (1980). A direct proof of strong universal consistency is given in Chapter 9.

The next theorem establishes universal consistency of certain cubic histogram rules.

Theorem 6.2. *If $h_n \to 0$ and $n h_n^d \to \infty$ as $n \to \infty$, then the cubic histogram rule is universally consistent.*

PROOF. We check the two simple conditions of Theorem 6.1. Clearly, the diameter of each cell is $\sqrt{d}h^d$. Therefore condition (1) follows trivially. To show condition (2), we need to prove that for any $M < \infty$, $\mathbf{P}\{N(X) \leq M\} \to 0$. Let S be an arbitrary ball centered at the origin. Then the number of cells intersecting S is not more than $c_1 + c_2/h^d$ for some positive constants c_1, c_2. Then

$$\mathbf{P}\{N(X) \leq M\}$$
$$\leq \sum_{j:A_{nj} \cap S \neq \emptyset} \mathbf{P}\{X \in A_{nj}, N(X) \leq M\} + \mathbf{P}\{X \in S^c\}$$
$$\leq \sum_{\substack{j:A_{nj} \cap S \neq \emptyset \\ \mu(A_{nj}) \leq 2M/n}} \mu(A_n j) + \sum_{\substack{j:A_{nj} \cap S \neq \emptyset \\ \mu(A_{nj}) > 2M/n}} \mu(A_{nj}) \mathbf{P}\{n\mu_n(A_{nj}) \leq M\} + \mu(S^c)$$

$$\leq \frac{2M}{n}\left(c_1 + \frac{c_2}{h^d}\right)$$

$$+ \sum_{\substack{j: A_{nj} \cap S \neq \emptyset \\ \mu(A_{nj}) > 2M/n}} \mu(A_{nj}) \mathbf{P}\{\mu_n(A_{nj}) - \mathbf{E}\{\mu_n(A_{nj})\} \leq M/n - \mu(A_{nj})\} + \mu(S^c)$$

$$\leq \frac{2M}{n}\left(c_1 + \frac{c_2}{h^d}\right)$$

$$+ \sum_{\substack{j: A_{nj} \cap S \neq \emptyset \\ \mu(A_{nj}) > 2M/n}} \mu(A_{nj}) \mathbf{P}\left\{\mu_n(A_{nj}) - \mathbf{E}\{\mu_n(A_{nj})\} \leq \frac{-\mu(A_{nj})}{2}\right\} + \mu(S^c)$$

$$\leq \frac{2M}{n}\left(c_1 + \frac{c_2}{h^d}\right) + \sum_{\substack{j: A_{nj} \cap S \neq \emptyset \\ \mu(A_{nj}) > 2M/n}} 4\mu(A_{nj}) \frac{\mathbf{Var}\{\mu_n(A_{nj})\}}{\left(\mu(A_{nj})\right)^2} + \mu(S^c)$$

(by Chebyshev's inequality)

$$\leq \frac{2M}{n}\left(c_1 + \frac{c_2}{h^d}\right) + \sum_{\substack{j: A_{nj} \cap S \neq \emptyset \\ \mu(A_{nj}) > 2M/n}} 4\mu(A_{nj}) \frac{1}{n\mu(A_{nj})} + \mu(S^c)$$

$$\leq \frac{2M+4}{n}\left(c_1 + \frac{c_2}{h^d}\right) + \mu(S^c)$$

$$\rightarrow \mu(S^c),$$

because $nh^d \rightarrow \infty$. Since S is arbitrary, the proof of the theorem is complete. $\square$

6.5 Stone's Theorem

A general theorem by Stone (1977) allows us to deduce universal consistency of several classification rules. Consider a rule based on an estimate of the a posteriori probability η of the form

$$\eta_n(x) = \sum_{i=1}^n I_{\{Y_i=1\}} W_{ni}(x) = \sum_{i=1}^n Y_i W_{ni}(x),$$

where the *weights* $W_{ni}(x) = W_{ni}(x, X_1, \ldots, X_n)$ are nonnegative and sum to one:

$$\sum_{i=1}^n W_{ni}(x) = 1.$$

The classification rule is defined as

$$g_n(x) = \begin{cases} 0 & \text{if } \sum_{i=1}^n I_{\{Y_i=1\}} W_{ni}(x) \leq \sum_{i=1}^n I_{\{Y_i=0\}} W_{ni}(x) \\ 1 & \text{otherwise,} \end{cases}$$

$$= \begin{cases} 0 & \text{if } \sum_{i=1}^n Y_i W_{ni}(x) \leq 1/2 \\ 1 & \text{otherwise.} \end{cases}$$

η_n is a *weighted average estimator* of η. It is intuitively clear that pairs (X_i, Y_i) such that X_i is close to x should provide more information about $\eta(x)$ than those far from x. Thus, the weights are typically much larger in the neighborhood of X, so η_n is roughly a (weighted) relative frequency of the X_i's that have label 1 among points in the neighborhood of X. Thus, η_n might be viewed as a *local average estimator*, and g_n a local (weighted) majority vote. Examples of such rules include the histogram, kernel, and nearest neighbor rules. These rules will be studied in depth later.

Theorem 6.3. (STONE (1977)). *Assume that for any distribution of X, the weights satisfy the following three conditions:*

(i) *There is a constant c such that, for every nonnegative measurable function f satisfying* $\mathbf{E} f(X) < \infty$,

$$\mathbf{E} \left\{ \sum_{i=1}^{n} W_{ni}(X) f(X_i) \right\} \leq c \mathbf{E} f(X).$$

(ii) *For all $a > 0$,*

$$\lim_{n \to \infty} \mathbf{E} \left\{ \sum_{i=1}^{n} W_{ni}(X) I_{\{\|X_i - X\| > a\}} \right\} = 0.$$

(iii)

$$\lim_{n \to \infty} \mathbf{E} \left\{ \max_{1 \leq i \leq n} W_{ni}(X) \right\} = 0.$$

Then g_n is universally consistent.

REMARK. Condition (ii) requires that the overall weight of X_i's outside of any ball of a fixed radius centered at X must go to zero. In other words, only points in a shrinking neighborhood of X should be taken into account in the averaging. Condition (iii) requires that no single X_i has too large a contribution to the estimate. Hence, the number of points encountered in the averaging must tend to infinity. Condition (i) is technical. $\square$

PROOF. By Corollary 6.2 it suffices to show that for every distribution of (X, Y)

$$\lim_{n \to \infty} \mathbf{E} \left\{ (\eta(X) - \eta_n(X))^2 \right\} = 0.$$

Introduce the notation

$$\widehat{\eta}_n(x) = \sum_{i=1}^{n} \eta(X_i) W_{ni}(x).$$

Then by the simple inequality $(a + b)^2 \leq 2(a^2 + b^2)$ we have

$$\mathbf{E}\left\{(\eta(X) - \eta_n(X))^2\right\}$$
$$= \mathbf{E}\left\{((\eta(X) - \widehat{\eta}_n(X)) + (\widehat{\eta}_n(X) - \eta_n(X)))^2\right\}$$
$$\leq 2\left(\mathbf{E}\left\{(\eta(X) - \widehat{\eta}_n(X))^2\right\} + \mathbf{E}\left\{(\widehat{\eta}_n(X) - \eta_n(X))^2\right\}\right). \qquad (6.1)$$

Therefore, it is enough to show that both terms on the right-hand side tend to zero. Since the W_{ni}'s are nonnegative and sum to one, by Jensen's inequality, the first term is

$$\mathbf{E}\left\{(\eta(X) - \widehat{\eta}_n(X))^2\right\} = \mathbf{E}\left\{\left(\sum_{i=1}^{n} W_{ni}(X)(\eta(X) - \eta(X_i))\right)^2\right\}$$
$$\leq \mathbf{E}\left\{\sum_{i=1}^{n} W_{ni}(X)(\eta(X) - \eta(X_i))^2\right\}.$$

If the function $0 \leq \eta^* \leq 1$ is continuous with bounded support, then it is uniformly continuous as well: for every $\epsilon > 0$, there is an $a > 0$ such that for $\|x_1 - x\| < a$, $|\eta^*(x_1) - \eta^*(x)|^2 < \epsilon$. Recall here that $\|x\|$ denotes the Euclidean norm of a vector $x \in \mathcal{R}^d$. Thus, since $|\eta^*(x_1) - \eta^*(x)| \leq 1$,

$$\mathbf{E}\left\{\sum_{i=1}^{n} W_{ni}(X)(\eta^*(X) - \eta^*(X_i))^2\right\}$$
$$\leq \mathbf{E}\left\{\sum_{i=1}^{n} W_{ni}(X)I_{\{\|X - X_i\| \geq a\}}\right\} + \mathbf{E}\left\{\sum_{i=1}^{n} W_{ni}(X)\epsilon\right\} \to \epsilon,$$

by (ii). Since the set of continuous functions with bounded support is dense in $L_2(\mu)$, for every $\epsilon > 0$ we can choose η^* such that

$$\mathbf{E}\left\{(\eta(X) - \eta^*(X))^2\right\} < \epsilon.$$

By this choice, using the inequality $(a + b + c)^2 \leq 3(a^2 + b^2 + c^2)$ (which follows from the Cauchy-Schwarz inequality),

$$\mathbf{E}\left\{(\eta(X) - \widehat{\eta}_n(X))^2\right\}$$
$$\leq \mathbf{E}\left\{\sum_{i=1}^{n} W_{ni}(X)(\eta(X) - \eta(X_i))^2\right\}$$
$$\leq 3\mathbf{E}\left\{\sum_{i=1}^{n} W_{ni}(X)\left((\eta(X) - \eta^*(X))^2 + (\eta^*(X) - \eta^*(X_i))^2\right.\right.$$
$$\left.\left. + (\eta^*(X_i) - \eta(X_i))^2\right)\right\}$$
$$\leq 3\mathbf{E}\left\{(\eta(X) - \eta^*(X))^2\right\}$$
$$+ 3\mathbf{E}\left\{\sum_{i=1}^{n} W_{ni}(X)(\eta^*(X) - \eta^*(X_i))^2\right\} + 3c\mathbf{E}\left\{(\eta(X) - \eta^*(X))^2\right\},$$

where we used (i). Therefore,

$$\limsup_{n \to \infty} \mathbf{E}\left\{(\eta(X) - \widehat{\eta}_n(X))^2\right\} \le 3\epsilon(1 + 1 + c).$$

To handle the second term of the right-hand side of (6.1), observe that

$$\mathbf{E}\left\{(Y_i - \eta(X_i))(Y_j - \eta(X_j))|X, X_1, \ldots, X_n\right\} = 0 \quad \text{for all } i \ne j$$

by independence. Therefore,

$$\begin{aligned}
&\mathbf{E}\left\{(\widehat{\eta}_n(X) - \eta_n(X))^2\right\} \\
&= \mathbf{E}\left\{\left(\sum_{i=1}^{n} W_{ni}(X)(\eta(X_i) - Y_i)\right)^2\right\} \\
&= \sum_{i=1}^{n}\sum_{j=1}^{n} \mathbf{E}\left\{W_{ni}(X)(\eta(X_i) - Y_i)W_{nj}(X)(\eta(X_j) - Y_j)\right\} \\
&= \sum_{i=1}^{n} \mathbf{E}\left\{W_{ni}^2(X)(\eta(X_i) - Y_i)^2\right\} \\
&\le \mathbf{E}\left\{\sum_{i=1}^{n} W_{ni}^2(X)\right\} \le \mathbf{E}\left\{\max_{1 \le i \le n} W_{ni}(X) \sum_{j=1}^{n} W_{nj}(X)\right\} \\
&= \mathbf{E}\left\{\max_{1 \le i \le n} W_{ni}(X)\right\} \to 0
\end{aligned}$$

by (iii), and the theorem is proved. □

6.6   The k-Nearest Neighbor Rule

In Chapter 5 we discussed asymptotic properties of the k-nearest neighbor rule when k remains fixed as the sample size n grows. In such cases the expected probability of error converges to a number between L^* and $2L^*$. In this section we show that if k is allowed to grow with n such that $k/n \to 0$, the rule is weakly universally consistent. The proof is a very simple application of Stone's theorem. This result, appearing in Stone's paper (1977), was the first universal consistency result for any rule. Strong consistency, and many other different aspects of the k-NN rule are studied in Chapters 11 and 26.

Recall the definition of the k-nearest neighbor rule: first the data are ordered according to increasing Euclidean distances of the X_j's to x:

$$(X_{(1)}(x), Y_{(1)}(x)), \ldots, (X_{(n)}(x), Y_{(n)}(x)),$$

that is, $X_{(i)}(x)$ is the i-th nearest neighbor of x among the points $X_1, \ldots, X_n$. Distance ties are broken by comparing indices, that is, in case of $\|X_i - x\| = \|X_j - x\|$, X_i is considered to be "closer" to x if $i < j$.

The k-NN classification rule is defined as

$$g_n(x) = \begin{cases} 0 & \text{if } \sum_{i=1}^{k} I_{\{Y_{(i)}(x)=1\}} \le \sum_{i=1}^{k} I_{\{Y_{(i)}(x)=0\}} \\ 1 & \text{otherwise.} \end{cases}$$

In other words, $g_n(x)$ is a majority vote among the labels of the k nearest neighbors of x.

Theorem 6.4. (STONE (1977)). *If $k \to \infty$ and $k/n \to 0$, then for all distributions* $EL_n \to L^*$.

PROOF. We proceed by checking the conditions of Stone's weak convergence theorem (Theorem 6.3). The weight $W_{ni}(X)$ in Theorem 6.3 equals $1/k$ iff X_i is among the k nearest neighbors of X, and equals 0 otherwise.

Condition (iii) is obvious since $k \to \infty$. For condition (ii) observe that

$$\mathbf{E} \left\{ \sum_{i=1}^{n} W_{ni}(X) I_{\{\|X_i - X\| > \epsilon\}} \right\} \to 0$$

holds whenever

$$\mathbf{P} \left\{ \|X_{(k)}(X) - X\| > \epsilon \right\} \to 0,$$

where $X_{(k)}(x)$ denotes the k-th nearest neighbor of x among $X_1, \ldots, X_n$. But we know from Lemma 5.1 that this is true for all $\epsilon > 0$ whenever $k/n \to 0$.

Finally, we consider condition (i). We have to show that for any nonnegative measurable function f with $\mathbf{E}\{f(X)\} < \infty$,

$$\mathbf{E} \left\{ \sum_{i=1}^{n} \frac{1}{k} I_{\{X_i \text{ is among the } k \text{ nearest neighbors of } X\}} f(X_i) \right\} \le \mathbf{E} \left\{ cf(X) \right\}$$

for some constant c. But we have shown in Lemma 5.3 that this inequality always holds with $c = \gamma_d$. Thus, condition (i) is verified. □

6.7 Classification Is Easier Than Regression Function Estimation

Once again assume that our decision is based on some estimate η_n of the a posteriori probability function η, that is,

$$g_n(x) = \begin{cases} 0 & \text{if } \eta_n(x) \le 1/2 \\ 1 & \text{otherwise.} \end{cases}$$

The bounds of Theorems 2.2, 2.3, and Corollary 6.2 point out that if η_n is a consistent estimate of η, then the resulting rule is also consistent. For example, writing $L_n = \mathbf{P}\{g_n(X) \neq Y | D_n\}$, we have

$$\mathbf{E}L_n - L^* \leq 2\sqrt{\mathbf{E}\left\{(\eta_n(X) - \eta(X))^2\right\}},$$

that is, L_2-consistent estimation of the regression function η leads to consistent classification, and in fact, this is the main tool used in the proof of Theorem 6.3. While the said bounds are useful for proving consistency, they are almost useless when it comes to studying rates of convergence. As Theorem 6.5 below shows, for consistent rules rates of convergence of $\mathbf{P}\{g_n(X) \neq Y\}$ to L^* are always orders of magnitude better than rates of convergence of $\sqrt{\mathbf{E}\left\{(\eta(X) - \eta_n(X))^2\right\}}$ to zero.

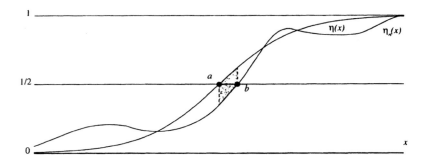

FIGURE 6.2. *The difference between the error probabilities grows roughly in proportion to the shaded area. Elsewhere $\eta_n(x)$ does not need to be close $\eta(x)$.*

Pattern recognition is thus easier than regression function estimation. This will be a recurring theme—to achieve acceptable results in pattern recognition, we can do more with smaller sample sizes than in regression function estimation. This is really just a consequence of the fact that less is required in pattern recognition. It also corroborates our belief that pattern recognition is dramatically different from regression function estimation, and that it deserves separate treatment in the statistical community.

Theorem 6.5. *Let η_n be a weakly consistent regression estimate, that is,*

$$\lim_{n \to \infty} \mathbf{E}\left\{(\eta_n(X) - \eta(X))^2\right\} = 0.$$

Define

$$g_n(x) = \begin{cases} 0 & \text{if } \eta_n(x) \leq 1/2 \\ 1 & \text{otherwise.} \end{cases}$$

Then

$$\lim_{n \to \infty} \frac{\mathbf{E}L_n - L^*}{\sqrt{\mathbf{E}\left\{(\eta_n(X) - \eta(X))^2\right\}}} = 0,$$

that is, $\mathbf{E}L_n - L^*$ converges to zero faster than the L_2-error of the regression estimate.

PROOF. We start with the equality of Theorem 2.2:

$$\mathbf{E}L_n - L^* = 2\,\mathbf{E}\left\{|\eta(X) - 1/2|I_{\{g_n(X) \neq g^*(X)\}}\right\}.$$

Fix $\epsilon > 0$. We may bound the last factor by

$$\mathbf{E}\left\{|\eta(X) - 1/2|I_{\{g_n(X) \neq g^*(X)\}}\right\}$$
$$\leq \mathbf{E}\left\{I_{\{\eta(X) \neq 1/2\}}|\eta(X) - \eta_n(X)|I_{\{g_n(X) \neq g^*(X)\}}\right\}$$
$$= \mathbf{E}\left\{|\eta(X) - \eta_n(X)|I_{\{g_n(X) \neq g^*(X)\}}I_{\{|\eta(X)-1/2|\leq \epsilon\}}I_{\{\eta(X) \neq 1/2\}}\right\}$$
$$\quad + \mathbf{E}\left\{|\eta(X) - \eta_n(X)|I_{\{g_n(X) \neq g^*(X)\}}I_{\{|\eta(X)-1/2|>\epsilon\}}\right\}$$
$$\leq \sqrt{\mathbf{E}\left\{(\eta_n(X) - \eta(X))^2\right\}}$$
$$\quad \times \left(\sqrt{\mathbf{P}\{|\eta(X) - 1/2| \leq \epsilon, \eta(X) \neq 1/2\}} \right.$$
$$\left. + \sqrt{\mathbf{P}\{g_n(X) \neq g^*(X), |\eta(X) - 1/2| > \epsilon\}}\right)$$

(by the Cauchy-Schwarz inequality).

Since $g_n(X) \neq g^*(X)$ and $|\eta(X) - 1/2| > \epsilon$ imply that $|\eta_n(X) - \eta(X)| > \epsilon$, consistency of the regression estimate implies that for any fixed $\epsilon > 0$,

$$\lim_{n \to \infty} \mathbf{P}\{g_n(X) \neq g^*(X), |\eta(X) - 1/2| > \epsilon\} = 0.$$

On the other hand,

$$\mathbf{P}\{|\eta(X) - 1/2| \leq \epsilon, \eta(X) \neq 1/2\} \to 0 \text{ as } \epsilon \to 0,$$

which completes the proof. $\square$

The actual value of the ratio

$$\rho_n = \frac{\mathbf{E}L_n - L^*}{\sqrt{\mathbf{E}\left\{(\eta_n(X) - \eta(X))^2\right\}}}$$

cannot be universally bounded. In fact, ρ_n may tend to zero arbitrarily slowly (see Problem 6.5). On the other hand, ρ_n may tend to zero extremely quickly. In Problems 6.6 and 6.7 and in the theorem below, upper bounds on ρ_n are given that may be used in deducing rate-of-convergence results. Theorem 6.6, in particular, states that $EL_n - L^*$ tends to zero as fast as the *square* of the L_2 error of the regression estimate, i.e., $E\left\{(\eta_n(X) - \eta(X))^2\right\}$, whenever $L^* = 0$. Just how slowly ρ_n tends to zero depends upon two things, basically: (1) the rate of convergence of η_n to η, and (2) the behavior of $P\{|\eta(X) - 1/2| \le \epsilon, \eta(X) \ne 1/2\}$ as a function of ϵ when $\epsilon \downarrow 0$ (i.e., the behavior of $\eta(x)$ at those x's where $\eta(x) \approx 1/2$).

Theorem 6.6. *Assume that* $L^* = 0$, *and consider the decision*

$$g_n(x) = \begin{cases} 0 & \text{if } \eta_n(x) \le 1/2 \\ 1 & \text{otherwise.} \end{cases}$$

Then

$$P\{g_n(X) \ne Y\} \le 4\,E\left\{(\eta_n(X) - \eta(X))^2\right\}.$$

PROOF. By Theorem 2.2,

$$\begin{aligned} P\{g_n(X) \ne Y\} &= 2\,E\left\{|\eta(X) - 1/2|I_{\{g_n(X) \ne g^*(X)\}}\right\} \\ &= 2\,E\left\{|\eta(X) - 1/2|I_{\{g_n(X) \ne Y\}}\right\} \\ &\quad \text{(since } g^*(X) = Y \text{ by the assumption } L^* = 0) \\ &\le 2\sqrt{E\left\{(\eta_n(X) - \eta(X))^2\right\}}\sqrt{P\{g_n(X) \ne Y\}} \\ &\quad \text{(by the Cauchy-Schwarz inequality).} \end{aligned}$$

Dividing both sides by $\sqrt{P\{g_n(X) \ne Y\}}$ yields the result. □

The results above show that the bounds of Theorems 2.2, 2.3, and Corollary 6.2 may be arbitrarily loose, and the error probability converges to L^* faster than the L_2-error of the regression estimate converges to zero. In some cases, consistency may even occur without convergence of $E|\eta_n(X) - \eta(X)|$ to zero. Consider for example a *strictly separable* distribution, that is, a distribution such that there exist two sets $A, B \subset \mathcal{R}^d$ with

$$\inf_{x \in A, y \in B} \|x - y\| \ge \delta > 0$$

for some $\delta > 0$, and having the property that

$$P\{X \in A|Y = 1\} = P\{X \in B|Y = 0\} = 1.$$

In such cases, there is a version of η that has $\eta(x) = 1$ on A and $\eta(x) = 0$ on B. We say version because η is not defined on sets of measure zero. For such strictly separable distributions, $L^* = 0$. Let $\bar{\eta}$ be $1/2 - \epsilon$ on B and $1/2 + \epsilon$ on A. Then, with

$$g(x) = \begin{cases} 0 & \text{if } \bar{\eta}(x) \le 1/2 \\ 1 & \text{otherwise,} \end{cases} = \begin{cases} 0 & \text{if } x \in B \\ 1 & \text{if } x \in A, \end{cases}$$

we have $\mathbf{P}\{g(X) \neq Y\} = L^* = 0$. Yet,

$$2\,\mathbf{E}|\eta(X) - \bar{\eta}(X)| = 1 - 2\epsilon$$

is arbitrarily close to one.

In a more realistic example, we consider the kernel rule (see Chapter 10),

$$g_n(x) = \begin{cases} 0 & \text{if } \eta_n(x) \leq 1/2 \\ 1 & \text{otherwise,} \end{cases}$$

in which

$$\eta_n(x) = \frac{\sum_{i=1}^n Y_i K(x - X_i)}{\sum_{i=1}^n K(x - X_i)},$$

where K is the standard normal density in $\mathcal{R}^d$:

$$K(u) = \frac{1}{(2\pi)^{d/2}} e^{-\|u\|^2/2}.$$

Assume that A and B consist of one point each, at distance δ from each other—that is, the distribution of X is concentrated on two points. If $\mathbf{P}\{Y = 0\} = \mathbf{P}\{Y = 1\} = 1/2$, we see that

$$\lim_{n \to \infty} \frac{1}{n} \sum_{i=1}^n K(x - X_i) = \frac{K(0) + K(\delta)}{2} \quad \text{with probability one}$$

at $x \in A \cup B$, by the law of large numbers. Also,

$$\lim_{n \to \infty} \frac{1}{n} \sum_{i=1}^n Y_i K(x - X_i) = \begin{cases} K(0)/2 & \text{if } x \in A \\ K(\delta)/2 & \text{if } x \in B \end{cases} \quad \text{with probability one.}$$

Thus,

$$\lim_{n \to \infty} \eta_n(x) = \begin{cases} \frac{K(0)}{K(0)+K(\delta)} & \text{if } x \in A \\ \frac{K(\delta)}{K(0)+K(\delta)} & \text{if } x \in B \end{cases} \quad \text{with probability one.}$$

Hence, as $\eta(x) = 1$ on A and $\eta(x) = 0$ on B,

$$\lim_{n \to \infty} 2\,\mathbf{E}|\eta(X) - \eta_n(X)|$$

$$= 2\frac{1}{2}\left(\left|1 - \frac{K(0)}{K(0) + K(\delta)}\right| + \left|0 - \frac{K(\delta)}{K(0) + K(\delta)}\right|\right)$$

$$= \frac{2K(\delta)}{K(0) + K(\delta)}.$$

Yet, $L^* = 0$ and $\mathbf{P}\{g_n(X) \neq Y\} \to 0$. In fact, if D_n denotes the training data,

$$\lim_{n \to \infty} \mathbf{P}\{g_n(X) \neq Y | D_n\} = L^* \quad \text{with probability one,}$$

and

$$\lim_{n \to \infty} 2\,\mathbf{E}\left\{ |\eta_n(X) - \eta(X)| \Big| D_n \right\} = \frac{2K(\delta)}{K(0) + K(\delta)} \quad \text{with probability one.}$$

This shows very strongly that for any $\delta > 0$, for many practical classification rules, we *do not need* convergence of η_n to η at all!

As all the consistency proofs in Chapters 6 through 11 rely on the convergence of η_n to η, we will create unnecessary conditions for some distributions, although it will always be possible to find distributions of (X, Y) for which the conditions are needed—in the latter sense, the conditions of these universal consistency results are not improvable.

6.8   Smart Rules

A rule is a sequence of mappings $g_n : \mathcal{R}^d \times \left(\mathcal{R}^d \times \{0, 1\} \right)^n \to \{0, 1\}$. Most rules are expected to perform better when n increases. So, we say that a rule is *smart* if for all distributions of (X, Y), $\mathbf{E}\{L(g_n)\}$ is nonincreasing, where

$$L(g_n) = \mathbf{P}\{g_n(X, D_n) \neq Y | D_n\}.$$

Some dumb rules are smart, such as the (useless) rule that, for each n, takes a majority over all Y_i's, ignoring the X_i's. This follows from the fact that

$$\mathbf{P}\left\{ \sum_{i=1}^{n}(2Y_i - 1) > 0,\, Y = 0 \text{ or } \sum_{i=1}^{n}(2Y_i - 1) \leq 0,\, Y = 1 \right\}$$

is monotone in n. This is a property of the binomial distribution (see Problem 6.12). A histogram rule with a fixed partition is smart (Problem 6.13). The 1-nearest neighbor rule is not smart. To see this, let (X, Y) be $(0, 1)$ and $(Z, 0)$ with probabilities p and $1 - p$, respectively, where Z is uniform on $[-1000, 1000]$. Verify that for $n = 1$, $\mathbf{E}L_n = 2p(1 - p)$, while for $n = 2$,

$$\mathbf{E}L_n = 2p(1 - p)^2 \left(\frac{1}{2} + \frac{\mathbf{E}|Z|}{4000} \right) + p^2(1 - p) + (1 - p)^2 p$$

$$= 2p(1 - p) \left(\frac{5(1 - p)}{8} + \frac{1}{2} \right),$$

which is larger than $2p(1 - p)$ whenever $p \in (0, 1/5)$. This shows that in all these cases it is better to have $n = 1$ than $n = 2$. Similarly, the standard kernel rule—discussed in Chapter 10—with fixed h is not smart (see Problems 6.14, 6.15).

The error probabilities of the above examples of smart rules do not change dramatically with n. However, change is necessary to guarantee Bayes risk consistency. At the places of change—for example when h_n jumps to a new value in the histogram rule—the monotonicity may be lost. This leads to the conjecture that *no* universally consistent rule can be smart.

Problems and Exercises

PROBLEM 6.1. Let the i.i.d. random variables $X_1, \ldots, X_n$ be distributed on $\mathcal{R}^d$ according to the density f. Estimate f by f_n, a function of x and $X_1, \ldots, X_n$, and assume that $\int |f_n(x) - f(x)| dx \to 0$ in probability (or with probability one). Then show that there exists a consistent (or strongly consistent) classification rule whenever the conditional densities f_0 and f_1 exist.

PROBLEM 6.2. HISTOGRAM DENSITY ESTIMATION. Let $X_1, \ldots, X_n$ be i.i.d. random variables in $\mathcal{R}^d$ with density f. Let $\mathcal{P}_n$ be a partition of $\mathcal{R}^d$ into cubes of size h_n, and define the histogram density estimate by

$$f_n(x) = \frac{1}{nh_n^d} \sum_{i=1}^{n} I_{\{X_i \in A_n(x)\}},$$

where $A_n(x)$ is the set in $\mathcal{P}_n$ that contains x. Prove that the estimate is *universally consistent* in L_1 if $h_n \to 0$ and $nh_n^d \to \infty$ as $n \to \infty$, that is, for any f the L_1 error of the estimate $\int |f_n(x) - f(x)| dx$ converges to zero in probability, or equivalently,

$$\lim_{n \to \infty} \mathbf{E} \left\{ \int |f_n(x) - f(x)| dx \right\} = 0.$$

HINT: The following suggestions may be helpful
  (1) $\mathbf{E} \left\{ \int |f_n - f| \right\} \leq \mathbf{E} \left\{ \int |f_n - \mathbf{E} f_n| \right\} + \int |\mathbf{E} f_n - f|$.
  (2) $\mathbf{E} \left\{ \int |f_n - \mathbf{E} f_n| \right\} = \sum_j |\mu(A_{nj}) - \mu_n(A_{nj})|$.
  (3) First show $\int |\mathbf{E} f_n - f| \to 0$ for uniformly continuous f, and then extend it to arbitrary densities.

PROBLEM 6.3. Let X be uniformly distributed on $[0, 1]$ with probability $1/2$, and let X be atomic on the rationals with probability $1/2$ (e.g., if the rationals are enumerated $r_1, r_2, r_3, \ldots$, then $\mathbf{P}\{X = r_i\} = 1/2^{i+1}$). Let $Y = 1$ if X is rational and $Y = 0$ if X is irrational. Give a direct proof of consistency of the 1-nearest neighbor rule. HINT: Given $Y = 1$, the conditional distribution of X is discrete. Thus, for every $\epsilon > 0$, there is an integer k such that given $Y = 1$, X equals one of k rationals with probability at least $1 - \epsilon$. Now, if n is large enough, every point in this set captures data points with label 1 with large probability. Also, for large n, the space between these points is filled with data points labeled with zeros.

PROBLEM 6.4. Prove the consistency of the cubic histogram rule by checking the conditions of Stone's theorem. HINT: To check (i), first bound $W_{ni}(x)$ by

$$I_{\{X_i \in A_n(x)\}} / \sum_{j=1}^{n} I_{\{X_j \in A_n(x)\}} + 1/n.$$

Since

$$\mathbf{E} \left\{ \sum_{i=1}^{n} \frac{1}{n} f(X_i) \right\} = \mathbf{E} f(X),$$

it suffices to show that there is a constant $c' > 0$ such that for any nonnegative function f with $\mathbf{E} f(X) < \infty$,

$$\mathbf{E} \left\{ \sum_{i=1}^{n} f(X_i) \frac{I_{\{X_i \in A_n(X)\}}}{\sum_{j=1}^{n} I_{\{X_j \in A_n(X)\}}} \right\} \leq c' \mathbf{E} f(X).$$

In doing so, you may need to use Lemma A.2 (i). To prove that condition (iii) holds, write

$$E\left\{\max_{1\le i\le n} W_{ni}(X)\right\}$$

$$\le \quad \frac{1}{n} + P\{X \in S^c\} + \sum_{j:A_{nj}\cap S\neq\emptyset} E\left\{I_{\{X\in A_{nj}\}}\frac{1}{n\mu_n(A_{nj})}I_{\{\mu_n(A_{nj})>0\}}\right\}$$

and use Lemma A.2 (ii).

PROBLEM 6.5. Let $\{a_n\}$ be a sequence of positive numbers converging to zero. Give an example of an a posteriori probability function η, and a sequence of functions $\{\eta_n\}$ approximating η such that

$$\liminf_{n\to\infty} \frac{P\{g_n(X) \neq Y\} - L^*}{a_n\sqrt{E\left\{(\eta_n(X) - \eta(X))^2\right\}}} > 0,$$

where

$$g_n(x) = \begin{cases} 0 & \text{if } \eta_n(x) \le 1/2 \\ 1 & \text{otherwise.} \end{cases}$$

Thus, the rate of convergence in Theorem 6.5 may be arbitrarily slow. HINT: Define $\eta = 1/2 + b(x)$, where $b(x)$ is a very slowly increasing nonnegative function.

PROBLEM 6.6. Let $\delta > 0$, and assume that $|\eta(x) - 1/2| \ge \delta$ for all x. Consider the decision

$$g_n(x) = \begin{cases} 0 & \text{if } \eta_n(x) \le 1/2 \\ 1 & \text{otherwise.} \end{cases}$$

Prove that

$$P\{g_n(X) \neq Y\} - L^* \le \frac{2E\left\{(\eta_n(X) - \eta(X))^2\right\}}{\delta}.$$

This shows that the rate of convergence implied by the inequality of Theorem 6.6 may be preserved for very general classes of distributions.

PROBLEM 6.7. Assume that $L^* = 0$, and consider the decision

$$g_n(x) = \begin{cases} 0 & \text{if } \eta_n(x) \le 1/2 \\ 1 & \text{otherwise.} \end{cases}$$

Show that for all $1 \le p < \infty$,

$$P\{g_n(X) \neq Y\} \le 2^p E\{|\eta_n(X) - \eta(X)|^p\}.$$

HINT: Proceed as in the proof of Theorem 6.6, but use Hölder's inequality.

PROBLEM 6.8. Theorem 6.5 cannot be generalized to the L_1 error. In particular, show by example that it is not always true that

$$\lim_{n\to\infty} \frac{EL_n - L^*}{E\{|\eta_n(X) - \eta(X)|\}} = 0$$

when $E\{|\eta_n(X) - \eta(X)|\} \to 0$ as $n \to \infty$ for some regression function estimate η_n. Thus, the inequality (Corollary 6.2)

$$EL_n - L^* \le 2E\{|\eta_n(X) - \eta(X)|\}$$

cannot be universally improved.

PROBLEM 6.9. Let $\eta' : \mathcal{R}^d \to [0, 1]$, and define $g(x) = I_{\{\eta'(x) > 1/2\}}$. Assume that the random variable X satisfies that $\mathbf{P}\{\eta'(X) = 1/2\} = 0$. Let $\eta_1, \eta_2, \ldots$ be a sequence of functions such that $\mathbf{E}\{|\eta'(X) - \eta_n(X)|\} \to 0$ as $n \to \infty$. Prove that $L(g_n) \to L(g)$ for all distributions of (X, Y) satisfying the condition on X above, where $g_n(x) = I_{\{\eta_n(x) > 1/2\}}$ (Lugosi (1992)).

PROBLEM 6.10. A LYING TEACHER. Sometimes the training labels $Y_1, \ldots, Y_n$ are not available, but can only be observed through a noisy binary channel. Still, we want to decide on Y. Consider the following model. Assume that the Y_i's in the training data are replaced by the i.i.d. binary-valued random variables Z_i, whose distribution is given by

$$\mathbf{P}\{Z_i = 1 | Y_i = 0, X_i = x\} \quad = \quad \mathbf{P}\{Z_i = 1 | Y_i = 0\} = p < 1/2,$$

$$\mathbf{P}\{Z_i = 0 | Y_i = 1, X_i = x\} \quad = \quad \mathbf{P}\{Z_i = 0 | Y_i = 1\} = q < 1/2.$$

Consider the decision

$$g(x) = \begin{cases} 0 & \text{if } \eta'(x) \leq 1/2 \\ 1 & \text{otherwise,} \end{cases}$$

where $\eta'(x) = \mathbf{P}\{Z_1 = 1 | X_1 = x\}$. Show that

$$\mathbf{P}\{g(X) \neq Y\} \leq L^* \left(1 + \frac{2|p - q|}{1 - 2\max(p, q)} \right).$$

Use Problem 6.9 to conclude that if the binary channel is symmetric (i.e., $p = q < 1/2$), and $\mathbf{P}\{\eta'(X) = 1/2\} = 0$, then L_1-consistent estimation leads to a consistent rule, in spite of the fact that the labels Y_i were not available in the training sequence (Lugosi (1992)).

PROBLEM 6.11. Develop a discrimination rule which has the property

$$\lim_{n \to \infty} \mathbf{E}L_n = \rho = \mathbf{E}\left\{ \sqrt{\eta(X)(1 - \eta(X))} \right\},$$

for all distributions such that X has a density. Note: clearly, since $\rho \geq L^*$, this rule is not universally consistent, but it will aid you in "visualizing" the Matushita error!

PROBLEM 6.12. If Z_n is binomial (n, p) and Z is Bernoulli (p), independent of Z_n, then show that $\mathbf{P}\{Z_n > n/2, Z = 0\} + \mathbf{P}\{Z_n \leq n/2, Z = 1\}$ is nonincreasing in n.

PROBLEM 6.13. Let g_n be the histogram rule based on a fixed partition $\mathcal{P}$. Show that g_n is smart.

PROBLEM 6.14. Show that the kernel rule with gaussian kernel and $h = 1$, $d = 1$, is not smart (kernel rules are discussed in Chapter 10). HINT: Consider $n = 1$ and $n = 2$ only.

PROBLEM 6.15. Show that the kernel rule on $\mathcal{R}$, with $K(x) = I_{[-1,1]}(x)$, and $h \downarrow 0$, such that $nh \to \infty$, is not smart.

PROBLEM 6.16. Conjecture: no universally consistent rule is smart.

PROBLEM 6.17. A rule $g_n : \mathcal{R}^d \times (\mathcal{R}^d \times \{0, 1\})^n$ is called symmetric if $g_n(x, D_n) = g_n(x, D'_n)$ for every x, and every training sequence D_n, where D'_n is an arbitrary permutation of the pairs (X_i, Y_i) in D_n. Any nonsymmetric rule g_n may be symmetrized by taking a majority vote at every $x \in \mathcal{R}^d$ over all $g_n(x, D'_n)$, obtained by the $n!$ permutations of D_n. It may intuitively be expected that symmetrized rules perform better. Prove that this is false, that is, exhibit a distribution and a nonsymmetric classifier g_n such that its expected probability of error is smaller than that of the symmetrized version of g_n. HINT: Take $g_3(x, D_3) = 1 - Y_1$.

7
Slow Rates of Convergence

In this chapter we consider the general pattern recognition problem: Given the observation X and the training data $D_n = ((X_1, Y_1), \ldots, (X_n, Y_n))$ of independent identically distributed random variable pairs, we estimate the label Y by the decision

$$g_n(X) = g_n(X, D_n).$$

The error probability is

$$L_n = \mathbf{P}\{Y \neq g_n(X)|D_n\}.$$

Obviously, the average error probability $\mathbf{E}L_n = \mathbf{P}\{Y \neq g_n(X)\}$ is completely determined by the distribution of the pair (X, Y), and the classifier g_n. We have seen in Chapter 6 that there exist classification rules such as the cubic histogram rule with properly chosen cube sizes such that $\lim_{n\to\infty} \mathbf{E}L_n = L^*$ for all possible distributions. The next question is whether there are classification rules with $\mathbf{E}L_n$ tending to the Bayes risk at a specified rate for all distributions. Disappointingly, such rules do not exist.

7.1 Finite Training Sequence

The first negative result shows that for any classification rule and for any *fixed n*, there exists a distribution such that the difference between the error probability of the rule and L^* is larger than $1/4$. To explain this, note that for fixed n, we can find a sufficiently complex distribution for which the sample size n is hopelessly small.

Theorem 7.1. (DEVROYE (1982B)). *Let $\epsilon > 0$ be an arbitrarily small number. For any integer n and classification rule g_n, there exists a distribution of (X, Y) with Bayes risk $L^* = 0$ such that*

$$EL_n \geq 1/2 - \epsilon.$$

PROOF. First we construct a family of distributions of (X, Y). Then we show that the error probability of any classifier is large for at least one member of the family. For every member of the family, X is uniformly distributed on the set $\{1, 2, \ldots, K\}$ of positive integers

$$p_i = P\{X = i\} = \begin{cases} 1/K & \text{if } i \in \{1, \ldots, K\} \\ 0 & \text{otherwise,} \end{cases}$$

where K is a large integer specified later. Now, the family of distributions of (X, Y) is parameterized by a number $b \in [0, 1)$, that is, every b determines a distribution as follows. Let $b \in [0, 1)$ have binary expansion $b = 0.b_0 b_1 b_2 \ldots$, and define $Y = b_X$. As the label Y is a function of X, there exists a perfect decision, and thus $L^* = 0$. We show that for any decision rule g_n there is a b such that if $Y = b_X$, then g_n has very poor performance. Denote the average error probability corresponding to the distribution determined by b, by $R_n(b) = EL_n$.

The proof of the existence of a bad distribution is based on the so-called probabilistic method. Here the key trick is the randomization of b. Define a random variable B which is uniformly distributed in $[0, 1)$ and independent of X and $X_1, \ldots, X_n$. Then we may compute the expected value of the random variable $R_n(B)$. Since for any decision rule g_n,

$$\sup_{b \in [0,1)} R_n(b) \geq E\{R_n(B)\},$$

a lower bound for $E\{R_n(B)\}$ proves the existence of a $b \in [0, 1)$ whose corresponding error probability exceeds the lower bound.

Since B is uniformly distributed in $[0, 1)$, its binary extension $B = 0.B_1 B_2 \ldots$ is a sequence of independent binary random variables with $P\{B_i = 0\} = P\{B_i = 1\} = 1/2$. But

$E\{R_n(B)\}$

$\quad = \quad P\{g_n(X, D_n) \neq Y\}$

$\quad = \quad P\{g_n(X, D_n) \neq B_X\}$

$\quad = \quad P\{g_n(X, X_1, B_{X_1}, \ldots, X_n, B_{X_n}) \neq B_X\}$

$\quad = \quad E\{P\{g_n(X, X_1, B_{X_1}, \ldots, X_n, B_{X_n}) \neq B_X \mid X, X_1, \ldots, X_n\}\}$

$\quad \geq \quad \dfrac{1}{2} P\{X \neq X_1, X \neq X_2, \ldots, X \neq X_n\},$

since if $X \neq X_i$ for all $i = 1, 2, \ldots, n$, then given $X, X_1, \ldots, X_n$, $Y = B_X$ is conditionally independent of $g_n(X, D_n)$ and Y takes values 0 and 1 with probability

1/2. But clearly,

$$\mathbf{P}\{X \neq X_1, X \neq X_2, \ldots, X \neq X_n \mid X\} = \mathbf{P}\{X \neq X_1 \mid X\}^n = (1 - 1/K)^n.$$

In summary,

$$\sup_{b \in [0,1)} R_n(b) \geq \frac{1}{2}(1 - 1/K)^n.$$

The lower bound tends to $1/2$ as $K \to \infty$. $\square$

Theorem 7.1 states that even though we have rules that are universally consistent, that is, they *asymptotically* provide the optimal performance for any distribution, their *finite sample* performance is *always* extremely bad for some distributions. This means that no classifier guarantees that with a sample size of (say) $n = 10^8$ we get within $1/4$ of the Bayes error probability for all distributions. However, as the bad distribution depends upon n, Theorem 7.1 does not allow us to conclude that there is one distribution for which the error probability is more than $L^* + 1/4$ for all n. Indeed, that would contradict the very existence of universally consistent rules.

7.2   Slow Rates

The next question is whether a certain universal rate of convergence to L^* is achievable for some classifier. For example, Theorem 7.1 does not exclude the existence of a classifier such that for every n, $\mathbf{E}L_n - L^* \leq c/n$ for all distributions, for some constant c depending upon the actual distribution. The next negative result is that this cannot be the case. Theorem 7.2 below states that the error probability $\mathbf{E}L_n$ of any classifier is larger than (say) $L^* + c/(\log\log\log n)$ *for every* n for some distribution, even if c depends on the distribution. (This can be seen by considering that by Theorem 7.2, there exists a distribution of (X, Y) such that $\mathbf{E}L_n \geq L^* + 1/\sqrt{\log\log\log n}$ for every n.) Moreover, there is no sequence of numbers a_n converging to zero such that there is a classification rule with error probability below L^* plus a_n for all distributions.

Thus, in practice, no classifier assures us that its error probability is close to L^*, unless the actual distribution is known to be a member of a restricted class of distributions. Now, it is easily seen that in the proof of both theorems we could take X to have uniform distribution on $[0, 1]$, or any other density (see Problem 7.2). Therefore, putting restrictions on the distribution of X alone does not suffice to obtain rate-of-convergence results. For such results, one needs conditions on the a posteriori probability $\eta(x)$ as well. However, if only training data give information about the joint distribution, then theorems with extra conditions on the distribution have little practical value, as it is impossible to detect whether, for example, the a posteriori probability $\eta(x)$ is twice differentiable or not.

Now, the situation may look hopeless, but this is not so. Simply put, the Bayes error is too difficult a target to shoot at.

Weaker versions of Theorem 7.2 appeared earlier in the literature. First Cover (1968b) showed that for any sequence of classification rules, for sequences $\{a_n\}$ converging to zero at arbitrarily slow algebraic rates (i.e., as $1/n^\delta$ for arbitrarily small $\delta > 0$), there exists a distribution such that $\mathbf{E}L_n \geq L^* + a_n$ infinitely often. Devroye (1982b) strengthened Cover's result allowing sequences tending to zero arbitrarily slowly. The next result asserts that $\mathbf{E}L_n > L^* + a_n$ for *every* n.

Theorem 7.2. *Let $\{a_n\}$ be a sequence of positive numbers converging to zero with $1/16 \geq a_1 \geq a_2 \geq \dots$. For every sequence of classification rules, there exists a distribution of (X, Y) with $L^* = 0$, such that*

$$\mathbf{E}L_n \geq a_n$$

for all n.

This result shows that universally good classification rules do not exist. Rate of convergence studies for particular rules must necessarily be accompanied by conditions on (X, Y). That these conditions too are necessarily restrictive follows from examples suggested in Problem 7.2. Under certain regularity conditions it is possible to obtain upper bounds for the rates of convergence for the probability of error of certain rules to L^*. Then it is natural to ask what the fastest achievable rate is for the given class of distributions. A theory for regression function estimation was worked out by Stone (1982). Related results for classification were obtained by Marron (1983). In the proof of Theorem 7.2 we will need the following simple lemma:

Lemma 7.1. *For any monotone decreasing sequence $\{a_n\}$ of positive numbers converging to zero with $a_1 \leq 1/16$, a probability distribution $(p_1, p_2, \dots)$ may be found such that $p_1 \geq p_2 \geq \dots$, and for all n*

$$\sum_{i=n+1}^{\infty} p_i \geq \max\left(8a_n, 32np_{n+1}\right).$$

PROOF. It suffices to look for p_i's such that

$$\sum_{i=n+1}^{\infty} p_i \geq \max\left(8a_n, 32np_n\right).$$

These conditions are easily satisfied. For positive integers $u < v$, define the function $H(v, u) = \sum_{i=u}^{v-1} 1/i$. First we find a sequence $1 = n_1 < n_2 < \dots$ of integers with the following properties:

    (a) $H(n_{k+1}, n_k)$ is monotonically increasing,

    (b) $H(n_2, n_1) \geq 32$,

    (c) $8a_{n_k} \leq 1/2^k$ for all $k \geq 1$.

Note that (c) may only be satisfied if $a_{n_1} = a_1 \leq 1/16$. To this end, define constants $c_1, c_2, \ldots$ by

$$c_k = \frac{32}{2^k H(n_{k+1}, n_k)}, \quad k \geq 1,$$

so that the c_k's are decreasing in k, and

$$\frac{1}{32} \sum_{k=1}^{\infty} c_k H(n_{k+1}, n_k) = \sum_{k=1}^{\infty} \frac{1}{2^k} = 1.$$

For $n \in [n_k, n_{k+1})$, we define $p_n = c_k/(32n)$. We claim that these numbers have the required properties. Indeed, $\{p_n\}$ is decreasing, and

$$\sum_{n=1}^{\infty} p_n = \sum_{k=1}^{\infty} \frac{c_k}{32} H(n_{k+1}, n_k) = 1.$$

Finally, if $n \in [n_k, n_{k+1})$, then

$$\sum_{i=n+1}^{\infty} p_i \geq \sum_{j=k+1}^{\infty} \frac{c_j}{32} H(n_{j+1}, n_j) = \sum_{j=k+1}^{\infty} \frac{1}{2^j} = \frac{1}{2^k}.$$

Clearly, on the one hand, by the monotonicity of $H(n_{k+1}, n_k)$, $1/2^k \geq c_k = 32np_n$. On the other hand, $1/2^k \geq 8a_{n_k} \geq 8a_n$. This concludes the proof. $\square$

PROOF OF THEOREM 7.2. We introduce some notation. Let $b = 0.b_1 b_2 b_3 \ldots$ be a real number on $[0, 1]$ with the shown binary expansion, and let B be a random variable uniformly distributed on $[0, 1]$ with expansion $B = 0.B_1 B_2 B_3 \ldots$. Let us restrict ourselves to a random variable X with

$$\mathbf{P}\{X = i\} = p_i, \quad i \geq 1,$$

where $p_1 \geq p_2 \geq \ldots > 0$, and $\sum_{i=n+1}^{\infty} p_i \geq \max(8a_n, 32np_{n+1})$ for every n. That such p_i's exist follows from Lemma 7.1. Set $Y = b_X$. As Y is a function of X, we see that $L^* = 0$. Each $b \in [0, 1)$ however describes a different distribution. With b replaced by B we have a random distribution. Introduce the short notation $\Delta_n = ((X_1, B_{X_1}), \ldots, (X_n, B_{X_n}))$, and define $G_{ni} = g_n(i, \Delta_n)$. If $L_n(B)$ denotes the probability of error $\mathbf{P}\{g_n(X, \Delta_n) \neq Y | B, X_1, \ldots, X_n\}$ for the random distribution, then we note that we may write

$$L_n(B) = \sum_{i=1}^{\infty} p_i I_{\{G_{ni} \neq B_i\}}.$$

If $L_n(b)$ is the probability of error for a distribution parametrized by b, then

$$\sup_b \inf_n \mathbf{E} \frac{L_n(b)}{2a_n} \geq \sup_b \mathbf{E} \left\{ \inf_n \frac{L_n(b)}{2a_n} \right\}$$

$$\geq \mathbf{E} \left\{ \inf_n \frac{L_n(B)}{2a_n} \right\}$$

$$= \mathbf{E} \left\{ \mathbf{E} \left\{ \inf_n \frac{L_n(B)}{2a_n} \middle| X_1, X_2, \ldots \right\} \right\}.$$

We consider only the conditional expectation for now. We have

$$\mathbf{E} \left\{ \inf_n \frac{L_n(B)}{2a_n} \middle| X_1, X_2, \ldots \right\}$$

$$\geq \mathbf{P} \left\{ \bigcap_{n=1}^{\infty} \{L_n(B) \geq 2a_n\} \middle| X_1, X_2, \ldots \right\}$$

$$\geq 1 - \sum_{n=1}^{\infty} \mathbf{P} \{ L_n(B) < 2a_n | X_1, X_2, \ldots \}$$

$$= 1 - \sum_{n=1}^{\infty} \mathbf{P} \{ L_n(B) < 2a_n | X_1, X_2, \ldots, X_n \}$$

$$= 1 - \sum_{n=1}^{\infty} \mathbf{E} \{ \mathbf{P} \{ L_n(B) < 2a_n | \Delta_n \} | X_1, X_2, \ldots, X_n \}.$$

We bound the conditional probabilities inside the sum:

$$\mathbf{P} \{ L_n(B) < 2a_n | \Delta_n \}$$

$$\leq \mathbf{P} \left\{ \sum_{i \notin \{X_1, \ldots, X_n\}} p_i I_{\{G_{ni} \neq B_i\}} < 2a_n \middle| \Delta_n \right\}$$

(and, noting that $G_{ni}, X_1, \ldots, X_n$

are all functions of Δ_n, we have:)

$$= \mathbf{P} \left\{ \sum_{i \notin \{X_1, \ldots, X_n\}} p_i I_{\{B_i = 1\}} < 2a_n \middle| \Delta_n \right\}$$

$$\leq \mathbf{P} \left\{ \sum_{i=n+1}^{\infty} p_i I_{\{B_i = 1\}} < 2a_n \right\}$$

(since the p_i's are decreasing, by stochastic dominance)

$$= \mathbf{P} \left\{ \sum_{i=n+1}^{\infty} p_i B_i < 2a_n \right\}.$$

Now everything boils down to bounding these probabilities from above. We proceed by *Chernoff's bounding technique*. The idea is the following: For any random variable X, and $s > 0$, by Markov's inequality,

$$P\{X \geq \epsilon\} = P\{e^{sX} \geq e^{s\epsilon}\} \leq \frac{E\{e^{sX}\}}{e^{s\epsilon}}.$$

By cleverly choosing s one can often obtain very sharp bounds. For more discussion and examples of Chernoff's method, refer to Chapter 8. In our case,

$$P\left\{\sum_{i=n+1}^{\infty} p_i B_i < 2a_n\right\}$$

$$\leq E\left\{e^{2sa_n - s\sum_{i=n+1}^{\infty} p_i B_i}\right\}$$

$$= e^{2sa_n} \prod_{i=n+1}^{\infty} \left(\frac{1}{2} + \frac{1}{2}e^{-sp_i}\right)$$

$$\leq e^{2sa_n} \prod_{i=n+1}^{\infty} \frac{1}{2}\left(2 - sp_i + \frac{s^2 p_i^2}{2}\right)$$

(since $e^{-x} \leq 1 - x + x^2/2$ for $x \geq 0$)

$$\leq \exp\left(2sa_n + \sum_{i=n+1}^{\infty}\left(-\frac{sp_i}{2} + \frac{s^2 p_i^2}{4}\right)\right)$$

(since $1 - x \leq e^{-x}$)

$$\leq \exp\left(2sa_n - \frac{s\Sigma}{2} + \frac{s^2 p_{n+1}\Sigma}{4}\right)$$

(where $\Sigma = \sum_{i=n+1}^{\infty} p_i$)

$$= \exp\left(-\frac{1}{4}\frac{(4a_n - \Sigma)^2}{\Sigma p_{n+1}}\right)$$

(by taking $s = \frac{\Sigma - 4a_n}{p_{n+1}\Sigma}$, and the fact that $\Sigma > 4a_n$)

$$\leq \exp\left(-\frac{1}{16}\frac{\Sigma}{p_{n+1}}\right) \quad \text{(since } \Sigma \geq 8a_n\text{)}$$

$$\leq e^{-2n} \quad \text{(since } \Sigma \geq 32p_{n+1}n\text{)}.$$

Thus, we conclude that

$$\sup_b \inf_n E\frac{L_n(b)}{2a_n} \geq 1 - \sum_{n=1}^{\infty} e^{-2n} = \frac{e^2 - 2}{e^2 - 1} > \frac{1}{2},$$

so that there exists a b for which $EL_n(b) \geq a_n$ for all n. $\square$

Problems and Exercises

PROBLEM 7.1. Extend Theorem 7.2 for distributions with $0 < L^* < 1/2$: show that if a_n is a sequence of positive numbers as in Theorem 7.2, then for any classification rule there is a distribution such that $\mathbf{E}L_n - L^* \geq a_n$ for every n for which $L^* + a_n < 1/2$.

PROBLEM 7.2. Prove Theorems 7.1 and 7.2, under one of the following additional assumptions, which make the case that one will need very restrictive conditions indeed to study rates of convergence.
  (1) X has a uniform density on $[0, 1)$.
  (2) X has a uniform density on $[0, 1)$ and η is infinitely many times continuously differentiable on $[0, 1)$.
  (3) η is unimodal in $x \in \mathcal{R}^2$, that is, $\eta(\lambda x)$ decreases as $\lambda > 0$ increases for any $x \in \mathcal{R}^2$.
  (4) η is $\{0, 1\}$-valued, X is $\mathcal{R}^2$-valued, and the set $\{x : \eta(x) = 1\}$ is a compact convex set containing the origin.

PROBLEM 7.3. THERE IS NO SUPER-CLASSIFIER. Show that for every sequence of classification rules $\{g_n\}$ there is a universally consistent sequence of rules $\{g'_n\}$, such that for some distribution of (X, Y),

$$\mathbf{P}\{g_n(X) \neq Y\} > \mathbf{P}\{g'_n(X) \neq Y\}$$

for all n.

PROBLEM 7.4. The next two exercises are intended to demonstrate that the weaponry of pattern recognition can often be successfully used for attacking other statistical problems. For example, a consequence of Theorem 7.2 is that estimating infinite discrete distributions is hard. Consider the problem of estimating a distribution $(p_1, p_2, \ldots)$ on the positive integers $\{1, 2, 3, \ldots\}$ from a sample $X_1, \ldots, X_n$ of i.i.d. random variables with $\mathbf{P}\{X_1 = i\} = p_i$, $i \geq 1$. Show that for any decreasing sequence $\{a_n\}$ of positive numbers converging to zero with $a_1 \leq 1/16$, and any estimate $\{p_{i,n}\}$, there exists a distribution such that

$$\mathbf{E}\left\{\sum_{i=1}^{\infty} |p_i - p_{i,n}|\right\} \geq a_n.$$

HINT: Consider a classification problem with $L^* = 0$, $\mathbf{P}\{Y = 0\} = 1/2$, and X concentrated on $\{1, 2, \ldots\}$. Assume that the class-conditional probabilities $p_i^{(0)} = \mathbf{P}\{X = i | Y = 0\}$ and $p_i^{(1)} = \mathbf{P}\{X = i | Y = 1\}$ are estimated from two i.i.d. samples $X_1^{(0)}, \ldots, X_n^{(0)}$ and $X_1^{(1)}, \ldots, X_n^{(1)}$, distributed according to $\{p_i^{(0)}\}$ and $\{p_i^{(1)}\}$, respectively. Use Theorem 2.3 to show that for the classification rule obtained from these estimates in a natural way,

$$\mathbf{E}L_n \leq \frac{1}{2}\mathbf{E}\left\{\sum_{i=1}^{\infty} |p_i^{(0)} - p_{i,n}^{(0)}| + \sum_{i=1}^{\infty} |p_i^{(1)} - p_{i,n}^{(1)}|\right\},$$

therefore the lower bound of Theorem 7.2 can be applied.

PROBLEM 7.5. A similar slow-rate result appears in density estimation. Consider the problem of estimating a density f on $\mathcal{R}$, from an i.i.d. sample $X_1, \ldots, X_n$ having density f. Show that for any decreasing sequence $\{a_n\}$ of positive numbers converging to zero with $a_1 \leq 1/16$, and any density estimate f_n, there exists a distribution such that

$$\mathbf{E}\left\{\int |f(x) - f_n(x)|dx\right\} \geq a_n.$$

This result was proved by Birgé (1986) using a different—and in our view much more complicated—argument. HINT: Put $p_i = \int_i^{i+1} f(x)dx$ and $p_{i,n} = \int_i^{i+1} f_n(x)dx$ and apply Problem 7.4.

8
Error Estimation

8.1 Error Counting

Estimating the error probability $L_n = \mathbf{P}\{g_n(X) \neq Y | D_n\}$ of a classification function g_n is of essential importance. The designer always wants to know what performance can be expected from a classifier. As the designer does not know the distribution of the data—otherwise there would not be any need to design a classifier—it is important to find error estimation methods that work well without any condition on the distribution of (X, Y). This motivates us to search for distribution-free performance bounds for error estimation methods.

Suppose that we want to estimate the error probability of a classifier g_n designed from the training sequence $D_n = ((X_1, Y_1), \ldots, (X_n, Y_n))$. Assume first that a *testing sequence*

$$T_m = ((X_{n+1}, Y_{n+1}), \ldots, (X_{n+m}, Y_{n+m}))$$

is available, which is a sequence of i.i.d. pairs that are independent of (X, Y) and D_n, and that are distributed as (X, Y). An obvious way to estimate L_n is to count the number of errors that g_n commits on T_m. The *error-counting estimator* $\widehat{L}_{n,m}$ is defined by the relative frequency

$$\widehat{L}_{n,m} = \frac{1}{m} \sum_{j=1}^{m} I_{\{g_n(X_{n+j}) \neq Y_{n+j}\}}.$$

The estimator is clearly *unbiased* in the sense that

$$\mathbf{E}\left\{\widehat{L}_{n,m} | D_n\right\} = L_n,$$

and the conditional distribution of $m\widehat{L}_{n,m}$, given the training data D_n, is binomial with parameters m and L_n. This makes analysis easy, for properties of the binomial distribution are well known. One main tool in the analysis is Hoeffding's inequality, which we will use many many times throughout this book.

8.2   Hoeffding's Inequality

The following inequality was proved for binomial random variables by Chernoff (1952) and Okamoto (1958). The general format is due to Hoeffding (1963):

Theorem 8.1. (HOEFFDING (1963)).   *Let $X_1, \ldots, X_n$ be independent bounded random variables such that X_i falls in the interval $[a_i, b_i]$ with probability one. Denote their sum by $S_n = \sum_{i=1}^{n} X_i$. Then for any $\epsilon > 0$ we have*

$$P\{S_n - ES_n \geq \epsilon\} \leq e^{-2\epsilon^2 / \sum_{i=1}^{n}(b_i - a_i)^2}$$

and

$$P\{S_n - ES_n \leq -\epsilon\} \leq e^{-2\epsilon^2 / \sum_{i=1}^{n}(b_i - a_i)^2}.$$

The proof uses a simple auxiliary inequality:

Lemma 8.1. *Let X be a random variable with $EX = 0$, $a \leq X \leq b$. Then for $s > 0$,*

$$E\left\{e^{sX}\right\} \leq e^{s^2(b-a)^2/8}.$$

PROOF. Note that by convexity of the exponential function

$$e^{sx} \leq \frac{x - a}{b - a} e^{sb} + \frac{b - x}{b - a} e^{sa} \quad \text{for } a \leq x \leq b.$$

Exploiting $EX = 0$, and introducing the notation $p = -a/(b - a)$ we get

$$\begin{aligned}
Ee^{sX} &\leq \frac{b}{b - a} e^{sa} - \frac{a}{b - a} e^{sb} \\
&= \left(1 - p + pe^{s(b-a)}\right) e^{-ps(b-a)} \\
&\stackrel{\text{def}}{=} e^{\phi(u)},
\end{aligned}$$

where $u = s(b - a)$, and $\phi(u) = -pu + \log(1 - p + pe^u)$. But by straightforward calculation it is easy to see that the derivative of ϕ is

$$\phi'(u) = -p + \frac{p}{p + (1 - p)e^{-u}},$$

therefore $\phi(0) = \phi'(0) = 0$. Moreover,

$$\phi''(u) = \frac{p(1 - p)e^{-u}}{(p + (1 - p)e^{-u})^2} \leq \frac{1}{4}.$$

Thus, by Taylor series expansion with remainder, for some $\theta \in [0, u]$,

$$\phi(u) = \phi(0) + u\phi'(0) + \frac{u^2}{2}\phi''(\theta) \leq \frac{u^2}{8} = \frac{s^2(b-a)^2}{8}. \quad \square$$

PROOF OF THEOREM 8.1.   The proof is based on *Chernoff's bounding method* (Chernoff (1952)): by Markov's inequality, for any nonnegative random variable X, and any $\epsilon > 0$,

$$P\{X \geq \epsilon\} \leq \frac{EX}{\epsilon}.$$

Therefore, if s is an arbitrary positive number, then for any random variable X,

$$P\{X \geq \epsilon\} = P\{e^{sX} \geq e^{s\epsilon}\} \leq \frac{Ee^{sX}}{e^{s\epsilon}}.$$

In Chernoff's method, we find an $s > 0$ that minimizes the upper bound or makes the upper bound small. In our case, we have

$$P\{S_n - ES_n \geq \epsilon\}$$

$$\leq e^{-s\epsilon} E\left\{\exp\left(s \sum_{i=1}^{n}(X_i - EX_i)\right)\right\}$$

$$= e^{-s\epsilon} \prod_{i=1}^{n} E\left\{e^{s(X_i - EX_i)}\right\} \quad \text{(by independence)}$$

$$\leq e^{-s\epsilon} \prod_{i=1}^{n} e^{s^2(b_i - a_i)^2/8} \quad \text{(by Lemma 8.1)}$$

$$= e^{-s\epsilon} e^{s^2 \sum_{i=1}^{n}(b_i - a_i)^2/8}$$

$$= e^{-2\epsilon^2 / \sum_{i=1}^{n}(b_i - a_i)^2} \quad \text{(by choosing } s = 4\epsilon / \sum_{i=1}^{n}(b_i - a_i)^2\text{)}.$$

The second inequality is proved analogously. $\square$

The two inequalities in Theorem 8.1 may be combined to get

$$P\{|S_n - ES_n| \geq \epsilon\} \leq 2e^{-2\epsilon^2 / \sum_{i=1}^{n}(b_i - a_i)^2}.$$

Now, we can apply this inequality to get a distribution-free performance bound for the counting error estimate:

COROLLARY 8.1.   *For every $\epsilon > 0$,*

$$P\left\{|\widehat{L}_{n,m} - L_n| > \epsilon \mid D_n\right\} \leq 2e^{-2m\epsilon^2}.$$

The variance of the estimate can easily be computed using the fact that, conditioned on the data D_n, $m\widehat{L}_{n,m}$ is binomially distributed:

$$\mathbf{E}\left\{\left(\widehat{L}_{n,m} - L_n\right)^2 \Big| D_n\right\} = \frac{L_n(1 - L_n)}{m} \leq \frac{1}{4m}.$$

These are just the types of inequalities we want, for these are valid for any distribution and data size, and the bounds do not even depend on g_n.

Consider a special case in which all the X_i's take values on $[-c, c]$ and have zero mean. Then Hoeffding's inequality states that

$$\mathbf{P}\left\{S_n/n > \epsilon\right\} \leq e^{-n\epsilon^2/(2c^2)}.$$

This bound, while useful for ϵ larger than $c/\sqrt{n}$, ignores variance information. When $\mathbf{Var}\{X_i\} \ll c^2$, it is indeed possible to outperform Hoeffding's inequality. In particular, we have:

Theorem 8.2. (BENNETT (1962) AND BERNSTEIN (1946)). *Let $X_1, \ldots, X_n$ be independent real-valued random variables with zero mean, and assume that $X_i \leq c$ with probability one. Let*

$$\sigma^2 = \frac{1}{n}\sum_{i=1}^{n}\mathbf{Var}\{X_i\}.$$

Then, for any $\epsilon > 0$,

$$\mathbf{P}\left\{\frac{1}{n}\sum_{i=1}^{n}X_i > \epsilon\right\} \leq \exp\left(-\frac{n\epsilon}{2c}\left(\left(1 + \frac{\sigma^2}{2c\epsilon}\right)\log\left(1 + \frac{2c\epsilon}{\sigma^2}\right) - 1\right)\right)$$

(Bennett (1962)), and

$$\mathbf{P}\left\{\frac{1}{n}\sum_{i=1}^{n}X_i > \epsilon\right\} \leq \exp\left(-\frac{n\epsilon^2}{2\sigma^2 + 2c\epsilon/3}\right)$$

(Bernstein (1946)).

The proofs are left as exercises (Problem 8.2). We note that Bernstein's inequality kicks in when ϵ is larger than about max $\left(\sigma/\sqrt{n}, c/\sqrt{n}\right)$. It is typically better than Hoeffding's inequality when $\sigma \ll c$.

8.3   Error Estimation Without Testing Data

A serious problem concerning the practical applicability of the estimate introduced above is that it requires a large, independent testing sequence. In practice, however, an additional sample is rarely available. One usually wants to incorporate all available (X_i, Y_i) pairs in the decision function. In such cases, to estimate L_n, we have to rely on the training data only. There are well-known methods that we will discuss later that are based on cross-validation (or leave-one-out) (Lunts and Brailovsky (1967); Stone (1974)); and holdout, resubstitution, rotation, smoothing, and bootstrapping (Efron (1979), (1983)), which may be employed to construct an empirical risk from the training sequence, thus obviating the need for a testing sequence. (See Kanal (1974), Cover and Wagner (1975), Toussaint (1974a), Glick (1978), Hand (1986), Jain, Dubes, and Chen (1987), and McLachlan (1992) for surveys, discussion, and empirical comparison.)

Analysis of these methods, in general, is clearly a much harder problem, as g_n can depend on D_n in a rather complicated way. If we construct some estimator $\widehat{L}_n$ from D_n, then it would be desirable to obtain distribution-free bounds on

$$\mathbf{P}\left\{|\widehat{L}_n - L_n| > \epsilon\right\},$$

or on

$$\mathbf{E}\left\{|\widehat{L}_n - L_n|^q\right\}$$

for some $q \geq 1$. Conditional probabilities and expectations given D_n are meaningless, since everything is a function of D_n. Here, however, we have to be much more careful as we do not want $\widehat{L}_n$ to be optimistically biased because the same data are used both for training and testing.

Distribution-free bounds for the above quantities would be extremely helpful, as we usually do not know the distribution of (X, Y). While for some rules such estimates exist—we will exhibit several avenues in Chapters 22, 23, 24, 25, 26, and 31—it is disappointing that a single error estimation method cannot possibly work for all discrimination rules. It is therefore important to point out that we have to consider $(g_n, \widehat{L}_n)$ pairs—for every rule one or more error estimates must be found if possible, and vice versa, for every error estimate, its limitations have to be stated. Secondly, rules for which no good error estimates are known should be avoided. Luckily, most popular rules do not fall into this category. On the other hand, proven distribution-free performance guarantees are rarely available—see Chapters 23 and 24 for examples.

8.4 Selecting Classifiers

Probably the most important application of error estimation is in the selection of a classification function from a class C of functions. If a class C of classifiers is given, then it is tempting to pick the one that minimizes an estimate of the error probability over the class. A good method should pick a classifier with an error probability that is close to the minimal error probability in the class. Here we require much more than distribution-free performance bounds of the error estimator for each of the classifiers in the class. Problem 8.8 demonstrates that it is not sufficient to be able to estimate the error probability of all classifiers in the class. Intuitively, if we can estimate the error probability for the classifiers in C *uniformly* well, then the classification function that minimizes the estimated error probability is likely to have an error probability that is close to the best in the class. To certify this intuition, consider the following situation: Let C be a class of classifiers, that is, a class of mappings of the form $\phi : \mathcal{R}^d \rightarrow \{0, 1\}$. Assume that the error count

$$\widehat{L}_n(\phi) = \frac{1}{n} \sum_{j=1}^{n} I_{\{\phi(X_j) \neq Y_j\}}$$

is used to estimate the error probability $L(\phi) = \mathbf{P}\{\phi(X) \neq Y\}$ of each classifier $\phi \in C$. Denote by ϕ_n^* the classifier that minimizes the estimated error probability over the class:

$$\widehat{L}_n(\phi_n^*) \leq \widehat{L}_n(\phi) \quad \text{for all } \phi \in C.$$

Then for the error probability

$$L(\phi_n^*) = \mathbf{P}\left\{ \phi_n^*(X) \neq Y \,\middle|\, D_n \right\}$$

of the selected rule we have:

Lemma 8.2. (VAPNIK AND CHERVONENKIS (1974C); SEE ALSO DEVROYE (1988B)).

$$L(\phi_n^*) - \inf_{\phi \in C} L(\phi) \leq 2 \sup_{\phi \in C} |\widehat{L}_n(\phi) - L(\phi)|,$$

$$|\widehat{L}_n(\phi_n^*) - L(\phi_n^*)| \leq \sup_{\phi \in C} |\widehat{L}_n(\phi) - L(\phi)|.$$

PROOF.

$$
\begin{aligned}
L(\phi_n^*) - \inf_{\phi \in C} L(\phi) &= L(\phi_n^*) - \widehat{L}_n(\phi_n^*) + \widehat{L}_n(\phi_n^*) - \inf_{\phi \in C} L(\phi) \\
&\leq L(\phi_n^*) - \widehat{L}_n(\phi_n^*) + \sup_{\phi \in C} |\widehat{L}_n(\phi) - L(\phi)| \\
&\leq 2 \sup_{\phi \in C} |\widehat{L}_n(\phi) - L(\phi)|.
\end{aligned}
$$

The second inequality is trivially true. $\square$

We see that upper bounds for $\sup_{\phi \in C} |\widehat{L}_n(\phi) - L(\phi)|$ provide us with upper bounds for two things simultaneously:

(1) An upper bound for the suboptimality of ϕ_n^* within C, that is, a bound for $L(\phi_n^*) - \inf_{\phi \in C} L(\phi)$.

(2) An upper bound for the error $|\widehat{L}_n(\phi_n^*) - L(\phi_n^*)|$ committed when $\widehat{L}_n(\phi_n^*)$ is used to estimate the probability of error $L(\phi_n^*)$ of the selected rule.

In other words, by bounding $\sup_{\phi \in C} |\widehat{L}_n(\phi) - L(\phi)|$, we kill two flies at once. It is particularly useful to know that even though $\widehat{L}_n(\phi_n^*)$ is usually optimistically biased, it is within given bounds of the unknown probability of error with ϕ_n^*, and that no other test sample is needed to estimate this probability of error. Whenever our bounds indicate that we are close to the optimum in C, we must at the same time have a good estimate of the probability of error, and vice versa.

As a simple, but interesting application of Lemma 8.2 we consider the case when the class C contains finitely many classifiers.

Theorem 8.3. *Assume that the cardinality of C is bounded by N. Then we have for all $\epsilon > 0$,*

$$\mathbf{P}\left\{\sup_{\phi \in C} |\widehat{L}_n(\phi) - L(\phi)| > \epsilon\right\} \leq 2Ne^{-2n\epsilon^2}.$$

PROOF.

$$\mathbf{P}\left\{\sup_{\phi \in C} |\widehat{L}_n(\phi) - L(\phi)| > \epsilon\right\} \leq \sum_{\phi \in C} \mathbf{P}\left\{|\widehat{L}_n(\phi) - L(\phi)| > \epsilon\right\}$$

$$\leq 2Ne^{-2n\epsilon^2},$$

where we used Hoeffding's inequality, and the fact that the random variable $n\widehat{L}_n(\phi)$ is binomially distributed with parameters n and $L(\phi)$. $\square$

REMARK. DISTRIBUTION-FREE PROPERTIES. Theorem 8.3 shows that the problem studied here is purely combinatorial. The actual distribution of the data does not play a role at all in the upper bounds. $\square$

REMARK. WITHOUT TESTING DATA. Very often, a class of rules C of the form $\phi_n(x) = \phi_n(x, D_n)$ is given, and the same data D_n are used to select a rule by minimizing some estimates $\widehat{L}_n(\phi_n)$ of the error probabilities $L(\phi_n) = \mathbf{P}\{\phi_n(X) \neq Y | D_n\}$. A similar analysis can be carried out in this case. In particular, if ϕ_n^* denotes the selected rule, then we have similar to Lemma 8.2:

Theorem 8.4.

$$L(\phi_n^*) - \inf_{\phi_n \in C} L(\phi_n) \leq 2 \sup_{\phi_n \in C} |\widehat{L}_n(\phi_n) - L(\phi_n)|,$$

and

$$|\widehat{L}_n(\phi_n^*) - L(\phi_n^*)| \leq \sup_{\phi_n \in C} |\widehat{L}_n(\phi_n) - L(\phi_n)|.$$

If C is finite, then again, similar to Theorem 8.3, we have for example

$$\mathbf{P}\left\{L(\phi_n^*) - \inf_{\phi_n \in C} L(\phi_n) > \epsilon\right\} \leq N \sup_{\phi_n \in C} \mathbf{P}\left\{|\widehat{L}_n(\phi_n) - L(\phi_n)| > \epsilon/2\right\}. \quad \square$$

8.5   Estimating the Bayes Error

It is also important to have a good estimate of the optimal error probability L^*. First of all, if L^* is large, we would know beforehand that any rule is going to perform poorly. Then perhaps the information might be used to return to the feature selection stage. Also, a comparison of estimates of L_n and L^* gives us an idea how much room is left for improvement. Typically, L^* is estimated by an estimate of the error probability of some consistent classification rule (see Fukunaga and Kessel (1971), Chen and Fu (1973), Fukunaga and Hummels (1987), and Garnett and Yau (1977)). Clearly, if the estimate $\widehat{L}_n$ we use is consistent in the sense that $\widehat{L}_n - L_n \to 0$ with probability one as $n \to \infty$, and the rule is strongly consistent, then

$$\widehat{L}_n \to L^*$$

with probability one. In other words, we have a consistent estimate of the Bayes error probability. There are two problems with this approach. The first problem is that if our purpose is comparing L^* with L_n, then using the same estimate for both of them does not give any information. The other problem is that even though for many classifiers, $L_n - L_n$ can be guaranteed to converge to zero rapidly, regardless what the distribution of (X, Y) is (see Chapters 23 and 24), in view of the results of Chapter 7, the rate of convergence of L_n to L^* using such a method may be arbitrarily slow. Thus, we cannot expect good performance for all distributions from such a method. The question is whether it is possible to come up with a method of estimating L^* such that the difference $\widehat{L}_n - L^*$ converges to zero rapidly for all distributions. Unfortunately, there is no method that guarantees a certain finite sample performance for all distributions. This disappointing fact is reflected in the following negative result:

Theorem 8.5. *For every n, for any estimate $\widehat{L}_n$ of the Bayes error probability L^*, and for every $\epsilon > 0$, there exists a distribution of (X, Y), such that*

$$\mathbf{E}\left\{|\widehat{L}_n - L^*|\right\} \geq \frac{1}{4} - \epsilon.$$

PROOF. For a fixed n, we construct a family $\mathcal{F}$ of distributions, and show that for at least one member of the family, $\mathbf{E}\left\{|\widehat{L}_n - L^*|\right\} \geq \frac{1}{4} - \epsilon$. The family contains $2^m + 1$ distributions, where m is a large integer specified later. In all cases, $X_1, \ldots, X_n$ are drawn independently by a uniform distribution from the set $\{1, \ldots, m\}$. Let $B_0, B_1, B_2, \ldots, B_n$ be i.i.d. Bernoulli random variables, independent of the X_i's, with $\mathbf{P}\{B_i = 0\} = \mathbf{P}\{B_i = 1\} = 1/2$. For the first member of the family $\mathcal{F}$, let $Y_i = B_i$ for $i = 1, \ldots, n$. Thus, for this distribution, $L^* = 1/2$. The Bayes error for the other 2^m members of the family is zero. These distributions are determined by m binary parameters $a_1, a_2, \ldots, a_m \in \{0, 1\}$ as follows:

$$\eta(i) = \mathbf{P}\{Y = 1 | X = i\} = a_i.$$

In other words, $Y_i = a_{X_i}$ for every $i = 1, \ldots, n$. Clearly, $L^* = 0$ for these distributions. Note also that all distributions with X distributed uniformly on $\{1, \ldots, m\}$

and $L^* = 0$ are members of the family. Just as in the proofs of Theorems 7.1 and 7.2, we randomize over the family $\mathcal{F}$ of distributions. However, the way of randomization is different here. The trick is to use $B_0, B_1, B_2, \ldots, B_n$ in randomly picking a distribution. (Recall that these random variables are just the labels $Y_1, \ldots, Y_n$ in the training sequence for the first distribution in the family.) We choose a distribution randomly, as follows: If $B_0 = 0$, then we choose the first member of $\mathcal{F}$ (the one with $L^* = 1/2$). If $B_0 = 1$, then the labels of the training sequence are given by

$$Y_i = \begin{cases} B_i & \text{if } X_i \neq X_1, X_i \neq X_2, \ldots, X_i \neq X_{i-1} \\ B_j & \text{if } j < i \text{ is the smallest index such that } X_i = X_j. \end{cases}$$

Note that in case of $B_0 = 1$, for any fixed realization $b_1, \ldots, b_n \in \{0, 1\}$ of $B_1, \ldots, B_n$, the Bayes risk is zero. Therefore, the distribution is in the family $\mathcal{F}$.

Now, let A be the event that all the X_i's are different. Observe that under A, $\widehat{L}_n$ is a function of $X_1, \ldots, X_n, B_1, \ldots, B_n$ only, but *not* B_0. Therefore,

$$\sup_{\mathcal{F}} \mathbf{E}\left\{|\widehat{L}_n - L^*|\right\} \geq \mathbf{E}\left\{|\widehat{L}_n - L^*|\right\}$$

$$\text{(with } B_0, B_1, \ldots, B_n \text{ random)}$$

$$\geq \mathbf{E}\left\{I_A|\widehat{L}_n - L^*|\right\}$$

$$= \mathbf{E}\left\{I_A\left(I_{\{B_0=0\}}\left|\widehat{L}_n - \frac{1}{2}\right| + I_{\{B_0=1\}}\left|\widehat{L}_n - 0\right|\right)\right\}$$

$$= \mathbf{E}\left\{I_A\frac{1}{2}\left(\left|\widehat{L}_n - \frac{1}{2}\right| + |\widehat{L}_n - 0|\right)\right\}$$

$$\geq \mathbf{E}\left\{I_A\frac{1}{4}\right\}$$

$$= \frac{1}{4}\mathbf{P}\{A\}.$$

Now, if we pick m large enough, $\mathbf{P}\{A\}$ can be as close to 1 as desired. Hence,

$$\sup \mathbf{E}\left\{|\widehat{L}_n - L^*|\right\} \geq \frac{1}{4},$$

where the supremum is taken over all distributions of (X, Y). $\square$

Problems and Exercises

PROBLEM 8.1. Let B be a binomial random variable with parameters n and p. Show that

$$\mathbf{P}\{B > \epsilon\} \leq e^{\epsilon - np - \epsilon \log(\epsilon/np)} \ (\epsilon > np)$$

and

$$\mathbf{P}\{B < \epsilon\} \leq e^{\epsilon - np - \epsilon \log(\epsilon/np)} \ (\epsilon < np).$$

(Chernoff (1952).) HINT: Proceed by Chernoff's bounding method.

PROBLEM 8.2. Prove the inequalities of Bennett and Bernstein given in Theorem 8.2. To help you, we will guide you through different stages:

(1) Show that for any $s > 0$, and any random variable X with $\mathbf{E}X = 0$, $\mathbf{E}X^2 = \sigma^2$, $X \le c$,

$$\mathbf{E}\left\{e^{sX}\right\} \le e^{f(\sigma^2/c^2)},$$

where

$$f(u) = \log\left(\frac{1}{1+u}e^{-csu} + \frac{u}{1+u}e^{cs}\right).$$

(2) Show that $f''(u) \le 0$ for $u \ge 0$.

(3) By Chernoff's bounding method, show that

$$\mathbf{P}\left\{\frac{1}{n}\sum_{i=1}^{n}X_i \ge \epsilon\right\} \le e^{-sn\epsilon + \sum_{i=1}^{n}f(\mathbf{Var}\{X_i\}/c^2)}.$$

(4) Show that $f(u) \le f(0) + uf'(0) = (e^{cs} - 1 - cs)u$.

(5) Using the bound of (4), find the optimal value of s and derive Bennett's inequality.

PROBLEM 8.3. Use Bernstein's inequality to show that if B is a binomial (n, p) random variable, then for $\epsilon > 0$,

$$\mathbf{P}\left\{\frac{B}{n} - p > \epsilon p\right\} \le e^{-3np\epsilon^2/8}$$

and

$$\mathbf{P}\left\{\frac{B}{n} - p < -\epsilon p\right\} \le e^{-3np\epsilon^2/8}.$$

PROBLEM 8.4. Let $X_1, \ldots, X_n$ be independent binary-valued random variables with $\mathbf{P}\{X_i = 1\} = 1 - \mathbf{P}\{X_i = 0\} = p_i$. Set $p = (1/n)\sum_{i=1}^{n}p_i$ and $S_n = \sum_{i=1}^{n}X_i$. Prove that

$$\mathbf{P}\{S_n - np \ge n\epsilon\} \le e^{-np\epsilon^2/3} \quad \text{and} \quad \mathbf{P}\{S_n - np \le -n\epsilon\} \le e^{-np\epsilon^2/2}$$

(Angluin and Valiant (1979), see also Hagerup and Rüb (1990)). Compare the results with Bernstein's inequality for this case. HINT: Put $s = \log(1 + \epsilon)$ and $s = -\log(1 - \epsilon)$ in the Chernoff bounding argument. Prove and exploit the elementary inequalities

$$-\frac{\epsilon^2}{2} \le \epsilon - (1 + \epsilon)\log(1 + \epsilon) \le -\frac{\epsilon^2}{3}, \quad \epsilon \in [0, 1],$$

and

$$-\frac{\epsilon^2}{2} \ge \epsilon - (1 + \epsilon)\log(1 + \epsilon), \quad \epsilon \in (-1, 0].$$

PROBLEM 8.5. Let B be a Binomial (n, p) random variable. Show that for $p \le a < 1$,

$$\mathbf{P}\{B > an\} \le \left(\left(\frac{p}{a}\right)^a \left(\frac{1-p}{1-a}\right)^{1-a}\right)^n \le \left(\left(\frac{p}{a}\right)^a e^{a-p}\right)^n.$$

Show that for $0 < a < p$ the same upper bounds hold for $\mathbf{P}\{B \le an\}$ (Karp (1988), see also Hagerup and Rüb (1990)). HINT: Use Chernoff's method with parameter s and set $s = \log(a(1 - p)/(p(1 - a)))$.

PROBLEM 8.6. Let B be a Binomial (n, p) random variable. Show that if $p \geq 1/2$,

$$\mathbf{P}\{B - np \geq n\epsilon\} < e^{-\frac{n\epsilon^2}{2p(1-p)}},$$

and if $p \leq 1/2$,

$$\mathbf{P}\{B - np \leq -n\epsilon\} < e^{-\frac{n\epsilon^2}{2p(1-p)}}$$

(Okamoto (1958)). HINT: Use Chernoff's method, and the inequality

$$x \log \frac{x}{p} + (1 - x) \log \frac{1 - x}{1 - p} \geq \frac{(x - p)^2}{2p(1 - p)}$$

for $1/2 \leq p \leq x \leq 1$, and for $0 \leq x \leq p \leq 1/2$.

PROBLEM 8.7. Let B be a Binomial (n, p) random variable. Prove that

$$\mathbf{P}\{\sqrt{B} - \sqrt{np} \geq \epsilon\sqrt{n}\} < e^{-2n\epsilon^2},$$

and

$$\mathbf{P}\{\sqrt{B} - \sqrt{np} \leq -\epsilon\sqrt{n}\} < e^{-n\epsilon^2}$$

(Okamoto (1958)). HINT: Use Chernoff's method, and the inequalities

$$x \log \frac{x}{p} + (1 - x) \log \frac{1 - x}{1 - p} \geq 2 \left(\sqrt{x} - \sqrt{p}\right)^2 \quad x \in [p, 1],$$

and

$$x \log \frac{x}{p} + (1 - x) \log \frac{1 - x}{1 - p} \geq \left(\sqrt{x} - \sqrt{p}\right)^2 \quad x \in [0, p].$$

PROBLEM 8.8. Give a class C of decision functions of the form $\phi : \mathcal{R}^d \to \{0, 1\}$ (i.e., the training data do not play any role in the decision) such that for every $\epsilon > 0$

$$\sup_{\phi \in C} \mathbf{P}\left\{|\widehat{L}_n(\phi) - L(\phi)| > \epsilon\right\} \leq 2e^{-2n\epsilon^2}$$

for every distribution, where $\widehat{L}_n(\phi)$ is the error-counting estimate of the error probability $L(\phi) = \mathbf{P}\{\phi(X) \neq Y\}$ of decision ϕ, and at the same time, if $\mathcal{F}_n$ is the class of mappings ϕ_n^* minimizing the error count $\widehat{L}_n(\phi)$ over the class C, then there exists one distribution such that

$$\mathbf{P}\left\{\sup_{\phi \in \mathcal{F}_n} L(\phi) - \inf_{\phi \in C} L(\phi) = 1\right\} = 1$$

for all n.

PROBLEM 8.9. Let C be a class of classifiers, that is, a class of mappings of the form $\phi_n(x, D_n) = \phi_n(x)$. Assume that an independent testing sequence T_m is given, and that the error count

$$\widehat{L}_{n,m}(\phi_n) = \frac{1}{m} \sum_{j=1}^{m} I_{\{\phi_n(X_{n+j}) \neq Y_{n+j}\}}$$

is used to estimate the error probability $L(\phi_n) = \mathbf{P}\{\phi_n(X) \neq Y | D_n\}$ of each classifier $\phi_n \in C$. Denote by $\phi_{n,m}^*$ the classifier that minimizes the estimated error probability over the class. Prove that for the error probability

$$L(\phi_{n,m}^*) = \mathbf{P}\left\{\phi_{n,m}^*(X) \neq Y | D_n\right\}$$

of the selected rule we have

$$L(\phi_{n,m}^*) - \inf_{\phi_n \in C} L(\phi_n) \le 2 \sup_{\phi_n \in C} |\widehat{L}_{n,m}(\phi_n) - L(\phi_n)|,$$

$$|\widehat{L}_{n,m}(\phi_{n,m}^*) - L(\phi_{n,m}^*)| \le \sup_{\phi_n \in C} |\widehat{L}_{n,m}(\phi_n) - L(\phi_n)|.$$

Also, if C is of finite cardinality with $|C| = N$, then

$$\mathbf{P}\left\{ \sup_{\phi_n \in C} |\widehat{L}_{n,m}(\phi_n) - L(\phi_n)| > \epsilon \,\middle|\, D_n \right\} \le 2Ne^{-2m\epsilon^2}.$$

PROBLEM 8.10. Show that if a rule g_n is consistent, then we can always find an estimate of the error such that $\mathbf{E}\{|\hat{L}_n - L_n|^q\} \to 0$ for all $q > 0$. HINT: Split the data sequence D_n and use the second half to estimate the error probability of $g_{\lceil n/2 \rceil}$.

PROBLEM 8.11. OPEN-ENDED PROBLEM. Is there a rule for which no error estimate works for all distributions? More specifically, is there a sequence of classification rules g_n such that for all n large enough,

$$\inf_{\hat{L}_n} \sup_{X,Y} \mathbf{E}\{(\hat{L}_n - L_n)^2\} \ge c$$

for some constant $c > 0$, where the infimum is taken over all possible error estimates? Are such rules necessarily inconsistent?

PROBLEM 8.12. Consider the problem of estimating the asymptotic probability of error of the nearest neighbor rule $L_{NN} = 2\mathbf{E}\{\eta(X)(1 - \eta(X))\}$. Show that for every n, for any estimate $\widehat{L}_n$ of L_{NN}, and for every $\epsilon > 0$, there exists a distribution of (X, Y), such that

$$\mathbf{E}\left\{|\widehat{L}_n - L_{NN}|\right\} \ge \frac{1}{4} - \epsilon.$$

9
The Regular Histogram Rule

In this chapter we study the cubic histogram rule. Recall that this rule partitions $\mathcal{R}^d$ into cubes of the same size, and gives the decision according to the number of zeros and ones among the Y_i's such that the corresponding X_i falls in the same cube as X. $\mathcal{P}_n = \{A_{n1}, A_{n2}, \ldots\}$ denotes a partition of $\mathcal{R}^d$ into cubes of size $h_n > 0$, that is, into sets of the type $\prod_{i=1}^{d}[k_i h_n, (k_i + 1)h_n)$, where the k_i's are integers, and the histogram rule is defined by

$$g_n(x) = \begin{cases} 0 & \text{if } \sum_{i=1}^{n} I_{\{Y_i=1\}} I_{\{X_i \in A_n(x)\}} \leq \sum_{i=1}^{n} I_{\{Y_i=0\}} I_{\{X_i \in A_n(x)\}} \\ 1 & \text{otherwise,} \end{cases}$$

where for every $x \in \mathcal{R}^d$, $A_n(x) = A_{ni}$ if $x \in A_{ni}$. That is, the decision is zero if the number of ones does not exceed the number of zeros in the cell in which x falls. Weak universal consistency of this rule was shown in Chapter 6 under the conditions $h_n \to 0$ and $nh_n^d \to \infty$ as $n \to \infty$. The purpose of this chapter is to introduce some techniques by proving *strong* universal consistency of this rule. These techniques will prove very useful in handling other problems as well. First we introduce the method of bounded differences.

9.1 The Method of Bounded Differences

In this section we present a generalization of Hoeffding's inequality, due to McDiarmid (1989). The result will equip us with a powerful tool to handle complicated functions of independent random variables. This inequality follows by results of Hoeffding (1963) and Azuma (1967) who observed that Theorem 8.1

can be generalized to bounded martingale difference sequences. The inequality has found many applications in combinatorics, as well as in nonparametric statistics (see McDiarmid (1989) and Devroye (1991a) for surveys).

Let us first recall the notion of martingales. Consider a probability space $(\Omega, \mathcal{F}, \mathbf{P})$.

DEFINITION 9.1. *A sequence of random variables* $Z_1, Z_2, \ldots$ *is called a* martingale *if*

$$\mathbf{E}\{Z_{i+1}|Z_1, \ldots, Z_i\} = Z_i \quad \text{with probability one}$$

for each $i > 0$.

Let $X_1, X_2, \ldots$ *be an arbitrary sequence of random variables.* $Z_1, Z_2, \ldots$ *is called a* martingale *with respect to the sequence* $X_1, X_2, \ldots$ *if for every* $i > 0$, Z_i *is a function of* $X_1, \ldots, X_i$, *and*

$$\mathbf{E}\{Z_{i+1}|X_1, \ldots, X_i\} = Z_i \quad \text{with probability one.}$$

Obviously, if $Z_1, Z_2, \ldots$ is a martingale with respect to $X_1, X_2, \ldots$, then Z_1, $Z_2, \ldots$ is a martingale, since

$$
\begin{aligned}
\mathbf{E}\{Z_{i+1}|Z_1, \ldots, Z_i\} &= \mathbf{E}\{\mathbf{E}\{Z_{i+1}|X_1, \ldots, X_i\}Z_1, \ldots, Z_i\} \\
&= \mathbf{E}\{Z_i|Z_1, \ldots, Z_i\} \\
&= Z_i.
\end{aligned}
$$

The most important examples of martingales are sums of independent zero-mean random variables. Let $U_1, U_2, \ldots$ be independent random variables with zero mean. Then the random variables

$$S_i = \sum_{j=1}^{i} U_j, \quad i > 0,$$

form a martingale (see Problem 9.1). Martingales share many properties of sums of independent variables. Our purpose here is to extend Hoeffding's inequality to martingales. The role of the independent random variables is played here by a so-called martingale difference sequence.

DEFINITION 9.2. *A sequence of random variables* $V_1, V_2, \ldots$ *is a* martingale difference sequence *if*

$$\mathbf{E}\{V_{i+1}|V_1, \ldots, V_i\} = 0 \quad \text{with probability one}$$

for every $i > 0$.

A sequence of random variables $V_1, V_2, \ldots$ *is called a* martingale difference sequence *with respect to a sequence of random variables* $X_1, X_2, \ldots$ *if for every* $i > 0$ V_i *is a function of* $X_1, \ldots, X_i$, *and*

$$\mathbf{E}\{V_{i+1}|X_1, \ldots, X_i\} = 0 \quad \text{with probability one.}$$

Again, it is easily seen that if V_1, V_2, ... is a martingale difference sequence with respect to a sequence X_1, X_2, ... of random variables, then it is a martingale difference sequence. Also, any martingale Z_1, Z_2, ... leads naturally to a martingale difference sequence by defining

$$V_i = Z_i - Z_{i-1}$$

for $i > 0$.

The key result in the method of bounded differences is the following inequality that relaxes the independence assumption in Theorem 8.1, allowing martingale difference sequences:

Theorem 9.1. (HOEFFDING (1963), AZUMA (1967)). *Let X_1, X_2, ... be a sequence of random variables, and assume that V_1, V_2, ... is a martingale difference sequence with respect to X_1, X_2, ... Assume furthermore that there exist random variables Z_1, Z_2, ... and nonnegative constants c_1, c_2, ... such that for every $i > 0$ Z_i is a function of $X_1, \ldots, X_{i-1}$, and*

$$Z_i \le V_i \le Z_i + c_i \quad \text{with probability one.}$$

Then for any $\epsilon > 0$ and n

$$\mathbf{P}\left\{\sum_{i=1}^{n} V_i \ge \epsilon\right\} \le e^{-2\epsilon^2 / \sum_{i=1}^{n} c_i^2}$$

and

$$\mathbf{P}\left\{\sum_{i=1}^{n} V_i \le -\epsilon\right\} \le e^{-2\epsilon^2 / \sum_{i=1}^{n} c_i^2}.$$

The proof is a rather straightforward extension of that of Hoeffding's inequality. First we need an analog of Lemma 8.1:

Lemma 9.1. *Assume that the random variables V and Z satisfy with probability one that $\mathbf{E}\{V|Z\} = 0$, and for some function f and constant $c \ge 0$*

$$f(Z) \le V \le f(Z) + c.$$

Then for every $s > 0$

$$\mathbf{E}\left\{e^{sV}|Z\right\} \le e^{s^2 c^2 / 8}.$$

The proof of the lemma is left as an exercise (Problem 9.2).

PROOF OF THEOREM 9.1. As in the proof of Hoeffding's inequality, we proceed by Chernoff's bounding method. Set $S_k = \sum_{i=1}^{k} V_i$. Then for any $s > 0$

$$\begin{aligned}
\mathbf{P}\{S_n \ge \epsilon\} &\le e^{-s\epsilon} \mathbf{E}\left\{e^{sS_n}\right\} \\
&= e^{-s\epsilon} \mathbf{E}\left\{e^{sS_{n-1}} \mathbf{E}\left\{e^{sV_n}|X_1, \ldots, X_{n-1}\right\}\right\}
\end{aligned}$$

$$\leq \quad e^{-s\epsilon} \mathbf{E}\left\{e^{sS_{n-1}}\right\} e^{s^2 c_n^2/8} \quad \text{(by Lemma 9.1)}$$

$$\leq \quad e^{-s\epsilon} e^{s^2 \sum_{i=1}^n c_i^2/8} \quad \text{(iterate previous argument)}$$

$$= \quad e^{-2\epsilon^2/\sum_{i=1}^n c_i^2} \quad \text{(choose } s = 4\epsilon/\sum_{i=1}^n c_i^2\text{)}.$$

The second inequality is proved analogously. $\square$

Now, we are ready to state the main inequality of this section. It is a large deviation-type inequality for functions of independent random variables such that the function is relatively robust to individual changes in the values of the random variables. The condition of the function requires that by changing the value of its i-th variable, the value of the function cannot change by more than a constant c_i.

Theorem 9.2. (MCDIARMID (1989)). *Let $X_1, \ldots, X_n$ be independent random variables taking values in a set A, and assume that $f : A^n \to R$ satisfies*

$$\sup_{\substack{x_1,\ldots,x_n, \\ x_i' \in A}} |f(x_1, \ldots, x_n) - f(x_1, \ldots, x_{i-1}, x_i', x_{i+1}, \ldots, x_n)| \leq c_i, \quad 1 \leq i \leq n.$$

Then for all $\epsilon > 0$

$$\mathbf{P}\left\{f(X_1, \ldots, X_n) - \mathbf{E}f(X_1, \ldots, X_n) \geq \epsilon\right\} \leq e^{-2\epsilon^2/\sum_{i=1}^n c_i^2},$$

and

$$\mathbf{P}\left\{\mathbf{E}f(X_1, \ldots, X_n) - f(X_1, \ldots, X_n) \geq \epsilon\right\} \leq e^{-2\epsilon^2/\sum_{i=1}^n c_i^2}.$$

PROOF. Define $V = f(X_1, \ldots, X_n) - \mathbf{E}f(X_1, \ldots, X_n)$. Introduce $V_1 = \mathbf{E}\{V|X_1\} - \mathbf{E}V$, and for $k > 1$,

$$V_k = \mathbf{E}\{V|X_1 \ldots, X_k\} - \mathbf{E}\{V|X_1, \ldots, X_{k-1}\},$$

so that $V = \sum_{k=1}^n V_k$. Clearly, $V_1, \ldots, V_n$ form a martingale difference sequence with respect to $X_1, \ldots, X_n$. Define the random variables

$$H_k(X_1, \ldots, X_k) = \mathbf{E}\{f(X_1, \ldots, X_n)|X_1, \ldots, X_k\}$$

and

$$V_k = H_k(X_1, \ldots, X_k) - \int H_k(X_1, \ldots, X_{k-1}, x)F_k(dx),$$

where the integration is with respect to F_k, the probability measure of X_k. Introduce the random variables

$$W_k = \sup_u \left(H_k(X_1, \ldots, X_{k-1}, u) - \int H_k(X_1, \ldots, X_{k-1}, x)F_k(dx) \right),$$

and

$$Z_k = \inf_v \left(H_k(X_1, \ldots, X_{k-1}, v) - \int H_k(X_1, \ldots, X_{k-1}, x)F_k(dx) \right).$$

Clearly, $Z_k \leq V_k \leq W_k$ with probability one. Since for every k Z_k is a function of $X_1, \ldots, X_{k-1}$, we can apply Theorem 9.1 directly to $V = \sum_{k=1}^{n} V_k$, if we can show that $W_k - Z_k \leq c_k$. But this follows from

$$
\begin{aligned}
W_k - Z_k &= \sup_u \sup_v \left(H_k(X_1, \ldots, X_{k-1}, u) - H_k(X_1, \ldots, X_{k-1}, v) \right) \\
&\leq c_k,
\end{aligned}
$$

by the condition of the theorem. $\square$

Clearly, if the X_i's are bounded, then the choice $f(x_1, \ldots, x_n) = \sum_{i=1}^{n} x_i$ yields Hoeffding's inequality. Many times the inequality can be used to handle very complicated functions of independent random variables with great elegance. For examples in nonparametric statistics, see Problems 9.3, 9.6, 10.3.

Similar methods to those used in the proof of Theorem 9.2 may be used to bound the variance $\mathbf{Var}\{f(X_1, \ldots, X_n)\}$. Other inequalities for the variance of general functions of independent random variables were derived by Efron and Stein (1981) and Steele (1986).

Theorem 9.3. (DEVROYE (1991A)). *Assume that the conditions of Theorem 9.2 hold. Then*

$$
\mathbf{Var}\{f(X_1, \ldots, X_n)\} \leq \frac{1}{4} \sum_{i=1}^{n} c_i^2.
$$

PROOF. Using the notations of the proof of Theorem 9.2, we have to show that

$$
\mathbf{Var}\{V\} \leq \frac{1}{4} \sum_{i=1}^{n} c_i^2.
$$

Observe that

$$
\begin{aligned}
\mathbf{Var}\{V\} &= \mathbf{E}V^2 \\
&= \mathbf{E}\left\{ \left(\sum_{i=1}^{n} V_i \right)^2 \right\} \\
&= \mathbf{E}\left\{ \sum_{i=1}^{n} V_i^2 \right\} + 2 \sum_{1 \leq i < j \leq n} \mathbf{E}\{V_i V_j\} \\
&= \mathbf{E}\left\{ \sum_{i=1}^{n} V_i^2 \right\},
\end{aligned}
$$

where in the last step we used the martingale property in the following way: for $i < j$ we have

$$
\begin{aligned}
\mathbf{E}\{V_i V_j | X_1, \ldots, X_{j-1}\} \\
= V_i \mathbf{E}\{V_j | X_1, \ldots, X_{j-1}\} \\
= 0 \quad \text{with probability one.}
\end{aligned}
$$

Thus, the theorem follows if we can show that

$$\mathbf{E}\{V_i^2 | X_1, \ldots, X_{i-1}\} \le \frac{c_i^2}{4}.$$

Introducing W_i and Z_i as in the proof of Theorem 9.2, we see that with probability one $Z_i \le V_i \le Z_i + c_i$. Since Z_i is a function of $X_1, \ldots, X_{i-1}$, therefore, conditioned on $X_1, \ldots, X_{i-1}$, V_i is a zero mean random variable taking values in the interval $[Z_i, Z_i + c_i]$. But an arbitrary random variable U taking values in an interval $[a, b]$ has variance not exceeding

$$\mathbf{E}\left\{(U - (a + b)/2)^2\right\} \le \frac{(b - a)^2}{4},$$

so that

$$\mathbf{E}\{V_i^2 | X_1, \ldots, X_{i-1}\} \le \frac{c_i^2}{4},$$

which concludes the proof. $\square$

9.2   Strong Universal Consistency

The purpose of this section is to prove *strong* universal consistency of the histogram rule. This is the first such result that we mention. Later we will prove the same property for other rules, too. The theorem, stated here for cubic partitions, is essentially due to Devroye and Györfi (1983). For more general sequences of partitions, see Problem 9.7. An alternative proof of the theorem based on the Vapnik-Chervonenkis inequality will be given later—see the remark following Theorem 17.2.

Theorem 9.4. *Assume that the sequence of partitions $\{\mathcal{P}_n\}$ satisfies the following two conditions as $n \to \infty$:*

$$h_n \to 0$$

and

$$nh_n^d \to \infty.$$

For any distribution of (X, Y), and for every $\epsilon > 0$ there is an integer n_0 such that for $n > n_0$, for the error probability L_n of the histogram rule

$$\mathbf{P}\{L_n - L^* > \epsilon\} \le 2e^{-n\epsilon^2/32}.$$

Thus, the cubic histogram rule is strongly universally consistent.

PROOF. Define

$$\eta_n^*(x) = \frac{\sum_{i=1}^n Y_i I_{\{X_i \in A_n(x)\}}}{n\mu(A_n(x))}.$$

Clearly, the decision based on η_n^*,

$$g_n(x) = \begin{cases} 0 & \text{if } \frac{\sum_{i=1}^n Y_i I_{\{X_i \in A_n(x)\}}}{n\mu(A_n(x))} \leq \frac{\sum_{i=1}^n (1-Y_i) I_{\{X_i \in A_n(x)\}}}{n\mu(A_n(x))} \\ 1 & \text{otherwise} \end{cases}$$

is just the histogram rule. Therefore, by Theorem 2.3, it suffices to prove that for n large enough,

$$\mathbf{P}\left\{ \int |\eta(x) - \eta_n^*(x)|\mu(dx) > \frac{\epsilon}{2} \right\} \leq e^{-n\epsilon^2/32}.$$

Decompose the difference as

$$|\eta(x) - \eta_n^*(x)| = \mathbf{E}|\eta(x) - \eta_n^*(x)| + \left(|\eta(x) - \eta_n^*(x)| - \mathbf{E}|\eta(x) - \eta_n^*(x)| \right). \quad (9.1)$$

The convergence of the first term on the right-hand side implies weak consistency of the histogram rule. The technique we use to bound this term is similar to that which we already saw in the proof of Theorem 6.3. For completeness, we give the details here. However, new ideas have to appear in our handling of the second term.

We begin with the first term. Since the set of continuous functions with bounded support is dense in $L_1(\mu)$, it is possible to find a continuous function of bounded support $r(x)$ such that

$$\int |\eta(x) - r(x)|\mu(dx) < \epsilon/16.$$

Note that $r(x)$ is uniformly continuous. Introduce the function

$$r_n^*(x) = \frac{\mathbf{E}\left\{ r(X)I_{\{X \in A_n(x)\}} \right\}}{\mu(A_n(x))}.$$

Then we can further decompose the first term on the right-hand side of (9.1) as

$$\begin{aligned} &\mathbf{E}|\eta(x) - \eta_n^*(x)| \\ &\leq \quad |\eta(x) - r(x)| + |r(x) - r_n^*(x)| \\ &\quad + |r_n^*(x) - \mathbf{E}\eta_n^*(x)| + \mathbf{E}|\mathbf{E}\eta_n^*(x) - \eta_n^*(x)|. \end{aligned} \quad (9.2)$$

We proceed term by term:

FIRST TERM: The integral of $|\eta(x) - r(x)|$ (with respect to μ) is smaller than $\epsilon/16$ by the definition of $r(x)$.

SECOND TERM: Using Fubini's theorem we have

$$\begin{aligned} &\int |r(x) - r_n^*(x)|\mu(dx) \\ &= \sum_{A_j : \mu(A_j) \neq 0} \int_{A_j} \left| r(x) - \frac{\mathbf{E}\left\{ r(X)I_{\{X \in A_j\}} \right\}}{\mu(A_j)} \right| \mu(dx) \end{aligned}$$

$$= \sum_{A_j:\mu(A_j)\neq 0} \frac{1}{\mu(A_j)} \int_{A_j} \left| r(x)\mu(A_j) - \mathbf{E}\left\{r(X)I_{\{X\in A_j\}}\right\} \right| \mu(dx)$$

$$= \sum_{A_j:\mu(A_j)\neq 0} \frac{1}{\mu(A_j)} \int_{A_j} \left| r(x)\int_{A_j}\mu(dy) - \int_{A_j} r(y)\mu(dy) \right| \mu(dx)$$

$$\leq \sum_{A_j:\mu(A_j)\neq 0} \frac{1}{\mu(A_j)} \int_{A_j}\int_{A_j} |r(x) - r(y)|\mu(dx)\mu(dy).$$

As $r(x)$ is uniformly continuous, if h_n is small enough, then $|r(x) - r(y)| < \epsilon/16$ for every $x, y \in A$ for any cell $A \in \mathcal{P}_n$. Then the double integral in the above expression can be bounded from above as follows:

$$\int_{A_j}\int_{A_j} |r(x) - r(y)|\mu(dx)\mu(dy) \leq \epsilon\mu^2(A_j)/16.$$

Note that we used the condition $h_n \to 0$ here. Summing over the cells we get

$$\int |r(x) - r_n^*(x)|\mu(dx) \leq \epsilon/16.$$

THIRD TERM: We have

$$\int |r_n^*(x) - \mathbf{E}\eta_n^*(x)|\mu(dx) = \sum_{A_j\in\mathcal{P}_n} \left| \mathbf{E}\left\{r(X)I_{\{X\in A_j\}} - YI_{\{X\in A_j\}}\right\}\right|$$

$$= \sum_{A_j\in\mathcal{P}_n} \left| \int_{A_j} r(x)\mu(dx) - \int_{A_j}\eta(x)\mu(dx) \right|$$

$$\leq \int |r(x) - \eta(x)|\mu(dx) < \epsilon/16.$$

FOURTH TERM: Our aim is to show that for n large enough,

$$\mathbf{E}\int |\mathbf{E}\eta_n^*(x) - \eta_n^*(x)|\mu(dx) < \epsilon/16.$$

To this end, let S be an arbitrary large ball centered at the origin. Denote by m_n the number of cells of the partition $\mathcal{P}_n$ that intersect S. Clearly, m_n is proportional to $1/h_n^d$ as $h_n \to 0$. Using the notation $\nu_n(A) = \frac{1}{n}\sum_{i=1}^n I_{\{Y_i=1,X_i\in A\}}$, it is clear that $\nu_n(A) = \int_A \eta_n^*(x)\mu(dx)$. Now, we can write

$$\mathbf{E}\int |\mathbf{E}\eta_n^*(x) - \eta_n^*(x)|\mu(dx)$$

$$= \mathbf{E}\sum_j \int_{A_{n,j}} |\mathbf{E}\eta_n^*(x) - \eta_n^*(x)|\mu(dx)$$

$$= \mathbf{E}\sum_j |\mathbf{E}\nu_n(A_{n,j}) - \nu_n(A_{n,j})|$$

$$\leq \quad \mathbf{E} \sum_{j:A_{n,j} \cap S \neq \emptyset} \left| \mathbf{E} v_n(A_{n,j}) - v_n(A_{n,j}) \right| + 2\mu(S^c)$$

(where S^c denotes the complement of S)

$$\leq \quad \sum_{j:A_{n,j} \cap S \neq \emptyset} \sqrt{\mathbf{E} \left| \mathbf{E} v_n(A_{n,j}) - v_n(A_{n,j}) \right|^2} + 2\mu(S^c)$$

(by the Cauchy-Schwarz inequality)

$$\leq \quad m_n \frac{1}{m_n} \sum_{j:A_{n,j} \cap S \neq \emptyset} \sqrt{\frac{\mu(A_{n,j})}{n}} + 2\mu(S^c)$$

$$\leq \quad m_n \sqrt{\frac{\frac{1}{m_n} \sum_{j:A_{n,j} \cap S \neq \emptyset} \mu(A_{n,j})}{n}} + 2\mu(S^c)$$

(by Jensen's inequality)

$$\leq \quad \sqrt{\frac{m_n}{n}} + 2\mu(S^c)$$

$$\leq \quad \epsilon/16$$

if n and the radius of S are large enough, since m_n/n converges to zero by the condition $n h_n^d \to \infty$, and $\mu(S^c)$ can be made arbitrarily small by choice of S.

We have proved for the first term on the right-hand side of (9.1) that for n large enough,

$$\mathbf{E} \int |\eta(x) - \eta_n^*(x)| \mu(dx) < \epsilon/4.$$

Finally, we handle the second term on the right-hand side of (9.1) by obtaining an exponential bound for

$$\int |\eta(x) - \eta_n^*(x)| \mu(dx) - \mathbf{E} \int |\eta(x) - \eta_n^*(x)| \mu(dx)$$

using Theorem 9.2. Fix the training data $(x_1, y_1), \ldots, (x_n, y_n) \in \mathcal{R}^d \times \{0, 1\}$, and replace (x_i, y_i) by $(\widehat{x}_i, \widehat{y}_i)$ changing the value of $\eta_n^*(x)$ to $\eta_{ni}^*(x)$. Then $\eta_n^*(x) - \eta_{ni}^*(x)$ differs from zero only on $A_n(x_i)$ and $A_n(\widehat{x}_i)$, and thus

$$\int |\eta(x) - \eta_n^*(x)| \mu(dx) - \int |\eta(x) - \eta_{ni}^*(x)| \mu(dx)$$

$$\leq \quad \int |\eta_n^*(x) - \eta_{ni}^*(x)| \mu(dx)$$

$$\leq \quad \left(\frac{1}{n\mu(A_n(x_i))} \mu(A_n(x_i)) + \frac{1}{n\mu(A_n(\widehat{x}_i))} \mu(A_n(\widehat{x}_i)) \right) \leq \frac{2}{n}.$$

So by Theorem 9.2, for sufficiently large n,

$$\mathbf{P}\left\{\int |\eta(x) - \eta_n^*(x)|\mu(dx) > \frac{\epsilon}{2}\right\}$$

$$\leq \mathbf{P}\left\{\int |\eta(x) - \eta_n^*(x)|\mu(dx) - \mathbf{E}\int |\eta(x) - \eta_n^*(x)|\mu(dx) > \frac{\epsilon}{4}\right\}$$

$$\leq e^{-n\epsilon^2/32}. \quad \square$$

REMARK. Strong universal consistency follows from the exponential bound on the probability $\mathbf{P}\{L_n - L^* > \epsilon\}$ via the Borel-Cantelli lemma. The inequality in Theorem 9.4 may seem universal in nature. However, it is distribution-dependent in a surreptitious way because its range of validity, $n \geq n_0$, depends heavily on ϵ, h_n, and the distribution. We know that distribution-free upper bounds could not exist anyway, in view of Theorem 7.2. $\square$

Problems and Exercises

PROBLEM 9.1. Let $U_1, U_2, \ldots$ be independent random variables with zero mean. Show that the random variables $S_i = \sum_{j=1}^{i} U_j \quad i > 0$ form a martingale.

PROBLEM 9.2. Prove Lemma 9.1.

PROBLEM 9.3. Let $X_1, \ldots, X_n$ be real valued i.i.d. random variables with distribution function $F(x)$, and corresponding empirical distribution function $F_n(x) = \frac{1}{n}\sum_{i=1}^{n} I_{\{X_i \leq x\}}$. Denote the *Kolmogorov-Smirnov statistic* by

$$V_n = \sup_{x \in \mathcal{R}} |F_n(x) - F(x)|.$$

Use Theorem 9.2 to show that

$$\mathbf{P}\{|V_n - \mathbf{E}V_n| \geq \epsilon\} \leq 2e^{-2n\epsilon^2}.$$

Compare this result with Theorem 12.9. (None of them implies the other.)

Also, consider a class $\mathcal{A}$ of subsets of $\mathcal{R}^d$. Let $Z_1, \ldots, Z_n$ be i.i.d. random variables in $\mathcal{R}^d$ with common distribution $\mathbf{P}\{Z_1 \in A\} = \nu(A)$, and consider the random variable

$$W_n = \sup_{A \in \mathcal{A}} |\nu_n(A) - \nu(A)|,$$

where $\nu_n(A) = \frac{1}{n}\sum_{i=1}^{n} I_{\{Z_i \in A\}}$ denotes the standard empirical measure of A. Prove that

$$\mathbf{P}\{|W_n - \mathbf{E}W_n| \geq \epsilon\} \leq 2e^{-2n\epsilon^2}.$$

Compare this result with Theorem 12.5, and note that this result is true even if $s(\mathcal{A}, n) = 2^n$ for all n.

PROBLEM 9.4. THE LAZY HISTOGRAM RULE. Let $\mathcal{P}_n = \{A_{n1}, A_{n2}, \ldots\}$ be a sequence of partitions satisfying the conditions of the convergence Theorem 9.4. Define the lazy histogram rule as follows:

$$g_n(x) = Y_j, \quad x \in A_{ni},$$

where X_j is the minimum-index point among $X_1, \ldots, X_n$ for which $X_j \in A_{ni}$. In other words, we ignore all but one point in each set of the partition. If L_n is the conditional probability of error for the lazy histogram rule, then show that for any distribution of (X, Y),

$$\limsup_{n \to \infty} \mathbf{E} L_n \leq 2L^*.$$

PROBLEM 9.5. Assume that $\mathcal{P}_n = \mathcal{P} = \{A_1, \ldots, A_k\}$ is a fixed partition into k sets. Consider the lazy histogram rule defined in Problem 9.4 based on $\mathcal{P}$. Show that for all distributions of (X, Y), $\lim_{n \to \infty} \mathbf{E} L_n$ exists and satisfies

$$\lim_{n \to \infty} \mathbf{E} L_n = \sum_{i=1}^{k} 2 p_i (1 - p_i) \mu(A_i),$$

where μ is the probability measure for X, and $p_i = \int_{A_i} \eta(X) \mu(dx) / \mu(A_i)$. Show that the limit of the probability of error $\mathbf{E} L'_n$ for the ordinary histogram rule is

$$\lim_{n \to \infty} \mathbf{E} L'_n = \sum_{i=1}^{k} \min(p_i, 1 - p_i) \mu(A_i),$$

and show that

$$\lim_{n \to \infty} \mathbf{E} L_n \leq 2 \lim_{n \to \infty} \mathbf{E} L'_n.$$

PROBLEM 9.6. HISTOGRAM DENSITY ESTIMATION. Let $X_1, \ldots, X_n$ be i.i.d. random variables in $\mathcal{R}^d$ with density f. Let $\mathcal{P}$ be a partition of $\mathcal{R}^d$, and define the histogram density estimate by

$$f_n(x) = \frac{1}{n \lambda(A(x))} \sum_{i=1}^{n} I_{\{X_i \in A(x)\}},$$

where $A(x)$ is the set in $\mathcal{P}$ that contains x and λ is the Lebesgue measure. Prove for the L_1-error of the estimate that

$$\mathbf{P}\left\{ \left| \int |f_n(x) - f(x)| dx - \mathbf{E} \int |f_n(x) - f(x)| dx \right| > \epsilon \right\} \leq 2 e^{-n\epsilon^2 / 2}.$$

(Devroye (1991a).) Conclude that weak L_1-consistency of the estimate implies strong consistency (Abou-Jaoude (1976a; 1976c), see also Problem 6.2).

PROBLEM 9.7. GENERAL PARTITIONS. Extend the consistency result of Theorem 9.4 for sequences of general, not necessarily cubic partitions. Actually, cells of the partitions need not even be hyperrectangles. Assume that the sequence of partitions $\{\mathcal{P}_n\}$ satisfies the following two conditions. For every ball S centered at the origin

$$\lim_{n \to \infty} \max_{i : A_{ni} \cap S \neq \emptyset} \left(\sup_{x, y \in A_{ni}} \|x - y\| \right) = 0$$

and

$$\lim_{n \to \infty} \frac{1}{n} \left| \{i : A_{ni} \cap S \neq \emptyset\} \right| = 0.$$

Prove that the corresponding histogram classification rule is strongly universally consistent.

PROBLEM 9.8. Show that for cubic histograms the conditions of Problem 9.7 on the partition are equivalent to the conditions $h_n \to 0$ and $nh_n^d \to \infty$, respectively.

PROBLEM 9.9. LINEAR SCALING. Partition $\mathcal{R}^d$ into congruent rectangles of the form

$$[k_1h_1, (k_1 + 1)h_1) \times \cdots \times [k_dh_d, (k_d + 1)h_d),$$

where $k_1, \ldots, k_d$ are integers, and $h_1, \ldots, h_d > 0$ denote the size of the edges of the rectangles. Prove that the corresponding histogram rule is strongly universally consistent if $h_i \to 0$ for every $i = 1, \ldots, d$, and $nh_1h_2 \ldots h_d \to \infty$ as $n \to \infty$. HINT: This is a Corollary of Problem 9.7.

PROBLEM 9.10. NONLINEAR SCALING. Let $F_1, \ldots, F_d : \mathcal{R} \to \mathcal{R}$ be invertible, strictly monotone increasing functions. Consider the partition of $\mathcal{R}^d$, whose cells are rectangles of the form

$$[F_1^{-1}(k_1h_1), F_1^{-1}((k_1 + 1)h_1)) \times \ldots \times [F_d^{-1}(k_dh_d), F_d^{-1}((k_d + 1)h_d)).$$

(See Problem 9.9.) Prove that the histogram rule corresponding to this partition is strongly universally consistent under the conditions of Problem 9.9. HINT: Use Problem 9.7.

PROBLEM 9.11. NECESSARY AND SUFFICIENT CONDITIONS FOR THE BIAS. A sequence of partitions $\{\mathcal{P}_n\}$ is called μ-approximating if for every measurable set A, for every $\epsilon > 0$, and for all sufficiently large n there is a set $A_n \in \sigma(\mathcal{P}_n)$, ($\sigma(\mathcal{P})$ denotes the σ-algebra generated by cells of the partition $\mathcal{P}$) such that $\mu(A_n \triangle A) < \epsilon$. Prove that the bias term $\int |\eta(x) - \mathbf{E}\eta_n^*(x)|\mu(dx)$ converges to zero for all distributions of (X, Y) if and only if the sequence of partitions $\{\mathcal{P}_n\}$ is μ-approximating for every probability measure μ on $\mathcal{R}^d$ (Abou-Jaoude (1976a)). Conclude that the first condition of Problem 9.7 implies that $\{\mathcal{P}_n\}$ is μ-approximating for every probability measure μ (Csiszár (1973)).

PROBLEM 9.12. NECESSARY AND SUFFICIENT CONDITIONS FOR THE VARIATION. Assume that for every probability measure μ on $\mathcal{R}^d$, every measurable set A, every $c > 0$, and every $\epsilon > 0$, there is an $N(\epsilon, c, A, \mu)$, such that for all $n > N(\epsilon, c, A, \mu)$,

$$\sum_{j:\mu(A_{n,j}\cap A)\leq c/n} \mu(A_{n,j} \cap A) < \epsilon.$$

Prove that the variation term $\int |\mathbf{E}\eta_n^*(x) - \eta_n^*(x)|\mu(dx)$ converges to zero for all distributions of (X, Y) if and only if the sequence of partitions $\{\mathcal{P}_n\}$ satisfies the condition above (Abou-Jaoude (1976a)).

PROBLEM 9.13. The ϵ-effective cardinality $m(\mathcal{P}, \mu, A, \epsilon)$ of a partition with respect to the probability measure μ, restricted to a set A is the minimum number of sets in $\mathcal{P}$ such that the union of the remaining sets intersected with A has μ-measure less than ϵ. Prove that the sequence of partitions $\{\mathcal{P}_n\}$ satisfies the condition of Problem 9.12 if and only if for every $\epsilon > 0$,

$$\lim_{n\to\infty} \frac{m(\mathcal{P}_n, \mu, A, \epsilon)}{n} = 0.$$

(Barron, Györfi, and van der Meulen (1992)).

PROBLEM 9.14. In $\mathcal{R}^2$, partition the plane by taking three fixed points not on a line, x, y and z. At each of these points, partition $\mathcal{R}^2$ by considering k equal sectors of angle $2\pi/k$ each. Sets in the histogram partition are obtained as intersections of cones. Is the induced histogram rule strongly universally consistent? If yes, state the conditions on k, and if no, provide a counterexample.

PROBLEM 9.15. Partition $\mathcal{R}^2$ into shells of size h each. The i-th shell contains all points at distance $d \in [(i - 1)h, ih)$ from the origin. Let $h \to 0$ and $nh \to \infty$ as $n \to \infty$. Consider the histogram rule. As $n \to \infty$, to what does $\mathbf{E}L_n$ converge?

10
Kernel Rules

Histogram rules have the somewhat undesirable property that the rule is less accurate at borders of cells of the partition than in the middle of cells. Looked at intuitively, this is because points near the border of a cell should have less weight in a decision regarding the cell's center. To remedy this problem, one might introduce the *moving window* rule, which is smoother than the histogram rule. This classifier simply takes the data points within a certain distance of the point to be classified, and decides according to majority vote. Working formally, let h be a positive number. Then the moving window rule is defined as

$$g_n(x) = \begin{cases} 0 & \text{if } \sum_{i=1}^n I_{\{Y_i=0, X_i \in S_{x,h}\}} \geq \sum_{i=1}^n I_{\{Y_i=1, X_i \in S_{x,h}\}} \\ 1 & \text{otherwise,} \end{cases}$$

where $S_{x,h}$ denotes the closed ball of radius h centered at x.

It is possible to make the decision even smoother by giving more weight to closer points than to more distant ones. Let $K : \mathcal{R}^d \to \mathcal{R}$ be a *kernel function*, which is usually nonnegative and monotone decreasing along rays starting from the origin. The *kernel classification rule* is given by

$$g_n(x) = \begin{cases} 0 & \text{if } \sum_{i=1}^n I_{\{Y_i=0\}} K\left(\frac{x-X_i}{h}\right) \geq \sum_{i=1}^n I_{\{Y_i=1\}} K\left(\frac{x-X_i}{h}\right) \\ 1 & \text{otherwise.} \end{cases}$$

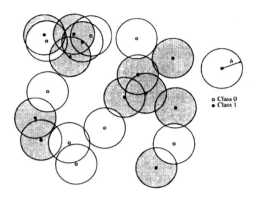

FIGURE 10.1. *The moving window rule in* $\mathcal{R}^2$. *The decision is* 1 *in the shaded area.*

o Class 0
• Class 1

The number h is called the *smoothing factor*, or *bandwidth*. It provides some form of distance weighting.

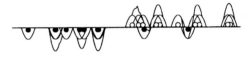

FIGURE 10.2. *Kernel rule on the real line. The figure shows* $\sum_{i=1}^{n}(2Y_i - 1)K((x - X_i)/h)$ *for* $n = 20$, $K(u) = (1 - u^2)I_{\{|u|\leq 1\}}$ *(the Epanechnikov kernel), and three smoothing factors* h. *One definitely undersmooths and one oversmooths. We took* $p = 1/2$, *and the class-conditional densities are* $f_0(x) = 2(1 - x)$ *and* $f_1(x) = 2x$ *on* [0, 1].

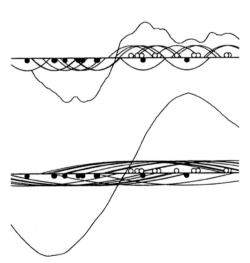

Clearly, the kernel rule is a generalization of the moving window rule, since taking the special kernel $K(x) = I_{\{x \in S_{0,1}\}}$ yields the moving window rule. This kernel is sometimes called the *naïve kernel*. Other popular kernels include the Gaussian kernel, $K(x) = e^{-\|x\|^2}$; the Cauchy kernel, $K(x) = 1/(1 + \|x\|^{d+1})$; and the

Epanechnikov kernel $K(x) = (1 - \|x\|^2)I_{\{\|x\| \leq 1\}}$, where $\| \cdot \|$ denotes Euclidean distance.

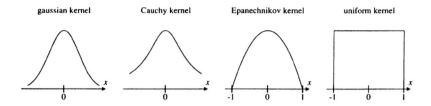

FIGURE 10.3. *Various kernels on $\mathcal{R}$.*

Kernel-based rules are derived from the kernel estimate in density estimation originally studied by Parzen (1962), Rosenblatt (1956), Akaike (1954), and Cacoullos (1965) (see Problems 10.2 and 10.3); and in regression estimation, introduced by Nadaraya (1964; 1970), and Watson (1964). For particular choices of K, rules of this sort have been proposed by Fix and Hodges (1951; 1952), Sebestyen (1962), Van Ryzin (1966), and Meisel (1969). Statistical analysis of these rules and/or the corresponding regression function estimate can be found in Nadaraya (1964; 1970), Rejtő and Révész (1973), Devroye and Wagner (1976b; 1980a; 1980b), Greblicki (1974; 1978b; 1978a), Krzyżak and Pawlak (1984b), and Devroye and Krzyżak (1989). Usage of Cauchy kernels in discrimination is investigated by Arkadjew and Braverman (1966), Hand (1981), and Coomans and Broeckaert (1986).

10.1   Consistency

In this section we demonstrate strong universal consistency of kernel-based rules under general conditions on h and K. Let $h > 0$ be a smoothing factor depending only on n, and let K be a kernel function. If the conditional densities f_0, f_1 exist, then weak and strong consistency follow from Problems 10.2 and 10.3, respectively, via Problem 2.11. We state the universal consistency theorem for a large class of kernel functions, namely, for all *regular* kernels.

DEFINITION 10.1. *The kernel K is called regular if it is nonnegative, and there is a ball $S_{0,r}$ of radius $r > 0$ centered at the origin, and constant $b > 0$ such that $K(x) \geq bI_{S_{0,r}}$ and $\int \sup_{y \in x + S_{0,r}} K(y)dx < \infty$.*

We provide three informative exercises on regular kernels (Problems 10.18, 10.19, 10.20). In all cases, regular kernels are bounded and integrable. The last condition holds whenever K is integrable and uniformly continuous. Introduce the short notation $K_h(x) = \frac{1}{h} K(\frac{x}{h})$. The next theorem states strong universal consistency of kernel rules. The theorem is essentially due to Devroye and Krzyżak

(1989). Under the assumption that X has a density, it was proven by Devroye and Györfi (1985) and Zhao (1989).

Theorem 10.1. (DEVROYE AND KRZYŻAK (1989)). *Assume that K is a regular kernel. If*

$$h \to 0 \quad and \quad nh^d \to \infty \quad as \ n \to \infty,$$

then for any distribution of (X, Y), and for every $\epsilon > 0$ there is an integer n_0 such that for $n > n_0$ for the error probability L_n of the kernel rule

$$\mathbf{P}\{L_n - L^* > \epsilon\} \leq 2e^{-n\epsilon^2/(32\rho^2)},$$

where the constant ρ depends on the kernel K and the dimension only. Thus, the kernel rule is strongly universally consistent.

Clearly, naïve kernels are regular, and moving window rules are thus strongly universally consistent. For the sake of readability, we give the proof for this special case only, and leave the extension to regular kernels to the reader—see Problems 10.14, 10.15, and 10.16. Before we embark on the proof in the next section, we should warn the reader that Theorem 10.1 is of no help whatsoever regarding the choice of K or h. One possible solution is to derive explicit upper bounds for the probability of error as a function of descriptors of the distribution of (X, Y), and of K, n and h. Minimizing such bounds with respect to K and h will lead to some expedient choices. Typically, such bounds would be based upon the inequality

$$\mathbf{E}L_n - L^*$$
$$\leq \quad \mathbf{E}\left\{\int |(1 - p)f_0(x) - p_{n0}f_{n0}(x)|dx + \int |pf_1(x) - p_{n1}f_{n1}(x)|dx\right\}$$

(see Chapter 6), where f_0, f_1 are the class densities, f_{n0}, f_{n1} are their kernel estimates (see Problem 10.2) $(1 - p)$ and p are the class probabilities, and p_{n0}, p_{n1} are their relative-frequency estimates. Bounds for the expected L_1-error in density estimation may be found in Devroye (1987) for $d = 1$ and Holmström and Klemelä (1992) for $d > 1$. Under regularity conditions on the distribution, the choice $h = cn^{-d/(d+4)}$ for some constant is asymptotically optimal in density estimation. However, c depends upon unknown distributional parameters. Rather than following this roundabout process, we ask the reader to be patient and to wait until Chapter 25, where we study automatic kernel rules, i.e., rules in which h, and sometimes K as well, is picked by the data without intervention from the statistician.

It is still too early to say meaningful things about the choice of a kernel. The kernel density estimate

$$f_n(x) = \frac{1}{nh^d} \sum_{i=1}^{n} K\left(\frac{x - X_i}{h}\right)$$

based upon an i.i.d. sample $X_1, \ldots, X_n$ drawn from an unknown density f is clearly a density in its own right if $K \geq 0$ and $\int K = 1$. Also, there are certain popular choices of K that are based upon various optimality criteria. In pattern recognition, the story is much more confused, as there is no compelling a priori reason to pick a function K that is nonnegative or integrable. Let us make a few points with the trivial case $n = 1$. Taking $h = 1$, the kernel rule is given by

$$g_1(x) = \begin{cases} 0 & \text{if } Y_1 = 0, K(x - X_1) \geq 0 \text{ or if } Y_1 = 1, K(x - X_1) \leq 0 \\ 1 & \text{otherwise.} \end{cases}$$

If $K \geq 0$, then $g_n(x) = 0$ if $Y_1 = 0$, or if $Y_1 = 1$ and $K(x - X_1) = 0$. As we would obviously like $g_n(x) = 0$ if and only if $Y_1 = 0$, it seems necessary to insist on $K > 0$ everywhere. However, this restriction makes the kernel estimate nonlocal in nature.

For $n = 1$ and $d = 1$, consider next a negative-valued kernel such as the Hermite kernel

$$K(x) = (1 - x^2)e^{-x^2}.$$

FIGURE 10.4. *Hermite kernel.*

It is easy to verify that $K(x) \geq 0$ if and only if $|x| \leq 1$. Also, $\int K = 0$. Nevertheless, we note that it yields a simple rule:

$$g_1(x) = \begin{cases} 0 & \text{if } Y_1 = 0, |x - X_1| \leq 1 \text{ or if } Y_1 = 1, |x - X_1| \geq 1 \\ 1 & \text{otherwise.} \end{cases}$$

If we have a biatomic distribution for X, with equally likely atoms at 0 and 2, and $\eta(0) = 0$ and $\eta(2) = 1$ (i.e., $Y = 0$ if $X = 0$ and $Y = 1$ if $X = 2$), then $L^* = 0$ and the probability of error for this kernel rule (L_1) is 0 as well. Note also that for all n, $g_n = g_1$ if we keep the same K. Consider now any positive kernel in the same example. If $X_1, \ldots, X_n$ are all zero, then the decision is $g_n(x) = 0$ for all x. Hence $L_n \geq \frac{1}{2}P\{X_1 = \cdots = X_n\} = 1/2^{n+1} > 0$. Our negative zero-integral kernel is *strictly better for all n* than any positive kernel! Such kernels should not be discarded without further thought. In density estimation, negative-valued kernels are used to reduce the bias under some smoothness conditions. Here, as shown above, there is an additional reason—negative weights given to points far away from the X_i's may actually be beneficial.

Staying with the same example, if $K > 0$ everywhere, then

$$\mathbf{E}L_1 = P\{Y_1 = 0, Y = 1\} + P\{Y_1 = 1, Y = 0\} = 2\mathbf{E}\eta(X)\mathbf{E}\{1 - \eta(X)\},$$

which may be $1/2$ (if $\mathbf{E}\eta(X) = 1/2$) even if $L^* = 0$ (which happens when $\eta \in \{0, 1\}$ everywhere). For this particular example, we would have obtained the same result even if $K \equiv 1$ everywhere. With $K \equiv 1$, we simply ignore the X_i's and take a majority vote among the Y_i's (with $K \equiv -1$, it would be a minority vote!):

$$g_n(x) = \begin{cases} 0 & \text{if } \sum_{i=1}^n I_{\{Y_i=0\}} \geq \sum_{i=1}^n I_{\{Y_i=1\}} \\ 1 & \text{otherwise.} \end{cases}$$

Let N_n be the number of Y_i's equal to zero. As N_n is binomial $(n, 1 - p)$ with $p = \mathbf{E}\eta(X) = \mathbf{P}\{Y = 1\}$, we see that

$$\mathbf{E}L_n = p\mathbf{P}\left\{N_n \geq \frac{n}{2}\right\} + (1 - p)\mathbf{P}\left\{N_n < \frac{n}{2}\right\} \to \min(p, 1 - p),$$

simply by invoking the law of large numbers. Thus, $\mathbf{E}L_n \to \min(p, 1 - p)$. As in the case with $n = 1$, the limit is $1/2$ when $p = 1/2$, even though $L^* = 0$ when $\eta \in \{0, 1\}$ everywhere. It is interesting to note the following though:

$$\begin{aligned} \mathbf{E}L_1 &= 2p(1 - p) \\ &= 2\min(p, 1 - p)(1 - \min(p, 1 - p)) \\ &\leq 2\min(p, 1 - p) \\ &= 2 \lim_{n \to \infty} \mathbf{E}L_n. \end{aligned}$$

The expected error with *one* observation is at most twice as bad as the expected error with an infinite sequence. We have seen various versions of this inequality at work in many instances such as the nearest neighbor rule.

Let us apply the inequality for $\mathbf{E}L_1$ to each part in a fixed partition $\mathcal{P}$ of $\mathcal{R}^d$. On each of the k sets $A_1, \ldots, A_k$ of $\mathcal{P}$, we apply a simple majority vote among the Y_i's, as in the histogram rule. If we define the *lazy histogram rule* as the one in which in each set A_i, we assign the class according to the Y_j for which $X_j \in A_i$ and j is the lowest such index ("the first point to fall in A_i"). It is clear (see Problems 9.4 and 9.5) that

$$\lim_{n \to \infty} \mathbf{E}L_{\text{LAZY},n} \leq 2 \lim_{n \to \infty} \mathbf{E}L_n$$

$$= 2\sum_{i=1}^k \frac{1}{\mu(A_i)}\int_{A_i} \eta(x)\mu(dx)\int_{A_i}(1 - \eta(x))\mu(dx),$$

where L_n is the probability of error for the ordinary histogram rule. Again, the vast majority of observations is barely needed to reach a good decision.

Just for fun, let us return to a majority vote rule, now applied to the first three observations only. With $p = \mathbf{P}\{Y = 1\}$, we see that

$$\mathbf{E}L_3 = p\left((1 - p)^3 + 3(1 - p)^2p\right) + (1 - p)\left(3(1 - p)p^2 + p^3\right)$$

by just writing down binomial probabilities. Observe that

$$\begin{aligned} \mathbf{E}L_3 &= p(1 - p)(1 + 4p(1 - p)) \\ &\leq \min(p, 1 - p)(1 + 4\min(p, 1 - p)) \\ &= \lim_{n \to \infty} \mathbf{E}L_n\left(1 + 4\lim_{n \to \infty} 4\mathbf{E}L_n\right). \end{aligned}$$

If $\lim_{n \to \infty} \mathbf{E} L_n$ is small to start with, e.g., $\lim_{n \to \infty} \mathbf{E} L_n = 0.01$, then $\mathbf{E} L_3 \leq 0.01 \times 1.04 = 0.0104$. In such cases, it just does not pay to take more than three observations.

Kernels with fixed smoothing factors have no local sensitivity and, except in some circumstances, have probabilities of error that do not converge to L^*. The universal consistency theorem makes a strong case for decreasing smoothing factors—there is no hope in general of approaching L^* unless decisions are asymptotically local.

The consistency theorem describes kernel rules with $h \to 0$: these rules become more and more local in nature as $n \to \infty$. The necessity of local rules is not apparent from the previous biatomic example. However, it is clear that if we consider a distribution in which given $Y = 0$, X is uniform on $\{\delta, 3\delta, \ldots, (2k+1)\delta\}$, and given $Y = 1$, X is uniform on $\{2\delta, 4\delta, \ldots, 2k\delta\}$, that is, with the two classes intimately interwoven, a kernel rule with $K \geq 0$ of compact support $[-1, 1]$, and $h < \delta < 1$ will have

$$L_n \leq I_{\left\{\bigcup_{i=1}^{2k} \{N_i = 0\}\right\}},$$

where N_i is the number of X_j's at the i-th atom. Hence $\mathbf{E} L_n$ goes to zero exponentially fast. If in the above example we assign X by a geometric distribution on $\delta, \delta^3, \delta^5, \ldots$ when $Y = 0$, and by a geometric distribution on $\delta^2, \delta^4, \delta^6, \ldots$ when $Y = 1$, then to obtain $\mathbf{E} L_n \to 0$, it is necessary that $h \to 0$ (see Problem 10.1).

REMARK. It is worthwhile to investigate what happens for negative-valued kernels K when $h \to 0$, $nh^d \to \infty$ and K has compact support. Every decision becomes an average over many local decisions. If μ has a density f, then at almost all points x, f may be approximated very nicely by $\int_{S_{x,\delta}} f / \lambda(S_{x,\delta})$ for small $\delta > 0$, where $S_{x,\delta}$ is the closed ball of radius δ about x. This implies, roughly speaking, that the number of weighted votes from class 0 observations in a neighborhood of x is about $(1 - \eta(x)) f(x) n h^d \int K$, while for class 1 the weight is about $\eta(x) f(x) n h^d \int K$. The correct decision is nearly always made for nh^d large enough provided that $\int K > 0$. See Problem 10.4 on why kernels with $\int K < 0$ should be avoided. $\square$

10.2 Proof of the Consistency Theorem

In the proof we can proceed as for the histogram. The crucial difference is captured in the following covering lemmas. Let β_d denote the minimum number of balls of radius $1/2$ that cover the ball $S_{0,1}$. If K is the naïve kernel, then $\rho = \beta_d$ in Theorem 10.1.

Lemma 10.1. (COVERING LEMMA). *If $K(x) = I_{\{x \in S_{0,1}\}}$, then for any $y \in \mathcal{R}^d$, $h > 0$, and probability measure μ,*

$$\int \frac{K_h(x - y)}{\int K_h(x - z)\mu(dz)} \mu(dx) \leq \beta_d.$$

PROOF. Cover the ball $S_{y,h}$ by β_d balls of radius $h/2$. Denote their centers by $x_1, \ldots, x_{\beta_d}$. Then $x \in S_{x_i,h/2}$ implies $S_{x_i,h/2} \subset S_{x,h}$ and thus

$$\mu(S_{x,h}) \geq \mu(S_{x_i,h/2}).$$

We may write

$$
\begin{aligned}
\int \frac{K_h(x-y)}{\int K_h(x-z)\mu(dz)}\mu(dx) &= \int \frac{I_{\{x \in S_{y,h}\}}}{\mu(S_{x,h})}\mu(dx) \\
&\leq \sum_{i=1}^{\beta_d} \int \frac{I_{\{x \in S_{x_i,h/2}\}}}{\mu(S_{x,h})}\mu(dx) \\
&\leq \sum_{i=1}^{\beta_d} \int \frac{I_{\{x \in S_{x_i,h/2}\}}}{\mu(S_{x_i,h/2})}\mu(dx) \\
&= \beta_d. \quad \square
\end{aligned}
$$

Lemma 10.2. *Let $0 < h \leq R < \infty$, and let $S \subset \mathcal{R}^d$ be a ball of radius R. Then for every probability measure μ,*

$$\int_S \frac{1}{\sqrt{\mu(S_{x,h})}}\mu(dx) \leq \left(1 + \frac{R}{h}\right)^{d/2} c_d,$$

where c_d depends upon the dimension d only.

PROOF. Cover S with balls of radius $h/2$, centered at center points of a regular grid of dimension $h/(2\sqrt{d}) \times \ldots \times h/(2\sqrt{d})$. Denote these centers by $x_1, \ldots, x_m$, where m is the number of balls that cover S. Clearly,

$$
\begin{aligned}
m &\leq \frac{\text{volume}(S_{0,R+h})}{\text{volume(grid cell)}} \\
&= \frac{V_d(R+h)^d}{\left(h/(2\sqrt{d})\right)^d} \quad (V_d \text{ is the volume of the unit ball in } \mathcal{R}^d) \\
&\leq \left(1 + \frac{R}{h}\right)^d c_d',
\end{aligned}
$$

where the constant c_d' depends upon the dimension only. Every x gets covered at most k_1 times where k_1 depends upon d only. Then we have

$$
\begin{aligned}
\int_S \frac{1}{\sqrt{\mu(S_{x,h})}}\mu(dx) &\leq \sum_{i=1}^m \int_{S_{x_i,h/2}} \frac{1}{\sqrt{\mu(S_{x,h})}}\mu(dx) \\
&= \sum_{i=1}^m \int \frac{I_{\{x \in S_{x_i,h/2}\}}}{\sqrt{\mu(S_{x,h})}}\mu(dx)
\end{aligned}
$$

$$\leq \quad \sum_{i=1}^{m} \int \frac{I_{\{x \in S_{x_i, h/2}\}}}{\sqrt{\mu(S_{x_i, h/2})}} \mu(dx)$$

(by the same argument as in Lemma 10.1)

$$\leq \quad \sum_{i=1}^{m} \sqrt{\mu(S_{x_i, h/2})}$$

$$\leq \quad \sqrt{m \sum_{i=1}^{m} \mu(S_{x_i, h/2})}$$

(by the Cauchy-Schwarz inequality)

$$\leq \quad \sqrt{k_1 m}$$

$$\leq \quad \sqrt{\left(1 + \frac{R}{h}\right)^d c_d},$$

where c_d depends upon the dimension only. $\square$

PROOF OF THEOREM 10.1. Define

$$\eta_n(x) = \frac{\sum_{j=1}^{n} Y_j K_h(x - X_j)}{n E K_h(x - X)}.$$

Since the decision rule can be written as

$$g_n(x) = \begin{cases} 0 & \text{if } \dfrac{\sum_{j=1}^{n} Y_j K_h(x - X_j)}{n E K_h(x - X)} \leq \dfrac{\sum_{j=1}^{n} (1 - Y_j) K_h(x - X_j)}{n E K_h(x - X)} \\ 1 & \text{otherwise,} \end{cases}$$

by Theorem 2.3, what we have to prove is that for n large enough

$$\mathbf{P}\left\{ \int |\eta(x) - \eta_n(x)| \mu(dx) > \frac{\epsilon}{2} \right\} \leq e^{-n\epsilon^2/(32\rho^2)}.$$

We use a decomposition as in the proof of strong consistency of the histogram rule:

$$|\eta(x) - \eta_n(x)|$$
$$= \quad \mathbf{E}|\eta(x) - \eta_n(x)| + \Big(|\eta(x) - \eta_n(x)| - \mathbf{E}|\eta(x) - \eta_n(x)| \Big). \quad (10.1)$$

To handle the first term on the right-hand side, fix $\epsilon' > 0$, and let $r : \mathcal{R}^d \to \mathcal{R}$ be a continuous function of bounded support satisfying

$$\int |\eta(x) - r(x)| \mu(dx) < \epsilon'.$$

Obviously, we can choose the function r such that $0 \le r(x) \le 1$ for all $x \in \mathcal{R}^d$. Then we have the following simple upper bound:

$$E|\eta(x) - \eta_n(x)|$$

$$\le |\eta(x) - r(x)| + \left| r(x) - \frac{E\{r(X)K_h(x - X)\}}{EK_h(x - X)} \right|$$

$$+ \left| \frac{E\{r(X)K_h(x - X)\}}{EK_h(x - X)} - E\eta_n(x) \right| + E|E\eta_n(x) - \eta_n(x)|. \quad (10.2)$$

Next we bound the integral of each term on the right-hand side of the inequality above.

FIRST TERM: By the definition of r,

$$\int |\eta(x) - r(x)|\mu(dx) < \epsilon'.$$

SECOND TERM: Since $r(x)$ is continuous and zero outside of a bounded set, it is also uniformly continuous, that is, there exists a $\delta > 0$ such that $\|x - y\| < \delta$ implies $|r(x) - r(y)| < \epsilon'$. Also, $r(x)$ is bounded. Thus,

$$\int \left| r(x) - \frac{E\{r(X)K_h(x - X)\}}{EK_h(x - X)} \right| \mu(dx)$$

$$= \int \left| r(x) - \frac{\int r(y)K_h(x - y)\mu(dy)}{EK_h(x - X)} \right| \mu(dx)$$

$$\le \int \int_{S_{x,\delta}} \frac{K_h(x - y)}{EK_h(x - X)} |r(x) - r(y)|\mu(dy)\mu(dx)$$

$$+ \int \int_{S^c_{x,\delta}} \frac{K_h(x - y)}{EK_h(x - X)} |r(x) - r(y)|\mu(dy)\mu(dx)$$

$$\le \int \left(\int_{S_{x,\delta}} \frac{K_h(x - y)}{EK_h(x - X)} \mu(dy) \sup_{z \in S_{x,\delta}} |r(x) - r(z)| \right) \mu(dx)$$

$$+ \int \left(\int_{S^c_{x,\delta}} \frac{K_h(x - y)}{EK_h(x - X)} \mu(dy) \right) \mu(dx).$$

In the last step we used the fact that $\sup_{x,y} |r(x) - r(y)| \le 1$. Clearly, we have $\int_{S_{x,\delta}} \frac{K_h(x-y)}{EK_h(x-X)}\mu(dy) \le 1$, and by the uniform continuity of $r(x)$, $\sup_{z \in S_{x,\delta}} |r(x) - r(z)| < \epsilon'$. Thus, the first term at the end of the chain of inequalities above is bounded by ϵ'. The second term converges to zero since $h < \delta$ for all n large enough, which in turn implies $\int_{S^c_{x,\delta}} \frac{K_h(x-y)}{EK_h(x-X)}\mu(dy) = 0$. (This is obvious for the naïve kernel. For regular kernels, convergence to zero follows from Problem 10.15.) The convergence of the integral (with respect to $\mu(dx)$) follows from the dominated convergence theorem. In summary, we have shown that

$$\limsup_{n \to \infty} \int \left| r(x) - \frac{\mathbf{E}\{r(X)K_h(x-X)\}}{\mathbf{E}K_h(x-X)} \right| \mu(dx) \le \epsilon'.$$

THIRD TERM:

$$\int \left| \frac{\mathbf{E}\{r(X)K_h(x-X)\}}{\mathbf{E}K_h(x-X)} - \mathbf{E}\eta_n(x) \right| \mu(dx)$$

$$= \int \left| \frac{\int (r(y) - \eta(y))K_h(x-y)\mu(dy)}{\mathbf{E}K_h(x-X)} \right| \mu(dx)$$

$$\le \int \int |r(y) - \eta(y)| \frac{K_h(x-y)}{\int K_h(x-z)\mu(dz)} \mu(dy)\mu(dx)$$

$$= \int \left(\int \frac{K_h(x-y)}{\int K_h(x-z)\mu(dz)} \mu(dx) \right) |r(y) - \eta(y)|\mu(dy)$$

(by Fubini's theorem)

$$\le \int \rho |r(y) - \eta(y)|\mu(dy) \le \rho\epsilon',$$

where in the last two steps we used the covering lemma (see Lemma 10.1 for the naïve kernel, and Problem 10.14 for general kernels), and the definition of $r(x)$. ρ is the constant appearing in the covering lemma:

FOURTH TERM: We show that

$$\mathbf{E}\left\{ \int |\mathbf{E}\eta_n(x) - \eta_n(x)|\mu(dx) \right\} \to 0.$$

For the naïve kernel, we have

$$\mathbf{E}\{|\mathbf{E}\eta_n(x) - \eta_n(x)|\}$$

$$\le \sqrt{\mathbf{E}\{|\mathbf{E}\eta_n(x) - \eta_n(x)|^2\}}$$

$$= \sqrt{\frac{\mathbf{E}\left\{ \left(\sum_{j=1}^{n}(Y_j K_h(x-X_j) - \mathbf{E}\{Y K_h(x-X)\}) \right)^2 \right\}}{n^2(\mathbf{E}K_h(x-X))^2}}$$

$$= \sqrt{\frac{\mathbf{E}\left\{ (Y K_h(x-X) - \mathbf{E}\{Y K_h(x-X)\})^2 \right\}}{n(\mathbf{E}K_h(x-X))^2}}$$

$$\le \sqrt{\frac{\mathbf{E}\left\{ (Y K_h(x-X))^2 \right\}}{n(\mathbf{E}K_h(x-X))^2}} \le \sqrt{\frac{\mathbf{E}\left\{ (K_h(x-X))^2 \right\}}{n(\mathbf{E}K_h(x-X))^2}}$$

$$\le \sqrt{\frac{\mathbf{E}K((x-X)/h)}{n(\mathbf{E}K((x-X)/h))^2}} \le \sqrt{\frac{1}{n\mu(S_{x,h})}},$$

where we used the Cauchy-Schwarz inequality, and properties of the naïve kernel. Extension to regular kernels is straightforward.

Next we use the inequality above to show that the integral converges to zero. Divide the integral over $\mathcal{R}^d$ into two terms, namely an integral over a large ball S centered at the origin, of radius $R > 0$, and an integral over S^c. For the integral outside of the ball we have

$$\int_{S^c} \mathbf{E} \{|\mathbf{E}\eta_n(x) - \eta_n(x)|\} \, \mu(dx) \le 2 \int_{S^c} \mathbf{E}\eta_n(x)\mu(dx) \to 2 \int_{S^c} \eta(x)\mu(dx)$$

with probability one as $n \to \infty$, which can be shown in the same way we proved

$$\int_{\mathcal{R}^d} \mathbf{E}\eta_n(x)\mu(dx) \to \int_{\mathcal{R}^d} \eta(x)\mu(dx)$$

(see the first, second, and third terms of (10.2)). Clearly, the radius R of the ball S can be chosen such that $2 \int_{S^c} \eta(x)\mu(dx) < \epsilon/4$. To bound the integral over S we employ Lemma 10.2:

$$\int_S \mathbf{E}\{|\mathbf{E}\eta_n(x) - \eta_n(x)|\} \, \mu(dx) \quad \le \quad \frac{1}{\sqrt{n}} \int_S \frac{1}{\sqrt{\mu(S_{x,h})}} \mu(dx)$$

$$\text{(by the inequality obtained above)}$$

$$\le \quad \frac{1}{\sqrt{n}} \left(1 + \frac{R}{h}\right)^{d/2} c_d$$

$$\to \quad 0 \quad \text{(since by assumption } nh^d \to \infty).$$

Therefore, if n is sufficiently large, then for the first term on the right-hand side of (10.1) we have

$$\mathbf{E}\left\{\int |\eta(x) - \eta_n(x)|\mu(dx)\right\} < \epsilon'(\rho + 3) = \epsilon/4$$

if we take $\epsilon' = \epsilon/(4\rho + 12)$.

It remains to show that the second term on the right-hand side of (10.1) is small with large probability. To do this, we use McDiarmid's inequality (Theorem 9.2) for

$$\int |\eta(x) - \eta_n(x)|\mu(dx) - \mathbf{E}\left\{\int |\eta(x) - \eta_n(x)|\mu(dx)\right\}.$$

Fix the training data at $((x_1, y_1), \ldots, (x_n, y_n))$ and replace the i-th pair (x_i, y_i) by $(\widehat{x}_i, \widehat{y}_i)$, changing the value of $\eta_n(x)$ to $\eta_{ni}^*(x)$. Clearly, by the covering lemma (Lemma 10.1),

$$\int |\eta(x) - \eta_n(x)|\mu(dx) - \int |\eta(x) - \eta_{ni}^*(x)|\mu(dx)$$

$$\le \quad \int |\eta_n(x) - \eta_{ni}^*(x)|\mu(dx) \le \sup_{y \in \mathcal{R}^d} \int \frac{2K_h(x - y)}{n\mathbf{E}K_h(x - X)}\mu(dx)$$

$$\le \quad \frac{2\rho}{n}.$$

So by Theorem 9.2,

$$\mathbf{P}\left\{\int |\eta(x) - \eta_n(x)|\mu(dx) > \frac{\epsilon}{2}\right\}$$

$$\leq \mathbf{P}\left\{\int |\eta(x) - \eta_n(x)|\mu(dx) - \mathbf{E}\left\{\int |\eta(x) - \eta_n(x)|\mu(dx)\right\} > \frac{\epsilon}{4}\right\}$$

$$\leq e^{-n\epsilon^2/(32\rho^2)}.$$

The proof is now completed. $\square$

10.3 Potential Function Rules

Kernel classification rules may be formulated in terms of the so-called *potential function rules*. These rules were originally introduced and studied by Bashkirov, Braverman and Muchnik (1964), Aizerman, Braverman and Rozonoer (1964c; 1964b; 1964a; 1970), Braverman (1965), and Braverman and Pyatniskii (1966). The original idea was the following: put a unit of positive electrical charge at every data point X_i, where $Y_i = 1$, and a unit of negative charge, at data points X_i where $Y_i = 0$. The resulting potential field defines an intuitively appealing rule: the decision at a point x is one if the potential at that point is positive, and zero if it is negative. This idea leads to a rule that can be generalized to obtain rules of the form

$$g_n(x) = \begin{cases} 0 & \text{if } f_n(x) \leq 0 \\ 1 & \text{otherwise,} \end{cases}$$

where

$$f_n(x) = \sum_{i=1}^{n} r_{n,i}(D_n)K_{n,i}(x, X_i),$$

where the $K_{n,i}$'s describe the potential field around X_i, and the $r_{n,i}$'s are their weights. Rules that can be put into this form are often called potential function rules. Here we give a brief survey of these rules.

KERNEL RULES. Clearly, kernel rules studied in the previous section are potential function rules with

$$K_{n,i}(x, y) = K\left(\frac{x - y}{h_n}\right), \quad \text{and} \quad r_{n,i}(D_n) = 2Y_i - 1.$$

Here K is a fixed kernel function, and $h_1, h_2, \ldots$ is a sequence of positive numbers.

HISTOGRAM RULES. Similarly, histogram rules (see Chapters 6 and 9) can be put in this form, by choosing

$$K_{n,i}(x, y) = I_{\{y \in A_n(x)\}}, \quad \text{and} \quad r_{n,i}(D_n) = 2Y_i - 1.$$

Recall that $A_n(x)$ denotes the cell of the partition in which x falls.

POLYNOMIAL DISCRIMINANT FUNCTIONS.   Specht (1967) suggested applying a polynomial expansion to the kernel $K\left(\frac{x-y}{h}\right)$. This led to the choice

$$K_{n,i}(x, y) = \sum_{j=1}^{k} \psi_j(x)\psi_j(y), \quad \text{and} \quad r_{n,i}(D_n) = 2Y_i - 1,$$

where $\psi_1, \ldots, \psi_k$ are fixed real-valued functions on $\mathcal{R}^d$. When these functions are polynomials, the corresponding classifier g_n is called a *polynomial discriminant function*. The potential function rule obtained this way is a generalized linear rule (see Chapter 17) with

$$f_n(x) = \sum_{j=1}^{k} a_{n,j}\psi_j(x),$$

where the coefficients $a_{n,j}$ depend on the data D_n only, through

$$a_{n,j} = \sum_{i=1}^{n} (2Y_i - 1)\psi_j(X_i).$$

This choice of the coefficients does not necessarily lead to a consistent rule, unless the functions $\psi_1, \ldots, \psi_k$ are allowed to change with n, or k is allowed to vary with n. Nevertheless, the rule has some computational advantages over kernel rules. In many practical situations there is enough time to preprocess the data D_n, but once the observation X becomes known, the decision has to be made very quickly. Clearly, the coefficients $a_{n,1}, \ldots, a_{n,n}$ can be computed by knowing the training data D_n only, and if the values $\psi_1(X), \ldots, \psi_k(X)$ are easily computable, then $f_n(X)$ can be computed much more quickly than in a kernel-based decision, where all n terms of the sum have to be computed in real time, if no preprocessing is done. However, using preprocessing of the data may also help with kernel rules, especially when $d = 1$. For a survey of computational speed-up with kernel methods, see Devroye and Machell (1985).

RECURSIVE KERNEL RULES.   Consider the choice

$$K_{n,i}(x, y) = K\left(\frac{x - y}{h_i}\right), \quad \text{and} \quad r_{n,i}(D_n) = 2Y_i - 1. \tag{10.3}$$

Observe that the only difference between this and the ordinary kernel rule is that in the expression of $K_{n,i}$, the smoothing parameter h_n is replaced with h_i. With this change, we can compute the rule recursively by observing that

$$f_{n+1}(x) = f_n(x) + (2Y_{n+1} - 1)K\left(\frac{x - X_{n+1}}{h_{n+1}}\right).$$

The computational advantage of this rule is that if one collects additional data, then the rule does not have to be entirely recomputed. It can be adjusted using the formula above. Consistency properties of this rule were studied by Devroye and Wagner (1980b), Krzyżak and Pawlak (1984a), Krzyżak (1986), and Greblicki and Pawlak (1987). Several similar recursive kernel rules have been studied in the literature. Wolverton and Wagner (1969b), Greblicki (1974), and Krzyżak and Pawlak (1983), studied the situation when

$$K_{n,i}(x, y) = \frac{1}{h_i^d} K\left(\frac{x - y}{h_i}\right) \quad \text{and} \quad r_{n,i}(D_n) = 2Y_i - 1. \tag{10.4}$$

The corresponding rule can be computed recursively by

$$f_{n+1}(x) = f_n(x) + (2Y_{n+1} - 1)\frac{1}{h_{n+1}^d} K\left(\frac{x - X_{n+1}}{h_{n+1}}\right).$$

Motivated by stochastic approximation methods (see Chapter 17), Révész (1973) suggested and studied the rule obtained from

$$f_{n+1}(x) = f_n(x) + \frac{1}{n + 1}(2Y_{n+1} - 1 - f_n(x))\frac{1}{h_{n+1}^d} K\left(\frac{x - X_{n+1}}{h_{n+1}}\right).$$

A similar rule was studied by Györfi (1981):

$$f_{n+1}(x) = f_n(x) + \frac{1}{n + 1}(2Y_{n+1} - 1 - f_n(x))K\left(\frac{x - X_{n+1}}{h_{n+1}}\right).$$

Problems and Exercises

PROBLEM 10.1. Let K be a nonnegative kernel with compact support on $[-1, 1]$. Show that for some distribution, $h \to 0$ is necessary for consistency of the kernel rule. To this end, consider the following example. Given $Y = 0$, X has a geometric distribution on $\delta, \delta^3, \delta^5, \ldots$, and given $Y = 1$, X has a geometric distribution on $\delta^2, \delta^4, \delta^6, \ldots$. Then show that to obtain $\mathbf{E}L_n \to L^* = 0$, it is necessary that $h \to 0$.

PROBLEM 10.2. KERNEL DENSITY ESTIMATION. Let $X_1, \ldots, X_n$ be i.i.d. random variables in $\mathcal{R}^d$ with density f. Let K be a kernel function integrating to one, and let $h_n > 0$ be a smoothing factor. The kernel density estimate is defined by

$$f_n(x) = \frac{1}{nh_n^d} \sum_{i=1}^{n} K\left(\frac{x - X_i}{h_n}\right)$$

(Rosenblatt (1956), Parzen (1962)). Prove that the estimate is *weakly universally consistent* in L_1 if $h_n \to 0$ and $nh_n^d \to \infty$ as $n \to \infty$. HINT: Proceed as in Problem 6.2.

PROBLEM 10.3. STRONG CONSISTENCY OF KERNEL DENSITY ESTIMATION. Let $X_1, \ldots, X_n$ be i.i.d. random variables in $\mathcal{R}^d$ with density f. Let K be a nonnegative function integrating to one (a kernel) and $h > 0$ a smoothing factor. As in the previous exercise, the kernel density estimate is defined by

$$f_n(x) = \frac{1}{nh^d} \sum_{i=1}^{n} K(\frac{x - X_i}{h}).$$

Prove for the L_1-error of the estimate that

$$\mathbf{P}\left\{\left|\int |f_n(x) - f(x)|dx - \mathbf{E}\int |f_n(x) - f(x)|dx\right| > \epsilon\right\} \leq 2e^{-n\epsilon^2/2}$$

(Devroye (1991a)). Conclude that weak L_1-consistency of the estimate implies strong consistency (see Problem 10.2). This is a way to show that weak and strong L_1-consistencies of the kernel density estimate are equivalent (Devroye (1983).) Also, for $d = 1$, since if K is nonnegative, then $\mathbf{E}\int |f_n(x) - f(x)|dx$ cannot converge to zero faster than $n^{-2/5}$ for any density (see Devroye and Györfi (1985)), therefore, the inequality above implies that for any density

$$\lim_{n \to \infty} \frac{\int |f_n(x) - f(x)|dx}{\mathbf{E}\int |f_n(x) - f(x)|dx} = 0$$

with probability one (Devroye (1988d)). This property is called the *relative stability* of the L_1 error. It means that the asymptotic behavior of the L_1-error is the same as that of its expected value. HINT: Use McDiarmid's inequality.

PROBLEM 10.4. If $\int K < 0$, show that under the assumption that μ has a density f, and that $h \to 0$, $nh^d \to \infty$, the kernel rule has

$$\lim_{n \to \infty} \mathbf{E}L_n = \mathbf{E}\{\max(\eta(X), 1 - \eta(X))\} = 1 - L^*.$$

Thus, the rule makes the wrong decisions, and such kernels should be avoided. HINT: You may use the fact that for the kernel density estimate with kernel L satisfying $\int L = 1$,

$$\int |f_n(x) - f(x)|dx \to 0$$

with probability one, if $h \to 0$ and $nh^d \to \infty$ (see Problems 10.2 and 10.3).

PROBLEM 10.5. Consider a devilish kernel that attaches counterproductive weight to the origin:

$$K(x) = \begin{cases} -1 & \text{if } \|x\| \leq 1/3 \\ 1 & \text{if } 1/3 < \|x\| \leq 1 \\ 0 & \text{if } \|x\| > 1. \end{cases}$$

Assume that $h \to 0$, yet $nh^d \to \infty$. Assume that $L^* = 0$. Show that $L_n \to 0$ with probability one. CONCESSION: if you find that you can't handle the universality, try first proving the statement for strictly separable distributions.

PROBLEM 10.6. Show that for the distribution depicted in Figure 10.2, the kernel rule with kernel $K(u) = (1 - u^2)I_{\{|u| \leq 1\}}$ is consistent whenever h, the smoothing factor, remains fixed and $0 < h \leq 1/2$.

PROBLEM 10.7. THE LIMIT FOR FIXED h. Consider a kernel rule with fixed $h \equiv 1$, and fixed kernel K. Find a simple argument that proves

$$\lim_{n \to \infty} \mathbf{E}L_n = L_\infty,$$

where L_∞ is the probability of error for the decision g_∞ defined by

$$g_\infty(x) = \begin{cases} 0 & \text{if } \mathbf{E}\{K(x - X)(2\eta(X) - 1)\} \le 0 \\ 1 & \text{if } \mathbf{E}\{K(x - X)(2\eta(X) - 1)\} > 0. \end{cases}$$

Find a distribution such that for the window kernel, $L_\infty = 1/2$, yet $L^* = 0$. Is there such a distribution for any kernel? HINT: Try proving a convergence result at each x by invoking the law of large numbers, and then replace x by X.

PROBLEM 10.8. Show that the conditions $h_n \to 0$ and $nh^d \to \infty$ of Theorem 10.1 are not necessary for consistency, that is, exhibit a distribution such that the kernel rule is consistent with $h_n = 1$, and exhibit another distribution for which the kernel rule is consistent with $h_n \sim 1/n^{1/d}$.

PROBLEM 10.9. Prove that the conditions $h_n \to 0$ and $nh^d \to \infty$ of Theorem 10.1 are necessary for universal consistency, that is, show that if one of these conditions is violated then there is a distribution for which the kernel rule is not consistent (Krzyżak (1991)).

PROBLEM 10.10. This exercise provides an argument in favor of monotonicity of the kernel K. In $\mathcal{R}^2$, find a nonatomic distribution for (X, Y), and a positive kernel with $\int K > 0$, K vanishing off $S_{0,\delta}$ for some $\delta > 0$, such that for all $h > 0$, and all n, the kernel rule has $\mathbf{E}L_n = 1/2$, while $L^* = 0$. This result says that the condition $K(x) \ge bI_{\{S_{0,\delta}\}}$ for some $b > 0$ in the universal consistency theorem cannot be abolished altogether.

PROBLEM 10.11. With K as in the previous problem, and taking $h = 1$, show that

$$\lim_{n \to \infty} \mathbf{E}L_n = L^*$$

under the following conditions:
  (1) K has compact support vanishing off $S_{0,\delta}$, $K \ge 0$, and $K \ge bI_{\{S_{0,\epsilon}\}}$ for some $\epsilon > 0$.
  (2) We say that we have *agreement* on $S_{x,\delta}$ when for all $z \in S_{x,\delta}$, either $\eta(z) \le 1/2$, or $\eta(z) \ge 1/2$. We ask that $\mathbf{P}\{\text{Agreement on } S_{X,\delta}\} = 1$.

PROBLEM 10.12. The previous exercise shows that at points where there is agreement, we make asymptotically the correct decision with kernels with fixed smoothing factor. Let D be the set $\{x : \eta(x) = 1/2\}$, and let the δ-neighborhood of D be defined by $D_\delta = \{y : \|y - x\| \le \delta \text{ for some } x \in D\}$. Let μ be the probability measure for X. Take K, h as in the previous exercise. Noting that $x \notin D_\delta$ means that we have agreement on $S_{x,\delta}$, show that for all distributions of (X, Y),

$$\limsup_{n \to \infty} \mathbf{E}L_n \le L^* + \mu(D_\delta).$$

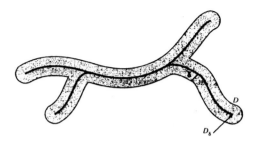

FIGURE 10.5. δ-neighborhood of a set D.

PROBLEM 10.13. CONTINUATION. Clearly, $\mu(D_\delta) \to 0$ as $\delta \to 0$ when $\mu(D) = 0$. Convince yourself that $\mu(D) = 0$ for most problems. If you knew how fast $\mu(D_\delta)$ tended to zero, then the previous exercise would enable you to pick h as a function of n such that $h \to 0$ and such that the upper bound for EL_n obtained by analogy from the previous exercise is approximately minimal. If in $\mathcal{R}^d$, D is the surface of the unit ball, X has a bounded density f, and η is Lipschitz, determine a bound for $\mu(D_\delta)$. By considering the proof of the universal consistency theorem, show how to choose h such that

$$EL_n - L^* = O\left(\frac{1}{n^{1/(2+d)}}\right).$$

PROBLEM 10.14. EXTENSION OF THE COVERING LEMMA (LEMMA 10.1) TO REGULAR KERNELS. Let K be a regular kernel, and let μ be an arbitrary probability measure. Prove that there exists a finite constant $\rho = \rho(K)$ only depending upon K such that for any y and h

$$\int \frac{K_h(x-y)}{\int K_h(x-z)\mu(dz)}\mu(dx) \le \rho$$

(Devroye and Krzyżak (1989)). HINT: Prove this by checking the following:
(1) First take a bounded overlap cover of $\mathcal{R}^d$ with translates of $S_{0,r/2}$, where $r > 0$ is the constant appearing in the definition of a regular kernel. This cover has an infinite number of member balls, but every x gets covered at most k_1 times where k_1 depends upon d only.
(2) The centers of the balls are called x_i, $i = 1, 2, \ldots$. The integral condition on K implies that

$$\sum_{i=1}^{\infty} \sup_{x \in x_i + S_{0,r/2}} K(x) \le \frac{k_1}{\int_{S_{0,r/2}} dx} \int \sup_{y \in x + S_{0,r}} K(y)dx \le k_2$$

for another finite constant k_2.
(3) Show that

$$K_h(x-y) \le \sum_{i=1}^{\infty} \sup_{x \in y + hx_i + S_{0,rh/2}} K_h(x-y)I_{[x \in y + hx_i + S_{0,rh/2}]},$$

and

$$\int K_h(x-z)\mu(dz) \ge b\mu(y + hx_i + S_{0,rh/2}) \quad (x \in y + hx_i + S_{0,rh/2}).$$

From (c) conclude

$$\int \frac{K_h(x-y)}{\int K_h(x-z)\mu(dz)}\mu(dx)$$

$$\leq \sum_{i=1}^{\infty}\int_{x\in y+hx_i+S_{0,rh/2}}\frac{\sup_{z\in hx_i+S_{0,rh/2}}K_h(z)}{b\mu(y+hx_i+S_{0,rh/2})}\mu(dx)$$

$$= \sum_{i=1}^{\infty}\frac{\mu(y+hx_i+S_{0,rh/2})\sup_{z\in hx_i+S_{0,rh/2}}K_h(z)}{b\mu(y+hx_i+S_{0,rh/2})}$$

$$= \frac{1}{b}\sum_{i=1}^{\infty}\sup_{z\in hx_i+S_{0,rh/2}}K_h(z) \leq \frac{k_2}{b},$$

where k_2 depends on K and d only.

PROBLEM 10.15. Let K be a regular kernel, and let μ be an arbitrary probability measure. Prove that for any $\delta > 0$

$$\lim_{h\to 0}\sup_{y}\int \frac{K_h(x-y)I_{\{\|x-y\|\geq\delta\}}}{\int K_h(x-z)\mu(dz)}\mu(dx) = 0.$$

HINT: Substitute $K_h(z)$ in the proof of Problem 10.15 by $K_h(z)I(\|z\| \geq \delta)$ and notice that

$$\sup_{y}\int \frac{K_h(x-y)I_{\{\|x-y\|\geq\delta\}}}{\int K_h(x-z)\mu(dz)}\mu(dx) \leq \sum_{i=1}^{\infty}\sup_{z\in hx_i+S_{0,rh/2}}K_h(z)I(\|z\| \geq \delta) \to 0$$

as $h \to 0$.

PROBLEM 10.16. Use Problems 10.14 and 10.15 to extend the proof of Theorem 10.1 for arbitrary regular kernels.

PROBLEM 10.17. Show that the constant β_d in Lemma 10.1 is never more than 4^d.

PROBLEM 10.18. Show that if $L \geq 0$ is a bounded function that is monotonically decreasing on $[0, \infty)$ with the property that $\int u^{d-1}L(u)du < \infty$, and if $K : \mathcal{R}^d \to [0, \infty)$ is a function with $K(x) \leq L(\|x\|)$, then K is regular.

PROBLEM 10.19. Find a kernel $K \geq 0$ that is monotonically decreasing along rays (i.e., $K(rx) \leq K(x)$ for all $x \in \mathcal{R}^d$ and all $r \geq 1$) such that K is not regular. (This exercise is intended to convince you that it is very difficult to find well-behaved kernels that are not regular.)

PROBLEM 10.20. Let $K(x) = L(\|x\|)$ for some bounded function $L \geq 0$. Show that K is regular if L is decreasing on $[0, \infty)$ and $\int K(x)dx < \infty$. Conclude that the Gaussian and Cauchy kernels are regular.

PROBLEM 10.21. Regularity of the kernel is not necessary for universal consistency. Investigate universal consistency with a nonintegrable kernel—that is, for which $\int K(x)dx = \infty$—such as $K(x) = 1/(1 + |x|)$. Greblicki, Krzyżak, and Pawlak (1984) proved consistency of the kernel rule with smoothing factor h_n satisfying $h_n \to 0$ and $nh_n^d \to \infty$ if the kernel K satisfies the following conditions: $K(x) \geq cI_{\{\|x\|\leq 1\}}$ for some $c > 0$ and for some $c_1, c_2 > 0, c_1H(\|x\|) \leq K(x) \leq c_2H(\|x\|)$, where H is a nonincreasing function on $[0, \infty)$ with $u^dH(u) \to 0$ as $u \to \infty$.

PROBLEM 10.22. Consider the kernel rule with kernel $K(x) = 1/\|x\|^r$, $r > 0$. Such kernels are useless for atomic distributions unless we take limits and define g_n as usual when $x \notin S$, the collection of points z with $X_i = X_j = z$ for some pair $(i \neq j)$. For $x \in S$, we take a majority vote over the Y_i's for which $X_i = x$. Discuss the weak universal consistency of this rule, which has the curious property that g_n is invariant to the smoothing factor h—so, we might as well set $h = 1$ without loss of generality. Note also that for $r \geq d$, $\int_{S_{0,1}} K(x)dx = \infty$, and for $r \leq d$, $\int_{S_{0,1}^c} K(x)dx = \infty$, where $S_{0,1}$ is the unit ball of $\mathcal{R}^d$ centered at the origin. In particular, if $r \leq d$, by considering X uniform on $S_{0,1}$ if $Y = 1$ and X uniform on the surface of $S_{0,1}$ if $Y = 0$, show that even though $L^* = 0$, the probability of error of the rule may tend to a nonzero limit for certain values of $P\{Y = 1\}$. Hence, the rule is not universally consistent. For $r \geq d$, prove or disprove the weak universal consistency, noting that the rules' decisions are by-and-large based on the few nearest neighbors. Prove the rule is weakly consistent for all $r \leq d$ whenever X has a density.

PROBLEM 10.23. Assume that the class densities coincide, that is, $f_0(x) = f_1(x)$ for every $x \in \mathcal{R}$, and assume $p = P\{Y = 1\} > 1/2$. Show that the expected probability of error of the kernel rule with $K \equiv 1$ is smaller than that with any unimodal regular kernel for every n and h small enough. Exhibit a distribution such that the kernel rule with a symmetric kernel such that $K(|x|)$ is monotone *increasing* has smaller expected error probability than that with any unimodal regular kernel.

PROBLEM 10.24. SCALING. Assume that the kernel K can be written into the following product form of one-dimensional kernels:

$$K(x) = K(x^{(1)}, \ldots, x^{(d)}) = \prod_{i=1}^{d} K_i(x^{(i)}).$$

Assume also that K is regular. One can use different smoothing factors along the different coordinate axes to define a kernel rule by

$$g_n(x) = \begin{cases} 0 & \text{if } \sum_{i=1}^{n}(2Y_i - 1) \prod_{j=1}^{d} K_j \left(\frac{x^{(j)} - X_i^{(j)}}{h_{jn}} \right) \leq 0 \\ 1 & \text{otherwise,} \end{cases}$$

where $X_i^{(j)}$ denotes the j-th component of X_i. Prove that g_n is strongly universally consistent if $h_{in} \to 0$ for all $i = 1, \ldots, d$, and $nh_{1n}h_{2n} \ldots h_{dn} \to \infty$.

PROBLEM 10.25. Let $K : [0, \infty) \to [0, \infty)$ be a function, and Σ a symmetric positive definite $d \times d$ matrix. For $x \in \mathcal{R}^d$ define $K'(x) = K\left(x^T \Sigma x\right)$. Find conditions on K such that the kernel rule with kernel K' is universally consistent.

PROBLEM 10.26. Prove that the recursive kernel rule defined by (10.4) is strongly universally consistent if K is a regular kernel, $h_n \to 0$, and $nh_n^d \to \infty$ as $n \to \infty$ (Krzyżak and Pawlak (1983)).

PROBLEM 10.27. Show that the recursive kernel rule of (10.3) is strongly universally consistent whenever K is a regular kernel, $\lim_{n\to\infty} h_n = 0$, and $\sum_{n=1}^{\infty} h_n^d = \infty$ (Greblicki and Pawlak (1987)). Note: Greblicki and Pawlak (1987) showed convergence under significantly weaker assumptions on the kernel. They assume that $K(x) \geq cI_{\{\|x\|\leq 1\}}$ for some $c > 0$ and that for some $c_1, c_2 > 0$, $c_1 H(\|x\|) \leq K(x) \leq c_2 H(\|x\|)$, where H is a nonincreasing function on $[0, \infty)$ with $u^d H(u) \to 0$ as $u \to \infty$. They also showed that under the additional assumption $\int K(x)dx < \infty$ the following conditions on h_n are necessary and sufficient for universal consistency:

$$\frac{\sum_{i=1}^{n} h_i^d I_{\{h_i > a\}}}{\sum_{i=1}^{n} h_i^d} \to 0 \text{ for all } a > 0 \quad \text{and} \quad \sum_{i=1}^{\infty} h_i^d = \infty.$$

PROBLEM 10.28. OPEN-ENDED PROBLEM. Let $\mathbf{P}\{Y = 1\} = 1/2$. Given $Y = 1$, let X be uniformly distributed on $[0, 1]$. Given $Y = 0$, let X be atomic on the rationals with the following distribution: let $X = V/W$, where V and W are independent identically distributed, and $\mathbf{P}\{V = i\} = 1/2^i$, $i \geq 1$. Consider the kernel rule with the window kernel. What is the behavior of the smoothing factor h_n^* that minimizes the expected probability of error $\mathbf{E}L_n$?

11
Consistency of the k-Nearest Neighbor Rule

In Chapter 5 we discuss results about the asymptotic behavior of k-nearest neighbor classification rules, where the value of k—the number of neighbors taken into account at the decision—is kept at a fixed number as the size of the training data n increases. This choice leads to asymptotic error probabilities smaller than $2L^*$, but no universal consistency. In Chapter 6 we showed that if we let k grow to infinity as $n \rightarrow \infty$ such that $k/n \rightarrow 0$, then the resulting rule is weakly consistent. The main purpose of this chapter is to demonstrate strong consistency, and to discuss various versions of the rule.

We are not concerned here with the data-based choice of k—that subject deserves a chapter of its own (Chapter 26). We are also not tackling the problem of the selection of a suitable—even data-based—metric. At the end of this chapter and in the exercises, we draw the attention to 1-nearest neighbor relabeling rules, which combine the computational comfort of the 1-nearest neighbor rule with the asymptotic performance of variable-k nearest neighbor rules.

Consistency of k-nearest neighbor classification, and corresponding regression and density estimation has been studied by many researchers. See Fix and Hodges (1951; 1952), Cover (1968a), Stone (1977), Beck (1979), Györfi and Györfi (1975), Devroye (1981a; 1982b), Collomb (1979; 1980; 1981), Bickel and Breiman (1983), Mack (1981), Stute (1984), Devroye and Györfi (1985), Bhattacharya and Mack (1987), Zhao (1987), and Devroye, Györfi, Krzyżak, and Lugosi (1994).

Recall the definition of the k-nearest neighbor rule: first reorder the data

$$(X_{(1)}(x), Y_{(1)}(x)), \ldots, (X_{(n)}(x), Y_{(n)}(x))$$

according to increasing Euclidean distances of the X_j's to x. In other words, $X_{(i)}(x)$ is the i-th nearest neighbor of x among the points $X_1, \ldots, X_n$. If distance ties occur, a tie-breaking strategy must be defined. If μ is absolutely continuous with respect to the Lebesgue measure, that is, it has a density, then no ties occur with probability one, so formally we break ties by comparing indices. However, for general μ, the problem of distance ties turns out to be important, and its solution is messy. The issue of tie breaking becomes important when one is concerned with convergence of L_n with probability one. For weak universal consistency, it suffices to break ties by comparing indices.

The k-NN classification rule is defined as

$$g_n(x) = \begin{cases} 0 & \text{if } \sum_{i=1}^{k} I_{\{Y_{(i)}(x)=1\}} \leq \sum_{i=1}^{k} I_{\{Y_{(i)}(x)=0\}} \\ 1 & \text{otherwise.} \end{cases}$$

In other words, $g_n(x)$ is a majority vote among the labels of the k nearest neighbors of x.

11.1 Strong Consistency

In this section we prove Theorem 11.1. We assume the existence of a density for μ, so that we can avoid messy technicalities necessary to handle distance ties. We discuss this issue briefly in the next section.

The following result implies strong consistency whenever X has an absolutely continuous distribution. The result was proved by Devroye and Györfi (1985), and Zhao (1987). The proof presented here basically appears in Devroye, Györfi, Krzyżak, and Lugosi (1994), where strong *universal* consistency is proved under an appropriate tie-breaking strategy (see discussion later). Some of the main ideas appeared in the proof of the strong universal consistency of the regular histogram rule (Theorem 9.4).

Theorem 11.1. (DEVROYE AND GYÖRFI (1985), ZHAO (1987)). *Assume that μ has a density. If $k \to \infty$ and $k/n \to 0$ then for every $\epsilon > 0$ there is an n_0 such that for $n > n_0$*

$$P\{L_n - L^* > \epsilon\} \leq 2e^{-n\epsilon^2/(72\gamma_d^2)},$$

where γ_d is the minimal number of cones centered at the origin of angle $\pi/6$ that cover $\mathcal{R}^d$. (For the definition of a cone, see Chapter 5.) Thus, the k-NN rule is strongly consistent.

REMARK. At first glance the upper bound in the theorem does not seem to depend on k. It is n_0 that depends on the sequence of k's. What we really prove is the following: for every $\epsilon > 0$ there exists a $\beta_0 \in (0, 1)$ such that for any $\beta < \beta_0$ there is an n_0 such that if $n > n_0$, $k > 1/\beta$, and $k/n < \beta$, the exponential inequality holds. $\square$

For the proof we need a generalization of Lemma 5.3. The role of this covering lemma is analogous to that of Lemma 10.1 in the proof of consistency of kernel rules.

Lemma 11.1. (DEVROYE AND GYÖRFI (1985)). *Let*

$$B_a(x') = \left\{ x : \mu(S_{x,\|x-x'\|}) \le a \right\}.$$

Then for all $x' \in \mathcal{R}^d$

$$\mu(B_a(x')) \le \gamma_d a.$$

PROOF. For $x \in \mathcal{R}^d$ let $C(x, s) \subset \mathcal{R}^d$ be a cone of angle $\pi/6$ centered at x. The cone consists of all y with the property that either $y = x$ or angle$(y - x, s) \le \pi/6$, where s is a fixed direction. If $y, y' \in C(x, s)$, and $\|x - y\| < \|x - y'\|$, then $\|y - y'\| < \|x - y'\|$. This follows from a simple geometric argument in the vector space spanned by x, y and y' (see the proof of Lemma 5.3).

Now, let $C_1, \ldots, C_{\gamma_d}$ be a collection of cones centered at x with different central direction covering $\mathcal{R}^d$. Then

$$\mu(B_a(x')) \le \sum_{i=1}^{\gamma_d} \mu(C_i \cap B_a(x')).$$

Let $x^* \in C_i \cap B_a(x')$. Then by the property of the cones mentioned above we have

$$\mu(C_i \cap S_{x',\|x'-x^*\|} \cap B_a(x')) \le \mu(S_{x^*,\|x'-x^*\|}) \le a,$$

where we use the fact that $x^* \in B_a(x')$. Since x^* is arbitrary,

$$\mu(C_i \cap B_a(x')) \le a,$$

which completes the proof of the lemma. $\square$

An immediate consequence of the lemma is that the number of points among $X_1, \ldots, X_n$ such that X is one of their k nearest neighbors is not more than a constant times k.

COROLLARY 11.1.

$$\sum_{i=1}^{n} I_{\{X \text{ is among the } k \text{ NN's of } X_i \text{ in } \{X_1,\ldots,X_n,X\}-\{X_i\}\}} \le k\gamma_d.$$

PROOF. Apply Lemma 11.1 with $a = k/n$ and let μ be the empirical measure μ_n of $X_1, \ldots, X_n$, that is, for each Borel set $A \subseteq \mathcal{R}^d$, $\mu_n(A) = (1/n)\sum_{i=1}^{n} I_{\{X_i \in A\}}$. $\square$

PROOF OF THEOREM 11.1. Since the decision rule g_n may be rewritten as

$$g_n(x) = \begin{cases} 0 & \text{if } \eta_n(x) \le 1/2 \\ 1 & \text{otherwise,} \end{cases}$$

where η_n is the corresponding regression function estimate

$$\eta_n(x) = \frac{1}{k} \sum_{i=1}^{k} Y_{(i)}(x),$$

the statement follows from Theorem 2.2 if we show that for sufficiently large n

$$\mathbf{P}\left\{\int |\eta(x) - \eta_n(x)|\mu(dx) > \frac{\epsilon}{2}\right\} \leq 2e^{-n\epsilon^2/(72\gamma_d^2)}.$$

Define $\rho_n(x)$ as the solution of the equation

$$\frac{k}{n} = \mu(S_{x,\rho_n(x)}).$$

Note that the absolute continuity of μ implies that the solution always exists. (This is the only point in the proof where we use this assumption.) Also define

$$\eta_n^*(x) = \frac{1}{k}\sum_{j=1}^{n} Y_j I_{\{\|X_j - x\| < \rho_n(x)\}}.$$

The basis of the proof is the following decomposition:

$$|\eta(x) - \eta_n(x)| \leq |\eta_n(x) - \eta_n^*(x)| + |\eta_n^*(x) - \eta(x)|.$$

For the first term on the right-hand side, observe that denoting $R_n(x) = \|X_{(k)}(x) - x\|$,

$$
\begin{aligned}
|\eta_n^*(x) - \eta_n(x)| &= \frac{1}{k}\left|\sum_{j=1}^{n} Y_j I_{\{X_j \in S_{x,\rho_n(x)}\}} - \sum_{j=1}^{n} Y_j I_{\{X_j \in S_{x,R_n(x)}\}}\right| \\
&\leq \frac{1}{k}\sum_{j=1}^{n}\left|I_{\{X_j \in S_{x,\rho_n(x)}\}} - I_{\{X_j \in S_{x,R_n(x)}\}}\right| \\
&= \left|\frac{1}{k}\sum_{j=1}^{n} I_{\{X_j \in S_{x,\rho_n(x)}\}} - 1\right| = |\widehat{\eta}_n^*(x) - \widehat{\eta}(x)|,
\end{aligned}
$$

where $\widehat{\eta}_n^*$ is defined as η_n^* with Y replaced by the constant random variable $\widehat{Y} = 1$, and $\widehat{\eta} \equiv 1$ is the corresponding regression function. Thus,

$$|\eta(x) - \eta_n(x)| \leq |\widehat{\eta}_n^*(x) - \widehat{\eta}(x)| + |\eta_n^*(x) - \eta(x)|. \tag{11.1}$$

First we show that the expected values of the integrals of both terms on the right-hand side converge to zero. Then we use McDiarmid's inequality to prove that both terms are very close to their expected values with large probability.

For the expected value of the first term on the right-hand side of (11.1), using the Cauchy-Schwarz inequality, we have

$$
\begin{aligned}
\mathbf{E}\int |\eta_n^*(x) - \eta_n(x)|\mu(dx) &\leq \mathbf{E}\int |\widehat{\eta}_n^*(x) - \widehat{\eta}(x)|\mu(dx) \\
&\leq \int \sqrt{\mathbf{E}\{|\widehat{\eta}_n^*(x) - \widehat{\eta}(x)|^2\}}\mu(dx)
\end{aligned}
$$

$$= \int \sqrt{\frac{1}{k^2} n \, \mathbf{Var}\{I_{\{X \in S_{x,\rho_n(x)}\}}\}} \mu(dx)$$

$$\leq \int \sqrt{\frac{1}{k^2} n \mu(S_{x,\rho_n(x)})} \mu(dx)$$

$$= \int \sqrt{\frac{n}{k^2} \frac{k}{n}} \mu(dx)$$

$$= \frac{1}{\sqrt{k}},$$

which converges to zero.

For the expected value of the second term on the right-hand side of (11.1), note that in the proof of Theorems 6.3 and 6.4 we already showed that

$$\lim_{n \to \infty} \mathbf{E} \int |\eta(x) - \eta_n(x)| \mu(dx) = 0.$$

Therefore,

$$\mathbf{E} \int |\eta_n^*(x) - \eta(x)| \mu(dx)$$

$$\leq \mathbf{E} \int |\eta_n^*(x) - \eta_n(x)| \mu(dx) + \mathbf{E} \int |\eta(x) - \eta_n(x)| \mu(dx) \to 0.$$

Assume now that n is so large that

$$\mathbf{E} \int |\widehat{\eta}_n^*(x) - \widehat{\eta}(x)| \mu(dx) + \mathbf{E} \int |\eta_n^*(x) - \eta(x)| \mu(dx) < \frac{\epsilon}{6}.$$

Then, by (11.1), we have

$$\mathbf{P}\left\{ \int |\eta(x) - \eta_n(x)| \mu(dx) > \frac{\epsilon}{2} \right\} \tag{11.2}$$

$$\leq \mathbf{P}\left\{ \int |\eta_n^*(x) - \eta(x)| \mu(dx) - \mathbf{E} \int |\eta_n^*(x) - \eta(x)| \mu(dx) > \frac{\epsilon}{6} \right\}$$

$$+ \mathbf{P}\left\{ \int |\widehat{\eta}_n^*(x) - \widehat{\eta}(x)| \mu(dx) - \mathbf{E} \int |\widehat{\eta}_n^*(x) - \widehat{\eta}(x)| \mu(dx) > \frac{\epsilon}{6} \right\}.$$

Next we get an exponential bound for the first probability on the right-hand side of (11.2) by McDiarmid's inequality (Theorem 9.2). Fix an arbitrary realization of the data $D_n = (x_1, y_1), \ldots, (x_n, y_n)$, and replace (x_i, y_i) by $(\widehat{x}_i, \widehat{y}_i)$, changing the value of $\eta_n^*(x)$ to $\eta_{ni}^*(x)$. Then

$$\left| \int |\eta_n^*(x) - \eta(x)| \mu(dx) - \int |\eta_{ni}^*(x) - \eta(x)| \mu(dx) \right| \leq \int |\eta_n^*(x) - \eta_{ni}^*(x)| \mu(dx).$$

But $|\eta_n^*(x) - \eta_{ni}^*(x)|$ is bounded by $2/k$ and can differ from zero only if $\|x - x_i\| < \rho_n(x)$ or $\|x - \widehat{x_i}\| < \rho_n(x)$. Observe that $\|x - x_i\| < \rho_n(x)$ if and only if $\mu(S_{x,\|x-x_i\|}) < k/n$. But the measure of such x's is bounded by $\gamma_d k/n$ by Lemma 11.1. Therefore

$$\sup_{x_1,y_1,\ldots,x_n,y_n,\widehat{x_i},\widehat{y_i}} \int |\eta_n^*(x) - \eta_{ni}^*(x)|\mu(dx) \le \frac{2}{k}\frac{\gamma_d k}{n} = \frac{2\gamma_d}{n}$$

and by Theorem 9.2

$$\mathbf{P}\left\{\left|\int |\eta(x) - \eta_n^*(x)|\mu(dx) - \mathbf{E}\int |\eta(x) - \eta_n^*(x)|\mu(dx)\right| > \frac{\epsilon}{6}\right\} \le 2e^{-n\epsilon^2/(72\gamma_d^2)}.$$

Finally, we need a bound for the second term on the right-hand side of (11.2). This probability may be bounded by McDiarmid's inequality exactly the same way as for the first term, obtaining

$$\mathbf{P}\left\{\left|\int |\widehat{\eta}_n^*(x) - \widehat{\eta}(x)|\mu(dx) - \mathbf{E}\int |\widehat{\eta}_n^*(x) - \widehat{\eta}(x)|\mu(dx)\right| > \frac{\epsilon}{6}\right\} \le 2e^{-n\epsilon^2/(72\gamma_d^2)},$$

and the proof is completed. $\square$

REMARK. The conditions $k \to \infty$ and $k/n \to 0$ are optimal in the sense that they are also necessary for consistency for some distributions with a density. However, for some distributions they are not necessary for consistency, and in fact, keeping $k = 1$ for all n may be a better choice. This latter property, dealt with in Problem 11.1, shows that the 1-nearest neighbor rule is admissible. $\square$

11.2   Breaking Distance Ties

Theorem 11.1 provides strong consistency under the assumption that X has a density. This assumption was needed to avoid problems caused by equal distances. Turning to the general case, we see that if μ does not have a density then distance ties can occur with nonzero probability, so we have to deal with the problem of breaking them. To see that the density assumption cannot be relaxed to the condition that μ is merely nonatomic without facing frequent distance ties, consider the following distribution on $\mathcal{R}^d \times \mathcal{R}^{d'}$ with $d, d' \ge 2$:

$$\mu = \frac{1}{2}(\tau_d \times \sigma_{d'}) + \frac{1}{2}(\sigma_d \times \tau_{d'}),$$

where τ_d denotes the uniform distribution on the surface of the unit sphere of $\mathcal{R}^d$ and σ_d denotes the unit point mass at the origin of $\mathcal{R}^d$. Observe that if X has distribution $\tau_d \times \sigma_{d'}$ and X' has distribution $\sigma_d \times \tau_{d'}$, then $\|X - X'\| = \sqrt{2}$.

Hence, if X_1, X_2, X_3, X_4 are independent with distribution μ, then $\mathbf{P}\{\|X_1 - X_2\| = \|X_3 - X_3\|\} = 1/4$.

Next we list some methods of breaking distance ties.

- TIE-BREAKING BY INDICES: If X_i and X_j are equidistant from x, then X_i is declared closer if $i < j$. This method has some undesirable properties. For example, if X is monoatomic, with $\eta < 1/2$, then X_1 is the nearest neighbor of all X_j's, $j > 1$, but X_j is only the $j - 1$-st nearest neighbor of X_1. The influence of X_1 in such a situation is too large, making the estimate very unstable and thus undesirable. In fact, in this monoatomic case, if $L^* + \epsilon < 1/2$,

$$\mathbf{P}\left\{L_n - L^* > \epsilon\right\} \geq e^{-ck}$$

for some $c > 0$ (see Problem 11.2). Thus, we cannot expect a distribution-free version of Theorem 11.1 with this tie-breaking method.

- STONE'S TIE-BREAKING: Stone (1977) introduced a version of the nearest neighbor rule, where the labels of the points having the same distance from x as the k-th nearest neighbor are averaged. If we denote the distance of the k-th nearest neighbor to x by $R_n(x)$, then Stone's rule is the following:

$$g_n(x) = \begin{cases} 0 & \text{if } \displaystyle\sum_{i:\|x - x_i\| < R_n(x)} I_{\{Y_i = 0\}} + \dfrac{k - |\{i : \|x - x_i\| < R_n(x)\}|}{|\{i : \|x - x_i\| = R_n(x)\}|} \\ & \qquad \times \displaystyle\sum_{i:\|x - x_i\| = R_n(x)} I_{\{Y_i = 0\}} \geq \dfrac{k}{2} \\ 1 & \text{otherwise.} \end{cases}$$

This is not a k-nearest neighbor rule in a strict sense, since this estimate, in general, uses more than k neighbors. Stone (1977) proved weak universal consistency of this rule.

- ADDING A RANDOM COMPONENT: To circumvent the aforementioned difficulties, we may artificially increase the dimension of the feature vector by one. Define the the $d + 1$-dimensional random vectors

$$X' = (X, U), X_1' = (X_1, U_1), \ldots, X_n' = (X_n, U_n),$$

where the randomizing variables $U, U_1, \ldots, U_n$ are real-valued i.i.d. random variables independent of X, Y, and D_n, and their common distribution has a density. Clearly, because of the independence of U, the Bayes error corresponding to the pair (X', Y) is the same as that of (X, Y). The algorithm performs the k-nearest neighbor rule on the modified data set

$$D_n' = ((X_1', Y_1), \ldots, (X_n', Y_n)).$$

It finds the k nearest neighbors of X', and uses a majority vote among these labels to guess Y. Since U has a density and is independent of X, distance ties occur with zero probability. Strong universal consistency of this rule can be seen by observing that the proof of Theorem 11.1 used the existence of the density in the definition of $\rho_n(x)$ only. With our randomization, $\rho_n(x)$ is well-defined, and the same proof yields *strong universal* consistency. Interestingly, this rule is consistent whenever U has a density and is independent of (X, Y). If, for example, the magnitude of U is much larger than that of $\|X\|$, then the rule defined this way will significantly differ from the k-nearest neighbor rule, though it still preserves universal consistency. One should expect however a dramatic decrease in the performance. Of course, if U is very small, then the rule remains intuitively appealing.

- TIE-BREAKING BY RANDOMIZATION: There is another, perhaps more natural way of breaking ties via randomization. We assume that (X, U) is a random vector independent of the data, where U is independent of X and uniformly distributed on $[0, 1]$. We also artificially enlarge the data by introducing $U_1, U_2, \ldots, U_n$, where the U_i's are i.i.d. uniform $[0, 1]$ as well. Thus, each (X_i, U_i) is distributed as (X, U). Let

$$(X_{(1)}(x, u), Y_{(1)}(x, u)), \ldots, (X_{(n)}(x, u), Y_{(n)}(x, u))$$

be a reordering of the data according to increasing values of $\|x - X_i\|$. In case of distance ties, we declare (X_i, U_i) closer to (x, u) than (X_j, U_j) provided that

$$|U_i - u| \le |U_j - u|.$$

Define the k-NN classification rule as

$$g_n(x) = \begin{cases} 0 & \text{if } \sum_{i=1}^k I_{\{Y_{(i)}(x,u)=1\}} \le \sum_{i=1}^k I_{\{Y_{(i)}(x,u)=0\}} \\ 1 & \text{otherwise,} \end{cases} \qquad (11.3)$$

and denote the error probability of g_n by

$$L_n = \mathbf{P}\{g_n(X, U) \ne Y | X_1, U_1, Y_1, \ldots, X_n, U_n, Y_n\}.$$

Devroye, Györfi, Krzyżak, and Lugosi (1994) proved that $L_n \to L^*$ with probability one for all distributions if $k \to \infty$ and $k/n \to 0$. The basic argument in (1994) is the same as that of Theorem 11.1, except that the covering lemma (Lemma 11.1) has to be appropriately modified.

It should be stressed again that if μ has a density, or just has an absolutely continuous component, then tie-breaking is needed with zero probability, and becomes therefore irrelevant.

11.3 Recursive Methods

To find the nearest neighbor of a point x among $X_1, \ldots, X_n$, we may preprocess the data in $O(n \log n)$ time, such that each query may be answered in $O(\log n)$

worst-case time—see, for example, Preparata and Shamos (1985). Other recent developments in computational geometry have made the nearest neighbor rules computationally feasible even when n is formidable. Without preprocessing however, one must resort to slow methods. If we need to find a decision at x and want to process the data file once when doing so, a simple rule was proposed by Devroye and Wise (1980). It is a fully recursive rule that may be updated as more observations become available.

Split the data sequence D_n into disjoint blocks of length $l_1, \ldots, l_N$, where $l_1, \ldots, l_N$ are positive integers satisfying $\sum_{i=1}^{N} l_i = n$. In each block find the nearest neighbor of x, and denote the nearest neighbor of x from the i-th block by $X_i^*(x)$. Let $Y_i^*(x)$ be the corresponding label. Ties are broken by comparing indices. The classification rule is defined as a majority vote among the nearest neighbors from each block:

$$g_n(x) = \begin{cases} 0 & \text{if } \sum_{i=1}^{N} I_{\{Y_i^*(x)=1\}} \leq \sum_{i=1}^{N} I_{\{Y_i^*(x)=0\}} \\ 1 & \text{otherwise.} \end{cases}$$

Note that we have only defined the rule g_n for n satisfying $\sum_{i=1}^{N} l_i = n$ for some N. A possible extension for all n's is given by $g_n(x) = g_m(x)$, where m is the largest integer not exceeding n that can be written as $\sum_{i=1}^{N} l_i$ for some N. The rule is weakly universally consistent if

$$\lim_{N \to \infty} l_N = \infty$$

(Devroye and Wise (1980), see Problem 11.3).

11.4 Scale-Invariant Rules

A scale-invariant rule is a rule that is invariant under rescalings of the components. It is motivated by the lack of a universal yardstick when components of a vector represent physically different quantities, such as temperature, blood pressure, alcohol, and the number of lost teeth. More formally, let $x^{(1)}, \ldots, x^{(d)}$ be the d components of a vector x. If $\psi_1, \ldots, \psi_d$ are strictly monotone mappings: $\mathcal{R} \to \mathcal{R}$, and if we define

$$\psi(x) = \left(\psi_1(x^{(1)}), \ldots, \psi_d(x^{(d)}) \right),$$

then g_n is scale-invariant if

$$g_n(x, D_n) = g_n(\psi(x), D_n'),$$

where $D_n' = ((\psi(X_1), Y_1), \ldots, (\psi(X_n), Y_n))$. In other words, if all the X_i's and x are transformed in the same manner, the decision does not change.

Some rules based on statistically equivalent blocks (discussed in Chapters 21 and 22) are scale-invariant, while the k-nearest neighbor rule clearly is not. Here we describe a scale-invariant modification of the k-nearest neighbor rule, suggested by Olshen (1977) and Devroye (1978).

The scale-invariant k-nearest neighbor rule is based upon *empirical distances* that are defined in terms of the order statistics along the d coordinate axes. First order the points $x, X_1, \ldots, X_n$ according to increasing values of their first components $x^{(1)}, X_1^{(1)}, \ldots, X_n^{(1)}$, breaking ties via randomization. Denote the rank of $X_i^{(1)}$ by $r_i^{(1)}$, and the rank of $x^{(1)}$ by $r^{(1)}$. Repeating the same procedure for the other coordinates, we obtain the ranks

$$r_i^{(j)}, \ r^{(j)}, \quad j = 1, \ldots, d, \ i = 1, \ldots, n.$$

Define the empirical distance between x and X_i by

$$\rho(x, X_i) = \max_{1 \le j \le d} |r_i^{(j)} - r^{(j)}|.$$

A k-NN rule can be defined based on these distances, by a majority vote among the Y_i's with the corresponding X_i's whose empirical distance from x are among the k smallest. Since these distances are integer-valued, ties frequently occur. These ties should be broken by randomization. Devroye (1978) proved that this rule (with randomized tie-breaking) is weakly universally consistent when $k \to \infty$ and $k/n \to 0$ (see Problem 11.5). For another consistent scale-invariant nearest neighbor rule we refer to Problem 11.6.

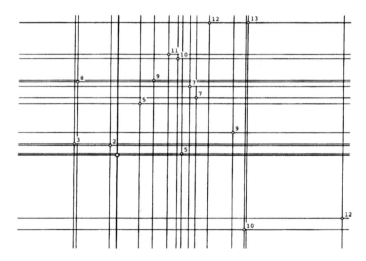

FIGURE 11.1. *Scale-invariant distances of 15 points from a fixed point are shown here.*

11.5   Weighted Nearest Neighbor Rules

In the k-NN rule, each of the k nearest neighbors of a point x plays an equally important role in the decision. However, intuitively speaking, nearer neighbors should provide more information than more distant ones. Royall (1966) first suggested using rules in which the labels Y_i are given unequal voting powers in the decision according to the distances of the X_i's from x: the i-th nearest neighbor receives weight w_{ni}, where usually $w_{n1} \geq w_{n2} \geq ... \geq w_{nn} \geq 0$ and $\sum_{i=1}^{n} w_{ni} = 1$. The rule is defined as

$$g_n(x) = \begin{cases} 0 & \text{if } \sum_{i=1}^{n} w_{ni} I_{\{Y_{(i)}(x)=1\}} \leq \sum_{i=1}^{n} w_{ni} I_{\{Y_{(i)}(x)=0\}} \\ 1 & \text{otherwise.} \end{cases}$$

We get the ordinary k-nearest neighbor rule back by the choice

$$w_{ni} = \begin{cases} 1/k & \text{if } i \leq k \\ 0 & \text{otherwise.} \end{cases}$$

The following conditions for consistency were established by Stone (1977):

$$\lim_{n \to \infty} \max_{1 \leq i \leq n} w_{ni} = 0,$$

and

$$\lim_{n \to \infty} \sum_{k \leq i \leq n} w_{ni} = 0$$

for some k with $k/n \to 0$ (see Problem 11.7). Weighted versions of the recursive and scale-invariant methods described above can also be defined similarly.

11.6   Rotation-Invariant Rules

Assume that an affine transformation T is applied to x and $X_1, \ldots, X_n$ (i.e., any number of combinations of rotations, translations, and linear rescalings), and that for any such linear transformation T,

$$g_n(x, D_n) = g_n(T(x), D'_n),$$

where $D'_n = ((T(X_1), Y_1), \ldots, (T(X_n), Y_n))$. Then we call g_n rotation-invariant. Rotation-invariance is indeed a very strong property. In $\mathcal{R}^d$, in the context of k-NN estimates, it suffices to be able to define a rotation-invariant distance measure. These are necessarily data-dependent. An example of this goes as follows. Any collection of d points in general position defines a polyhedron in a hyperplane of $\mathcal{R}^d$. For points $(X_{i_1}, \ldots, X_{i_d})$, we denote this polyhedron by $P(i_1, \ldots, i_d)$. Then we define the distance

$$\rho(X_i, x) = \sum_{\substack{i_1, \ldots, i_d, \\ i \notin \{i_1, \ldots, i_d\}}} I_{\{\text{segment } (X_i, x) \text{ intersects } P(i_1, \ldots, i_d)\}}.$$

Near points have few intersections. Using $\rho(\cdot, \cdot)$ in a k-NN rule with $k \to \infty$ and $k/n \to 0$, we expect weak universal consistency under an appropriate scheme of tie-breaking. The answer to this is left as an open problem for the scholars.

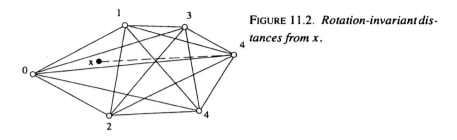

FIGURE 11.2. *Rotation-invariant distances from* x.

11.7   Relabeling Rules

The 1-NN rule appeals to the masses who crave simplicity and attracts the programmers who want to write short understandable code. Nearest neighbors may be found efficiently if the data are preprocessed (see Chapter 5 for references). Can we make the 1-NN rule universally consistent as well? In this section we introduce a tool called *relabeling*, which works as follows. Assume that we have a classification rule $\{g_n(x, D_n), x \in \mathcal{R}^d, n \geq 1\}$, where D_n is the data $(X_1, Y_1), \ldots, (X_n, Y_n)$. This rule will be called the ancestral rule. Define the *labels*

$$Z_i = g_n(X_i, D_n), \quad 1 \leq i \leq n.$$

These are the decisions for the X_i's themselves obtained by mere resubstitution. In the *relabeling method*, we apply the 1-NN rule to the new data $(X_1, Z_1), \ldots, (X_n, Z_n)$. If all goes well, when the ancestral rule g_n is universally consistent, so should the relabeling rule. We will show this by example, starting from a consistent k-NN rule as ancestral rule (with $k \rightarrow \infty, k/n \rightarrow 0$).

Unfortunately, relabeling rules do not always inherit consistency from their ancestral rules, so that a more general theorem is more difficult to obtain, unless one adds in a lot of regularity conditions—this does not seem to be the right time for that sort of effort. To see that universal consistency of the ancestral rule does not imply consistency of the relabeling rule, consider the following rule h_n:

$$h_n(x) = \begin{cases} 1 - Y_i & \text{if } x = X_i \text{ and } x \neq X_j, \text{ all } j \neq i \\ g_n(x, D_n) & \text{otherwise,} \end{cases}$$

where g_n is a weakly universally consistent rule. It is easy to show (see Problem 11.15) that h_n is universally consistent as well. Changing a rule on a set of measure zero indeed does not affect L_n. Also, if $x = X_i$ is at an atom of the distribution of X, we only change g_n to $1 - Y_i$ if X_i is the sole occurrence of that atom in the data. This has asymptotically no impact on L_n. However, if h_n is used as an ancestral rule, and X is nonatomic, then $h_n(X_i, D_n) = 1 - Y_i$ for all i, and therefore, the relabeling rule is a 1-NN rule based on the data $(X_1, 1 - Y_1), \ldots, (X_n, 1 - Y_n)$. If $L^* = 0$ for the distribution of (X, Y), then the relabeling rule has probability of error converging to one!

For most nonpathological ancestral rules, relabeling does indeed preserve universal consistency. We offer a prototype proof for the k-NN rule.

Theorem 11.2. *Let g_n be the k-NN rule in which tie-breaking is done by randomization as in (11.3). Assume that $k \to \infty$ and $k/n \to 0$ (so that g_n is weakly universally consistent). Then the relabeling rule based upon g_n is weakly universally consistent as well.*

PROOF. We verify the conditions of Stone's weak convergence theorem (see Theorem 6.3). To keep things simple, we assume that the distribution of X has a density so that distance ties happen with probability zero.

In our case, the weight $W_{ni}(X)$ of Theorem 6.3 equals $1/k$ iff X_i is among the k nearest neighbors of $X_{(1)}(X)$, where $X_{(1)}(X)$ is the nearest neighbor of X. It is zero otherwise.

Condition 6.3 (iii) is trivially satisfied since $k \to \infty$. For condition 6.3 (ii), we note that if $X_{(i)}(x)$ denotes the i-th nearest neighbor of X among $X_1, \ldots, X_n$, then

$$\| X_{(1)}(x) - X_{(k)}(X_{(1)}(x)) \| \leq 2 \| x - X_{(k)} \|.$$

Just note that $S_{X_{(1)}(x), 2\|x - X_{(k)}\|} \supseteq S_{x, \|x - X_{(k)}\|}$ and that the latter sphere contains k data points. But we already know from the proof of weak consistency of the k-NN rule that if $k/n \to 0$, then for all $\epsilon > 0$,

$$\mathbf{P} \left\{ \| X_{(k)}(X) - X \| > \epsilon \right\} \to 0.$$

Finally, we consider condition 6.3 (i). Here we have, arguing partially as in Stone (1977),

$$\mathbf{E} \left\{ \sum_{i=1}^{n} \frac{1}{k} I_{\{X_i \text{ is among the } k \text{ NN's of } X_{(1)}(X)\}} f(X_i) \right\}$$

$$= \mathbf{E} \left\{ \frac{1}{k} \sum_{i=1}^{n} I_{\{X \text{ is among the } k \text{ NN's of } X_{(1)}(X_i) \text{ in } \{X_1, \ldots, X_n, X\} - \{X_i\}\}} f(X) \right\}$$

(reverse the roles of X_i and X)

$$= \mathbf{E} \left\{ \frac{1}{k} \sum_{i=1}^{n} \sum_{j=1}^{n} I_{\{X_j \text{ is the NN of } X_i \text{ in } \{X_1, \ldots, X_n, X\} - \{X_i\}\}} \right.$$

$$\left. \times I_{\{X \text{ is among the } k \text{ NN's of } X_j \text{ in } \{X_1, \ldots, X_n, X\} - \{X_i\}\}} f(X) \right\}.$$

However, by Lemma 11.1,

$$\sum_{i=1}^{n} I_{\{X_j \text{ is the NN of } X_i \text{ in } \{X_1, \ldots, X_n, X\} - \{X_i\}\}} \leq \gamma_d.$$

Also,

$$\sum_{j=1}^{n} I_{\{X \text{ is among the } k \text{ NN's of } X_j \text{ in } \{X_1,\ldots,X_n,X\}-\{X_i\}\}} \le k\gamma_d.$$

Therefore, by a double application of Lemma 11.1,

$$\mathbf{E}\left\{\sum_{i=1}^{n}\frac{1}{k}I_{\{X_i \text{ is among the } k \text{ NN's of } X_{(1)}(X)\}}f(X_i)\right\} \le \gamma_d^2\,\mathbf{E}f(X),$$

and condition 6.3 (i) is verified. $\square$

Problems and Exercises

PROBLEM 11.1. Show that the conditions $k \to \infty$ and $k/n \to 0$ are necessary for universal consistency of the k-nearest neighbor rule. That is, exhibit a distribution such that if k remains bounded, $\liminf_{n\to\infty} \mathbf{E}L_n > L^*$. Exhibit a second distribution such that if $k/n \ge \epsilon > 0$ for all n, and some ϵ, then $\liminf_{n\to\infty} \mathbf{E}L_n > L^*$.

PROBLEM 11.2. Let X be monoatomic, with $\eta < 1/2$. Show that for $\epsilon < 1/2 - \eta$,

$$\mathbf{P}\{L_n - L^* > \epsilon\} \ge e^{-ck}$$

for some $c > 0$, where L_n is the error probability of the k-NN rule with tie-breaking by indices.

PROBLEM 11.3. Prove that the recursive nearest neighbor rule is universally consistent provided that $\lim_{N\to\infty} l_N = \infty$ (Devroye and Wise (1980)). HINT: Check the conditions of Theorem 6.3.

PROBLEM 11.4. Prove that the nearest neighbor rule defined by any L_p-distance measure $0 < p \le \infty$ is universally consistent under the usual conditions on k. The L_p-distance between x, $y \in \mathcal{R}^d$ is defined by $\left(\sum_{i=1}^{d} |x^{(i)} - y^{(i)}|^p\right)^{1/p}$ for $0 < p < \infty$, and by $\sup_i |x^{(i)} - y^{(i)}|$ for $p = \infty$, where $x = (x^{(1)}, \ldots, x^{(d)})$. HINT: Check the conditions of Theorem 6.3.

PROBLEM 11.5. Let $\sigma(x, z, X_i, Z_i) = \rho(x, X_i) + |Z_i - z|$ be a generalized distance between (x, z) and (X_i, Z_i), where $x, X_1, \ldots, X_n$ are as in the description of the scale-invariant k-NN rule, $\rho(x, X_i)$ is the empirical distance defined there, and $z, Z_i \in [0, 1]$ are real numbers added to break ties at random. The sequence $Z_1, \ldots, Z_n$ is i.i.d. uniform $[0, 1]$ and is independent of the data D_n. With the k-NN rule based on the artificial distances σ, show that the rule is universally consistent by verifying the conditions of Theorem 6.3, when $k \to \infty$ and $k/n \to 0$. In particular, show first that if Z is uniform $[0, 1]$ and independent of X, Y, D_n and $Z_1, \ldots, Z_n$, and if $W_{ni}(X, Z)$ is the weight of (X_i, Z_i) in this k-NN rule (i.e., it is $1/k$ iff (X_i, Z_i) is among the k nearest neighbors of (X, Z) according to σ), then

(1) $$\mathbf{E}\left\{\sum_{i=1}^{n} W_{ni}(X, Z)f(X_i)\right\} \le 2^d\mathbf{E}f(X)$$

for all nonnegative measurable f with $\mathbf{E}f(X) < \infty$.

(2) If $k/n \to 0$, then

$$E\left\{ \sum_{i=1}^{n} W_{ni}(X, Z)I_{\{\|X_i - X\| > a\}} \right\}$$

for all $a > 0$.

HINT: Check the conditions of Theorem 6.3 (Devroye (1978)).

PROBLEM 11.6. The *layered nearest neighbor rule* partitions the space at x into 2^d quadrants. In each quadrant, the outer-layer points are marked, that is, those X_i for which the hyperrectangle defined by x and X_i contains no other data point. Then it takes a majority vote over the Y_i's for the marked points. Observe that this rule is scale-invariant. Show that whenever X has nonatomic marginals (to avoid ties), $E\{L_n\} \to L^*$ in probability.

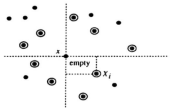

FIGURE 11.3. *The layered nearest neighbor rule takes a majority vote over the marked points.*

HINT: It suffices to show that the number of marked points increases unboundedly in probability, and that its proportion to unmarked points tends to zero in probability.

PROBLEM 11.7. Prove weak universal consistency of the weighted nearest neighbor rule if the weights satisfy

$$\lim_{n \to \infty} \max_{1 \le i \le n} w_{ni} = 0,$$

and

$$\lim_{n \to \infty} \sum_{k \le i \le n} w_{ni} = 0$$

for some k with $k/n \to 0$ (Stone (1977)). HINT: Check the conditions of Theorem 6.3.

PROBLEM 11.8. If $(w_{n1}, \ldots, w_{nn})$ is a probability vector, then $\lim_{n \to \infty} \sum_{i > n\delta} w_{ni} = 0$ for all $\delta > 0$ if and only if there exists a sequence of integers $k = k_n$ such that $k = o(n)$, $k \to \infty$, and $\sum_{i > k} w_{ni} = o(1)$. Show this. Conclude that the conditions of Problem 11.7 are equivalent to

(i) $$\lim_{n \to \infty} \max_{1 \le i \le n} w_{ni} = 0,$$

(ii) $$\lim_{n \to \infty} \sum_{i > n\delta} w_{ni} = 0 \quad \text{for all } \delta > 0.$$

PROBLEM 11.9. Verify the conditions of Problems 11.7 and 11.8 for weight vectors of the form $w_{ni} = c_n/i^\alpha$, where $\alpha > 0$ is a constant and c_n is a normalizing constant. In particular, check that they do not hold for $\alpha > 1$ but that they do hold for $0 < \alpha \le 1$.

PROBLEM 11.10. Consider

$$w_{ni} = \frac{\rho_n}{(1 + \rho_n)^i} \times \frac{1}{1 - (1 + \rho_n)^{-n}}, \quad 1 \le i \le n,$$

as weight vector. Show that the conditions of Problem 11.8 hold if $\rho_n \to 0$ and $n\rho_n \to \infty$.

PROBLEM 11.11. Let $w_{ni} = \mathbf{P}\{Z = i\}/\mathbf{P}\{Z \leq n\}$, $1 \leq i \leq n$, where Z is a Poisson (λ_n) random variable. Show that there is no choice of λ_n such that $\{w_{ni}, 1 \leq i \leq n\}$ is a consistent weight sequence following the conditions of Problem 11.8.

PROBLEM 11.12. Let $w_{ni} = \mathbf{P}\{\text{Binomial}(n, p_n) = i\} = \binom{n}{i}p_n^i(1 - p_n)^{n-i}$. Derive conditions on p_n for this choice of weight sequence to be consistent in the sense of Problem 11.8.

PROBLEM 11.13. k-NN DENSITY ESTIMATION. We recall from Problem 2.11 that if the conditional densities f_0, f_1 exist, then L_1-consistent density estimation leads to a consistent classification rule. Consider now the k-nearest neighbor density estimate introduced by Loftsgaarden and Quesenberry (1965). Let $X_1, \ldots, X_n$ be independent, identically distributed random variables in $\mathcal{R}^d$, with common density f. The k-NN estimate of f is defined by

$$f_n(x) = \frac{k}{\lambda\left(S_{x, \|x - X_{(k)}(x)\|}\right)},$$

where $X_{(k)}(x)$ is the k-th nearest neighbor of x among $X_1, \ldots, X_n$. Show that for every n, $\int |f(x) - f_n(x)|dx = \infty$, so that the density estimate is never consistent in L_1. On the other hand, according to Problem 11.14, the corresponding classification rule is consistent, so read on.

PROBLEM 11.14. Assume that the conditional densities f_0 and f_1 exist. Then we can use a rule suggested by Patrick and Fischer (1970): let $N_0 = \sum_{i=1}^n I_{\{Y_i = 0\}}$ and $N_1 = n - N_0$ be the number of zeros and ones in the training data. Denote by $X_0^{(k)}(x)$ the k-th nearest neighbor of x among the X_i's with $Y_i = 0$. Define $X_1^{(k)}(x)$ similarly. If $\lambda(A)$ denotes the volume of a set $A \subset \mathcal{R}^d$, then the rule is defined as

$$g_n(x) = \begin{cases} 0 & \text{if } N_0/\lambda\left(S_{x, \|x - X_0^{(k)}(x)\|}\right) \geq N_1/\lambda\left(S_{x, \|x - X_1^{(k)}(x)\|}\right) \\ 1 & \text{otherwise.} \end{cases}$$

This estimate is based on the k-nearest neighbor density estimate introduced by Loftsgaarden and Quesenberry (1965). Their estimate of f_0 is

$$\widehat{f}_{0,n}(x) = \frac{k}{\lambda\left(S_{x, \|x - X_0^{(k)}(x)\|}\right)}.$$

Then the rule g_n can be re-written as

$$g_n(x) = \begin{cases} 0 & \text{if } p_{0,n}\widehat{f}_{0,n} \geq p_{1,n}\widehat{f}_{1,n} \\ 1 & \text{otherwise,} \end{cases}$$

where $p_{0,n} = N_0/n$ and $p_{1,n} = N_1/n$ are the obvious estimates of the class probabilities. Show that g_n is weakly consistent if $k \to \infty$ and $k/n \to 0$, whenever the conditional densities exist.

PROBLEM 11.15. Consider a weakly universally consistent rule g_n, and define the rule

$$h_n(x) = \begin{cases} 1 - Y_i & \text{if } x = X_i \text{ and } x \neq X_j, \text{ all } j \neq i \\ g_n(x, D_n) & \text{otherwise.} \end{cases}$$

Show that h_n too is weakly universally consistent. Note: it is the atomic (or partially atomic) distributions of X that make this exercise interesting. HINT: The next exercise may help.

PROBLEM 11.16. Let X have an atomic distribution which puts probability p_i at atom i. Let $X_1, \ldots, X_n$ be an i.i.d. sample drawn from this distribution. Then show the following.

(1) $P\{|N - EN| > \epsilon\} \leq 2e^{-2\epsilon^2/n}$, where $\epsilon > 0$, and N is the number of "occupied" atoms (the number of different values in the data sequence).

(2) $EN/n \to 0$.

(3) $N/n \to 0$ almost surely.

(4) $\sum_{i:X_j \neq i \text{ for all } j \leq n} p_i \to 0$ almost surely.

PROBLEM 11.17. ROYALL'S RULE. Royall (1966) proposes the regression function estimate

$$\eta_n(x) = \frac{1}{nh_n} \sum_{i=1}^{\lfloor nh_n \rfloor} J\left(\frac{i}{nh_n}\right) Y_{(i)}(x),$$

where $J(u)$ is a smooth kernel-like function on $[0, 1]$ with $\int_0^1 J(u)du = 1$ and $h_n > 0$ is a smoothing factor. Suggestions included

(i) $J(u) \equiv 1$;

(ii) $J(u) = \frac{(d+2)^2}{4}\left(1 - \frac{d+4}{d+2}u^{2/d}\right)$

(note that this function becomes negative).

Define

$$g_n(x) = \begin{cases} 1 & \text{if } \eta_n(x) > 1/2 \\ 0 & \text{otherwise.} \end{cases}$$

Assume that $h_n \to 0$, $nh_n \to \infty$. Derive sufficient conditions on J that guarantee the weak universal consistency of Royall's rule. In particular, insure that choice (ii) is weakly universally consistent. HINT: Try adding an appropriate smoothness condition to J.

PROBLEM 11.18. Let K be a kernel and let $R_n(x)$ denote the distance between x and its k-th nearest neighbor, $X_{(k)}(x)$, among $X_1, \ldots, X_n$. The discrimination rule that corresponds to a kernel-type nearest neighbor regression function estimate of Mack (1981) is

$$g_n(x) = \begin{cases} 0 & \text{if } \sum_{i=1}^n (2Y_i - 1)K\left(\frac{x-X_i}{R_n(x)}\right) \leq 0 \\ 1 & \text{otherwise.} \end{cases}$$

(The idea of replacing the smoothing factor in the kernel estimate by a local rank-based value such as $R_n(x)$ is due to Breiman, Meisel, and Purcell (1977).) For the kernel $K = I_{S_{0,1}}$, this rule coincides with the k-NN rule. For regular kernels (see Chapter 10), show that the rule remains weakly universally consistent whenever $k \to \infty$ and $k/n \to 0$ by verifying Stone's conditions of Theorem 6.3.

PROBLEM 11.19. Let $\{g_n\}$ be a weakly universally consistent sequence of classifiers. Split the data sequence D_n into two parts: $D_m = ((X_1, Y_1), \ldots, (X_m, Y_m))$ and $T_{n-m} = ((X_{m+1}, Y_{m+1}), \ldots, (X_n, Y_n))$. Use g_{n-m} and the second part T_{n-m} to relabel the first part, i.e., define $Y_i' = g_{n-m}(X_i, T_{n-m})$ for $i = 1, \ldots, m$. Prove that the 1-NN rule based on the data (X_1, Y_1'), $\ldots, (X_m, Y_m')$ is weakly universally consistent whenever $m \to \infty$ and $n - m \to \infty$. HINT: Use Problem 5.40.

PROBLEM 11.20. Consider the k-NN rule with a fixed k as the ancestral rule, and apply the 1-NN rule using the relabeled data. Investigate the convergence of EL_n. Is the limit L_{kNN} or something else?

12
Vapnik-Chervonenkis Theory

12.1 Empirical Error Minimization

In this chapter we select a decision rule from a class of rules with the help of training data. Working formally, let C be a class of functions $\phi : \mathcal{R}^d \to \{0, 1\}$. One wishes to select a function from C with small error probability. Assume that the training data $D_n = ((X_1, Y_1), \ldots, (X_n, Y_n))$ are given to pick one of the functions from C to be used as a classifier. Perhaps the most natural way of selecting a function is to minimize the *empirical error probability*

$$\widehat{L}_n(\phi) = \frac{1}{n} \sum_{i=1}^{n} I_{\{\phi(X_i) \neq Y_i\}}$$

over the class C. Denote the empirically optimal rule by ϕ_n^*:

$$\phi_n^* = \arg\min_{\phi \in C} \widehat{L}_n(\phi).$$

Thus, ϕ_n^* is the classifier that, according to the data D_n, "looks best" among the classifiers in C. This idea of minimizing the empirical risk in the construction of a rule was developed to great extent by Vapnik and Chervonenkis (1971; 1974c; 1974a; 1974b).

Intuitively, the selected classifier ϕ_n^* should be good in the sense that its *true* error probability $L(\phi_n^*) = \mathbf{P}\{\phi_n^*(X) \neq Y | D_n\}$ is expected to be close to the optimal error probability within the class. Their difference is the quantity that primarily interests us in this chapter:

$$L(\phi_n^*) - \inf_{\phi \in C} L(\phi).$$

The latter difference may be bounded in a distribution-free manner, and a rate of convergence results that only depends on the structure of C. While this is very exciting, we must add that $L(\phi_n^*)$ may be far away from the Bayes error L^*. Note that

$$L(\phi_n^*) - L^* = \left(L(\phi_n^*) - \inf_{\phi \in C} L(\phi) \right) + \left(\inf_{\phi \in C} L(\phi) - L^* \right).$$

The size of C is a compromise: when C is large, $\inf_{\phi \in C} L(\phi)$ may be close to L^*, but the former error, the *estimation error*, is probably large as well. If C is too small, there is no hope to make the *approximation error* $\inf_{\phi \in C} L(\phi) - L^*$ small. For example, if C is the class of all (measurable) decision functions, then we can always find a classifier in C with zero empirical error, but it may have arbitrary values outside of the points $X_1, \ldots, X_n$. For example, an empirically optimal classifier is

$$\phi_n^*(x) = \begin{cases} Y_i & \text{if } x = X_i, i = 1, \ldots, n \\ 0 & \text{otherwise.} \end{cases}$$

This is clearly not what we are looking for. This phenomenon is called *overfitting*, as the overly large class C overfits the data. We will give precise conditions on C that allow us to avoid this anomaly. The choice of C such that $\inf_{\phi \in C} L(\phi)$ is close to L^* has been the subject of various chapters on consistency—just assume that C is allowed to grow with n in some manner. Here we take the point of view that C is fixed, and that we have to live with the functions in C. The best we may then hope for is to minimize $L(\phi_n^*) - \inf_{\phi \in C} L(\phi)$. A typical situation is shown in Figure 12.1.

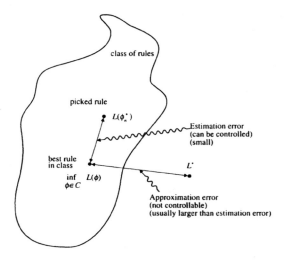

FIGURE 12.1. *Various errors in empirical classifier selection.*

Consider first a finite collection $\mathcal{C}$, and assume that one of the classifiers in $\mathcal{C}$ has zero error probability, that is, $\min_{\phi \in \mathcal{C}} L(\phi) = 0$. Then clearly, $\widehat{L}_n(\phi_n^*) = 0$ with probability one. We then have the following performance bound:

Theorem 12.1. (VAPNIK AND CHERVONENKIS (1974c)). *Assume* $|\mathcal{C}| < \infty$ *and* $\min_{\phi \in \mathcal{C}} L(\phi) = 0$. *Then for every n and* $\epsilon > 0$,

$$\mathbf{P}\{L(\phi_n^*) > \epsilon\} \leq |\mathcal{C}| e^{-n\epsilon},$$

and

$$\mathbf{E}\{L(\phi_n^*)\} \leq \frac{1 + \log |\mathcal{C}|}{n}.$$

PROOF. Clearly,

$$
\begin{aligned}
\mathbf{P}\{L(\phi_n^*) > \epsilon\} &\leq \mathbf{P}\left\{\max_{\phi \in \mathcal{C}: \widehat{L}_n(\phi)=0} L(\phi) > \epsilon\right\} \\
&= \mathbf{E}\left\{I_{\left\{\max_{\phi \in \mathcal{C}: \widehat{L}_n(\phi)=0} L(\phi) > \epsilon\right\}}\right\} \\
&= \mathbf{E}\left\{\max_{\phi \in \mathcal{C}} I_{\{\widehat{L}_n(\phi)=0\}} I_{\{L(\phi) > \epsilon\}}\right\} \\
&\leq \sum_{\phi \in \mathcal{C}: L(\phi) > \epsilon} \mathbf{P}\left\{\widehat{L}_n(\phi) = 0\right\} \\
&\leq |\mathcal{C}|(1 - \epsilon)^n,
\end{aligned}
$$

since the probability that no (X_i, Y_i) pair falls in the set $\{(x, y) : \phi(x) \neq y\}$ is less than $(1 - \epsilon)^n$ if the probability of the set is larger than ϵ. The probability inequality of the theorem follows from the simple inequality $1 - x \leq e^{-x}$.

To bound the expected error probability, note that for any $u > 0$,

$$
\begin{aligned}
\mathbf{E}\{L(\phi_n^*)\} &= \int_0^\infty \mathbf{P}\{L(\phi_n^*) > t\} dt \\
&\leq u + \int_u^\infty \mathbf{P}\{L(\phi_n^*) > t\} dt \\
&\leq u + |\mathcal{C}| \int_u^\infty e^{-nt} dt \\
&= u + \frac{|\mathcal{C}|}{n} e^{-nu}.
\end{aligned}
$$

Since u was arbitrary, we may choose it to minimize the obtained upper bound. The optimal choice is $u = \log |\mathcal{C}|/n$, which yields the desired inequality. $\square$

Theorem 12.1 shows that empirical selection works very well if the sample size n is much larger than the logarithm of the size of the family $\mathcal{C}$. Unfortunately, the assumption on the distribution of (X, Y), that is, that $\min_{\phi \in C} L(\phi) = 0$, is very restrictive. In the sequel we drop this assumption, and deal with the distribution-free problem.

One of our main tools is taken from Lemma 8.2:

$$L(\phi_n^*) - \inf_{\phi \in C} L(\phi) \leq 2 \sup_{\phi \in C} \left| \widehat{L}_n(\phi) - L(\phi) \right|.$$

This leads to the study of *uniform* deviations of relative frequencies from their probabilities by the following simple observation: let ν be a probability measure of (X, Y) on $\mathcal{R}^d \times \{0, 1\}$, and let ν_n be the empirical measure based upon D_n. That is, for any fixed measurable set $A \subset \mathcal{R}^d \times \{0, 1\}$, $\nu(A) = \mathbf{P}\{(X, Y) \in A\}$, and $\nu_n(A) = \frac{1}{n} \sum_{i=1}^{n} I_{\{(X_i, Y_i) \in A\}}$. Then

$$L(\phi) = \nu(\{(x, y) : \phi(x) \neq y\})$$

is just the ν-measure of the set of pairs $(x, y) \in \mathcal{R}^d \times \{0, 1\}$, where $\phi(x) \neq y$. Formally, $L(\phi)$ is the ν-measure of the set

$$\{\{x : \phi(x) = 1\} \times \{0\}\} \bigcup \{\{x : \phi(x) = 0\} \times \{1\}\}.$$

Similarly, $\widehat{L}_n(\phi) = \nu_n(\{(x, y) : \phi(x) \neq y\})$. Thus,

$$\sup_{\phi \in C} |\widehat{L}_n(\phi) - L(\phi)| = \sup_{A \in \mathcal{A}} |\nu_n(A) - \nu(A)|,$$

where $\mathcal{A}$ is the collection of all sets

$$\{\{x : \phi(x) = 1\} \times \{0\}\} \bigcup \{\{x : \phi(x) = 0\} \times \{1\}\}, \quad \phi \in \mathcal{C}.$$

For a fixed set A, for any probability measure ν, by the law of large numbers $\nu_n(A) - \nu(A) \to 0$ almost surely as $n \to \infty$. Moreover, by Hoeffding's inequality (Theorem 8.1),

$$\mathbf{P}\{|\nu_n(A) - \nu(A)| > \epsilon\} \leq 2e^{-2n\epsilon^2}.$$

However, it is a much harder problem to obtain such results for $\sup_{A \in \mathcal{A}} |\nu_n(A) - \nu(A)|$. If the class of sets $\mathcal{A}$ (or, analogously, in the pattern recognition context, $\mathcal{C}$) is of finite cardinality, then the union bound trivially gives

$$\mathbf{P} \left\{ \sup_{A \in \mathcal{A}} |\nu_n(A) - \nu(A)| > \epsilon \right\} \leq 2|\mathcal{A}|e^{-2n\epsilon^2}.$$

However, if $\mathcal{A}$ contains infinitely many sets (as in many of the interesting cases) then the problem becomes nontrivial, spawning a vast literature. The most powerful weapons to attack these problems are distribution-free large deviation-type inequalities first proved by Vapnik and Chervonenkis (1971) in their pioneering work. However, in some situations, we can handle the problem in a much simpler way. We have already seen such an example in Section 4.5.

12.2   Fingering

Recall that in Section 4.5 we studied a specific rule that selects a linear classifier by minimizing the empirical error. The performance bounds provided by Theorems 4.5 and 4.6 show that the selected rule performs very closely to the best possible linear rule. These bounds apply only to the specific algorithm used to find the empirical minima—we have not showed that any classifier minimizing the empirical error performs well. This matter will be dealt with in later sections. In this section we extend Theorems 4.5 and 4.6 to classes other than linear classifiers.

Let C be the class of classifiers assigning 1 to those x contained in a closed hyperrectangle, and 0 to all other points. Then a classifier minimizing the empirical error $\widehat{L}_n(\phi)$ over all $\phi \in C$ may be obtained by the following algorithm: to each $2d$-tuple $(X_{i_1}, \ldots, X_{i_{2d}})$ of points from $X_1, \ldots, X_n$, assign the smallest hyperrectangle containing these points. If we assume that X has a density, then the points $X_1, \ldots, X_n$ are in general position with probability one. This way we obtain at most $\binom{n}{2d}$ sets. Let ϕ_i be the classifier corresponding to the i-th such hyperrectangle, that is, the one assigning 1 to those x contained in the hyperrectangle, and 0 to other points. Clearly, for each $\phi \in C$, there exists a ϕ_i, $i = 1, \ldots, \binom{n}{2d}$, such that

$$\phi(X_j) = \phi_i(X_j)$$

for all X_j, except possibly for those on the boundary of the hyperrectangle. Since the points are in general position, there are at most $2d$ such exceptional points. Therefore, if we select a classifier $\widehat{\phi}$ among $\phi_1, \ldots, \phi_{\binom{n}{2d}}$ to minimize the empirical error, then it approximately minimizes the empirical error over the whole class C as well. A quick scan through the proof of Theorem 4.5 reveals that by similar arguments we may obtain the performance bound

$$\mathbf{P}\left\{ L(\widehat{\phi}) - \inf_{\phi \in C} L(\phi) > \epsilon \right\} \le e^{4d\epsilon}\left(\binom{n}{2d} + 1 \right) e^{-n\epsilon^2/2},$$

for $n \ge 2d$ and $\epsilon \ge 4d/n$.

The idea may be generalized. It always works, if for some k, k-tuples of points determine classifiers from C such that no matter where the other data points fall, the minimal empirical error over these sets coincides with the overall minimum. Then we may fix these sets—"put our finger on them"—and look for the empirical minima over this finite collection. The next theorems, whose proofs are left as an exercise (Problem 12.2), show that if C has this property, then "fingering" works extremely well whenever $n \gg k$.

Theorem 12.2. *Assume that the class C of classifiers has the following property: for some integer k there exists a function $\Psi : \left(\mathcal{R}^d \right)^k \to C$ such that for all $x_1, \ldots, x_n \in \mathcal{R}^d$ and all $\phi \in C$, there exists a k-tuple $i_1, \ldots, i_k \in \{1, \ldots, n\}$ of different indices such that*

$$\Psi(x_{i_1}, \ldots, x_{i_k})(x_j) = \phi(x_j) \quad \text{for all } j = 1, \ldots, n \text{ with } j \ne i_l, l = 1, \ldots, k$$

with probability one. Let $\widehat{\phi}$ be found by fingering, that is, by empirical error mini-mization over the collection of $n!/(n-k)!$ classifiers of the form

$$\Psi(X_{i_1}, \ldots, X_{i_k}), \qquad i_1, \ldots, i_k \in \{1, \ldots, n\}, \qquad different.$$

Then for $n \geq k$ and $2k/n \leq \epsilon \leq 1$,

$$\mathbf{P}\left\{L(\widehat{\phi}) - \inf_{\phi \in C} L(\phi) > \epsilon\right\} \leq e^{2k\epsilon}\left(n^k + 1\right)e^{-n\epsilon^2/2}.$$

Moreover, if $n \geq k$, then

$$\mathbf{E}\left\{L(\widehat{\phi}) - \inf_{\phi \in C} L(\phi)\right\} \leq \sqrt{\frac{2(k+1)\log n + (2k+2)}{n}}.$$

The smallest k for which C has the property described in the theorem may be called the *fingering dimension* of C. In most interesting cases, it is independent of n. Problem 12.3 offers a few such classes. We will see later in this chapter that the fingering dimension is closely related the so-called VC dimension of C (see also Problem 12.4).

Again, we get much smaller errors if $\inf_{\phi \in C} L(\phi) = 0$. The next inequality generalizes Theorem 4.6.

Theorem 12.3. *Assume that C has the property described in Theorem 12.2 with fingering dimension k. Assume, in addition, that $\inf_{\phi \in C} L(\phi) = 0$. Then for all n and ϵ,*

$$\mathbf{P}\left\{L(\widehat{\phi}) > \epsilon\right\} \leq n^k e^{-(n-k)\epsilon},$$

and

$$\mathbf{E}\left\{L(\widehat{\phi})\right\} \leq \frac{k\log n + 2}{n-k}.$$

REMARK. Even though the results in the next few sections based on the Vapnik-Chervonenkis theory supersede those of this section (by requiring less from the class C and being able to bound the error of any classifier minimizing the empirical risk), we must remark that the exponents in the above probability inequalities are the best possible, and bounds of the same type for the general case can only be obtained with significantly more effort. □

12.3   The Glivenko-Cantelli Theorem

In the next two sections, we prove the Vapnik-Chervonenkis inequality, a powerful generalization of the classical Glivenko-Cantelli theorem. It provides upper bounds on random variables of the type

$$\sup_{A \in \mathcal{A}} |\nu_n(A) - \nu(A)|.$$

As we noted in Section 12.1, such bounds yield performance bounds for any classifier selected by minimizing the empirical error. To make the material more digestible, we first present the main ideas in a simple one-dimensional setting, and then prove the general theorem in the next section.

We drop the pattern recognition setting momentarily, and return to probability theory. The following theorem is sometimes referred to as the fundamental theorem of mathematical statistics, stating uniform almost sure convergence of the empirical distribution function to the true one:

Theorem 12.4. (GLIVENKO-CANTELLI THEOREM). *Let* $Z_1, \ldots, Z_n$ *be i.i.d. real-valued random variab'$_\bullet$' with distribution function* $F(z) = P\{Z_1 \leq z\}$. *Denote the standard empirical distribution function by*

$$F_n(z) = \frac{1}{n} \sum_{i=1}^{n} I_{\{Z_i \leq z\}}.$$

Then

$$P\left\{\sup_{z \in \mathcal{R}} |F(z) - F_n(z)| > \epsilon\right\} \leq 8(n+1)e^{-n\epsilon^2/32},$$

and, in particular, by the Borel-Cantelli lemma,

$$\lim_{n \to \infty} \sup_{z \in \mathcal{R}} |F(z) - F_n(z)| = 0 \quad \text{with probability one.}$$

PROOF. The proof presented here is not the simplest possible, but it contains the main ideas leading to a powerful generalization. Introduce the notation $\nu(A) = P\{Z_1 \in A\}$ and $\nu_n(A) = (1/n) \sum_{j=1}^{n} I_{\{Z_j \in A\}}$ for all measurable sets $A \subset \mathcal{R}$. Let $\mathcal{A}$ denote the class of sets of form $(-\infty, z]$ for $z \in \mathcal{R}$. With these notations,

$$\sup_{z \in \mathcal{R}} |F(z) - F_n(z)| = \sup_{A \in \mathcal{A}} |\nu_n(A) - \nu(A)|.$$

We prove the theorem in several steps, following symmetrization ideas of Dudley (1978), and Pollard (1984). We assume that $n\epsilon^2 \geq 2$, since otherwise the bound is trivial. In the first step we introduce a symmetrization.

STEP 1. FIRST SYMMETRIZATION BY A GHOST SAMPLE. Define the random variables $Z_1', \ldots, Z_n' \in \mathcal{R}$ such that $Z_1, \ldots, Z_n, Z_1', \ldots, Z_n'$ are all independent and identically distributed. Denote by ν_n' the empirical measure corresponding to the new sample:

$$\nu_n'(A) = \frac{1}{n} \sum_{i=1}^{n} I_{\{Z_i' \in A\}}.$$

Then for $n\epsilon^2 \geq 2$ we have

$$P\left\{\sup_{A \in \mathcal{A}} |\nu_n(A) - \nu(A)| > \epsilon\right\} \leq 2P\left\{\sup_{A \in \mathcal{A}} |\nu_n(A) - \nu_n'(A)| > \frac{\epsilon}{2}\right\}.$$

To see this, let $A^* \in \mathcal{A}$ be a set for which $|v_n(A^*) - v(A^*)| > \epsilon$ if such a set exists, and let A^* be a fixed set in $\mathcal{A}$ otherwise. Then

$$
\begin{aligned}
\mathbf{P} &\left\{ \sup_{A \in \mathcal{A}} |v_n(A) - v_n'(A)| > \epsilon/2 \right\} \\
&\geq \quad \mathbf{P}\left\{ |v_n(A^*) - v_n'(A^*)| > \epsilon/2 \right\} \\
&\geq \quad \mathbf{P}\left\{ |v_n(A^*) - v(A^*)| > \epsilon, \, |v_n'(A^*) - v(A^*)| < \frac{\epsilon}{2} \right\} \\
&= \quad \mathbf{E}\left\{ I_{\{|v_n(A^*)-v(A^*)|>\epsilon\}} \mathbf{P}\left\{ |v_n'(A^*) - v(A^*)| < \frac{\epsilon}{2} \middle| Z_1, \ldots, Z_n \right\} \right\}.
\end{aligned}
$$

The conditional probability inside may be bounded by Chebyshev's inequality as follows:

$$
\begin{aligned}
\mathbf{P}\left\{ |v_n'(A^*) - v(A^*)| < \frac{\epsilon}{2} \middle| Z_1, \ldots, Z_n \right\} &\geq \quad 1 - \frac{v(A^*)(1 - v(A^*))}{n\epsilon^2/4} \\
&\geq \quad 1 - \frac{1}{n\epsilon^2} \geq \frac{1}{2}
\end{aligned}
$$

whenever $n\epsilon^2 \geq 2$. In summary,

$$
\begin{aligned}
\mathbf{P}\left\{ \sup_{A \in \mathcal{A}} |v_n(A) - v_n'(A)| > \epsilon/2 \right\} &\geq \quad \frac{1}{2}\mathbf{P}\{|v_n(A^*) - v(A^*)| > \epsilon\} \\
&\geq \quad \frac{1}{2}\mathbf{P}\left\{ \sup_{A \in \mathcal{A}} |v_n(A) - v(A)| > \epsilon \right\}.
\end{aligned}
$$

STEP 2. SECOND SYMMETRIZATION BY RANDOM SIGNS. Let $\sigma_1, \ldots, \sigma_n$ be i.i.d. sign variables, independent of $Z_1, \ldots, Z_n$ and $Z_1', \ldots, Z_n'$, with $\mathbf{P}\{\sigma_i = -1\} = \mathbf{P}\{\sigma_i = 1\} = 1/2$. Clearly, because $Z_1, Z_1', \ldots, Z_n, Z_n'$ are all independent and identically distributed, the distribution of

$$
\sup_{A \in \mathcal{A}} \left| \sum_{i=1}^{n} (I_A(Z_i) - I_A(Z_i')) \right|
$$

is the same as the distribution of

$$
\sup_{A \in \mathcal{A}} \left| \sum_{i=1}^{n} \sigma_i(I_A(Z_i) - I_A(Z_i')) \right|.
$$

Thus, by Step 1,

$$
\begin{aligned}
\mathbf{P}&\left\{ \sup_{A \in \mathcal{A}} |v_n(A) - v(A)| > \epsilon \right\} \\
&\leq \quad 2\mathbf{P}\left\{ \sup_{A \in \mathcal{A}} \frac{1}{n} \left| \sum_{i=1}^{n} (I_A(Z_i) - I_A(Z_i')) \right| > \frac{\epsilon}{2} \right\} \\
&= \quad 2\mathbf{P}\left\{ \sup_{A \in \mathcal{A}} \frac{1}{n} \left| \sum_{i=1}^{n} \sigma_i(I_A(Z_i) - I_A(Z_i')) \right| > \frac{\epsilon}{2} \right\}.
\end{aligned}
$$

Simply applying the union bound, we can remove the auxiliary random variables $Z'_1, \ldots, Z'_n$:

$$
\mathbf{P}\left\{\sup_{A \in \mathcal{A}} \frac{1}{n}\left|\sum_{i=1}^{n} \sigma_i(I_A(Z_i) - I_A(Z'_i))\right| > \frac{\epsilon}{2}\right\}
$$

$$
\leq \quad \mathbf{P}\left\{\sup_{A \in \mathcal{A}} \frac{1}{n}\left|\sum_{i=1}^{n} \sigma_i I_A(Z_i)\right| > \frac{\epsilon}{4}\right\} + \mathbf{P}\left\{\sup_{A \in \mathcal{A}} \frac{1}{n}\left|\sum_{i=1}^{n} \sigma_i I_A(Z'_i)\right| > \frac{\epsilon}{4}\right\}
$$

$$
= \quad 2\mathbf{P}\left\{\sup_{A \in \mathcal{A}} \frac{1}{n}\left|\sum_{i=1}^{n} \sigma_i I_A(Z_i)\right| > \frac{\epsilon}{4}\right\}.
$$

STEP 3. CONDITIONING.   To bound the probability

$$
\mathbf{P}\left\{\sup_{A \in \mathcal{A}} \frac{1}{n}\left|\sum_{i=1}^{n} \sigma_i I_A(Z_i)\right| > \frac{\epsilon}{4}\right\} = \mathbf{P}\left\{\sup_{z \in \mathcal{R}} \frac{1}{n}\left|\sum_{i=1}^{n} \sigma_i I_{\{Z_i \leq z\}}\right| > \frac{\epsilon}{4}\right\},
$$

we condition on $Z_1, \ldots, Z_n$. Fix $z_1, \ldots, z_n \in \mathcal{R}^d$, and note that as z ranges over $\mathcal{R}$, the number of different vectors $\left(I_{\{z_1 \leq z\}}, \ldots, I_{\{z_n \leq z\}}\right)$ is at most $n + 1$. Thus, conditional on $Z_1, \ldots, Z_n$, the supremum in the probability above is just a maximum taken over at most $n + 1$ random variables. Thus, applying the union bound gives

$$
\mathbf{P}\left\{\sup_{A \in \mathcal{A}} \frac{1}{n}\left|\sum_{i=1}^{n} \sigma_i I_A(Z_i)\right| > \frac{\epsilon}{4}\middle| Z_1, \ldots, Z_n\right\}
$$

$$
\leq \quad (n + 1)\sup_{A \in \mathcal{A}} \mathbf{P}\left\{\frac{1}{n}\left|\sum_{i=1}^{n} \sigma_i I_A(Z_i)\right| > \frac{\epsilon}{4}\middle| Z_1, \ldots, Z_n\right\}.
$$

With the supremum now outside the probability, it suffices to find an exponential bound on the conditional probability

$$
\mathbf{P}\left\{\frac{1}{n}\left|\sum_{i=1}^{n} \sigma_i I_A(Z_i)\right| > \frac{\epsilon}{4}\middle| Z_1, \ldots, Z_n\right\}.
$$

STEP 4. HOEFFDING'S INEQUALITY.   With $z_1, \ldots, z_n$ fixed, $\sum_{i=1}^{n} \sigma_i I_A(z_i)$ is the sum of n independent zero mean random variables bounded between -1 and 1. Therefore, Theorem 8.1 applies in a straightforward manner:

$$
\mathbf{P}\left\{\frac{1}{n}\left|\sum_{i=1}^{n} \sigma_i I_A(Z_i)\right| > \frac{\epsilon}{4}\middle| Z_1, \ldots, Z_n\right\} \leq 2e^{-n\epsilon^2/32}.
$$

Thus,

$$
\mathbf{P}\left\{\sup_{A \in \mathcal{A}} \frac{1}{n}\left|\sum_{i=1}^{n} \sigma_i I_A(Z_i)\right| > \frac{\epsilon}{4}\middle| Z_1, \ldots, Z_n\right\} \leq 2(n + 1)e^{-n\epsilon^2/32}.
$$

Taking the expected value on both sides we have

$$\mathbf{P}\left\{\sup_{A\in\mathcal{A}}\frac{1}{n}\left|\sum_{i=1}^{n}\sigma_i I_A(Z_i)\right| > \frac{\epsilon}{4}\right\} \le 2(n+1)e^{-n\epsilon^2/32}.$$

In summary,

$$\mathbf{P}\left\{\sup_{A\in\mathcal{A}}|\nu_n(A) - \nu(A)| > \epsilon\right\} \le 8(n+1)e^{-n\epsilon^2/32}. \quad \square$$

12.4   Uniform Deviations of Relative Frequencies from Probabilities

In this section we prove the Vapnik-Chervonenkis inequality, a mighty generalization of Theorem 12.4. In the proof we need only a slight adjustment of the proof above. In the general setting, let the independent identically distributed random variables $Z_1, \ldots, Z_n$ take their values from $\mathcal{R}^d$. Again, we use the notation $\nu(A) = \mathbf{P}\{Z_1 \in A\}$ and $\nu_n(A) = (1/n)\sum_{j=1}^{n} I_{\{Z_j \in A\}}$ for all measurable sets $A \subset \mathcal{R}^d$. The Vapnik-Chervonenkis theory begins with the concepts of *shatter coefficient* and *Vapnik-Chervonenkis (or VC) dimension*:

DEFINITION 12.1. *Let $\mathcal{A}$ be a collection of measurable sets. For $(z_1, \ldots, z_n) \in \{\mathcal{R}^d\}^n$, let $N_\mathcal{A}(z_1, \ldots, z_n)$ be the number of different sets in*

$$\{\{z_1, \ldots, z_n\} \cap A; A \in \mathcal{A}\}.$$

The n-th shatter coefficient of $\mathcal{A}$ is

$$s(\mathcal{A}, n) = \max_{(z_1,\ldots,z_n)\in\{\mathcal{R}^d\}^n} N_\mathcal{A}(z_1, \ldots, z_n).$$

That is, the shatter coefficient is the maximal number of different subsets of n points that can be picked out by the class of sets $\mathcal{A}$.

The shatter coefficients measure the richness of the class $\mathcal{A}$. Clearly, $s(\mathcal{A}, n) \le 2^n$, as there are 2^n subsets of a set with n elements. If $N_\mathcal{A}(z_1, \ldots, z_n) = 2^n$ for some $(z_1, \ldots, z_n)$, then we say that $\mathcal{A}$ shatters $\{z_1, \ldots, z_n\}$. If $s(\mathcal{A}, n) < 2^n$, then any set of n points has a subset such that there is no set in $\mathcal{A}$ that contains exactly that subset of the n points. Clearly, if $s(\mathcal{A}, k) < 2^k$ for some integer k, then $s(\mathcal{A}, n) < 2^n$ for all $n > k$. The first time when this happens is important:

DEFINITION 12.2. *Let $\mathcal{A}$ be a collection of sets with $|\mathcal{A}| \ge 2$. The largest integer $k \ge 1$ for which $s(\mathcal{A}, k) = 2^k$ is denoted by $V_\mathcal{A}$, and it is called the Vapnik-Chervonenkis dimension (or VC dimension) of the class $\mathcal{A}$. If $s(\mathcal{A}, n) = 2^n$ for all n, then by definition, $V_\mathcal{A} = \infty$.*

For example, if A contains all halflines of form $(-\infty, x]$, $x \in \mathcal{R}$, then $s(A, 2) = 3 < 2^2$, and $V_A = 1$. This is easily seen by observing that for any two different points $z_1 < z_2$ there is no set of the form $(-\infty, x]$ that contains z_2, but not z_1. A class of sets A for which $V_A < \infty$ is called a Vapnik-Chervonenkis (or VC) class. In a sense, V_A may be considered as the complexity, or size, of A. Several properties of the shatter coefficients and the VC dimension will be shown in Chapter 13. The main purpose of this section is to prove the following important result by Vapnik and Chervonenkis (1971):

Theorem 12.5. (VAPNIK AND CHERVONENKIS (1971)). *For any probability measure v and class of sets A, and for any n and $\epsilon > 0$,*

$$\mathbf{P}\left\{\sup_{A \in A} |v_n(A) - v(A)| > \epsilon\right\} \leq 8s(A, n)e^{-n\epsilon^2/32}.$$

PROOF. The proof parallels that of Theorem 12.4. We may again assume that $n\epsilon^2 \geq 2$. In the first two steps we prove that

$$\mathbf{P}\left\{\sup_{A \in A} |v_n(A) - v(A)| > \epsilon\right\} \leq 4\mathbf{P}\left\{\sup_{A \in A} \frac{1}{n}\left|\sum_{i=1}^{n} \sigma_i I_A(Z_i)\right| > \frac{\epsilon}{4}\right\}.$$

This may be done exactly the same way as in Theorem 12.4; we do not repeat the argument. The only difference appears in Step 3:

STEP 3. CONDITIONING. To bound the probability

$$\mathbf{P}\left\{\sup_{A \in A} \frac{1}{n}\left|\sum_{i=1}^{n} \sigma_i I_A(Z_i)\right| > \frac{\epsilon}{4}\right\},$$

again we condition on $Z_1, \ldots, Z_n$. Fix $z_1, \ldots, z_n \in \mathcal{R}^d$, and observe that as A ranges over A, the number of different vectors $(I_A(z_1), \ldots, I_A(z_n))$ is just the number of different subsets of $\{z_1, \ldots, z_n\}$ produced by intersecting it with sets in A, which, by definition, cannot exceed $s(A, n)$. Therefore, with $Z_1, \ldots, Z_n$ fixed, the supremum in the above probability is a maximum of at most $N_A(Z_1, \ldots, Z_n)$ random variables. This number, by definition, is bounded from above by $s(A, n)$. By the union bound we get

$$\mathbf{P}\left\{\sup_{A \in A} \frac{1}{n}\left|\sum_{i=1}^{n} \sigma_i I_A(Z_i)\right| > \frac{\epsilon}{4}\bigg| Z_1, \ldots, Z_n\right\}$$

$$\leq \quad s(A, n) \sup_{A \in A} \mathbf{P}\left\{\frac{1}{n}\left|\sum_{i=1}^{n} \sigma_i I_A(Z_i)\right| > \frac{\epsilon}{4}\bigg| Z_1, \ldots, Z_n\right\}.$$

Therefore, as before, it suffices to bound the conditional probability

$$\mathbf{P}\left\{\frac{1}{n}\left|\sum_{i=1}^{n} \sigma_i I_A(Z_i)\right| > \frac{\epsilon}{4}\bigg| Z_1, \ldots, Z_n\right\}.$$

This may be done by Hoeffding's inequality exactly as in Step 4 of the proof of Theorem 12.4. Finally, we obtain

$$\mathbf{P}\left\{\sup_{A\in\mathcal{A}}|\nu_n(A) - \nu(A)| > \epsilon\right\} \le 8s(\mathcal{A}, n)e^{-n\epsilon^2/32}. \quad \Box$$

The bound of Theorem 12.5 is useful when the shatter coefficients do not increase too quickly with n. For example, if $\mathcal{A}$ contains all Borel sets of $\mathcal{R}^d$, then we can shatter any collection of n different points at will, and obtain $s(\mathcal{A}, n) = 2^n$. This would be useless, of course. The smaller $\mathcal{A}$, the smaller the shatter coefficient is. To apply the VC bound, it suffices to compute shatter coefficients for certain families of sets. Examples may be found in Cover (1965), Vapnik and Chervonenkis (1971), Devroye and Wagner (1979a), Feinholz (1979), Devroye (1982a), Massart (1983), Dudley (1984), Simon (1991), and Stengle and Yukich (1989). This list of references is far from exhaustive. More information about shatter coefficients is given in Chapter 13.

REMARK. MEASURABILITY. The supremum in Theorem 12.5 is not always measurable. Measurability must be verified for every family $\mathcal{A}$. For all our examples, the quantities are indeed measurable. For more on the measurability question, see Dudley (1978; 1984), Massart (1983), and Gaenssler (1983). Giné and Zinn (1984) and Yukich (1985) provide further work on suprema of the type shown in Theorem 12.5. $\Box$

REMARK. OPTIMAL EXPONENT. For the sake of readability we followed the line of Pollard's proof (1984) instead of the original by Vapnik and Chervonenkis. In particular, the exponent $-n\epsilon^2/32$ in Theorem 12.5 is worse than the $-n\epsilon^2/8$ established in the original paper. The best known exponents together with some other related results are mentioned in Section 12.8. The basic ideas of the original proof by Vapnik and Chervonenkis appear in the proof of Theorem 12.7 below. $\Box$

REMARK. NECESSARY AND SUFFICIENT CONDITIONS. It is clear from the proof of the theorem that it can be strengthened to

$$\mathbf{P}\left\{\sup_{A\in\mathcal{A}}|\nu_n(A) - \nu(A)| > \epsilon\right\} \le 8\mathbf{E}\left\{N_{\mathcal{A}}(Z_1, \dots, Z_n)\right\}e^{-n\epsilon^2/32},$$

where $Z_1, \dots, Z_n$ are i.i.d. random variables with probability measure ν. Although this upper bound is tighter than that in the stated inequality, it is usually more difficult to handle, since the coefficient in front of the exponential term depends on the distribution of Z_1, while $s(\mathcal{A}, n)$ is purely combinatorial in nature. However, this form is important in a different setting: we say that the *uniform law of large numbers* holds if

$$\sup_{A\in\mathcal{A}}|\nu_n(A) - \nu(A)| \to 0 \quad \text{in probability.}$$

It follows from this form of Theorem 12.5 that the uniform law of large numbers holds if

$$\frac{\mathbf{E}\{\log(N_A(Z_1, \ldots, Z_n))\}}{n} \to 0.$$

Vapnik and Chervonenkis showed (1971; 1981) that this condition is also *necessary* for the uniform law of large numbers. Another characterization of the uniform law of large numbers is given by Talagrand (1987), who showed that the uniform law of large numbers holds if and only if there does not exist a set $A \subset \mathcal{R}^d$ with $\nu(A) > 0$ such that, with probability one, the set $\{Z_1, \ldots, Z_n\} \cap A$ is shattered by $\mathcal{A}$. □

12.5   Classifier Selection

The following theorem relates the results of the previous sections to empirical classifier selection, that is, when the empirical error probability $\widehat{L}_n(\phi)$ is minimized over a class of classifiers $\mathcal{C}$. We emphasize that unlike in Section 12.2, here we allow *any* classifier with the property that has minimal empirical error $\widehat{L}_n(\phi)$ in $\mathcal{C}$. First introduce the shatter coefficients and VC dimension of $\mathcal{C}$:

DEFINITION 12.3. *Let $\mathcal{C}$ be a class of decision functions of the form $\phi : \mathcal{R}^d \to \{0, 1\}$. Define $\mathcal{A}$ as the collection of all sets*

$$\{\{x : \phi(x) = 1\} \times \{0\}\} \bigcup \{\{x : \phi(x) = 0\} \times \{1\}\}, \quad \phi \in \mathcal{C}.$$

Define the n-th shatter coefficient $S(\mathcal{C}, n)$ of the class of classifiers $\mathcal{C}$ as

$$S(\mathcal{C}, n) = s(\mathcal{A}, n).$$

Furthermore, define the VC dimension $V_{\mathcal{C}}$ of $\mathcal{C}$ as

$$V_{\mathcal{C}} = V_{\mathcal{A}}.$$

For the performance of the empirically selected decision ϕ_n^*, we have the following:

Theorem 12.6. *Let $\mathcal{C}$ be a class of decision functions of the form $\phi : \mathcal{R}^d \to \{0, 1\}$. Then using the notation $\widehat{L}_n(\phi) = \frac{1}{n} \sum_{i=1}^n I_{\{\phi(X_i) \neq Y_i\}}$ and $L(\phi) = \mathbf{P}\{\phi(X) \neq Y\}$, we have*

$$\mathbf{P}\left\{\sup_{\phi \in \mathcal{C}} |\widehat{L}_n(\phi) - L(\phi)| > \epsilon\right\} \leq 8S(\mathcal{C}, n)e^{-n\epsilon^2/32},$$

and therefore

$$\mathbf{P}\left\{L(\phi_n^*) - \inf_{\phi \in \mathcal{C}} L(\phi) > \epsilon\right\} \leq 8S(\mathcal{C}, n)e^{-n\epsilon^2/128},$$

where ϕ_n^* denotes the classifier minimizing $\widehat{L}_n(\phi)$ over the class $\mathcal{C}$.

PROOF. The statements are immediate consequences of Theorem 12.5 and Lemma 8.2. □

The next corollary, an easy application of Theorem 12.6 (see Problem 12.1), makes things a little more transparent.

COROLLARY 12.1. *In the notation of Theorem 12.6,*

$$\mathbf{E}\left\{L(\phi_n^*)\right\} - \inf_{\phi \in C} L(\phi) \le 16\sqrt{\frac{\log(8eS(C, n))}{2n}}.$$

If $S(C, n)$ increases polynomially with n, then the average error probability of the selected classifier is within $O\left(\sqrt{\log n / n}\right)$ of the error of the best rule in the class. We point out that this result is completely distribution-free. Furthermore, note the nonasymptotic nature of these inequalities: they hold for *every* n. From here on the problem is purely combinatorial—one has to estimate the shatter coefficients. Many properties are given in Chapter 13. In particular, if $V_C > 2$, then $S(C, n) \le n^{V_C}$, that is, if the class C has finite VC dimension, then $S(C, n)$ increases at a polynomial rate, and

$$\mathbf{E}\left\{L(\phi_n^*)\right\} - \inf_{\phi \in C} L(\phi) \le 16\sqrt{\frac{V_C \log n + 4}{2n}}.$$

REMARK. Theorem 12.6 provides a bound for the behavior of the empirically optimal classifier. In practice, finding an empirically optimal classifier is often computationally very expensive. In such cases, the designer is often forced to put up with algorithms yielding suboptimal classifiers. Assume for example, that we have an algorithm which selects a classifier g_n such that its empirical error is not too far from the optimum with large probability:

$$\mathbf{P}\left\{\widehat{L}_n(g_n) \le \inf_{\phi \in C} \widehat{L}_n(\phi) + \epsilon_n\right\} \ge 1 - \delta_n,$$

where $\{\epsilon_n\}$ and $\{\delta_n\}$ are sequences of positive numbers converging to zero. Then it is easy to see (see Problem 12.6) that

$$\mathbf{P}\left\{L(g_n) - \inf_{\phi \in C} L(\phi) > \epsilon\right\} \le \delta_n + \mathbf{P}\left\{2\sup_{\phi \in C}\left|\widehat{L}_n(\phi) - L(\phi)\right| > \epsilon - \epsilon_n\right\},$$

and Theorem 12.6 may be used to obtain bounds for the error probability of g_n. □

Interestingly, the empirical error probability of the empirically optimal classifier is always close to its expected value, as may be seen from the following example:

COROLLARY 12.2. *Let C be an arbitrary class of classification rules (that is, functions of the form $\phi : \mathcal{R}^d \to \{0, 1\}$). Let $\phi_n^* \in C$ be the rule that minimizes the* *number of errors committed on the training sequence* D_n, *among the classifiers in* C. *In other words,*

$$\widehat{L}_n(\phi_n^*) \leq \widehat{L}_n(\phi) \quad \text{for every } \phi \in C,$$

where $\widehat{L}_n(\phi) = \frac{1}{n} \sum_{i=1}^{n} I_{\{\phi(X_i) \neq Y_i\}}$. *Then for every* n *and* $\epsilon > 0$,

$$\mathbf{P}\left\{\left|\widehat{L}_n(\phi_n^*) - \mathbf{E}\left\{\widehat{L}_n(\phi_n^*)\right\}\right| > \epsilon\right\} \leq 2e^{-2n\epsilon^2}.$$

The corollary follows immediately from Theorem 9.1 by observing that changing the value of one (X_i, Y_i) pair in the training sequence results in a change of at most $1/n$ in the value of $\widehat{L}_n(\phi_n^*)$. The corollary is true even if the VC dimension of C is infinite (!). The result shows that $\widehat{L}_n(\phi_n^*)$ is always very close to its expected value with large probability, even if $\mathbf{E}\left\{\widehat{L}_n(\phi_n^*)\right\}$ is far from $\inf_{\phi \in C} L(\phi)$ (see also Problem 12.12).

12.6 Sample Complexity

In his theory of learning, Valiant (1984) rephrases the empirical classifier selection problem as follows. For $\epsilon, \delta > 0$, define an (ϵ, δ) *learning algorithm* as a method that selects a classifier g_n from C using the data D_n such that for the selected rule

$$\sup_{(X,Y)} \mathbf{P}\left\{L(g_n) - \inf_{\phi \in C} L(\phi) > \epsilon\right\} \leq \delta,$$

whenever $n \geq N(\epsilon, \delta)$. Here $N(\epsilon, \delta)$ is the *sample complexity* of the algorithm, defined as the smallest integer with the above property. Since the supremum is taken over all possible distributions of (X, Y), the integer $N(\epsilon, \delta)$ is the number of data pairs that guarantees ϵ *accuracy* with δ *confidence* for any distribution. Note that we use the notation g_n and not ϕ_n^* in the definition above, as the definition does not force us to take the empirical risk minimizer ϕ_n^*.

We may use Theorem 12.6 to get an upper bound on the sample complexity of the selection algorithm based on empirical error minimization (i.e., the classifier ϕ_n^*).

COROLLARY 12.3. *The sample complexity of the method based on empirical error minimization is bounded from above by*

$$N(\epsilon, \delta) \leq \max\left(\frac{512V_C}{\epsilon^2} \log \frac{256V_C}{\epsilon^2}, \frac{256}{\epsilon^2} \log \frac{8}{\delta}\right).$$

The corollary is a direct consequence of Theorem 12.6. The details are left as an exercise (Problem 12.5). The constants may be improved by using refined versions of Theorem 12.5 (see, e.g., Theorem 12.8). The sample size that guarantees the prescribed accuracy and confidence is proportional to the maximum of

$$\frac{V_C}{\epsilon^2} \log \frac{V_C}{\epsilon^2} \quad \text{and} \quad \frac{1}{\epsilon^2} \log \frac{1}{\delta}.$$

Here we have our first practical interpretation of the VC dimension. Doubling the VC dimension requires that we basically double the sample size to obtain the same accuracy and confidence. Doubling the accuracy however, forces us to quadruple the sample size. On the other hand, the confidence level has little influence on the sample size, as it is hidden behind a logarithmic term, thanks to the exponential nature of the Vapnik-Chervonenkis inequality.

The Vapnik-Chervonenkis bound and the sample complexity $N(\epsilon, \delta)$ also allow us to compare different classes in a unified manner. For example, if we pick ϕ_n^* by minimizing the empirical error over all hyperrectangles of $\mathcal{R}^{18}$, will we need a sample size that exceeds that of the rule that minimizes the empirical error over all linear halfspaces of $\mathcal{R}^{29}$? With the sample complexity in hand, it is just a matter of comparing VC dimensions.

As a function of ϵ, the above bound grows as $O\big((1/\epsilon^2)\log(1/\epsilon^2)\big)$. It is possible, interestingly, to get rid of the "log" term, at the expense of increasing the linearity in the VC dimension (see Problem 12.11).

12.7   The Zero-Error Case

Theorem 12.5 is completely general, as it applies to any class of classifiers and all distributions. In some cases, however, when we have some additional information about the distribution, it is possible to obtain even better bounds. For example, in the theory of concept learning one commonly assumes that $L^* = 0$, and that the Bayes decision is contained in C (see, e.g., Valiant (1984), Blumer, Ehrenfeucht, Haussler, and Warmuth (1989), Natarajan (1991)). The following theorem provides significant improvement. Its various forms have been proved by Devroye and Wagner (1976b), Vapnik (1982), and Blumer, Ehrenfeucht, Haussler, and Warmuth (1989). For a sharper result, see Problem 12.9.

Theorem 12.7. *Let C be a class of decision functions mapping $\mathcal{R}^d$ to $\{0, 1\}$, and let ϕ_n^* be a function in C that minimizes the empirical error based on the training sample D_n. Suppose that $\inf_{\phi \in C} L(\phi) = 0$, i.e., the Bayes decision is contained in C, and $L^* = 0$. Then*

$$\mathbf{P}\left\{L(\phi_n^*) > \epsilon\right\} \le 2S(C, 2n)2^{-n\epsilon/2}.$$

To contrast this with Theorem 12.6, observe that the exponent in the upper bound for the empirically optimal rule is proportional to $-n\epsilon$ instead of $-n\epsilon^2$. To see the significance of this difference, note that Theorem 12.7 implies that the error probability of the selected classifier is within $O(\log n/n)$ of the optimal rule in the class (which equals zero in this case, see Problem 12.8), as opposed to $O\left(\sqrt{\log n/n}\right)$ from Theorem 12.6. We show in Chapter 14 that this is not a technical coincidence, but since both bounds are essentially tight, it is a mathematical witness to the fact that it is remarkably easier to select a good classifier when $L^* = 0$. The proof is based on the random permutation argument developed in the original proof of the Vapnik-Chervonenkis inequality (1971).

PROOF. For $n\epsilon \leq 2$, the inequality is clearly true. So, we assume that $n\epsilon > 2$. First observe that since $\inf_{\phi\in C} L(\phi) = 0$, $\widehat{L}_n(\phi_n^*) = 0$ with probability one. It is easily seen that

$$L(\phi_n^*) \leq \sup_{\phi:\widehat{L}_n(\phi)=0} |L(\phi) - \widehat{L}_n(\phi)|.$$

Now, we return to the notation of the previous sections, that is, Z_i denotes the pair (X_i, Y_i), ν denotes its measure, and ν_n is the empirical measure based on $Z_1, \ldots, Z_n$. Also, $\mathcal{A}$ consists of all sets of the form $A = \{(x, y) : \phi(x) \neq y\}$ for $\phi \in C$. With these notations,

$$\sup_{\phi:\widehat{L}_n(\phi)=0} |L(\phi) - \widehat{L}_n(\phi)| = \sup_{A:\nu_n(A)=0} |\nu_n(A) - \nu(A)|.$$

STEP 1. SYMMETRIZATION BY A GHOST SAMPLE. The first step of the proof is similar to that of Theorem 12.5. Introduce the auxiliary sample $Z_1', \ldots, Z_n'$ such that the random variables $Z_1, \ldots, Z_n, Z_1', \ldots, Z_n'$ are i.i.d., and let ν_n' be the empirical measure for $Z_1', \ldots, Z_n'$. Then for $n\epsilon > 2$

$$\mathbf{P}\left\{\sup_{A:\nu_n(A)=0} |\nu_n(A) - \nu(A)| > \epsilon\right\} \leq 2\mathbf{P}\left\{\sup_{A:\nu_n(A)=0} |\nu_n(A) - \nu_n'(A)| > \frac{\epsilon}{2}\right\}.$$

The proof of this inequality parallels that of the corresponding one in the proof of Theorem 12.5. However, an important difference is that the condition $n\epsilon^2 > 2$ there is more restrictive than the condition $n\epsilon > 2$ here. The details of the proof are left to the reader (see Problem 12.7).

STEP 2. SYMMETRIZATION BY PERMUTING. Note that the distribution of

$$\sup_{A:\nu_n(A)=0} |\nu_n(A) - \nu_n'(A)| = \sup_{A:\sum_{i=1}^n I_A(Z_i)=0} \left|\frac{1}{n}\sum_{i=1}^n I_A(Z_i) - \frac{1}{n}\sum_{i=1}^n I_A(Z_i')\right|$$

is the same as the distribution of

$$\beta(\Pi) \overset{\text{def}}{=} \sup_{A:\sum_{i=1}^n I_A(\Pi(Z_i))=0} \left|\frac{1}{n}\sum_{i=1}^n I_A(\Pi(Z_i)) - \frac{1}{n}\sum_{i=1}^n I_A(\Pi(Z_i'))\right|,$$

where $\Pi(Z_1), \ldots, \Pi(Z_n), \Pi(Z_1'), \ldots, \Pi(Z_n')$ is an arbitrary permutation of the random variables $Z_1, \ldots, Z_n, Z_1', \ldots, Z_n'$. The $(2n)!$ possible permutations are denoted by

$$\Pi_1, \Pi_2, \ldots, \Pi_{(2n)!}.$$

Therefore,

$$\mathbf{P}\left\{\sup_{A:\nu_n(A)=0} |\nu_n(A) - \nu_n'(A)| > \frac{\epsilon}{2}\right\}$$

$$= \mathbf{E}\left\{\frac{1}{(2n)!}\sum_{j=1}^{(2n)!} I_{\{\beta(\Pi_j)>\frac{\epsilon}{2}\}}\right\}$$

$$= \mathbf{E}\left\{\frac{1}{(2n)!}\sum_{j=1}^{(2n)!}\sup_{A:\sum_{i=1}^n I_A(\Pi_j(Z_i))=0} I_{\{|\frac{1}{n}\sum_{i=1}^n I_A(\Pi_j(Z_i))-\frac{1}{n}\sum_{i=1}^n I_A(\Pi_j(Z_i'))|>\frac{\epsilon}{2}\}}\right\}.$$

STEP 3. CONDITIONING. Next we fix $z_1, \ldots, z_n, z_1', \ldots, z_n'$ and bound the value of the random variable above. Let $\bar{A} \subset \mathcal{A}$ be a collection of sets such that any two sets in $\bar{A}$ pick different subsets of $\{z_1, \ldots, z_n, z_1', \ldots, z_n'\}$, and its cardinality is $N_A(z_1, \ldots, z_n, z_1', \ldots, z_n')$, that is, all possible subsets are represented exactly once. Then it suffices to take the supremum over $\bar{A}$ instead of $\mathcal{A}$:

$$\frac{1}{(2n)!}\sum_{j=1}^{(2n)!}\sup_{A:\sum_{i=1}^n I_A(\Pi_j(z_i))=0} I_{\{|\frac{1}{n}\sum_{i=1}^n I_A(\Pi_j(z_i))-\frac{1}{n}\sum_{i=1}^n I_A(\Pi_j(z_i'))|>\frac{\epsilon}{2}\}}$$

$$= \frac{1}{(2n)!}\sum_{j=1}^{(2n)!}\sup_{A\in\bar{A}:\sum_{i=1}^n I_A(\Pi_j(z_i))=0} I_{\{\frac{1}{n}\sum_{i=1}^n I_A(\Pi_j(z_i'))>\frac{\epsilon}{2}\}}$$

$$\leq \frac{1}{(2n)!}\sum_{j=1}^{(2n)!}\sum_{A\in\bar{A}:\sum_{i=1}^n I_A(\Pi_j(z_i))=0} I_{\{\frac{1}{n}\sum_{i=1}^n I_A(\Pi_j(z_i'))>\frac{\epsilon}{2}\}}$$

$$= \frac{1}{(2n)!}\sum_{j=1}^{(2n)!}\sum_{A\in\bar{A}} I_{\{\sum_{i=1}^n I_A(\Pi_j(z_i))=0\}} I_{\{\frac{1}{n}\sum_{i=1}^n I_A(\Pi_j(z_i'))>\frac{\epsilon}{2}\}}$$

$$= \sum_{A\in\bar{A}}\frac{1}{(2n)!}\sum_{j=1}^{(2n)!} I_{\{\sum_{i=1}^n I_A(\Pi_j(z_i))=0\}} I_{\{\frac{1}{n}\sum_{i=1}^n I_A(\Pi_j(z_i'))>\frac{\epsilon}{2}\}}.$$

STEP 4. COUNTING. Clearly, the expression behind the first summation sign is just the number of permutations of the $2n$ points $z_1, z_1', \ldots, z_n, z_n'$ with the property

$$I_{\{\sum_{i=1}^n I_A(\Pi_j(z_i))=0\}} I_{\{\frac{1}{n}\sum_{i=1}^n I_A(\Pi_j(z_i'))>\frac{\epsilon}{2}\}} = 1,$$

divided by $(2n)!$, the total number of permutations. Observe that if $l = \sum_{i=1}^n (I_A(z_i)+I_A(z_i'))$ is the total number of points in A among $z_1, z_1', \ldots, z_n, z_n'$ for a fixed set A, then the number of permutations such that

$$I_{\{\sum_{i=1}^n I_A(\Pi_j(z_i))=0\}} I_{\{\frac{1}{n}\sum_{i=1}^n I_A(\Pi_j(z_i'))>\frac{\epsilon}{2}\}} = 1$$

is zero if $l \leq n\epsilon/2$. If $l > n\epsilon/2$, then the fraction of the number of permutations with the above property and the number of all permutations can not exceed

$$\frac{\binom{n}{l}}{\binom{2n}{l}}.$$

To see this, note that for the above product of indicators to be 1, all the points falling in A have to be in the second half of the permuted sample. Now clearly,

$$\frac{\binom{n}{l}}{\binom{2n}{l}} = \frac{n(n-1)\cdots(n-l+1)}{2n(2n-1)\cdots(2n-l+1)} \leq 2^{-l} \leq 2^{-n\epsilon/2}.$$

Summarizing, we have

$$\mathbf{P}\left\{\sup_{A:\nu_n(A)=0} |\nu_n(A) - \nu'_n(A)| > \frac{\epsilon}{2}\right\}$$

$$\leq \quad \mathbf{E}\left\{\frac{1}{(2n)!}\sum_{j=1}^{(2n)!} \sup_{A:\sum_{i=1}^{n} I_A(\Pi_j(Z_i))=0} I_{\left\{\left|\frac{1}{n}\sum_{i=1}^{n} I_A(\Pi_j(Z_i)) - \frac{1}{n}\sum_{i=1}^{n} I_A(\Pi_j(Z'_i))\right| > \frac{\epsilon}{2}\right\}}\right\}$$

$$\leq \quad \mathbf{E}\left\{\sum_{A\in\bar{A}} 2^{-n\epsilon/2}\right\}$$

$$= \quad \mathbf{E}\left\{N_A(Z_1,\ldots,Z_n,Z'_1,\ldots,Z'_n)2^{-n\epsilon/2}\right\}$$

$$\leq \quad s(A,2n)2^{-n\epsilon/2},$$

and the theorem is proved. $\square$

Again, we can bound the sample complexity $N(\epsilon,\delta)$ restricted to the class of distributions with $\inf_{\phi\in C} L(\phi) = 0$. Just as Theorem 12.6 implies Corollary 12.3, Theorem 12.7 yields

COROLLARY 12.4. *The sample complexity $N(\epsilon,\delta)$ that guarantees*

$$\sup_{(X,Y):\inf_{\phi\in C} L(\phi)=0} \mathbf{P}\left\{L(g_n) - \inf_{\phi\in C} L(\phi) > \epsilon\right\} \leq \delta$$

for $n \geq N(\epsilon,\delta)$, is bounded by

$$N(\epsilon,\delta) \leq \max\left(\frac{8V_C}{\epsilon}\log_2\frac{13}{\epsilon}, \frac{4}{\epsilon}\log_2\frac{2}{\delta}\right).$$

A quick comparison with Corollary 12.3 shows that the ϵ^2 factors in the denominators there are now replaced by ϵ. For the same accuracy, much smaller samples suffice if we know that $\inf_{\phi\in C} L(\phi) = 0$. Interestingly, the sample complexity is still roughly linear in the VC dimension.

REMARK. We also note that under the condition of Theorem 12.7,

$$\mathbf{P}\left\{L(\phi_n^*) > \epsilon\right\} \leq 2\mathbf{E}\left\{N_A(Z_1,\ldots,Z_n,Z'_1,\ldots,Z'_n)\right\}2^{-n\epsilon/2}$$

$$= 2\mathbf{E}\{N_{\mathcal{A}}(Z_1, \ldots, Z_{2n})\}\, 2^{-n\epsilon/2},$$

where the class $\mathcal{A}$ of sets is defined in the proof of Theorem 12.7. $\square$

REMARK. As Theorem 12.7 shows, the difference between the error probability of an empirically optimal classifier and that of the optimal in the class is much smaller if the latter quantity is known to be zero than if no restriction is imposed on the distribution. To bridge the gap between the two bounds, one may put the restriction $\inf_{\phi \in C} L(\phi) \leq L$ on the distribution, where $L \in (0, 1/2)$ is a fixed number. Devroye and Wagner (1976b) and Vapnik (1982) obtained such bounds. For example, it follows from a result by Vapnik (1982) that

$$\mathbf{P}\left\{L(\phi_n^*) - \inf_{\phi \in C} L(\phi) > \epsilon\right\} \leq 8S(C, n)e^{-n\epsilon^2/8(L+\epsilon)}.$$

As expected, the bound becomes smaller as L decreases. We face the same phenomenon in Chapter 14, where lower bounds are obtained for the probability above. $\square$

12.8 Extensions

We mentioned earlier that the constant in the exponent in Theorem 12.5 can be improved at the expense of a more complicated argument. The best possible exponent appears in the following result, whose proof is left as an exercise (Problem 12.15):

Theorem 12.8. (DEVROYE (1982A)).

$$\mathbf{P}\left\{\sup_{A \in \mathcal{A}} |\nu_n(A) - \nu(A)| > \epsilon\right\} \leq cs(\mathcal{A}, n^2)e^{-2n\epsilon^2},$$

where the constant c does not exceed $4e^{4\epsilon + 4\epsilon^2} \leq 4e^8$, $\epsilon \leq 1$.

Even though the coefficient in front is larger than in Theorem 12.5, it becomes very quickly absorbed by the exponential term. We will see in Chapter 13 that for $V_{\mathcal{A}} > 2$, $s(\mathcal{A}, n) \leq n^{V_{\mathcal{A}}}$, so $s(\mathcal{A}, n^2) \leq n^{2V_{\mathcal{A}}}$. This difference is negligible compared to the difference between the exponential terms, even for moderately large values of $n\epsilon^2$.

Both Theorem 12.5 and Theorem 12.8 imply that

$$\mathbf{E}\left\{\sup_{A \in \mathcal{A}} |\nu_n(A) - \nu(A)|\right\} = O\left(\sqrt{\log n/n}\right)$$

(see Problem 12.1). However, it is possible to get rid of the logarithmic term to obtain $O(1/\sqrt{n})$. For example, for the Kolmogorov-Smirnov statistic we have the following result by Dvoretzky, Kiefer and Wolfowitz (1956), sharpened by Massart (1990):

Theorem 12.9. (DVORETZKY, KIEFER, AND WOLFOWITZ (1956); MASSART (1990)). *Using the notation of Theorem 12.4, we have for every n and $\epsilon > 0$,*

$$P\left\{\sup_{z \in \mathcal{R}} |F(z) - F_n(z)| > \epsilon\right\} \le 2e^{-2n\epsilon^2}.$$

For the general case, we also have Alexander's bound:

Theorem 12.10. (ALEXANDER (1984)). *For $n\epsilon^2 \ge 64$,*

$$P\left\{\sup_{A \in \mathcal{A}} |\nu_n(A) - \nu(A)| > \epsilon\right\} \le 16\left(\sqrt{n}\epsilon\right)^{4096 V_A} e^{-2n\epsilon^2}.$$

The theorem implies the following (Problem 12.10):

COROLLARY 12.5.

$$E\left\{\sup_{A \in \mathcal{A}} |\nu_n(A) - \nu(A)|\right\} \le \frac{8 + \sqrt{2048 V_A \log(4096 V_A)}}{\sqrt{n}}.$$

The bound in Theorem 12.10 is theoretically interesting, since it implies (see Problem 12.10) that for fixed V_A the expected value of the supremum decreases as $a/\sqrt{n}$ instead of $\sqrt{\log n / n}$. However, a quick comparison reveals that Alexander's bound is larger than that of Theorem 12.8, unless $n > 2^{6144}$, an astronomically large value. Recently, Talagrand (1994) obtained a very strong result. He proved that there exists a universal constant c such that

$$P\left\{\sup_{A \in \mathcal{A}} |\nu_n(A) - \nu(A)| > \epsilon\right\} \le \frac{c}{\sqrt{n}\epsilon} \left(\frac{cn\epsilon^2}{V_A}\right)^{V_A} e^{-2n\epsilon^2}.$$

For more information about these inequalities, see also Vapnik (1982), Gaenssler (1983), Gaenssler and Stute (1979), and Massart (1983).

It is only natural to ask whether the uniform law of large numbers

$$\sup_{A \in \mathcal{A}} |\nu_n(A) - \nu(A)| \to 0$$

holds if we allow $\mathcal{A}$ to be the class of *all measurable subsets* of $\mathcal{R}^d$. In this case the supremum above is called the *total variation* between the measures ν_n and ν. The convergence clearly can not hold if ν_n is the standard empirical measure

$$\nu_n(A) = \frac{1}{n} \sum_{i=1}^{n} I_{\{Z_i \in A\}}.$$

But is there another empirical measure such that the convergence holds? The somewhat amusing answer is no. As Devroye and Györfi (1992) proved, for any empirical measure $\tilde{\nu}_n$—that is, a function depending on $Z_1, \ldots, Z_n$ assigning a nonnegative number to any measurable set—there exists a distribution of Z such that for *all* n

$$\sup_{A \in \mathcal{A}} |\tilde{\nu}_n(A) - \nu(A)| > 1/4$$

almost surely. Thus, in this generality, the problem is hopeless. For meaningful results, either $\mathcal{A}$ or ν must be restricted. For example, if we assume that ν is absolutely continuous with density f, and that $\tilde{\nu}_n$ is absolutely continuous too (with density f_n), then by Scheffé's theorem (1947),

$$\sup_{A \in \mathcal{A}} |\tilde{\nu}_n(A) - \nu(A)| = \frac{1}{2} \int_{\mathcal{R}^d} |f_n(x) - f(x)| dx$$

(see Problem 12.13). But as we see from Problems 6.2, 10.2, 9.6, and 10.3, there exist density estimators (such as histogram and kernel estimates) such that the L_1-error converges to zero almost surely for *all possible densities*. Therefore, the total variation between the empirical measures derived from these density estimates and the true measure converges to zero almost surely for all distributions with a density. For other large classes of distributions that can be estimated consistently in total variation, we refer to Barron, Györfi, and van der Meulen (1992).

Problems and Exercises

PROBLEM 12.1. Prove that if a nonnegative random variable Z satisfies $\mathbf{P}\{Z > t\} \le ce^{-2nt^2}$ for all $t > 0$ and some $c > 0$, then,

$$\mathbf{E}\{Z^2\} \le \frac{\log(ce)}{2n}.$$

Furthermore,

$$\mathbf{E}Z \le \sqrt{\mathbf{E}\{Z^2\}} \le \sqrt{\frac{\log(ce)}{2n}}.$$

HINT: Use the identity $\mathbf{E}\{Z^2\} = \int_0^\infty \mathbf{P}\{Z^2 > t\}dt$, and set $\int_0^\infty = \int_0^u + \int_u^\infty$. Bound the first integral by u, and the second by the exponential inequality. Find the value of u that minimizes the obtained upper bound.

PROBLEM 12.2. Generalize the arguments of Theorems 4.5 and 4.6 to prove Theorems 12.2 and 12.3.

PROBLEM 12.3. Determine the fingering dimension of classes of classifiers $\mathcal{C} = \{\phi : \phi(x) = I_{\{x \in A\}}; A \in \mathcal{A}\}$ if the class $\mathcal{A}$ is
   (1) the class of all closed intervals in $\mathcal{R}$,
   (2) the class of all sets obtained as the union of m closed intervals in $\mathcal{R}$,
   (3) the class of balls in $\mathcal{R}^d$ centered at the origin,
   (4) the class of all balls in $\mathcal{R}^d$,

(5) the class of sets of the form $(-\infty, x_1] \times \cdots \times (-\infty, x_d]$ in $\mathcal{R}^d$, and
(6) the class of all convex polygons in $\mathcal{R}^2$.

PROBLEM 12.4. Let C be a class of classifiers with fingering dimension $k > 4$ (independently of n). Show that $V_C \leq k \log_2^2 k$.

PROBLEM 12.5. Prove that Theorem 12.6 implies Corollary 12.3. HINT: Find $N(\epsilon, \delta)$ such that $8n^{V_C} e^{-n\epsilon^2/128} \leq \delta$ whenever $n > N(\epsilon, \delta)$. To see this, first show that $n^{V_C} \leq e^{n\epsilon^2/256}$ is satisfied for $n \geq \frac{512 V_C}{\epsilon^2} \log \frac{256 V_C}{\epsilon^2}$, which follows from the fact that $2 \log x \leq x$ if $x \geq e^2$. But in this case $8n^{V_C} e^{-n\epsilon^2/128} \leq 8e^{-n\epsilon^2/256}$. The upper bound does not exceed δ if $n \geq \frac{256}{\epsilon^2} \log \frac{8}{\delta}$.

PROBLEM 12.6. Let C be a class of classifiers $\phi : \mathcal{R}^d \to \{0, 1\}$, and let ϕ_n^* be a classifier minimizing the empirical error probability measured on D_n. Assume that we have an algorithm which selects a classifier g_n such that

$$\mathbf{P}\left\{\widehat{L}_n(g_n) \leq \inf_{\phi \in C} \widehat{L}_n(\phi) + \epsilon_n\right\} \geq 1 - \delta_n,$$

where $\{\epsilon_n\}$ and $\{\delta_n\}$ are sequences of positive numbers converging to zero. Show that

$$\mathbf{P}\left\{L(g_n) - \inf_{\phi \in C} L(\phi) > \epsilon\right\} \leq \delta_n + \mathbf{P}\left\{2 \sup_{\phi \in C} |\widehat{L}_n(\phi) - L(\phi)| > \epsilon - \epsilon_n\right\}.$$

Find conditions on $\{\epsilon_n\}$ and $\{\delta_n\}$ so that

$$\mathbf{E}L(g_n) - \inf_{\phi \in C} L(\phi) = O\left(\sqrt{\frac{\log n}{n}}\right),$$

that is, $\mathbf{E}L(g_n)$ converges to the optimum at the same order as the error probability of ϕ_n^*.

PROBLEM 12.7. Prove that

$$\mathbf{P}\left\{\sup_{A : \nu_n(A)=0} |\nu_n(A) - \nu(A)| > \epsilon\right\} \leq 2\mathbf{P}\left\{\sup_{A : \nu_n(A)=0} |\nu_n(A) - \nu_n'(A)| > \frac{\epsilon}{2}\right\}$$

holds if $n\epsilon > 2$. This inequality is needed to complete the proof of Theorem 12.7. HINT: Proceed as in the proof of Theorem 12.5. Introduce A^* with $\nu_n(A^*) = 0$ and justify the validity of the steps of the following chain of inequalities:

$$\mathbf{P}\left\{\sup_{A : \nu_n(A)=0} |\nu_n(A) - \nu_n'(A)| > \epsilon/2\right\}$$

$$\geq \quad \mathbf{E}\left\{I_{\{\nu(A^*)>\epsilon\}}\mathbf{P}\left\{\nu_n'(A^*) \geq \frac{\epsilon}{2}\bigg| Z_1, \ldots, Z_n\right\}\right\}$$

$$\geq \quad \mathbf{P}\left\{B(n, \epsilon) > \frac{n\epsilon}{2}\right\}\mathbf{P}\{|\nu_n(A^*) - \nu(A^*)| > \epsilon\},$$

where $B(n, \epsilon)$ is a binomial random variable with parameters n and ϵ. Finish the proof by showing that the probability on the right-hand side is greater than or equal to $1/2$ if $n\epsilon > 2$. (Under the slightly more restrictive condition $n\epsilon > 8$, this follows from Chebyshev's inequality.)

PROBLEM 12.8. Prove that Theorem 12.7 implies that if $\inf_{\phi \in C} L(\phi) = 0$, then

$$EL(\phi_n^*) \leq \frac{2V_C \log(2n) + 4}{n \log 2}.$$

We note here that Haussler, Littlestone, and Warmuth (1988) demonstrated the existence of a classifier ϕ_n^* with $EL(\phi_n^*) \leq 2V_C/n$ when $\inf_{\phi \in C} L(\phi) = 0$. HINT: Use the identity $EX = \int_0^\infty P\{X > t\}dt$ for nonnegative random variables X, and employ the fact that $S(C, n) \leq n^{V_C}$ (see Theorem 13.3).

PROBLEM 12.9. Prove the following version of Theorem 12.7. Let ϕ_n^* be a function that minimizes the empirical error over a class in C. Assume that $\inf_{\phi \in C} L(\phi) = 0$. Then

$$P\left\{L(\phi_n^*) > \epsilon\right\} \leq 2S(C, n^2)e^{-n\epsilon}$$

(Shawe-Taylor, Anthony, and Biggs (1993)). HINT: Modify the proof of Theorem 12.7 by introducing a ghost sample $Z_1', \ldots, Z_m'$ with size m (to be specified after optimization). Only the first symmetrization step (Problem 12.7) needs adjusting: show that for any $\alpha \in (0, 1)$,

$$P\left\{\sup_{A:v_n(A)=0} v(A) > \epsilon\right\} \leq \frac{1}{1 - e^{-m\alpha\epsilon}} P\left\{\sup_{A:v_n(A)=0} v_m'(A) > (1 - \alpha)\epsilon\right\},$$

where v_m' is the empirical measure based on the ghost sample (use Bernstein's inequality—Theorem 8.2). The rest of the proof is similar to that of Theorem 12.7. Choose $m = n^2 - n$ and $\alpha = n/(n + m)$.

PROBLEM 12.10. Prove that Alexander's bound (Theorem 12.10) implies that if A is a Vapnik-Chervonenkis class with vc dimension $V = V_A$, then

$$E\left\{\sup_{A \in A} |v_n(A) - v(A)|\right\} \leq \frac{8 + \sqrt{2048 V \log(4096 V)}}{\sqrt{n}}.$$

HINT: Justify the following steps:

(1)   If ψ is a negative decreasing concave function, then

$$\int_u^\infty e^{\psi(t)}dt \leq \frac{e^{\psi(u)}}{-\psi'(u)}.$$

HINT: Bound ψ by using its Taylor series expansion.

(2)   Let $b > 0$ be fixed. Then for $u \geq \sqrt{b/2}$,

$$\int_u^\infty t^b e^{-2t^2} dt \leq \frac{u^{b-1}e^{-2u^2}}{2}.$$

HINT: Use the previous step.

(3)   Let X be a positive random variable for which

$$P\{X > u\} \leq au^b e^{-2u^2}, \quad u \geq \sqrt{c},$$

where a, b, c are positive constants. Then, if $b \geq 2c \geq e$,

$$EX \leq \frac{a}{2} + \sqrt{\frac{b \log b}{2}}.$$

HINT: Use $EX = \int_0^\infty P\{X > u\}du$, and bound the probability either by one, or by the bound of the previous step.

PROBLEM 12.11. Use Alexander's inequality to obtain the following sample size bound for empirical error minimization:

$$N(\epsilon, \delta) \leq \max\left(\frac{2^{14}V_C \log\left(2^{14}V_C\right)}{\epsilon^2}, \frac{4}{\epsilon^2}\log\frac{16}{\delta}\right).$$

For what values of ϵ does this bound beat Corollary 12.3?

PROBLEM 12.12. Let $Z_1, \ldots, Z_n$ be i.i.d. random variables in $\mathcal{R}^d$, with measure ν, and standard empirical measure ν_n. Let $\mathcal{A}$ be an arbitrary class of subsets of $\mathcal{R}^d$. Show that as $n \to \infty$,

$$\sup_{A \in \mathcal{A}} |\nu_n(A) - \nu(A)| \to 0 \quad \text{in probability}$$

if and only if

$$\sup_{A \in \mathcal{A}} |\nu_n(A) - \nu(A)| \to 0 \quad \text{with probability one.}$$

HINT: Use McDiarmid's inequality.

PROBLEM 12.13. Prove Scheffé's theorem (1947): let μ and ν be absolute continuous probability measures on $\mathcal{R}^d$ with densities f and g, respectively. Prove that

$$\sup_{A \in \mathcal{A}} |\mu(A) - \nu(A)| = \frac{1}{2}\int |f(x) - g(x)|dx,$$

where $\mathcal{A}$ is the class of all Borel-measurable sets. HINT: Show that the supremum is achieved for the set $\{x : f(x) > g(x)\}$.

PROBLEM 12.14. LEARNING BASED ON EMPIRICAL COVERING. This problem demonstrates an alternative method of picking a classifier which works as well as empirical error minimization. The method, based on empirical covering of the class of classifiers, was introduced by Buescher and Kumar (1996a). The idea of covering the class goes back to Vapnik (1982). See also Benedek and Itai (1988), Kulkarni (1991), and Dudley, Kulkarni, Richardson, and Zeitouni (1994). Let C be a class of classifiers $\phi : \mathcal{R}^d \to \{0, 1\}$. The data set D_n is split into two parts, $D_m = ((X_1, Y_1), \ldots, (X_m, Y_m))$, and $T_l = ((X_{m+1}, Y_{m+1}), \ldots, (X_n, Y_n))$, where $n = m + l$. We use the first part D_m to cover C as follows. Define the random variable N as the number of different values the binary vector $\mathbf{b}_m(\phi) = (\phi(X_1), \ldots, \phi(X_m))$ takes as ϕ is varied over C. Clearly, $N \leq S(C, m)$. Take N classifiers from C, such that all N possible values of the binary vector $\mathbf{b}_m(\phi)$ are represented exactly once. Denote these classifiers by $\tilde{\phi}_1, \ldots, \tilde{\phi}_N$. Among these functions, pick one that minimizes the empirical error on the second part of the data set T_l:

$$\widehat{L}_l(\tilde{\phi}_i) = \frac{1}{l}\sum_{j=1}^{l} I_{\{\tilde{\phi}_i(X_{m+j}) \neq Y_{m+j}\}}.$$

Denote the selected classifier by $\widehat{\phi}_n$. Show that for every n, m and $\epsilon > 0$, the difference between the error probability $L(\widehat{\phi}_n) = \mathbf{P}\{\widehat{\phi}_n(X) \neq Y | D_n\}$ and the minimal error probability in the class satisfies

$$\mathbf{P}\left\{L(\widehat{\phi}_n) - \inf_{\phi \in C} L(\phi) > \epsilon\right\} \leq 2S(C, m)e^{-(n-m)\epsilon^2/8} + 2S^4(C, 2m)e^{-m\epsilon \log 2/4}$$

(Buescher and Kumar (1996a)). For example, by taking $m \sim \sqrt{n}$, we get

$$\mathbf{E}\left\{L(\widehat{\phi}_n) - \inf_{\phi \in C} L(\phi)\right\} \leq c\sqrt{\frac{V_C \log n}{n}},$$

where c is a universal constant. The fact that the number of samples m used for covering C is very small compared to n, may make the algorithm computationally more attractive than the method of empirical error minimization. HINT: Use the decomposition

$$\mathbf{P}\left\{L(\widehat{\phi}_n) - \inf_{\phi \in C} L(\phi) > \epsilon\right\}$$

$$\leq \quad \mathbf{P}\left\{L(\widehat{\phi}_n) - \inf_{i=1,\ldots,N} L(\tilde{\phi}_i) > \frac{\epsilon}{2}\right\} + \mathbf{P}\left\{\inf_{i=1,\ldots,N} L(\tilde{\phi}_i) - \inf_{\phi \in C} L(\phi) > \frac{\epsilon}{2}\right\}.$$

Bound the first term on the right-hand side by using Lemma 8.2 and Hoeffding's inequality:

$$\mathbf{P}\left\{L(\widehat{\phi}_n) - \inf_{i=1,\ldots,N} L(\tilde{\phi}_i) > \frac{\epsilon}{2}\right\} \leq 2S(C, m)e^{-l\epsilon^2/8}.$$

To bound the second term of the decomposition, observe that

$$\inf_{i=1,\ldots,N} L(\tilde{\phi}_i) - \inf_{\phi \in C} L(\phi) \quad \leq \quad \sup_{\phi,\phi' \in C: \mathbf{b}_m(\phi) = \mathbf{b}_m(\phi')} |L(\phi) - L(\phi')|$$

$$\leq \quad \sup_{\phi,\phi' \in C: \mathbf{b}_m(\phi) = \mathbf{b}_m(\phi')} \mathbf{P}\left\{\phi(X) \neq \phi'(X)\right\}$$

$$= \quad \sup_{A \in \mathcal{A}: \nu_m(A) = 0} \nu(A),$$

where

$$\mathcal{A} = \{\{x : \phi(x) = 1\} : \phi(x) = |\phi_1(x) - \phi_2(x)|, \phi_1, \phi_2 \in C\},$$

and $\nu_m(A) = \frac{1}{m}\sum_{i=1}^{m} I_{\{X_i \in A\}}$. Bound the latter quantity by applying Theorem 12.7. To do this, you will need to bound the shatter coefficients $s(\mathcal{A}, 2m)$. In Chapter 13 we introduce simple tools for this. For example, it is easy to deduce from parts (ii), (iii), and (iv) of Theorem 13.5, that $s(\mathcal{A}, 2m) \leq S^4(C, 2m)$.

PROBLEM 12.15. Prove that for all $\epsilon \in (0, 1)$,

$$\mathbf{P}\left\{\sup_{A \in \mathcal{A}} |\nu_n(A) - \nu(A)| > \epsilon\right\} \leq cs(\mathcal{A}, n^2)e^{-2n\epsilon^2},$$

where $c \leq 4e^{4\epsilon + 4\epsilon^2}$ (Devroye (1982a)). HINT: Proceed as indicated by the following steps:
  (1) Introduce an i.i.d. ghost sample $Z'_1, \ldots, Z'_m$ of size $m = n^2 - n$, where Z'_1 is distributed as Z_1. Denote the corresponding empirical measure by ν'_m. As in the proof of the first step of Theorem 12.4, prove that for $\alpha, \epsilon \in (0, 1)$,

$$\mathbf{P}\left\{\sup_{A \in \mathcal{A}} |\nu_n(A) - \nu'_m(A)| > (1 - \alpha)\epsilon\right\}$$

$$\geq \quad \left(1 - \frac{1}{4\alpha^2\epsilon^2 m}\right)\mathbf{P}\left\{\sup_{A \in \mathcal{A}} |\nu_n(A) - \nu(A)| > \epsilon\right\}.$$

(2) Introduce $n^2!$ permutations $\Pi_1, \ldots, \Pi_{n+m}$ of the $n + m$ random variables as in Step 2 of the proof of Theorem 12.5. Show that

$$\frac{1}{n^2!} \sum_{j=1}^{n^2!} \sup_{A \in \mathcal{A}} I_{\left\{\left|\frac{1}{n}\sum_{i=1}^{n} I_A(\Pi_j(Z_i)) - \frac{1}{m}\sum_{i=1}^{m} I_A(\Pi_j(Z_i'))\right| > (1-\alpha)\epsilon\right\}}$$

$$\leq \mathcal{S}(\mathcal{A}, n^2) \max_{A \in \mathcal{A}} \frac{1}{n^2!} \sum_{j=1}^{n^2!} I_{\left\{\left|\frac{1}{n}\sum_{i=1}^{n} I_A(\Pi_j(Z_i)) - \frac{1}{m}\sum_{i=1}^{m} I_A(\Pi_j(Z_i'))\right| > (1-\alpha)\epsilon\right\}}.$$

(3) Show that for each $A \in \mathcal{A}$,

$$\frac{1}{n^2!} \sum_{j=1}^{n^2!} I_{\left\{\left|\frac{1}{n}\sum_{i=1}^{n} I_A(\Pi_j(Z_i)) - \frac{1}{m}\sum_{i=1}^{m} I_A(\Pi_j(Z_i'))\right| > (1-\alpha)\epsilon\right\}} \leq 2e^{-2n\epsilon^2 + 4\alpha n\epsilon^2 + 4\epsilon^2}$$

by using Hoeffding's inequality for sampling without replacement from n^2 binary-valued elements (see Theorem A.25). Choose $\alpha = 1/(n\epsilon)$.

13
Combinatorial Aspects of Vapnik-Chervonenkis Theory

13.1 Shatter Coefficients and VC Dimension

In this section we list a few interesting properties of shatter coefficients $s(\mathcal{A}, n)$ and of the VC dimension $V_{\mathcal{A}}$ of a class of sets $\mathcal{A}$. We begin with a property that makes things easier. In Chapter 12 we noted the importance of classes of the form

$$\bar{\mathcal{A}} = \left\{ A \times \{0\} \cup A^c \times \{1\}; A \in \mathcal{A} \right\}.$$

(The sets A are of the form $\{x : \phi(x) = 1\}$, and the sets in $\bar{\mathcal{A}}$ are sets of pairs (x, y) for which $\phi(x) \neq y$.) Recall that if $\mathcal{C}$ is a class of classifiers $\phi : \mathcal{R}^d \to \{0, 1\}$, then by definition, $S(\mathcal{C}, n) = s(\bar{\mathcal{A}}, n)$ and $V_{\mathcal{C}} = V_{\bar{\mathcal{A}}}$. The first result states that $S(\mathcal{C}, n) = s(\mathcal{A}, n)$, so it suffices to investigate properties of $\mathcal{A}$, a class of subsets of $\mathcal{R}^d$.

Theorem 13.1. *For every n we have $s(\bar{\mathcal{A}}, n) = s(\mathcal{A}, n)$, and therefore $V_{\bar{\mathcal{A}}} = V_{\mathcal{A}}$.*

PROOF. Let N be a positive integer. We show that for any n pairs from $\mathcal{R}^d \times \{0, 1\}$, if N sets from $\bar{\mathcal{A}}$ pick N different subsets of the n pairs, then there are N corresponding sets in $\mathcal{A}$ that pick N different subsets of n points in $\mathcal{R}^d$, and vice versa. Fix n pairs $(x_1, 0), \ldots, (x_m, 0), (x_{m+1}, 1), \ldots, (x_n, 1)$. Note that since ordering does not matter, we may arrange any n pairs in this manner. Assume that for a certain set $A \in \mathcal{A}$, the corresponding set $\bar{A} = A \times \{0\} \bigcup A^c \times \{1\} \in \bar{\mathcal{A}}$ picks out the pairs $(x_1, 0), \ldots, (x_k, 0), (x_{m+1}, 1), \ldots, (x_{m+l}, 1)$, that is, the set of these pairs is the intersection of $\bar{A}$ and the n pairs. Again, we can assume without loss of generality that the pairs are ordered in this way. This means that A picks from the set

$\{x_1, \ldots, x_n\}$ the subset $\{x_1, \ldots, x_k, x_{m+l+1}, \ldots, x_n\}$, and the two subsets uniquely determine each other. This proves $s(\bar{A}, n) \leq s(A, n)$. To prove the other direction, notice that if A picks a subset of k points $x_1, \ldots, x_k$, then the corresponding set $\bar{A} \in \bar{A}$ picks the pairs with the same indices from $\{(x_1, 0), \ldots, (x_k, 0)\}$. Equality of the VC dimensions follows from the equality of the shatter coefficients. $\square$

The following theorem, attributed to Vapnik and Chervonenkis (1971) and Sauer (1972), describes the relationship between the VC dimension and shatter coefficients of a class of sets. This is the most important tool for obtaining useful upper bounds on the shatter coefficients in terms of the VC dimension.

Theorem 13.2. *If A is a class of sets with VC dimension V_A, then for every n*

$$s(A, n) \leq \sum_{i=0}^{V_A} \binom{n}{i}.$$

PROOF. Recall the definition of the shatter coefficients

$$s(A, n) = \max_{(x_1, \ldots, x_n)} N_A(x_1, \ldots, x_n),$$

where

$$N_A(x_1, \ldots, x_n) = \left| \left\{ \{x_1, \ldots, x_n\} \bigcap A; A \in A \right\} \right|.$$

Clearly, it suffices to prove that for every $x_1, \ldots, x_n$,

$$N_A(x_1, \ldots, x_n) \leq \sum_{i=0}^{V_A} \binom{n}{i}.$$

But since $N_A(x_1, \ldots, x_n)$ is just the shatter coefficient of the class of finite sets

$$\left\{ \{x_1, \ldots, x_n\} \bigcap A; A \in A \right\},$$

we need only to prove the theorem for finite sets. We assume without loss of generality that A is a class of subsets of $\{x_1, \ldots, x_n\}$ with VC dimension V_A. Note that in this case $s(A, n) = |A|$.

We prove the theorem by induction with respect to n and V_A. The statement is obviously true for $n = 1$ for any class with $V_A \geq 1$. It is also true for any $n \geq 1$ if $V_A = 0$, since $s(A, n) = 1$ for all n in this case. Thus, we assume $V_A \geq 1$. Assume that the statement is true for all $k < n$ for all classes of subsets of $\{x_1, \ldots, x_k\}$ with VC dimension not exceeding V_A, and for n for all classes with VC dimension smaller than V_A. Define the following two classes of subsets of $\{x_1, \ldots, x_n\}$:

$$A' = \{A - \{x_n\}; A \in A\},$$

and

$$\hat{A} = \left\{ \hat{A} \in A : x_n \notin \hat{A}, \hat{A} \cup \{x_n\} \in A \right\}.$$

Note that both $\mathcal{A}'$ and $\widehat{\mathcal{A}}$ contain subsets of $\{x_1, \ldots, x_{n-1}\}$. $\widehat{\mathcal{A}}$ contains all sets $\widehat{A}$ that are members of $\mathcal{A}$ such that $\widehat{A} \cup \{x_n\}$ is also in $\mathcal{A}$, but $x_n \notin \widehat{A}$.

Then $|\mathcal{A}| = |\mathcal{A}'| + |\widehat{\mathcal{A}}|$. To see this, write

$$\mathcal{A}' = \{A - \{x_n\} : x_n \in A, A \in \mathcal{A}\} \bigcup \{A - \{x_n\} : x_n \notin A, A \in \mathcal{A}\} = \mathcal{B}_1 \cup \mathcal{B}_2.$$

Thus,

$$\begin{aligned}
|\mathcal{A}'| &= |\mathcal{B}_1| + |\mathcal{B}_2| - |\mathcal{B}_1 \cap \mathcal{B}_2| \\
&= |\{A - \{x_n\} : x_n \in A, A \in \mathcal{A}\}| + |\{A - \{x_n\} : x_n \notin A, A \in \mathcal{A}\}| - |\widehat{\mathcal{A}}| \\
&= |\{A : x_n \in A, A \in \mathcal{A}\}| + |\{A : x_n \notin A, A \in \mathcal{A}\}| - |\widehat{\mathcal{A}}| \\
&= |\mathcal{A}| - |\widehat{\mathcal{A}}|.
\end{aligned}$$

Since $|\mathcal{A}'| \leq |\mathcal{A}|$, and $\mathcal{A}'$ is a class of subsets of $\{x_1, \ldots, x_{n-1}\}$, the induction hypothesis implies that

$$|\mathcal{A}'| = s(\mathcal{A}', n - 1) \leq \sum_{i=0}^{V_\mathcal{A}} \binom{n - 1}{i}.$$

Next we show that $V_{\widehat{\mathcal{A}}} \leq V_\mathcal{A} - 1$, which will imply

$$|\widehat{\mathcal{A}}| = s(\widehat{\mathcal{A}}, n - 1) \leq \sum_{i=0}^{V_\mathcal{A}-1} \binom{n - 1}{i}$$

by the induction hypothesis. To see this, consider a set $S \subset \{x_1, \ldots, x_{n-1}\}$ that is shattered by $\widehat{\mathcal{A}}$. Then $S \cup \{x_n\}$ is shattered by $\mathcal{A}$. To prove this we have to show that any set $S' \subset S$ and $S' \cup \{x_n\}$ is the intersection of $S \cup \{x_n\}$ and a set from $\mathcal{A}$. Since S is shattered by $\widehat{\mathcal{A}}$, if $S' \subset S$, then there exists a set $\widehat{A} \in \widehat{\mathcal{A}}$ such that $S' = S \cap \widehat{A}$. But since by definition $x_n \notin \widehat{A}$, we must have

$$S' = (S \cup \{x_n\}) \cap \widehat{A}$$

and

$$S' \cup \{x_n\} = (S \cup \{x_n\}) \cap \left(\widehat{A} \cup \{x_n\}\right).$$

Since by the definition of $\widehat{\mathcal{A}}$ both $\widehat{A}$ and $\widehat{A} \cup \{x_n\}$ are in $\mathcal{A}$, we see that $S \cup \{x_n\}$ is indeed shattered by $\mathcal{A}$. But any set that is shattered by $\mathcal{A}$ must have cardinality not exceeding $V_\mathcal{A}$, therefore $|S| \leq V_\mathcal{A} - 1$. But S was an arbitrary set shattered by $\widehat{\mathcal{A}}$, which means $V_{\widehat{\mathcal{A}}} \leq V_\mathcal{A} - 1$. Thus, we have shown that

$$s(\mathcal{A}, n) = |\mathcal{A}| = |\mathcal{A}'| + |\widehat{\mathcal{A}}| \leq \sum_{i=0}^{V_\mathcal{A}} \binom{n - 1}{i} + \sum_{i=0}^{V_\mathcal{A}-1} \binom{n - 1}{i}.$$

Straightforward application of the identity $\binom{n}{i} = \binom{n-1}{i} + \binom{n-1}{i-1}$ shows that

$$\sum_{i=0}^{V_\mathcal{A}} \binom{n}{i} = \sum_{i=0}^{V_\mathcal{A}} \binom{n - 1}{i} + \sum_{i=0}^{V_\mathcal{A}-1} \binom{n - 1}{i}. \quad \square$$

Theorem 13.2 has some very surprising implications. For example, it follows immediately from the binomial theorem that $s(\mathcal{A}, n) \leq (n + 1)^{V_{\mathcal{A}}}$. This means that a shatter coefficient falls in one of two categories: either $s(\mathcal{A}, n) = 2^n$ for all n, or $s(\mathcal{A}, n) \leq (n + 1)^{V_{\mathcal{A}}}$, which happens if the vc dimension of $\mathcal{A}$ is finite. We cannot have $s(\mathcal{A}, n) \sim 2^{\sqrt{n}}$, for example. If $V_{\mathcal{A}} < \infty$, the upper bound in Theorem 12.5 decreases exponentially quickly with n. Other sharper bounds are given below.

Theorem 13.3. *For all $n > 2V$,*

$$s(\mathcal{A}, n) \leq \sum_{i=0}^{V_{\mathcal{A}}} \binom{n}{i} \leq \left(\frac{en}{V_{\mathcal{A}}}\right)^{V_{\mathcal{A}}}.$$

For $V_{\mathcal{A}} > 2$, $s(\mathcal{A}, n) \leq n^{V_{\mathcal{A}}}$, and for all $V_{\mathcal{A}}$, $s(\mathcal{A}, n) \leq n^{V_{\mathcal{A}}} + 1$.

Theorem 13.3 follows from Theorem 13.4 below. We leave the details as an exercise (see Problem 13.2).

Theorem 13.4. *For all $n \geq 1$ and $V_{\mathcal{A}} < n/2$,*

$$s(\mathcal{A}, n) \leq e^{n\mathcal{H}\left(\frac{V_{\mathcal{A}}}{n}\right)},$$

where $\mathcal{H}(x) = -x \log x - (1 - x)\log(1 - x)$ for $x \in (0, 1)$, and $\mathcal{H}(0) = \mathcal{H}(1) = 0$ is the binary entropy function.

Theorem 13.4 is a consequence of Theorem 13.2, and the inequality below. A different, probabilistic proof is sketched in Problem 13.3 (see also Problem 13.4).

Lemma 13.1. *For $k < n/2$,*

$$\sum_{i=0}^{k} \binom{n}{i} \leq e^{n\mathcal{H}\left(\frac{k}{n}\right)}.$$

PROOF. Introduce $\lambda = k/n \leq 1/2$. By the binomial theorem,

$$
\begin{aligned}
1 &= (\lambda + (1 - \lambda))^n \\
&\geq \sum_{i=1}^{\lambda n} \binom{n}{i} \lambda^i (1 - \lambda)^{n-i} \\
&\geq \sum_{i=1}^{\lambda n} \binom{n}{i} \left(\frac{\lambda}{1 - \lambda}\right)^{\lambda n} (1 - \lambda)^n \quad \text{(since } \lambda/(1 - \lambda) \leq 1\text{)} \\
&= e^{-n\mathcal{H}(\lambda)} \sum_{i=0}^{k} \binom{n}{i},
\end{aligned}
$$

the desired inequality. $\square$

REMARK. The binary entropy function $\mathcal{H}(x)$ plays a central role in information theory (see, e.g., Csiszár and Körner (1981), Cover and Thomas (1991)). Its main properties are the following: $\mathcal{H}(x)$ is symmetric around $1/2$, where it takes its maximum. It is continuous, concave, strictly monotone increasing in $[0, 1/2]$, decreasing in $[1/2, 1]$, and equals zero for $x = 0$, and $x = 1$. $\square$

Next we present some simple results about shatter coefficients of classes that are obtained by combinations of classes of sets.

Theorem 13.5.

   (i)   *If $\mathcal{A} = \mathcal{A}_1 \cup \mathcal{A}_2$, then $s(\mathcal{A}, n) \leq s(\mathcal{A}_1, n) + s(\mathcal{A}_2, n)$.*

   (ii)   *Given a class $\mathcal{A}$ define $\mathcal{A}_c = \{A^c; A \in \mathcal{A}\}$. Then $s(\mathcal{A}_c, n) = s(\mathcal{A}, n)$.*

   (iii)   *For $\mathcal{A} = \left\{ \widehat{A} \cap \tilde{A}; \widehat{A} \in \widehat{\mathcal{A}}, \tilde{A} \in \tilde{\mathcal{A}} \right\}$, $s(\mathcal{A}, n) \leq s(\widehat{\mathcal{A}}, n)s(\tilde{\mathcal{A}}, n)$.*

   (iv)   *For $\mathcal{A} = \left\{ \widehat{A} \cup \tilde{A}; \widehat{A} \in \widehat{\mathcal{A}}, \tilde{A} \in \tilde{\mathcal{A}} \right\}$, $s(\mathcal{A}, n) \leq s(\widehat{\mathcal{A}}, n)s(\tilde{\mathcal{A}}, n)$.*

   (v)   *For $\mathcal{A} = \left\{ \widehat{A} \times \tilde{A}; \widehat{A} \in \widehat{\mathcal{A}}, \tilde{A} \in \tilde{\mathcal{A}} \right\}$, $s(\mathcal{A}, n) \leq s(\widehat{\mathcal{A}}, n)s(\tilde{\mathcal{A}}, n)$.*

PROOF. (i), (ii), and (v) are trivial. To prove (iii), fix n points $x_1, \ldots, x_n$, and assume that $\widehat{\mathcal{A}}$ picks $N \leq s(\widehat{\mathcal{A}}, n)$ subsets $C_1, \ldots, C_N$. Then $\tilde{\mathcal{A}}$ picks from C_i at most $s(\tilde{\mathcal{A}}, |C_i|)$ subsets. Therefore, sets of the form $\widehat{A} \cap \tilde{A}$ pick at most

$$\sum_{i=1}^{N} s(\tilde{\mathcal{A}}, |C_i|) \leq s(\widehat{\mathcal{A}}, n)s(\tilde{\mathcal{A}}, n)$$

subsets. Here we used the obvious monotonicity property $s(\mathcal{A}, n) \leq s(\mathcal{A}, n + m)$. To prove (iv), observe that

$$\left\{ \widehat{A} \cup \tilde{A}; \widehat{A} \in \widehat{\mathcal{A}}, \tilde{A} \in \tilde{\mathcal{A}} \right\} = \left\{ \left(\widehat{A}^c \cap \tilde{A}^c \right)^c ; \widehat{A} \in \widehat{\mathcal{A}}, \tilde{A} \in \tilde{\mathcal{A}} \right\}.$$

The statement now follows from (ii) and (iii). $\square$

13.2   Shatter Coefficients of Some Classes

Here we calculate shatter coefficients of some simple but important examples of classes of subsets of $\mathcal{R}^d$. We begin with a simple observation.

Theorem 13.6. *If $\mathcal{A}$ contains finitely many sets, then $V_{\mathcal{A}} \leq \log_2 |\mathcal{A}|$, and $s(\mathcal{A}, n) \leq |\mathcal{A}|$ for every n.*

PROOF. The first inequality follows from the fact that at least 2^n sets are necessary to shatter n points. The second inequality is trivial. $\square$

In the next example, it is interesting to observe that the bound of Theorem 13.2 is tight.

Theorem 13.7.

(i) *If A is the class of all half lines: $A = \{(-\infty, x]; x \in \mathcal{R}\}$, then $V_A = 1$, and*

$$s(A, n) = n + 1 = \binom{n}{0} + \binom{n}{1}.$$

(ii) *If A is the class of all intervals in $\mathcal{R}$, then $V_A = 2$, and*

$$s(A, n) = \frac{n(n+1)}{2} + 1 = \binom{n}{0} + \binom{n}{1} + \binom{n}{2}.$$

PROOF. (i) is easy. To see that $V_A = 2$ in (ii), observe that if we fix three different points in $\mathcal{R}$, then there is no interval that does not contain the middle point, but does contain the other two. The shatter coefficient can be calculated by counting that there are at most $n - k + 1$ sets in $\{A \cap \{x_1, \ldots, x_n\}; A \in A\}$ such that $|A \cap \{x_1, \ldots, x_n\}| = k$ for $k = 1, \ldots, n$, and one set (namely $\emptyset$) such that $|A \cap \{x_1, \ldots, x_n\}| = 0$. This gives altogether

$$1 + \sum_{k=1}^{n}(n - k + 1) = \frac{n(n+1)}{2} + 1. \quad \square$$

Now we can generalize the result above for classes of intervals and rectangles in $\mathcal{R}^d$:

Theorem 13.8.

(i) *If $A = \{(-\infty, x_1] \times \cdots \times (-\infty, x_d]\}$, then $V_A = d$.*
(ii) *If A is the class of all rectangles in $\mathcal{R}^d$, then $V_A = 2d$.*

PROOF. We prove (ii). The first part is left as an exercise (Problem 13.5). We have to show that there are $2d$ points that can be shattered by A, but for any set of $2d + 1$ points there is a subset of it that can not be picked by sets in A. To see the first part just consider the following $2d$ points:

$$(1, 0, 0, \ldots, 0), (0, 1, 0, 0, \ldots, 0), \ldots, (0, 0, \ldots, 0, 1),$$
$$(-1, 0, 0, \ldots, 0), (0, -1, 0, 0, \ldots, 0), \ldots, (0, 0, \ldots, 0, -1),$$

(see Figure 13.1).

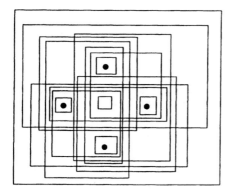

FIGURE 13.1. $2^4 = 16$ *rectangles shatter 4 points in* $\mathcal{R}^2$.

On the other hand, for any given set of $2d + 1$ points we can choose a subset of at most $2d$ points with the property that it contains a point with largest first coordinate, a point with smallest first coordinate, a point with largest second coordinate, and so forth. Clearly, there is no set in $\mathcal{A}$ that contains these points, but does not contain the others (Figure 13.2). □

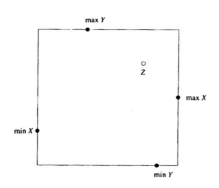

FIGURE 13.2. *No 5 points can be shattered by rectangles in* $\mathcal{R}^2$.

Theorem 13.9. (STEELE (1975), DUDLEY (1978)). *Let* $\mathcal{G}$ *be a finite-dimensional vector space of real functions on* $\mathcal{R}^d$. *The class of sets*

$$\mathcal{A} = \{\{x : g(x) \geq 0\} : g \in \mathcal{G}\}$$

has VC *dimension* $V_{\mathcal{A}} \leq r$, *where* $r = dimension(\mathcal{G})$.

PROOF. It suffices to show that no set of size $m = 1 + r$ can be shattered by sets of the form $\{x : g(x) \geq 0\}$. Fix m arbitrary points $x_1, \ldots, x_m$, and define the linear mapping $L : \mathcal{G} \to \mathcal{R}^m$ as

$$L(g) = (g(x_1), \ldots, g(x_m)).$$

Then the image of $\mathcal{G}$, $L(\mathcal{G})$, is a linear subspace of $\mathcal{R}^m$ of dimension not exceeding the dimension of $\mathcal{G}$, that is, $m - 1$. Then there exists a nonzero vector $\gamma = (\gamma_1, \ldots, \gamma_m) \in \mathcal{R}^m$, that is orthogonal to $L(\mathcal{G})$, that is, for every $g \in \mathcal{G}$

$$\gamma_1 g(x_1) + \ldots + \gamma_m g(x_m) = 0.$$

We can assume that at least one of the γ_i's is negative. Rearrange this equality so that terms with nonnegative γ_i stay on the left-hand side:

$$\sum_{i:\gamma_i \geq 0} \gamma_i g(x_i) = \sum_{i:\gamma_i < 0} -\gamma_i g(x_i).$$

Now, suppose that there exists a $g \in \mathcal{G}$ such that the set $\{x : g(x) \geq 0\}$ picks exactly the x_i's on the left-hand side. Then all terms on the left-hand side are nonnegative, while the terms on the right-hand side must be negative, which is a contradiction, so $x_1, \ldots, x_m$ cannot be shattered, and the proof is completed. $\square$

REMARK. Theorem 13.2 implies that the shatter coefficients of the class of sets in Theorem 13.9 are bounded as follows:

$$s(\mathcal{A}, n) \leq \sum_{i=0}^{r} \binom{n}{i}.$$

In many cases it is possible to get sharper estimates. Let

$$\mathcal{G} = \left\{ \sum_{i=1}^{r} a_i \psi_i : a_1, \ldots, a_r \in \mathcal{R} \right\}$$

be the linear space of functions spanned by some fixed functions $\psi_1, \ldots, \psi_r$: $\mathcal{R}^d \to \mathcal{R}$, and define $\Psi(x) = (\psi_1(x), \ldots, \psi_r(x))$. Cover (1965) showed that if for some $x_1, \ldots, x_n \in \mathcal{R}^d$, every r-element subset of $\Psi(x_1), \ldots, \Psi(x_n)$ is linearly independent, then the n-th shatter coefficient of the class of sets $\mathcal{A} = \{\{x : g(x) \geq 0\} : g \in \mathcal{G}\}$ actually equals

$$s(\mathcal{A}, n) = 2 \sum_{i=0}^{r-1} \binom{n-1}{i}$$

(see Problem 13.6). By using the last identity in the proof of Theorem 13.2, it is easily seen that the difference between the bound obtained from Theorem 13.9 and the true value is $\binom{n-1}{r}$. Using Cover's result for the shatter coefficients we can actually improve Theorem 13.9, in that the VC dimension of the class $\mathcal{A}$ *equals* r. To see this, note that

$$s(\mathcal{A}, r) = 2 \sum_{i=0}^{r-1} \binom{r-1}{i} = 2 \cdot 2^{r-1} = 2^r,$$

while

$$s(\mathcal{A}, r+1) = 2 \sum_{i=0}^{r-1} \binom{r}{i} = 2 \sum_{i=0}^{r} \binom{r}{i} - 2\binom{r}{r} = 2 \cdot 2^r - 2 < 2^{r+1}.$$

Therefore no $r + 1$ points are shattered. It is interesting that Theorem 13.9 above combined with Theorem 13.2 and Theorem 13.3 gives the bound

$$s(\mathcal{A}, n) \le n^r + 1$$

when $r > 2$. Cover's result, however, improves it to

$$s(\mathcal{A}, n) \le 2(n-1)^{r-1} + 2. \quad \square$$

Perhaps the most important class of sets is the class of halfspaces in $\mathcal{R}^d$, that is, sets containing points falling on one side of a hyperplane. The shatter coefficients of this class can be obtained from the results above:

COROLLARY 13.1. *Let $\mathcal{A}$ be the class of halfspaces, that is, subsets of $\mathcal{R}^d$ of the form $\{x : ax \ge b\}$, where $a \in \mathcal{R}^d$, $b \in \mathcal{R}$ take all possible values. Then $V_{\mathcal{A}} = d+1$, and*

$$s(\mathcal{A}, n) = 2 \sum_{i=0}^{d} \binom{n-1}{i} \le 2(n-1)^d + 2.$$

PROOF. This is an immediate consequence of the remark above if we take $\mathcal{G}$ to be the linear space spanned by the functions

$$\phi_1(x) = x^{(1)}, \quad \phi_2(x) = x^{(2)}, \quad \ldots, \quad \phi_d(x) = x^{(d)}, \quad \text{and } \phi_{d+1}(x) = 1,$$

where $x^{(1)}, \ldots, x^{(d)}$ denote the d components of the vector x. $\square$

It is equally simple now to obtain an upper bound on the VC dimension of the class of all closed balls in $\mathcal{R}^d$ (see Cover (1965), Devroye (1978), or Dudley (1979) for more information).

COROLLARY 13.2. *Let $\mathcal{A}$ be the class of all closed balls in $\mathcal{R}^d$, that is, subsets of $\mathcal{R}^d$ of the form*

$$\left\{ x = (x^{(1)}, \ldots, x^{(d)}) : \sum_{i=1}^{d} |x^{(i)} - a_i|^2 \le b \right\},$$

where $a_1, \ldots, a_d, b \in \mathcal{R}$ take all possible values. Then $V_{\mathcal{A}} \le d + 2$.

PROOF. If we write

$$\sum_{i=1}^{d} |x^{(i)} - a_i|^2 - b = \sum_{i=1}^{d} |x^{(i)}|^2 - 2 \sum_{i=1}^{d} x^{(i)} a_i + \sum_{i=1}^{d} a_i^2 - b,$$

then it is clear that Theorem 13.9 yields the result by setting $\mathcal{G}$ to be the linear space spanned by

$$\phi_1(x) = \sum_{i=1}^{d} |x^{(i)}|^2, \quad \phi_2(x) = x^{(1)}, \quad \ldots, \quad \phi_{d+1}(x) = x^{(d)}, \quad \text{and } \phi_{d+2}(x) = 1. \quad \Box$$

It follows from Theorems 13.9 and 13.5 (iii) that the class of all polytopes with a bounded number of faces has finite VC dimension. The next negative example demonstrates that this boundedness is necessary.

Theorem 13.10. *If $\mathcal{A}$ is the class of all convex polygons in $\mathcal{R}^2$, then $V_{\mathcal{A}} = \infty$.*

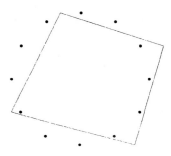

FIGURE 13.3. *Any subset of n points on the unit circle can be picked by a convex polygon.*

PROOF. Let $x_1, \ldots, x_n \in \mathcal{R}^2$ lie on the unit circle. Then it is easy to see that for any subset of these (different) points there is a polygon that picks that subset. $\Box$

13.3   Linear and Generalized Linear Discrimination Rules

Recall from Chapter 4 that a linear classification rule classifies x into one of the two classes according to whether

$$a_0 + \sum_{i=1}^{d} a_i x^{(i)}$$

is positive or negative, where $x^{(1)}, \ldots, x^{(d)}$ denote the components of $x \in \mathcal{R}^d$. The coefficients a_i are determined by the training sequence. These decisions dichotomize the space $\mathcal{R}^d$ by virtue of a halfspace, and assign class 1 to one halfspace, and class 0 to the other. Points on the border are treated as belonging to class 0. Consider a classifier that adjusts the coefficients by minimizing the number of errors committed on D_n. In the terminology of Chapter 12, C is the collection of all linear classifiers.

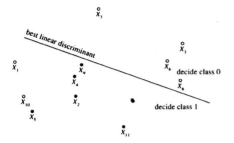

FIGURE 13.4. *An empirically optimal linear classifier.*

Glick (1976) pointed out that for the error probability $L(\phi_n^*)$ of this classifier, $L(\phi_n^*) - \inf_{\phi \in C} L(\phi) \to 0$ almost surely. However, from Theorems 12.6, 13.1 and Corollary 13.1, we can now provide more details:

Theorem 13.11. *For all n and $\epsilon > 0$, the error probability $L(\phi_n^*)$ of the empirically optimal linear classifier satisfies*

$$\mathbf{P}\left\{ L(\phi_n^*) - \inf_{\phi \in C} L(\phi) > \epsilon \right\} \le 8n^{d+1} e^{-n\epsilon^2/128}.$$

Comparing the above inequality with Theorem 4.5, note that there we selected ϕ_n^* by a specific algorithm, while this result holds for any linear classifier whose empirical error is minimal.

Generalized linear classification rules (see Duda and Hart (1973)) are defined by

$$g_n(x) = \begin{cases} 0 & \text{if } a_0 + \sum_{i=1}^{d^*} a_i \psi_i(x) \ge 0 \\ 1 & \text{otherwise,} \end{cases}$$

where d^* is a positive integer, the functions $\psi_1, \ldots, \psi_{d^*}$ are fixed, and the coefficients $a_0, \ldots, a_{d^*}$ are functions of the data D_n. These include for example all quadratic discrimination rules in $\mathcal{R}^d$ when we choose all functions that are either components of x, or squares of components of x, or products of two components of x. That is, the functions $\psi_i(x)$ are of the form either $x^{(j)}$, or $x^{(j)}x^{(k)}$. In all, $d^* = 2d + d(d-1)/2$. The argument used for linear discriminants remain valid, and we obtain

Theorem 13.12. *Let C be the class of generalized linear discriminants (i.e., the coefficients vary, the basis functions ψ_i are fixed). For the error probability $L(\phi_n^*)$ of the empirically optimal classifier, for all $d^* > 1, n$ and $\epsilon > 0$, we have*

$$\mathbf{P}\left\{ L(\phi_n^*) - \inf_{\phi \in C} L(\phi) > \epsilon \right\} \le 8n^{d^*+1} e^{-n\epsilon^2/128}.$$

Also, for $n > 2d^* + 1$,

$$\mathbf{P}\left\{L(\phi_n^*) - \inf_{\phi \in C} L(\phi) > \epsilon\right\} \leq 8e^{n\mathcal{H}(\frac{d^*+1}{n})}e^{-n\epsilon^2/128}.$$

The second inequality is obtained by using the bound of Theorem 13.4 for the shatter coefficients. Note nevertheless that unless d^* (and therefore C) is allowed to increase with n, there is no hope of obtaining universal consistency. The question of universal consistency will be addressed in Chapters 17 and 18.

13.4   Convex Sets and Monotone Layers

Classes of infinite VC dimension are not hopeless by any means. In this section, we offer examples that will show how they may be useful in pattern recognition. The classes of interest to us for now are

$$C = \left\{\text{all convex sets of } \mathcal{R}^2\right\}$$

$$\mathcal{L} = \left\{\text{all monotone layers of } \mathcal{R}^2, \text{ i.e., all sets of the form}\right.$$

$$\left. \{(x_1, x_2) : x_2 \leq \psi(x_1)\} \text{ for some nonincreasing function } \psi\right\}.$$

In discrimination, this corresponds to making decisions of the form $\phi(x) = I_{\{x \in C\}}$, $C \in C$, or $\phi(x) = I_{\{x \notin C\}}$, $C \in C$, and similarly for $\mathcal{L}$. Decisions of these forms are important in many situations. For example, if $\eta(x)$ is monotone decreasing in both components of $x \in \mathcal{R}^2$, then the Bayes rule is of the form $g^*(x) = I_{\{x \in L\}}$ for some $L \in \mathcal{L}$. We have pointed out elsewhere (Theorem 13.10, and Problem 13.19) that $V_C = V_{\mathcal{L}} = \infty$. To see this, note that any set of n points on the unit circle is shattered by C, while any set of n points on the antidiagonal $x_2 = -x_1$ is shattered by $\mathcal{L}$. Nevertheless, shattering becomes unlikely if X has a density. Our starting point here is the bound obtained while proving Theorem 12.5:

$$\mathbf{P}\left\{\sup_{A \in \mathcal{A}} |\mu_n(A) - \mu(A)| > \epsilon\right\} \leq 8\mathbf{E}\{N_A(X_1, \ldots, X_n)\}e^{-n\epsilon^2/32},$$

where $N_A(X_1, \ldots, X_n)$ is the number of sets in $\{A \cap \{X_1, \ldots, X_n\} : A \in \mathcal{A}\}$. The following theorem is essential:

Theorem 13.13. *If X has a density f on $\mathcal{R}^2$, then*

$$\mathbf{E}\{N_A(X_1, \ldots, X_n)\} = 2^{o(n)}$$

when $\mathcal{A}$ is either C or $\mathcal{L}$.

This theorem, a proof of which is a must for the reader, implies the following:

COROLLARY 13.3. *Let X have a density f on $\mathcal{R}^2$. Let ϕ_n^* be picked by minimizing the empirical error over all classifiers of the form*

$$\phi(x) = \begin{cases} 1 & \text{if } x \in A \\ 0 & \text{if } x \notin A, \end{cases}$$

where A or A^c is in C (or $\mathcal{L}$). Then

$$L(\phi_n^*) \rightarrow \inf_{\phi = I_C \text{ for } C \text{ or } C^c \text{ in } C} L(\phi)$$

with probability one (and similarly for $\mathcal{L}$).

PROOF. This follows from the inequality of Lemma 8.2, Theorem 13.13, and the Borel-Cantelli lemma. □

REMARK. Theorem 13.13 and the corollary may be extended to $\mathcal{R}^d$, but this generalization holds nothing new and will only result in tedious notations. □

PROOF OF THEOREM 13.13.   We show the theorem for $\mathcal{L}$, and indicate the proof for C. Take two sequences of integers, m and r, where $m \sim \sqrt{n}$ and $r \sim m^{1/3}$, so that $m \rightarrow \infty$, yet $r^2/m \rightarrow 0$, as $n \rightarrow \infty$. Consider the set $[-r, r]^2$ and partition each side into m equal intervals, thus obtaining an $m \times m$ grid of square cells $C_1, \ldots, C_{m^2}$. Denote $C_0 = \mathcal{R}^2 - [-r, r]^2$. Let $N_0, N_1, \ldots, N_{m^2}$ be the number of points among $X_1, \ldots, X_n$ that belong to these cells. The vector $(N_0, N_1, \ldots, N_{m^2})$ is clearly multinomially distributed.

Let ψ be a nonincreasing function $\mathcal{R} \rightarrow \mathcal{R}$ defining a set in $\mathcal{L}$ by $L = \{(x_1, x_2) : x_2 \leq \psi(x_1)\}$. Let $C(\psi)$ be the collection of all cell sets cut by ψ, that is, all cells with a nonempty intersection with both L and L^c. The collection $C(\psi)$ is shaded in Figure 13.5.

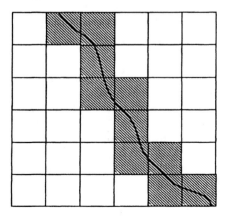

FIGURE 13.5. *A monotone layer and bordering cells.*

We bound $N_{\mathcal{L}}(X_1, \ldots, X_n)$ from above, conservatively, as follows:

$$\sum_{\text{all possible } C(\psi)} \left(\prod_{C_i \in C(\psi)} 2^{N_i} \right) 2^{N_0}. \tag{13.1}$$

The number of different collections $C(\psi)$ cannot exceed 2^{2m} because each cell in $C(\psi)$ may be obtained from its predecessor cell by either moving right on the same row or moving down one cell in the same column. For a particular collection, denoting $p_i = \mathbf{P}\{X \in C_i\}$, we have

$$\mathbf{E}\left\{ \left(\prod_{C_i \in C(\psi)} 2^{N_i} \right) 2^{N_0} \right\}$$

$$= \left(\sum_{C_i \in C(\psi)} 2p_i + 2p_0 + 1 - \sum_{C_i \in C(\psi)} p_i - p_0 \right)^n$$

(by applying Lemma A.7)

$$\leq \exp\left(n \left(\sum_{C_i \in C(\psi)} p_i + p_0 \right) \right)$$

$$\leq \exp\left(n \left(\sup_{\text{all sets } A \text{ with } \lambda(A) \leq 8r^2/m} \int_A f + \int_{C_0} f \right) \right)$$

$$\left(\text{since } \lambda \left(\bigcup_{i:C_i \in C(\psi)} C_i \right) \leq 2m(2r/m)^2 = 8r^2/m \right)$$

$$= e^{o(n)},$$

because $r^2/m \to 0$ and $r \to \infty$ and by the absolute continuity of X. As this estimate is uniform over all collections $C(\psi)$, we see that the expected value of (13.1) is not more than $2^{2m} e^{o(n)} = e^{o(n)}$. The argument for the collection $\mathcal{C}$ of convex sets is analogous. $\square$

REMARK. The theorem implies that if $\mathcal{A}$ is the class of all convex sets, then $\sup_{A \in \mathcal{A}} |\mu_n(A) - \mu(A)| \to 0$ with probability one whenever μ has a density. This is a special case of a result of Ranga Rao (1962). Assume now that the density f of X is bounded and of bounded support. Then in the proof above we may take r fixed so that $[-r, r]^2$ contains the support of f. Then the estimate of Theorem 13.13 is

$$\mathbf{E}\{N_A(X_1, \ldots, X_n)\} \leq 2^{2m} \cdot e^{8n\|f\|_\infty r^2/m}$$

(where $\|f\|_\infty$ denotes the bound on f)

$$= \quad 2^{4r\sqrt{n}\|f\|_\infty} e^{4r\sqrt{n}\|f\|_\infty}$$

$$(\text{if we take } m = 2r\sqrt{n\|f\|_\infty})$$

$$\leq \quad e^{\alpha\sqrt{n}}$$

for a constant α. This implies by Corollary 12.1 that

$$\mathbf{E}\left\{ L(\phi_n^*) - \inf_{\phi = I_C \text{ for } C \text{ or } C^c \text{ in } C} L(\phi) \right\} = O\left(n^{-1/4}\right).$$

This latter inequality was proved by Steele (1975). To see that it cannot be extended to arbitrary densities, observe that the data points falling on the convex hull (or upper layer) of the points $X_1, \ldots, X_n$ can always be shattered by convex sets (or monotone layers, respectively). Thus, $N_A(X_1, \ldots, X_n)$ is at least 2^{M_n}, where M_n is the number of points among $X_1, \ldots, X_n$, falling on the convex hull (or upper layer) of $X_1, \ldots, X_n$. Thus,

$$\mathbf{E}\{N_A(X_1, \ldots, X_n)\} \geq \mathbf{E}\left\{2^{M_n}\right\} \geq 2^{\mathbf{E}\{M_n\}}.$$

But it follows from results of Carnal (1970) and Devroye (1991b) that for each $a < 1$, there exists a density such that

$$\limsup_{n \to \infty} \frac{\mathbf{E}\{M_n\}}{n^a} > 1. \quad \square$$

The important point of this interlude is that with infinite VC dimension, we may under some circumstances get expected error rates that are $o(1)$ but larger than $1/\sqrt{n}$. However, the bounds are sometimes rather loose. The reason is the looseness of the Vapnik-Chervonenkis inequality when the collections A become very big. To get such results for classes with infinite VC dimension it is necessary to impose some conditions on the distribution. We will prove this in Chapter 14.

Problems and Exercises

PROBLEM 13.1. Show that the inequality of Theorem 13.2 is tight, that is, exhibit a class A of sets such that for each n, $s(A, n) = \sum_{i=0}^{V_A} \binom{n}{i}$.

PROBLEM 13.2. Show that for all $n > 2V_A$

$$\sum_{i=0}^{V_A} \binom{n}{i} \leq 2\left(\frac{n^{V_A}}{V_A!}\right) \leq \left(\frac{en}{V_A}\right)^{V_A}$$

and that

$$\sum_{i=0}^{V_A} \binom{n}{i} \leq n^{V_A}.$$

if $V_A > 2$. HINT: There are several ways to prove the statements. One can proceed directly by using the recurrence $\sum_{i=0}^{V_A} \binom{n}{i} = \sum_{i=0}^{V_A} \binom{n-1}{i} + \sum_{i=0}^{V_A-1} \binom{n-1}{i}$. A simpler way to prove $\sum_{i=0}^{V_A} \binom{n}{i} \leq \left(\frac{en}{V_A}\right)^{V_A}$ is by using Theorem 13.4. The third inequality is an immediate consequence of the first two.

PROBLEM 13.3. Give an alternative proof of Lemma 13.1 by completing the following probabilistic argument. Observe that for $k' = n - k$,

$$\sum_{i=0}^{k} \binom{n}{i} = \sum_{i=k'}^{n} \binom{n}{i} = 2^n P\{B_n \geq k'\},$$

where B_n is a binomial random variable with parameters n and $1/2$. Then Chernoff's bounding technique (see the proof of Theorem 8.1) may be used to bound this probability: for all $s > 0$,

$$P\{B_n \geq k'\} \leq e^{-sk'} E\{e^{sB_n}\} = \exp\left(-n\left(sk'/n - \log\left(\frac{e^s + 1}{2}\right)\right)\right).$$

Take the derivative of the exponent with respect to s to minimize the upper bound. Substitute the obtained value into the bound to get the desired inequality.

PROBLEM 13.4. Let B be a binomial random variable with parameters n and p. Prove that for $k > np$

$$P\{B \geq k\} \leq \exp\left(-n\left(\mathcal{H}(k/n) + \frac{k}{n}\log p + \frac{n-k}{n}\log(1-p)\right)\right).$$

HINT: Use Chernoff's bounding technique as in the previous problem.

PROBLEM 13.5. Prove part (i) of Theorem 13.8.

PROBLEM 13.6. Prove that for the class of sets defined in the remark following Theorem 13.9,

$$s(\mathcal{A}, n) = 2\sum_{i=0}^{r-1} \binom{n-1}{i}$$

(Cover (1965)). HINT: Proceed by induction with respect to n and r. In particular, show that the recurrence $s(\mathcal{A}_r, n) = s(\mathcal{A}_r, n-1) + s(\mathcal{A}_{r-1}, n-1)$ holds, where $\mathcal{A}_r$ denotes the class of sets defined as $\{x : g(x) \geq 0\}$ where g runs through a vector space spanned by the first r of the sequence of functions $\psi_1, \psi_2, \ldots$.

PROBLEM 13.7. Let $\mathcal{A}$ and $\mathcal{B}$ be two families of subsets of $\mathcal{R}^d$. Assume that for some $r \geq 2$,

$$s(\mathcal{A}, n) \leq n^r s(\mathcal{B}, n)$$

for all $n \geq 1$. Show that if $V_B > 2$, then $V_A \leq 2(V_B + r - 1)\log(V_B + r - 1)$. HINT: By Theorem 13.3, $s(\mathcal{A}, n) \leq n^{V_B+r-1}$. Clearly, V_A is not larger than any k for which $k^{V_B+r-1} < 2^k$.

PROBLEM 13.8. Determine the VC dimension of the class of subsets of the real line such that each set in the class can be written as a union of k intervals.

PROBLEM 13.9. Determine the VC dimension of the collection of all polygons with k vertices in the plane.

PROBLEM 13.10. What is the VC dimension of the collection of all ellipsoids of $\mathcal{R}^d$?

PROBLEM 13.11. Determine the VC dimension of the collection of all subsets of $\{1, \ldots, k\}^d$, where k and d are fixed. How does the answer change if we restrict the subsets to those of cardinality $l \leq k^d$?

PROBLEM 13.12. Let $\mathcal{A}$ consist of all simplices of $\mathcal{R}^d$, that is, all sets of the form

$$\left\{ x : x = \sum_{i=1}^{d+1} \lambda_i x_i \text{ for some } \lambda_1, \ldots, \lambda_{d+1} : \sum_{i=1}^{d+1} \lambda_i = 1, \ \lambda_i \geq 0 \right\},$$

where $x_1, \ldots, x_{d+1}$ are fixed points of $\mathcal{R}^d$. Determine the VC dimension of $\mathcal{A}$.

PROBLEM 13.13. Let $A(x_1, \ldots, x_k)$ be the set of all $x \in \mathcal{R}$ that are of the form

$$x = \psi_1(i)x_1 + \psi_2(i)x_2 + \cdots + \psi_k(i)x_k, \quad i = 1, 2, \ldots,$$

where $x_1, \ldots, x_k$ are fixed numbers and $\psi_1, \ldots, \psi_k$ are fixed functions on the integers. Let $\mathcal{A} = \{A(x_1, \ldots, x_k) : x_1, \ldots, x_k \in \mathcal{R}\}$. Determine the VC dimension of $\mathcal{A}$.

PROBLEM 13.14. In some sense, VC dimension measures the "size" of a class of sets. However it has little to do with cardinalities, as this exercise demonstrates. Exhibit a class of subsets of the integers with uncountably many sets, yet VC dimension 1. (This property was pointed out to us by András Faragó.) Note: On the other hand, the class of all subsets of integers cannot be written as a countable union of classes with finite VC dimension (Theorem 18.6). HINT: Find a class of subsets of the reals with the desired properties, and make a proper correspondence between sets of integers and sets in the class.

PROBLEM 13.15. Show that if a class of sets $\mathcal{A}$ is linearly ordered by inclusion, that is, for any pair of sets $A, B \in \mathcal{A}$ either $A \subset B$ or $B \subset A$ and $|\mathcal{A}| \geq 2$, then $V_A = 1$. Conversely, assume that $V_A = 1$ and for every set B with $|B| = 2$,

$$\emptyset, B \in \{A \cap B : A \in \mathcal{A}\}.$$

Prove that then $\mathcal{A}$ is linearly ordered by inclusion (Dudley (1984)).

PROBLEM 13.16. We say that four sets A, B, C, D form a *diamond* if $A \subset B \subset C, A \subset D \subset C$, but $B \not\subset D$ and $D \not\subset C$. Let $\mathcal{A}$ be a class of sets. Show that $V_A \geq 2$ if and only if for some set R, the class $\{A \cap R : A \in \mathcal{A}\}$ includes a diamond (Dudley (1984)).

PROBLEM 13.17. Let $\mathcal{A}$ be a class of sets, and define its *density* by

$$D_A = \inf \left\{ r > 0 : \sup_{n \geq 1} \frac{s(\mathcal{A}, n)}{n^r} < \infty \right\}.$$

Verify the following properties:
   (1) $D_A \leq V_A$;
   (2) For each positive integer k, there exists a class $\mathcal{A}$ of sets such that $V_A = k$, yet $D_A = 0$; and

(3) $D_A < \infty$ if and only if $V_A < \infty$
(Assouad (1983a)).

PROBLEM 13.18. CONTINUED. Let A and A' be classes of sets, and define $B = A \cup A'$. Show that

    (1) $D_B = \max(D_A, D_{A'})$;
    (2) $V_B \leq V_A + V_{A'} + 1$; and
    (3)   For every pair of positive integers k, m there exist classes A and A' such that $V_A = k$, $V_{A'} = m$, and $V_B = k + m + 1$

(Assouad (1983a)).

PROBLEM 13.19. A set $A \subset \mathcal{R}^d$ is called a monotone layer if $x \in A$ implies that $y \in A$ for all $y \leq x$ (i.e., each component of x is not larger than the corresponding component of y). Show that the class of all monotone layers has infinite VC dimension.

PROBLEM 13.20. Let (X, Y) be a pair of random variables in $\mathcal{R}^2 \times \{0, 1\}$ such that $Y = I_{\{X \in A\}}$, where A is a convex set. Let $D_n = ((X_1, Y_1), \ldots, (X_n, Y_n))$ be an i.i.d. training sequence, and consider the classifier

$$g_n(x) = \begin{cases} 1 & \text{if } x \text{ is in the convex hull of the } X_i\text{'s with } Y = 1 \\ 0 & \text{otherwise.} \end{cases}$$

Find a distribution for which g_n is not consistent, and find conditions for consistency. HINT: Recall Theorem 13.10 and its proof.

14

Lower Bounds for Empirical Classifier Selection

In Chapter 12 a classifier was selected by minimizing the empirical error over a class of classifiers $\mathcal{C}$. With the help of the Vapnik-Chervonenkis theory we have been able to obtain distribution-free performance guarantees for the selected rule. For example, it was shown that the difference between the expected error probability of the selected rule and the best error probability in the class behaves at least as well as $O(\sqrt{V_{\mathcal{C}} \log n / n})$, where $V_{\mathcal{C}}$ is the Vapnik-Chervonenkis dimension of $\mathcal{C}$, and n is the size of the training data D_n. (This upper bound is obtained from Theorem 12.5. Corollary 12.5 may be used to replace the $\log n$ term with $\log V_{\mathcal{C}}$.) Two questions arise immediately: Are these upper bounds (at least up to the order of magnitude) tight? Is there a much better way of selecting a classifier than minimizing the empirical error? This chapter attempts to answer these questions. As it turns out, the answer is essentially affirmative for the first question, and negative for the second.

These questions were also asked in the learning theory setup, where it is usually assumed that the error probability of the best classifier in the class is zero (see Blumer, Ehrenfeucht, Haussler, and Warmuth (1989), Haussler, Littlestone, and Warmuth (1988), and Ehrenfeucht, Haussler, Kearns, and Valiant (1989)). In this case, as the bound of Theorem 12.7 implies, the error of the rule selected by minimizing the empirical error is within $O(V_{\mathcal{C}} \log n / n)$ of that of the best in the class (which equals zero, by assumption). We will see that essentially there is no way to beat this upper bound either.

14.1   Minimax Lower Bounds

Let us formulate exactly what we are interested in. Let C be a class of decision functions $\phi : \mathcal{R}^d \to \{0, 1\}$. The training sequence $D_n = ((X_1, Y_1), \ldots, (X_n, Y_n))$ is used to select the classifier $g_n(X) = g_n(X, D_n)$ from C, where the selection is based on the data D_n. We emphasize here that g_n can be an arbitrary function of the data, we do not restrict our attention to empirical error minimization, where g_n is a classifier in C that minimizes the number errors committed on the data D_n.

As before, we measure the performance of the selected classifier by the difference between the error probability $L(g_n) = \mathbf{P}\{g_n(X) \neq Y | D_n\}$ of the selected classifier and that of the best in the class. To save space further on, denote this optimum by

$$L_C \overset{\text{def}}{=} \inf_{\phi \in C} \mathbf{P}\{\phi(X) \neq Y\}.$$

In particular, we seek lower bounds for

$$\sup \mathbf{P}\{L(g_n) - L_C > \epsilon\},$$

and

$$\sup \mathbf{E}L(g_n) - L_C,$$

where the supremum is taken over all possible distributions of the pair (X, Y). A lower bound for one of these quantities means that no matter what our method of picking a rule from C is, we may face a distribution such that our method performs worse than the bound. This view may be criticized as too pessimistic. However, it is clearly a perfectly meaningful question to pursue, as typically we have no other information available than the training data, so we have to be prepared for the worst situation.

Actually, we investigate a stronger problem, in that the supremum is taken over all distributions with L_C kept at a fixed value between zero and $1/2$. We will see that the bounds depend on n, V_C, and L_C jointly. As it turns out, the situations for $L_C > 0$ and $L_C = 0$ are quite different. Because of its relative simplicity, we first treat the case $L_C = 0$. All the proofs are based on a technique called "the probabilistic method." The basic idea here is that the existence of a "bad" distribution is proved by considering a large class of distributions, and bounding the average behavior over the class.

Lower bounds on the probabilities $\mathbf{P}\{L(g_n) - L_C > \epsilon\}$ may be translated into lower bounds on the sample complexity $N(\epsilon, \delta)$. We obtain lower bounds for the size of the training sequence such that for any classifier, $\mathbf{P}\{L(g_n) - L_C > \epsilon\}$ cannot be smaller than δ for all distributions if n is smaller than this bound.

14.2   The Case $L_C = 0$

In this section we obtain lower bounds under the assumption that the best classifier in the class has zero error probability. In view of Theorem 12.7 we see that the situation here is different from when $L_C > 0$; there exist methods of picking a classifier from C (e.g., minimization of the empirical error) such that the error probability decreases to zero at a rate of $O(V_C \log n / n)$. We obtain minimax lower bounds close to the upper bounds obtained for empirical error minimization. For example, Theorem 14.1 shows that if $L_C = 0$, then the expected error probability cannot decrease faster than a sequence proportional to V_C / n for some distributions.

Theorem 14.1. (VAPNIK AND CHERVONENKIS (1974C); HAUSSLER, LITTLESTONE, AND WARMUTH (1988)). *Let C be a class of discrimination functions with VC dimension V. Let $\mathcal{X}$ be the set of all random variables (X, Y) for which $L_C = 0$. Then, for every discrimination rule g_n based upon $X_1, Y_1, \ldots, X_n, Y_n$, and $n \geq V - 1$,*

$$\sup_{(X,Y) \in \mathcal{X}} \mathbf{E} L_n \geq \frac{V-1}{2en} \left(1 - \frac{1}{n} \right) .$$

PROOF. The idea is to construct a family $\mathcal{F}$ of 2^{V-1} distributions within the distributions with $L_C = 0$ as follows: first find points $x_1, \ldots, x_V$ that are shattered by C. Each distribution in $\mathcal{F}$ is concentrated on the set of these points. A member in $\mathcal{F}$ is described by $V - 1$ bits, $b_1, \ldots, b_{V-1}$. For convenience, this is represented as a bit vector b. Assume $V - 1 \leq n$. For a particular bit vector, we let $X = x_i$ $(i < V)$ with probability $1/n$ each, while $X = x_V$ with probability $1 - (V - 1)/n$. Then set $Y = f_b(X)$, where f_b is defined as follows:

$$f_b(x) = \begin{cases} b_i & \text{if } x = x_i, i < V \\ 0 & \text{if } x = x_V. \end{cases}$$

Note that since Y is a function of X, we must have $L^* = 0$. Also, $L_C = 0$, as the set $\{x_1, \ldots, x_V\}$ is shattered by C, i.e., there is a $g \in C$ with $g(x_i) = f_b(x_i)$ for $1 \leq i \leq V$. Clearly,

$$\sup_{(X,Y):L_C=0} \mathbf{E}\{L_n - L_C\}$$

$$\geq \sup_{(X,Y) \in \mathcal{F}} \mathbf{E}\{L_n - L_C\}$$

$$= \sup_b \mathbf{E}\{L_n - L_C\}$$

$$\geq \mathbf{E}\{L_n - L_C\}$$

(where b is replaced by B, uniformly distributed over $\{0, 1\}^{V-1}$)

$$= \mathbf{E}\{L_n\},$$

$$= \mathbf{P}\{g_n(X, X_1, Y_1, \ldots, X_n, Y_n) \neq f_B(X)\} .$$

The last probability may be viewed as the error probability of the decision function $g_n : \mathcal{R}^d \times (\mathcal{R}^d \times \{0, 1\})^n \to \{0, 1\}$ in predicting the value of the random variable $f_B(X)$ based on the observation $Z_n = (X, X_1, Y_1, \ldots, X_n, Y_n)$. Naturally, this probability is bounded from below by the Bayes probability of error

$$L^*(Z_n, f_B(X)) = \inf_{g_n} \mathbf{P}\{g_n(Z_n) \neq f_B(X)\}$$

corresponding to the decision problem $(Z_n, f_B(X))$. By the results of Chapter 2,

$$L^*(Z_n, f_B(X)) = \mathbf{E}\left\{\min(\eta^*(Z_n), 1 - \eta^*(Z_n))\right\},$$

where $\eta^*(Z_n) = \mathbf{P}\{f_B(X) = 1 | Z_n\}$. Observe that

$$\eta^*(Z_n) = \begin{cases} 1/2 & \text{if } X \neq X_1, \ldots, X \neq X_n, X \neq x_V \\ 0 \text{ or } 1 & \text{otherwise.} \end{cases}$$

Thus, we see that

$$\begin{aligned}
\sup_{(X,Y):L_C=0} \mathbf{E}\{L_n - L_C\} &\geq L^*(Z_n, f_B(X)) \\
&= \frac{1}{2}\mathbf{P}\{X \neq X_1, \ldots, X \neq X_n, X \neq x_V\} \\
&= \frac{1}{2}\sum_{i=1}^{V-1}\mathbf{P}\{X = x_i\}(1 - \mathbf{P}\{X = x_i\})^n \\
&= \frac{V-1}{2n}(1 - 1/n)^n \\
&\geq \frac{V-1}{2en}\left(1 - \frac{1}{n}\right) \quad (\text{since } (1 - 1/n)^{n-1} \downarrow 1/e).
\end{aligned}$$

This concludes the proof. $\square$

Minimax lower bounds on the probability $\mathbf{P}\{L_n \geq \epsilon\}$ can also be obtained. These bounds have evolved through several papers: see Ehrenfeucht, Haussler, Kearns and Valiant (1989); and Blumer, Ehrenfeucht, Haussler and Warmuth (1989). The tightest bounds we are aware of thus far are given by the next theorem.

Theorem 14.2. (DEVROYE AND LUGOSI (1995)). *Let C be a class of discrimination functions with VC dimension $V \geq 2$. Let $\mathcal{X}$ be the set of all random variables (X, Y) for which $L_C = 0$. Assume $\epsilon \leq 1/4$. Assume $n \geq V - 1$. Then for every discrimination rule g_n based upon $X_1, Y_1, \ldots, X_n, Y_n$,*

$$\sup_{(X,Y)\in\mathcal{X}} \mathbf{P}\{L_n \geq \epsilon\} \geq \frac{1}{e\sqrt{\pi V}}\left(\frac{2ne\epsilon}{V-1}\right)^{(V-1)/2} e^{-4n\epsilon/(1-4\epsilon)}.$$

If on the other hand $n \geq 15$ and $n \leq (V - 1)/(12\epsilon)$, then

$$\sup_{(X,Y)\in\mathcal{X}} \mathbf{P}\{L_n \geq \epsilon\} \geq \frac{1}{10}.$$

PROOF. We randomize as in the proof of Theorem 14.1. The difference now is that we pick $x_1, \ldots, x_{V-1}$ with probability p each. Thus, $\mathbf{P}\{X = x_V\} = 1 - p(V - 1)$.

We inherit the notation from the proof of Theorem 14.1. For a fixed b, denote the error probability by

$$L_n(b) = P\{g_n(X, X_1, f_b(X_1), \ldots, X_n, f_b(X_n)) \neq f_b(X)|X_1, \ldots, X_n\}.$$

We now randomize and replace b by B. Clearly,

$$\sup_{(X,Y):L_C=0} P\{L_n \geq \epsilon\} \geq \sup_b P\{L_n(b) \geq \epsilon\}$$
$$\geq E\{P\{L_n(B) \geq \epsilon|B\}\}$$
$$= P\{L_n(B) \geq \epsilon\}.$$

As in the proof of Theorem 14.1, observe that $L_n(B)$ cannot be smaller than the Bayes risk corresponding to the decision problem $(Z_n, f_B(X))$, where

$$Z_n = (X, X_1, f_B(X_1), \ldots, X_n, f_B(X_n)).$$

Thus,

$$L_n(B) \geq E\left\{\min(\eta^*(Z_n), 1 - \eta^*(Z_n))|X_1, \ldots, X_n\right\}.$$

As in the proof of Theorem 14.1, we see that

$$E\left\{\min(\eta^*(Z_n), 1 - \eta^*(Z_n))|X_1, \ldots, X_n\right\}$$
$$= \frac{1}{2}P\{X \neq X_1, \ldots, X \neq X_n, X \neq x_V|X_1, \ldots, X_n\}$$
$$\geq \frac{1}{2}p\sum_{i=1}^{V-1} I_{\{x_i \neq X_1, \ldots, x_i \neq X_n\}}.$$

For fixed $X_1, \ldots, X_n$, we denote by J the collection $\{j : 1 \leq j \leq V-1, \cap_{i=1}^n \{X_i \neq x_j\}\}$. This is the collection of empty cells x_i. We summarize:

$$\sup_{(X,Y):L_C=0} P\{L_n \geq \epsilon\} \geq P\left\{\frac{1}{2}p\sum_{i=1}^{V-1} I_{\{x_i \neq X_1, \ldots, x_i \neq X_n\}} \geq \epsilon\right\}$$
$$\geq P\{|J| \geq 2\epsilon/p\}.$$

We consider two choices for p:

CHOICE A.   Take $p = 1/n$, and assume $12n\epsilon \leq V - 1$, $\epsilon < 1/2$. Note that for $n \geq 15$, $E|J| = (V-1)(1-p)^n \geq (V-1)/3$. Also, since $0 \leq |J| \leq V - 1$, we have $\text{Var}|J| \leq (V-1)^2/4$. By the Chebyshev-Cantelli inequality (Theorem A.17),

$$P\{|J| \geq 2n\epsilon\} = 1 - P\{|J| < 2n\epsilon\}$$
$$\geq 1 - P\{|J| < (V-1)/6\}$$
$$= 1 - P\{|J| - E|J| \leq -(V-1)/6\}$$

$$\geq \; 1 - \frac{\text{Var}\,|J|}{\text{Var}\,|J| + (V-1)^2/36}$$

$$\geq \; 1 - \frac{(V-1)^2/4}{(V-1)^2/4 + (V-1)^2/36}$$

$$= \; \frac{1}{10}\,.$$

This proves the second inequality for $\sup \mathbf{P}\{L_n \geq \epsilon\}$.

CHOICE B. Assume that $\epsilon \leq 1/4$. By the pigeonhole principle, $|J| \geq 2\epsilon/p$ if the number of points X_i, $1 \leq i \leq n$, that are not equal to x_V does not exceed $V - 1 - 2\epsilon/p$. Therefore, we have a further lower bound:

$$\mathbf{P}\{|J| \geq 2\epsilon/p\} \geq \mathbf{P}\{\text{Binomial}(n, (V-1)p) \leq V - 1 - 2\epsilon/p\}\,.$$

Define $v = \lceil (V-1)/2 \rceil$. Take $p = 2\epsilon/v$. By assumption, $n \geq 2v - 1$. Then the lower bound is

$$\mathbf{P}\{\text{Binomial}(n, 4\epsilon) \leq v\}$$

$$\geq \; \binom{n}{v}(4\epsilon)^v(1 - 4\epsilon)^{n-v}$$

$$\geq \; \frac{1}{e\sqrt{2\pi v}}\left(\frac{4e\epsilon(n - v + 1)}{v(1 - 4\epsilon)}\right)^v (1 - 4\epsilon)^n$$

$$\left(\text{since } \binom{n}{v} \geq \left(\frac{(n-v+1)e}{v}\right)^v \frac{1}{e\sqrt{2\pi v}} \text{ by Stirling's formula}\right)$$

$$\geq \; \frac{1}{e\sqrt{2\pi v}}\left(\frac{4e\epsilon(n - v + 1)}{v}\right)^v (1 - 4\epsilon)^n$$

$$\geq \; \frac{1}{e\sqrt{2\pi v}}\left(\frac{4en\epsilon}{v}\right)^v (1 - 4\epsilon)^n \left(1 - \frac{v - 1}{n}\right)^v$$

$$\geq \; \frac{1}{e\sqrt{2\pi v}}\left(\frac{2en\epsilon}{v}\right)^v e^{-4n\epsilon/(1-4\epsilon)} \qquad (\text{since } n \geq 2(v - 1))$$

$$(\text{use } 1 - x \geq \exp(-x/(1 - x))).$$

This concludes the proof. $\square$

14.3   Classes with Infinite VC Dimension

The results presented in the previous section may also be applied to classes with infinite VC dimension. For example, it is not hard to derive the following result from Theorem 14.1:

Theorem 14.3. *Assume that $V_C = \infty$. For every n, $\delta > 0$ and classification rule g_n, there is a distribution with $L_C = 0$ such that*

$$\mathbf{E}L(g_n) > \frac{1}{2e} - \delta.$$

For the proof, see Problem 14.2. Thus, when $V_C = \infty$, distribution-free nontrivial performance guarantees for $L(g_n) - L_C$ do not exist. This generalizes Theorem 7.1, where a similar result is shown if C is the class of all measurable discrimination functions. We have also seen in Theorem 7.2, that if C is the class of all measurable classifiers, then no universal rate of convergence exists. However, we will see in Chapter 18 that for some classes with infinite VC dimension, it is possible to find a classification rule such that $L(g_n) - L^* \leq c\sqrt{\log n / n}$ for any distribution such that the Bayes classifier is in C. The constant c, however, necessarily depends on the distribution, as is apparent from Theorem 14.3.

Infinite VC dimension means that the class C shatters finite sets of any size. On the other hand, if C shatters *infinitely many* points, then no universal rate of convergence exists. This may be seen by an appropriate modification of Theorem 7.2, as follows. See Problem 14.3 for the proof.

Theorem 14.4. *Let $\{a_n\}$ be a sequence of positive numbers converging to zero with $1/16 \geq a_1 \geq a_2 \geq \ldots$. Let C be a class of classifiers with the property that there exists a set $A \subset \mathcal{R}^d$ of infinite cardinality such that for any subset B of A, there exists $\phi \in C$ such that $\phi(x) = 1$ if $x \in B$ and $\phi(x) = 0$ if $x \in A - B$. Then for every sequence of classification rules, there exists a distribution of (X, Y) with $L_C = 0$, such that*

$$\mathbf{E}L_n \geq a_n$$

for all n.

Note that the basic difference between this result and all others in this chapter is that in Theorem 14.4 the "bad" distribution does not vary with n. This theorem shows that selecting a classifier from a class shattering infinitely many points is essentially as hard as selecting one from the class of all classifiers.

14.4   The Case $L_C > 0$

In the more general case, when the best decision in the class C has positive error probability, the upper bounds derived in Chapter 12 for the expected error probability of the classifier obtained by minimizing the empirical risk are much larger than when $L_C = 0$. In this section we show that these upper bounds are necessarily large, and they may be tight for some distributions. Moreover, there is no other classifier that performs significantly better than empirical error minimization.

Theorem 14.5 below gives a lower bound for $\sup_{(X,Y):L_C \text{ fixed}} \mathbf{E}L(g_n) - L_C$. As a function of n and V_C, the bound decreases basically as in the upper bound obtained from Theorem 12.6. Interestingly, the lower bound becomes smaller as L_C

decreases, as should be expected. The bound is largest when L_C is close to $1/2$. The constants in the bound may be tightened at the expense of more complicated expressions. The theorem is essentially due to Devroye and Lugosi (1995), though the proof given here is different. Similar bounds—without making the dependence on L_C explicit—have been proved by Vapnik and Chervonenkis (1974c) and Simon (1993).

Theorem 14.5. *Let C be a class of discrimination functions with VC dimension $V \geq 2$. Let $\mathcal{X}$ be the set of all random variables (X, Y) for which for fixed $L \in (0, 1/2)$,*

$$L = \inf_{g \in C} \mathbf{P}\{g(X) \neq Y\} \, .$$

Then, for every discrimination rule g_n based upon $X_1, Y_1, \ldots, X_n, Y_n$,

$$\sup_{(X,Y) \in \mathcal{X}} \mathbf{E}(L_n - L) \geq \sqrt{\frac{L(V - 1)}{24n}} e^{-8} \quad \text{if } n \geq \frac{V-1}{2L} \max(9, 1/(1 - 2L)^2).$$

PROOF. Again we consider the finite family $\mathcal{F}$ from the previous section. The notation b and B is also as above. X now puts mass p at x_i, $i < V$, and mass $1 - (V - 1)p$ at x_V. This imposes the condition $(V - 1)p \leq 1$, which will be satisfied. Next introduce the constant $c \in (0, 1/2)$. We no longer have Y as a function of X. Instead, we have a uniform $[0, 1]$ random variable U independent of X and define

$$Y = \begin{cases} 1 & \text{if } U \leq \frac{1}{2} - c + 2cb_i, X = x_i, i < V \\ 0 & \text{otherwise.} \end{cases}$$

Thus, when $X = x_i$, $i < V$, Y is 1 with probability $1/2 - c$ or $1/2 + c$. A simple argument shows that the best rule for b is the one which sets

$$f_b(x) = \begin{cases} 1 & \text{if } x = x_i, i < V, b_i = 1 \\ 0 & \text{otherwise.} \end{cases}$$

Also, observe that

$$L = (V - 1)p(1/2 - c) \, .$$

Noting that $|2\eta(x_i) - 1| = c$ for $i < V$, for fixed b, by the equality in Theorem 2.2, we may write

$$L_n - L \geq \sum_{i=1}^{V-1} 2pc I_{\{g_n(x_i, X_1, Y_1, \ldots, X_n, Y_n) = 1 - f_b(x_i)\}} \, .$$

It is sometimes convenient to make the dependence of g_n upon b explicit by considering $g_n(x_i)$ as a function of x_i, $X_1, \ldots, X_n$, $U_1, \ldots, U_n$ (an i.i.d. sequence of uniform $[0, 1]$ random variables), and b_i. The proof given here is based on the ideas used in the proofs of Theorems 14.1 and 14.2. We replace b by a uniformly distributed random B over $\{0, 1\}^{V-1}$. After this randomization, denote $Z_n = (X, X_1, Y_1, \ldots, X_n, Y_n)$. Thus,

$$
\begin{aligned}
\sup_{(X,Y)\in\mathcal{F}} \mathbf{E}\{L_n - L\} &= \sup_b \mathbf{E}\{L_n - L\} \\
&\geq \mathbf{E}\{L_n - L\} \quad \text{(with random B)} \\
&\geq \sum_{i=1}^{V-1} 2pc\mathbf{E}I_{\{g_n(x_i,X_1,\ldots,Y_n)=1-f_B(x_i)\}} \\
&= 2c\mathbf{P}\{g_n(Z_n) \neq f_B(X)\} \\
&\geq 2cL^*(Z_n, f_B(X)),
\end{aligned}
$$

where, as before, $L^*(Z_n, f_B(X))$ denotes the Bayes probability of error of predicting the value of $f_B(X)$ based on observing Z_n. All we have to do is to find a suitable lower bound for

$$
L^*(Z_n, f_B(X)) = \mathbf{E}\left\{\min(\eta^*(Z_n), 1 - \eta^*(Z_n))\right\},
$$

where $\eta^*(Z_n) = \mathbf{P}\{f_B(X) = 1|Z_n\}$. Observe that

$$
\eta^*(Z_n) = \begin{cases} 1/2 & \text{if $X \neq X_1, \ldots, X \neq X_n$ and $X \neq x_V$} \\ \mathbf{P}\{B_i = 1|Y_{i_1}, \ldots, Y_{i_k}\} & \text{if $X = X_{i_1} = \cdots = X_{i_k} = x_i, i < V$.} \end{cases}
$$

Next we compute $\mathbf{P}\{B_i = 1|Y_{i_1} = y_1, \ldots, Y_{i_k} = y_k\}$ for $y_1, \ldots, y_k \in \{0, 1\}$. Denoting the numbers of zeros and ones by $k_0 = |\{j \leq k : y_j = 0\}|$ and $k_1 = |\{j \leq k : y_j = 1\}|$, we see that

$$
\begin{aligned}
&\mathbf{P}\{B_i = 1|Y_{i_1} = y_1, \ldots, Y_{i_k} = y_k\} \\
&= \frac{(1 - 2c)^{k_1}(1 + 2c)^{k_0}}{(1 - 2c)^{k_1}(1 + 2c)^{k_0} + (1 + 2c)^{k_1}(1 - 2c)^{k_0}}.
\end{aligned}
$$

Therefore, if $X = X_{i_1} = \cdots = X_{i_k} = x_i, i < V$, then

$$
\begin{aligned}
&\min(\eta^*(Z_n), 1 - \eta^*(Z_n)) \\
&= \frac{\min\left((1 - 2c)^{k_1}(1 + 2c)^{k_0}, (1 + 2c)^{k_1}(1 - 2c)^{k_0}\right)}{(1 - 2c)^{k_1}(1 + 2c)^{k_0} + (1 + 2c)^{k_1}(1 - 2c)^{k_0}} \\
&= \frac{\min\left(1, \left(\frac{1+2c}{1-2c}\right)^{k_1-k_0}\right)}{1 + \left(\frac{1+2c}{1-2c}\right)^{k_1-k_0}} \\
&= \frac{1}{1 + \left(\frac{1+2c}{1-2c}\right)^{|k_1-k_0|}}.
\end{aligned}
$$

In summary, denoting $a = (1 + 2c)/(1 - 2c)$, we have

$$
\begin{aligned}
L^*(Z_n, f_B(X)) &= \mathbf{E}\left\{\frac{1}{1 + a^{\left|\sum_j x_j = x(2Y_j - 1)\right|}}\right\} \\
&\geq \mathbf{E}\left\{\frac{1}{2a^{\left|\sum_j x_j = x(2Y_j - 1)\right|}}\right\} \\
&\geq \frac{1}{2}\sum_{i=1}^{V-1} \mathbf{P}\{X = x_i\}\mathbf{E}\left\{a^{-\left|\sum_{j:x_j = x_i}(2Y_j - 1)\right|}\right\} \\
&\geq \frac{1}{2}(V - 1)pa^{-\mathbf{E}\left\{\left|\sum_{j:x_j = x_i}(2Y_j - 1)\right|\right\}}
\end{aligned}
$$

(by Jensen's inequality).

Next we bound $\mathbf{E}\left\{\left|\sum_{j:X_j = x_i}(2Y_j - 1)\right|\right\}$. Clearly, if $B(k, q)$ denotes a binomial random variable with parameters k and q,

$$
\mathbf{E}\left\{\left|\sum_{j:X_j = x_i}(2Y_j - 1)\right|\right\} = \sum_{k=0}^{n}\binom{n}{k}p^k(1 - p)^{n-k}\mathbf{E}\{|2B(k, 1/2 - c) - k|\}.
$$

However, by straightforward calculation we see that

$$
\begin{aligned}
\mathbf{E}\{|2B(k, 1/2 - c) - k|\} &\leq \sqrt{\mathbf{E}\{(2B(k, 1/2 - c) - k)^2\}} \\
&= \sqrt{k(1 - 4c^2) + 4k^2c^2} \\
&\leq 2kc + \sqrt{k}.
\end{aligned}
$$

Therefore, applying Jensen's inequality once again, we get

$$
\sum_{k=0}^{n}\binom{n}{k}p^k(1 - p)^{n-k}\mathbf{E}\{|2B(k, 1/2 - c) - k|\} \leq 2npc + \sqrt{np}.
$$

Summarizing what we have obtained so far, we have

$$
\begin{aligned}
\sup_b \mathbf{E}\{L_n - L\} &\geq 2cL^*(Z_n, f_B(X)) \\
&\geq 2c\frac{1}{2}(V - 1)pa^{-2npc - \sqrt{np}} \\
&\geq c(V - 1)pe^{-2npc(a-1) - (a-1)\sqrt{np}}
\end{aligned}
$$

(by the inequality $1 + x \leq e^x$)

$$
= c(V - 1)pe^{-8npc^2/(1-2c) - 4c\sqrt{np}/(1-2c)}.
$$

A rough asymptotic analysis shows that the best asymptotic choice for c is given by

$$
c = \frac{1}{\sqrt{4np}}.
$$

Then the constraint $L = (V - 1)p(1/2 - c)$ leaves us with a quadratic equation in c. Instead of solving this equation, it is more convenient to take $c = \sqrt{(V - 1)/(8nL)}$. If $2nL/(V - 1) \geq 9$, then $c \leq 1/6$. With this choice for c, using $L = (V - 1)p(1/2 - c)$, straightforward calculation provides

$$\sup_{(X,Y)\in\mathcal{F}} \mathbf{E}(L_n - L) \geq \sqrt{\frac{(V - 1)L}{24n}} e^{-8}.$$

The condition $p(V - 1) \leq 1$ implies that we need to ask that $n \geq (V - 1)/(2L(1 - 2L)^2)$. This concludes the proof of Theorem 14.5. $\square$

Next we obtain a probabilistic bound. Its proof below is based upon Hellinger distances, and its methodology is essentially due to Assouad (1983b). For refinements and applications, we refer to Birgé (1983; 1986) and Devroye (1987).

Theorem 14.6. (DEVROYE AND LUGOSI (1995)). *Let C be a class of discrimination functions with VC dimension $V \geq 2$. Let $\mathcal{X}$ be the set of all random variables (X, Y) for which for fixed $L \in (0, 1/4]$,*

$$L = \inf_{g \in C} \mathbf{P}\{g(X) \neq Y\} .$$

Then, for every discrimination rule g_n based upon $X_1, Y_1, \ldots, X_n, Y_n$, and any $\epsilon \leq L$,

$$\sup_{(X,Y)\in\mathcal{X}} \mathbf{P}\{L_n - L \geq \epsilon\} \geq \frac{1}{4} e^{-4n\epsilon^2/L} .$$

PROOF. The method of randomization here is similar to that in the proof of Theorem 14.5. Using the same notation as there, it is clear that

$$\sup_{(X,Y)\in\mathcal{X}} \mathbf{P}\{L_n - L \geq \epsilon\}$$

$$\geq \mathbf{E} I_{\{\sum_{i=1}^{V-1} 2pc I_{\{g_n(x_i, X_1, \ldots, Y_n)=1-f_\beta(x_i)\}} \geq \epsilon\}}$$

$$= 2^{-(V-1)} \sum_{\substack{(x'_1, \ldots, x'_n, y_1, \ldots, y_n) \\ \in (\{x_1, \ldots, x_V\} \times \{0,1\})^n}}$$

$$\sum_b I_{\{\sum_{i=1}^{V-1} 2pc I_{\{g_n(x_i, x'_1, y_1, \ldots, x'_n, y_n)=1-f_b(x_i)\}} \geq \epsilon\}} \prod_{j=1}^n p_b(x'_j, y_j) .$$

First observe that if

$$\epsilon/(2pc) \leq (V - 1)/2, \tag{14.1}$$

then

$$I_{\{\sum_{i=1}^{V-1} 2pc I_{\{g_n(x_i, x'_1, y_1, \ldots, x'_n, y_n)=1-f_b(x_i)\}} \geq \epsilon\}}$$

$$+ \ I_{\{\sum_{i=1}^{V-1} 2pc I_{\{g_n(x_i, x'_1, y_1, \ldots, x'_n, y_n)=1-f_{b^c}(x_i)\}} \geq \epsilon\}} \geq 1 ,$$

where b^c denotes the binary vector $(1 - b_1, \ldots, 1 - b_{V-1})$, that is, the complement of b. Therefore, for $\epsilon \leq pc(V - 1)$, the last expression in the lower bound above is bounded from below by

$$\frac{1}{2^{V-1}} \sum_{\substack{(x_1', \ldots, x_n', y_1, \ldots, y_n) \\ \in (\{x_1, \ldots, x_V\} \times \{0,1\})^n}} \sum_b \frac{1}{2} \min \left(\prod_{j=1}^n p_b(x_j', y_j) , \prod_{j=1}^n p_{b^c}(x_j', y_j) \right)$$

$$\geq \frac{1}{2^{V+1}} \sum_b \left(\sum_{\substack{(x_1', \ldots, x_n', y_1, \ldots, y_n) \\ \in (\{x_1, \ldots, x_V\} \times \{0,1\})^n}} \sqrt{\prod_{j=1}^n p_b(x_j', y_j) \times \prod_{j=1}^n p_{b^c}(x_j', y_j)} \right)^2$$

(by LeCam's inequality, Lemma 3.1)

$$= \frac{1}{2^{V+1}} \sum_b \left(\sum_{(x,y)} \sqrt{p_b(x, y) p_{b^c}(x, y)} \right)^{2n} .$$

It is easy to see that for $x = x_V$,

$$p_b(x, y) = p_{b^c}(x, y) = \frac{1 - (V - 1)p}{2} ,$$

and for $x = x_i, i < V$,

$$p_b(x, y) p_{b^c}(x, y) = p^2 \left(\frac{1}{4} - c^2 \right) .$$

Thus, we have the equality

$$\sum_{(x,y)} \sqrt{p_b(x, y) p_{b^c}(x, y)} = 1 - (V - 1)p + 2(V - 1)p \sqrt{\frac{1}{4} - c^2} .$$

Summarizing, since $L = p(V - 1)(1/2 - c)$, we have

$$\sup_{(X,Y) \in \mathcal{X}} \mathbf{P}\{L_n - L \geq \epsilon\} \geq \frac{1}{4} \left(1 - \frac{L}{\frac{1}{2} - c} \left(1 - \sqrt{1 - 4c^2} \right) \right)^{2n}$$

$$\geq \frac{1}{4} \left(1 - \frac{L}{\frac{1}{2} - c} 4c^2 \right)^{2n}$$

$$\geq \frac{1}{4} \exp \left(- \frac{16nLc^2}{1 - 2c} \middle/ \left(1 - \frac{8Lc^2}{1 - 2c} \right) \right) ,$$

where we used the inequality $1 - x \geq e^{-x/(1-x)}$ again. We may choose c as $\frac{\epsilon}{2L + 2\epsilon}$. It is easy to verify that condition (14.1) holds. Also, $p(V - 1) \leq 1$. From the condition $L \geq \epsilon$ we deduce that $c \leq 1/4$. The exponent in the expression above may be bounded as

$$\frac{\frac{16nLc^2}{1-2c}}{1 - \frac{8Lc^2}{1-2c}} = \frac{16nLc^2}{1 - 2c - 8Lc^2}$$

$$= \frac{\frac{4n\epsilon^2}{L+\epsilon}}{1 - \frac{2\epsilon^2}{L+\epsilon}} \qquad \text{(by substituting } c = \epsilon/(2L + 2\epsilon))$$

$$\leq \bullet \frac{4n\epsilon^2}{L}.$$

Thus,

$$\sup_{(X,Y)\in\mathcal{X}} \mathbf{P}\{L_n - L \geq \epsilon\} \geq \frac{1}{4} \exp\left(-4n\epsilon^2/L\right),$$

as desired. $\Box$

14.5   Sample Complexity

We may rephrase the probability bounds above in terms of the sample complexity of algorithms for selecting a classifier from a class. Recall that for given $\epsilon, \delta > 0$, the sample complexity of a selection rule g_n is the smallest integer $N(\epsilon, \delta)$ such that

$$\sup \mathbf{P}\{L(g_n) - L_C \geq \epsilon\} \leq \delta$$

for all $n \geq N(\epsilon, \delta)$. The supremum is taken over a class of distributions of (X, Y). Here we are interested in distributions such that L_C is fixed.

We start with the case $L_C = 0$, by checking the implications of Theorem 14.2 for the sample complexity $N(\epsilon, \delta)$. First Blumer, Ehrenfeucht, Haussler, and Warmuth (1989) showed that for any algorithm,

$$N(\epsilon, \delta) \geq C \left(\frac{1}{\epsilon} \log\left(\frac{1}{\delta}\right) + V_C\right),$$

where C is a universal constant. In Ehrenfeucht, Haussler, Kearns, and Valiant (1989), the lower bound was partially improved to

$$N(\epsilon, \delta) \geq \frac{V_C - 1}{32\epsilon}$$

when $\epsilon \leq 1/8$ and $\delta \leq 1/100$. It may be combined with the previous bound. Theorem 14.2 provides the following bounds:

COROLLARY 14.1. *Let C be a class of discrimination functions with VC dimension $V \geq 2$. Let $\mathcal{X}$ be the set of all random variables (X, Y) for which $L_C = 0$. Assume*

$\epsilon \leq 1/4$, and denote $v = \lceil (V - 1)/2 \rceil$. Then for every discrimination rule g_n based upon $X_1, Y_1, \ldots, X_n, Y_n$, when $\epsilon \leq 1/8$ and

$$\log\left(\frac{1}{\delta}\right) \geq \left(\frac{4v}{e}\right)\left(e\sqrt{2\pi v}\right)^{1/v} ,$$

then

$$N(\epsilon, \delta) \geq \frac{1}{8\epsilon} \log\left(\frac{1}{\delta}\right) .$$

Finally, for $\delta \leq 1/10$, and $\epsilon < 1/2$,

$$N(\epsilon, \delta) \geq \frac{V - 1}{12\epsilon} .$$

PROOF. The second bound follows trivially from the second inequality of Theorem 14.2. By the first inequality there,

$$\sup_{(X,Y)\in\mathcal{X}} P\{L_n \geq \epsilon\} \geq \frac{1}{e\sqrt{2\pi v}}\left(\frac{2ne\epsilon}{V - 1}\right)^{(V-1)/2} e^{-4n\epsilon/(1-4\epsilon)}$$

$$\geq \frac{1}{e\sqrt{2\pi v}}\left(\frac{2en\epsilon}{v}\right)^{v} e^{-8n\epsilon} \quad \text{(since } \epsilon \leq 1/8\text{)}$$

$$\geq \frac{(8\epsilon)^v}{\log^v(1/\delta)} n^v e^{-8n\epsilon}$$

(since we assume $\log\left(\frac{1}{\delta}\right) \geq \left(\frac{4v}{e}\right)\left(e\sqrt{2\pi v}\right)^{1/v}$).

The function $n^v e^{-8n\epsilon}$ varies unimodally in n, and achieves a peak at $n = v/(8\epsilon)$. For n below this threshold, by monotonicity, we apply the bound at $n = v/(8\epsilon)$. It is easy to verify that the value of the bound at $v/(8\epsilon)$ is always at least δ. If on the other hand, $(1/8\epsilon)\log(1/\delta) \geq n \geq v/(8\epsilon)$, the lower bound achieves its minimal value at $(1/8\epsilon)\log(1/\delta)$, and the value there is δ. This proves the first bound. $\square$

Corollary 14.1 shows that for any classifier, at least

$$\max\left(\frac{1}{8\epsilon}\log\left(\frac{1}{\delta}\right), \frac{V_C - 1}{4\epsilon}\right)$$

training samples are necessary to achieve ϵ accuracy with δ confidence for all distributions. Apart from a $\log\left(\frac{1}{\epsilon}\right)$ factor, the order of magnitude of this expression is the same as that of the upper bound for empirical error minimization, obtained in Corollary 12.4. That the upper and lower bounds are very close, has two important messages. On the one hand it gives a very good estimate for the number of training samples needed to achieve a certain performance. On the other hand it shows that there is essentially no better method than minimizing the empirical error probability.

In the case $L_C > 0$, we may derive lower bounds for $N(\epsilon, \delta)$ from Theorems 14.5 and 14.6:

COROLLARY 14.2. *Let $\mathcal{G}$ be a class of discrimination functions with VC dimension $V \geq 2$. Let $\mathcal{X}$ be the set of all random variables (X, Y) for which for fixed $L \in (0, 1/2)$,*

$$L = \inf_{g \in \mathcal{G}} P\{g(X) \neq Y\} \ .$$

Then, for every discrimination rule g_n based upon $X_1, Y_1, \ldots, X_n, Y_n$,

$$N(\epsilon, \delta) \geq \frac{L(V - 1)e^{-10}}{32} \times \min\left(\frac{1}{\delta^2}, \frac{1}{\epsilon^2}\right) \ .$$

Also, and in particular, for $\epsilon \leq L \leq 1/4$,

$$N(\epsilon, \delta) \geq \frac{L}{4\epsilon^2} \log \frac{1}{4\delta} \ .$$

PROOF. The first bound may be obtained easily from the expectation-bound of Theorem 14.5 (see Problem 14.1). Setting the bound of Theorem 14.6 equal to δ provides the second bound on $N(\epsilon, \delta)$. □

These bounds may of course be combined. They show that $N(\epsilon, \delta)$ is bounded from below by terms like $(1/\epsilon^2)\log(1/\delta)$ (independent of V_C) and $(V_C - 1)/\epsilon^2$, as δ is typically much smaller than ϵ. By comparing these bounds with the upper bounds of Corollary 12.3, we see that the only difference between the orders of magnitude is a $\log(1/\epsilon)$-factor, so all remarks made for the case $L_C = 0$ remain valid.

Interestingly, all bounds depend on the class C only through its VC dimension. This fact suggests that when studying distribution-free properties of $L(g_n) - L_C$, the VC dimension is the most important characteristic of the class. Also, all bounds are linear in the VC dimension, which links it conveniently to sample size.

REMARK. It is easy to see from the proofs that all results remain valid if we allow randomization in the rules g_n. □

Problems and Exercises

PROBLEM 14.1. Show that Theorem 14.5 implies that for every discrimination rule g_n based upon D_n,

$$N(\epsilon, \delta) \geq \frac{L(V - 1)e^{-10}}{32} \times \min\left(\frac{1}{\delta^2}, \frac{1}{\epsilon^2}\right) \ .$$

HINT: Assume that $P\{L_n - L > \epsilon\} < \delta$. Then clearly, $E\{L_n - L\} \leq \epsilon + \delta$.

PROBLEM 14.2. Prove Theorem 14.3: First apply the proof method of Theorem 14.1 to show that for every n, and g_n, there is a distribution with $L_C = 0$ such that $EL(g_n) \geq (1 - 1/n)/(2e)$. Use a monotonicity argument to finish the proof.

PROBLEM 14.3. Prove Theorem 14.4 by modifying the proof of Theorem 7.2.

15
The Maximum Likelihood Principle

In this chapter we explore the various uses of the maximum likelihood principle in discrimination. In general, the principle is only applicable if we have some a priori knowledge of the problem at hand. We offer definitions, consistency results, and examples that highlight the advantages and shortcomings.

15.1 Maximum Likelihood: The Formats

Sometimes, advance information takes a very specific form (e.g., "if $Y = 1$, X is normal (μ, σ^2))"). Often, it is rather vague (e.g., "we believe that X has a density," or "$\eta(x)$ is thought to be a monotone function of $x \in \mathcal{R}$").

If we have information in *set format*, the maximum likelihood principle is less appropriate. Here we know that the Bayes rule $g^*(x)$ is of the form $g(x) = I_{\{x \in A\}}$, where $A \in \mathcal{A}$ and $\mathcal{A}$ is a class of sets of $\mathcal{R}^d$. We refer to the chapters on empirical risk minimization (see Chapter 12 and also Chapter 18) for this situation.

If we know that the true (unknown) η belongs to a class $\mathcal{F}$ of functions that map $\mathcal{R}^d$ to $[0, 1]$, then we say that we are given information in *regression format*. With each $\eta' \in \mathcal{F}$ we associate a set $A = \{x : \eta'(x) > 1/2\}$ and a discrimination rule $g(x) = I_{\{x \in A\}}$. The class of these rules is denoted by $\mathcal{C}$. Assume that we somehow could estimate η by η_n. Then it makes sense to use the associated rule $g_n(x) = I_{\{\eta_n(x) > 1/2\}}$. The maximum likelihood method suggests a way of picking the η_n from $\mathcal{F}$ that in some sense is most likely given the data. It is fully automatic—the user does not have to pick any parameters—but it does require a serious implementation effort in many cases. In a sense, the regression format is more powerful than the set format, as there is more information in knowing a function η than in knowing the indicator function $I_{\{\eta > 1/2\}}$. Still, no structure is assumed on the part of X, and none is needed to obtain consistency results.

A third format, even more detailed, is that in which we know that the distribution of (X, Y) belongs to a class $\mathcal{D}$ of distributions on $\mathcal{R}^d \times \{0, 1\}$. For a given distribution, we know η, so we may once again deduce a rule g by setting $g(x) = I_{\{\eta(x) > 1/2\}}$. This *distribution format* is even more powerful, as the positions $X_1, \ldots, X_n$ alone in some cases may determine the unknown parameters in the model. This situation fits in squarely with classical parameter estimation in mathematical statistics. Once again, we may apply the maximum likelihood principle to select a distribution from $\mathcal{D}$. Unfortunately, as we move to more restrictive and stronger formats, the number of conditions under which the maximum likelihood principle is consistent increases as well. We will only superficially deal with the distribution format (see Chapter 16 for more detail).

15.2   The Maximum Likelihood Method: Regression Format

Given $X_1, \ldots, X_n$, the probability of observing $Y_1 = y_1, \ldots, Y_n = y_n$ is

$$\prod_{i=1}^{n} \eta(X_i)^{y_i} (1 - \eta(X_i))^{1-y_i}.$$

If η is unknown but belongs to a family $\mathcal{F}$ of functions, we may wish to select that η' from $\mathcal{F}$ (if it exists) for which that likelihood product is maximal. More formally, we select η' so that the logarithm

$$\mathcal{L}_n(\eta') = \frac{1}{n} \sum_{i=1}^{n} \left(Y_i \log \eta'(X_i) + (1 - Y_i) \log(1 - \eta'(X_i)) \right)$$

is maximal. If the family $\mathcal{F}$ is too rich, this will overfit, and consequently, the selected function η_n has a probability of error

$$L(\eta_n) = \mathbf{P}\left\{ I_{\{\eta_n(X) > 1/2\}} \neq Y | D_n \right\}$$

that does not tend to L^*. For convenience, we assume here that there exists an element of $\mathcal{F}$ maximizing $\mathcal{L}_n$.

We do not assume here that the class $\mathcal{F}$ is very small. Classes in which each η' in $\mathcal{F}$ is known up to one or a few parameters are loosely called parametric. Sometimes $\mathcal{F}$ is defined via a generic description such as: $\mathcal{F}$ is the class of all $\eta' : \mathcal{R}^d \to [0, 1]$ that are Lipschitz with constant c. Such classes are called nonparametric. In certain cases, the boundary between parametric and nonparametric is unclear. We will be occupied with the consistency question: does $L(\eta_n) \to \inf_{\eta' \in \mathcal{F}} L(\eta')$ with probability one for all distribution of (X, Y)? (Here $L(\eta') = \mathbf{P}\left\{ I_{\{\eta'(X) > 1/2\}} \neq Y \right\}$ is the probability of error of the natural rule that corresponds to η'.) If, additionally, $\mathcal{F}$ is rich enough or our prior information is good enough, we may have $\inf_{\eta' \in \mathcal{F}} L(\eta') = L^*$, but that is not our concern here, as $\mathcal{F}$ is given to us.

We will not be concerned with the computational problems related to the maximization of $\mathcal{L}_n(\eta')$ over $\mathcal{F}$. Gradient methods or variations of them are sometimes used—refer to McLachlan (1992) for a bibliography. It suffices to say that in simple cases, an explicit form for η_n may be available. An example follows.

Our first lemma illustrates that the maximum likelihood method should only be used when the true (but unknown) η indeed belongs to $\mathcal{F}$. Recall that the same was not true for empirical risk minimization over VC classes (see Chapter 12).

Lemma 15.1. *Consider the class $\mathcal{F}$ with two functions $\eta_A \equiv 0.45$, $\eta_B \equiv 0.95$. Let η_n be the function picked by the maximum likelihood method. There exists a distribution for (X, Y) with $\eta \notin \mathcal{F}$ such that with probability one, as $n \to \infty$,*

$$L(\eta_n) \to \max_{\eta' \in \mathcal{F}} L(\eta') > \min_{\eta' \in \mathcal{F}} L(\eta').$$

Thus, maximum likelihood picks the wrong classifier.

PROOF. Define the distribution of (X, Y) on $\{0, 1\} \times \{0, 1\}$ by $\mathbf{P}\{X = 0, Y = 0\} = \mathbf{P}\{X = 1, Y = 0\} = 2/9$, $\mathbf{P}\{X = 0, Y = 1\} = 1/9$, and $\mathbf{P}\{X = 1, Y = 1\} = 4/9$. Then one may quickly verify that

$$L(\eta_A) = \frac{5}{9}, \quad L(\eta_B) = \frac{4}{9}.$$

(Note that $L^* = 1/3$, but this is irrelevant here.) Within $\mathcal{F}$, η_B is the better for our distribution. By the strong law of large numbers, we have

$$\mathcal{L}_n(\eta_A) \to \mathbf{E}\{Y \log \eta_A(X) + (1 - Y) \log(1 - \eta_A(X))\}$$

with probability one (and similarly for η_B). If one works out the values, it is seen that with probability one, $\eta_n \equiv \eta_A$ for all n large enough. Hence, $L(\eta_n) \to \max_{\eta' \in \mathcal{F}} L(\eta')$ with probability one. $\square$

REMARK. Besides the clear theoretical hazard of not capturing η in $\mathcal{F}$, the maximum likelihood method runs into a practical problem with "infinity." For example, take $\mathcal{F} = \{\eta_A \equiv 0, \eta_B \equiv 1\}$, and assume $\eta \equiv 1/3$. For all n large enough, both classes are represented in the data sample with probability one. This implies that $\mathcal{L}_n(\eta_A) = \mathcal{L}_n(\eta_B) = -\infty$. The maximum likelihood estimate η_n is ill-defined, while any reasonable rule should quickly be able to pick η_A over η_B. $\square$

The lemma shows that when $\eta \notin \mathcal{F}$, the maximum likelihood method is not even capable of selecting one of two choices. The situation changes dramatically when $\eta \in \mathcal{F}$. For finite classes $\mathcal{F}$, nothing can go wrong. Noting that whenever $\eta \in \mathcal{F}$, we have $L^* = \inf_{\eta' \in \mathcal{F}} L(\eta')$, we may now expect that $L(\eta_n) \to L^*$ in probability, or with probability one.

Theorem 15.1. *If $|\mathcal{F}| = k < \infty$, and $\eta \in \mathcal{F}$, then the maximum likelihood method is consistent, that is,*

$$L(\eta_n) \to L^* \quad \text{in probability.}$$

PROOF. For a fixed distribution of (X, Y), we rank the members $\eta^{(1)}, \ldots, \eta^{(k)}$ of $\mathcal{F}$ by increasing values of $L(\eta^{(i)})$. Put $\eta^{(1)} \equiv \eta$. Let i_0 be the largest index for which $L(\eta^{(i)}) = L^*$. Let η_n be the maximum likelihood choice from $\mathcal{F}$. For any α, we have

$$\mathbf{P}\{L(\eta_n) \neq L^*\}$$
$$\leq \mathbf{P}\left\{ \mathcal{L}_n(\eta^{(1)}) \leq \max_{i > i_0} \mathcal{L}_n(\eta^{(i)}) \right\}$$
$$\leq \mathbf{P}\{\mathcal{L}_n(\eta^{(1)}) \leq \alpha\} + \sum_{i > i_0} \mathbf{P}\{\mathcal{L}_n(\eta^{(i)}) \geq \alpha\}.$$

Define the entropy of (X, Y) by

$$\mathcal{E} = \mathcal{E}(\eta) = -\mathbf{E}\{\eta(X) \log \eta(X) + (1 - \eta(X)) \log(1 - \eta(X))\}$$

(see Chapter 3). Recall that $0 \leq \mathcal{E} \leq \log 2$. We also need the negative divergences

$$D_i = \mathbf{E}\left\{ \eta(X) \log \frac{\eta^{(i)}(X)}{\eta(X)} + (1 - \eta(X)) \log \frac{1 - \eta^{(i)}(X)}{1 - \eta(X)} \right\},$$

which are easily seen to be nonpositive for all i (by Jensen's inequality). Furthermore, $D_i = 0$ if and only if $\eta^{(i)}(X) = \eta(X)$ with probability one. Observe that for $i > i_0$, we cannot have this. Let $\theta = \max_{i > i_0} D_i$. (If $i_0 = k$, this set is empty, but then the theorem is trivially true.)

It is advantageous to take $\alpha = -\mathcal{E} + \theta/2$. Observe that

$$\mathbf{E}\{\mathcal{L}_n(\eta^{(i)})\} = -\mathcal{E} + D_i \begin{cases} = -\mathcal{E} & \text{if } i = 1 \\ \leq -\mathcal{E} + \theta & \text{if } i > i_0. \end{cases}$$

Thus,

$$\mathbf{P}\{L(\eta_n) \neq L^*\} \leq \mathbf{P}\left\{ \mathcal{L}_n(\eta^{(1)}) \leq -\mathcal{E} + \frac{\theta}{2} \right\} + \sum_{i > i_0} \mathbf{P}\left\{ \mathcal{L}_n(\eta^{(i)}) \geq -\mathcal{E} + \frac{\theta}{2} \right\}.$$

By the law of large numbers, we see that both terms converge to zero. Note, in particular, that it is true even if for some $i > i_0$, $D_i = -\infty$. $\square$

For infinite classes, many things can go wrong. Assume that $\mathcal{F}$ is the class of all $\eta' : \mathcal{R}^2 \to [0, 1]$ with $\eta' = I_A$ and A is a convex set containing the origin. Pick X uniformly on the perimeter of the unit circle. Then the maximum likelihood estimate η_n matches the data, as we may always find a closed polygon P_n with vertices at the X_i's with $Y_i = 1$. For $\eta_n = I_{P_n}$, we have $\mathcal{L}_n(\eta_n) = 0$ (its maximal value), yet $L(\eta_n) = \mathbf{P}\{Y = 1\} = p$ and $L^* = 0$. The class $\mathcal{F}$ is plainly too rich.

For distributions of X on the positive integers, maximum likelihood does not behave as poorly, even though it must pick among infinitely many possibilities. Assume $\mathcal{F}$ is the class of all η', but we know that X puts all its mass on the positive integers. Then maximum likelihood tries to maximize

$$\prod_{i=1}^{\infty} (\eta'(i))^{N_{1,i}} (1 - \eta'(i))^{N_{0,i}}$$

over $\eta' \in \mathcal{F}$, where $N_{1,i} = \sum_{j=1}^{n} I_{\{X_j=i, Y_j=1\}}$ and $N_{0,i} = \sum_{j=1}^{n} I_{\{X_j=i, Y_j=0\}}$. The maximization is to be done over *all* $(\eta'(1), \eta'(2), \ldots)$ from $\bigotimes_{i=1}^{\infty}[0, 1]$. Fortunately, this is turned into a maximization for each individual i—usually we are not so lucky. Thus, if $N_{0,i} + N_{1,i} = 0$, we set $\eta_n(i) = 0$ (arbitrarily), while if $N_{0,i} + N_{1,i} > 0$, we pick $\eta_n(i)$ as the value u that maximizes

$$N_{1,i} \log u + N_{0,i} \log(1 - u).$$

Setting the derivative with respect to u equal to zero shows that

$$\eta_n(i) = \frac{N_{1,i}}{N_{1,i} + N_{0,i}}.$$

In other words, η_n is the familiar histogram estimate with bin width less than one half. It is known to be universally consistent (see Theorem 6.2). Thus, maximum likelihood may work for large $\mathcal{F}$ if we restrict the distribution of (X, Y) a bit.

15.3 Consistency

Finally, we are ready for the main consistency result for η_n when $\mathcal{F}$ may have infinitely many elements. The conditions of the theorem involve the *bracketing metric entropy* of $\mathcal{F}$, defined as follows: for every distribution of X, and $\epsilon > 0$, let $\mathcal{F}_{X,\epsilon}$ be a set of functions such that for each $\eta' \in \mathcal{F}$, there exist functions $\eta'_L, \eta'_U \in \mathcal{F}_{X,\epsilon}$ such that for all $x \in \mathcal{R}^d$

$$\eta'_L(x) \le \eta'(x) \le \eta'_U(x),$$

and

$$\mathbf{E}\{\eta'_U(X) - \eta'_L(X)\} \le \epsilon.$$

That is, every $\eta' \in \mathcal{F}$ is "bracketed" between two members of $\mathcal{F}_{X,\epsilon}$ whose $L_1(\mu)$ distance is not larger than ϵ. Let $N(X, \epsilon)$ denote the cardinality of the smallest such $\mathcal{F}_{X,\epsilon}$. If $N(X, \epsilon) < \infty$,

$$\log N(X, \epsilon)$$

is called the *bracketing ϵ-entropy* of $\mathcal{F}$, corresponding to X.

Theorem 15.2. *Let $\mathcal{F}$ be a class of regression functions $\mathcal{R}^d \to [0, 1]$. Assume that for every distribution of X and $\epsilon > 0$, $N(X, \epsilon) < \infty$. Then the maximum likelihood choice η_n satisfies*

$$\lim_{n \to \infty} L(\eta_n) = L^* \quad \text{in probability}$$

for all distributions of (X, Y) with $\eta \in \mathcal{F}$.

Thus, consistency is guaranteed if $\mathcal{F}$ has a finite bracketing ϵ-entropy for every X and ϵ. We provide examples in the next section. For the proof, first we need a simple corollary of Lemma 3.2:

COROLLARY 15.1. *Let $\eta, \eta' : \mathcal{R}^d \to [0, 1]$, and let (X, Y) be an $\mathcal{R}^d \times \{0, 1\}$-valued random variable pair with $P\{Y = 1 | X = x\} = \eta(x)$. Define*

$$L^* = E\{\min(\eta(X), 1 - \eta(X))\}, \quad L(\eta') = P\left\{I_{\{\eta'(X) > 1/2\}} \neq Y\right\}.$$

Then

$$
\begin{aligned}
E\left\{\sqrt{\eta'(X)\eta(X)} + \sqrt{(1 - \eta'(X))(1 - \eta(X))}\right\} &\leq 1 - \frac{\left(2E|\eta(X) - \eta'(X)|\right)^2}{8} \\
&\leq 1 - \frac{1}{8}(L(\eta') - L^*)^2.
\end{aligned}
$$

PROOF. The first inequality follows by Lemma 3.2, and the second by Theorem 2.2. $\square$

Now, we are ready to prove the theorem. Some of the ideas used here appear in Wong and Shen (1992).

PROOF OF THEOREM 15.2.   Look again at the proof for the case $|\mathcal{F}| < \infty$ (Theorem 15.1). Define $\mathcal{E}$ as there. Let $\mathcal{F}^{(\epsilon)}$ be the collection of those $\eta' \in \mathcal{F}$ with $L(\eta') > L^* + \epsilon$, recalling that $L(\eta') = P\left\{I_{\{\eta'(X) > 1/2\}} \neq Y\right\}$. For every α,

$$
\begin{aligned}
P\{L(\eta_n) - L^* > \epsilon\} &\leq P\left\{\sup_{\eta' \in \mathcal{F}^{(\epsilon)}} \mathcal{L}_n(\eta') \geq \mathcal{L}_n(\eta)\right\} \\
&\leq P\{\mathcal{L}_n(\eta) \leq \alpha\} + P\left\{\sup_{\eta' \in \mathcal{F}^{(\epsilon)}} \mathcal{L}_n(\eta') \geq \alpha\right\}.
\end{aligned}
$$

For reasons that will be obvious later, we take $\alpha = -\mathcal{E} - \epsilon^2/16$. Thus,

$$
\begin{aligned}
&P\{L(\eta_n) - L^* > \epsilon\} \\
&\leq P\left\{\mathcal{L}_n(\eta) \leq -\mathcal{E} - \frac{\epsilon^2}{16}\right\} + P\left\{\sup_{\eta' \in \mathcal{F}^{(\epsilon)}} \mathcal{L}_n(\eta') \geq -\mathcal{E} - \frac{\epsilon^2}{16}\right\}
\end{aligned}
$$

$$= \mathbf{P}\left\{\mathcal{L}_n(\eta) \le -\mathcal{E} - \frac{\epsilon^2}{16}\right\}$$

$$+ \mathbf{P}\left\{\left(\sup_{\eta' \in \mathcal{F}^{(\epsilon)}} \mathcal{L}_n(\eta') - \mathcal{L}_n(\eta)\right) + (\mathcal{L}_n(\eta) + \mathcal{E}) \ge \frac{\epsilon^2}{16} - \frac{\epsilon^2}{8}\right\}$$

$$\le \mathbf{P}\left\{|\mathcal{L}_n(\eta) + \mathcal{E}| \ge \frac{\epsilon^2}{16}\right\} + \mathbf{P}\left\{\sup_{\eta' \in \mathcal{F}^{(\epsilon)}} \mathcal{L}_n(\eta') - \mathcal{L}_n(\eta) \ge -\frac{\epsilon^2}{8}\right\}.$$

Noting that $\mathbf{E}\{\mathcal{L}_n(\eta)\} = -\mathcal{E}$, the law of large numbers implies that the first term on the right-hand side converges to zero for every ϵ. (See Problem 15.4 for more information.)

Next we bound the second term. For a fixed distribution, let $\mathcal{F}_{X,\delta}$ be the smallest set of functions such that for each $\eta' \in \mathcal{F}$, there exist $\eta_L', \eta_U' \in \mathcal{F}_{X,\delta}$ with

$$\eta_L'(x) \le \eta'(x) \le \eta_U'(x), \quad x \in \mathcal{R}^d$$

and

$$\mathbf{E}\{\eta_U'(X) - \eta_L'(X)\} \le \delta,$$

where $\delta > 0$ will be specified later. By assumption, $N(X, \delta) = |\mathcal{F}_{X,\delta}| < \infty$. We have,

$$\mathbf{P}\left\{\sup_{\eta' \in \mathcal{F}^{(\epsilon)}} \mathcal{L}_n(\eta') - \mathcal{L}_n(\eta) \ge -\frac{\epsilon^2}{8}\right\}$$

$$= \mathbf{P}\left\{\sup_{\eta' \in \mathcal{F}^{(\epsilon)}} \prod_{i=1}^n \frac{\eta'(X_i)^{Y_i}(1 - \eta'(X_i))^{1-Y_i}}{\eta(X_i)^{Y_i}(1 - \eta(X_i))^{1-Y_i}} \ge e^{-n\epsilon^2/8}\right\}$$

$$\le \mathbf{P}\left\{\sup_{\eta' \in \mathcal{F}^{(\epsilon)}} \prod_{i=1}^n \frac{\eta_U'(X_i)^{Y_i}(1 - \eta_L'(X_i))^{1-Y_i}}{\eta(X_i)^{Y_i}(1 - \eta(X_i))^{1-Y_i}} \ge e^{-n\epsilon^2/8}\right\}$$

(where $\eta_L', \eta_U' \in \mathcal{F}_{X,\delta}$, $\eta_L' \le \eta' \le \eta_U'$ and $\mathbf{E}\{\eta_U'(X) - \eta_L'(X)\} < \delta$)

$$\le N^2(X, \delta) \sup_{\eta' \in \mathcal{F}^{(\epsilon)}} \mathbf{P}\left\{\prod_{i=1}^n \frac{\eta_U'(X_i)^{Y_i}(1 - \eta_L'(X_i))^{1-Y_i}}{\eta(X_i)^{Y_i}(1 - \eta(X_i))^{1-Y_i}} \ge e^{-n\epsilon^2/8}\right\}$$

(by the union bound)

$$= N^2(X, \delta) \sup_{\eta' \in \mathcal{F}^{(\epsilon)}} \mathbf{P}\left\{\prod_{i=1}^n \sqrt{\frac{\eta_U'(X_i)^{Y_i}(1 - \eta_L'(X_i))^{1-Y_i}}{\eta(X_i)^{Y_i}(1 - \eta(X_i))^{1-Y_i}}} \ge e^{-n\epsilon^2/16}\right\}$$

$$\le N^2(X, \delta)$$

$$\times \sup_{\eta' \in \mathcal{F}^{(\epsilon)}} \left(\mathbf{E}\left\{\sqrt{\eta_U'(X)\eta(X)} + \sqrt{(1 - \eta_L'(X))(1 - \eta(X))}\right\}\right)^n e^{n\epsilon^2/16},$$

where in the last step we used Markov's inequality and independence. We now find a good bound for

$$\mathbf{E}\left\{\sqrt{\eta'_U(X)\eta(X)} + \sqrt{(1 - \eta'_L(X))(1 - \eta(X))}\right\}$$

for each $\eta' \in \mathcal{F}^{(\epsilon)}$ as follows:

$$\mathbf{E}\left\{\sqrt{\eta'_U(X)\eta(X)} + \sqrt{(1 - \eta'_L(X))(1 - \eta(X))}\right\}$$

$$\leq \quad \mathbf{E}\left\{\sqrt{\eta'(X)\eta(X)} + \sqrt{(1 - \eta'(X))(1 - \eta(X))}\right.$$

$$\left. + \sqrt{\eta(X)\left(\eta'_U(X) - \eta'(X)\right)} + \sqrt{(1 - \eta(X))\left(\eta'(X) - \eta'_L(X)\right)}\right\}$$

(here we used $\sqrt{a + b} \leq \sqrt{a} + \sqrt{b}$)

$$\leq \quad \mathbf{E}\left\{\sqrt{\eta'(X)\eta(X)} + \sqrt{(1 - \eta'(X))(1 - \eta(X))}\right\}$$

$$+ \sqrt{\mathbf{E}\{\eta(X)\}}\sqrt{\mathbf{E}\left\{\eta'_U(X) - \eta'(X)\right\}}$$

$$+ \sqrt{\mathbf{E}\{1 - \eta(X)\}}\sqrt{\mathbf{E}\left\{\eta'(X) - \eta'_L(X)\right\}}$$

(by the Cauchy-Schwarz inequality)

$$\leq \quad \mathbf{E}\left\{\sqrt{\eta'(X)\eta(X)} + \sqrt{(1 - \eta'(X))(1 - \eta(X))}\right\} + 2\sqrt{\delta}$$

$$\leq \quad 1 - \frac{\epsilon^2}{8} + 2\sqrt{\delta} \quad \text{(by Corollary 15.1).}$$

Summarizing, we obtain

$$\mathbf{P}\left\{\sup_{\eta' \in \mathcal{F}^{(\epsilon)}} \mathcal{L}_n(\eta') - \mathcal{L}_n(\eta) \geq -\frac{\epsilon^2}{8}\right\}$$

$$\leq \quad N^2(X, \delta)\left(1 - \frac{\epsilon^2}{8} + 2\sqrt{\delta}\right)^n e^{n\epsilon^2/16}$$

$$\leq \quad N^2(X, \delta)e^{-n\epsilon^2/16 + 2n\sqrt{\delta}} \quad \text{(since } 1 + x \leq e^x\text{).}$$

By taking $\delta < \left(\frac{\epsilon^2}{32}\right)^2$, we see that the probability converges to zero exponentially rapidly, which concludes the proof. $\square$

15.4 Examples

In this section we show by example how to apply the previous consistency results. In all cases, we assume that $\eta \in \mathcal{F}$ and we are concerned with the weak convergence of $L(\eta_n)$ to L^* for all such distributions of (X, Y). The classes are as follows:

$$\mathcal{F}_1 = \left\{\eta = I_{[a,b]}, -\infty \le a \le b \le \infty\right\}$$

$$\mathcal{F}_1' = \left\{\eta = cI_{[a,b]}, c \in [0, 1], -\infty \le a \le b \le \infty\right\}$$

$$\mathcal{F}_2 = \left\{\eta = I_{[a_1,b_1] \times \cdots \times [a_d,b_d]}, -\infty \le a_i \le b_i \le \infty, 1 \le i \le d\right\}$$

$$\mathcal{F}_3 = \left\{\eta : \eta(x) = cx/(1 + cx) : x \ge 0, c \ge 0\right\}$$

$$\mathcal{F}_4 = \left\{\eta \text{ is monotone decreasing on } [0, 1]\right\}$$

$$\mathcal{F}_5 = \left\{\eta : \eta(x) = \frac{1}{1 + \|x - m\|^2}, m \in \mathcal{R}^d\right\}$$

$$\mathcal{F}_6 = \left\{\eta : \eta(x) = \sin^2(\theta x), \theta \in \mathcal{R}\right\}$$

$$\mathcal{F}_7 = \left\{\eta : \eta(x) = \frac{e^{\alpha_0 + \alpha^T x}}{1 + e^{\alpha_0 + \alpha^T x}}, \alpha_0 \in \mathcal{R}, x, \alpha \in \mathcal{R}^d\right\}.$$

These are all rather simple, yet they will illustrate various points. Of these classes, $\mathcal{F}_4$ is nonparametric; yet, it behaves "better" than the one-parameter class $\mathcal{F}_6$, for example. We emphasize again that we are interested in consistency for all distributions of X.

For $\mathcal{F}_1$, η_n will agree with the samples, that is, $\eta_n(X_i) = Y_i$ for all i, and therefore, $\eta_n \in \mathcal{F}_1$ is any function of the form $I_{[a,b]}$ with

$$X^{(l-1)} < a \le X^{(l)}, \quad X^{(r)} \le b < X^{(r+1)},$$

where $X^{(1)} \le \cdots \le X^{(n)}$ are the order statistics for $X_1, \ldots, X_n$ (with $X^{(0)} = -\infty$, $X^{(n+1)} = \infty$), $X^{(l)}$ is the smallest data point with $Y^{(l)} = 1$, and $X^{(r)}$ is the largest data point with $Y^{(r)} = 1$. As $L^* = 0$, we claim that

$$\mathbf{E}\{L(\eta_n)\} \le \mathbf{P}\left\{X^{(l-1)} < X < X^{(l)}\right\} + \mathbf{P}\left\{X^{(r)} < X < X^{(r+1)}\right\}$$

$$\le \frac{4}{n+1}.$$

The rule is simply excellent, and has universal performance guarantees.

REMARK. Note that in this case, maximum likelihood minimizes the empirical risk over the class of classifiers $\mathcal{C} = \{\phi = I_{[a,b]}, a \le b\}$. As $\mathcal{C}$ has VC dimension $V_{\mathcal{C}} = 2$ (Theorem 13.7), and $\inf_{\phi \in \mathcal{C}} L(\phi) = 0$, Theorem 12.7 implies that for all $\epsilon > 0$,

$$\mathbf{P}\{L(\eta_n) > \epsilon\} \le 8n^2 2^{-n\epsilon/2},$$

and that

$$\mathbf{E}\{L(\eta_n)\} \le \frac{4 \log n + 4}{n \log 2}$$

(see Problem 12.8). With the analysis given here, we have gotten rid of the $\log n$ factor. □

The class $\mathcal{F}_1'$ is much more interesting. Here you will observe a dramatic difference with empirical risk minimization, as the parameter c plays a key role that will aid a lot in the selection. Note that the likelihood product is zero if $N_1 = 0$ or if $Y_i = 1$ for some i with $X_i \notin [a, b]$, and it is

$$c^{N_1}(1 - c)^{N_0} \text{ otherwise,}$$

where N_0 is the number of (X_i, Y_i) pairs with $a \leq X_i \leq b$, $Y_i = 0$, and N_1 is the number of (X_i, Y_i) pairs with $a \leq X_i \leq b$, $Y_i = 1$. For fixed N_0, N_1, this is maximal when

$$c = \frac{N_1}{N_0 + N_1}.$$

Resubstitution yields that the likelihood product is zero if $N_1 = 0$ or if $Y_i = 1$ for some i with $X_i \notin [a, b]$, and equals

$$\exp\left\{ N_1 \log \frac{N_1}{N_0 + N_1} + N_0 \log \frac{N_0}{N_0 + N_1} \right\} \text{ if otherwise.}$$

Thus, we should pick $\eta_n = c I_{[a,b]}$ such that $c = N_1/(N_0+N_1)$, and $[a, b]$ maximizes the divergence

$$N_1 \log \frac{N_1}{N_0 + N_1} + N_0 \log \frac{N_0}{N_0 + N_1}.$$

See Problem 15.5.

For $\mathcal{F}_2$, by a similar argument as for $\mathcal{F}_1$, let $[A_1, B_1] \times \cdots \times [A_d, B_d]$ be the smallest rectangle of $\mathcal{R}^d$ that encloses all X_i's for which $Y_i = 1$. Then, we know that $\eta_n = I_{[A_1, B_1] \times \cdots \times [A_d, B_d]}$ agrees with the data. Furthermore, $L^* = 0$, and

$$\mathbf{E}\{L(\eta_n)\} \leq \frac{4d}{n + 1}$$

(see Problem 15.6).

The logarithm of the likelihood product for $\mathcal{F}_3$ is

$$\sum_{i:Y_i=1} \log(cX_i) - \sum_{i=1}^{n} \log(1 + cX_i).$$

Setting the derivative with respect to c equal to zero in the hope of obtaining an equation that must be satisfied by the maximum, we see that c must satisfy

$$N_1 = \sum_{i=1}^{n} \frac{cX_i}{1 + cX_i} = \sum_{i=1}^{n} \frac{X_i}{1/c + X_i}$$

unless $X_1 = \cdots = X_n = 0$, where $N_1 = \sum_{i=1}^{n} I_{\{Y_i=1\}}$. This equation has a unique solution for c. The rule corresponding to η_n is of the form $g(x) = I_{\{cx > 1\}} = I_{\{x > 1/c\}}$. Equivalently,

$$g(x) = \begin{cases} 1 & \text{if } N_1 > \sum_{i=1}^{n} \frac{X_i}{x + X_i} \\ 0 & \text{otherwise.} \end{cases}$$

This surprising example shows that we do not even have to know the parameter c in order to describe the maximum likelihood rule. In quite a few cases, this shortcut makes such rules very appealing indeed. Furthermore, as the condition of Theorem 15.2 is fulfilled, $\eta \in \mathcal{F}_3$ implies that the rule is consistent as well.

For $\mathcal{F}_4$, it is convenient to order the X_i's from small to large and to identify k consecutive groups, each group consisting of any number of X_i's with $Y_i = 0$, and one X_i with $Y_i = 1$. Thus, $k = \sum_{i=1}^{n} I_{\{Y_i=1\}}$. Then, a moment's thought shows that η_n must be piecewise constant, taking values $1 \geq a_1 \geq \cdots \geq a_k \geq 0$ on the consecutive groups, and the value zero past the k-th group (which can only consist of X_i's with $Y_i = 0$. The likelihood product thus is of the form

$$a_1(1 - a_1)^{n_1} a_2(1 - a_2)^{n_2} \cdots a_k(1 - a_k)^{n_k},$$

where $n_1, \ldots, n_k$ are the cardinalities of the groups minus one (i.e., the number of $Y_j = 0$-elements in the i-th group). Finally, we have

$$(a_1, \ldots, a_k) = \operatorname*{arg\,max}_{1 \geq a_1 \geq \cdots \geq a_k \geq 0} \prod_{j=1}^{k} a_j(1 - a_j)^{n_j}.$$

To check consistency, we see that for every X and $\epsilon > 0$,

$$N(X, \epsilon) \leq 4^{\lceil \frac{1}{\epsilon} \rceil}$$

(see Problem 15.7). Thus, the condition of Theorem 15.2 is satisfied, and $L(\eta_n) \to L^*$ in probability, whenever $\eta \in \mathcal{F}_4$.

In $\mathcal{F}_5$, the maximum likelihood method will attempt to place m at the center of the highest concentration in $\mathcal{R}^d$ of X_i's with $Y_i = 1$, while staying away from X_i's with $Y_i = 0$. Certainly, there are computational problems, but it takes little thought to verify that the conditions of Theorem 15.2 are satisfied. Explicit description of the rule is not necessary for some theoretical analysis!

Class $\mathcal{F}_6$ is a simple one-parameter class that does *not* satisfy the condition of Theorem 15.2. In fact, maximum likelihood fails here for the following reason: Let X be uniform on $[0, 1]$. Then the likelihood product is

$$\prod_{i:Y_i=1} \sin^2(\theta X_i) \times \prod_{i:Y_i=0} \cos^2(\theta X_i).$$

This product reaches a degenerate global maximum (1) as $\theta \to \infty$, regardless of the true (unknown) value of θ that gave rise to the data. See Problem 15.9.

Class $\mathcal{F}_7$ is used in the popular logistic discrimination problem, reviewed and studied by Anderson (1982), see also McLachlan (1992, Chapter 8). It is particularly important to observe that with this model,

$$\{\eta > 1/2\} \equiv \left\{ e^{\alpha_0 + \alpha^T x} > 1/2 \right\} \equiv \left\{ \alpha^T x > \beta \right\},$$

where $\beta = -\alpha_0 - \log 2$. Thus, $\mathcal{F}_7$ subsumes linear discriminants. It does also force a bit more structure on the problem, making η approach zero or one as we move away from the separating hyperplane. Day and Kerridge (1967) point out that model $\mathcal{F}_7$ is appropriate if the class-conditional densities take the form

$$cf(x) \exp\left\{-\frac{1}{2}(x - m)^T \Sigma^{-1}(x - m)\right\},$$

where c is a normalizing constant, f is a density, m is a vector, and Σ is a positive definite matrix; f and Σ must be the same for the two densities, but c and m may be different. Unfortunately, obtaining the best values for α_0 and α by maximum likelihood takes a serious computational effort. Had we tried to estimate f, m, and Σ in the last example, we would have done more than what is needed, as both f and Σ drop out of the picture. In this respect, the regression format is both parsimonious and lightweight.

15.5   Classical Maximum Likelihood: Distribution Format

In a more classical approach, we assume that given $Y = 1$, X has a density f_1, and given $Y = 0$, X has a density f_0, where both f_0 and f_1 belong to a given family $\mathcal{F}$ of densities. A similar setup may be used for atomic distributions, but this will not add anything new here, and is rather routine. The likelihood product for the data $(X_1, Y_1), \ldots, (X_n, Y_n)$ is

$$\prod_{i=1}^{n}(pf_1(X_i))^{Y_i}((1 - p)f_0(X_i))^{1-Y_i},$$

where $p = \mathbf{P}\{Y = 1\}$ is assumed to be unknown as well. The maximum likelihood choices for p, f_0, f_1 are given by

$$\left(p^*, f_0^*, f_1^*\right) = \underset{p\in[0,1],f_0,f_1\in\mathcal{F}}{\arg\max}\ \mathcal{L}_n(p, f_0, f_1),$$

where

$$\mathcal{L}_n(p, f_0, f_1) = \frac{1}{n}\sum_{i=1}^{n}(Y_i \log(pf_1(X_i)) + (1 - Y_i)\log((1 - p)f_0(X_i))).$$

Having determined p^*, f_0^*, f_1^* (recall that the solution is not necessarily unique), the maximum likelihood rule is

$$g_n(x) = \begin{cases} 0 & \text{if } (1 - p^*)f_0^*(x) \geq p^* f_1^*(x) \\ 1 & \text{otherwise.} \end{cases}$$

Generally speaking, the distribution format is more sensitive than the regression format. It may work better under the correct circumstances. However, we give up our universality, as the nature of the distribution of X must be known. For example, if X is distributed on a lower dimensional nonlinear manifold of $\mathcal{R}^d$, the distribution format is particularly inconvenient. Consistency results and examples are provided in a few exercises.

Problems and Exercises

PROBLEM 15.1. Show that if $|\mathcal{F}| = k < \infty$ and $\eta \in \mathcal{F}$, then $L(\eta_n) \to L^*$ with probability one, where η_n is obtained by the maximum likelihood method. Also, prove that convergence with probability one holds in Theorem 15.2.

PROBLEM 15.2. Prove that if η, η' are $[0, 1]$-valued regression functions, then the divergence

$$D(\eta') = \mathbf{E}\left\{\eta(X)\log\frac{\eta'(X)}{\eta(X)} + (1 - \eta(X))\log\frac{1 - \eta'(X)}{1 - \eta(X)}\right\}$$

is nonpositive, and that $D = 0$ if and only if $\eta(X) = \eta'(X)$ with probability one. In this sense, D measures the distance between η and η'.

PROBLEM 15.3. Let

$$L^* = \mathbf{E}\{\min(\eta(X), 1 - \eta(X))\}, \quad L(\eta') = \mathbf{P}\left\{I_{\{\eta'(X) > 1/2\}} \neq Y\right\},$$

and

$$D(\eta') = \mathbf{E}\left\{\eta(X)\log\frac{\eta'(X)}{\eta(X)} + (1 - \eta(X))\log\frac{1 - \eta'(X)}{1 - \eta(X)}\right\},$$

where $\eta(x) = \mathbf{P}\{Y = 1 | X = x\}$, and $\eta' : \mathcal{R}^d \to [0, 1]$ is arbitrary. Prove that

$$D(\eta') \leq -\frac{\mathbf{E}\left\{(\eta(X) - \eta'(X))^2\right\}}{2} \leq -\frac{1}{2}(L(\eta') - L^*)^2$$

(see also Problem 3.22). HINT: First prove that for $p, q \in [0, 1]$,

$$p\log\frac{q}{p} + (1 - p)\log\frac{1 - q}{1 - p} \leq -\frac{(p - q)^2}{2}$$

(use Taylor's series with remainder term for $\mathcal{H}(\cdot)$).

PROBLEM 15.4. Show that for each η there exists an $\epsilon_0 > 0$ such that for all $\epsilon \in (0, \epsilon_0)$,

$$\mathbf{P}\{\mathcal{L}_n(\eta) > \mathbf{E}\{\mathcal{L}_n(\eta)\} + \epsilon\} \leq e^{-n\epsilon^2/16}.$$

HINT: Proceed by Chernoff's bounding technique.

PROBLEM 15.5. Consider $\eta \in \mathcal{F}'_1$ and let η_n be a maximum likelihood estimate over $\mathcal{F}'_1$. Let $p = \mathbf{P}\{Y = 1\}$. Show that

$$L^* = p\min\left(1, \frac{1 - c}{c}\right).$$

Derive an upper bound for

$$\mathbf{E}\{L(\eta_n) - L^*\},$$

where in case of multiple choices for η_n, you take the smallest η_n in the equivalence class. This is an important problem, as $\mathcal{F}_1'$ picks a histogram cell in a data-based manner. $\mathcal{F}_1'$ may be generalized to the automatic selection of the best k-cell histogram: let $\mathcal{F}$ be the collection of all η's that are constant on the k intervals determined by breakpoints $-\infty < a_1 < \cdots < a_{k-1} < \infty$.

PROBLEM 15.6. Show that for the class $\mathcal{F}_2$,

$$\mathbf{E}\{L(\eta_n)\} \le \frac{4d}{n+1}$$

when $\eta \in \mathcal{F}_2$ and η_n is the maximum likelihood estimate.

PROBLEM 15.7. Show that for the class $\mathcal{F}_4$,

$$N(X, \epsilon) \le 4^{\lceil 3/\epsilon \rceil}$$

holds for all X and $\epsilon > 0$, that is, the bracketing ϵ-entropy of $\mathcal{F}_4$ is bounded by $\lceil 3/\epsilon \rceil$. HINT: Cover the class $\mathcal{F}_4$ by a class of monotone decreasing, piecewise constant functions, whose values are multiples of $\epsilon/3$, and whose breakpoints are at the $k\epsilon/3$-quantiles of the distribution of X ($k = 1, 2, \ldots, \lceil 3/\epsilon \rceil$).

PROBLEM 15.8. Discuss the maximum likelihood method for the class

$$\mathcal{F} = \left\{ \eta : \eta(x) = \begin{cases} c & \text{if } \alpha^T x > \beta \\ c' & \text{otherwise;} \end{cases} \quad c, c' \in [0, 1], \alpha \in \mathcal{R}^d, \beta \in \mathcal{R} \right\}.$$

What do the discrimination rules look like? If $\eta \in \mathcal{F}$, is the rule consistent? Can you guarantee a certain rate of convergence for $\mathbf{E}\{L(\eta_n)\}$? If $\eta \notin \mathcal{F}$, can you prove that $L(\eta_n)$ does not converge in probability to $\inf_{\eta' \in \mathcal{F}} L(\eta')$ for some distribution of (X, Y) with $\eta(x) = \mathbf{P}\{Y = 1 | X = x\}$? How would you obtain the values of c, c', α, β for the maximum likelihood choice η_n?

PROBLEM 15.9. Let $X_1, \ldots, X_n$ be i.i.d. uniform $[0, 1]$ random variables. Let $y_1, \ldots, y_n$ be arbitrary $\{0, 1\}$-valued numbers. Show that with probability one,

$$\limsup_{\theta \to \infty} \prod_{i:y_i=1} \sin^2(\theta X_i) \times \prod_{i:y_i=0} \cos^2(\theta X_i) = 1,$$

while for any $t < \infty$, with probability one,

$$\sup_{0 \le \theta \le t} \prod_{i:y_i=1} \sin^2(\theta X_i) \times \prod_{i:y_i=0} \cos^2(\theta X_i) < 1.$$

16

Parametric Classification

What do you do if you believe (or someone tells you) that the conditional distributions of X given $Y = 0$ and $Y = 1$ are members of a given family of distributions, described by finitely many real-valued parameters? Of course, it does not make sense to say that there are, say, six parameters. By interleaving the bits of binary expansions, we can always make one parameter out of six, and by splitting binary expressions, we may make a countable number of parameters out of one parameter (by writing the bits down in triangular fashion as shown below).

$$
\begin{array}{llll}
b_1 & b_2 & b_4 & b_7 \quad \cdots \\
b_3 & b_5 & b_8 \quad \cdots \\
b_6 & b_9 \quad \cdots \\
b_{10} \quad \cdots \\
\vdots
\end{array}
$$

Thus, we must proceed with care. The number of parameters of a family really is measured more by the sheer size or vastness of the family than by mere representation of numbers. If the family is relatively small, we will call it parametric but we will not give you a formal definition of "parametric." For now, we let Θ, the set of all possible values of the parameter θ, be a subset of a finite-dimensional Euclidean space. Formally, let

$$
\mathcal{P}_\Theta = \{P_\theta : \theta \in \Theta\},
$$

be a class of probability distributions on the Borel sets of $\mathcal{R}^d$. Typically, the family $\mathcal{P}_\Theta$ is parametrized in a smooth way. That is, two distributions, corresponding to two parameter vectors close to each other are in some sense close to each other, as well. Assume that the class-conditional densities f_0 and f_1 exist, and that both belong to the class of densities

$$\mathcal{F}_\Theta = \{f_\theta : \theta \in \Theta\}.$$

Discrete examples may be handled similarly. Take for example all gaussian distributions on $\mathcal{R}^d$, in which

$$f_\theta(x) = \frac{1}{\sqrt{(2\pi)^d \det(\Sigma)}} e^{-\frac{1}{2}(x-m)^T \Sigma^{-1}(x-m)},$$

where $m \in \mathcal{R}^d$ is the vector of means, and Σ is the covariance matrix. Recall that x^T denotes the transposition of the column vector x, and $\det(\Sigma)$ is the determinant of Σ. This class is conveniently parametrized by $\theta = (m, \Sigma)$, that is, a vector of $d + d(d + 1)/2$ real numbers.

Knowing that the class-conditional distributions are in $\mathcal{P}_\Theta$ makes discrimination so much easier—rates of convergence to L^* are excellent. Take $\mathcal{F}_\Theta$ as the class of uniform densities on hyperrectangles of $\mathcal{R}^d$: this has $2d$ natural parameters, the coordinates of the lower left and upper right vertices.

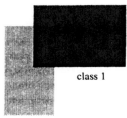

FIGURE 16.1. *The class-conditional densities are uniform on hyperrectangles.*

class 1

class 0

Given $(X_1, Y_1), \ldots, (X_n, Y_n)$, a child could not do things wrong—for class 1, estimate the upper right vertex by

$$\left(\max_{1 \le i \le n : Y_i = 1} X_i^{(1)}, \max_{1 \le i \le n : Y_i = 1} X_i^{(2)}, \ldots, \max_{1 \le i \le n : Y_i = 1} X_i^{(d)}\right)$$

and similarly for the upper right vertex of the class 0 density. Lower left vertices are estimated by considering minima. If A_0, A_1 are the two unknown hyperrectangles and $p = \mathbf{P}\{Y = 1\}$, the Bayes rule is simply

$$g^*(x) = \begin{cases} 1 & \text{if } x \in A_1 - A_0 \\ 0 & \text{if } x \in A_0 - A_1 \\ 1 & \text{if } x \in A_1 \cap A_0, \ \dfrac{p}{\lambda(A_1)} > \dfrac{1-p}{\lambda(A_0)} \\ 0 & \text{if } x \in A_1 \cap A_0, \ \dfrac{p}{\lambda(A_1)} \le \dfrac{1-p}{\lambda(A_0)}. \end{cases}$$

In reality, replace A_0, A_1, and p by the sample estimates $\widehat{A}_0$, $\widehat{A}_1$ (described above) and $\widehat{p} = (1/n) \sum_{i=1}^{n} I_{\{Y_i = 1\}}$. This way of doing things works very well, and we will pick up the example a bit further on. However, it is a bit ad hoc. There are indeed a few main principles that may be used in the design of classifiers under the additional information given here. In no particular order, here are a few methodologies:

(A) As the Bayes classifiers belong to the class

$$ C = \left\{ \phi = I_{\{pf_{\theta_1} > (1-p)f_{\theta_0}\}} : p \in [0, 1], \theta_0, \theta_1 \in \Theta \right\}, $$

it suffices to consider classifiers in C. For example, if $\mathcal{F}_\Theta$ is the normal family, then C coincides with indicators of functions in the set

$$ \left\{ \left\{ x : x^T A x + b^T x + c > 0 : A \text{ is a } d \times d \text{ matrix}, b \in \mathcal{R}^d, c \in \mathcal{R} \right\} \right\}, $$

that is, the family of quadratic decisions. In the hyperrectangular example above, every ϕ is of the form $I_{A_1 - A_2}$ where A_1 and A_2 are hyperrectangles of $\mathcal{R}^d$. Finding the best classifier of the form $\phi = I_A$ where $A \in \mathcal{A}$ is something we can do in a variety of ways: one such way, empirical risk minimization, is dealt with in Chapter 12 for example.

(B) Plug-in rules estimate (θ_0, θ_1) by $(\widehat{\theta}_0, \widehat{\theta}_1)$ and $p = \mathbf{P}\{Y = 1\}$ by $\widehat{p}$ from the data, and form the rule

$$ g_n(x) = \begin{cases} 1 & \text{if } \widehat{p} f_{\widehat{\theta}_1}(x) > (1 - \widehat{p}) f_{\widehat{\theta}_0}(x) \\ 0 & \text{otherwise.} \end{cases} $$

The rule here is within the class C described in the previous paragraph. We are hopeful that the performance with g_n is close to the performance with the Bayes rule g^* when $(\widehat{p}, \widehat{\theta}_0, \widehat{\theta}_1)$ is close to (p, θ_0, θ_1). For this strategy to work it is absolutely essential that $L(p, \theta_0, \theta_1)$ (the probability of error when p, θ_0, θ_1 are the parameters) be continuous in (p, θ_0, θ_1). Robustness is a key ingredient. If the continuity can be captured in an inequality, then we may get performance guarantees for $\mathbf{E}\{L_n\} - L^*$ in terms of the distance between $(\widehat{p}, \widehat{\theta}_0, \widehat{\theta}_1)$ and (p, θ_0, θ_1). Methods of estimating the parameters include maximum likelihood. This methodology is dealt with in Chapter 15. This method is rather sensitive to incorrect hypotheses (what if we were wrong about our assumption that the class-conditional distributions were in $\mathcal{P}_\Theta$?). Another strategy, minimum distance estimation, picks that member from $\mathcal{P}_\Theta$ that is closest in some sense to the raw empirical measure that puts mass $1/n$ at each if the n data points. See Section 16.3 This approach does *not* care about continuity of $L(p, \theta_0, \theta_1)$, as it judges members of $\mathcal{P}_\Theta$ by closeness in some space under a metric that is directly related to the probability of error. Robustness will drop out naturally.

A general approach should not have to know whether Θ can be described by a finite number of parameters. For example, it should equally well handle descriptions as $\mathcal{P}_\Theta$ is the class of all distributions of (X, Y) in which $\eta(x) = \mathbf{P}\{Y = 1 | X = x\}$ is monotonically increasing in all the components of x. Universal paradigms such as maximum likelihood, minimum distance estimation, and empirical error minimization are all applicable here. This particular $\mathcal{P}_\Theta$ is dealt with in Chapter 15, just to show you that the description of the class does invalidate the underlying principles.

16.1   Example: Exponential Families

A class $\mathcal{P}_\Theta$ is exponential, if every class-conditional density f_θ can be written in the form

$$f_\theta(x) = c\alpha(\theta)\beta(x)e^{\sum_{i=1}^k \pi_i(\theta)\psi_i(x)},$$

where $\beta, \psi_1, \ldots, \psi_k : \mathcal{R}^d \to \mathcal{R}, \beta \geq 0, \alpha, \pi_1, \ldots, \pi_k : \Theta \to \mathcal{R}$ are fixed functions, and c is a normalizing constant. Examples of exponential families include the gaussian, gamma, beta, Rayleigh, and Maxwell densities (see Problem 16.4). The Bayes-rule can be rewritten as

$$g^*(x) = \begin{cases} 1 & \text{if } \log\left(\frac{pf_{\theta_1^*}(x)}{(1-p)f_{\theta_0^*}(x)}\right) > 0 \\ 0 & \text{otherwise.} \end{cases}$$

This is equivalent to

$$g^*(x) = \begin{cases} 1 & \text{if } \sum_{i=1}^k \psi_i(x)(\pi_i(\theta_0^*) - \pi_i(\theta_1^*)) < \log\left(\frac{p\alpha(\theta_1^*)}{(1-p)\alpha(\theta_0^*)}\right) \\ 0 & \text{otherwise.} \end{cases}$$

The Bayes-rule is a so-called generalized linear rule with $1, \psi_1, \ldots, \psi_k$ as basis functions. Such rules are easily dealt with by empirical risk minimization and related methods such as complexity regularization (Chapters 12, 17, 18, 22).

Another important point is that g^* does *not* involve the function β. For all we know, β may be some esoteric ill-behaved function that would make estimating $f_\theta(x)$ all but impossible if β were unknown. Even if $\mathcal{P}_\Theta$ is the huge family in which $\beta \geq 0$ is left undetermined, but it is known to be identical for the two class-conditional densities (and $\alpha(\theta)$ is just a normalization factor), we would still only have to look at the same small class of generalized linear discrimination rules! So, densities do not matter—ratios of densities do. Pattern recognition should be, and is, easier than density estimation. Those who first estimate (f_0, f_1) by $(\widehat{f_0}, \widehat{f_1})$ and then construct rules based on the sign of $\widehat{p}\widehat{f_1}(x) - (1 - \widehat{p})\widehat{f_0}(x)$ do themselves a disservice.

16.2   Standard Plug-In Rules

In standard plug-in methods, we construct estimates $\widehat{\theta}_0$, $\widehat{\theta}_1$ and $\widehat{p}$ from the data and use them to form a classifier

$$g_n(x) = \begin{cases} 1 & \text{if } \widehat{p} f_{\widehat{\theta}_1}(x) > (1 - \widehat{p}) f_{\widehat{\theta}_0}(x) \\ 0 & \text{otherwise.} \end{cases}$$

It is generally *not* true that if $\widehat{p} \to p$, $\widehat{\theta}_0 \to \theta_0$, and $\widehat{\theta}_1 \to \theta_1$ in probability, then $L(g_n) \to L^*$ in probability, where $L(g_n)$ is the probability of error with g_n. Consider the following simple example:

EXAMPLE. Let $\mathcal{F}_\Theta$ be the class of all uniform densities on $[-\theta, 0]$ if $\theta \neq 1$ and on $[0, \theta]$ if $\theta = 1$. Let $\theta_0 = 1$, $\theta_1 = 2$, $p = 1/2$. Then a reasonable estimate of θ_i would be $\widehat{\theta}_i = \max_{i:Y_i=1} |X_i|$. Clearly, $\widehat{\theta}_i \to \theta_i$ in probability. However, as $\widehat{\theta}_i \neq 1$ with probability one, we note that $g_n(x) = 0$ for $x > 0$ and thus, even though $L^* = 0$ (as the supports of f_{θ_0} and f_{θ_1} are not overlapping), $L(g_n) \geq P\{Y = 1\} = p = 1/2$. The problem with this is that there is no continuity with respect to θ in $\mathcal{F}_\Theta$. $\square$

Basic consistency based upon continuity considerations is indeed easy to establish. As ratios of densities matter, it helps to introduce

$$\eta_\theta(x) = \frac{p f_{\theta_1}(x)}{(1 - p) f_{\theta_0}(x) + p f_{\theta_1}(x)} = P\{Y = 1 | X = x\},$$

where $\theta = (p, \theta_0, \theta_1)$ or (θ_0, θ_1) as the case may be. We recall that if $g_n(x) = I_{\{\eta_{\widehat{\theta}}(x) > 1/2\}}$ where $\widehat{\theta}$ is an estimate of θ, then

$$L(g_n) - L^* \leq 2E\left\{ |\eta_{\widehat{\theta}}(X) - \eta_\theta(X)| \big| D_n \right\},$$

where D_n is the data sequence. Thus,

$$E\left\{ L(g_n) - L^* \right\} \leq 2E\left\{ |\eta_{\widehat{\theta}}(X) - \eta_\theta(X)| \right\}.$$

By the Lebesgue dominated convergence theorem, we have, without further ado:

Theorem 16.1. *If η_θ is continuous in θ in the $L_1(\mu)$ sense, where μ is the measure of X, and $\widehat{\theta} \to \theta$ in probability, then $E\{L(g_n)\} \to L^*$ for the standard plug-in rule.*

In some cases, we can do better and derive rates of convergence by examining the local behavior of $\eta_\theta(x)$. For example, if $\eta_\theta(x) = e^{-\theta|x|}$, $x \in \mathcal{R}$, then

$$|\eta_\theta(x) - \eta_{\theta'}(x)| \leq |x||\theta - \theta'|,$$

and

$$\begin{aligned} E\{L(g_n)\} - L^* &\leq 2E\left\{ |X| |\theta - \widehat{\theta}| \right\} \\ &\leq 2\sqrt{E\{X^2\}} \sqrt{E\left\{ (\theta - \widehat{\theta})^2 \right\}}, \end{aligned}$$

yielding an explicit bound. In general, θ is multivariate, consisting at least of the triple (p, θ_0, θ_1), but the above example shows the way to happy analysis. For the simple example given here, $\mathbf{E}\{L(g_n)\} \to L^*$ if $\mathbf{E}\left\{(\theta - \widehat{\theta})^2\right\} \to 0$. In fact, this seems to suggest that for this family, $\widehat{\theta}$ should be found to minimize $\mathbf{E}\left\{(\theta - \widehat{\theta})^2\right\}$. This is false. One is always best off minimizing the probability of error. Other criteria may be relevant via continuity, but should be considered with care.

How certain parameters are estimated for given families of distributions is what mathematical statistics is all about. The maximum likelihood principle looms large: θ_0 is estimated for densities f_θ by

$$\widehat{\theta}_0 = \arg\max_\theta \prod_{i:Y_i=0} f_\theta(X_i)$$

for example. If you work out this (likelihood) product, you will often discover a simple form for the estimate of the data. In discrimination, only η matters, not the class-conditional densities. Maximum likelihood in function of the η's was studied in Chapter 15. We saw there that this is often consistent, but that maximum likelihood behaves poorly when the true distribution is not in $\mathcal{P}_\theta$. We will work out two simple examples:

As an example of the maximum likelihood method in discrimination, we assume that

$$\mathcal{F}_\Theta = \left\{ f_\theta(x) = \frac{x^{\alpha-1}e^{-x/\beta}}{\beta^\alpha \Gamma(\alpha)} I_{\{x>0\}} : \alpha, \beta > 0 \right\}$$

is the class of all gamma densities with $\theta = (\alpha, \beta)$. The likelihood product given $(X_1, Y_1), \ldots, (X_n, Y_n)$, is, if θ_0, θ_1 are the unknown parameters,

$$\prod_{i=1}^n \left(p f_{\theta_1}(X_i) \right)^{Y_i} \left((1-p) f_{\theta_0}(X_i) \right)^{1-Y_i}.$$

This is the probability of observing the data sequence if the f_θ's were in fact discrete probabilities. This product is simply

$$\left(\frac{p}{\beta_1^{\alpha_1} \Gamma(\alpha_1)} \right)^{\sum_{i=1}^n Y_i} \left(\frac{1-p}{\beta_0^{\alpha_0} \Gamma(\alpha_0)} \right)^{n-\sum_{i=1}^n Y_i}$$

$$e^{-\frac{\sum Y_i X_i}{\beta_1} - \frac{\sum(1-Y_i)X_i}{\beta_0} + (\alpha_1-1)\sum Y_i \log X_i + (\alpha_0-1)\sum(1-Y_i)\log X_i},$$

where $\theta_0 = (\alpha_0, \beta_0)$, $\theta_1 = (\alpha_1, \beta_1)$. The first thing we notice is that this expression depends only on certain functions of the data, notably $\sum Y_i X_i$, $\sum(1 - Y_i)X_i$, $\sum Y_i \log X_i$, $\sum(1 - Y_i)\log X_i$, and $\sum Y_i$. These are called the sufficient statistics for the problem at hand. We may in fact throw away the data and just store the sufficient statistics. The likelihood product has to be maximized. Even in this rather simple univariate example, this is a nontrivial task. Luckily, we immediately note that p occurs in the factor $p^N(1 - p)^{n-N}$, where $N = \sum Y_i$. This is maximal at $\widehat{p} = N/n$, a well-known result. For fixed α_0, α_1, we can also get β_0, β_1; but for variable $\alpha_0, \alpha_1, \beta_0, \beta_1$, the optimization is difficult. In d-dimensional cases, one has nearly always to resort to specialized algorithms.

As a last example, we return to $\mathcal{F}_\Theta = \{$uniform densities on rectangles of $\mathcal{R}^d\}$. Here the likelihood product once again has $p^N(1-p)^{n-N}$ as a factor, leading to $\widehat{p} = N/n$. The other factor (if (a_1, b_1), (a_0, b_0) are the lower left, upper right vertices of the rectangles for θ_1, θ_0) is zero if for some i, $Y_i = 1$, $X_i \notin$ rectangle(a_1, b_1), or $Y_i = 0$, $X_i \notin$ rectangle(a_0, b_0). Otherwise, it is

$$\frac{1}{\|b_1 - a_1\|^N} \cdot \frac{1}{\|b_0 - a_0\|^{n-N}},$$

where $\|b_1 - a_1\| = \prod_{j=1}^d (b_1^{(j)} - a_1^{(j)})$ denotes the volume of the rectangle (a_1, b_1), and similarly for $\|b_0 - a_0\|$. This is maximal if $\|b_1 - a_1\|$ and $\|b_0 - a_0\|$ are minimal. Thus, the maximum likelihood estimates are

$$\widehat{a}_i^{(k)} = \min_{j:Y_j=i} X_i^{(k)}, \quad 1 \le k \le d, \ i = 0, 1,$$

$$\widehat{b}_i^{(k)} = \max_{j:Y_j=i} X_i^{(k)}, \quad 1 \le k \le d, \ i = 0, 1,$$

where $\widehat{a}_i = (\widehat{a}_i^{(1)}, \ldots, \widehat{a}_i^{(d)})$, $\widehat{b}_i = (\widehat{b}_i^{(1)}, \ldots, \widehat{b}_i^{(d)})$, and $X_i = (X_i^{(1)}, \ldots, X_i^{(d)})$.

Rates of convergence may be obtained via some of the (in)equalities of Chapter 6, such as

(1) $\mathbf{E}\{L(g_n)\} - L^* \le 2\mathbf{E}\{|\eta_n(X) - \eta(X)|\}$

   (where $\eta_n = \widehat{p} f_{\widehat{\theta}_1} / (\widehat{p} f_{\widehat{\theta}_1} + (1 - \widehat{p}) f_{\widehat{\theta}_0})$),

(2) $\mathbf{E}\{L(g_n)\} - L^* \le 2\sqrt{\mathbf{E}\left\{(\eta_n(X) - \eta(X))^2\right\}}$,

(3) $\mathbf{E}\{L(g_n)\} - L^* = 2\mathbf{E}\left\{\left|\eta(X) - \frac{1}{2}\right| I_{\{g_n(X) \ne g^*(X)\}}\right\}$.

The rate with which $\widehat{\theta}$ approaches θ in Θ (measured with some metric) may be very different from that with which $L(g_n)$ approaches L^*. As shown in Theorem 6.5, the inequality (2) is always loose, yet it is this inequality that is often used to derive rates of convergence by authors and researchers. Let us take a simple example to illustrate this point. Assume $\mathcal{F}_\Theta$ is the family of normal densities on the real line. If $\theta_0 = (m_0, \sigma_0)$, $\theta_1 = (m_1, \sigma_1)$ are the unknown means and standard deviations of the class-conditional densities, and $p = \mathbf{P}\{Y = 1\}$ is also unknown, then we may estimate p by $\widehat{p} = \sum_{j=1}^n Y_j/n$, and m_i and σ_i by

$$\widehat{m}_i = \frac{\sum_{j=1}^n X_j I_{\{Y_j=i\}}}{\sum_{j=1}^n I_{\{Y_j=i\}}} \quad \text{and} \quad \widehat{\sigma}_i^2 = \frac{\sum_{j=1}^n (X_j - \widehat{m}_i)^2 I_{\{Y_j=i\}}}{\sum_{j=1}^n I_{\{Y_j=i\}}}, \quad i = 0, 1,$$

respectively, when denominators are positive. If a denominator is 0, set $\widehat{m}_i = 0$, $\widehat{\sigma}_i^2 = 1$. From Chebyshev's inequality, we can verify that for fixed $p \in (0, 1)$,

$E\{|p - \widehat{p}|\} = O\left(1/\sqrt{n}\right)$, $E\{|m_i - \widehat{m}_i|\} = O\left(1/\sqrt{n}\right)$, $E\{|\sigma_i - \widehat{\sigma}_i|\} = O\left(1/\sqrt{n}\right)$ (Problem 16.3).

If we compute $E\{|\eta_n(X) - \eta(X)|\}$, we will discover that $E\{L(g_n)\} - L^* = O\left(1/\sqrt{n}\right)$. However, if we compute $E\left\{\left|\eta(X) - \frac{1}{2}\right| I_{\{g_n(X) \neq g^*(X)\}}\right\}$, we will find that $E\{L(g_n)\} - L^* = O(1/n)$. Thus, while the parameters converge at a rate $O\left(1/\sqrt{n}\right)$ dictated by the central limit theorem, and while η_n converges to η in $L_1(\mu)$ with the same rate, the error rate in discrimination is much smaller. See Problems 16.7 to 16.9 for some practice in this respect.

BIBLIOGRAPHIC REMARKS. McLachlan (1992) has a comprehensive treatment on parametric classification. Duda and Hart (1973) have many good introductory examples and a nice discussion on sufficient statistics, a topic we do not deal with in this text. For maximum likelihood estimation, see Hjort (1986a; 1986b). □

16.3   Minimum Distance Estimates

Here we describe a general parameter estimation principle that appears to be more suitable for plug-in classification rules than the maximum likelihood method. The estimated parameter is obtained by the projection of the empirical measure on the parametric family.

The principle of minimum distance estimation may be described as follows. Let $\mathcal{P}_\Theta = \{P_\theta : \theta \in \Theta\}$ be a parametric family of distributions, and assume that P_{θ^*} is the unknown distribution of the i.i.d. observations $Z_1, \ldots, Z_n$. Denote by ν_n the empirical measure

$$\nu_n(A) = \frac{1}{n} \sum_{i=1}^{n} I_{\{Z_i \in A\}}, \quad A \subset \mathcal{R}^d.$$

Let $D(\cdot, \cdot)$ be a metric on the set of all probability distributions on $\mathcal{R}^d$. The *minimum distance estimate* of θ^* is defined as

$$\theta_n = \arg\min_{\theta \in \Theta} D(\nu_n, P_\theta),$$

if it exists and is unique. If it is not unique, select one candidate for which the minimum is attained.

Consider for example the Kolmogorov-Smirnov distance

$$D_{KS}(P, Q) = \sup_{z \in \mathcal{R}^d} |F(z) - G(z)|,$$

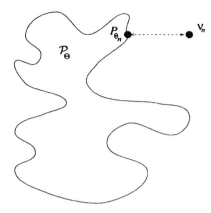

FIGURE 16.2. *The member of $\mathcal{P}_\Theta$ closest to the empirical measure ν_n is chosen by the minimum distance estimate.*

where F and G are the distribution functions of the measures P and Q on $\mathcal{R}^d$, respectively. It is easy to see that D_{ks} is a metric. Note that

$$\sup_{z \in \mathcal{R}^d} |F(z) - G(z)| = \sup_{A \in \mathcal{A}} |P(A) - Q(A)|,$$

where $\mathcal{A}$ is the class of sets of the form $(-\infty, z^{(1)}) \times \cdots \times (-\infty, z^{(d)})$, for $z = (z^{(1)}, \ldots, z^{(d)}) \in \mathcal{R}^d$. For the Kolmogorov-Smirnov distance between the estimated and the true distributions, by the triangle inequality, we have

$$\begin{aligned} D_{ks}(P_{\theta_n}, P_{\theta^*}) &\leq D_{ks}(P_{\theta_n}, \nu_n) + D_{ks}(\nu_n, P_{\theta^*}) \\ &\leq 2D_{ks}(\nu_n, P_{\theta^*}), \end{aligned}$$

where in the second inequality we used the definition of θ_n. Now notice that the upper bound is just twice the Kolmogorov-Smirnov distance between an empirical distribution and the true distribution. By a straightforward application of Theorem 12.5, for every n and $\epsilon > 0$,

$$\mathbf{P}\{D_{ks}(\nu_n, P_{\theta^*}) > \epsilon\} = \mathbf{P}\left\{\sup_{A \in \mathcal{A}} |\nu_n(A) - P_{\theta^*}(A)| > \epsilon\right\} \leq 8 \left(\frac{ne}{d}\right)^d e^{-n\epsilon^2/32},$$

since the n-th shatter coefficient $s(\mathcal{A}, n)$ of the class of sets

$$\mathcal{A} = \left\{(-\infty, z^{(1)}) \times \cdots \times (-\infty, z^{(d)}), z = (z^{(1)}, \ldots, z^{(d)}) \in \mathcal{R}^d\right\}$$

cannot exceed $(ne/d)^d$. This can easily be seen by an argument similar to that in the proof of Theorem 13.8. From the inequalities above, we see that the Kolmogorov-Smirnov distance between the estimated and the true distributions is always $O\left(\sqrt{\log n/n}\right)$. The only condition we require is that θ_n be well defined.

Of course, rather than the Kolmogorov-Smirnov distance, it is the error probability of the plug-in classification rule in which we are primarily interested. In order to make the connection, we adapt the notions of minimum distance estimation and Kolmogorov-Smirnov distance to better suit the classification problem we are after. Every parametric family of distributions defines a class of sets in $\mathcal{R}^d$ as the collection $\mathcal{A}$ of sets of the form $\{x : \phi(x) = 1\}$, where the classifiers ϕ are the possible plug-in rules defined by the parametric family. The idea here is to perform minimum distance estimation with the generalized Kolmogorov-Smirnov distance with respect to a class of sets closely related to $\mathcal{A}$.

Assume that both class-conditional distributions P_{θ_0}, P_{θ_1} belong to a parametric family $\mathcal{P}_\Theta$, and the class-conditional densities f_{θ_0}, f_{θ_1} exist. Then the Bayes rule may be written as

$$g^*(x) = \begin{cases} 1 & \text{if } \alpha_\theta(x) > 0 \\ 0 & \text{otherwise}, \end{cases}$$

where $\alpha_\theta(x) = p f_{\theta_1}(x) - (1 - p) f_{\theta_0}(x)$ and $p = \mathbf{P}\{Y = 1\}$. We use the short notation $\theta = (p, \theta_0, \theta_1)$. The function α_θ may be thought of as the Radon-Nikodym derivative of the signed measure $Q_\theta = p P_{\theta_1} - (1 - p) P_{\theta_0}$. In other words, to each Borel set $A \subset \mathcal{R}^d$, Q_θ assigns the real number $Q_\theta(A) = p P_{\theta_1}(A) - (1 - p) P_{\theta_0}(A)$. Given the data $D_n = ((X_1, Y_1), \ldots, (X_n, Y_n))$, we define the empirical counterpart of Q_θ by

$$\nu_n(A) = \frac{1}{n} \sum_{i=1}^n (2Y_i - 1) I_{\{X_i \in A\}}.$$

The minimum distance classification rule we propose projects the empirical signed measure ν_n on the set of measures Q_θ. The metric we use is also specifically fitted to the given pattern recognition problem: define the class of sets

$$\mathcal{A} = \left\{ \{x \in \mathcal{R}^d : \alpha_\theta(x) > 0\} : p \in [0, 1], \theta_0, \theta_1 \in \Theta \right\}.$$

$\mathcal{A}$ is just the class of sets $A \subset \mathcal{R}^d$ such that $I_{\{x \in A\}}$ is the Bayes rule for some $\theta = (p, \theta_0, \theta_1)$. Also introduce

$$\mathcal{B} = \left\{ A \cap B^c : A, B \in \mathcal{A} \right\}.$$

Given two signed measures Q, Q', we define their *generalized Kolmogorov-Smirnov distance* by

$$D_\mathcal{B}(Q, Q') = \sup_{A \in \mathcal{B}} |Q(A) - Q'(A)|,$$

that is, instead of the class of half-infinite intervals as in the definition of the ordinary Kolmogorov-Smirnov distance, here we take the supremum over $\mathcal{B}$, a class tailored to our discrimination problem. Now, we are ready to define our minimum distance estimate

$$\widehat{\theta} = \arg\min_\theta D_\mathcal{B}(Q_\theta, \nu_n),$$

where the minimum is taken over all triples $\theta = (p, \theta_0, \theta_1)$ with $p \in [0, 1]$, $\theta_0, \theta_1 \in \Theta$. The corresponding classification rule is

$$g_{\hat{\theta}}(x) = \begin{cases} 1 & \text{if } \alpha_{\hat{\theta}}(x) > 0 \\ 0 & \text{otherwise.} \end{cases}$$

The next theorem shows that if the parametric assumption is valid, then $g_{\hat{\theta}}$ performs extremely well. The theorem shows that if $\mathcal{A}$ has finite VC dimension $V_{\mathcal{A}}$, then without any additional conditions imposed on the parametric class, the corresponding error probability $L(g_{\hat{\theta}})$ is not large: $L(g_{\hat{\theta}}) - L^* = O\left(\sqrt{V_{\mathcal{A}} \log n / n}\right)$.

Theorem 16.2. *Assume that both conditional densities $f_{\theta_0}, f_{\theta_1}$ are in the parametric class $\mathcal{F}_{\Theta}$. Then for the classification rule defined above, we have for every n and $\epsilon > 0$ that*

$$\mathbf{P}\left\{L(g_{\hat{\theta}}) - L^* > \epsilon\right\} \leq 8s(\mathcal{B}, n)e^{-n\epsilon^2/512},$$

where $s(\mathcal{B}, n)$ is the n-th shatter coefficient of $\mathcal{B}$. Furthermore,

$$\mathbf{E}\left\{L(g_{\hat{\theta}}) - L^*\right\} \leq 32\sqrt{\frac{\log(8es(\mathcal{B}, n))}{2n}}.$$

REMARK. Recalling from Chapter 13 that $s(\mathcal{B}, n) \leq s^2(\mathcal{A}, n)$ and that $s(\mathcal{A}, n) \leq (n + 1)^{V_{\mathcal{A}}}$, where $V_{\mathcal{A}}$ is the VC dimension of $\mathcal{A}$, we obtain the bound

$$\mathbf{P}\left\{L(g_{\hat{\theta}}) - L^* > \epsilon\right\} \leq 8(n + 1)^{2V_{\mathcal{A}}}e^{-n\epsilon^2/512}. \quad \square$$

The proof of the theorem is based upon two key observations. The first lemma provides a bound on $L(g_{\hat{\theta}}) - L^*$ in terms of the generalized Kolmogorov-Smirnov distance between the estimated and the true parameters. The second lemma is a straightforward extension of the Vapnik-Chervonenkis inequality to signed measures.

Lemma 16.1.

$$L(g_{\hat{\theta}}) - L^* \leq 2D_{\mathcal{B}}(Q_\theta, Q_{\hat{\theta}}).$$

PROOF. Denote

$$A_\theta = \{x : \alpha_\theta(x) > 0\} \quad \text{and} \quad A_{\hat{\theta}} = \{x : \alpha_{\hat{\theta}}(x) > 0\},$$

that is, $g_{\hat{\theta}}(x) = I_{\{\alpha_{\hat{\theta}}(x) > 0\}}$, $g^*(x) = I_{\{\alpha_\theta(x) > 0\}}$, and $A_\theta, A_{\hat{\theta}} \in \mathcal{A}$. At the first crucial step, we use the equality of Theorem 2.2:

$$L(g_{\hat{\theta}}) - L^* = \int I_{\{g_{\hat{\theta}}(x) \neq g_H(x)\}} |\alpha_\theta(x)| dx$$

$$= -\int_{A_{\hat{\theta}} \cap A_H^c} \alpha_\theta(x) dx + \int_{A_{\hat{\theta}}^c \cap A_\theta} \alpha_\theta(x) dx$$

$$\leq \int_{A_{\widehat{\theta}} \cap A_{\theta}^c} \alpha_{\widehat{\theta}}(x)dx - \int_{A_{\widehat{\theta}} \cap A_{\theta}^c} \alpha_{\theta}(x)dx$$

$$+ \int_{A_{\widehat{\theta}}^c \cap A_{\theta}} \alpha_{\theta}(x)dx - \int_{A_{\widehat{\theta}}^c \cap A_{\theta}} \alpha_{\widehat{\theta}}(x)dx$$

$$= Q_{\widehat{\theta}}(A_{\widehat{\theta}} \cap A_{\theta}^c) - Q_{\theta}(A_{\widehat{\theta}} \cap A_{\theta}^c) + Q_{\theta}(A_{\widehat{\theta}}^c \cap A_{\theta}) - Q_{\widehat{\theta}}(A_{\widehat{\theta}}^c \cap A_{\theta})$$

$$\leq 2D_{\mathcal{B}}(Q_{\theta}, Q_{\widehat{\theta}}),$$

since both $A_{\widehat{\theta}} \cap A_{\theta}^c$ and $A_{\widehat{\theta}}^c \cap A_{\theta}$ are members of $\mathcal{B}_{\Theta}$. □

The next lemma is an extension of the Vapnik-Chervonenkis inequality. The proof is left to the reader (Problem 16.11).

Lemma 16.2. *For every n and* $\epsilon > 0$,

$$\mathbf{P}\{D_{\mathcal{B}}(Q_{\theta}, \nu_n) > \epsilon\} \leq 8s(\mathcal{B}, n)e^{-n\epsilon^2/32}.$$

The rest of the proof of Theorem 16.2 shows that the generalized Kolmogorov-Smirnov distance with respect to $\mathcal{B}$ between the estimated and true distributions is small with large probability. This can be done as we proved for the Kolmogorov-Smirnov distance at the beginning of the section.

PROOF OF THEOREM 16.2. By Lemma 16.1,

$$\begin{aligned}
L(g_{\widehat{\theta}}) - L^* &\leq 2D_{\mathcal{B}}(Q_{\theta}, Q_{\widehat{\theta}}) \\
&\leq 2D_{\mathcal{B}}(Q_{\theta}, \nu_n) + 2D_{\mathcal{B}}(Q_{\widehat{\theta}}, \nu_n) \quad \text{(by the triangle inequality)} \\
&\leq 4D_{\mathcal{B}}(Q_{\theta}, \nu_n) \quad \text{(by the definition of } \widehat{\theta}\text{)}.
\end{aligned}$$

The theorem now follows by Lemma 16.2. □

Finally, we examine robustness of the minimum distance rule against modeling errors, that is, what happens if the distributions are not in $\mathcal{P}_{\Theta}$. A good rule should still work reasonably well if the distributions are in some sense close to the modeled parametric class $\mathcal{P}_{\Theta}$. Observe that if for some $\theta = (p, \theta_0, \theta_1)$ the Bayes rule can be written as

$$g^*(x) = \begin{cases} 1 & \text{if } \alpha_{\theta}(x) > 0 \\ 0 & \text{otherwise,} \end{cases}$$

then Lemma 16.1 remains valid even when the class-conditional distributions are not in $\mathcal{P}_{\Theta}$. Denote the true class-conditional distributions by P_0^*, P_1^*, let $p^* = \mathbf{P}\{Y = 1\}$, and introduce $Q^* = p^*P_1^* - (1 - p)P_0^*$. Thus,

$$\begin{aligned}
L(g_{\widehat{\theta}}) &- L^* \\
&\leq 2D_{\mathcal{B}}(Q^*, Q_{\widehat{\theta}}) \quad \text{(by Lemma 16.1)} \\
&\leq 2D_{\mathcal{B}}(Q^*, \nu_n) + 2D_{\mathcal{B}}(\nu_n, Q_{\widehat{\theta}}) \quad \text{(by the triangle inequality)} \\
&\leq 2D_{\mathcal{B}}(Q^*, \nu_n) + 2D_{\mathcal{B}}(\nu_n, Q_{\theta}) \quad \text{(by the definition of } Q_{\widehat{\theta}}\text{)} \\
&\leq 4D_{\mathcal{B}}(Q^*, \nu_n) + 2D_{\mathcal{B}}(Q^*, Q_{\theta}) \quad \text{(again by the triangle inequality)}.
\end{aligned}$$

Lemma 16.2 now applies to the first term on the right-hand side. Thus, we conclude that if the Bayes rule is in $\mathcal{A}$, then for all $\epsilon \geq 4 \inf_{Q_\theta} D_B(Q^*, Q_\theta)$,

$$\mathbf{P}\left\{L(g_{\hat{\theta}}) - L^* > \epsilon\right\} \leq 8(n+1)^{2V_A} e^{-n\epsilon^2/2^{11}}.$$

The constant in the exponent may be improved significantly by more careful analysis. In other words, if the Bayes rule is in $\mathcal{A}$ and the true distribution is close to the parametric family in the generalized Kolmogorov-Smirnov distance specified above, then the minimum distance rule still performs close to the Bayes error. Unfortunately, we cannot say the same if $\mathcal{A}$ does not contain the Bayes rule. Empirical error minimization, discussed in the next section, is however very robust in all situations.

16.4 Empirical Error Minimization

In this section we explore the connection between parametric classification and rule selection by minimizing the empirical error, studied in Chapter 12.

Consider the class $\mathcal{C}$ of classifiers of the form

$$\phi(x) = \begin{cases} 0 & \text{if } (1-p)f_{\theta_0}(x) \geq pf_{\theta_1}(x) \\ 1 & \text{otherwise,} \end{cases}$$

where $p \in [0, 1]$, and $\theta_0, \theta_1 \in \Theta$. The parametric assumption means that the Bayes rule is contained in $\mathcal{C}$. Then it is a very natural approach to minimize the empirical error probability

$$\widehat{L}_n(\phi) = \frac{1}{n} \sum_{i=1}^{n} I_{\{\phi(X_i) \neq Y_i\}},$$

measured on the training data D_n over classifiers ϕ in the class $\mathcal{C}$. Denote the empirically selected rule (i.e., the one minimizing $\widehat{L}_n(\phi)$) by ϕ_n^*. For most typical parametric classes Θ, the VC dimension $V_\mathcal{C}$ is finite. Therefore, as a straightforward consequence of Theorem 12.6, we have

COROLLARY 16.1. *If both conditional distributions are contained in the parametric family $\mathcal{P}_\Theta$, then for the error probability $L(\phi_n^*) = \mathbf{P}\{\phi_n^*(X) \neq Y | D_n\}$ of the empirically optimal rule ϕ_n^*, we have for every n and $\epsilon > 0$*

$$\mathbf{P}\left\{L(\phi_n^*) - L^* > \epsilon\right\} \leq 8S(\mathcal{C}, n)e^{-n\epsilon^2/128}.$$

The result above means that $O\left(\sqrt{\log n/n}\right)$ rate of convergence to the Bayes rule is guaranteed for the empirically optimal rule, whenever the VC dimension $V_\mathcal{C}$ is finite. This is the case, for example, for exponential families. If $\mathcal{P}_\Theta$ is an exponential family, with densities of the form

$$f_\theta(x) = c\alpha(\theta)\beta(x) \exp\left\{\sum_{i=1}^{k} \pi_i(\theta)\psi_i(x)\right\},$$

then by results from Chapter 13, $S(C, n) \leq (n+1)^{k+1}$. Observe that in this approach, nothing but properties of the class C are used to derive the $O\left(\sqrt{\log n/n}\right)$ rate of convergence.

REMARK. ROBUSTNESS. The method of empirical minimization is clearly extremely robust against errors in the parametric model. Obviously, (see Theorem 12.6) if the true conditional distributions are not contained in the class $\mathcal{P}_\Theta$, then L^* can be replaced in Corollary 16.1 by $\inf_{\phi \in C} \mathbf{P}\{\phi(X) \neq Y\}$. If the model is fairly good, then this number should be very close to the Bayes risk L^*. □

REMARK. LOWER BOUNDS. The results of Chapter 14 imply that for some distributions the error probability of the selected rule is about $O(1/\sqrt{n})$ away from the Bayes risk. In the parametric case, however, since the class of distributions is restricted by the parametric model, this is not necessarily true. In some cases, a much faster rate of convergence is possible than the $O(\sqrt{\log n/n})$ rate guaranteed by Corollary 16.1. See, for example, Problem 16.3, where an example is given in which the error rate is $O(1/n)$. □

Problems and Exercises

PROBLEM 16.1. Show that if both conditional distributions of X, given $Y = 0$ and $Y = 1$, are gaussian, then the Bayes decision is quadratic, that is, it can be written as

$$g^*(x) = \begin{cases} 0 & \text{if } \sum_{i=1}^{d+d(d+1)/2} a_i \psi_i(x) + a_0 \geq 0 \\ 1 & \text{otherwise,} \end{cases}$$

where the functions $\psi_i(x)$ are either of the form $x^{(i)}$ (the i-th component of the vector x), or $x^{(i)} x^{(j)}$, and $a_0, \ldots, a_{d+d(d+1)/2} \in \mathcal{R}$.

PROBLEM 16.2. Let f be the normal density on the real line with mean m and standard deviation σ^2, and we draw an i.i.d. sample $X_1, \ldots, X_n$ from f, and set

$$\widehat{m} = \frac{1}{n} \sum_{i=1}^n X_i \quad \text{and} \quad \widehat{\sigma}^2 = \frac{1}{n} \sum_{i=1}^n (X_i - \widehat{m})^2.$$

Show that $\mathbf{E}\{|m - \widehat{m}|\} = O\left(1/\sqrt{n}\right)$ and $\mathbf{E}\{|\sigma - \widehat{\sigma}|\} = O\left(1/\sqrt{n}\right)$ by using Chebyshev's inequality. Show that this rate is in fact tight. Prove also that the result remains true if n is replaced by N, a binomial (n, p) random variable independent of $X_1, \ldots, X_n$, where $p \in (0, 1)$. That is, $\widehat{m}$ becomes $(1/N) \sum_{i=1}^N X_i$ if $N > 0$ and 0 otherwise, and similarly for $\widehat{\sigma}$.

PROBLEM 16.3. Assume that $p = 1/2$, and that both class-conditional densities f_0 and f_1 are gaussian on $\mathcal{R}$ with unit variance, but different means. We use the maximum likelihood estimates $\widehat{m}_0, \widehat{m}_1$ of the conditional means $m_0 = \mathbf{E}\{X|Y = 0\}$ and $m_1 = \mathbf{E}\{X|Y = 1\}$ to obtain the plug-in classifier $g_{\widehat{\theta}}$. Show that $\mathbf{E}\left\{(\widehat{m}_0 - m_0)^2\right\} = O(1/n)$. Then go on to show that $\mathbf{E}\left\{L(g_{\widehat{\theta}})\right\} - L^* \leq O(1/n)$.

PROBLEM 16.4. Show that the following classes of densities on $\mathcal{R}$ constitute exponential families:

(1)   gaussian family:

$$\left\{ \frac{e^{-\frac{(x-m)^2}{2\sigma^2}}}{\sqrt{2\pi}\,\sigma} : m \in \mathcal{R}, \sigma > 0 \right\};$$

(2)   gamma family:

$$\left\{ \frac{x^{\alpha-1}e^{-x/\beta}}{\beta^\alpha \Gamma(\alpha)} I_{\{x>0\}} : \alpha, \beta > 0 \right\};$$

(3)   beta family:

$$\left\{ \frac{\Gamma(\alpha+\beta)}{\Gamma(\alpha)\Gamma(\beta)} x^{\alpha-1}(1-x)^{\beta-1} I_{\{x\in 0,1\}} : \alpha, \beta > 0 \right\};$$

(4)   Rayleigh family:

$$\left\{ \theta^\lambda \beta^{1-\lambda} e^{-\beta^2/(2\theta)} x^\lambda e^{-\theta x^2/2} \sum_{j=0}^{\infty} \frac{1}{j!\Gamma(j+\lambda)} \left(\frac{x}{2}\right)^{2j+\lambda-1} : \theta, \beta > 0, \lambda \geq 0 \right\}.$$

PROBLEM 16.5. This exercise shows that one-parameter classes may be incredibly rich. Let C be the class of rules of the form

$$g_\theta(x) = \begin{cases} 1 & \text{if } x \in A + \theta \\ 0 & \text{if } x \notin A + \theta, \end{cases}$$

where $x \in \mathcal{R}$, $\theta \in \mathcal{R}$ is a parameter, and A is a union of intervals on the real line. Equivalently, $g_\theta = I_{A+\theta}$. Let

$$A = \bigcup_{i:b_i=1} \left[i - \frac{1}{2}, i + \frac{1}{2} \right),$$

where $b_1, b_2, \ldots$ are bits, obtained as follows: first list all sequences of length 1, then those of length 2, et cetera, so that $(b_1, b_2, \ldots)$ is a concatenation of $(1, 0), (1, 1, 1, 0, 0, 1, 0, 0)$, et cetera.

(1)   Show that for all n, there exists a set $\{x_1, \ldots, x_n\} \subset \mathcal{R}$ that can be shattered by a set from $\{A + \theta : \theta \in \mathcal{R}\}$. Conclude that the VC dimension of C is infinite.

(2)   If we use C for empirical error minimization and $L^* = 0$, what can you say about $E\{L_n\}$, the probability of error of the selected rule?

PROBLEM 16.6. CONTINUATION. Let X be uniform on $\theta + [i - 1/2, i + 1/2)$ with probability $1/2^i$, $i \geq 1$. Set $Y = b_i$ if $X \in \theta + [i - 1/2, i + 1/2)$, so that $L^* = 0$.

(1)   Derive the class of Bayes rules.

(2)   Work out the details of the generalized Kolmogorov-Smirnov distance minimizer based on the class of (1).

(3)   Provide the best upper bound on $E\{L_n\}$ you can get with any method.

(4)   Regardless of discrimination, how would you estimate θ? Derive the asymptotic behavior of $E\{L_n\}$ for the plug-in rule based on your estimate of θ.

PROBLEM 16.7. If $\mathcal{F}_\Theta = \{$all uniform densities on rectangles of $\mathcal{R}^d\}$ and if we use the maximum likelihood estimates of p, θ_0, θ_1 derived in the text, show that $E\{L_n\} - L^* = O(1/n)$.

PROBLEM 16.8. CONTINUED. Assume that the true class-conditional densities are f_0, f_1, and f_0, $f_1 \notin \mathcal{F}_\theta$. With the same maximum likelihood estimates given above, find f_0, f_1 for which $\mathbf{E}\{L_n\} \to 1$, yet $L^* = 0$.

PROBLEM 16.9. CONTINUED. Show that the $O(1/n)$ rate above can, in general, not be improved.

PROBLEM 16.10. Show that

$$\mathbf{P}\{D_{KS}(\nu_n, P_{\theta^*}) > \epsilon\} \leq 4n^d e^{-2(n-d)\epsilon^2}$$

by noting that the supremum in the definition of D_{KS} may be replaced by a maximum over n^d carefully selected points of $\mathcal{R}^d$. HINT: Use the idea of fingering from Chapters 4 and 12.

PROBLEM 16.11. Prove Lemma 16.2 by following the line of the proof of the Vapnik-Chervonenkis inequality (Theorem 12.5). HINT: In the second symmetrization step observe that

$$\sup_{A \in B} \left| \frac{1}{n} \sum_{i=1}^n \sigma_i (2Y_i - 1) I_{\{X_i \in A\}} \right|$$

has the same distribution as that of

$$\sup_{A \in B} \left| \frac{1}{n} \sum_{i=1}^n \sigma_i I_{\{X_i \in A\}} \right|.$$

PROBLEM 16.12. MINIMUM DISTANCE DENSITY ESTIMATION. Let $\mathcal{F}_\Theta = \{f_\theta : \theta \in \Theta\}$ be a parametric class of densities on $\mathcal{R}^d$, and assume that the i.i.d. sample $X_1, \ldots, X_n$ is drawn from the density $f_\theta \in \mathcal{F}_\Theta$. Define the class of sets

$$A = \left\{ \{x \in \mathcal{R}^d : f_{\theta_1}(x) > f_{\theta_2}(x)\} : \theta_1, \theta_2 \in \Theta \right\},$$

and define the minimum distance estimate of θ by

$$\widehat{\theta} = \arg \min_{\theta' \in \Theta} D_A(P_{\theta'}, \mu_n),$$

where $P_{\theta'}$ is the distribution belonging to the density $f_{\theta'}$, μ_n is the empirical distribution defined by $X_1, \ldots, X_n$, and D_A is the generalized Kolmogorov-Smirnov distance defined by

$$D_A(P, Q) = \sup_{A \in A} |P(A) - Q(A)|.$$

Prove that if A has finite VC dimension, then

$$\mathbf{E}\left\{ \int |f_{\widehat{\theta}}(x) - f_\theta(x)| dx \right\} = O\left(1/\sqrt{n}\right).$$

For more information on minimum distance density estimation, see Yatracos (1985) and Devroye (1987). HINT: Follow the steps listed below:

(1) $\int |f_{\widehat{\theta}} - f_\theta| \leq 2D_A(P_{\widehat{\theta}}, P_\theta)$ (use Scheffé's theorem).
(2) $D_A(P_{\widehat{\theta}}, P_\theta) \leq 2D_A(P_\theta, \mu_n)$ (proceed as in the text).
(3) Finish the proof by applying Alexander's improvement of the Vapnik-Chervonenkis inequality (Theorem 12.10).

17
Generalized Linear Discrimination

The classifiers we study here have their roots in the Fourier series estimate or other series estimates of an unknown density, potential function methods (see Chapter 10), and generalized linear classifiers. All these estimators can be put into the following form: classify x as belonging to class 0 if

$$\sum_{j=1}^{k} a_{n,j} \psi_j(x) \le 0,$$

where the ψ_j's are fixed functions, forming a base for the series estimate, $a_{n,j}$ is a fixed function of the training data, and k controls the amount of smoothing. When the ψ_j's are the usual trigonometric basis, then this leads to the Fourier series classifier studied by Greblicki and Pawlak (1981; 1982). When the ψ_j's form an orthonormal system based upon Hermite polynomials, we obtain the classifiers studied by Greblicki (1981), and Greblicki and Pawlak (1983; 1985). When $\{\psi_j(x)\}$ is the collection of all products of components of x (such as 1, $(x^{(i)})^k$, $(x^{(i)})^k(x^{(j)})^l$, etcetera), we obtain the polynomial method of Specht (1971).

We start with classification based on Fourier series expansion, which has its origins in Fourier series density estimation, which, in turn, was developed by Cencov (1962), Schwartz (1967), Kronmal and Tarter (1968), Tarter and Kronmal (1970), and Specht (1971). Its use in classification was considered by Greblicki (1981), Specht (1971), Greblicki and Pawlak (1981; 1982; 1983), and others.

17.1   Fourier Series Classification

Let f be the density of X, and assume that $f \in L_2(\lambda)$, that is, $f \geq 0$, $\int f(x)dx = 1$ and $\int f^2(x)dx < \infty$, and recall that λ denotes the Lebesgue measure on $\mathcal{R}^d$. Let $\psi_1, \psi_2, \ldots$ be a complete orthonormal set of bounded functions in $L_2(\lambda)$ with $\sup_{j,x} |\psi_j(x)| \leq B < \infty$. An orthonormal set of functions $\{\psi_j\}$ in $L_2(\mu)$ is such that $\int \psi_j \psi_i = I_{\{i=j\}}$ for all i, j. It is complete if $\int \alpha \psi_i = 0$ for all i for some function $\alpha \in L_2(\lambda)$ implies that $\alpha \equiv 0$ almost everywhere (with respect to λ). If $f \in L_2(\lambda)$, then the class-conditional densities f_0 and f_1 exist, and are in $L_2(\lambda)$. Then the function

$$\alpha(x) = pf_1(x) - (1 - p)f_0(x) = (2\eta(x) - 1)f(x)$$

is in $L_2(\lambda)$ as well. Recall that the Bayes decision is given by

$$g^*(x) = \begin{cases} 0 & \text{if } \alpha(x) \leq 0 \\ 1 & \text{otherwise.} \end{cases}$$

Let the bounded functions $\psi_1, \psi_2, \ldots$ form a complete orthonormal basis, and let the Fourier representation of α be given by

$$\alpha = \sum_{j=1}^{\infty} \alpha_j \psi_j.$$

The above convergence is understood in $L_2(\lambda)$, that is, the infinite sum means that

$$\lim_{k \to \infty} \int \left(\alpha(x) - \sum_{j=1}^{k} \alpha_j \psi_j(x) \right)^2 dx = 0.$$

The coefficients are given by

$$\alpha_j = \int \psi_j(x)\alpha(x)dx.$$

We approximate $\alpha(x)$ by a truncation of its Fourier representation to finitely many terms, and use the data D_n to estimate the coefficients appearing in this sum. Formally, consider the classification rule

$$g_n(x) = \begin{cases} 0 & \text{if } \sum_{j=1}^{k_n} \alpha_{n,j} \psi_j(x) \leq 0 \\ 1 & \text{otherwise,} \end{cases} \tag{17.1}$$

where the estimates $\alpha_{n,j}$ of α_j are very easy to compute:

$$\alpha_{n,j} = \frac{1}{n} \sum_{i=1}^{n} (2Y_i - 1)\psi_j(X_i).$$

Before discussing the properties of these rules, we list a few examples of complete orthonormal systems on the real line. Some of these systems contain functions on a bounded interval. These, of course, can only be used if the distribution of X is concentrated on an interval. The completeness of these systems may typically be checked via the Stone-Weierstrass theorem (Theorem A.9). A general way of producing complete orthonormal systems on $\mathcal{R}^d$ is taking products of one-dimensional functions of the d components, as sketched in Problem 17.1. For more information on orthogonal series, we refer to Sansone (1969), Szegő (1959), and Zygmund (1959).

(1) The *standard trigonometric basis* on the interval $[-\pi, \pi]$ is formed by the functions

$$\psi_0(x) = \frac{1}{\sqrt{2\pi}}, \quad \psi_{2i-1}(x) = \frac{\cos(ix)}{\sqrt{\pi}}, \quad \psi_{2i}(x) = \frac{\sin(ix)}{\sqrt{\pi}}, \quad i = 1, 2, \ldots.$$

This is a complete orthonormal system in $L_2([-\pi, \pi])$.

(2) The *Legendre polynomials* form a complete orthonormal basis over the interval $[-1, 1]$:

$$\psi_i(x) = \sqrt{\frac{2i + 1}{2}} \frac{1}{2^i i!} \frac{d^i}{dx^i} \left((x^2 - 1)^i \right), \quad i = 0, 1, 2, \ldots.$$

(3) The set of *Hermite functions* is complete and orthonormal over the whole real line:

$$\psi_i(x) = \frac{e^{x^2/2}}{\sqrt{2^i i! \sqrt{\pi}}} \frac{d^i}{dx^i} \left(e^{-x^2} \right), \quad i = 0, 1, 2, \ldots.$$

(4) Functions of the *Laguerre basis* are defined on $[0, \infty)$ by

$$\psi_i(x) = \left(\frac{\Gamma(i + 1)}{\Gamma(i + \alpha + 1)} x^{-\alpha} e^x \right)^{1/2} \frac{1}{i!} \frac{d^i}{dx^i} \left(x^{i+\alpha} e^{-x} \right), \quad i = 0, 1, 2, \ldots,$$

where $\alpha > -1$ is a parameter of the system. A complete orthonormal system over the whole real line may be obtained by

$$\psi'_{2i}(x) = \psi_i(x) I_{\{x \geq 0\}}, \quad \psi'_{2i+1}(x) = \psi_i(-x) I_{\{x < 0\}}, \quad i = 0, 1, 2, \ldots.$$

(5) The *Haar basis* contains functions on $[0, 1]$ that take three different values. It is orthonormal and complete. Functions with finitely many values have computational advantages in pattern recognition. Write the integer $i \geq 0$ as $i = 2^k + j$, where $k = \lfloor \log_2 i \rfloor$ (i.e., $0 \leq j < 2^k$). Then

$$\psi_i(x) = \begin{cases} 2^{k/2} & \text{if } x \in \left((j - 1)/2^k, (j - 1/2)/2^k \right) \\ 2^{-k/2} & \text{if } x \in \left((j - 1/2)/2^k, j/2^k \right) \\ 0 & \text{otherwise.} \end{cases}$$

(6) Functions on $[0, 1]$ of the *Rademacher basis* take only two values, -1 and 1. It is an orthonormal system, but it is not complete:

$$\psi_0(x) \equiv 1, \quad \psi_i(x) = (-1)^{\lfloor 2^i x \rfloor}, \quad i = 1, 2, \ldots.$$

The basis may be completed as follows: write $i = \sum_{j=1}^{k} 2^{i_j}$ such that $0 \leq i_1 < i_2 < \cdots < i_k$. This form is unique. Define

$$\psi'(x) = \psi_{i_1}(x) \cdots \psi_{i_k}(x),$$

where the ψ_j's are the Rademacher functions. The resulting basis ψ_0', ψ_1', ... is orthonormal and complete, and is called the *Walsh basis*.

There is no system of basis functions that is better than another for all distributions. In selecting the basis of a Fourier series rule, the designer must use a mixture of intuition, error estimation, and computational concerns. We have the following consistency theorem:

Theorem 17.1. *Let $\{\psi_0, \psi_1, \ldots\}$ be a complete orthonormal basis on $\mathcal{R}^d$ such that for some $B < \infty$, $|\psi_i(x)| \leq B$ for each i and x. Assume that the class-conditional densities f_0 and f_1 exist and are in $L_2(\lambda)$. If*

$$k_n \to \infty \quad and \quad \frac{k_n}{n} \to 0 \quad as \quad n \to \infty,$$

then the Fourier series classification rule defined in (17.1) is consistent:

$$\lim_{n \to \infty} \mathbf{E}L(g_n) = L^*.$$

If

$$k_n \to \infty \quad and \quad \frac{k_n \log n}{n} \to 0 \quad as \quad n \to \infty,$$

then

$$\lim_{n \to \infty} L(g_n) = L^*$$

with probability one, that is, the rule is strongly consistent.

PROOF. First observe that the $\alpha_{n,j}$'s are unbiased estimates of the α_j's:

$$
\begin{aligned}
\mathbf{E}\{\alpha_{n,j}\} &= \mathbf{E}\{(2Y-1)\psi_j(X)\} = \mathbf{E}\{\mathbf{E}\{(2Y-1)\psi_j(X)|X\}\} \\
&= \mathbf{E}\{\psi_j(X)\mathbf{E}\{(2Y-1)|X\}\} = \mathbf{E}\{(2\eta(X)-1)\psi_j(X)\} \\
&= \int (2\eta(x)-1)\psi_j(x)f(x)dx = \int \psi_j(x)\alpha(x)dx \\
&= \alpha_j,
\end{aligned}
$$

and that their variance can be bounded from above as follows:

$$
\begin{aligned}
\mathbf{Var}\{\alpha_{n,j}\} &= \frac{\mathbf{Var}\{\psi_j(X_1)(2\eta(X_1) - 1)\}}{n} \\
&= \frac{\int \psi_j^2(x)(2\eta(x) - 1)^2 f(x)dx - \alpha_j^2}{n} \\
&\leq \frac{B^2 - \alpha_j^2}{n} \leq \frac{B^2}{n},
\end{aligned}
$$

where we used the boundedness of the ψ_j's. By Parseval's identity,

$$
\int \alpha^2(x)dx = \sum_{j=1}^{\infty} \alpha_j^2.
$$

Therefore, exploiting orthonormality of the ψ_j's, we have

$$
\begin{aligned}
&\int \left(\alpha(x) - \sum_{j=1}^{k_n} \alpha_{n,j}\psi_j(x) \right)^2 dx \\
&= \int \alpha^2(x)dx + \int \left(\sum_{j=1}^{k_n} \alpha_{n,j}\psi_j(x) \right)^2 dx \\
&\quad - 2\int \alpha(x) \sum_{j=1}^{k_n} \alpha_{n,j}\psi_j(x)dx \\
&= \sum_{j=1}^{\infty} \alpha_j^2 + \sum_{j=1}^{k_n} \alpha_{n,j}^2 - 2\sum_{j=1}^{k_n} \alpha_j \alpha_{n,j} \\
&= \sum_{j=1}^{k_n} \left(\alpha_{n,j} - \alpha_j \right)^2 + \sum_{j=k_n+1}^{\infty} \alpha_j^2.
\end{aligned}
$$

Thus, the expected L_2-error is bounded as follows:

$$
\begin{aligned}
&\mathbf{E}\left\{ \int \left(\alpha(x) - \sum_{j=1}^{k_n} \alpha_{n,j}\psi_j(x) \right)^2 dx \right\} \\
&= \mathbf{E}\left\{ \sum_{j=1}^{k_n} \left(\alpha_{n,j} - \alpha_j \right)^2 \right\} + \sum_{j=k_n+1}^{\infty} \alpha_j^2 \\
&= \sum_{j=1}^{k_n} \mathbf{Var}\{\alpha_{n,j}\} + \sum_{j=k_n+1}^{\infty} \alpha_j^2 \\
&\leq \frac{k_n B^2}{n} + \sum_{j=k_n+1}^{\infty} \alpha_j^2.
\end{aligned}
$$

Since $\sum_{j=1}^{\infty} \alpha_j^2 < \infty$, the second term tends to zero if $k_n \to \infty$. If at the same time $k_n/n \to 0$, then the expected L_2-error converges to zero, that is, the estimate is consistent in L_2. Now, convergence of the error probability follows from Problem 2.11.

To prove strong consistency (i.e., convergence with probability one), fix $\epsilon > 0$, and assume that n is so large that

$$\sum_{j=k_n+1}^{\infty} \alpha_j^2 < \epsilon/2.$$

Then

$$\mathbf{P}\left\{ \int \left(\alpha(x) - \sum_{j=1}^{k_n} \alpha_{n,j} \psi_j(x) \right)^2 dx > \epsilon \right\}$$

$$= \mathbf{P}\left\{ \sum_{j=1}^{k_n} (\alpha_{n,j} - \alpha_j)^2 + \sum_{j=k_n+1}^{\infty} \alpha_j^2 > \epsilon \right\}$$

$$\leq \mathbf{P}\left\{ \sum_{j=1}^{k_n} (\alpha_{n,j} - \alpha_j)^2 > \epsilon/2 \right\}$$

$$\leq \sum_{j=1}^{k_n} \mathbf{P}\left\{ (\alpha_{n,j} - \alpha_j)^2 > \epsilon/(2k_n) \right\}$$

(by the union bound)

$$= \sum_{j=1}^{k_n} \mathbf{P}\left\{ \left| \frac{1}{n} \sum_{i=1}^{n} \left(\psi_j(X_i)(2Y_i - 1) - \mathbf{E}\{\psi_j(X)(2Y_i - 1)\} \right) \right| > \sqrt{\frac{\epsilon}{2k_n}} \right\}$$

$$\leq 2k_n e^{-n\epsilon/(4k_n B^2)},$$

where we used Hoeffding's inequality (Theorem 8.1). Because $k_n \log n/n \to 0$, the upper bound is eventually smaller than $e^{-2\log n} = n^{-2}$, which is summable. The Borel-Cantelli lemma yields strong L_2 consistency. Strong consistency of the classifier then follows from Problem 2.11. □

REMARK. It is clear from the inequality

$$\mathbf{E}\left\{ \int \left(\alpha(x) - \sum_{j=1}^{k_n} \alpha_{n,j} \psi_j(x) \right)^2 dx \right\} \leq \frac{k_n B^2}{n} + \sum_{j=k_n+1}^{\infty} \alpha_j^2,$$

that the rate of convergence is determined by the choice of k_n. If k_n is small, then the first term, which corresponds to the *estimation error*, is small, but the *approximation error*, expressed by the second term, is larger. For large k_n, the situation is reversed. The optimal choice of k_n depends on how fast the second term goes to zero as k_n grows. This depends on other properties of f, such as the size of its tail and its smoothness. $\square$

REMARK. Consistency in Theorem 17.1 is not universal, since we needed to assume the existence of square integrable conditional densities. This, however, is not a restrictive assumption in some practical situations. For example, if the observations are corrupted by additive gaussian noise, then the conditions of the theorem hold (Problem 17.2). However, if μ does not have a density, the method may be inconsistent (see Problem 17.4). $\square$

Fourier series classifiers have two rather unattractive features in general:

(i) They are not invariant under translations of the coordinate space.

(ii) Most series classifiers are not local in nature—points at arbitrary distances from x affect the decision at x. In kernel and nearest neighbor rules, the locality is easily controlled.

17.2   Generalized Linear Classification

When X is purely atomic or singular continuous, Theorem 17.1 is not applicable. A theme of this book is that pattern recognition can be developed in a distribution-free manner since, after all, the distribution of (X, Y) is not known. Besides, even if we had an i.i.d. sample $(X_1, Y_1), \ldots, (X_n, Y_n)$ at our disposal, we do not know of any test for verifying whether X has a square integrable density. So, we proceed a bit differently to develop universally consistent rules. To generalize Fourier series classifiers, let $\psi_1, \psi_2, \ldots$ be bounded functions on $\mathcal{R}^d$. These functions do not necessarily form an orthonormal basis of L_2. Consider the classifier

$$g_n(x) = \begin{cases} 0 & \text{if } \sum_{j=1}^{k_n} \alpha_{n,j} \psi_j(x) \leq 0 \\ 1 & \text{otherwise,} \end{cases}$$

where the coefficients $\alpha_{n,j}$ are some functions of the data D_n. This may be viewed as a generalization of Fourier series rules, where the coefficients were unbiased estimates of the Fourier coefficients of $\alpha(x) = (2\eta(x) - 1)f(x)$. Here we will consider some other choices of the coefficients. Observe that g_n is a *generalized linear classifier*, as defined in Chapter 13. An intuitively appealing way to determine the coefficients is to pick them to minimize the empirical error committed on D_n, as in empirical risk minimization. The critical parameter here is k_n, the number of basis functions used in the linear combination. If we keep k_n fixed, then as we saw in Chapter 13, the error probability of the selected rule converges quickly to that of the best classifier of the above form. However, for some distributions, this infimum may be far from the Bayes risk, so it is useful to let k_n grow as n becomes larger. However, choosing k_n too large may result in overfitting the data. Using the terminology introduced in Chapter 12, let $\mathcal{C}^{(k_n)}$ be the class of classifiers of the form

$$\phi(x) = \begin{cases} 0 & \text{if } \sum_{j=1}^{k_n} a_j \psi_j(x) \leq 0 \\ 1 & \text{otherwise,} \end{cases}$$

where the a_j's range through $\mathcal{R}$. Choose the coefficients to minimize the empirical error

$$\widehat{L}_n(\phi) = \frac{1}{n} \sum_{i=1}^{n} I_{\{\phi(X_i) \neq Y_i\}}.$$

Let $g_n = \phi_n^*$ be the corresponding classifier. We recall from Chapter 13 that the vc dimension of $\mathcal{C}^{(k_n)}$ is not more than k_n. Therefore, by Theorem 13.12, for every $n > 2k_n + 1$,

$$\mathbf{P}\left\{ L(g_n) - \inf_{\phi \in \mathcal{C}^{(k_n)}} L(\phi) > \epsilon \right\} \leq 8 e^{n\mathcal{H}(\frac{k_n}{n})} e^{-n\epsilon^2/128},$$

where $\mathcal{H}$ is the binary entropy function. The right-hand side is $e^{-n\epsilon^2/128+o(n)}$ if $k_n = o(n)$. However, to obtain consistency, we need to know how close $\inf_{\phi \in \mathcal{C}^{(k_n)}} L(\phi)$ is to L^*. This obviously depends on the choice of the ψ_j's, as well as on the distribution. If k_n is not allowed to grow with n, and is bounded by a number K, then universal consistency eludes us, as for some distribution $\inf_{\phi \in \mathcal{C}^{(K)}} L(\phi) - L^* > 0$. It follows from Theorem 2.2 that for every $B > 0$

$$\inf_{\phi \in \mathcal{C}^{(k_n)}} L(\phi) - L^*$$

$$\leq \inf_{a_1,\ldots,a_{k_n}} \int_{\|x\| \leq B} \left| (2\eta(x) - 1) - \sum_{j=1}^{k_n} a_j \psi_j(x) \right| \mu(dx) + \int_{\|x\| > B} \mu(dx).$$

The limit supremum of the right-hand side can be arbitrarily close to zero for all distributions if $k_n \to \infty$ as $n \to \infty$ and the set of functions

$$\bigcup_{k=1}^{\infty} \left\{ \sum_{j=1}^{k} a_j \psi_j(x); a_1, a_2, \ldots \in \mathcal{R} \right\}$$

is dense in $L_1(\mu)$ on balls of the form $\{x : \|x\| \leq B\}$ for all μ. This means that given any probability measure μ, and function f with $\int |f| d\mu < \infty$, for every $\epsilon > 0$, there exists an integer k and coefficients $a_1, \ldots, a_k$ such that

$$\int \left| f(x) - \sum_{j=1}^{k} a_j \psi_j(x) \right| \mu(dx) < \epsilon.$$

Thus, we have obtained the following consistency result:

Theorem 17.2. *Let* $\psi_1, \psi_2, \ldots$ *be a sequence of functions such that the set of all finite linear combinations of the* ψ_j's *is dense in* $L_1(\mu)$ *on balls of the form* $\{x : \|x\| \leq B\}$ *for any probability measure* μ. *Then then* g_n *is strongly universally consistent when*

$$k_n \to \infty \quad and \quad \frac{k_n}{n} \to 0 \quad as \quad n \to \infty.$$

REMARK. To see that the statement of Theorem 17.2 is not vacuous, note that denseness in $L_1(\mu)$ on balls follows from denseness with respect to the supremum norm on balls. Denseness in the L_∞ sense may be checked by the Stone-Weierstrass theorem (Theorem A.9). !• example, the class of all polynomial classifiers satisfies the conditions of the theorem. This class can be obtained by choosing the functions $\psi_1, \psi_2, \ldots$ as the simple polynomials $x^{(1)}, \ldots, x^{(d)}, x^{(1)}x^{(2)}, \ldots$. Similarly, the functions ψ_j can be chosen as bases of a trigonometric series. $\square$

REMARK. THE HISTOGRAM RULE. It is clear that Theorem 17.2 can be modified in the following way: let $\psi_{j,k}, j = 1, \ldots, k; k = 1, 2, \ldots$, be functions such that for every $f \in L_1(\mu)$ (with μ concentrated on a bounded ball) and $\epsilon > 0$ there is a k_0 such that for all $k > k_0$ there is a function of the form $\sum_{j=1}^{k} a_j \psi_{j,k}$ whose distance from f in $L_1(\mu)$ is smaller than ϵ. Let $C^{(k_n)}$ be the class of generalized linear classifiers based on the functions $\psi_{1,k_n}, \ldots, \psi_{k_n,k_n}$. If $k_n \to \infty$ and $k_n/n \to 0$, then the classifier g_n that minimizes the empirical error over $C^{(k_n)}$ is strongly universally consistent. This modification has an interesting implication. Consider for example functions $\psi_{1,k_n}, \ldots, \psi_{k_n,k_n}$ that are indicators of sets of a partition of $\mathcal{R}^d$. Then it is easy to see that the classifier that minimizes the empirical error is just the histogram classifier based on the same partition. The denseness assumption requires that the partition becomes infinitesimally fine as $n \to \infty$. In fact, we have obtained an alternative proof for the strong universal consistency of the regular histogram rule (Theorem 9.4). The details are left as an exercise (Problem 17.3). $\square$

Problems and Exercises

PROBLEM 17.1. Let $\psi_1, \psi_2, \ldots$ be a sequence of real-valued functions on a bounded interval $[a, b]$ such that $\int_a^b \psi_i(x)\psi_j(x)dx = I_{\{i=j\}}$, and the set of finite linear combinations $\sum_{i=1}^{k} a_i \psi_i(x)$ is dense in the class of continuous functions on $[a, b]$ with respect to the supremum norm. Define the functions $\Psi_{i_1,\ldots,i_d} : [a, b]^d \to \mathcal{R}$ by

$$\Psi_{i_1,\ldots,i_d}(x) = \psi_{i_1}(x^{(1)}) \cdots \psi_{i_d}(x^{(d)}), \quad i_1, \ldots, i_d \in \{1, 2, \ldots\}.$$

Show that these functions form a complete orthonormal set of functions on $[a, b]^d$.

PROBLEM 17.2. Let Z be an arbitrary random variable on $\mathcal{R}$, and V be a real random variable, independent of Z, that has a density $h \in L_2(\lambda)$. Show that the density f of the random variable $X = Z + V$ exists, and $\int f^2(x)dx < \infty$. HINT: Use Jensen's inequality to prove that $\int f^2(x)dx \le \int h^2(x)dx$.

PROBLEM 17.3. Derive the strong universal consistency of the regular histogram rule from Theorem 17.2, as indicated in the remark following it.

PROBLEM 17.4. Let $\{\psi_0, \psi_1, \psi_2, \ldots\}$ be the standard trigonometric basis, and consider the classification rule defined in (17.1). Show that the rule is not consistent for the following distribution: given $Y = 1$, let X be 0, and given $Y = 0$, let X be uniformly distributed on $[-\pi, \pi]$. Assume furthermore that $\mathbf{P}\{Y = 1\} = 1/2$.

PROBLEM 17.5. The Haar basis is not bounded. Determine whether or not the Laguerre, Hermite, and Legendre bases are bounded.

18

Complexity Regularization

This chapter offers key theoretical results that confirm the existence of certain "good" rules. Although the proofs *are* constructive—we *do* tell you how you may design such rules—the computational requirements are often prohibitive. Many of these rules are thus not likely to filter down to the software packages and pattern recognition implementations. An attempt at reducing the computational complexity somewhat is described in the section entitled "Simple empirical covering." Nevertheless, we feel that much more serious work on discovering practical algorithms for empirical risk minimization is sorely needed.

In Chapter 12, the empirical error probability was minimized over a class C of decision rules $\phi : \mathcal{R}^d \to \{0, 1\}$. We saw that if the vc dimension V_C of the class is finite, then the error probability of the selected rule is within constant times $\sqrt{V_C \log n / n}$ of that of the best rule in C. Unfortunately, classes with finite vc dimension are nearly always too small, and thus the error probability of the best rule in C is typically far from the Bayes risk L^* for some distribution (see Theorem 18.4). In Chapter 17 we investigated the special classes of generalized linear rules. Theorem 17.2, for example, shows that if we increase the size of the class in a controlled fashion as the sample size n increases, the error probability of the selected rule approaches L^* for any distribution. Thus, V_C increases with n!

Theorem 17.2 may be generalized straightforwardly for other types of classifiers. Consider a sequence of classes $C^{(1)}, C^{(2)}, \ldots$, containing classifiers of the form $\phi : \mathcal{R}^d \to \{0, 1\}$. The training data $D_n = ((X_1, Y_1), \ldots, (X_n, Y_n))$ are used to select a classifier ϕ_n^* by minimizing the empirical error probability $\hat{L}_n(\phi)$ over the class $C^{(k_n)}$, where the integer k_n depends on n in a specified way. The following generalization is based on the proof of Theorem 17.2 (see Problem 18.1):

Theorem 18.1. *Assume that $C^{(1)}, C^{(2)}, \ldots$ is a sequence of classes of decision rules such that for any distribution of (X, Y),*

$$\lim_{i \to \infty} \inf_{\phi \in C^{(i)}} L(\phi) = L^*,$$

and that the VC dimensions $V_{C^{(1)}}, V_{C^{(2)}}, \ldots$ are all finite. If

$$k_n \to \infty \quad and \quad \frac{V_{C^{(k_n)}} \log n}{n} \to 0 \quad as \ n \to \infty,$$

then the classifier ϕ_n^* that minimizes the empirical error over the class $C^{(k_n)}$ is strongly universally consistent.

Theorem 18.1 is the missing link—we are now ready to apply the rich theory of Vapnik-Chervonenkis classes in the construction of universally consistent rules. The theorem does not help us, however, with the choice of the classes $C^{(i)}$, or the choice of the sequence $\{k_n\}$. Examples for sequences of classes satisfying the condition of Theorem 18.1 include generalized linear decisions with properly chosen basis functions (Chapter 17), or neural networks (Chapter 30).

A word of warning here. The universally consistent rules obtained via Theorem 18.1 may come at a tremendous computational price. As we will see further on, we will often have exponential complexities in n instead of polynomial time that would be obtained if we just minimized the empirical risk over a fixed VC class. The computational complexity of these rules are often incomparably larger than that of some simple universally consistent rules as k-nearest neighbor, kernel, or partitioning rules.

18.1   Structural Risk Minimization

Let $C^{(1)}, C^{(2)}, \ldots$ be a sequence of classes of classifiers, from which we wish to select a sequence of classifiers with the help of the training data D_n. Previously, we picked ϕ_n^* from the k_n-th class $C^{(k_n)}$, where the integer k_n is some prespecified function of the sample size n only. The integer k_n basically determines the complexity of the class from which the decision rule is selected. Theorem 18.1 shows that under mild conditions on the sequence of classes, it is p· ible to find sequences $\{k_n\}$ such that the rule is universally consistent. Typically, k_n should grow with n in order to assure convergence of the approximation error $\inf_{\phi \in C^{(k_n)}} L(\phi) - L^*$, but it cannot grow too rapidly, for otherwise the estimation error $L(\phi_n^*) - \inf_{\phi \in C^{(k_n)}} L(\phi)$ might fail to converge to zero. Ideally, to get best performance, the two types of error should be about the same order of magnitude. Clearly, a prespecified choice of the complexity k_n cannot balance the two sides of the trade-off for all distributions. Therefore, it is important to find methods such that the classifier is selected from a class whose index is automatically determined by the data D_n. This section deals with such methods.

The most obvious method would be based on selecting a candidate decision rule $\tilde{\phi}_{n,j}$ from every class $C^{(j)}$ (for example, by minimizing the empirical error over $C^{(j)}$), and then minimizing the empirical error over these rules. However, typically, the VC dimension of

$$C^* = \bigcup_{j=1}^{\infty} C^{(j)}$$

equals infinity, which, in view of results in Chapter 14, implies that this approach will not work. In fact, in order to guarantee

$$\inf_{j\geq 1} \inf_{\phi\in C^{(j)}} L(\phi) = L^*,$$

it is necessary that

$$V_{C^*} = \infty$$

(see Theorems 18.4 and 18.5).

A possible solution for the problem is a method proposed by Vapnik and Chervonenkis (1974c) and Vapnik (1982), called *structural risk minimization*. First we select a classifier $\tilde{\phi}_{n,j}$ from every class $C^{(j)}$ which minimizes the empirical error over the class. Then we know from Theorem 12.5 that for every j, with very large probability,

$$L(\tilde{\phi}_{n,j}) \leq \widehat{L}_n(\tilde{\phi}_{n,j}) + O\left(\sqrt{\frac{V_{C^{(j)}}\log n}{n}}\right).$$

Now, pick a classifier that minimizes the upper bound over $j \geq 1$. To make things more precise, for every n and j, we introduce the complexity penalty

$$r(j,n) = \sqrt{\frac{32}{n} V_{C^{(j)}}\log(en)}.$$

Let $\tilde{\phi}_{n,1}, \tilde{\phi}_{n,2}, \ldots$ be classifiers minimizing the empirical error $\widehat{L}_n(\phi)$ over the classes $C^{(1)}, C^{(2)}, \ldots$, respectively. For $\phi \in C^{(j)}$, define the complexity-penalized error estimate

$$\tilde{L}_n(\phi) = \widehat{L}_n(\phi) + r(j,n).$$

Finally, select the classifier ϕ_n^* minimizing the complexity penalized error estimate $\tilde{L}_n(\tilde{\phi}_{n,j})$ over $j \geq 1$. The next theorem states that this method avoids overfitting the data. The only condition is that each class in the sequence has finite VC dimension.

Theorem 18.2. *Let $C^{(1)}, C^{(2)}, \ldots$ be a sequence of classes of classifiers such that for any distribution of (X, Y),*

$$\lim_{j\to\infty} \inf_{\phi\in C^{(j)}} L(\phi) = L^*.$$

Assume also that the VC dimensions $V_{C^{(1)}}, V_{C^{(2)}}, \ldots$ are finite and satisfy

$$\Delta = \sum_{j=1}^{\infty} e^{-V_{C^{(j)}}} < \infty.$$

Then the classification rule ϕ_n^* based on structural risk minimization, as defined above, is strongly universally consistent.

REMARK. Note that the condition on the VC dimensions is satisfied if we insist that $V_{C^{(j+1)}} > V_{C^{(j)}}$ for all j, a natural assumption. See also Problem 18.3. $\square$

Instead of minimizing the empirical error $\widehat{L}_n(\phi)$ over the set of candidates C^*, the method of structural risk minimization minimizes $\tilde{L}_n(\phi)$, the sum of the empirical error, and a term $\sqrt{\frac{32}{n} V_{C^{(j)}} \log(en)}$, which increases as the VC dimension of the class $C^{(j)}$ containing ϕ increases. Since classes with larger VC dimension can be considered as more complex than those with smaller VC dimension, the term added to the empirical error may be considered as a penalty for complexity. The idea of minimizing the sum of the empirical error and a term penalizing the complexity has been investigated in various statistical problems by, for example, Rissanen (1983), Akaike (1974), Barron (1985; 1991), Barron and Cover (1991), and Lugosi and Zeger (1996). Barron (1991) minimizes the penalized empirical risk over a suitably chosen countably infinite list of candidates. This approach is close in spirit to the skeleton estimates discussed in Chapter 28. He makes the connection between structural risk minimization and the *minimum description length* principle, and obtains results similar to those discussed in this section. The theorems presented here were essentially developed in Lugosi and Zeger (1996).

PROOF OF THEOREM 18.2.   We show that both terms on the right-hand side of the following decomposition converge to zero with probability one:

$$L(\phi_n^*) - L^* = \left(L(\phi_n^*) - \inf_{j \geq 1} \tilde{L}_n(\tilde{\phi}_{n,j}) \right) + \left(\inf_{j \geq 1} \tilde{L}_n(\tilde{\phi}_{n,j}) - L^* \right).$$

For the first term we have

$$
\begin{aligned}
&\mathbf{P}\left\{ L(\phi_n^*) - \inf_{j \geq 1} \tilde{L}_n(\tilde{\phi}_{n,j}) > \epsilon \right\} \\
&= \mathbf{P}\left\{ L(\phi_n^*) - \tilde{L}_n(\phi_n^*) > \epsilon \right\} \\
&\leq \mathbf{P}\left\{ \sup_{j \geq 1} \left(L(\tilde{\phi}_{n,j}) - \tilde{L}_n(\tilde{\phi}_{n,j}) \right) > \epsilon \right\} \\
&= \mathbf{P}\left\{ \sup_{j \geq 1} \left(L(\tilde{\phi}_{n,j}) - \widehat{L}_n(\tilde{\phi}_{n,j}) - r(j,n) \right) > \epsilon \right\} \\
&\leq \sum_{j=1}^{\infty} \mathbf{P}\left\{ \left| L(\tilde{\phi}_{n,j}) - \widehat{L}_n(\tilde{\phi}_{n,j}) \right| > \epsilon + r(j,n) \right\} \\
&\leq \sum_{j=1}^{\infty} 8n^{V_{C^{(j)}}} e^{-n(\epsilon + r(j,n))^2/32} \quad \text{(by Theorem 12.5)}
\end{aligned}
$$

$$\leq \sum_{j=1}^{\infty} 8n^{V_{C^{(j)}}} e^{-nr^2(j,n)/32} e^{-n\epsilon^2/32}$$

$$= 8e^{-n\epsilon^2/32} \sum_{j=1}^{\infty} e^{-V_{C(j)}} = \Delta e^{-n\epsilon^2/32}$$

using the defining expression for $r(j,n)$, where $\Delta = 8\sum_{j=1}^{\infty} e^{-V_{C(j)}} < \infty$, by assumption. Thus, it follows from the Borel-Cantelli lemma that

$$L(\phi_n^*) - \inf_{j \geq 1} \tilde{L}_n(\tilde{\phi}_{n,j}) \to 0$$

with probability one as $n \to \infty$. Now, we can turn to investigating the second term $\inf_{j \geq 1} \tilde{L}_n(\tilde{\phi}_{n,j}) - L^*$. By our assumptions, for every $\epsilon > 0$, there exists an integer k such that

$$\inf_{\phi \in C^{(k)}} L(\phi) - L^* \leq \epsilon.$$

Fix such a k. Thus, it suffices to prove that

$$\limsup_{n \to \infty} \inf_{j \geq 1} \tilde{L}_n(\tilde{\phi}_{n,j}) - \inf_{\phi \in C^{(k)}} L(\phi) \leq 0 \quad \text{with probability one.}$$

Clearly, for any fixed k, if n is large enough, then

$$r(k,n) = \sqrt{\frac{32}{n} V_{C^{(k)}} \log(en)} \leq \frac{\epsilon}{2}.$$

Thus, for such large n,

$$\mathbf{P}\left\{ \inf_{j \geq 1} \tilde{L}_n(\tilde{\phi}_{n,j}) - \inf_{\phi \in C^{(k)}} L(\phi) > \epsilon \right\}$$

$$\leq \mathbf{P}\left\{ \tilde{L}_n(\tilde{\phi}_{n,k}) - \inf_{\phi \in C^{(k)}} L(\phi) > \epsilon \right\}$$

$$= \mathbf{P}\left\{ \widehat{L}_n(\tilde{\phi}_{n,k}) + r(k,n) - \inf_{\phi \in C^{(k)}} L(\phi) > \epsilon \right\}$$

$$\leq \mathbf{P}\left\{ \widehat{L}_n(\tilde{\phi}_{n,k}) - \inf_{\phi \in C^{(k)}} L(\phi) > \frac{\epsilon}{2} \right\}$$

$$\leq \mathbf{P}\left\{ \sup_{\phi \in C^{(k)}} \left| \widehat{L}_n(\phi) - L(\phi) \right| > \frac{\epsilon}{2} \right\}$$

$$\leq 8n^{V_{C^{(k)}}} e^{-n\epsilon^2/128}$$

by Theorem 12.5. Therefore, since $V_{C^{(k)}} < \infty$, the proof is completed. $\square$

Theorem 18.2 shows that the method of structural risk minimization is universally consistent under very mild conditions on the sequence of classes $C^{(1)}$, $C^{(2)}$, ...

This property, however, is shared with the minimizer of the empirical error over the class $C^{(k_n)}$, where k_n is a properly chosen function of the sample size n (Theorem 18.1). The real strength, then, of structural risk minimization is seen from the next result.

Theorem 18.3. *Let* $C^{(1)}, C^{(2)}, \dots$ *be a sequence of classes of classifiers such that the* VC *dimensions* $V_{C^{(1)}}, V_{C^{(2)}}, \dots$ *are all finite. Assume further that the Bayes rule*

$$g^* \in C^* = \bigcup_{j=1}^{\infty} C^{(j)},$$

that is, a Bayes rule is contained in one of the classes. Let k be the smallest integer such that $g^* \in C^{(k)}$. Then for every n and $\epsilon > 0$ satisfying

$$V_{C^{(k)}} \log(en) \le \frac{n\epsilon^2}{512},$$

the error probability of the classification rule based on structural risk minimization ϕ_n^* satisfies

$$\mathbf{P}\left\{ L(\phi_n^*) - L^* > \epsilon \right\} \le \Delta e^{-n\epsilon^2/128} + 8n^{V_{C^{(k)}}} e^{-n\epsilon^2/512},$$

where $\Delta = \sum_{j=1}^{\infty} e^{-V_{C^{(j)}}}$.

PROOF. Theorem 18.3 follows by examining the proof of Theorem 18.2. The only difference is that by assumption, there is an integer k such that $\inf_{\phi \in C^{(k)}} L(\phi) = L^*$. Therefore, for this k,

$$L(\phi_n^*) - L^* = \left(L(\phi_n^*) - \inf_{j \ge 1} \tilde{L}_n(\tilde{\phi}_{n,j}) \right) + \left(\inf_{j \ge 1} \tilde{L}_n(\tilde{\phi}_{n,j}) - \inf_{\phi \in C^{(k)}} L(\phi) \right),$$

and the two terms on the right-hand side may be bounded as in the proof of Theorem 18.2. $\square$

Theorem 18.3 implies that if $g^* \in C^*$, there is a universal constant c_0 and a finite k such that

$$\mathbf{E}L(\phi_n^*) - L^* \le c_0 \sqrt{\frac{V_{C^{(k)}} \log n}{n}},$$

that is, the rate of convergence is always of the order of $\sqrt{\log n / n}$, and the constant factor $V_{C^{(k)}}$ depends on the distribution. The number $V_{C^{(k)}}$ may be viewed as the inherent complexity of the Bayes rule for the distribution. The intuition is that the simplest rules are contained in $C^{(1)}$, and more and more complex rules are added to the class as the index of the class increases. The size of the error is about the same as if we had known k beforehand, and minimized the empirical error over $C^{(k)}$. In view of the results of Chapter 14 it is clear that the classifier described in Theorem 18.1 does not share this property, since if $L^* > 0$, then the error probability of the rule selected from $C^{(k_n)}$ is larger than a constant times $\sqrt{V_{C^{(k_n)}} \log n / n}$ for some distribution—even if $g^* \in C^{(k)}$ for some fixed k. Just as in designs based upon minimum description length, automatic model selection, and other complexity regularization methods (Rissanen (1983), Akaike (1974), Barron (1985; 1991), and Barron and Cover (1991)), structural risk minimization automatically finds where to look for the optimal classifier. The constants appearing in Theorem 18.3 may be improved by using refined versions of the Vapnik-Chervonenkis inequality.

The condition $g^* \in C^*$ in Theorem 18.3 is not very severe, as C^* can be a large class with infinite VC dimension. The only requirement is that it should be written as a countable union of classes of finite VC dimension. Note however that the class of all decision rules can not be decomposed as such (see Theorem 18.6). We also emphasize that in order to achieve the $O\left(\sqrt{\log n / n}\right)$ rate of convergence, we do not have to assume that the distribution is a member of a known finite-dimensional parametric family (see Chapter 16). The condition is imposed solely on the form of the Bayes classifier g^*.

By inspecting the proof of Theorem 18.2 we see that for every k,

$$\mathbf{E}L(\phi_n^*) - L^* \leq c_0 \sqrt{\frac{V_{C^{(k)}} \log n}{n}} + \left(\inf_{\phi \in C(k)} L(\phi) - L^*\right).$$

In fact, Theorem 18.3 is the consequence of this inequality. The first term on the right-hand side, which may be called estimation error, usually increases with growing k, while the second, the approximation error, usually decreases with it. Importantly, the above inequality is true for *every* k, so that

$$\mathbf{E}L(\phi_n^*) - L^* \leq \inf_k \left(c_0 \sqrt{\frac{V_{C^{(k)}} \log n}{n}} + \left(\inf_{\phi \in C(k)} L(\phi) - L^*\right)\right).$$

Thus structural risk minimization finds a nearly optimal balance between the two terms. See also Problem 18.6.

REMARK. It is worthwhile mentioning that under the conditions of Theorem 18.3, an even faster, $O\left(\sqrt{1/n}\right)$, rate of convergence is achievable at the expense of magnifying the constant factor. More precisely, it is possible to define the complexity penalties $r(j, n)$ such that the resulting classifier satisfies

$$\mathbf{E}L(\phi_n^*) - L^* \leq \frac{c_1}{\sqrt{n}},$$

where the constant c_1 depends on the distribution. These penalties may be defined by exploiting Alexander's inequality (Theorem 12.10), and the inequality above can be proved by using the bound in Problem 12.10, see Problem 18.6. □

REMARK. We see from the proof of Theorem 18.2 that

$$\mathbf{P}\left\{L(\phi_n^*) > \bar{L}_n(\phi_n^*) + \epsilon\right\} \leq \Delta e^{-n\epsilon^2/32}.$$

This means that $\tilde{L}_n(\phi_n^*)$ does not underestimate the true error probability of ϕ_n^* by much. This is a very attractive feature of $\tilde{L}_n$ as an error estimate, as the designer can be confident about the performance of the rule ϕ_n^*. □

The initial excitement over a consistent rule with a guaranteed $O(\sqrt{\log n/n})$ rate of convergence to the Bayes error is tempered by a few sobering facts:

(1) The user needs to choose the sequence $C^{(i)}$ and has to know $V_{C^{(i)}}$ (see, however, Problems 18.3, 18.4, and the method of simple empirical covering below).

(2) It is difficult to find the structural risk minimizer ϕ_n^* efficiently. After all, the minimization is done over an infinite sequence of infinite sets.

The second concern above deserves more attention. Consider the following simple example: let $d = 1$, and let $C^{(j)}$ be the class of classifiers ϕ for which

$$\{x : \phi(x) = 1\} = \bigcup_{i=1}^{j} A_i,$$

where each A_i is an interval of $\mathcal{R}$. Then, from Theorem 13.7, $V_{C^{(j)}} = 2j$, and we may take

$$r(j, n) = \sqrt{\frac{64j}{n} \log(en)}.$$

In structural risk minimization, we find those j (possibly empty) intervals that minimize

$$\widehat{L}_n(\phi) + r(j, n),$$

and call the corresponding classifier $\tilde{\phi}_{n,j}$. For $j = 1$ we have $r(j, n) = \sqrt{\frac{64}{n} \log(en)}$. As $r(n, n) > 1 + r(1, n)$, and $r(j, n)$ is monotone in j, it is easily seen that to pick the best j as well, we need only consider $1 \leq j \leq n$. Still, this is a formidable exercise. For fixed j, the best j intervals may be found by considering all possible insertions of $2j$ interval boundaries among the n X_i's. This brute force method takes computation time bounded from below by $\binom{n+2j}{2j}$. If we let j run up to n, then we have as a lower bound

$$\sum_{j=1}^{n} \binom{n + 2j}{2j}.$$

Just the last term alone, $\binom{3n}{2n}$, grows as $\sqrt{\frac{3}{4\pi n}} \left(\frac{27}{4}\right)^n$, and is prohibitively large for $n \geq 20$. Fortunately, in this particular case, there is an algorithm which finds a classifier minimizing $\widehat{L}_n(\phi) + r(j, n)$ over $C^* = \bigcup_{j=1}^{\infty} C^{(j)}$ in computational time polynomial in n, see Problem 18.5. Another example when the structural risk minimizer ϕ_n^* is easy to find is described in Problem 18.6. However, such fast algorithms are not always available, and exponential running time prohibits the use of structural risk minimization even for relatively small values of n.

18.2   Poor Approximation Properties of VC Classes

We pause here for a moment to summarize some interesting by-products that readily follow from Theorem 18.3 and the slow convergence results of Chapter 7. Theorem 18.3 states that for a large class of distributions an $O\left(\sqrt{\log n/n}\right)$ rate of convergence to the Bayes error L^* is achievable. On the other hand, Theorem 7.2 asserts that no universal rates of convergence to L^* exist. Therefore, the class of distributions for which Theorem 18.3 is valid cannot be that large, after all. The combination of these facts results in the following three theorems, which say that VC classes—classes of subsets of $\mathcal{R}^d$ with finite VC dimension—have necessarily very poor approximation properties. The proofs are left to the reader as easy exercises (see Problems 18.7 to 18.9). For direct proofs, see, for example, Benedek and Itai (1994).

Theorem 18.4. Let $\mathcal{C}$ be any class of classifiers of the form $\phi : \mathcal{R}^d \to \{0, 1\}$, with VC dimension $V_\mathcal{C} < \infty$. Then for every $\epsilon > 0$ there exists a distribution such that

$$\inf_{\phi \in \mathcal{C}} L(\phi) - L^* > \frac{1}{2} - \epsilon.$$

Theorem 18.5. Let $C^{(1)}, C^{(2)}, \ldots$ be a sequence of classifiers such that the VC dimensions $V_{C^{(1)}}, V_{C^{(2)}}, \ldots$ are all finite. Then for any sequence $\{a_k\}$ of positive numbers converging to zero arbitrarily slowly, there exists a distribution such that for every k large enough,

$$\inf_{\phi \in C^{(k)}} L(\phi) - L^* > a_k.$$

Theorem 18.6. The class C^* of all (Borel measurable) decision rules of form $\phi : \mathcal{R}^d \to \{0, 1\}$ cannot be written as

$$C^* = \bigcup_{j=1}^{\infty} C^{(j)},$$

where the VC dimension of each class $C^{(j)}$ is finite. In other words, the class $\mathcal{B}$ of all Borel subsets of $\mathcal{R}^d$ cannot be written as a union of countably many VC classes. In fact, the same is true for the class of all subsets of the set of positive integers.

18.3   Simple Empirical Covering

As Theorem 18.2 shows, the method of structural risk minimization provides automatic protection against the danger of overfitting the data, by penalizing complex candidate classifiers. One of the disadvantages of the method is that the penalty terms $r(j, n)$ require knowledge of the VC dimensions of the classes $C^{(j)}$ or upper bounds of these dimensions. Next we discuss a method proposed by

Buescher and Kumar (1996b) which is applicable even if the VC dimensions of the classes are completely unknown. The method, called *simple empirical covering*, is closely related to the method based on empirical covering studied in Problem 12.14. In simple empirical covering, we first split the data sequence D_n into two parts. The first part is $D_m = ((X_1, Y_1), \ldots, (X_m, Y_m))$ and the second part is $T_l = ((X_{m+1}, Y_{m+1}), \ldots, (X_n, Y_n))$. The positive integers m and $l = n - m$ will be specified later. The first part D_m is used to cover $C^* = \bigcup_{j=1}^{\infty} C^{(j)}$ as follows. For every $\phi \in C^*$, define the binary m-vector $b_m(\phi)$ by

$$b_m(\phi) = (\phi(X_1), \ldots, \phi(X_m)) .$$

There are $N \le 2^m$ different values of $b_m(\phi)$. Usually, as $V_{C^*} = \infty$, $N = 2^m$, that is, all possible values of $b_m(\phi)$ occur as ϕ is varied over C^*, but of course, N depends on the values of $X_1, \ldots, X_m$. We call a classifier ϕ *simpler* than another classifier ϕ', if the smallest index i such that $\phi \in C^{(i)}$ is smaller than or equal to the smallest index j such that $\phi' \in C^{(j)}$. For every binary m-vector $b \in \{0, 1\}^m$ that can be written as $b = b_m(\phi)$ for some $\phi \in C^*$, we pick a candidate classifier $\tilde{\phi}_{m,k}$ with $k \in \{1, \ldots, N\}$ such that $b_m(\tilde{\phi}_{m,k}) = b$, and it is the simplest such classifier, that is, there is no $\phi \in C^*$ such that simultaneously $b_m(\tilde{\phi}_{m,k}) = b(\phi)$ and ϕ is simpler than $\tilde{\phi}_{m,k}$. This yields $N \le 2^m$ candidates $\tilde{\phi}_{m,1}, \ldots, \tilde{\phi}_{m,N}$. Among these, we select one that minimizes the empirical error

$$\widehat{L}_l(\tilde{\phi}_{m,k}) = \frac{1}{l} \sum_{i=1}^{l} I_{\{\tilde{\phi}_{m,k}(X_{m+i}) \ne Y_{m+i}\}} ,$$

measured on the independent testing sequence T_l. Denote the selected classifier by ϕ_n^*. The next theorem asserts that the method works under circumstances similar to structural risk minimization.

Theorem 18.7. (BUESCHER AND KUMAR (1996B)). *Let $C^{(1)} \subseteq C^{(2)} \subseteq \ldots$ be a nested sequence of classes of classifiers such that for any distribution of (X, Y),*

$$\lim_{j \to \infty} \inf_{\phi \in C^{(j)}} L(\phi) = L^* .$$

Assume also that the VC dimensions $V_{C^{(1)}}, V_{C^{(2)}}, \ldots$ are all finite. If $m / \log n \to \infty$ and $m/n \to 0$, then the classification rule ϕ_n^* based on simple empirical covering as defined above is strongly universally consistent.

PROOF. We decompose the difference between the error probability of the selected rule ϕ_n^* and the Bayes risk as follows:

$$L(\phi_n^*) - L^* = \left(L(\phi_n^*) - \inf_{1 \le k \le N} L(\tilde{\phi}_{m,k}) \right) + \left(\inf_{1 \le k \le N} L(\tilde{\phi}_{m,k}) - L^* \right) .$$

The first term can be handled by Lemma 8.2 and Hoeffding's inequality:

$$\mathbf{P}\left\{L(\phi_n^*) - \inf_{1\leq k\leq N} L(\tilde{\phi}_{m,k}) > \epsilon\right\}$$

$$\leq \quad \mathbf{P}\left\{2\sup_{1\leq k\leq N}\left|L(\tilde{\phi}_{m,k}) - \widehat{L}_l(\tilde{\phi}_{m,k})\right| > \epsilon\right\}$$

$$= \quad \mathbf{E}\left\{\mathbf{P}\left\{2\sup_{1\leq k\leq N}\left|L(\tilde{\phi}_{m,k}) - \widehat{L}_l(\tilde{\phi}_{m,k})\right| > \epsilon\,\bigg|\,D_m\right\}\right\}$$

$$\leq \quad \mathbf{E}\left\{2Ne^{-l\epsilon^2/2}\right\}$$

$$\leq \quad 2^{m+1}e^{-l\epsilon^2/2} = 2^{m+1}e^{-(n-m)\epsilon^2/2}.$$

Because $m = o(n)$, the latter expression converges to zero exponentially rapidly. Thus, it remains to show that

$$\inf_{1\leq k\leq N} L(\tilde{\phi}_{m,k}) - L^* \to 0$$

with probability one. By our assumptions, for every $\epsilon > 0$, there is an integer k such that

$$\inf_{\phi\in C^{(k)}} L(\phi) - L^* < \epsilon.$$

Then there exists a classifier $\phi^{(\epsilon)} \in C^{(k)}$ with $L(\phi^{(\epsilon)}) - L^* \leq \epsilon$. Therefore, we are done if we prove that

$$\limsup_{n\to\infty} \inf_{1\leq k\leq N} L(\tilde{\phi}_{m,k}) - L(\phi^{(\epsilon)}) \leq 0 \quad \text{with probability one.}$$

Clearly, there is a classifier $\tilde{\phi}_{m,j}$ among the candidates $\tilde{\phi}_{m,1}, \ldots, \tilde{\phi}_{m,N}$, such that

$$b_m(\tilde{\phi}_{m,j}) = b_m(\phi^{(\epsilon)}).$$

Since by definition, $\tilde{\phi}_{m,j}$ is simpler than $\phi^{(\epsilon)}$, and the classes $C^{(1)}, C^{(2)}, \ldots$ are nested, it follows that $\tilde{\phi}_{m,j} \in C^{(k)}$. Therefore,

$$\inf_{1\leq k\leq N} L(\tilde{\phi}_{m,k}) - L(\phi^{(\epsilon)}) \quad\leq\quad L(\tilde{\phi}_{m,j}) - L(\phi^{(\epsilon)})$$

$$\leq \sup_{\phi,\phi'\in C^{(k)}:b_m(\phi)=b_m(\phi')} |L(\phi) - L(\phi')|,$$

where the last supremum is taken over all pairs of classifiers such that their corresponding binary vectors $b_m(\phi)$ and $b_m(\phi')$ are equal. But from Problem 12.14,

$$P\left\{\sup_{\phi,\phi'\in C^{(k)}:b_m(\phi)=b_m(\phi')}|L(\phi)-L(\phi')|>\epsilon\right\}\le 2S^4(C^{(k)},2m)e^{-m\epsilon\log 2/4},$$

which is summable if $m/\log n\to\infty$. $\square$

REMARK. As in Theorem 18.3, we may assume again that there is an integer k such that $\inf_{\phi\in C^{(k)}}L(\phi)=L^*$. Then from the proof of the theorem above we see that the error probability of the classifier ϕ_n^*, obtained by the method of simple empirical covering, satisfies

$$P\left\{L(\phi_n^*)-L^*>\epsilon\right\}\le 2^{m+1}e^{-(n-m)\epsilon^2/8}+2S^4(C^{(k)},2m)e^{-m\epsilon\log 2/8}.$$

Unfortunately, for any choice of m, this bound is much larger than the analogous bound obtained for structural risk minimization. In particular, it does not yield an $O(\sqrt{\log n/n})$ rate of convergence. Thus, it appears that the price paid for the advantages of simple empirical covering—no knowledge of the VC dimensions is required, and the implementation may require significantly less computational time in general—is a slower rate of convergence. See Problem 18.10. $\square$

Problems and Exercises

PROBLEM 18.1. Prove Theorem 18.1.

PROBLEM 18.2. Define the complexity penalties $r(j,n)$ so that under the conditions of Theorem 18.3, the classification rule ϕ_n^* based upon structural risk minimization satisfies

$$EL(\phi_n^*)-L^*\le\frac{c_1}{\sqrt{n}},$$

where the constant c_1 depends on the distribution. HINT: Use Alexander's bound (Theorem 12.10), and the inequality of Problem 12.10.

PROBLEM 18.3. Let $C^{(1)},C^{(2)},\ldots$ be a sequence of classes of decision rules with finite VC dimensions. Assume that only upper bounds $\alpha_j\ge V_{C^{(j)}}$ on the VC dimensions are known. Define the complexity penalties by

$$r(j,n)=\sqrt{\frac{32}{n}\alpha_j\log(en)}.$$

Show that if $\sum_{j=1}^{\infty}e^{-\alpha_j}<\infty$, then Theorems 18.2 and 18.3 carry over to the classifier based on structural risk minimization defined by these penalties. This points out that knowledge of relatively rough estimates of the VC dimensions suffice.

PROBLEM 18.4. Let $C^{(1)},C^{(2)},\ldots$ be a sequence of classes of classifiers such that the VC dimensions $V_{C^{(1)}},V_{C^{(2)}},\ldots$ are all finite. Assume furthermore that the Bayes rule is contained in one of the classes and that $L^*=0$. Let ϕ_n^* be the rule obtained by structural risk minimization using the positive penalties $r(j,n)$, satisfying

(1) For each n, $r(j, n)$ is strictly monotone increasing in j.

(2) For each j, $\lim_{n \to \infty} r(j, n) = 0$.

Show that $E\{L(\phi_n^*)\} = O(\log n/n)$ (Lugosi and Zeger (1996)). For related work, see Benedek and Itai (1994).

PROBLEM 18.5. Let $C^{(j)}$ be the class of classifiers $\phi : \mathcal{R}^d \to \{0, 1\}$ satisfying

$$\{x : \phi(x) = 1\} = \bigcup_{i=1}^{j} A_i,$$

where the A_i's are bounded intervals in $\mathcal{R}$. The purpose of this exercise is to point out that there is a fast algorithm to find the structural risk minimizer ϕ_n^* over $C^* = \bigcup_{j=1}^{\infty} C^{(j)}$, that is, which minimizes $\tilde{L}_n = \hat{L}_n + r(j, n)$ over C^*, where the penalties $r(j, n)$ are defined as in Theorems 18.2 and 18.3. The property below was pointed out to us by Miklós Csűrös and Réka Szabó.

(1) Let $A_{1,j}^*, \ldots, A_{j,j}^*$ be the j intervals defining the classifier $\tilde{\phi}_{n,j}$ minimizing $\hat{L}_n$ over $C^{(j)}$. Show that the optimal intervals for $C^{(j+1)}$, $A_{1,j+1}^*, \ldots, A_{j+1,j+1}^*$ satisfy the following property: either j of the intervals coincide with $A_{1,j}^*, \ldots, A_{j,j}^*$, or $j - 1$ of them are among $A_{1,j}^*, \ldots, A_{j,j}^*$, and the remaining two intervals are subsets of the j-th interval.

(2) Use property (1) to define an algorithm that finds ϕ_n^* in running time polynomial in the sample size n.

PROBLEM 18.6. Assume that the distribution of X is concentrated on the unit cube, that is, $P\{X \in [0, 1]^d\} = 1$. Let $\mathcal{P}_j$ be a partition of $[0, 1]^d$ into cubes of size $1/j$, that is, $\mathcal{P}_j$ contains j^d cubic cells. Let $C^{(j)}$ be the class of all histogram classifiers $\phi : [0, 1]^d \to \{0, 1\}$ based on $\mathcal{P}_j$. In other words, $\mathcal{P}_j$ contains all 2^{j^d} classifiers which assign the same label to points falling in the same cell of $\mathcal{P}_j$. What is the VC dimension $V_{C^{(j)}}$ of $C^{(j)}$? Point out that the classifier minimizing $\hat{L}_n$ over $C^{(j)}$ is just the regular histogram rule based on $\mathcal{P}_j$. (See Chapter 9.) Thus, we have another example in which the empirically optimal classifier is computationally inexpensive. The structural risk minimizer ϕ_n^* based on $C^* = \bigcup_{j=1}^{\infty} C^{(j)}$ is also easy to find. Assume that the a posteriori probability $\eta(x)$ is uniformly Lipschitz, that is, for any $x, y \in [0, 1]^d$,

$$|\eta(x) - \eta(y)| \le c\|x - y\|,$$

where c is some constant. Find upper bounds for the rate of convergence of $EL(\phi_n^*)$ to L^*.

PROBLEM 18.7. Prove Theorem 18.4. HINT: Use Theorem 7.1.

PROBLEM 18.8. Prove Theorem 18.5. HINT: Use Theorem 7.2.

PROBLEM 18.9. Prove Theorem 18.6. HINT: Use Theorems 7.2 and 18.3.

PROBLEM 18.10. Assume that the Bayes rule g^* is contained in $C^* = \bigcup_{j=1}^{\infty} C^{(j)}$. Let ϕ_n^* be the classifier obtained by simple empirical covering. Determine the value of the design parameter m that minimizes the bounds obtained in the proof of Theorem 18.7. Obtain a tight upper bound for $EL(\phi_n^*) - L^*$. Compare your results with Theorem 18.3.

19
Condensed and Edited Nearest Neighbor Rules

19.1   Condensed Nearest Neighbor Rules

Condensing is the process by which we eliminate data points, yet keep the same behavior. For example, in the nearest neighbor rule, by condensing we might mean the reduction of $(X_1, Y_1), \ldots, (X_n, Y_n)$ to $(X'_1, Y'_1), \ldots, (X'_m, Y'_m)$ such that for all $x \in \mathcal{R}^d$, the 1-NN rule is identical based on the two samples. This will be called *pure condensing*. This operation has no effect on L_n, and therefore is recommended whenever space is at a premium. The space savings should be substantial whenever the classes are separated. Unfortunately, pure condensing is computationally expensive, and offers no hope of improving upon the performance of the ordinary 1-NN rule.

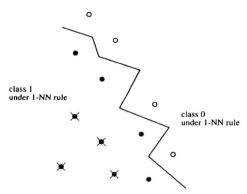

FIGURE 19.1. *Pure condensing: Eliminating the marked points does not change the decision.*

Hart (1968) has the first simple algorithm for condensing. He picks a subset that correctly classifies the remaining data by the 1-NN rule. Finding a minimal such subset is computationally difficult, but heuristics may do the job. Hart's heuristic is also discussed in Devijver and Kittler (1982, p.120). For probabilistic analysis, we take a more abstract setting. Let $(X_1', Y_1'), \ldots, (X_m', Y_m')$ be a sequence that depends in an arbitrary fashion on the data D_n, and let g_n be the 1-nearest neighbor rule with $(X_1', Y_1'), \ldots, (X_m', Y_m')$, where for simplicity, m is fixed beforehand. The data might, for example, be obtained by finding the subset of the data of size m for which the error with the 1-NN rule committed on the remaining $n - m$ data is minimal (this will be called *Hart's rule*). Regardless, if $\widehat{L}_n = (1/n) \sum_{i=1}^{n} I_{\{g_n(X_i) \neq Y_i\}}$ and $L_n = \mathbf{P}\{g_n(X) \neq Y | D_n\}$, then we have the following:

Theorem 19.1. (DEVROYE AND WAGNER (1979C)). *For all $\epsilon > 0$ and all distributions,*

$$\mathbf{P}\left\{|L_n - \widehat{L}_n| \geq \epsilon\right\} \leq 8 \left(\frac{ne}{d+1}\right)^{(d+1)m(m-1)} e^{-n\epsilon^2/32}.$$

REMARK. The estimate $\widehat{L}_n$ is called the *resubstitution estimate* of the error probability. It is thoroughly studied in Chapter 23, where several results of the aforementioned kind are stated. $\square$

PROOF. Observe that

$$\widehat{L}_n = \frac{1}{n} \sum_{j=1}^{n} I_{\{(X_j, Y_j) \notin \bigcup_{i=1}^{m} B_i \times \{Y_i'\}\}},$$

where B_i is the *Voronoi cell* of X_i' in the Voronoi partition corresponding to $X_1', \ldots, X_m'$, that is, B_i is the set of points of $\mathcal{R}^d$ closer to X_i' than to any other X_j' (with appropriate distance-tie breaking). Similarly,

$$L_n = \mathbf{P}\left\{(X, Y) \notin \bigcup_{i=1}^{m} B_i \times \{Y_i'\} \middle| D_n\right\}.$$

We use the simple upper bound

$$|L_n - \widehat{L}_n| \leq \sup_{A \in \mathcal{A}_m} |\nu_n(A) - \nu(A)|,$$

where ν denotes the measure of (X, Y), ν_n is the corresponding empirical measure, and $\mathcal{A}_m$ is the family of all subsets of $\mathcal{R}^d \times \{0, 1\}$ of the form $\bigcup_{i=1}^{m} B_i \times \{y_i\}$, where $B_1, \ldots, B_m$ are Voronoi cells corresponding to $x_1, \ldots, x_m$, $x_i \in \mathcal{R}^d$, $y_i \in \{0, 1\}$. We use the Vapnik-Chervonenkis inequality to bound the above supremum. By Theorem 13.5 (iv),

$$s(\mathcal{A}_m, n) \leq s(\mathcal{A}, n)^m,$$

where $\mathcal{A}$ is the class of sets $B_1 \times \{y_1\}$. But each set in $\mathcal{A}$ is an intersection of at most $m - 1$ hyperplanes. Therefore, by Theorems 13.5 (iii), 13.9, and 13.3,

$$s(\mathcal{A}, n) \leq \sup_{n_0, n_1 : n_0 + n_1 = n} \left(\prod_{j=0}^{1} \left(\frac{n_j e}{d+1} \right)^{(d+1)(k-1)} \right) \leq \left(\frac{ne}{d+1} \right)^{(d+1)(k-1)},$$

where n_j denotes the number of points in $\mathcal{R}^d \times \{j\}$. The result now follows from Theorem 12.5. $\square$

REMARK. With Hart's rule, at least m data points are correctly classified by the 1-NN rule (if we handle distance ties satisfactorily). Therefore, $\widehat{L}_n \leq 1 - m/n$. $\square$

The following is a special case: Let $m < n$ be fixed and let D_m be an arbitrary (possibly random) subset of m pairs from $(X_1, Y_1), \ldots, (X_n, Y_n)$, used in g_n. Let the remaining $n - m$ points be denoted by T_{n-m}. We write $L_n(D_m)$ for the probability of error with the 1-NN based upon D_m, and we define

$$\widehat{L}_{n,m}(D_m, T_{n-m}) = \frac{1}{n-m} \sum_{(X_i, Y_i) \in T_{n-m}} I_{\{g_n(X_i) \neq Y_i\}}.$$

In Hart's rule, $\widehat{L}_{n,m}$ would be zero, for example. Then we have:

Theorem 19.2. *For all $\epsilon > 0$,*

$$\mathbf{P}\left\{ |L_n - \widehat{L}_{n,m}| > \epsilon \right\} \leq 2\binom{n}{m} e^{-2(n-m)\epsilon^2}$$

$$\leq 2n^m e^{-2(n-m)\epsilon^2},$$

where L_n is the probability of error with g_n (note that D_m depends in an arbitrary fashion upon D_n), and $\widehat{L}_{n,m}$ is $\widehat{L}_{n,m}(D_m, T_{n-m})$ with the data set D_m.

PROOF. List the m-element subsets $\{i_1, \ldots, i_m\}$ of $\{1, 2, \ldots, n\}$, and define $D_m^{(i)}$ as the sequence of m pairs from D_n indexed by $i = \{i_1, \ldots, i_m\}$, $1 \leq i \leq \binom{n}{m}$. Accordingly, denote $T_{n-m}^{(i)} = D_n - D_m^{(i)}$. Then

$$\mathbf{P}\left\{ |L_n - \widehat{L}_{n,m}| > \epsilon \right\}$$

$$\leq \sum_{i=1}^{\binom{n}{m}} \mathbf{P}\left\{ |L_n(D_m^{(i)}) - \widehat{L}_{n,m}(D_m^{(i)}, T_{n-m}^{(i)})| > \epsilon \right\}$$

$$= \mathbf{E}\left\{ \sum_{i=1}^{\binom{n}{m}} \mathbf{P}\left\{ |L_n(D_m^{(i)}) - \widehat{L}_{n,m}(D_m^{(i)}, T_{n-m}^{(i)})| > \epsilon | D_m^{(i)} \right\} \right\}$$

$$\leq \quad E\left\{\sum_{i=1}^{\binom{n}{m}} 2e^{-2(n-m)\epsilon^2}\right\}$$

(by Hoeffding's inequality, because given $D_m^{(i)}$,

$(n-m)\widehat{L}_{n,m}(D_n^{(i)}, T_{n-m}^{(i)})$ is binomial $((n-m), L_n(D_m^{(i)})))$

$$= \quad 2\binom{n}{m} e^{-2(n-m)\epsilon^2}. \quad \square$$

By checking the proof, we also see that if D_m is selected to minimize the error estimate $\widehat{L}_{n,m}(D_m, T_{n-m})$, then the error probability L_n of the obtained rule satisfies

$$P\left\{|L_n - \inf_{D_m} L_n(D_m)| > \epsilon\right\}$$

$$\leq \quad P\left\{2\sup_{D_m} |L_n(D_m) - \widehat{L}_{n,m}(D_m, T_{n-m})| > \epsilon\right\} \quad \text{(by Theorem 8.4)}$$

$$\leq \quad 2\binom{n}{m} e^{-(n-m)\epsilon^2/2} \quad \text{(as in the proof of Theorem 19.2).} \quad (19.1)$$

Thus, for the particular rule that mimics Hart's rule (with the exception that m is fixed), if m is not too large—it must be much smaller than $n/\log n$—L_n is likely to be close to the best possible we can hope for with a 1-NN rule based upon a subsample of size m. With some work (see Problem 12.1), we see that

$$E\left\{\left|L_n - \inf_{D_m} L_n(D_m)\right|\right\} = O\left(\sqrt{\frac{m\log m}{n}}\right).$$

By Theorem 5.1,

$$E\left\{\inf_{D_m} L_n(D_m)\right\} \leq E\{L_m\} \rightarrow L_{NN} \quad \text{as} \quad m \rightarrow \infty,$$

where L_m is the probability of error with the 1-NN rule based upon a sample of m data pairs. Hence, if $m = o(n/\log n)$, $m \rightarrow \infty$,

$$\limsup_{n\rightarrow\infty} E\{L_n\} \leq L_{NN}$$

for the 1-NN rule based on m data pairs D_m selected to minimize the error estimate $\widehat{L}_{n,m}(D_m, T_{n-m})$. However, this is very pessimistic indeed. It reassures us that with only a small fraction of the original data, we obtain at least as good a performance as with the full data set—so, this method of condensing is worthwhile. This is not very surprising. Interestingly, however, the following much stronger result is true.

Theorem 19.3. *If $m = o(n/\log n)$ and $m \to \infty$, and if L_n is the probability of error for the condensed nearest neighbor rule in which $\widehat{L}_{n,m}(D_m, T_{n-m})$ is minimized over all data sets D_m, then*

$$\lim_{n \to \infty} \mathbf{E}\{L_n\} = L^*.$$

PROOF. By (19.1), it suffices to establish that

$$\mathbf{E}\left\{\inf_{D_m} L_n(D_m)\right\} \to L^*$$

as $m \to \infty$ such that $m = o(n)$, where $L_n(D_m)$ is the probability of error of the 1-NN rule with D_m. As this is one of the fundamental properties of nearest neighbor rules not previously found in texts, we offer a thorough analysis and proof of this result in the remainder of this section, culminating in Theorem 19.4. $\square$

The distribution-free result of Theorem 19.3 sets the stage for many consistency proofs for rules that use condensing (or editing, or prototyping, as defined in the next two sections). It states that inherently, partitions of the space by 1-NN rules are rich.

HISTORICAL REMARK.   Other condensed nearest neighbor rules are presented by Gates (1972), Ullmann (1974), Ritter, Woodruff, Lowry, and Isenhour (1975), Tomek (1976b), Swonger (1972), Gowda and Krishna (1979), and Fukunaga and Mantock (1984). $\square$

Define $Z = I_{\{\eta(X) > 1/2\}}$. Let (X_i', Y_i, Z_i), $i = 1, 2, \ldots, n$, be i.i.d. tuples, independent of (X, Y, Z), where X_i' may have a distribution different from X, but the support set of the distribution μ' of X_i' is identical to that of μ, the distribution of X. Furthermore, $\mathbf{P}\{Y_i = 1 | X_i' = x\} = \eta(x)$, and $Z_i = I_{\{\eta(X_i') > 1/2\}}$ is the Bayes decision at X_i'.

Lemma 19.1. *Let $L_n = \mathbf{P}\left\{Z'_{(1)}(X) \neq Z | X_1', Z_1, \ldots, X_n', Z_n\right\}$ be the probability of error for the 1-NN rule based on (X_i', Z_i), $1 \leq i \leq n$, that is, $Z'_{(1)}(X) = Z_i$ if X_i' is the nearest neighbor of X among $X_1', \ldots, X_n'$. Then*

$$\lim_{n \to \infty} \mathbf{E}\{L_n\} = 0.$$

PROOF. Denote by $X'_{(1)}(X)$ the nearest neighbor of X among $X_1', \ldots, X_n'$. Notice that the proof of Lemma 5.1 may be extended in a straightforward way to show that $\|X'_{(1)}(X) - X\| \to 0$ with probability one. Since this is the only property of the nearest neighbor of X that we used in deriving the asymptotic formula for the ordinary 1-NN error, $\lim_{n \to \infty} \mathbf{E}\{L_n\}$ equals L_{NN} corresponding to the pair (X, Z). But we have $\mathbf{P}\{Z = 1 | X = x\} = I_{\{\eta(x) > 1/2\}}$. Thus, the Bayes probability of error L^*

for the pattern recognition problem with (X, Z) is zero. Hence, for this distribution, the 1-NN rule is consistent as $L_{NN} = L^* = 0$. $\square$

Lemma 19.2. *Let* $Z'_{(1)}(X)$ *be as in the previous lemma. Let*

$$L_n = \mathbf{P}\left\{Z'_{(1)}(X) \neq Y \,|\, X'_1, Y_1, \dots, X'_n, Y_n\right\}$$

be the probability of error for the discrimination problem for (X, Y) *(not* (X, Z)*). Then*

$$\lim_{n \to \infty} \mathbf{E}\{L_n\} = L^*,$$

where L^* *is the Bayes error corresponding to* (X, Y).

PROOF.

$$
\begin{aligned}
\mathbf{E}\{L_n\} \;&=\; \mathbf{P}\left\{Z'_{(1)}(X) \neq Y\right\} \\
&\leq\; \mathbf{P}\left\{Z'_{(1)}(X) \neq Z\right\} + \mathbf{P}\{Y \neq Z\} \\
&=\; o(1) + L^*
\end{aligned}
$$

by Lemma 19.1. $\square$

Theorem 19.4. *Let* D_m *be a subset of size* m *drawn from* D_n. *If* $m \to \infty$ *and* $m/n \to 0$ *as* $n \to \infty$, *then*

$$\lim_{n \to \infty} \mathbf{P}\left\{\inf_{D_m} L_n(D_m) > L^* + \epsilon\right\} = 0$$

for all $\epsilon > 0$, *where* $L_n(D_m)$ *denotes the conditional probability of error of the nearest neighbor rule with* D_m, *and the infimum ranges over all* $\binom{n}{m}$ *subsets.*

PROOF. Let D be the subset of D_n consisting of those pairs (X_i, Y_i) for which $Y_i = I_{\{\eta(X_i) > 1/2\}} = Z_i$. If $|D| \geq m$, let D^* be the first m pairs of D, and if $|D| < m$, let $D^* = \{(X_1, Y_1), \dots, (X_m, Y_m)\}$. Then

$$\inf_{D_m} L_n(D_m) \leq L_n(D^*).$$

If $N = |D| \geq m$, then we know that the pairs in D^* are i.i.d. and drawn from the distribution of (X', Z), where X' has the same support set as X; see Problem 19.2 for properties of X'. In particular,

$$\mathbf{P}\left\{\inf_{D_m} L_n(D_m) > L^* + \epsilon\right\}$$

$$\leq\; \mathbf{P}\{N < m\} + \mathbf{P}\left\{N \geq m, L_n(D^*) > L^* + \epsilon\right\}$$

$$\leq \quad \mathbf{P}\{N < m\} + \mathbf{P}\left\{L_n(D^*) > L^* + \epsilon\right\}$$

$$\leq \quad \mathbf{P}\{\text{Binomial}(n, p) < m\} + \frac{\mathbf{E}\{L_n(D^*)\} - L^*}{\epsilon}$$

(where $p = \mathbf{P}\{Y = Z\} = 1 - L^* \geq 1/2$, and by Markov's inequality)

$$= \quad o(1),$$

by the law of large numbers (here we use $m/n \to 0$), and by Lemma 19.2 (here we use $m \to \infty$). $\square$

19.2   Edited Nearest Neighbor Rules

Edited nearest neighbor rules are 1-NN rules that are based upon carefully selected subsets $(X_1', Y_1'), \ldots, (X_m', Y_m')$. This situation is partially dealt with in the previous section, as the frontier between condensed and edited nearest neighbor rules is ill-defined. The idea of editing based upon the k-NN rule was first suggested by Wilson (1972) and later studied by Wagner (1973) and Penrod and Wagner (1977). Wilson suggests the following scheme: compute (X_i, Y_i, Z_i), where Z_i is the k-NN decision at X_i based on the full data set with (X_i, Y_i) deleted. Then eliminate all data pairs for which $Y_i \neq Z_i$. The remaining data pairs are used with the 1-NN rule (not the k-NN rule). Another rule, based upon data splitting is dealt with by Devijver and Kittler (1982). A survey is given by Dasarathy (1991), Devijver (1980), and Devijver and Kittler (1980). Repeated editing was investigated by Tomek (1976a). Devijver and Kittler (1982) propose a modification of Wilson's leave-one-out method of editing based upon data splitting.

19.3   Sieves and Prototypes

Let g_n be a rule that uses the 1-NN classification based upon *prototype data pairs* $(X_1', Y_1'), \ldots, (X_m', Y_m')$ that depend in some fashion on the original data. If the pairs form a subset of the data pairs (and thus, $m \leq n$), we have edited or condensed nearest neighbor rules. However, the (X_i', Y_i') pairs may be strategically picked outside the original data set. For example, in relabeling (see Section 11.7), $m = n$, $X_i' = X_i$ and $Y_i' = g_n'(X_i)$, where $g_n'(X_i)$ is the k-NN decision at X_i. Under some conditions, the relabeling rule is consistent (see Theorem 11.2). The true objective of *prototyping* is to extract information from the data by insisting that m be much smaller than n.

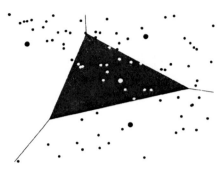

FIGURE 19.2. *A 1-NN rule based upon 4 prototypes. In this example, all the data points are correctly classified based upon the prototype 1-NN rule.*

Chang (1974) describes a rule in which we iterate the following step until a given stopping rule is satisfied: merge the two closest nearest neighbors of the same class and replace both pairs by a new average prototype pair. Kohonen (1988; 1990) recognizes the advantages of such prototyping in general as a device for partitioning $\mathcal{R}^d$—he calls this *learning vector quantization.* This theme was picked up again by Geva and Sitte (1991), who pick $X'_1, \ldots, X'_m$ as a random subset of $X_1, \ldots, X_n$ and allow Y'_i to be different from Y_i. Diverging a bit from Geva and Sitte, we might minimize the empirical error with the prototyped 1-NN rule over all $Y'_1, \ldots, Y'_m$, where the empirical error is that committed on the remaining data. We show that this simple strategy leads to a Bayes-risk consistent rule whenever $m \to \infty$ and $m/n \to 0$. Note, in particular, that we may take $(X'_1, \ldots, X'_m) = (X_1, \ldots, X_m)$, and that we "throw away" $Y_1, \ldots, Y_m$, as these are not used. We may, in fact, use additional data with missing Y_i-values for this purpose—the unclassified data are thus efficiently used to partition the space.

Let

$$\widehat{L}_n(g_n) = \frac{1}{n-m} \sum_{i=m+1}^{n} I_{\{g_n(X_i) \neq Y_i\}}$$

be the empirical risk on the remaining data, where g_n is the 1-NN rule based upon $(X_1, Y'_1), \ldots, (X_m, Y'_m)$. Let g_n^* be the 1-NN rule with the choice of $Y'_1, \ldots, Y'_m$ that minimizes $\widehat{L}_n(g_n)$. Let $L(g_n^*)$ denote its probability of error.

Theorem 19.5. $L(g_n^*) \to L^*$ *in probability for all distributions whenever $m \to \infty$ and $m/n \to 0$.*

PROOF. There are 2^m different possible functions g_n. Thus,

$$\mathbf{P}\left\{ \sup_{g_n} |\widehat{L}_n(g_n) - L(g_n)| > \epsilon \,\Big|\, X_1, \ldots, X_n \right\}$$

$$\leq 2^m \sup_{g_n} \mathbf{P}\left\{ |\widehat{L}_n(g_n) - L(g_n)| > \epsilon \,\Big|\, X_1, \ldots, X_m \right\}$$

$$\leq 2^{m+1} e^{-2(n-m)\epsilon^2}$$

by Hoeffding's inequality. Also,

$$\mathbf{P}\left\{L(g_n^*) > L^* + 3\epsilon\right\}$$

$$\leq \quad \mathbf{P}\left\{L(g_n^*) - \widehat{L}_n(g_n^*) > \epsilon\right\}$$

$$+\mathbf{P}\left\{\widehat{L}_n(\tilde{g}_n) - L(\tilde{g}_n) > \epsilon\right\} \quad (\tilde{g}_n \text{ minimizes } L(g_n))$$

$$+\mathbf{P}\left\{L(\tilde{g}_n) > L^* + \epsilon\right\} \quad (\text{here we used } \widehat{L}_n(\tilde{g}_n) \geq \widehat{L}_n(g_n^*))$$

$$\leq \quad 2\mathbf{E}\left\{\mathbf{P}\left\{\sup_{g_n} |\widehat{L}_n(g_n) - L(g_n)| > \epsilon \Big| X_1, \ldots, X_n\right\}\right\}$$

$$+\mathbf{P}\left\{L(g_n^+) > L^* + \epsilon\right\}$$

(where g_n^+ is the 1-NN rule based on $(X_1, Z_1), \ldots, (X_m, Z_m)$

with $Z_i = I_{\{\eta(X_i)>1/2\}}$ as in Lemma 19.1)

$$\leq \quad 2^{m+2}e^{-(n-m)\epsilon^2} + o(1)$$

(if $m \to \infty$, by Lemma 19.1)

$$= \quad o(1)$$

if $m \to \infty$ and $m/n \to 0$. $\square$

If we let $X_1', \ldots, X_m'$ have arbitrary values—not only among those taken by $X_1, \ldots, X_n$—then we get a much larger, more flexible class of classifiers. For example, every linear discriminant is nothing but a prototype 1-NN rule with $m = 2$—just take $(X_1', 1), (X_2', 0)$ and place X_1' and X_2' in the right places. In this sense, prototype 1-NN rules generalize a vast class of rules. The most promising strategy of choosing prototypes is to minimize the empirical error committed on the training sequence D_n. Finding this optimum may be computationally very expensive. Nevertheless, the theoretical properties provided in the next result may provide useful guidance.

Theorem 19.6. *Let C be the class of nearest neighbor rules based on prototype pairs $(x_1, y_1), \ldots, (x_m, y_m), m \geq 3$, where the (x_i, y_i)'s range through $\mathcal{R}^d \times \{0, 1\}$. Given the training data $D_n = (X_1, Y_1), \ldots, (X_n, Y_n)$, let g_n be the nearest neighbor rule from C minimizing the error estimate*

$$\widehat{L}_n(\phi) = \frac{1}{n} \sum_{i=1}^{n} I_{\{\phi(X_i) \neq Y_i\}}, \quad \phi \in C.$$

Then for each $\epsilon > 0$,

$$\mathbf{P}\left\{L(g_n) - \inf_{\phi \in C} L(\phi) > \epsilon\right\} \leq 8 \cdot 2^m n^{(d+1)m(m-1)/2} e^{-n\epsilon^2/128}.$$

The rule is consistent if $m \to \infty$ such that $m^2 \log m = o(n)$. For $d = 1$ and $d = 2$

the probability bound may be improved significantly. For $d = 1$,

$$\mathbf{P}\left\{L(g_n) - \inf_{\phi \in C} L(\phi) > \epsilon\right\} \leq 8 \cdot (2n)^m e^{-n\epsilon^2/128},$$

and for $d = 2$,

$$\mathbf{P}\left\{L(g_n) - \inf_{\phi \in C} L(\phi) > \epsilon\right\} \leq 8 \cdot 2^m n^{9m-18} e^{-n\epsilon^2/128}.$$

In both cases, the rule is consistent if $m \to \infty$ and $m \log m = o(n)$.

PROOF. It follows from Theorem 12.6 that

$$\mathbf{P}\left\{L(g_n) - \inf_{\phi \in C} L(\phi) > \epsilon\right\} \leq 8S(C, n)e^{-n\epsilon^2/128},$$

where $S(C, n)$ is the n-th shatter coefficient of the class of sets $\{x : \phi(x) = 1\}$, $\phi \in C$. All we need is to find suitable upper bounds for $S(C, n)$. Each classifier ϕ is a partitioning rule based on the m Voronoi cells defined by $x_1, \ldots, x_m$. Therefore, $S(C, n)$ is not more than 2^m times the number of different ways n points in $\mathcal{R}^d$ can be partitioned by Voronoi partitions defined by m points. In each partition, there are at most $m(m - 1)/2$ cell boundaries that are subsets of $d - 1$-dimensional hyperplanes. Thus, the sought number is not greater than the number of different ways $m(m - 1)/2$ hyperplanes can partition n points. By results of Chapter 13, this is at most $n^{(d+1)m(m-1)/2}$, proving the first inequality.

The other two inequalities follow by sharper bounds on the number of cell boundaries. For $d = 1$, this is clearly at most m. To prove the third inequality, for each Voronoi partition construct a graph whose vertices are $x_1, \ldots, x_m$, and two vertices are connected with an edge if and only if their corresponding Voronoi cells are connected. It is easy to see that this graph is planar. But the number of edges of a planar graph with m vertices cannot exceed $3m - 6$ (see Nishizeki and Chiba (1988, p.10)) which proves the inequality.

The consistency results follow from the stated inequalities and from the fact that $\inf_{\phi \in C} L(\phi)$ tends to L^* as $m \to \infty$ (check the proof of Theorem 5.15 again). $\square$

Problems and Exercises

PROBLEM 19.1. Let $N \leq n$ be the size of the data set after pure condensing. Show that $\lim \inf_{n \to \infty} \mathbf{E}\{N\}/n > 0$ whenever $L^* > 0$. True or false: if $L^* = 0$, then $\mathbf{E}\{N\} = o(n)$. HINT: Consider the real line, and note that all points whose right and left neighbors are of the same class are eliminated.

PROBLEM 19.2. Let $(X_1, Y_1), (X_2, Y_2), \ldots$ be an i.i.d. sequence of pairs of random variables in $\mathcal{R}^d \times \{0, 1\}$ with $\mathbf{P}\{Y_1 = 1|X_1 = x\} = \eta(x)$. Let (X', Z) be the first pair (X_i, Y_i) in the sequence such that $Y_i = I_{\{\eta(X_i) > 1/2\}}$. Show that the distribution μ' of X' is absolutely continuous with respect to the common distribution μ of the X_i's, with density (i.e., Radon-Nikodym derivative)

$$\frac{d\mu'}{d\mu}(x) = \frac{1 - \min(\eta(x), 1 - \eta(x))}{1 - L^*},$$

where L^* is the Bayes error corresponding to (X_1, Y_1). Let Y be a $\{0, 1\}$-valued random variable with $P\{Y = 1 | X' = x\} = \eta(x)$. If L' denotes the Bayes error corresponding to (X', Y), then show that $L' \leq L^*$.

PROBLEM 19.3. Consider the following edited NN rule. The pair (X_i, Y_i) is eliminated from the training sequence if the k-NN rule (based on the remaining $n - 1$ pairs) incorrectly classifies X_i. The 1-NN rule is used with the edited data. Show that this rule is consistent whenever X has a density if $k \to \infty$ and $k/n \to 0$. Related papers: Wilson (1972), Wagner (1973), Penrod and Wagner (1977), and Devijver and Kittler (1980).

20
Tree Classifiers

Classification trees partition $\mathcal{R}^d$ into regions, often hyperrectangles parallel to the axes. Among these, the most important are the binary classification trees, since they have just two children per node and are thus easiest to manipulate and update. We recall the simple terminology of books on data structures. The top of a binary tree is called the *root*. Each *node* has either no child (in that case it is called a *terminal node* or *leaf*), a *left child*, a *right child*, or a left child and a right child. Each node is the root of a tree itself. The trees rooted at the children of a node are called the left and right *subtrees* of that node. The *depth* of a node is the length of the *path* from the node to the root. The *height* of a tree is the maximal depth of any node.

Trees with more than two children per node can be reduced to binary trees by

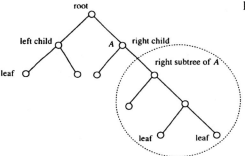

FIGURE 20.1. *A binary tree.*

a simple device—just associate a left child with each node by selecting the oldest child in the list of children. Call the right child of a node its next sibling (see Figures 20.2 and 20.3). The new binary tree is called the oldest-child/next-sibling binary tree (see, e.g., Cormen, Leiserson, and Rivest (1990) for a general introduction). We only mention this particular mapping because it enables us to only consider binary trees for simplicity.

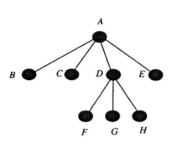
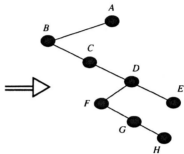

FIGURE 20.2. *Ordered tree: The children are ordered from oldest to youngest.*

FIGURE 20.3. *The corresponding binary tree.*

In a *classification tree*, each node represents a set in the space $\mathcal{R}^d$. Also, each node has exactly two or zero children. If a node u represents the set A and its children u', u'' represent A' and A'', then we require that $A = A' \cup A''$ and $A' \cap A'' = \emptyset$. The root represents $\mathcal{R}^d$, and the leaves, taken together, form a *partition* of $\mathcal{R}^d$.

Assume that we know $x \in A$. Then the question "is $x \in A$?" should be answered in a computationally simple manner so as to conserve time. Therefore, if $x = (x^{(1)}, \ldots, x^{(d)})$, we may just limit ourselves to questions of the following forms:

(i) Is $x^{(i)} \leq \alpha$? This leads to *ordinary binary classification trees* with partitions into hyperrectangles.

(ii) Is $a_1 x^{(1)} + \cdots + a_d x^{(d)} \leq \alpha$? This leads to BSP trees (*binary space partition trees*). Each decision is more time consuming, but the space is more flexibly cut up into convex polyhedral cells.

(iii) Is $\|x - z\| \leq \alpha$? (Here z is a point of $\mathcal{R}^d$, to be picked for each node.) This induces a partition into pieces of spheres. Such trees are called *sphere trees*.

(iv) Is $\psi(x) \geq 0$? Here, ψ is a nonlinear function, different for each node. Every classifier can be thought of as being described in this format—decide class one if $\psi(x) \geq 0$. However, this misses the point, as tree classifiers should really be built up from fundamental atomic operations and queries such as those listed in (i)–(iii). We will not consider such trees any further.

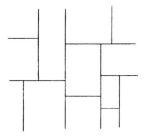

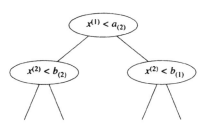

FIGURE 20.4. *Partition induced by an ordinary binary tree.*

FIGURE 20.5. *Corresponding tree.*

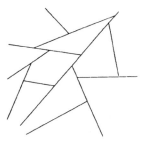

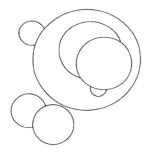

FIGURE 20.6. *Partition induced by a* BSP *tree.*

FIGURE 20.7. *Partition induced by a sphere tree.*

We associate a class in some manner with each leaf in a classification tree. The tree structure is usually data dependent, as well, and indeed, it is in the construction itself where methods differ. If a leaf represents region A, then we say that the classifier g_n is *natural* if

$$g_n(x) = \begin{cases} 1 & \text{if } \sum_{i:X_i \in A} Y_i > \sum_{i:X_i \in A} (1 - Y_i), x \in A \\ 0 & \text{otherwise.} \end{cases}$$

That is, in every leaf region, we take a majority vote over all (X_i, Y_i)'s with X_i in the same region. Ties are broken, as usual, in favor of class 0. In this set-up, natural tree classifiers are but special cases of data-dependent partitioning rules. The latter are further described in detail in Chapter 21.

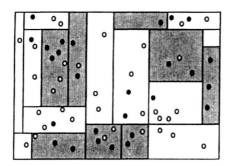

FIGURE 20.8. *A natural classifier based on an ordinary binary tree. The decision is 1 in regions where points with label 1 form a majority. These areas are shaded.*

Regular histograms can also be thought of as natural binary tree classifiers—the construction and relationship is obvious. However, as $n \to \infty$, histograms change size, and usually, histogram partitions are not nested as n grows. Trees offer the exciting perspective of fully dynamic classification—as data are added, we may update the tree slightly, say, by splitting a leaf or so, to obtain an updated classifier.

The most compelling reason for using binary tree classifiers is to explain complicated data and to have a classifier that is easy to analyze and understand. In fact, expert system design is based nearly exclusively upon decisions obtained by going down a binary classification tree. Some argue that binary classification trees are preferable over BSP trees for this simple reason. As argued in Breiman, Friedman, Olshen, and Stone (1984), trees allow mixing component variables that are heterogeneous—some components may be of a nonnumerical nature, others may represent integers, and still others may be real numbers.

20.1   Invariance

Nearly all rules in this chapter and in Chapters 21 and 30 show some sort of invariance with respect to certain transformations of the input. This is often a major asset in pattern recognition methods. We say a rule g_n is *invariant* under transformation T if

$$g_n(T(x); T(X_1), Y_1, \ldots, T(X_n), Y_n) = g_n(x; X_1, Y_1, \ldots, X_n, Y_n)$$

for all values of the arguments. In this sense, we may require *translation invariance*, *rotation invariance*, *linear translation invariance*, and *monotone transformation invariance* ($T(\cdot)$ maps each coordinate separately by a strictly increasing but possibly nonlinear function).

Monotone transformation invariance frees us from worries about the kind of measuring unit. For example, it would not matter whether earthquakes were measured on a logarithmic (Richter) scale or a linear scale. Rotation invariance matters of course in situations in which input data have no natural coordinate axis system. In many cases, data are of the ordinal form—colors and names spring to mind—and ordinal values may be translated into numeric values by creating bit vectors. Here, distance loses its physical meaning, and any rule that uses ordinal data perhaps mixed in with numerical data should be monotone transformation invariant.

Tree methods that are based upon perpendicular splits are usually (but not always) monotone transformation invariant and translation invariant. Tree methods based upon linear hyperplane splits are sometimes linear transformation invariant.

The partitions of space cause some problems if the data points can line up along hyperplanes. This is just a matter of housekeeping, of course, but the fact that some projections of X to a line have atoms or some components of X have atoms will make the proofs heavier to digest. For this reason only, we assume throughout this chapter that X has a density f. As typically no conditions are put on f in our consistency theorems, it will be relatively easy to generalize them to all distributions. The density assumption affords us the luxury of being able to say that, with probability one, no $d + 1$ points fall in a hyperplane, no d points fall in a hyperplane perpendicular to one axis, no $d - 1$ points fall in a hyperplane perpendicular to two axes, etcetera. If a rule is monotone transformation invariant, we can without harm transform all the data as follows for the purpose of analysis only. Let $f_1, \ldots, f_d$ be the marginal densities of X (see Problem 20.1), with corresponding distribution functions $F_1, \ldots, F_d$. Then replace in the data each X_i by $T(X_i)$, where

$$T(x) = \left(F_1(x^{(1)}), \ldots, F_d(x^{(d)}) \right).$$

Each component of $T(X_i)$ is now uniformly distributed on $[0, 1]$. Of course, as we do not know T beforehand, this device could only be used in the analysis. The transformation T will be called the *uniform marginal transformation*. Observe that the original density is now transformed into another density.

20.2   Trees with the X-Property

It is possible to prove the convergence of many tree classifiers all at once. What is needed, clearly, is a partition into small regions, yet, most majority votes should be over sufficiently large sample. In many of the cases considered, the form of the tree is determined by the X_i's only, that is, the labels Y_i do not play a role in constructing the partition, but they are used in voting. This is of course rather simplistic, but as a start, it is very convenient. We will say that the classification tree has the X-property, for lack of a better mnemonic. Let the leaf regions be $\{A_1, \ldots, A_N\}$ (with N possibly random). Define N_j as the number of X_i's falling in A_j. As the leaf regions form a partition, we have $\sum_{j=1}^{N} N_j = n$. By diam(A_j) we mean the diameter of the cell A_j, that is, the maximal distance between two points of A_j. Finally, decisions are taken by majority vote, so for $x \in A_j$, $1 \le j \le N$, the rule is

$$g_n(x) = \begin{cases} 1 & \text{if } \sum_{i:X_i \in A_j} Y_i > \sum_{i:X_i \in A_j} (1 - Y_i), x \in A_j \\ 0 & \text{otherwise.} \end{cases}$$

$A(x)$ denotes the set of the partition $\{A_1, \ldots, A_N\}$ into which x falls, and $N(x)$ is the number of data points falling in this set. Recall that the general consistency result given in Theorem 6.1 is applicable in such cases. Consider a natural classification tree as defined above and assume the X-property. Theorem 6.1 states that then $E\{L_n\} \to L^*$ if

(1) $\text{diam}(A(X)) \to 0$ in probability,
(2) $N(X) \to \infty$ in probability.

A more general, but also more complicated consistency theorem is proved in Chapter 21. Let us start with the simplest possible example. We verify the conditions of Theorem 6.1 for the k-spacing rule in one dimension. This rule partitions the real line by using the k-th, $2k$-th (and so on) order statistics (Mahalanobis, (1961); see also Parthasarathy and Bhattacharya (1961)).

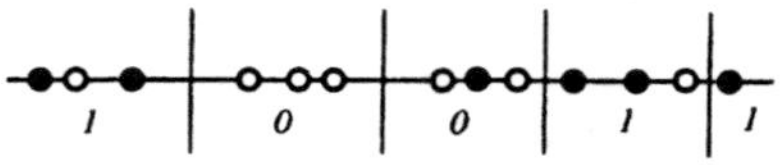

FIGURE 20.9. *A 3-spacing classifier.*

Formally, let $k < n$ be a positive integer, and let $X_{(1)}, X_{(2)}, \ldots, X_{(n)}$ be the *order statistics* of the data points. Recall that $X_{(1)}, X_{(2)}, \ldots, X_{(n)}$ are obtained by permuting $X_1, \ldots, X_n$ in such a way that $X_{(1)} \le X_{(2)} \le \cdots \le X_{(n)}$. Note that this ordering is unique with probability one as X has a density. We partition $\mathcal{R}$ into N intervals $A_1, \ldots, A_N$, where $N = \lceil n/k \rceil$, such that for $j = 1, \ldots, N - 1$, A_j satisfies

$$X_{(k(j-1)+1)}, \ldots, X_{(kj)} \in A_j,$$

and the rightmost cell A_N satisfies

$$X_{(k(N-1)+1)}, \ldots, X_{(n)} \in A_N.$$

We have not specified the endpoints of each cell of the partition. For simplicity, let the borders between A_j and A_{j+1} be put halfway between the rightmost data point in A_j and leftmost data point in A_{j+1}, $j = 1, \ldots, N - 1$.

The classification rule g_n is defined in the usual way:

$$g_n(x) = \begin{cases} 1 & \text{if } \sum_{i=1}^n I_{\{X_i \in A(x), Y_i = 1\}} > \sum_{i=1}^n I_{\{X_i \in A(x), Y_i = 0\}} \\ 0 & \text{otherwise.} \end{cases}$$

Theorem 20.1. *Let g_n be the k-spacing classifier given above. Assume that the distribution of X has a density f on $\mathcal{R}$. Then the classification rule g_n is consistent, that is, $\lim_{n \to \infty} E\{L_n\} = L^*$, if $k \to \infty$ and $k/n \to 0$ as n tends to infinity.*

REMARK. We discuss various generalizations of this rule in Chapter 21. $\square$

PROOF. We check the conditions of Theorem 6.1, as the partition has the X-property. Condition (2) is obvious from $k \to \infty$.

To establish condition (1), fix $\epsilon > 0$. Note that by the invariance of the rule under monotone transformations, we may assume without loss of generality that f is the uniform density on $[0, 1]$. Among the intervals $A_1, \ldots, A_N$, there are at most $1/\epsilon$ disjoint intervals of length greater than ϵ in $[0, 1]$. Thus,

$$P\{\text{diam}(A(X)) > \epsilon\}$$
$$\leq \quad \frac{1}{\epsilon} E \left\{ \max_{1 \leq j \leq N} \mu(A_j) \right\}$$
$$\leq \quad \frac{1}{\epsilon} E \left\{ \left(\max_{1 \leq j \leq N} \mu_n(A_j) + \max_{1 \leq j \leq N} |\mu(A_j) - \mu_n(A_j)| \right) \right\}$$
$$\leq \quad \frac{1}{\epsilon} \left(\frac{k}{n} + E \left\{ \sup_A |\mu(A) - \mu_n(A)| \right\} \right),$$

where the supremum is taken over all intervals in $\mathcal{R}$. The first term within the parentheses converges to zero by the second condition of the theorem, while the second term goes to zero by an obvious extension of the classical Glivenko-Cantelli theorem (Theorem 12.4). This completes the proof. $\square$

We will encounter several trees in which the partition is determined by a small fraction of the data, such as binary k-d trees and quadtrees. In these cases, condition (2) of Theorem 6.1 may be verified with the help of the following lemma:

Lemma 20.1. *Let $p_1, \ldots, p_k$ be a probability vector. Let $N_1, \ldots, N_k$ be multinomially distributed random variables with parameters n and $p_1, \ldots, p_k$. Then if the random variable X is independent of $N_1, \ldots, N_k$, and $P\{X = i\} = p_i$, we have for any M,*

$$P\{N_X \leq M\} \leq \frac{(2M + 4)k}{n}.$$

(Note: this probability goes to 0 if $k/n \to 0$ uniformly over all probability vectors with k components!)

PROOF. Let Z_i be binomial with parameters n and p_i. Then

$$P\{N_X \leq M\} \quad \leq \quad \sum_{i:np_i \leq 2M} p_i + \sum_{i:np_i > 2M} p_i P\{Z_i \leq M\}$$
$$\leq \quad \frac{2Mk}{n} + \sum_{i:np_i > 2M} p_i P\{Z_i - EZ_i \leq M - np_i\}$$
$$\leq \quad \frac{2Mk}{n} + \sum_{i:np_i > 2M} p_i P\left\{ Z_i - EZ_i \leq -\frac{EZ_i}{2} \right\}$$

$$\leq \frac{2Mk}{n} + \sum_{i:np_i > 2M} 4p_i \frac{\text{Var}\{Z_i\}}{(\text{E}\{Z_i\})^2}$$

(by Chebyshev's inequality)

$$\leq \frac{2Mk}{n} + \sum_{i:np_i > 2M} 4p_i \frac{1}{np_i}$$

$$\leq \frac{(2M + 4)k}{n}. \quad \square$$

The previous lemma implies that for any binary tree classifier constructed on the basis of $X_1, \ldots, X_k$ with $k + 1$ regions, $N(X) \to \infty$ in probability whenever $k/(n - k) \to 0$ (i.e., $k/n \to 0$). It suffices to note that we may take M arbitrarily large but fixed in Lemma 20.1. This remark saves us the trouble of having to verify just how large or small the probability mass of the region is. In fact, it also implies that we should not worry so much about regions with few data points. What matters more than anything else is the number of regions. Stopping rules based upon cardinalities of regions can effectively be dropped in many cases!

20.3   Balanced Search Trees

Balanced multidimensional search trees are computationally attractive. Binary trees with n leaves have $O(\log n)$ height, for example, when at each node, the size of every subtree is at least α times the size of the other subtree rooted at the parent, for some constant $\alpha > 0$. It is thus important to verify the consistency of balanced search trees used in classification. We again consider binary classification trees with the X-property and majority votes over the leaf regions. Take for example a tree in which we split every node perfectly, that is, if there are n points, we find the median according to one coordinate, and create two subtrees of sizes $\lfloor (n - 1)/2 \rfloor$ and $\lceil (n-1)/2 \rceil$. The median itself stays behind and is not sent down to the subtrees. Repeat this for k levels of nodes, at each level cutting along the next coordinate axe in a rotational manner. This leads to 2^k leaf regions, each having at least $n/2^k - k$ points and at most $n/2^k$ points. Such a tree will be called a *median tree.*

25%

25%

25%

25%

FIGURE 20.10. *Median tree with four leaf regions in* $\mathcal{R}^2$.

Setting up such a tree is very easy, and hence such trees may appeal to certain programmers. In hypothesis testing, median trees were studied by Anderson (1966).

Theorem 20.2. *Natural classifiers based upon median trees with k levels (2^k leaf regions) are consistent ($E\{L_n\} \to L^*$) whenever X has a density, if*

$$\frac{n}{k2^k} \to \infty \quad and \quad k \to \infty.$$

(Note: the conditions of k are fulfilled if $k \le \log_2 n - 2 \log_2 \log_2 n$, $k \to \infty$.)

We may prove the theorem by checking the conditions of Theorem 6.1. Condition (2) follows trivially by the fact that each leaf region contains at least $n/2^k - k$ points and the condition $n/(k2^k) \to \infty$. Thus, we need only verify the first condition of Theorem 6.1. To make the proof more transparent, we first analyze a closely related hypothetical tree, the *theoretical median tree*. Also, we restrict the analysis to $d = 2$. The multidimensional extension is straightforward. The theoretical median tree rotates through the coordinates and cuts each hyperrectangle precisely so that the two new hyperrectangles have equal μ-measure.

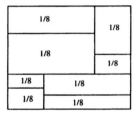

FIGURE 20.11. *Theoretical median tree with three levels of cuts.*

Observe that the rule is invariant under monotone transformations of the coordinate axes. Recall that in such cases there is no harm in assuming that the marginal distributions are all uniform on [0, 1]. We let $\{H_i, V_i\}$ denote the horizontal and vertical sizes of the rectangles after k levels of cuts. Of course, we begin with $H_1 = V_1 = 1$ when $k = 0$. We now show that, for the theoretical median tree, $\text{diam}(A(X)) \to 0$ in probability, as $k \to \infty$. Note that $\text{diam}(A(X)) \le H(X) + V(X)$, where $H(X)$ and $V(X)$ are the horizontal and vertical sizes of the rectangle $A(X)$. We show that if k is even,

$$E\{H(X) + V(X)\} = \frac{2}{2^{k/2}},$$

from which the claim follows. After the k-th round of splits, since all 2^k rectangles have equal probability measure, we have

$$E\{H(X) + V(X)\} = \sum_{i=1}^{2^k} \frac{1}{2^k}(H_i + V_i).$$

Apply another round of splits, all vertical. Then each term $\frac{1}{2^k}(H_i + V_i)$ spawns, so to speak, two new rectangles with horizontal and vertical sizes (H_i', V_i) and

(H_i''', V_i) with $H_i = H_i' + H_i'''$ that contribute

$$\frac{1}{2^{k+1}}(H_i' + V_i) + \frac{1}{2^{k+1}}(H_i''' + V_i) = \frac{1}{2^{k+1}}H_i + \frac{1}{2^k}V_i.$$

The next round yields horizontal splits, with total contribution now (see Figure 20.12)

$$\frac{1}{2^{k+2}}(H_i' + V_i' + H_i' + V_i'' + H_i''' + V_i''' + H_i''' + V_i'''') \;=\; \frac{1}{2^{k+2}}(2H_i + 2V_i)$$

$$=\; \frac{1}{2^{k+1}}(H_i + V_i).$$

Thus, over two iterations of splits, we see that $E\{H(X) + V(X)\}$ is halved, and the claim follows by simple induction.

We show now what happens in the real median tree when cuts are based upon a random sample. We deviate of course from the theoretical median tree, but consistency is preserved. The reason, seen intuitively, is that if the number of points in a cell is large, then the sample median will be close to the theoretical median, so that the shrinking-diameter property is preserved. The methodology followed here shows how one may approach the analysis in general by separating the theoretical model from the sample-based model.

PROOF OF THEOREM 20.2.   As we noted before, all we have to show is that $\mathrm{diam}(A(X)) \to 0$ in probability. Again, we assume without loss of generality that the marginals of X are uniform $[0, 1]$, and that $d = 2$. Again, we show that $E\{H(X) + V(X)\} \to 0$.

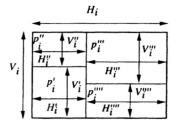

FIGURE 20.12.   *A rectangle after two rounds of splits.*

If a rectangle of probability mass p_i and sizes H_i, V_i is split into four rectangles as in Figure 20.12, with probability masses p_i', p_i'', p_i''', p_i'''', then the contribution $p_i(H_i + V_i)$ to $E\{H(X) + V(X)\}$ becomes

$$p_i'(H_i' + V_i') + p_i''(H_i'' + V_i'') + p_i'''(H_i''' + V_i''') + p_i''''(H_i'''' + V_i'''')$$

after two levels of cuts. This does not exceed

$$\frac{p_i}{2}(1 + \epsilon)(H_i + V_i),$$

if

$$\max(p_i' + p_i'', p_i''' + p_i'''') \le \frac{1}{2}\sqrt{1 + \epsilon}\, p_i,$$

$$\max(p_i', p_i'') \le \frac{1}{2}\sqrt{1 + \epsilon}(p_i' + p_i''),$$

and

$$\max(p_i''', p_i'''') \le \frac{1}{2}\sqrt{1 + \epsilon}(p_i''' + p_i''''),$$

that is, when all three cuts are within $(1/2)\sqrt{1 + \epsilon}$ of the true median. We call such "$(1/2)\sqrt{1 + \epsilon}$" cuts good. If *all* cuts are good, we thus note that in two levels of cuts, $E\{H(X) + V(X)\}$ is reduced by $(1+\epsilon)/2$. Also, all p_i's decrease at a controlled rate. Let G be the event that all $1 + 2 + \cdots + 2^{k-1}$ cuts in a median tree with k levels are good. Then, at level k, all p_i's are at most $\left(\sqrt{1 + \epsilon}/2\right)^k$. Thus,

$$\sum_{i=1}^{2^k} p_i(H_i + V_i) \le \left(\frac{\sqrt{1 + \epsilon}}{2}\right)^k \sum_{i=1}^{2^k}(H_i + V_i)$$

$$\le (3 + 2^{k/2+1})\left(\frac{\sqrt{1 + \epsilon}}{2}\right)^k,$$

since $\sum_{i=1}^{2^k}(H_i + V_i) \le 4 + 2 + 4 + 8 + \cdots + 2^{k/2} \le 2^{k/2+1} + 3$ if k is even. Hence, after k levels of cuts,

$$E\{H(X) + V(X)\} \le (3 + 2^{k/2+1})\, P\{G^c\} + (3 + 2^{k/2+1})\left(\frac{\sqrt{1 + \epsilon}}{2}\right)^k.$$

The last term tends to zero if ϵ is small enough. We bound $P\{G^c\}$ by 2^k times the probability that one cut is bad. Let us cut a cell with N points and probability content p in a given direction. A quick check of the median tree shows that given the position and size of the cell, the N points inside the cell are distributed in an i.i.d. manner according to the restriction of μ to the cell. After the cut, we have $\lfloor (N - 1)/2 \rfloor$ and $\lceil (N - 1)/2 \rceil$ points in the new cells, and probability contents p' and p''. It is clear that we may assume without loss of generality that $p = 1$. Thus, if all points are projected down in the direction of the cut, and F and F_N denote the distribution function and empirical distribution function of the obtained one-dimensional data, then

$$P\{\text{cut is not good}\,|\,N\}$$

$$\le P\left\{p' > \frac{\sqrt{1 + \epsilon}}{2} \text{ or } p'' > \frac{\sqrt{1 + \epsilon}}{2}\,\middle|\,N\right\}$$

$$\le P\left\{\text{for some } x,\ F(x) > \frac{\sqrt{1 + \epsilon}}{2},\ \text{and } F_N(x) \le \frac{1}{2}\,\middle|\,N\right\}$$

$$\leq \ \mathbf{P}\left\{\sup_x(F(x) - F_N(x)) > \frac{1}{2}\left(\sqrt{1+\epsilon} - 1\right)\Big| N\right\}$$

$$\leq \ 2\exp\left(-\frac{1}{2}N\left(\sqrt{1+\epsilon} - 1\right)^2\right)$$

(by Theorem 12.9)

$$\leq \ 2\exp\left(-\left(\frac{n}{2^{k+1}} - \frac{k}{2}\right)\left(\sqrt{1+\epsilon} - 1\right)^2\right)$$

(as $N \geq n/(2^k) - k$).

Hence, for n large enough,

$$\mathbf{P}\{G^c\} \leq 2^{k+1}e^{-(n/2^{k+2})\left(\sqrt{1+\epsilon} - 1\right)^2}$$

and $2^{k/2}\mathbf{P}\{G^c\} \to 0$ if $n/(k2^k) \to \infty$. $\square$

20.4   Binary Search Trees

The simplest trees to analyze are those whose structure depends in a straightforward way on the data. To make this point, we begin with the binary search tree and its multivariate extension, the k-d tree (see Cormen, Leiserson, and Rivest (1990) for the binary search tree; for multivariate binary trees, we refer to Samet (1990b)). A full-fledged k-d tree is defined as follows: we promote the first data point X_1 to the root and partition $\{X_2, \ldots, X_n\}$ into two sets: those whose first coordinate exceeds that of X_1, and the remaining points. Within each set, points are ordered by original index. The former set is used to build the right subtree of X_1 and the latter to construct the left subtree of X_1. For each subtree, the same construction is applied recursively with only one variant: at depth l in the tree, the $(l \bmod d + 1)$-st coordinate is used to split the data. In this manner, we rotate through the coordinate axes periodically.

Attach to each leaf two new nodes, and to each node with one child a second child. Call these new nodes *external nodes*. Each of the $n + 1$ external nodes correspond to a region of $\mathcal{R}^d$, and collectively the external nodes define a partition of $\mathcal{R}^d$ (if we define exactly what happens on the boundaries between regions).

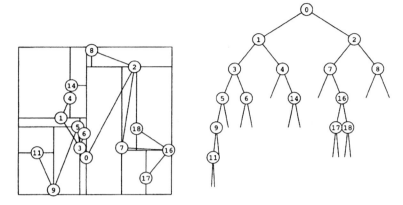

FIGURE 20.13. *A k-d tree of 15 random points on the plane and the induced partition.*

Put differently, we may look at the external nodes as the leaves of a new tree with $2n + 1$ nodes and declare this new tree to be our new binary classification tree. As there are $n + 1$ leaf regions and n data points, the natural binary tree classifier it induces is degenerate—indeed, all external regions contain very few points. Clearly, we must have a mechanism for trimming the tree to insure better populated leaves. Let us look at just three naïve strategies. For convenience, we assume that the data points determining the tree are not counted when taking a majority vote over the cells. As the number of these points is typically much smaller than n, this restriction does not make a significant difference.

(1) Fix $k < n$, and construct a k-d tree with k internal nodes and $k + 1$ external nodes, based on the first k data points $X_1, \ldots, X_k$. Classify by majority vote over all $k + 1$ regions as in natural classification tees (taking the data pairs $(X_{k+1}, Y_{k+1}), \ldots, (X_n, Y_n)$ into account). Call this the *chronological k-d tree*.

(2) Fix k and truncate the k-d tree to k levels. All nodes at level k are declared leaves and classification is again by majority vote over the leaf regions. Call this the *deep k-d tree*.

(3) Fix k and trim the tree so that each node represents at least k points in the original construction. Consider the maximal such tree. The number of regions here is random, with between 1 and n/k regions. Call this the *well-populated k-d tree*.

Let the leaf regions be $\{A_1, \ldots, A_N\}$ (with N possibly random), and denote the leaf nodes by $u_1, \ldots, u_N$. The strict *descendants* of u_i in the full k-d tree have indices that we will collect in an index set I_i. Define $|I_i| = N_i$. As the leaf regions form a partition, we have

$$\sum_{i=1}^{N} N_i = n - N,$$

because the leaf nodes themselves are not counted in I_i. Voting is by majority vote, so the rule is

$$g_n(x) = \begin{cases} 1 & \text{if } \sum_{i \in I_i} Y_i > \sum_{i \in I_i} (1 - Y_i), x \in A_i, \\ 0 & \text{otherwise.} \end{cases}$$

20.5   The Chronological k-d Tree

Here we have $N = k + 1$. Also, $(\mu(A_1), \ldots, \mu(A_{k+1}))$ are distributed as uniform spacings. That is, if $U_1, \ldots, U_k$ are i.i.d. uniform [0, 1] random variables defining $k + 1$ spacings $U_{(1)}, U_{(2)} - U_{(1)}, \ldots, U_{(k)} - U_{(k-1)}, 1 - U_{(k)}$ by their order statistics $U_{(1)} \leq U_{(2)} \leq \cdots \leq U_{(k)}$, then these spacings are jointly distributed as $(\mu(A_1), \ldots, \mu(A_{k+1}))$. This can be shown by induction. When U_{k+1} is added, U_{k+1} first picks a spacing with probability equal to the size of the spacing. Then it cuts that spacing in a uniform manner. As the same is true when the chronological k-d tree grows by one leaf, the property follows by induction on k.

Theorem 20.3. *We have* $E\{L_n\} \to L^*$ *for all distributions of X with a density for the chronological k-d tree classifier whenever*

$$k \to \infty, \qquad \text{and} \qquad k/n \to 0.$$

PROOF. We verify the conditions of Theorem 6.1. As the number of regions is $k + 1$, and the partition is determined by the first k data points, condition (2) immediately follows from Lemma 20.1 and the remark following it.

Condition (1) of Theorem 6.1 requires significantly more work. We verify condition (1) for $d = 2$, leaving the straightforward extension to $\mathcal{R}^d$, $d > 2$, to the reader. Throughout, we assume without loss of generality that the marginal distributions are uniform [0, 1]. We may do so by the invariance properties discussed earlier. Fix a point $x \in \mathcal{R}^d$. We insert points $X_1, \ldots, X_k$ into an initially empty k-d tree and let $R_1, \ldots, R_k$ be the rectangles containing x just after these points were inserted. Note that $R_1 \supseteq R_2 \supseteq \cdots \supseteq R_k$. Assume for simplicity that the integer k is a perfect cube and set $l = k^{1/3}$, $m = k^{2/3}$. Define the distances from x to the sides of R_i by H_i, H_i', V_i, V_i' (see Figure 20.14).

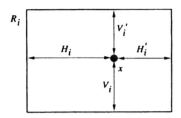

FIGURE 20.14. *A rectangle R_i containing x with its distances to the sides.*

We construct a sequence of events that forces the diameter of R_k to be small with high probability. l⦁t ϵ be a small positive number to be specified later.

Denote the four squares with opposite vertices x, $x + (\pm\epsilon, \pm\epsilon)$ by C_1, C_2, C_3, C_4. Then define the following events:

$$E_1 = \bigcap_{i=1}^{4} \{C_i \cap \{X_1, \ldots, X_l\} \neq \emptyset\},$$

$$E_2 = \{\max(V_l, V_l', H_l, H_l') \leq \epsilon\},$$

$$E_3 = \{\min(\max(V_l, V_l'), \max(H_l, H_l')) \leq \epsilon < \max(V_l, V_l', H_l, H_l')\},$$

$$E_4 = \{\text{at least three of } V_m, V_m', H_m, H_m' \text{ are } \leq \epsilon\},$$

$$E_5 = \{\max(V_m, V_m', H_m, H_m') \leq \epsilon\},$$

$$E_6 = \{\max(V_k, V_k', H_k, H_k') \leq \epsilon\}.$$

If E_2, E_5, or E_6 hold, then $\mathrm{diam}(R_k) \leq \epsilon\sqrt{8}$. Assume that we find a set $B \subset \mathcal{R}^d$ such that $\mathbf{P}\{X \in B\} = 1$, and for all $x \in B$, and sufficiently small $\epsilon > 0$, $\mathbf{P}\{E_2 \cup E_5 \cup E_6\} \to 1$ as $k \to \infty$. Then, by the Lebesgue dominated convergence theorem, $\mathrm{diam}(A(X)) \to 0$ in probability, and condition (1) of Theorem 6.1 would follow. In the remaining part of the proof we define such a set B, and show that $\mathbf{P}\{E_2 \cup E_5 \cup E_6\} \to 1$. This will require some work.

We define the set B in terms of the density f of X. $x \in B$ if and only if

(1) $\min_{1 \leq i \leq 4} \int_{C_i} f(z)dz > 0$ for all $\epsilon > 0$ small enough;

(2) $\displaystyle\inf_{\substack{\text{rectangles } R \text{ containing } x, \\ \text{of diameter } \leq \epsilon\sqrt{8}}} \frac{\int_R f(z)dz}{\lambda(R)} \geq \frac{f(x)}{2}$ for all $\epsilon > 0$ small enough;

(3) $f(x) \geq 0$.

That $\mathbf{P}\{X \in B\} = 1$ follows from a property of the support (Problem 20.6), a corollary of the Jessen-Marcinkiewicz-Zygmund theorem (Corollary A.2; this implies (2) for almost all x), and the fact that for μ-almost all x, $f(x) > 0$.

It is easy to verify the following:

$$\mathbf{P}\{E_6^c\} = \mathbf{P}\{E_1^c\}$$
$$+ \mathbf{P}\{E_1 \cap E_2^c \cap E_3^c\}$$

$$+ \, \mathbf{P}\{E_1 \cap E_2^c \cap E_3 \cap E_4^c\}$$
$$+ \, \mathbf{P}\{E_1 \cap E_2^c \cap E_3 \cap E_4 \cap E_5^c \cap E_6^c\}.$$

We show that each term tends to 0 at $x \in B$.

TERM 1. By the union bound,

$$
\begin{aligned}
\mathbf{P}\{E_1^c\} \; &\leq \; \sum_{i=1}^{4} \mathbf{P}\{X_1 \notin C_i, \ldots, X_l \notin C_i\} \\[2mm]
&\leq \; \sum_{i=1}^{4} (1 - \mu(C_i))^l \\[2mm]
&\leq \; \exp\left(-l \min_{1 \leq i \leq 4} \mu(C_i)\right) \\[2mm]
&\rightarrow \; 0,
\end{aligned}
$$

by part (1) of the definition of B.

TERM 2. $E_1 \subseteq E_3$ by a simple geometric argument. Hence $\mathbf{P}\{E_1 \cap E_3^c\} = 0$.

TERM 3. To show that $\mathbf{P}\{E_1 \cap E_2^c \cap E_3 \cap E_4^c\} \rightarrow 0$, we assume without loss of generality that $\max(V_l, V_l') \leq \epsilon$ while $\max(H_l, H_l') = a > \epsilon$. Let $\{X_i'\}$ be a subset of $\{X_i, l < i \leq m\}$ consisting of those X_i's that fall in R_l. We introduce three notions in this sequence: first, Z_i is the absolute value of the difference of the $x^{(2)}$-coordinates of x and X_i'. Let W_i be the absolute value of the difference of the $x^{(1)}$-coordinates of x and X_i'. We re-index the sequence X_i' (and W_i and Z_i) so that i runs from 1 to N, where

$$N = \sum_{i=l+1}^{m} I_{\{X_i \in R_l\}}.$$

To avoid trivialities, assume that $N \geq 1$ (this will be shown to happen with probability tending to one). Call X_i' a *record* if $Z_i = \min(Z_1, \ldots, Z_i)$. Call X_i' a *good point* if $W_i \leq \epsilon$. An X_i' causes $\min(H_m, H_m') \leq \epsilon$ if that X_i' is a good point and a record, and if it defines a vertical cut. The alternating nature of the cuts makes our analysis a bit heavier than needed. We show here what happens when all directions are picked independently of each other, leaving the rotating-cuts case to the reader (Problem 20.8). Thus, if we set

$$S_i = I_{\{X_i' \text{ is a record, } X_i' \text{ defines a vertical cut}\}},$$

we have,

$$E_1 \cap E_2^c \cap E_3 \cap E_4^c \subseteq \{N = 0\} \cup \left\{\left\{\sum_{i=1}^{N} S_i I_{\{W_i \leq \epsilon\}} = 0\right\} \cap \{N = 0\}\right\}.$$

Re-index again and let $X'_1, X'_2, \ldots$ all be records. Note that given X'_i, X'_{i+1} is distributed according to f restricted to the rectangle R' of

height $\min(V_l, Z_i)$ above x,

height $\min(V'_l, Z_i)$ below x,

width H_l to the left of x,

width H'_l to the right of x.

Call these four quantities v, v', h, h', respectively. Then

$$\mathbf{P}\{W_{i+1} \le \epsilon | X'_i\} \ge \frac{(v + v')\epsilon f(x)/2}{v + v'} = \frac{\epsilon f(x)}{2}$$

because the marginal distribution of an independent X_1 is uniform and thus, $\mathbf{P}\{X_1 \in R'|R'\} \le v + v'$, while, by property (2) of B, $\mathbf{P}\{X_1 \in R', W_1 \le \epsilon|R'\} \ge (v + v')\epsilon f(x)/2$.

Recall the re-indexing. Let M be the number of records (thus, M is the length of our sequence X'_i). Then

$$E_1 \cap E_2^c \cap E_3 \cap E_4^c \cap \{N > 0\} \subseteq \left\{ \sum_{i=1}^{M} I_{\{W_i \le \epsilon\}} I_{\{X'_i \text{ is a vertical cut}\}} = 0 \right\}.$$

But as cuts have independently picked directions, and since $\mathbf{P}\{W_{i+1} \le \epsilon | X'_i\} \ge \epsilon f(x)/2$, we see that

$$\mathbf{P}\{E_1 \cap E_2^c \cap E_3 \cap E_4^c, N > 0\} \le \mathbf{E}\left\{ \left(1 - \frac{\epsilon f(x)}{4}\right)^M I_{\{N>0\}} \right\}.$$

We rewrite $M = \sum_{i=1}^{N} I_{\{X'_i \text{ is a record}\}}$ and recall that the indicator variables in this sum are independent and are of mean $1/i$ (Problem 20.9). Hence, for $c > 0$,

$$
\begin{aligned}
\mathbf{E}\{c^M\} &= \mathbf{E}\left\{ \prod_{i=1}^{N} \left(\frac{c}{i} + 1 - \frac{1}{i} \right) \right\} \\
&\le \mathbf{E}\left\{ e^{-(1-c)\sum_{i=1}^{N} 1/i} \right\} \quad \text{(use } 1 - x \le e^{-x}) \\
&\le \mathbf{E}\left\{ \frac{1}{(N + 1)^{1-c}} \right\} \quad \text{(use } \sum_{i=1}^{N} 1/i \ge \log(N + 1)).
\end{aligned}
$$

The latter formula remains valid even if $N = 0$. Thus, with $c = 1 - \epsilon f(x)/4$,

$$
\begin{aligned}
\mathbf{P}\{E_1 \cap E_2^c \cap E_3 \cap E_4^c\} &\le \mathbf{E}\left\{ \left(1 - \frac{\epsilon f(x)}{4}\right)^M \right\} \\
&\le \mathbf{E}\left\{ \frac{1}{(N + 1)^{1-c}} \right\}.
\end{aligned}
$$

N is binomial with parameters $(m - l, \mu(R_l))$. We know from the introduction of this section that $\mu(R_l)$ is distributed as the minimum of l i.i.d. uniform $[0, 1]$ random variables. Thus, for $\delta > 0$, $\mathbf{E}\{\mu(R_l)\} = 1/(l+1)$, and $\mathbf{P}\{\mu(R_l) < \delta/l\} \le \delta$. Hence,

$$\mathbf{E}\left\{\frac{1}{(N+1)^{1-c}}\right\}$$

$$\le \quad \mathbf{P}\{\mu(R_l) < \delta/l\} + \mathbf{E}\left\{\frac{1}{(\text{Binomial}(m-l, \delta/l) + 1)^{1-c}}\right\}$$

$$\le \quad \delta + \left(\frac{l}{2(m-l)\delta}\right)^{1-c} + \mathbf{P}\left\{\text{Binomial}(m-l, \delta/l) < \frac{(m-l)\delta}{2l}\right\}.$$

The first term is small by choice of δ. The second one is $O\left(k^{-(1-c)/3}\right)$. The third one is bounded from above, by Chebyshev's inequality, by

$$\frac{(m-l)(\delta/l)(1 - \delta/l)}{((m-l)\delta/2l)^2} \le \frac{4l}{(m-l)\delta} \to 0.$$

TERM 4. This term is handled exactly as Term 3. Note, however, that l and m now become m and k respectively. The convergence to 0 requires now $m/(k - m) \to 0$, which is still the case.

This concludes the proof of Theorem 20.3. $\square$

20.6   The Deep k-d Tree

Theorem 20.4. *The deep k-d tree classifier is consistent (i.e., $\mathbf{E}\{L_n\} \to L^*$) for all distributions such that X has a density, whenever*

$$\lim_{n \to \infty} k = \infty \quad and \quad \limsup_{n \to \infty} \frac{k}{\log n} < 2.$$

PROOF. In Problem 20.10, you are asked to show that $k \to \infty$ implies $\text{diam}(A(X)) \to 0$ in probability. Theorem 20.3 may be invoked here. We now show that the assumption $\limsup_{n \to \infty} k/\log n < 2$ implies that $N(X) \to \infty$ in probability. Let D be the depth (distance from the root) of X when X is inserted into a k-d tree having n elements. Clearly, $N(X) \ge D - k$, so it suffices to show that $D - k \to \infty$ in probability. We know that $D/(2 \log n) \to 1$ in probability (see, e.g., Devroye (1988a) and Problem 20.10). This concludes the proof of the theorem. $\square$

20.7   Quadtrees

Quadtrees or hyperquadtrees are unquestionably the most prominent trees in computer graphics. Easy to manipulate and compact to store, they have found their way into mainstream computer science. Discovered by Finkel and Bentley (1974) and surveyed by Samet (1984), (1990a), they take several forms. We are given d-dimensional data $X_1, \ldots, X_n$. The tree is constructed as the k-d tree. In particular, X_1 becomes the root of the tree. It partitions $X_2, \ldots, X_n$ into 2^d (possibly empty) sets according to membership in one of the 2^d (hyper-) quadrants centered at X_1 (see Figure 20.15).

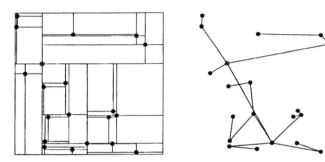

FIGURE 20.15. *Quadtree and the induced partition of* $\mathcal{R}^2$. *The points on the right are shown in the position in space. The root is specially marked.*

The partitioning process is repeated at the 2^d child nodes until a certain stopping rule is satisfied. In analogy with the k-d tree, we may define the chronological quadtree (only k splits are allowed, defined by the first k points $X_1, \ldots, X_k$), and the deep quadtree (k levels of splits are allowed). Other, more balanced versions may also be introduced. Classification is by majority vote over all (X_i, Y_i)'s—with $k < i \le n$ in the chronological quadtree—that fall in the same region as x. Ties are broken in favor of class 0. We will refer to this as the (chronological, deep) quadtree classifier.

Theorem 20.5. *Whenever X has a density, the chronological quadtree classifier is consistent ($\mathbf{E}\{L_n\} \to L^*$) provided that $k \to \infty$ and $k/n \to 0$.*

PROOF. Assume without loss of generality that all marginal distributions are uniform $[0, 1]$. As the X-property holds, Theorem 6.1 applies. By Lemma 20.1, since we have $k(2^d - 1) + 1$ external regions,

$$\mathbf{P}\{N(X) \le M\} \le \frac{(2M + 4)(k(2^d - 1) + 1)}{n - k} \to 0$$

for all $M > 0$, provided that $k/n \to 0$. Hence, we need only verify the condition diam$(A(X)) \to 0$ in probability. This is a bit easier than in the proof of Theorem 20.3 for the k-d tree and is thus left to the reader (Problem 20.18). $\square$

REMARK. Full-fledged random quadtrees with n nodes have expected height $O(\log n)$ whenever X has a density (see, e.g., Devroye and Laforest (1990)). With k nodes, every region is thus reached in only $O(\log k)$ steps on the average. Furthermore, quadtrees enjoy the same monotone transformation invariance that we observed for k-d trees and median trees. $\square$

20.8   Best Possible Perpendicular Splits

For computational reasons, classification trees are most often produced by determining the splits recursively. At a given stage of the tree-growing algorithm, some criterion is used to determine which node of the tree should be split next, and where the split should be made. As these criteria typically use all the data, the resulting trees no longer have the X-property. In this section we examine perhaps the most natural criterion. In the following sections we introduce some alternative splitting criteria.

A binary classification tree can be obtained by associating with each node a splitting function $\phi(x)$ obtained in a top-down fashion from the data. For example, at the root, we may select the function

$$\phi(x) = s x^{(i)} - \alpha,$$

where i, the component cut, $\alpha \in \mathcal{R}$, the threshold, and $s \in \{-1, +1\}$, a polarization, are all dependent upon the data. The root then splits the data $D_n = \{(X_1, Y_1), \ldots, (X_n, Y_n)\}$ into two sets, D'_n, D''_n, with $D'_n \cup D''_n = D_n$, $|D'_n| + |D''_n| = n$, such that

$$D'_n = \{(x, y) \in D_n : \phi(x) \geq 0\},$$
$$D''_n = \{(x, y) \in D_n : \phi(x) < 0\}.$$

A decision is made whether to split a node or not, and the procedure is applied recursively to the subtrees. Natural majority vote decisions are taken at the leaf level. All such trees will be called *perpendicular splitting trees*.

In Chapter 4, we introduced univariate Stoller splits, that is, splits that minimize the empirical error. This could be at the basis of a perpendicular splitting tree. One realizes immediately that the number of possibilities for stopping is endless. To name two, we could stop after k splitting nodes have been defined, or we could make a tree with k full levels of splits (so that all leaves are at distance k from the root). We first show that for $d > 1$, any such strategy is virtually doomed to fail. To make this case, we will argue on the basis of distribution functions only. For convenience, we consider a two-dimensional problem. Given a rectangle R now assigned to one class $y \in \{0, 1\}$, we see that the current probability of error in R, before splitting is $P\{X \in R, Y \neq y\}$. Let R' range over all rectangles of the form $R \cap ((-\infty, \alpha] \times \mathcal{R})$, $R \cap ([\alpha, \infty) \times \mathcal{R})$, $R \cap (\mathcal{R} \times (-\infty, \alpha])$, or $R \cap (R \times [\alpha, \infty))$, and let $R'' = R - R'$. Then after a split based upon (R', R''), the probability of error over the rectangle R is

$$P\{X \in R', Y = 1\} + P\{X \in R'', Y = 0\}$$

as we assign class 0 to R' and class 1 to R''. Call $\Delta(R)$ the decrease in probability of error if we minimize over all R'. Compute $\Delta(R)$ for all leaf rectangles and then proceed to split that rectangle (or leaf) for which $\Delta(R)$ is maximal. The data-based rule based upon this would proceed similarly, if $P\{A\}$ is replaced everywhere by the empirical estimate $(1/n) \sum_{i=1}^{n} I_{\{(X_i, Y_i) \in A\}}$, where A is of the form $R' \times \{1\}$, $R'' \times \{0\}$, or $R \times \{1 - y\}$, as the case may be.

Let us denote by $L_0, L_1, L_2, \ldots$ the sequence of the overall probabilities of error for the theoretically optimal sequence of cuts described above. Here we start with $R = \mathcal{R}^2$ and $y = 0$, for example. For fixed $\epsilon > 0$, we now construct a simple example in which

$$L^* = 0,$$

and

$$L_k \downarrow \frac{1 - \epsilon}{2} \quad \text{as } k \to \infty.$$

Thus, applying the best split incrementally, even if we use the true probability of error as our criterion for splitting, is not advisable.

The example is very simple: X has uniform distribution on $[0, 1]^2$ with probability ϵ and on $[1, 2]^2$ with probability $1 - \epsilon$. Also, Y is a deterministic function of X, so that $L^* = 0$.

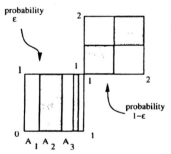

FIGURE 20.16. *Repeated Stoller splits are not consistent in this two-dimensional example. Cuts will always be made in the leftmost square.*

The way Y depends on X is shown in Figure 20.16: $Y = 1$ if

$$X \in [1, 3/2]^2 \cup [3/2, 2]^2 \cup (A_2 \cup A_4 \cup A_6 \ldots) \times [0, 1],$$

where

$$A_1 = [0, 1/4),$$
$$A_2 = [1/4, 1/4 + 3/8),$$
$$A_3 = [1/4 + 3/8, 1/4 + 3/8 + 3/16),$$
$$\vdots$$
$$A_k = [1/4 + 3/8 + \cdots + 3/2^k, 1/4 + 3/8 + \cdots + 3/2^{k+1}),$$

and so forth. We verify easily that $P\{Y = 1\} = 1/2$. Also, the error probability before any cut is made is $L_0 = 1/2$. The best split has $R' = (-\infty, 1/4) \times \mathcal{R}$ so that A_1 is cut off. Therefore, $L_1 = \epsilon/4 + (1 - \epsilon)/2$. We continue and split off A_2 and so forth, leading to the tree of Figure 20.17.

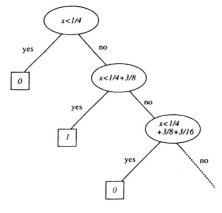

FIGURE 20.17. *The tree obtained by repeated Stoller splits.*

Verify that $L_2 = \epsilon/8 + (1 - \epsilon)/2$ and in general that

$$L_k = \frac{\epsilon}{2^{k+1}} + \frac{1 - \epsilon}{2} \downarrow \frac{1 - \epsilon}{2},$$

as claimed.

20.9   Splitting Criteria Based on Impurity Functions

In 1984, Breiman, Friedman, Olshen, and Stone presented their CART program for constructing classification trees with perpendicular splits. One of the key ideas in their approach is the notion that trees should be constructed from the bottom up, by combining small subtrees. The starting point is a tree with $n + 1$ leaf regions defined by a partition of the space based on the n data points. Such a tree is much too large and is pruned by some methods that will not be explored here. When constructing a starting tree, a certain splitting criterion is applied recursively. The criterion determines which rectangle should be split, and where the cut should be made. To keep the classifier invariant under monotone transformation of the coordinate axes, the criterion should only depend on the coordinatewise ranks of the points, and their labels. Typically the criterion is a function of the numbers of points labeled by 0 and 1 in the rectangles after the cut is made. One such class of criteria is described here.

Let $\alpha \in \mathcal{R}$, and let i be a given coordinate ($1 \le i \le d$). Let R be a hyperrectangle to be cut. Define the following quantities for a split at α, perpendicular to the i-th coordinate:

$$\mathcal{X}_i(R, \alpha) = \left\{ (X_j, Y_j) : X_j \in R, X_j^{(i)} \le \alpha \right\},$$

$$\mathcal{X}_i^c(R, \alpha) = \left\{ (X_j, Y_j) : X_j \in R, X_j^{(i)} > \alpha \right\}$$

are the sets of pairs falling to the left and to the right of the cut, respectively.

$$N_i(R, \alpha) = |\mathcal{X}_i(R, \alpha)|, \quad N_i'(R, \alpha) = |\mathcal{X}_i^c(R, \alpha)|$$

are the numbers of such pairs. Finally, the numbers of points with label 0 and label 1 in these sets are denoted, respectively, by

$$
\begin{aligned}
N_{i,0}(R, \alpha) &= \left| \mathcal{X}_i(R, \alpha) \cap \{(X_j, Y_j) : Y_j = 0\} \right|, \\
N_{i,1}(R, \alpha) &= \left| \mathcal{X}_i(R, \alpha) \cap \{(X_j, Y_j) : Y_j = 1\} \right|, \\
N_{i,0}'(R, \alpha) &= \left| \mathcal{X}_i^c(R, \alpha) \cap \{(X_j, Y_j) : Y_j = 0\} \right|, \\
N_{i,1}'(R, \alpha) &= \left| \mathcal{X}_i^c(R, \alpha) \cap \{(X_j, Y_j) : Y_j = 1\} \right|.
\end{aligned}
$$

Following Breiman, Friedman, Olshen, and Stone (1984), we define an *impurity function* for a possible split (i, α) by

$$I_i(R, \alpha) = \psi \left(\frac{N_{i0}}{N_{i0} + N_{i1}}, \frac{N_{i1}}{N_{i0} + N_{i1}} \right) N_i + \psi \left(\frac{N_{i0}'}{N_{i0}' + N_{i1}'}, \frac{N_{i1}'}{N_{i0}' + N_{i1}'} \right) N_i',$$

where we dropped the argument (R, α) throughout. Here ψ is a nonnegative function with the following properties:

(1) $\psi \left(\frac{1}{2}, \frac{1}{2} \right) \ge \psi(p, 1 - p)$ for any $p \in [0, 1]$;

(2) $\psi(0, 1) = \psi(1, 0) = 0$;

(3) $\psi(p, 1 - p)$ increases in p on $[0, 1/2]$ and decreases in p on $[1/2, 1]$.

A rectangle R is split at α along the i-th coordinate if $I_i(R, \alpha)$ is minimal. I_i penalizes splits in which the subregions have about equal proportions from both classes. Examples of such functions ψ include

(1) The *entropy function* $\psi(p, 1 - p) = -p \log p - (1 - p) \log(1 - p)$ (Breiman et al. (1984, pp.25,103)).

(2) The *Gini function* $\psi(p, 1 - p) = 2p(1 - p)$, leading to the Gini index of diversity I_i. (Breiman et al. (1984, pp.38,103)).

(3) The *probability of misclassification* $\psi(p, 1 - p) = \min(p, 1 - p)$. In this case the splitting criterion leads to the empirical Stoller splits studied in the previous section.

We have two kinds of splits:

(A) The *forced split*: force a split along the i-th coordinate, but minimize $I_i(\alpha, R)$ over all α and rectangles R.

(B) The *free split*: choose the most advantageous coordinate for splitting, that is, minimize $I_i(\alpha, R)$ over all α, i, and R.

Unfortunately, regardless of which kind of policy we choose, there are distributions for which no splitting based on an impurity function leads to a consistent classifier. To see this, note that the two-dimensional example of the previous section applies to all impurity functions. Assume that we had infinite sample size. Then $I_i(\alpha, R)$ would approach $a\psi(p, 1 - p) + b\psi(q, 1 - q)$, where p is the probability content of R' (one of the rectangles obtained after the cut is made), b is that of R'' (the other rectangle), $p = P\{Y = 1 | X \in R'\}$ and $q = P\{Y = 1 | X \in R''\}$. If X is uniformly distributed on the checkerboard shown in Figure 20.18, regardless where we try to cut, $p = q = 1/2$, and every cut seems to be undesirable.

FIGURE 20.18. *No cut decreases the value of the impurity function.*

This simple example may be made more interesting by mixing it with a distribution with much less weight in which $x^{(1)}$- and $x^{(2)}$-direction splits are alternatingly encouraged all the time. Therefore, impurity functions should be avoided in their raw form for splitting. This may explain partially why in CART, the original tree is undesirable and must be pruned from the bottom up. See Problems 20.22 to 20.24 for more information. In the next section and in the last section of this chapter we propose other ways of growing trees with much more desirable properties.

The derivation shown above does not indicate that the empirical version will not work properly, but simple versions of it will certainly not. See Problem 20.22.

REMARK. MALICIOUS SPLITTING. The impurity functions suggested above all avoid leaving the proportions of zeros and ones intact through splitting. They push towards more homogeneous regions. Assume now that we do the opposite. Through such splits, we can in fact create hyperplane classification trees that are globally poor, that is, that are such that every trimmed version of the tree is also a poor classifier. Such splitting methods must of course use the Y_i's. Our example shows that any general consistency theorem for hyperplane classification trees must come with certain restrictions on the splitting process—the X property is good; sometimes it is necessary to force cells to shrink; sometimes the position of the split is restricted by empirical error minimization or some other criterion.

The ham-sandwich theorem (see Edelsbrunner (1987)) states that given $2n$ class-0 points and $2m$ class-1 points in $\mathcal{R}^d$, $d > 1$, there exists a hyperplane cut that leaves n class-0 points and m class-1 points in each halfspace. So, assume that X has a density and that $p = P\{Y = 1\} = 1/2$. In a sample of size n, let y be the majority class (ties are broken in favor of class 0).

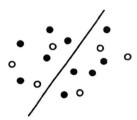

FIGURE 20.19. *Ham-sandwich cut: Each halfspace contains exactly half the points from each class.*

Regardless of the sample make-up, if $y = 0$, we may construct a hyperplane-tree classifier in which, during the construction, every node represents a region in which the majority vote would be 0. This property has nothing to do with the distribution of (X, Y), and therefore, for any trimmed version of the tree classifier,

$$L_n \geq p I_{\{y=0\}}$$

and $P\{L_n \geq p\} \geq P\{y = 0\} \geq 1/2$ if $p = 1/2$. Obviously, as we may take $L^* = 0$, these classifiers are hopeless. $\square$

BIBLIOGRAPHIC REMARKS. Empirical Stoller splits without forced rotation were recommended by Payne and Meisel (1977) and Rounds (1980), but their failure to be universally good was noted by Gordon and Olshen (1978). The last two authors recommended a splitting scheme that combined several ideas, but roughly speaking, they perform empirical Stoller splits with forced rotation through the coordinate axes (Gordon and Olshen (1978), (1980)). Other splitting criteria include the AID criterion of Morgan and Sonquist (1963), which is a predecessor of the Gini index of diversity used in CART (Breiman, Friedman, Olshen, and Stone (1984), see also Gelfand and Delp (1991), Guo and Gelfand (1992) Gelfand, Ravishankar, and Delp (1989), (1991), and Ciampi (1991)). Michel-Briand and Milhaud (1994) also observed the failure of multivariate classification trees based on the AID criterion.

The Shannon entropy or modifications of it are recommended by Talmon (1986), Sethi and Sarvarayudu (1982), Wang and Suen (1984), Goodman and Smyth (1988), and Chou (1991). Permutation statistics are used in Li and Dubes (1986), still without forced rotations through the coordinate axes. Quinlan (1993) has a more involved splitting criterion. A general discussion on tree splitting may be found in the paper by Sethi (1991). A class of impurity functions is studied in Burshtein, Della Pietra, Kanevsky, and Nádas (1992). Among the pioneers of tree splitting (with perpendicular cuts) are Sebestyen (1962), and Henrichon and Fu (1969). For related work, we refer to Stoffel (1974), Sethi and Chatterjee (1977), Argentiero, Chin and Beaudet (1982), You and Fu (1976), Anderson and Fu (1979), Qing-Yun and Fu (1983), Hartmann, Varshney, Mehrotra, and Gerberich (1982), and Casey and Nagy (1984). References on nonperpendicular splitting methods are given below in the section on BSP trees.

20.10   A Consistent Splitting Criterion

There is no reason for pessimism after the previous sections. Rest assured, there are several consistent splitting strategies that are fully automatic and depend only upon the populations of the regions. In this section we provide a solution for the simple case when X is univariate and nonatomic. It is possible to generalize the method for $d > 1$ if we force cuts to alternate directions. We omit here the detailed analysis for multidimensional cases for two reasons. First, it is significantly heavier than for $d = 1$. Secondly, in the last section of this chapter we introduce a fully automatic way of building up consistent trees, that is, without forcing the directions of the splits.

To a partition $A_1, \ldots, A_N$ of $\mathcal{R}$, we assign the quantity

$$Q = \sum_{i=1}^{N} N_0(A_i) N_1(A_i),$$

where

$$N_0(A) = \sum_{j=1}^{n} I_{\{X_j \in A, Y_j = 0\}}, \qquad N_1(A) = \sum_{j=1}^{n} I_{\{X_j \in A, Y_j = 1\}}$$

are the respective numbers of points labeled with 0 and 1 falling in the region A. The tree-growing algorithm starts with the trivial partition $\{\mathcal{R}\}$, and at each step it makes a cut that yields the minimal value of Q. It proceeds recursively until the improvement in the value of Q falls below a threshold Δ_n.

REMARK. Notice that this criterion always splits a cell that has many points from both classes (see the proof of the theorem below). Thus, it avoids the anomalies of impurity-function criteria described in the previous section. On the other hand, it does not necessarily split large cells, if they are almost homogeneous. For a comparison, recall that the Gini criterion minimizes the quantity

$$Q' = \sum_{i=1}^{N} \frac{N_0(A_i) N_1(A_i)}{N_0(A_i) + N_1(A_i)},$$

thus favoring cutting cells with very few points. We realize that the criterion Q introduced here is just one of many with similarly good properties, and albeit probably imperfect, it is certainly one of the simplest. $\square$

Theorem 20.6. *Let X have a nonatomic distribution on the real line, and consider the tree classifier obtained by the algorithm described above. If the threshold satisfies*

$$\frac{\Delta_n}{n} \to \infty \quad and \quad \frac{\Delta_n}{n^2} \to 0,$$

then the classification rule is strongly consistent.

PROOF. There are two key properties of the algorithm that we exploit:

PROPERTY 1. If $\min(N_0(A), N_1(A)) > \sqrt{2\Delta_n}$ for a cell A, then A gets cut by the algorithm. To see this, observe that (dropping the argument A from the notation) if $N_0 \geq N_1$, and we cut A so that the number of 0-labeled points in the two child regions differ by at most one, then the contribution of these two new regions to Q is

$$\lceil N_0/2 \rceil N_1' + \lfloor N_0/2 \rfloor N_1'' \leq \lceil N_0 N_1/2 \rceil,$$

where N_1' and N_1'' are the numbers of class-1 points in the two child regions. Thus, the decrease of Q if A is split is at least $\lfloor N_0 N_1/2 \rfloor$. If $\min(N_0, N_1) > \sqrt{2\Delta_n}$, then $\Delta_n \leq \lfloor N_0 N_1/2 \rfloor$, and a cut is made.

PROPERTY 2. No leaf node has less than Δ_n/n points. Assume that after a region is cut, in one of the child regions, the total number of points is $N_0' + N_1' \leq k$. Then the improvement in Q caused by the split is bounded by

$$N_0 N_1 - (N_0' N_1' + N_0'' N_1'') \leq N_0 N_1 - (N_0 - k)(N_1 - k) \leq (N_0 + N_1)k \leq nk.$$

Therefore, if $k < \Delta_n/n$, then the improvement is smaller than Δ_n. Thus, no cut is made that leaves behind a child region with fewer than Δ_n/n points.

It follows from Property 1 that if a leaf region has more than $4\sqrt{2\Delta_n}$ points, then since the class in minority has less than $\sqrt{2\Delta_n}$ points in it, it may be cut into intervals containing between $2\sqrt{2\Delta_n}$ and $4\sqrt{2\Delta_n}$ points without altering the decision, since the majority vote within each region remains the same.

Summarizing, we see that the tree classifier is equivalent to a classifier that partitions the real line into intervals, each containing at least Δ_n/n, and at most $4\sqrt{2\Delta_n}$ data points. Thus, in this partition, each interval has a number of points growing to infinity as $o(n)$. We emphasize that the number of points in a region of the studied tree classifier may be large, but such regions are almost homogeneous, and therefore the classifier is equivalent to another classifier which has $o(n)$ points in each region. Consistency of such partitioning classifiers is proved in the next chapter—see Theorem 21.3. □

20.11   BSP Trees

Binary space partition trees (or BSP trees) partition $\mathcal{R}^d$ by hyperplanes. Trees of this nature have evolved in the computer graphics literature via the work of Fuchs,

Kedem, and Naylor (1980) and Fuchs, Abram, and Grant (1983) (see also Samet (1990b), Kaplan (1985), and Sung and Shirley (1992)). In discrimination they are at the same time generalizations of linear discriminants, of histograms, and of binary tree classifiers. BSP trees were recommended for use in discrimination by Henrichon and Fu (1969), Mizoguchi, Kizawa, and Shimura (1977) and Friedman (1977). Further studies include Sklansky and Michelotti (1980), Argentiero, Chin, and Beaudet (1982), Qing-Yun and Fu (1983), Breiman, Friedman, Olshen, and Stone (1984), Loh and Vanichsetakul (1988), and Park and Sklansky (1990).

There are numerous ways of constructing BSP trees. Most methods of course use the Y-values to determine good splits. Nevertheless, we should mention first simple splits with the X-property, if only to better understand the BSP trees.

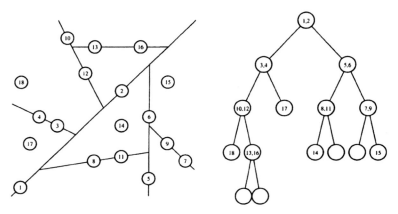

FIGURE 20.20. *A raw* BSP *tree and its induced partition in the plane. Every region is split by a line determined by the two data points with smallest index in the region.*

We call the *raw* BSP *tree* the tree obtained by letting $X_1, \ldots, X_d$ determine the first hyperplane. The d data points remain associated with the root, and the others $(X_{d+1}, \ldots, X_k)$ are sent down to the subtrees, where the process is repeated as far as possible. The remaining points $X_{k+1}, \ldots, X_n$ are used in a majority vote in the external regions. Note that the number of regions is not more than k/d. Thus, by Lemma 20.1, if $k/n \to 0$, we have $N(X) \to \infty$ in probability. Combining this with Problem 20.25, we have our first result:

Theorem 20.7. *The natural binary tree classifier based upon a raw* BSP *tree with $k \to \infty$ and $k/n \to 0$ is consistent whenever X has a density.*

Hyperplanes may also be selected by optimization of a criterion. Typically, this would involve separating the classes in some way. All that was said for perpendicular splitting remains valid here. It is worthwhile recalling therefore that there are distributions for which the best empirical Stoller split does not improve the probability of error. Take for example the uniform distribution in the unit ball of $\mathcal{R}^d$ in which

$$Y = \begin{cases} 1 & \text{if } \|X\| \geq 1/2 \\ 0 & \text{if } \|X\| < 1/2. \end{cases}$$

FIGURE 20.21. *No single split improves on the error probability for this distribution.*

Here, no linear split would be helpful as the 1's would always hold a strong majority. Minimizing other impurity functions such as the Gini criterion may be helpful, however (Problem 20.27).

BIBLIOGRAPHIC REMARKS. Hyperplane splits may be generalized to include quadratic splits (Henrichon and Fu (1969)). For example, Mui and Fu (1980) suggest taking $d' < d$ and forming quadratic classifiers as in normal discrimination (see Chapter 4) based upon vectors in $\mathcal{R}^{d'}$. The cuts are thus perpendicular to $d - d'$ axes but quadratic in the subspace $\mathcal{R}^{d'}$. Lin and Fu (1983) employ the Bhattacharyya distance or 2-means clustering for determining splits. As a novelty, within each leaf region, the decision is not by majority vote, but rather by a slightly more sophisticated rule such as the k-NN rule or linear discrimination. Here, no optimization is required along the way. Loh and Vanichsetakul (1988) allow linear splits but use F ratios to select desirable hyperplanes.

20.12   Primitive Selection

There are two reasons for optimizing a tree configuration. First of all, it just does not make sense to ignore the class labels when constructing a tree classifier, so the Y_i's must be used to help in the selection of a best tree. Secondly, some trees may not be consistent (or provably consistent), yet when optimized over a family of trees, consistency drops out. We take the following example: let $\mathcal{G}_k$ be a class of binary tree classifiers with the X-property, with the space partitioned into $k + 1$ regions determined by $X_1, \ldots, X_k$ only. Examples include the chronological k-d tree and some kinds of BSP trees. We estimate the error for $g \in \mathcal{G}_k$ by

$$\widehat{L}_n(g) = \frac{1}{n-k} \sum_{i=k+1}^{n} I_{\{g(X_i) \neq Y_i\}},$$

realizing the danger of using the same data that were used to obtain majority votes to estimate the error. An optimistic bias will be introduced. (For more on such error estimates, see Chapter 23.) Let g_n^* be the classifier (or one of the classifiers) in $\mathcal{G}_k$ for which $\widehat{L}_n$ is minimum. Assume that $|\mathcal{G}_k| < \infty$—for example, $\mathcal{G}_k$ could be all $k!$ chronological k-d trees obtained by permuting $X_1, \ldots, X_k$. We call g_n^* then the permutation-optimized chronological k-d classifier. When $k = \lfloor \sqrt{\log n} \rfloor$, one can verify that $k! = o(n^\epsilon)$ for any $\epsilon > 0$, so that the computational burden—at least on paper—is not out of sight. We assume that $\mathcal{G}_k$ has a consistent classifier sequence, that is, as $n \to \infty$, k usually grows unbounded, and $\mathbf{E}\{L_n(g_k)\} \to L^*$ for a sequence $g_k \in \mathcal{G}_k$.

EXAMPLE. Among the chronological k-d trees modulo permutations, the first one (i.e., the one corresponding to the identity permutation) was shown to be consistent in Theorem 20.3, if X has a density, $k \to \infty$, and $k/n \to 0$. □

EXAMPLE. Let $\mathcal{G}_k$ be the class of BSP trees in which we take as possible hyperplanes for splitting the root:

  (1) the hyperplane through $X_1, \ldots, X_d$;
  (2) the d hyperplanes through $X_1, \ldots, X_{d-1}$ that are parallel to one axis;
  (3) the $\binom{d}{2}$ hyperplanes through $X_1, \ldots, X_{d-2}$ that are parallel to two axes;

  $\vdots$

  (d) the $\binom{d}{d-1} = d$ hyperplanes through X_1 that are parallel to $d - 1$ axes.

Thus, conservatively estimated, $|\mathcal{G}_k| \leq k! 2^d$ because there are at most 2^d possible choices at each node and there are $k!$ permutations of $X_1, \ldots, X_k$. Granted, the number of external regions is very variable, but it remains bounded by $k + 1$ in any case. As $\mathcal{G}_k$ contains the chronological k-d tree, it has a consistent sequence of classifiers when $k \to \infty$ and $n/(k \log k) \to \infty$. □

Theorem 20.8. *Let* $g_n^* = \arg\min_{g \in \mathcal{G}_k} \widehat{L}_n(g)$. *Then, if* $\mathcal{G}_k$ *has a consistent sequence of classifiers, if the number of regions in the partitions for all* $g \in \mathcal{G}_k$ *are at most* $k + 1$, *and if* $k \to \infty$ *and* $n/\log|\mathcal{G}_k| \to \infty$, *then* $\mathbf{E}\{L_n(g_n^*)\} \to L^*$, *where* $L_n(g_n^*)$ *is the conditional probability of error of* g_n^*.

In the examples cited above, we must take $k \to \infty$, $n/k \to \infty$. Furthermore, $\log|\mathcal{G}_k| = O(\log k!) = O(k \log k)$ in both cases. Thus, g_n^* is consistent whenever X has a density, $k \to \infty$ and $k = o(n/\log n)$. This is a simple way of constructing a basically universally consistent BSP tree.

PROOF. Let $g^+ = \arg\min_{g \in \mathcal{G}_k} L_n(g)$.

$$
\begin{aligned}
L_n(g_n^*) - L^* &= L_n(g_n^*) - \widehat{L}_n(g_n^*) \\
&\quad + \widehat{L}_n(g_n^*) - \widehat{L}_n(g^+) \\
&\quad + \widehat{L}_n(g^+) - L_n(g^+) \\
&\quad + L_n(g^+) - L^*.
\end{aligned}
$$

Clearly,

$$
L_n(g_n^*) - L^* \leq L_n(g^+) - L^* + 2 \max_{g \in \mathcal{G}_k} |L_n(g) - \widehat{L}_n(g)|
$$

$$
\stackrel{\text{def}}{=} I + II.
$$

Obviously, $I \to 0$ in the mean by our assumption, and for $\epsilon > 0$,

$$\mathbf{P}\left\{2 \max_{g \in \mathcal{G}_k} |L_n(g) - \widehat{L}_n(g)| > \epsilon\right\} \leq |\mathcal{G}_k| \max_{g \in \mathcal{G}_k} \mathbf{P}\left\{|L_n(g) - \widehat{L}_n(g)| > \frac{\epsilon}{2}\right\}.$$

Next we bound the probabilities on the right-hand side. Let p_{i0}, p_{i1} denote $\mathbf{P}\{X \in$ region i, $Y = 0\}$ and $\mathbf{P}\{X \in$ region i, $Y = 1\}$ respectively, with regions determined by g, $1 \leq i \leq k + 1$. Let

$$N_{i0} = \sum_{j=k+1}^{n} I_{\{X_j \in \text{region } i, Y_j = 0\}} \quad \text{and} \quad N_{i1} = \sum_{j=k+1}^{n} I_{\{X_j \in \text{region } i, Y_j = 1\}}.$$

Then

$$L_n(g) = \sum_{i=1}^{k+1} p_{i1} I_{\{N_{i0} \geq N_{i1}\}} + \sum_{i=1}^{k+1} p_{i0} I_{\{N_{i0} < N_{i1}\}},$$

and

$$\widehat{L}_n(g) = \sum_{i=1}^{k+1} \frac{N_{i1}}{n-k} I_{\{N_{i0} \geq N_{i1}\}} + \sum_{i=1}^{k+1} \frac{N_{i0}}{n-k} I_{\{N_{i0} < N_{i1}\}}.$$

Thus,

$$|L_n(g) - \widehat{L}_n(g)| \leq \sum_{i=1}^{k+1} \left|p_{i1} - \frac{N_{i1}}{n-k}\right| + \sum_{i=1}^{k+1} \left|p_{i0} - \frac{N_{i0}}{n-k}\right|.$$

Introduce the notation Z for the random variable on the right-hand side of the above inequality. By the Cauchy-Schwarz inequality,

$$\mathbf{E}\{Z|X_1, \ldots, X_k\}$$

$$\leq \sum_{i=1}^{k+1} \sqrt{\mathbf{E}\left\{\left(p_{i1} - \frac{N_{i1}}{n-k}\right)^2 | X_1, \ldots, X_k\right\}}$$

$$+ \sum_{i=1}^{k+1} \sqrt{\mathbf{E}\left\{\left(p_{i0} - \frac{N_{i0}}{n-k}\right)^2 | X_1, \ldots, X_k\right\}}$$

(as given $X_1, \ldots, X_k$, N_{i1} is binomial $(n - k, p_{i1})$)

$$\leq \sum_{i=1}^{k+1} \left(\sqrt{\frac{p_{i1}}{n-k}} + \sqrt{\frac{p_{i0}}{n-k}}\right)$$

$$\leq \frac{2k+2}{\sqrt{n-k}} \sqrt{\frac{1}{2k+2} \sum_{i=1}^{k+1} (p_{i1} + p_{i0})}$$

(by another use of the Cauchy-Schwarz inequality)

$$= \sqrt{\frac{2k+2}{n-k}}.$$

Thus, $E\{Z|X_1, \ldots, X_k\} \to 0$ as $k/n \to 0$. Note that

$$P\{Z > \epsilon | X_1, \ldots, X_k\}$$

$$\leq \quad P\left\{Z - E\{Z|X_1, \ldots, X_k\} > \frac{\epsilon}{2} \Big| X_1, \ldots, X_k\right\}$$

(if $E\{Z|X_1, \ldots, X_k\} < \epsilon/2$, which happens when $\sqrt{\dfrac{2k+2}{n-k}} \leq \dfrac{\epsilon}{2}$)

$$\leq \quad \exp\left(-\left(\frac{\epsilon}{2}\right)^2 \Big/ 2(n-k)\frac{4}{(n-k)^2}\right)$$

(by McDiarmid's inequality, as changing the value of an X_j, $j \geq k$, changes Z by at most $2/(n-k)$; see Theorem 9.2)

$$= \quad \exp\left(-\frac{(n-k)\epsilon^2}{32}\right).$$

Thus, taking expected values, we see that

$$P\{II > \epsilon\} \leq |\mathcal{G}_k| e^{-(n-k)\epsilon^2/64}$$

if $\sqrt{(2k+2)/(n-k)} \leq \epsilon/2$. This tends to 0 for all $\epsilon > 0$ if $k/n \to 0$ and $n/\log|\mathcal{G}_k| \to \infty$. $\square$

20.13   Constructing Consistent Tree Classifiers

Thus far, we have taken you through a forest of beautiful trees and we have shown you a few tricks of the trade. When you read (or write) a research paper on tree classifiers, and try to directly apply a consistency theorem, you will get frustrated however—most real-life tree classifiers use the data in intricate ways to suit a certain application. It really helps to have a few truly general results that have universal impact. In this section we will point you to three different places in the book where you may find useful results in this respect. First of all, there is a consistency theorem—Theorem 21.2—that applies to rules that partition the space and decide by majority vote. The partition is arbitrary and may thus be generated by using some or all of $X_1, Y_1, \ldots, X_n, Y_n$. If a rule satisfies the two (weak) conditions of Theorem 21.2, it must be universally consistent. To put it differently, even the worst rule within the boundaries of the theorem's conditions must perform well asymptotically.

Second, we will briefly discuss the design of tree classifiers obtained by minimizing the empirical error estimate $(\widehat{L}_n)$ over possibly infinite classes of classifiers. Such classifiers, however hard to find by an algorithm, have asymptotic properties that are related to the VC dimension of the class of rules. Consistency follows almost without work if one can calculate or bound the VC dimension appropriately.

While Chapters 13 and 21 deal in more detail with the VC dimension, it is necessary to give a few examples here.

Third, we point the reader to Chapter 22 on data splitting, where the previous approach is applied to the minimization of the holdout estimate, obtained by trees based upon part of the sample, and using another part to select the best tree in the bunch. Here too the VC dimension plays a crucial role.

Theorem 21.2 allows space partitions that depend quite arbitrarily on *all* the data and extends earlier universally applicable results of Gordon and Olshen (1978), (1980), (1984), and Breiman, Friedman, Olshen, and Stone (1984). A particularly useful format is given in Theorem 21.8. If the partition is by recursive hyperplane splits as in BSP trees and the number of splits is at most m_n, if $m_n \log n/n \to 0$, and if

$$\text{diam}(A(X) \cap S_B) \to 0$$

with probability one for all S_B (where S_B is the ball of radius B centered at the origin), then the classification rule is strongly consistent. The last condition forces a randomly picked region in the partition to be small. However, $m_n \log n/n \to 0$ guarantees that no devilish partition can be inconsistent. The latter condition is certainly satisfied if each region contains at least k_n points, where $k_n/\log n \to \infty$.

Next we take a look at full-fledged minimization of $\widehat{L}_n$, the empirical error, over certain classes of tree classifiers. Here we are not concerned with the (often unacceptable) computational effort. For example, let $\mathcal{G}_k$ be the class of all binary tree classifiers based upon a tree consisting of k internal nodes, each representing a hyperplane cut (as in the BSP tree), and all possible 2^{k+1} labelings of the $k + 1$ leaf regions. Pick such a classifier for which

$$\widehat{L}_n(g) = \frac{1}{n} \sum_{i=1}^{n} I_{\{g(X_i) \neq Y_i\}}$$

is minimal. Observe that the chosen tree is always natural; that is, it takes majority votes over the leaf regions. Thus, the minimization is equivalent to the minimization of the *resubstitution* error estimate (defined in Chapter 23) over the corresponding class of natural tree classifiers. We say that a sequence $\{\mathcal{G}_k\}$ of classes is *rich* if we can find a sequence $g_k \in \mathcal{G}_k$ such that $L(g_k) \to L^*$. For hyperplanes, this is the case if $k \to \infty$ as $n \to \infty$—just make the hyperplane cuts form a regular histogram grid and recall Theorem 9.4. Let $S(\mathcal{G}_k, n)$ be the shatter coefficient of $\mathcal{G}_k$ (for a definition, see Chapter 12, Definition 12.3). For example, for the hyperplane family,

$$S(\mathcal{G}_k, n) \leq n^{k(d+1)}.$$

Then by Corollary 12.1 we have $E\{L_n(g_n^*)\} \to L^*$ for the selected classifier g_n^* when $\mathcal{G}_k$ is rich ($k \to \infty$ here) and $\log S(\mathcal{G}_k, n) = o(n)$, that is,

$$k = o\left(\frac{n}{\log n}\right).$$

Observe that no conditions are placed on the distribution of (X, Y) here! Consistency follows from basic notions—one combinatorial to keep us from overfitting, and one approximation-theoretical (the richness).

The above result remains valid under the same conditions on k in the following classes:

(1) All trees based upon k internal nodes each representing a perpendicular split. (Note: $S(\mathcal{G}_k, n) \leq ((n + 1)d)^k$.)

(2) All trees based upon k internal nodes, each representing a quadtree split. (Note: $S(\mathcal{G}_k, n) \leq (n^d + 1)^k$.)

20.14   A Greedy Classifier

In this section, we define simply a binary tree classifier that is grown via optimization of a simple criterion. It has the remarkable property that it does not require a forced rotation through the coordinate axes or special safeguards against small or large regions or the like. It remains entirely parameter-free (nothing is picked by the user), is monotone transformation invariant, and fully automatic. We show that in $\mathcal{R}^d$ it is always consistent. It serves as a prototype for teaching about such rules and should not be considered as more than that. For fully practical methods, we believe, one will have to tinker with the approach.

The space is partitioned into rectangles as shown below:

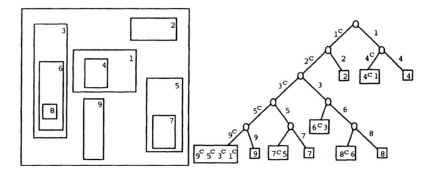

FIGURE 20.22. *A tree based on partitioning the plane into rectangles. The right subtree of each internal node belongs to the inside of a rectangle, and the left subtree belongs to the complement of the same rectangle (i^c denotes the complement of i). Rectangles are not allowed to overlap.*

A hyperrectangle defines a split in a natural way. The theory presented here applies for many other types of cuts. These will be discussed after the main consistency theorem is stated.

A partition is denoted by $\mathcal{P}$, and a decision on a set $A \in \mathcal{P}$ is by majority vote. We write $g_\mathcal{P}$ for such a rule:

$$
g_\mathcal{P}(x) = \begin{cases} 1 & \text{if } \sum_{i:X_i \in A} Y_i > \sum_{i:X_i \in A} (1 - Y_i), x \in A \\ 0 & \text{otherwise.} \end{cases}
$$

Given a partition $\mathcal{P}$, a legal rectangle A is one for which $A \cap B = \emptyset$ or $A \subseteq B$ for all sets $B \in \mathcal{P}$. If we refine $\mathcal{P}$ by adding a legal rectangle T somewhere, then we obtain the partition $\mathcal{T}$. The decision $g_\mathcal{T}$ agrees with $g_\mathcal{P}$ except on the set $B \in \mathcal{P}$ that contains T.

We introduce the convenient notation

$$
\begin{aligned}
\nu_j(A) &= \mathbf{P}\{X \in A, Y = j\}, \quad j \in \{0, 1\}, \\
\nu_{j,n}(A) &= \frac{1}{n} \sum_{i=1}^{n} I_{\{X_i \in A, Y_i = j\}}, \quad j \in \{0, 1\}.
\end{aligned}
$$

An estimate of the quality of $g_\mathcal{P}$ is

$$
\widehat{L}_n(\mathcal{P}) \overset{\text{def}}{=} \sum_{R \in \mathcal{P}} \widehat{L}_n(R),
$$

where

$$
\begin{aligned}
\widehat{L}_n(R) &= \frac{1}{n} \sum_{i=1}^{n} I_{\{X_i \in R, g_\mathcal{P}(X_i) \neq Y_i\}} \\
&= \min(\nu_{0,n}(R), \nu_{1,n}(R)).
\end{aligned}
$$

Here we use two different arguments for $\widehat{L}_n$ (R and $\mathcal{P}$), but the distinction should be clear. We may similarly define $\widehat{L}_n(\mathcal{T})$. Given a partition $\mathcal{P}$, the greedy classifier selects that legal rectangle T for which $\widehat{L}_n(\mathcal{T})$ is minimal (with any appropriate policy for breaking ties). Let R be the set of $\mathcal{P}$ containing T. Then the greedy classifier picks that T for which

$$
\widehat{L}_n(T) + \widehat{L}_n(R - T) - \widehat{L}_n(R)
$$

is minimal. Starting with the trivial partition $\mathcal{P}_0 = \{\mathcal{R}^d\}$, we repeat the previous step k times, leading thus to $k + 1$ regions. The sequence of partitions is denoted by $\mathcal{P}_0, \mathcal{P}_1, \ldots, \mathcal{P}_k$.

We put *no* safeguards in place—the rectangles are not forced to shrink. And in fact, it is easy to construct examples in which most rectangles do not shrink. The main result of the section, and indeed of this chapter, is that the obtained classifier is consistent:

Theorem 20.9. *For the greedy classifier with $k \to \infty$ and $k = o\left(\sqrt[3]{n/\log n}\right)$, assuming that X has nonatomic marginals, we have $L_n \to L^*$ with probability one.*

REMARK. We note that with techniques presented in the next chapter, it is possible to improve the second condition on k to $k = o\left(\sqrt{n/\log n}\right)$ (see Problem 20.31). □

Before proving the theorem, we mention that the same argument may be used to establish consistency of greedily grown trees with many other types of cuts. We have seen in Section 20.8 that repeated Stoller splits do not result in good classifiers. The reason is that optimization is over a collection of sets (halfspaces) that is not guaranteed to improve matters—witness the examples provided in previous sections. A good cutting method is one that includes somehow many (but not too many) small sets. For example, let us split at the root making $d + 1$ hyperplane cuts at once, that is, by finding the $d + 1$ cuts that together produce the largest decrease in the empirical error probability. Then repeat this step recursively in each region k times. The procedure is consistent under the same conditions on k as in Theorem 6.1, whenever X has a density (see Problem 20.30). The $d + 1$ hyperplane cuts may be considered as an *elementary cut* which is repeated in a greedy manner. In Figures 20.23 to 20.25 we show a few elementary cuts that may be repeated greedily for a consistent classifier. The straightforward proofs of consistency are left to the reader in Problem 20.30.

PROOF OF THEOREM 20.9.   We restrict ourselves to $\mathcal{R}^2$, but the proof remains similar in $\mathcal{R}^d$ (Problem 20.29). The notation $\widehat{L}_n(\cdot)$ was introduced above, where the argument is allowed to be a partition or a set in a partition. We similarly define

$$
\begin{aligned}
L(R) &= \min_{y \in \{0,1\}} \mathbf{P}\{X \in R, Y \neq y\} \\
&= \min(\nu_0(R), \nu_1(R)), \\
L(\mathcal{P}) &= \sum_{R \in \mathcal{P}} L(R).
\end{aligned}
$$

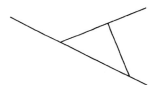

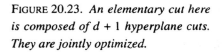

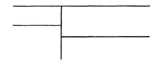

FIGURE 20.23. *An elementary cut here is composed of $d + 1$ hyperplane cuts. They are jointly optimized.*

FIGURE 20.24. *2d rectangular cuts determine an elementary cut. All 2d cuts have arbitrary directions; there are no forced directions.*

FIGURE 20.25. *Simplex cuts. A cut is determined by a polyhedron with d + 1 vertices. The simplices are not allowed to overlap, just as for rectangular cuts in the greedy classifier. Three consecutive simplex cuts in $\mathcal{R}^2$ are shown here.*

FIGURE 20.26. *At every iteration we refine the partition by selecting that rectangle R in the partition and that 3 × 3 × ··· × 3 rectangular grid cut of R for which the empirical error is minimal. Two consecutive rectangular grid cuts are shown.*

For example, let $\mathcal{G}_l$ denote a partition of $\mathcal{R}^2$ into a rectangular $l \times l$ grid.

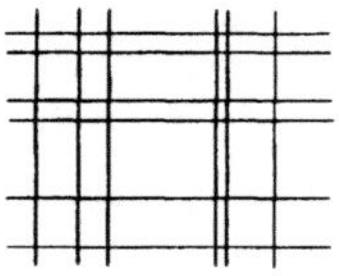

FIGURE 20.27. $\mathcal{G}_l$: *An $l \times l$ grid (with $l = 7$ here).*

It is clear that for all $\epsilon > 0$, there exists an $l = l(\epsilon)$ and an $l \times l$ grid $\mathcal{G}_l$ such that

$$L(\mathcal{G}_l) \leq L^* + \epsilon.$$

If Q is another finer partition into rectangles (i.e., each set of Q is a rectangle and intersects at most one rectangle of $\mathcal{G}_l$), then necessarily

$$L(\mathcal{Q}) \leq L(\mathcal{G}_l) \leq L^* + \epsilon.$$

We will call Q a *refinement* of $\mathcal{G}_l$. The next lemma is a key property of partitioning classifiers. In our eyes, it is the main technical property of this entire chapter. We say that the partition T is an *extension* of $\mathcal{P}$ by a set Q—where $Q \subseteq R \in \mathcal{P}$—if T contains all cells of $\mathcal{P}$ other than R, plus Q and $R - Q$.

Lemma 20.2. *Let $\mathcal{G}_l$ be a finite partition with $L(\mathcal{G}_l) \leq L^* + \epsilon$. Let $\mathcal{P}$ be a finite partition of $\mathcal{R}^d$, and let $\mathcal{Q}$ be a refinement of both $\mathcal{P}$ and $\mathcal{G}_l$. Then there exists a set $Q \in \mathcal{Q}$ (note: Q is contained in one set of $\mathcal{P}$ only) and an extension of $\mathcal{P}$ by Q to $\mathcal{T}_Q$ such that, if $L(\mathcal{P}) \geq L^* + \epsilon$,*

$$L(\mathcal{T}_Q) - (L^* + \epsilon) \leq \left(1 - \frac{1}{|\mathcal{Q}|}\right)\left(L(\mathcal{P}) - (L^* + \epsilon)\right).$$

PROOF. First fix $R \in \mathcal{P}$ and let $Q_1, \ldots, Q_N$ be the sets of $\mathcal{Q}$ contained in R. Define

$$p_i = \nu_0(Q_i), \qquad q_i = \nu_1(Q_i), \qquad L(Q_i) = \min(p_i, q_i),$$

$$p = \sum_{i=1}^{N} p_i, \qquad q = \sum_{i=1}^{N} q_i, \qquad L(R) = \min(p, q),$$

$$L(R - Q_i) = \min(p - p_i, q - q_i).$$

First we show that there exists an integer i such that

$$L(R) - L(Q_i) - L(R - Q_i) \geq \frac{L(R) - \sum_{i=1}^{N} L(Q_i)}{N},$$

or equivalently,

$$\max_{1 \leq i \leq N} \Delta_i \geq \frac{\min(p, q) - \sum_{i=1}^{N} \min(p_i, q_i)}{N},$$

where $\Delta_i = \min(p, q) - \min(p_i, q_i) - \min(p - p_i, q - q_i)$. To see this, assume without loss of generality that $p \leq q$. If $p_i \leq q_i$ for all i, then

$$\min(p, q) - \sum_{i=1}^{N} \min(p_i, q_i) = p - \sum_{i=1}^{N} p_i = 0,$$

so we are done. Assume therefore that $p_i > q_i$ for $i \in A$, where A is a set of indices with $|A| \geq 1$. For such i,

$$\Delta_i = p - q_i - (p - p_i) = p_i - q_i,$$

and thus,

$$\sum_{i \in A} \Delta_i = \sum_{i \in A}(p_i - q_i) = p - \sum_{i \notin A} p_i - \sum_{i \in A} q_i = p - \sum_{i=1}^{N} \min(p_i, q_i).$$

But then,

$$\max_{1 \leq i \leq N} \Delta_i \geq \frac{1}{|A|} \sum_{i \in A} \Delta_i \geq \frac{p - \sum_{i=1}^{N} \min(p_i, q_i)}{|A|} \geq \frac{p - \sum_{i=1}^{N} \min(p_i, q_i)}{N},$$

and the claim follows. To prove the lemma, notice that since $L(\mathcal{Q}) \leq L^* + \epsilon$, it suffices to show that

$$\max_{Q \in \mathcal{Q}}(L(\mathcal{P}) - L(T_Q)) \geq \frac{L(\mathcal{P}) - L(\mathcal{Q})}{|\mathcal{Q}|},$$

or equivalently, that

$$\max_{Q \in \mathcal{Q}}(L(R_Q) - L(Q) - L(R_Q - Q)) \geq \frac{L(\mathcal{P}) - L(\mathcal{Q})}{|\mathcal{Q}|},$$

where R_Q is the unique cell of $\mathcal{P}$ containing Q. However,

$$\max_{Q \in \mathcal{Q}}(L(R_Q) - L(Q) - L(R_Q - Q))$$

$$\geq \max_{R \in \mathcal{P}, Q \subseteq R}(L(R) - L(Q) - L(R - Q))$$

$$\geq \max_{R \in \mathcal{P}} \frac{L(R) - \sum_{Q \subseteq R} L(Q)}{|R|}$$

(by the inequality shown above, where $|R|$ denotes the number of sets of $\mathcal{Q}$ contained in R)

$$\geq \frac{\sum_{R \in \mathcal{P}}\left(L(R) - \sum_{Q \subseteq R} L(Q)\right)}{\sum_{R \in \mathcal{P}} |R|}$$

$$= \frac{\sum_{R \in \mathcal{P}} L(R) - \sum_{Q \in \mathcal{Q}} L(Q)}{|\mathcal{Q}|}$$

$$= \frac{L(\mathcal{P}) - L(\mathcal{Q})}{|\mathcal{Q}|},$$

and the lemma is proved. $\square$

Let us return to the proof of Theorem 20.9. The previous lemma applied to our situation ($\mathcal{P} = \mathcal{P}_i$) shows that we may extend $\mathcal{P}_i$ by a *rectangle* $Q \in \mathcal{Q}$ (as Q will be a collection of rectangles refining both $\mathcal{G}_l$ and $\mathcal{P}_i$) such that $L(T_Q) - (L^* + \epsilon)$ is smaller by a guaranteed amount than $L(\mathcal{P}_i) - (L^* + \epsilon)$. ($T_Q$ is the partition obtained by extending $\mathcal{P}_i$.) To describe $\mathcal{Q}$, we do the following (note that $\mathcal{Q}$ must consist entirely of rectangles for otherwise the lemma is useless): take the rectangles of $\mathcal{P}_i$ and extend all four sides (in the order of birth of the rectangles) until they hit a side of another rectangle or an extended border of a previous rectangle if they hit anything at all. Figure 20.28 illustrates the partition into rectangles. Note that this partition consists of $4i$ line segments, and at most $9i$ rectangles (this can be seen by noting that we can write in each nonoriginal rectangle of $\mathcal{P}_i$ the number of an original neighboring rectangle, each number appearing at most 8 times).

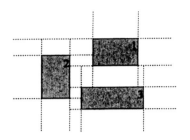

FIGURE 20.28. *Extensions of rectangles to get a rectangular grid.*

The rectangular grid partition thus obtained is intersected with $\mathcal{G}_l$ to yield $\mathcal{Q}$. To apply Lemma 20.2, we only need a bound on $|\mathcal{Q}|$. To the rectangular partition just created, we add each of the $2l - 2$ lines of the grid $\mathcal{G}_l$ one by one. Start with the horizontal lines. Each time one is added, it creates at most $9i$ new rectangles. Then, each vertical line adds at most $9i + l$ new rectangles for a total of not more than $9i + 9i(l - 1) + (l - 1)(9i + l) \leq l^2 + 18li$ (assuming $l \geq 1$). Hence,

$$|\mathcal{Q}| \leq l^2 + 18li.$$

Apply the lemma, and observe that there is a rectangle $Q \in \mathcal{Q}$ such that the extension of $\mathcal{P}_i$ by Q to $\mathcal{P}'_i$ would yield

$$L(\mathcal{P}'_i) - (L^* + \epsilon) \leq \left(L(\mathcal{P}_i) - (L^* + \epsilon) \right) \left(1 - \frac{1}{l^2 + 18li} \right).$$

At this point in the proof, the reader can safely forget $\mathcal{Q}$. It has done what it was supposed to do.

Of course, we cannot hope to find $\mathcal{P}'_i$ because we do not know the distribution. Let us denote the *actual* new partition by $\mathcal{P}_{i+1}$ and let R_{i+1} be the selected rectangle by the empirical minimization. Observe that

$$
\begin{aligned}
&L(\mathcal{P}_{i+1}) - L(\mathcal{P}'_i) \\
&= \left(L(\mathcal{P}_{i+1}) - \widehat{L}_n(\mathcal{P}_{i+1}) \right) + \left(\widehat{L}_n(\mathcal{P}_{i+1}) - \widehat{L}_n(\mathcal{P}'_i) \right) + \left(\widehat{L}_n(\mathcal{P}'_i) - L(\mathcal{P}'_i) \right) \\
&\leq \left(L(\mathcal{P}_{i+1}) - \widehat{L}_n(\mathcal{P}_{i+1}) \right) + \left(\widehat{L}_n(\mathcal{P}'_i) - L(\mathcal{P}'_i) \right).
\end{aligned}
$$

As $\mathcal{P}_{i+1}$ and $\mathcal{R}'_i$ have most sets in common, many terms cancel in the last double sum. We are left with just $L(\cdot)$ and $\widehat{L}_n(\cdot)$ terms for sets of the form $R - R_{i+1}$, R_{i+1}, $R - Q$, Q with $R \in \mathcal{P}_i$. Thus,

$$L(\mathcal{P}_{i+1}) - L(\mathcal{P}'_i) \leq 2 \sup_{R'} |\widehat{L}_n(R') - L(R')|,$$

where the supremum is taken over all rectangles of the above described form. These are sets of the form obtainable in $\mathcal{P}_{i+1}$. Every such set can be written as the difference of a (possibly infinite) rectangle and at most $i + 1$ nonoverlapping contained rectangles. As $i + 1 \leq k$ in our analysis, we call $\mathcal{Z}_k$ the class of all sets that may be described in this form: $Z_0 - Z_1 - \cdots - Z_k$, where $Z_0, Z_1, \ldots, Z_k$ are rectangles, each $Z_i \subseteq Z_0, Z_1, \ldots, Z_k$ are mutually disjoint. Rectangles may be infinite or half infinite. Hence, for $i < k$,

$$L(\mathcal{P}_{i+1}) - L(\mathcal{P}_i') \le 2 \sup_{Z \in Z_k} |\widehat{L}_n(Z) - L(Z)|.$$

For a fixed $Z \in Z_k$,

$$
\begin{aligned}
|\widehat{L}_n(Z) - L(Z)| &= |\min(\nu_{0,n}(Z), \nu_{1,n}(Z)) - \min(\nu_0(Z), \nu_1(Z))| \\
&\le |\nu_{0,n}(Z) - \nu_0(Z)| + |\nu_{1,n}(Z) - \nu_1(Z)|.
\end{aligned}
$$

Thus,

$$L(\mathcal{P}_{i+1}) - L(\mathcal{P}_i') \le V_n \stackrel{\text{def}}{=} 2 \sup_{Z \in Z_k} \Big(|\nu_{0,n}(Z) - \nu_0(Z)| + |\nu_{1,n}(Z) - \nu_1(Z)| \Big).$$

Define the (good) event

$$G = \{V_n < \delta\},$$

where $\delta > 0$ will be picked later. On G, we see that for *all* $i < k$,

$$
\begin{aligned}
L(\mathcal{P}_{i+1}) &- (L^* + \epsilon) \\
&\le L(\mathcal{P}_i') - (L^* + \epsilon) + \delta \\
&\le \big(L(\mathcal{P}_i) - (L^* + \epsilon) \big) \left(1 - \frac{1}{l^2 + 18li} \right) + \delta \quad \text{(Lemma 20.2).}
\end{aligned}
$$

We now introduce a convenient lemma.

Lemma 20.3. *Let a_n, b_n be sequences of positive numbers with $b_n \downarrow 0$, $b_0 < 1$, and let $\delta > 0$ be fixed. If $a_{n+1} \le a_n(1 - b_n) + \delta$, then*

$$a_{n+1} \le a_0 e^{-\sum_{j=0}^n b_j} + \delta(n+1).$$

PROOF. We have

$$
\begin{aligned}
a_{n+1} &\le a_0 \prod_{j=0}^n (1 - b_j) + \delta \left(\sum_{m=1}^{n-1} \prod_{j=m}^n (1 - b_j) \right) + \delta \\
&= I + II + III.
\end{aligned}
$$

Clearly, $I \le a_0 \exp\left(-\sum_{j=0}^n b_j \right)$. A trivial bound for $II + III$ is $\delta(n+1)$. $\square$

Conclude that on G,

$$
\begin{aligned}
L(\mathcal{P}_k) - (L^* + \epsilon) &\le \big(L(\mathcal{P}_0) - (L^* + \epsilon) \big) e^{-\sum_{j=0}^{k-1} 1/(l^2 + 18lj)} + \delta k \\
&\le e^{-\int_0^{k-1} 1/(l^2 + 18lu) du} + \delta k \\
&\le e^{-\frac{1}{18l} \log((l^2 + 18l(k-1))/(l^2))} + \delta k \\
&= \frac{1}{\left(1 + \frac{18(k-1)}{l} \right)^{1/(18l)}} + \delta k.
\end{aligned}
$$

Thus,

$$\mathbf{P}\{L(\mathcal{P}_k) > L^* + 2\epsilon\} \leq \mathbf{P}\{G^c\} = \mathbf{P}\{V_n \geq \delta\}$$

when $\delta k < \epsilon/2$ and

$$\frac{1}{\left(1 + \frac{18(k-1)}{l}\right)^{1/(18l)}} < \frac{\epsilon}{2}.$$

Finally, introduce the notation

$$
\begin{aligned}
L_n(R) &= \mathbf{P}\{X \in R, Y \neq g_{\mathcal{P}}(X)\}, \\
L_n(\mathcal{P}) &= \sum_{R \in \mathcal{P}} L_n(R).
\end{aligned}
$$

As $L(\mathcal{P}_k)$ takes the partition into account, but not the majority vote, and $L_n(\mathcal{P}_k)$ does, we need a small additional argument:

$$
\begin{aligned}
L_n(\mathcal{P}_k) &= \left(L_n(\mathcal{P}_k) - \widehat{L}_n(\mathcal{P}_k)\right) + \left(\widehat{L}_n(\mathcal{P}_k) - L(\mathcal{P}_k)\right) + L(\mathcal{P}_k) \\
&\leq 2 \sum_{R \in \mathcal{P}_k} \left(|\nu_0(R) - \nu_{0,n}(R)| + |\nu_1(R) - \nu_{1,n}(R)|\right) + L(\mathcal{P}_k) \\
&\stackrel{\text{def}}{=} 2W_n + L(\mathcal{P}_k).
\end{aligned}
$$

Putting things together,

$$
\begin{aligned}
&\mathbf{P}\left\{L_n(\mathcal{P}_k) > L^* + 4\epsilon\right\} \\
&\leq \mathbf{P}\{W_n > \epsilon\} + \mathbf{P}\left\{L(\mathcal{P}_k) > L^* + 2\epsilon\right\} \\
&\leq \mathbf{P}\{W_n > \epsilon\} + \mathbf{P}\{V_n \geq \delta\}
\end{aligned}
$$

under the given conditions on δ and k. Observe that if for a set Z,

$$U_n(Z) \stackrel{\text{def}}{=} |\nu_0(Z) - \nu_{0,n}(Z)| + |\nu_1(Z) - \nu_{1,n}(Z)|,$$

then

$$
\begin{aligned}
V_n &\leq \sup_{\text{rectangles } Z} U_n(Z) + \sup_{\substack{\text{disjoint sets of } k \\ \text{rectangles } Z_1,\dots,Z_k}} \sum_{i=1}^{k} U_n(Z_i) \\
&\leq (k+1) \sup_{\text{rectangles } Z} U_n(Z), \\
W_n &\leq (k+1) \sup_{\text{rectangles } Z} U_n(Z).
\end{aligned}
$$

We may now use the Vapnik-Chervonenkis inequality (Theorems 12.5 and 13.8) to conclude that for all $\epsilon > 0$,

$$\mathbf{P}\left\{ \sup_{\text{rectangles } Z} U_n(Z) > \epsilon \right\} \leq 16n^{2d} e^{-n\epsilon^2/32}.$$

Armed with this, we have

$$P\{W_n > \epsilon\} \;\leq\; 16n^{2d}e^{-n\left(\frac{\epsilon}{2(k+1)}\right)^2/32},$$

$$P\{V_n \geq \delta\} \;\leq\; 16n^{2d}e^{-n\left(\frac{\delta}{2(k+1)}\right)^2/32}.$$

Both terms tend to 0 with n if $n/(k^2 \log n) \to \infty$ and $n\delta^2/(k^2 \log n) \to \infty$. Take $\delta = \epsilon/(3k)$ to satisfy our earlier side condition, and note that we need $n/(k^3 \log n) \to \infty$. We, in fact, have strong convergence to 0 for $L_n(\mathcal{P}_k) - L^*$ by the Borel-Cantelli lemma. $\square$

The crucial element in the proof of Theorem 20.9 is the fact that the number of sets in the partition $\mathcal{Q}$ grows at most linearly in i, the iteration number. Had it grown quadratically, say, it would not have been good enough—we would not have had guaranteed improvements of large enough sizes to push the error probability towards L^*. In the multidimensional version and in extension to other types of cuts (Problems 20.29, 20.30) this is virtually the only thing that must be verified. For hyperplane cuts, an additional inequality of the VC type is needed, extending that for classes of hyperplanes.

Our proof is entirely combinatorial and geometric and comes with explicit bounds. The only reference to the Bayes error we need is the quantity $l(\epsilon)$, which is the smallest value l for which

$$\inf_{\text{all } l \times l \text{ grids } \mathcal{G}_l} L(\mathcal{G}_l) \leq L^* + \epsilon.$$

It depends heavily on the distribution of course. Call $l(\epsilon)$ the *grid complexity* of the distribution, for lack of a better term. For example, if X is discrete and takes values on the hypercube $\{0, 1\}^d$, then $l(\epsilon) \leq 2^d$. If X takes values on $\{a_1, \ldots, a_M\}^d$, then $l(\epsilon) \leq (M + 1)^d$. If X has an arbitrary distribution on the real line and $\eta(x)$ is monotone, then $l(\epsilon) \leq 2$. However, if η is unimodal, $l(\epsilon) \leq 3$. In one dimension, $l(\epsilon)$ is sensitive to the number of places where the Bayes decision changes. In two dimensions, however, $l(\epsilon)$ measures the complexity of the distribution, especially near regions with $\eta(x) \approx 1/2$. When X is uniform on $[0, 1]^2$ and $Y = 1$ if and only if the components of X sum to less than one, then $l(\epsilon) = \Theta(1/\sqrt{\epsilon})$ as $\epsilon \downarrow 0$ (Problem 20.32), for example. The grid complexity is eminently suited to study rates of convergence, as it is explicitly featured in the inequalities of our proof. There is no room in this book for following this thread, however.

Problems and Exercises

PROBLEM 20.1. Let X be a random vector with density f on $\mathcal{R}^d$. Show that each component has a density as well.

PROBLEM 20.2. Show that both condition (1) and condition (2) of Theorem 6.1 are necessary for consistency of trees with the X-property.

PROBLEM 20.3. For a theoretical median tree in $\mathcal{R}^d$ with uniform marginals and k levels of splitting, show that $E\{H(X) + V(X)\} = 1/(2^{k/d})$ when k is a multiple of d.

PROBLEM 20.4. α-BALANCED TREES. Consider the following generalization of the median tree: at a node in the tree that represents n points, if a split occurs, both subtrees must be of size at least $\alpha(n - 1)$ for some fixed $\alpha > 0$. Repeat the splits for k levels, resulting in 2^k leaf regions. The tree has the X-property and the classifier is natural. However, the points at which cuts are made are picked among the data points in an arbitrary fashion. The splits rotate through the coordinate axes. Generalize the consistency theorem for median trees.

PROBLEM 20.5. Consider the following tree classifier. First we find the median according to one coordinate, and create two subtrees of sizes $\lfloor (n - 1)/2 \rfloor$ and $\lceil (n - 1)/2 \rceil$. Repeat this at each level, cutting the next coordinate axis in a rotational manner. Do not cut a node any further if either all data points in the corresponding region have identical Y_i values or the region contains less than k points. Prove that the obtained natural classifier is consistent whenever X has a density if $k \to \infty$ and $\log n/k \to 0$.

PROBLEM 20.6. Let μ be a probability measure with a density f on $\mathcal{R}^d$, and define

$$C = \left\{ x = (x^{(1)}, \ldots, x^{(d)}) : \mu\left([x^{(1)}, x^{(1)} + \epsilon] \times \cdots \times [x^{(d)}, x^{(d)} + \epsilon]\right) > 0 \text{ all } \epsilon > 0 \right\}.$$

Show that $\mu(C) = 1$. HINT: Proceed as in the proof of Lemma A.1 in the Appendix.

PROBLEM 20.7. Let $U_{(1)}, \ldots, U_{(n)}$ be uniform order statistics defining spacings $S_1, \ldots, S_{n+1}$ with $S_i = U_{(i)} - U_{(i-1)}$, if $U_{(n+1)} = 1$ and $U_{(0)} = 0$. Show that
  (1) $S_1, \ldots, S_{n+1}$ are identically distributed;
  (2) $P\{S_1 > x\} = (1 - x)^n$, $x \in [0, 1]$;
  (3) $E\{S_1\} = 1/(n + 1)$.

PROBLEM 20.8. In the proof of Theorem 20.3, we assumed in the second part that horizontal and vertical cuts were meted out independently. Return to the proof and see how you can modify it to take care of the forced alternating cuts.

PROBLEM 20.9. Let $X_1, X_2, \ldots, X_n$ be i.i.d. with common density f on the real line. Call X_i a record if $X_i = \min(X_1, \ldots, X_i)$. Let $R_i = I_{\{X_i \text{ is a record}\}}$. Show that $R_1, \ldots, R_n$ are independent and that $P\{R_i = 1\} = 1/i$. Conclude that the expected number of records is $\sim \log n$.

PROBLEM 20.10. THE DEEP K-D TREE. Assume that X has a density f and that all marginal densities are uniform.
  (1) Show that $k \to \infty$ implies that $\text{diam}(A(X)) \to 0$ in probability. Do this by arguing that $\text{diam}(A(X)) \to 0$ in probability for the chronological k-d tree with the same parameter k.
  (2) In a random k-d tree with n elements, show that the depth D of the last inserted node satisfies $D/(2 \log n) \to 1$ in probability. Argue first that you may restrict yourself to $d = 1$. Then write D as a sum of indicators of records. Conclude by computing $E\{D\}$ and $\text{Var}\{D\}$ or at least bounding these quantities in an appropriate manner.
  (3) Improve the second condition of Theorem 20.4 to $(k - 2 \log n)/\sqrt{\log n} \to -\infty$. (Note that it is possible that $\lim \sup_{n \to \infty} k/\log n = 2$.) HINT: Show that $|D - 2 \log n| = O\left(\sqrt{\log n}\right)$ in probability by referring to the previous part of this exercise.

PROBLEM 20.11. Consider a full-fledged k-d tree with $n + 1$ external node regions. Let L_n be the probability of error if classification is based upon this tree and if external node regions are assigned the labels (classes) of their immediate parents (see Figure 20.29).

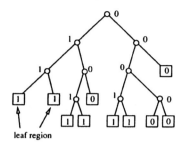

FIGURE 20.29. *The external nodes are assigned the classes of their parents.*

leaf region

Show that whenever X has a density, $E\{L_n\} \to L_{NN} \le 2L^*$ just as for the 1-nearest neighbor rule.

PROBLEM 20.12. CONTINUATION. Show that $L_n \to L_{NN}$ in probability.

PROBLEM 20.13. Meisel and Michalopoulos (1973) propose a binary tree classifier with perpendicular cuts in which all leaf regions are homogeneous, that is, all Y_i's for the X_i's in the same region are identical.

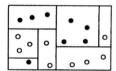

FIGURE 20.30. *Example of a tree partition into homogeneous regions.*

(1) If $L^* = 0$, give a stopping rule for constructing a consistent rule whenever X has a density.
(2) If $L^* = 0$, show that there exists a consistent homogeneous partition classifier with $o(n)$ expected number of cells.
(3) If $L^* > 0$, show a (more complicated) way of constructing a homogeneous partition that yields a tree classifier with $E\{L_n\} \to L^*$ whenever X has a density. HINT: First make a consistent nonhomogeneous binary tree classifier, and refine it to make it homogeneous.
(4) If $L^* > 0$, then show that the expected number of homogeneous regions is at least cn for some $c > 0$. Such rules are therefore not practical, unless $L^* = 0$. Overfitting will occur.

PROBLEM 20.14. Let X be uniform on $[0, 1]$ and let Y be independent of X with $P\{Y = 1\} = p$. Find a tree classifier based upon simple interval splitting for which each region has one data point, and

$$\liminf_{p \to 0} \frac{\liminf_{n \to \infty} E L_n}{p} \ge 3.$$

We know that for the nearest neighbor rule,

$$\lim_{p \to 0} \frac{\limsup_{n \to \infty} \mathbf{E}L_n}{p} = 2,$$

so that the given interval splitting method is worse than the nearest neighbor method. (This provides a counterexample to a conjecture of Breiman, Friedman, Olshen, and Stone (1984, pp.89–90).) HINT: Sort the points, and adjust the sizes of the intervals given to the class 0 and class 1 points in favor of the 1's. By doing so, the asymptotic probability of error can be made as close as desired to $(p^2 + 3p(1 - p))(1 - p)$.

PROBLEM 20.15. CONTINUATION. Let X be uniform on $[0, 1]^2$ and let Y be independent of X with $\mathbf{P}\{Y = 1\} = p$. Find a data-based tree classifier based upon perpendicular cuts for which each region has one data point, $\mathrm{diam}(A(X)) \to 0$ in probability (recall Theorems 6.1 and 21.2), and

$$\liminf_{p \to 0} \frac{\liminf_{n \to \infty} \mathbf{E}L_n}{p} = \infty.$$

Conclude that in $\mathcal{R}^d$, we can construct tree classifiers with one point per cell that are much worse than the nearest neighbor rule. HINT: The next two problems may help you with the construction and analysis.

PROBLEM 20.16. CONTINUATION. Draw an i.i.d. sample $X_1, \ldots, X_n$ from the uniform distribution on $[0, 1]^2$, and let $Y_1, \ldots, Y_n$ be i.i.d. and independent of the X_i's with $\mathbf{P}\{Y = 1\} = p$. Construct a binary tree partition with perpendicular cuts for $\{X_i : Y_i = 1\}$ such that every leaf region has one and only one point and $\mathrm{diam}(A(X)) \to 0$ in probability.
  (1) How would you proceed, avoiding putting X_i's on borders of regions?
  (2) Prove $\mathrm{diam}(A(X)) \to 0$ in probability.
  (3) Add the (X_i, Y_i) pairs with $Y_i = 0$ to the leaf regions, so that every region has one class-1 observation and zero or more class-0 observations. Give the class-1 observation the largest area containing no class-0 points, as shown in Figure 20.31. Show that this can always be done by adding perpendicular cuts and keeping at least one observation per region.

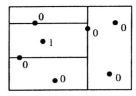

FIGURE 20.31.   *Cutting a rectangle by giving a large area to the single class-1 point.*

  (4) Partition all rectangles with more than one point further to finally obtain a one-point-per-leaf-region partition. If there are N points in a region of the tree before the class-1 and class-0 points are separated (thus, $N - 1$ class-0 points and one class-1 point), then show that the expected proportion of the region's area given to class 1, given N, times N tends to ∞. (An explicit lower bound will be helpful.) HINT: Use the next problem.
  (5) Write the probability of error for the rule in terms of areas of rectangles, and use part (4) to get an asymptotic lower bound.

(6)   Now let p tend to zero and get an asymptotic expression for your lower bound in terms of p. Compare this with $2p(1 - p)$, the asymptotic error probability of the 1-nearest neighbor rule.

PROBLEM 20.17. CONTINUATION. Draw a sample $X_1, \ldots, X_n$ of n i.i.d. observations uniformly distributed on $[0, 1]^d$. The rectangles defined by the origin and $X_1, \ldots, X_n$ as opposite vertices are denoted by $R_1, \ldots, R_n$, respectively. The probability content of R_i is clearly $\mu(R_i) = \prod_{i=1}^d X_i^{(j)}$. Study

$$M_n = \max\{\mu(R_i) : 1 \leq i \leq n, R_i \text{ does not contain } R_j \text{ for any } j \neq i\},$$

the probability content of the largest empty rectangle with a vertex at the origin. For $d = 1$, M_n is just the minimum of the X_i's, and thus $nM_n \xrightarrow{\mathcal{L}} \mathcal{E}$, where $\mathcal{E}$ is the exponential distribution. Also $E\{M_n\} = 1/(n + 1)$. For $d > 1$, M_n is larger. Show that $nM_n \to \infty$ in probability and try obtaining the first term in the rate of increase.

PROBLEM 20.18. Show that $\text{diam}(A(X)) \to 0$ in probability for the chronological quadtree whenever $k \to 0$ and X has a density. HINT: Mimic the proof for the chronological quadtree.

PROBLEM 20.19. Show that the deep quadtree is consistent if X has a density and k levels of splits are applied, where $k \to \infty$ and $k/\log n \to 0$.

PROBLEM 20.20. Consider a full-fledged quadtree with n nodes (and thus $n(2^d - 1) + 1$ leaf regions). Assign to each region the Y-label (class) of its parent in the quadtree. With this simple classifier, show that whenever X has a density, $E\{L_n\} \to 2E\{\eta(X)(1-\eta(X))\} = L_{NN}$. In particular, $\limsup_{n\to\infty} E\{L_n\} \leq 2L^*$ and the classifier is consistent when $L^* = 0$.

PROBLEM 20.21. In $\mathcal{R}^2$, partition the space as follows: X_1, X_2 define nine regions by vertical and horizontal lines through them. $X_3, \ldots, X_k$ are sent down to the appropriate subtrees in the 9-ary tree, and within each subtree with at least two points, the process is repeated recursively. A decision at x is by a majority vote (among $(X_{k+1}, Y_{k+1}), \ldots, (X_n, Y_n)$) among those X_i's in the same rectangle of the partition as x Show that if $k \to \infty, k/n \to 0$, the rule is consistent whenever X has a density.

PROBLEM 20.22. On the two-dimensional counterexample shown in the text for multivariate Stoller splits, prove that if splits are performed based upon a sample drawn from the distribution, and if we stop after k splits with k depending on n in such a way that $k/\log n \to 0$, then L_n, the conditional probability of error, satisfies $\liminf_{n\to\infty} E\{L_n\} \geq (1 - \epsilon)/2$. HINT: Bound the probability of ever splitting $[1, 2]^2$ anywhere by noting that the maximal difference between the empirical distribution functions for the first coordinate of X given $Y = 0$ and $Y = 1$ is $O\left(1/\sqrt{n}\right)$ when restricted to $[1, 2]^2$.

PROBLEM 20.23. Let X have the uniform distribution on $[0, 5]$ and let $Y = I_{\{2 < X < 3\}}$, so that $L^* = 0$. Construct a binary classification tree by selecting at each iteration the split that minimizes the impurity function I, where ψ is the Gini criterion. Consider just the first three splits made in this manner. Let L_n be the probability of error with the given rule (use a majority vote over the leaf regions). Show that $L_n \to 0$ in probability. Analyze the algorithm when the Gini criterion is replaced by the probability-of-error criterion.

PROBLEM 20.24. Let X be uniform on $[0, 1]$ and let Y be independent of X, with $P\{Y = 1\} = 2/3$. Draw a sample of size n from this distribution. Investigate where the first cut might take place, based upon minimizing the impurity function with $\psi(p, 1 - p)$ (the Gini criterion). (Once this is established, you will have discovered the nature of the classification tree, roughly speaking.)

PROBLEM 20.25. Complete the consistency proof of Theorem 20.7 for the raw BSP tree for $\mathcal{R}^2$ by showing that $\text{diam}(A(X)) \to 0$ in probability.

PROBLEM 20.26. BALANCED BSP TREES. We generalize median trees to allow splitting the space along hyperplanes.

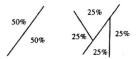

FIGURE 20.32. *A balanced BSP tree: Each hyperplane cut splits a region into two cells of the same cardinality.*

Assume that X has a density and that the tree possesses the X-property. Keep splitting until there are 2^k leaf regions, as with median trees. Call the trees balanced BSP trees.
  (1) Show that there are ways of splitting in $\mathcal{R}^2$ that lead to nonconsistent rules, regardless how k varies with n.
  (2) If every splitting hyperplane is *forced* to contain d data points (in $\mathcal{R}^d$) and these data points stay with the splitting node (they are not sent down to subtrees), then show that once again, there exists a splitting method that leads to nonconsistent rules, regardless of how k varies with n.

PROBLEM 20.27. Let X have a uniform distribution on the unit ball of $\mathcal{R}^d$. Let $Y = I_{\{\|X\| \geq 1/2\}}$, so that $L^* = 0$. Assume that we split the space by a hyperplane by minimizing an impurity function based upon the Gini criterion. If n is very large, where approximately would the cut take place (modulo a rotation, of course)?

PROBLEM 20.28. There exist singular continuous distributions that admit uniform $[0, 1]$ marginals in $\mathcal{R}^d$. Show, for example, that if X is uniformly distributed on the surface of the unit sphere of $\mathcal{R}^3$, then its three components are all uniformly distributed on $[-1, 1]$.

PROBLEM 20.29. Verify that Theorem 20.9 remains valid in $\mathcal{R}^d$.

PROBLEM 20.30. Prove that Theorem 20.9 remains valid if rectangular cuts are replaced by any of the elementary cuts shown on Figures 20.23 to 20.25, and such cuts are performed recursively k times, always by maximizing the decrease of the empirical error.

PROBLEM 20.31. Show that Theorem 20.11 remains valid if $k \to \infty$ and $k = o\left(\sqrt{n/\log n}\right)$. HINT: In the proof of the theorem, the bound on V_n and W_n is loose. You may get more efficient bounds by applying Theorem 21.1 from the next chapter.

PROBLEM 20.32. Study the behavior of the grid complexity as $\epsilon \downarrow 0$ for the following cases:
  (1) X is uniform on the perimeter of the unit circle of $\mathcal{R}^2$ with probability $1/2$ and X is uniform on $[0, 1]^2$ otherwise. Let $Y = 1$ if and only if X is on the perimeter of that circle (so that $L^* = 0$).
  (2) $X = \left(X^{(1)}, X^{(2)}\right)$ is uniform on $[0, 1]^2$ and $Y = 1$ if and only if $X^{(1)} + X^{(2)} \leq 1$.

21
Data-Dependent Partitioning

21.1 Introduction

In Chapter 9 we investigated properties of the regular histogram rule. Histogram classifiers partition the observation space $\mathcal{R}^d$ and classify the input vector X according to a majority vote among the Y_i's whose corresponding X_i's fall in the same cell of the partition as X. Partitions discussed in Chapter 9 could depend on the sample size n, but were not allowed to depend on the data D_n itself. We dealt mostly with grid partitions, but will now allow other partitions as well. Just consider clustered training observations $X_1, \ldots, X_n$. Near the cluster's center finer partitions are called for. Similarly, when the components have different physical dimensions, the scale of one coordinate axis is not related at all to the other scales, and some data-adaptive stretching is necessary. Sometimes the data are concentrated on or around a hyperplane. In all these cases, although consistent, the regular histogram method behaves rather poorly, especially if the dimension of the space is large. Therefore, it is useful to allow data-dependent partitions, while keeping a majority voting scheme within each cell.

The simplest data-dependent partitioning methods are based on *statistically equivalent blocks* in which each cell contains the same number of points. In one-dimensional problems statistically equivalent blocks reduce to *k-spacing estimates* where the k-th, $2k$-th, etc. order statistics determine the partition of the real line.

Sometimes, it makes sense to cluster the data points into groups such that points in a group are close to each other, and define the partition so that each group is in a different cell.

Many other data-dependent partitioning schemes have been introduced in the literature. In most of these algorithms the cells of the partition correspond to the leaves of a *binary tree*, which makes computation of the corresponding classification rule fast and convenient. Tree classifiers were dealt with in Chapter 20. Analysis of universal consistency for these algorithms and corresponding density estimates was begun by Abou-Jaoudé (1976b) and Gordon and Olshen (1984), (1978), (1980) in a general framework, and was extended, for example, by Breiman, Friedman, Olshen, and Stone (1984), Chen and Zhao (1987), Zhao, Krishnaiah, and Chen (1990), Lugosi and Nobel (1996), and Nobel (1994).

This chapter is more general than the chapter on tree classifiers, as every partition induced by a tree classifier is a valid partition of space, but not vice versa. The example below shows a rectangular partition of the plane that cannot be obtained by consecutive perpendicular cuts in a binary classification tree.

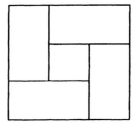

FIGURE 21.1. *A rectangular partition of* $[0, 1]^2$ *that cannot be achieved by a tree of consecutive cuts.*

In this chapter we first establish general sufficient conditions for the consistency of data-dependent histogram classifiers. Because of the complicated dependence of the partition on the data, methods useful for handling regular histograms have to be significantly refined. The main tool is a large deviation inequality for families of partitions that is related to the Vapnik-Chervonenkis inequality for families of sets. The reader is asked for forgiveness—we want to present a very generally applicable theorem and have to sacrifice (temporarily) by increasing the density of the text. However, as you will discover, the rewards will be sweet.

21.2   A Vapnik-Chervonenkis Inequality for Partitions

This is a technical section. We will use its results in the n xt section to establish a general consistency theorem for histogram rules with data-dependent partitions. The main goal of this section is to extend the basic Vapnik-Chervonenkis inequality (Theorem 12.5) to families of partitions from families of sets. The line of thought followed here essentially appears in Zhao, Krishnaiah, and Chen (1990) for rectangular partitions, and more generally in Lugosi and Nobel (1996). A substantial simplification in the proof was pointed out to us by Andrew Barron.

By a *partition* of $\mathcal{R}^d$ we mean a countable collection $\mathcal{P} = \{A_1, A_2, \ldots\}$ of subsets of $\mathcal{R}^d$ such that $\cup_{j=1}^{\infty} A_j = \mathcal{R}^d$ and $A_i \cap A_j = \emptyset$ if $i \neq j$. Each set A_j is called a *cell* of the partition $\mathcal{P}$.

Let M be a positive number. For each partition $\mathcal{P}$, we define $\mathcal{P}^{(M)}$ as the restriction of $\mathcal{P}$ to the ball S_M (recall that $S_M \subset \mathcal{R}^d$ denotes the closed ball of radius M centered at the origin). In other words, $\mathcal{P}^{(M)}$ is a partition of S_M, whose cells are obtained by intersecting the cells of $\mathcal{P}$ with S_M. We assume throughout that $\mathcal{P}$ is such that $|\mathcal{P}^{(M)}| < \infty$ for each $M < \infty$. We denote by $\mathcal{B}(\mathcal{P}^{(M)})$ the collection of all $2^{|\mathcal{P}^{(M)}|}$ sets obtained by unions of cells of $\mathcal{P}^{(M)}$.

Just as we dealt with classes of sets in Chapter 12, here we introduce families of partitions. Let $\mathcal{F}$ be a (possibly infinite) collection of partitions of $\mathcal{R}^d$. $\mathcal{F}^{(M)} = \{\mathcal{P}^{(M)} : \mathcal{P} \in \mathcal{F}\}$ denotes the family of partitions of S_M obtained by restricting members of $\mathcal{F}$ to S_M. For each M, we will measure the complexity of a family of partitions $\mathcal{F}^{(M)}$ by the shatter coefficients of the class of sets obtained as unions of cells of partitions taken from the family $\mathcal{F}^{(M)}$. Formally, we define the combinatorial quantity $\Delta_n(\mathcal{F}^{(M)})$ as follows: introduce the class $\mathcal{A}^{(M)}$ of subsets of $\mathcal{R}^d$ by

$$\mathcal{A}^{(M)} = \left\{ A \in \mathcal{B}(\mathcal{P}^{(M)}) \text{ for some } \mathcal{P}^{(M)} \in \mathcal{F}^{(M)} \right\},$$

and define

$$\Delta_n(\mathcal{F}^{(M)}) = s(\mathcal{A}^{(M)}, n),$$

the shatter coefficient of $\mathcal{A}^{(M)}$. $\mathcal{A}^{(M)}$ is thus the class of all sets that can be obtained as unions of cells of some partition of S_M in the collection $\mathcal{F}^{(M)}$. For example, if all members of $\mathcal{F}^{(M)}$ partition S_M into two sets, then $\Delta_n(\mathcal{F}^{(M)})$ is just the shatter coefficient of all sets in these partitions (with $\emptyset$ and S_M included in the collection of sets).

Let μ be a probability measure on $\mathcal{R}^d$ and let $X_1, X_2, \ldots$ be i.i.d. random vectors in $\mathcal{R}^d$ with distribution μ. For $n = 1, 2, \ldots$ let μ_n denote the empirical distribution of $X_1, \ldots, X_n$, which places mass $1/n$ at each of $X_1, \ldots, X_n$. To establish the consistency of data-driven histogram methods, we require information about the large deviations of random variables of the form

$$\sup_{\mathcal{P}^{(M)} \in \mathcal{F}^{(M)}} \sum_{A \in \mathcal{P}^{(M)}} |\mu_n(A) - \mu(A)|,$$

where $\mathcal{F}$ is an appropriate family of partitions.

REMARK. Just as in Chapter 12, the supremum above is not guaranteed to be measurable. In order to insure measurability, it is necessary to impose regularity conditions on uncountable collections of partitions. It suffices to mention that in all our applications, the measurability can be verified by checking conditions given, e.g., in Dudley (1978), or Pollard (1984). □

The following theorem is a consequence of the Vapnik-Chervonenkis inequality:

Theorem 21.1. (LUGOSI AND NOBEL (1996)). *Let $X_1, \ldots, X_n$ be i.i.d. random vectors in $\mathcal{R}^d$ with measure μ and empirical measure μ_n. Let $\mathcal{F}$ be a collection of partitions of $\mathcal{R}^d$. Then for each $M < \infty$ and $\epsilon > 0$,*

$$\mathbf{P}\left\{ \sup_{\mathcal{P}^{(M)} \in \mathcal{F}^{(M)}} \sum_{A \in \mathcal{P}^{(M)}} |\mu_n(A) - \mu(A)| > \epsilon \right\} \leq 8\Delta_n(\mathcal{F}^{(M)}) e^{-n\epsilon^2/512} + e^{-n\epsilon^2/2}.$$

PROOF. For a fixed partition $\mathcal{P}$, define A' as the set

$$A' = \bigcup_{A \in \mathcal{P}^{(M)}: \mu_n(A) \geq \mu(A)} A.$$

Then

$$\sum_{A \in \mathcal{P}^{(M)}} |\mu_n(A) - \mu(A)|$$

$$= \sum_{A \in \mathcal{P}^{(M)}: \mu_n(A) \geq \mu(A)} (\mu_n(A) - \mu(A))$$

$$+ \sum_{A \in \mathcal{P}^{(M)}: \mu_n(A) < \mu(A)} (\mu(A) - \mu_n(A))$$

$$= 2 \left(\mu_n(A') - \mu(A') \right) + \mu(S_M) - \mu_n(S_M)$$

$$\leq 2 \sup_{A \in \mathcal{B}(\mathcal{P}^{(M)})} |\mu_n(A) - \mu(A)| + \mu(S_M) - \mu_n(S_M).$$

Recall that the class of sets $\mathcal{B}(\mathcal{P}^{(M)})$ contains all $2^{|\mathcal{P}^{(M)}|}$ sets obtained by unions of cells of $\mathcal{P}^{(M)}$. Therefore,

$$\sup_{\mathcal{P}^{(M)} \in \mathcal{F}^{(M)}} \sum_{A \in \mathcal{P}^{(M)}} |\mu_n(A) - \mu(A)|$$

$$\leq 2 \sup_{\mathcal{P}^{(M)} \in \mathcal{F}^{(M)}} \sup_{A \in \mathcal{B}(\mathcal{P}^{(M)}) \cup \{S_M\}} |\mu_n(A) - \mu(A)| + \mu(S_M) - \mu_n(S_M).$$

Observe that the first term on the right-hand side of the inequality is a uniform deviation of the empirical measure μ_n from μ over a specific class of sets. The class contains all sets that can be written as unions of cells of a partition $\mathcal{P}^{(M)}$ in the class of partitions $\mathcal{F}^{(M)}$. This class of sets is just $\mathcal{A}^{(M)}$, defined above. The theorem now follows from the Vapnik-Chervonenkis inequality (Theorem 12.5), the definition of $\Delta_n(\mathcal{F}^{(M)})$, and Hoeffding's inequality (Theorem 8.1). $\square$

We will use a special application of Theorem 21.1, summarized in the following corollary:

COROLLARY 21.1. *Let $(X_1, Y_1), (X_2, Y_2) \ldots$ be a sequence of i.i.d. random pairs in $\mathcal{R}^d \times \{0, 1\}$ and let $\mathcal{F}_1, \mathcal{F}_2, \ldots$ be a sequence of families of partitions of $\mathcal{R}^d$. If for $M < \infty$*

$$\lim_{n \to \infty} \frac{\log \left(\Delta_n(\mathcal{F}_n^{(M)}) \right)}{n} = 0,$$

then

$$\sup_{\mathcal{P}^{(M)} \in \mathcal{F}_n^{(M)}} \sum_{A \in \mathcal{P}^{(M)}} \left| \frac{1}{n} \sum_{i=1}^{n} Y_i I_{\{X_i \in A\}} - \mathbf{E}\left\{ Y I_{\{X \in A\}} \right\} \right| \to 0$$

with probability one as n tends to infinity.

PROOF. Let ν be the measure of (X, Y) on $\mathcal{R}^d \times \{0, 1\}$, and let ν_n be the empirical measure corresponding to the sequence $(X_1, Y_1), \ldots, (X_n, Y_n)$. Using $\mathcal{F}_n^{(M)}$ we define a family $\mathcal{G}_n^{(M)}$ of partitions of $\mathcal{R}^d \times \{0, 1\}$ by

$$\mathcal{G}_n^{(M)} = \{\mathcal{P}^{(M)} \times \{0\} : \mathcal{P}^{(M)} \in \mathcal{F}_n^{(M)}\} \cup \{\mathcal{P}^{(M)} \times \{1\} : \mathcal{P}^{(M)} \in \mathcal{F}_n^{(M)}\},$$

where $\mathcal{P} \times \{0\} = \{A_1, A_2, \ldots\} \times \{0\} \overset{\text{def}}{=} \{A_1 \times \{0\}, A_2 \times \{0\}, \ldots\}$. We apply Theorem 21.1 for these families of partitions.

$$\sup_{\mathcal{P}^{(M)} \in \mathcal{F}_n^{(M)}} \sum_{A \in \mathcal{P}^{(M)}} \left| \frac{1}{n} \sum_{i=1}^n Y_i I_{\{X_i \in A\}} - \mathbf{E}\left\{ Y I_{\{X \in A\}} \right\} \right|$$

$$= \sup_{\mathcal{P}^{(M)} \in \mathcal{F}_n^{(M)}} \sum_{A \in \mathcal{P}^{(M)}} \left| \nu_n(A \times \{1\}) - \nu(A \times \{1\}) \right|$$

$$\leq \sup_{\bar{\mathcal{P}}^{(M)} \in \mathcal{G}_n^{(M)}} \sum_{\bar{A} \in \bar{\mathcal{P}}^{(M)}} \left| \nu_n(\bar{A}) - \nu(\bar{A}) \right|.$$

It is easy to see that $\Delta_n(\mathcal{G}_n^{(M)}) = \Delta_n(\mathcal{F}_n^{(M)})$. Therefore the stated convergence follows from Theorem 21.1 and the Borel-Cantelli lemma.□

Lemma 21.1. *Assume that the family $\mathcal{F}^{(M)}$ is such that the number of cells of the partitions in the family are uniformly bounded, that is, there is a constant N such that $|\mathcal{P}^{(M)}| \leq N$ for each $\mathcal{P}^{(M)} \in \mathcal{F}^{(M)}$. Then*

$$\Delta_n(\mathcal{F}^{(M)}) \leq 2^N \Delta_n^*(\mathcal{F}^{(M)}),$$

where $\Delta_n^*(\mathcal{F}^{(M)})$ is maximal number of different ways n points can be partitioned by members of $\mathcal{F}^{(M)}$.

EXAMPLE. FLEXIBLE GRIDS. As a first simple example, let us take in $\mathcal{F}_n$ all partitions into d-dimensional grids (called flexible grids as they may be visualized as chicken-wire fences with unequally spaced vertical and horizontal wires) in which cells are made up as Cartesian products of d intervals, and each coordinate axis contributes one of m_n intervals to these products. Clearly, if $\mathcal{P}$ is a member partition of $\mathcal{F}_n$, then $|\mathcal{P}| = m_n^d$. Nevertheless, there are uncountably many $\mathcal{P}$'s, as there are uncountably many intervals of the real line. This is why the finite quantity Δ_n comes in so handy. We will verify later that for each M,

$$\Delta_n(\mathcal{F}^{(M)}) \leq 2^{m_n^d} \binom{n + m_n}{n}^d,$$

so that the condition of Corollary 21.1 is fulfilled when

$$\lim_{n \to \infty} \frac{m_n^d}{n} = 0. \quad \square$$

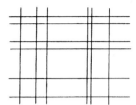

FIGURE 21.2. *A flexible grid partition.*

21.3   Consistency

In this section we establish a general consistency theorem for a large class of data-based partitioning rules. Using the training data D_n we produce a partition $\mathcal{P}_n = \pi_n(X_1, Y_1, \ldots, X_n, Y_n)$ according to a prescribed rule π_n. We then use the partition $\mathcal{P}_n$ in conjunction with D_n to produce a classification rule based on a majority vote within the cells of the (random) partition. That is, the training set is used twice and it is this feature of data-dependent histogram methods that distinguishes them from regular histogram methods.

Formally, an *n-sample partitioning rule* for $\mathcal{R}^d$ is a function π_n that associates every n-tuple of pairs $(x_1, y_1), \ldots, (x_n, y_n) \in \mathcal{R}^d \times \{0, 1\}$ with a measurable partition of $\mathcal{R}^d$. Associated with every partitioning rule π_n there is a fixed, non-random family of partitions

$$\mathcal{F}_n = \{\pi_n(x_1, y_1, \ldots, x_n, y_n) : (x_1, y_1) \ldots, (x_n, y_n) \in \mathcal{R}^d \times \{0, 1\}\}.$$

$\mathcal{F}_n$ is the family of partitions produced by the partitioning rule π_n for all possible realizations of the training sequence D_n. When a partitioning rule π_n is applied to the sequence $D_n = (X_1, Y_1), \ldots, (X_n, Y_n)$, it produces a random partition $\mathcal{P}_n = \pi_n(D_n) \in \mathcal{F}_n$. In what follows we suppress the dependence of $\mathcal{P}_n$ on D_n for notational simplicity. For every $x \in \mathcal{R}^d$ let $A_n(x)$ be the unique cell of $\mathcal{P}_n$ that contains the point x.

Now let $\{\pi_1, \pi_2, \ldots\}$ be a fixed sequence of partitioning rules. The classification rule $g_n(\cdot) = g_n(\cdot, D_n)$ is defined by taking a majority vote among the classes appearing in a given cell of $\mathcal{P}_n$, that is,

$$g_n(x) = \begin{cases} 0 & \text{if } \sum_{i=1}^{n} I_{\{X_i \in A_n(x), Y_i = 1\}} \leq \sum_{i=1}^{n} I_{\{X_i \in A_n(x), Y_i = 0\}} \\ 1 & \text{otherwise.} \end{cases}$$

We emphasize here that the partition $\mathcal{P}_n$ can depend on the vectors X_i and the labels Y_i. First we establish the strong consistency of the rules $\{g_n\}$ for a wide class of partitioning rules. As always, diam(A) denotes the *diameter* of a set A, that is, the maximum Euclidean distance between any two points of A:

$$\text{diam}(A) = \sup_{x, y \in A} \|x - y\|.$$

Theorem 21.2. (LUGOSI AND NOBEL (1996)). *Let* $\{\pi_1, \pi_2, \ldots\}$ *be a fixed sequence of partitioning rules, and for each n let* $\mathcal{F}_n$ *be the collection of partitions associated with the n-sample partitioning rule* π_n. *If*

(i) *for each* $M < \infty$

$$\lim_{n \to \infty} \frac{\log\left(\Delta_n(\mathcal{F}_n^{(M)})\right)}{n} = 0,$$

and

(ii) *for all balls* S_B *and all* $\gamma > 0$,

$$\lim_{n \to \infty} \mu\left(\{x : \text{diam}(A_n(x) \cap S_M) > \gamma\}\right) = 0$$

with probability one,
then the classification rule $\{g_n\}$ *corresponding to* π_n *satisfies*

$$L(g_n) \to L^*$$

with probability one. In other words, the rule $\{g_n\}$ *is strongly consistent.*

REMARK. In some applications we need a weaker condition to replace (ii) in the theorem. The following condition will do: for every $\gamma > 0$ and $\delta \in (0, 1)$

$$\lim_{n \to \infty} \inf_{T \subset \mathcal{R}^d : \mu(T) \geq 1 - \delta} \mu(\{x : \text{diam}(A_n(x) \cap T) > \gamma\}) = 0 \quad \text{with probability one.}$$

The verification of this is left to the reader (Problem 21.2). □

The proof, given below, requires quite some effort. The utility of the theorem is not immediately apparent. The length of the proof is indicative of the generality of the conditions in the theorem. Given a data-dependent partitioning rule, we must verify two things: condition (i) merely relates to the richness of the class of partitions of $\mathcal{R}^d$ that may possibly occur, such as flexible grids. Condition (ii) tells us that the rule should eventually make local decisions. From examples in earlier chapters, it should be obvious that (ii) is not necessary. Finite partitions of $\mathcal{R}^d$ necessarily have component sets with infinite diameter, hence we need a condition that states that such sets have small μ-measure. Condition (ii) requires that a randomly chosen cell have a small diameter. Thus, it may be viewed as the "with-probability-one" version of condition (1) of Theorem 6.1. However, the weaker version of condition (ii) stated in the above remark is more subtle. By considering examples in which μ has bounded support, more than just balls S_M are needed, as the sets of the partition near the boundary of the support may all have infinite diameter as well. Hence we introduced an infimum with respect to sets T over all T with $\mu(T) \geq 1 - \delta$.

It suffices to mention that (ii) is satisfied for all distributions in some of the examples that follow. Theorem 21.2 then allows us to conclude that such rules are strongly universally consistent. The theorem has done most of the digestion of such proofs, so we are left with virtually no work at all. Consistency results will follow like dominos falling.

PROOF OF THEOREM 21.2. Observe that the partitioning classifier g_n can be rewritten in the form

$$g_n(x) = \begin{cases} 0 & \text{if } \frac{\sum_{i=1}^n I_{\{Y_i=1\}} I_{\{X_i \in A_n(x)\}}}{\mu(A_n(x))} \leq \frac{\sum_{i=1}^n I_{\{Y_i=0\}} I_{\{X_i \in A_n(x)\}}}{\mu(A_n(x))} \\ 1 & \text{otherwise.} \end{cases}$$

Introduce the notation

$$\eta_n(x) = \frac{\sum_{i=1}^n Y_i I_{\{X_i \in A_n(x)\}}}{n\mu(A_n(x))}.$$

For any $\epsilon > 0$, there is an $M \in (0, \infty)$ such that $\mathbf{P}\{X \notin S_M\} < \epsilon$. Thus, by an application of Theorem 2.3 we see that

$$L(g_n) - L^*$$
$$\leq \quad \mathbf{P}\{g_n(X) \neq Y, X \in S_M | D_n\} - \mathbf{P}\{g^*(X) \neq Y, X \in S_M\} + \epsilon$$
$$\leq \quad \int_{S_M} |\eta(x) - \eta_n(x)|\mu(dx) + \int_{S_M} |(1 - \eta(x)) - \eta_n^{(0)}(x)|\mu(dx) + \epsilon,$$

where

$$\eta_n^{(0)}(x) = \frac{\sum_{i=1}^n (1 - Y_i) I_{\{X_i \in A_n(x)\}}}{n\mu(A_n(x))}.$$

By symmetry, since $\epsilon > 0$ is arbitrary, it suffices to show that for each $M < \infty$,

$$\int_{S_M} |\eta(x) - \eta_n(x)|\mu(dx) \to 0 \quad \text{with probability one.}$$

Fix $\epsilon > 0$ and let $r : \mathcal{R}^d \to \mathcal{R}$ be a continuous function with bounded support such that $\int_{S_M} |\eta(x) - r(x)|\mu(dx) < \epsilon$. Such r exists by Theorem A.8. Now define the auxiliary functions

$$\tilde{\eta}_n(x) = \frac{\mathbf{E}\left\{ Y I_{\{X \in A_n(x)\}} \middle| D_n \right\}}{\mu(A_n(x))}$$

and

$$\tilde{r}_n(x) = \frac{\mathbf{E}\left\{ r(X) I_{\{X \in A_n(x)\}} \middle| D_n \right\}}{\mu(A_n(x))}.$$

Note that both $\tilde{\eta}_n$ and $\tilde{r}_n$ are piecewise constant on the cells of the random partition $\mathcal{P}_n$. We may decompose the error as follows:

$$|\eta(x) - \eta_n(x)| \tag{21.1}$$
$$\leq \quad |\eta(x) - r(x)| + |r(x) - \tilde{r}_n(x)| + |\tilde{r}_n(x) - \tilde{\eta}_n(x)| + |\tilde{\eta}_n(x) - \eta_n(x)|.$$

The integral of the first term on the right-hand side is smaller than ϵ by the definition of r. For the third term we have

$$\int_{S_M} |\tilde{r}_n(x) - \tilde{\eta}_n(x)| \mu(dx)$$

$$= \sum_{A \in \mathcal{P}_n^{(M)}} |\mathbf{E}\{ Y I_{\{X \in A\}} | D_n \} - \mathbf{E}\{ r(X) I_{\{X \in A\}} | D_n \}|$$

$$= \sum_{A \in \mathcal{P}_n^{(M)}} \left| \int_A \eta(x) \mu(dx) - \int_A r(x) \mu(dx) \right|$$

$$\leq \int_{S_M} |\eta(x) - r(x)| \mu(dx) < \epsilon.$$

Consider the fourth term on the right-hand side of (21.1). Clearly,

$$\int_{S_M} |\tilde{\eta}_n(x) - \eta_n(x)| \mu(dx)$$

$$= \sum_{A \in \mathcal{P}_n^{(M)}} \left| \frac{1}{n} \sum_{i=1}^n Y_i I_{\{X_i \in A\}} - \mathbf{E}\{ Y I_{\{X \in A\}} | D_n \} \right|$$

$$\leq \sup_{\mathcal{P}^{(M)} \in \mathcal{F}_n^{(M)}} \sum_{A \in \mathcal{P}^{(M)}} \left| \frac{1}{n} \sum_{i=1}^n Y_i I_{\{X_i \in A\}} - \mathbf{E}\{ Y I_{\{X \in A\}} \} \right|.$$

It follows from the first condition of the theorem and Corollary 21.1 of Theorem 21.1 that

$$\int_{S_M} |\tilde{\eta}_n(x) - \eta_n(x)| \mu(dx) \to 0 \quad \text{with probability one.}$$

Finally, we consider the second term on the right-hand side of (21.1). Using Fubini's theorem we have

$$\int_{S_M} |r(x) - \tilde{r}_n(x)| \mu(dx)$$

$$= \sum_{A \in \mathcal{P}_n^{(M)}: \mu(A) \neq 0} \int_A \left| r(x) - \frac{\mathbf{E}\{ r(X) I_{\{X \in A\}} | D_n \}}{\mu(A)} \right| \mu(dx)$$

$$= \sum_{A \in \mathcal{P}_n^{(M)}: \mu(A) \neq 0} \frac{1}{\mu(A)} \int_A |r(x) \mu(A) - \mathbf{E}\{ r(X) I_{\{X \in A\}} | D_n \}| \mu(dx)$$

$$= \sum_{A \in \mathcal{P}_n^{(M)}: \mu(A) \neq 0} \frac{1}{\mu(A)} \int_A \left| r(x) \int_A \mu(dy) - \int_A r(y) \mu(dy) \right| \mu(dx)$$

$$\leq \sum_{A \in \mathcal{P}_n^{(M)}: \mu(A) \neq 0} \frac{1}{\mu(A)} \int_A \int_A |r(x) - r(y)| \mu(dx) \mu(dy).$$

Fix $\delta \in (0, 1)$. As r is uniformly continuous, there exists a number $\gamma > 0$ such that if $\text{diam}(A) < \gamma$, then $|r(x) - r(y)| < \delta$ for every $x, y \in A$. In addition, there is a constant $M < \infty$ such that $|r(x)| \le M$ for every $x \in \mathcal{R}^d$. Fix n now and consider the integrals

$$\frac{1}{\mu(A)} \int_A \int_A |r(x) - r(y)| \mu(dx)\mu(dy)$$

appearing in the sum above. We always have the upper bound

$$\frac{1}{\mu(A)} \int_A \int_A |r(x) - r(y)| \mu(dx)\mu(dy) \le 2M\mu(A).$$

Assume now that $A \in \mathcal{P}_n^{(M)}$ has $\text{diam}(A) < \gamma$. Then we can write

$$\frac{1}{\mu(A)} \int_A \int_A |r(x) - r(y)| \mu(dx)\mu(dy) \le \delta\mu(A).$$

Summing over the cells $A \in \mathcal{P}_n^{(M)}$ with $\mu(A) > 0$, these bounds give

$$\int_{S_M} |r(x) - \bar{r}_n(x)| \mu(dx)$$

$$\le \sum_{A \in \mathcal{P}_n^{(M)} : \text{diam}(A) \ge \gamma} 2M\mu(A) + \sum_{A \in \mathcal{P}_n^{(M)} : \text{diam}(A) < \gamma} \delta\mu(A)$$

$$\le 2M\mu(\{x : \text{diam}(A_n(x))\}) \ge \gamma) + \delta.$$

Letting n tend to infinity gives

$$\limsup_{n \to \infty} \int_{S_M} |r(x) - \bar{r}_n(x)| \mu(dx) \le \delta \quad \text{with probability one}$$

by the second condition of the theorem. Summarizing,

$$\limsup_{n \to \infty} \int_{S_M} |\eta(x) - \eta_n(x)| \mu(dx) \le 2\epsilon + \delta \quad \text{with probability one}.$$

Since ϵ and δ are arbitrary, the theorem is proved. $\square$

21.4   Statistically Equivalent Blocks

In this section we apply Theorem 21.2 both to classifiers based on uniform spacings in one dimension, and to their extension to multidimensional problems. We refer to these as rules based on statistically equivalent blocks. The order statistics of the components of the training data are used to construct a partition into rectangles. All such classifiers are invariant with respect to all strictly monotone transformations of the coordinate axes. The simplest such rule is the k-spacing rule studied in Chapter 20 (see Theorem 20.1). Generalizations are possible in several ways. Theorem 21.2 allows us to handle partitions depending on the whole data sequence—and not only on the X_i's. The next simple result is sometimes useful.

Theorem 21.3. *Consider a data-dependent partitioning classifier on the real line that partitions $\mathcal{R}$ into intervals in such a way that each interval contains at least a_n and at most b_n points. Assume that X has a nonatomic distribution. Then the classifier is strongly consistent whenever $a_n \to \infty$ and $b_n/n \to 0$ as $n \to \infty$.*

PROOF. We check the conditions of Theorem 21.2. Let $\mathcal{F}_n$ contain all partitions of $\mathcal{R}$ into $m = \lceil n/a_n \rceil$ intervals. Since for each M, all partitions in $\mathcal{F}_n^{(M)}$ have at most m cells, we can bound $\Delta_n(\mathcal{F}_n^{(M)})$ according to the Lemma 21.1. By the lemma, $\Delta_n(\mathcal{F}_n^{(M)})$ does not exceed 2^m times the number of different ways n points can be partitioned into m intervals. A little thought confirms that this number is

$$\binom{n+m}{n},$$

and therefore,

$$\Delta_n(\mathcal{F}_n^{(M)}) \leq 2^m \binom{n+m}{n}.$$

Let $\mathcal{H}$ denote the binary entropy function, $\mathcal{H}(x) = -x \log(x) - (1-x)\log(1-x)$ for $x \in (0,1)$. Note that $\mathcal{H}$ is symmetric about $1/2$ and that $\mathcal{H}$ is increasing for $0 < x \leq 1/2$. By the inequality of Theorem 13.4, $\log \binom{s}{t} \leq s\mathcal{H}(t/s)$. Therefore, it is easy to see that

$$\begin{aligned}
\log \Delta_n(\mathcal{F}_n^{(M)}) &\leq m + (n+m)\mathcal{H}\left(\frac{m}{n+m}\right) \\
&\leq n/a_n + 2n\mathcal{H}(1/a_n) + 1.
\end{aligned}$$

As $\mathcal{H}(x) \to 0$ when $x \to 0$, the condition $a_n \to \infty$ implies that

$$\frac{1}{n} \log \Delta_n(\mathcal{F}_n^{(M)}) \to 0,$$

which establishes condition (i).

To establish condition (ii) of Theorem 21.2, we proceed similarly to the proof of Theorem 20.1. Fix $\gamma, \epsilon > 0$. There exists an interval $[-M, M]$ such that $\mu([-M, M]^c) < \epsilon$, and consequently

$$\mu(\{x : \text{diam}(A_n(x)) > \gamma\}) \leq \epsilon + \mu(\{x : \text{diam}(A_n(x)) > \gamma\} \cap [-M, M]),$$

where $A_n(x)$ denotes the cell of the partition $\mathcal{P}_n$ containing x. Among the intervals of $\mathcal{P}_n$, there can be at most $2 + 2M/\gamma$ disjoint intervals of length greater than γ in

$[-M, M]$. Thus we may bound the second term on the right-hand side above by

$$\mu(\{x : \text{diam}(A_n(x)) > \gamma\} \cap [-M, M])$$

$$\leq \left(2 + \frac{2M}{\gamma}\right) \max_{A \in \mathcal{P}_n} \mu(A)$$

$$\leq \left(2 + \frac{2M}{\gamma}\right) \left(\max_{A \in \mathcal{P}_n} \mu_n(A) + \max_{A \in \mathcal{P}_n} |\mu(A) - \mu_n(A)|\right)$$

$$\leq \left(2 + \frac{2M}{\gamma}\right) \left(\frac{b_n}{n} + \sup_{A \in \mathcal{A}} |\mu(A) - \mu_n(A)|\right),$$

where $\mathcal{A}$ is the set of all intervals in $\mathcal{R}$. The first term in the parenthesis converges to zero by the second condition of the theorem, while the second term goes to zero with probability one by an obvious extension of the classical Glivenko-Cantelli theorem (Theorem 12.4). Summarizing, we have shown that for any $\gamma, \epsilon > 0$

$$\limsup_{n \to \infty} \mu(\{x : \text{diam}(A_n(x)) > \gamma\}) \leq \epsilon \quad \text{with probability one.}$$

This completes the proof. $\square$

The d-dimensional generalizations of k-spacing rules include rules based upon statistically equivalent blocks, that is, partitions with sets that contain k points each. It is obvious that one can proceed in many ways, see, for example, Anderson (1966), Patrick (1966), Patrick and Fisher (1967), Quesenberry and Gessaman (1968) and Gessaman and Gessaman (1972).

As a first example, consider the following algorithm: the k-th smallest $x^{(1)}$-coordinate among the training data defines the first cut. The (infinite) rectangle with $n - k$ points is cut according to the $x^{(2)}$-axis, isolating another k points. This can be repeated on a rotational basis for all coordinate axes. Unfortunately, the classifier obtained this way is not consistent. To see this, observe that if k is much smaller than n—a clearly necessary requirement for consistency—then almost all cells produced by the cuts are long and thin. We sketched a distribution in Figure 21.3 for which the error probability of this classifier fails to converge to L^*. The details are left to the reader (Problem 21.3). This example highlights the importance of condition (ii) of Theorem 21.2, that is, that the diameters of the cells should shrink in some sense as $n \to \infty$.

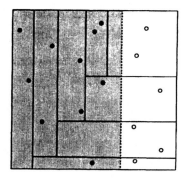

FIGURE 21.3. *A nonconsistent k-block algorithm (with k = 2 in the picture). $\eta(x) = 1$ in the shaded area and $\eta(x) = 0$ in the white area.*

Rules have been developed in which the rectangular partition depends not only upon the X_i's in the training sequence, but also upon the Y_i's (see, e.g., Henrichon and Fu (1969), Meisel and Michalopoulos (1973) and Friedman (1977)). For example, Friedman cuts the axes at the places where the absolute differences between the marginal empirical distribution functions are largest, to insure minimal empirical error after the cut. His procedure is based upon Stoller splits (see Chapters 4 and 20).

Rules depending on the coordinatewise ranks of data points are interesting because they are invariant under monotone transformations of the coordinate axes. This is particularly important in practice when the components are not physically comparable. "Distribution-free" is an adjective often used to point out a property that is universally valid. For such methods in statistics, see the survey of Das Gupta (1964). Statistically equivalent sets in partitions are called "distribution-free" because the measure $\mu(A)$ of a set in the partition does not depend upon the distribution of X. We already noted a similar distribution-free behavior for k-d trees and median trees (Chapter 20). There is no reason to stay with rectangular-shaped sets (Anderson and Benning (1970), Beakley and Tuteur (1972)) but doing so greatly simplifies the interpretation of a classifier. In this book, to avoid confusion, we reserve the term "distribution-free" for consistency results or other theoretical properties that hold for all distributions of (X, Y).

It is possible to define consistent partitions that have statistically equivalent sets. To fix the ideas, we take Gessaman's rule (1970) as our prototype rule for further study (note: for hypothesis testing, this partition was already noted by Anderson (1966)). For each n, let $m = \lceil (n/k_n)^{1/d} \rceil$. Project the vectors $X_1, \ldots, X_n$ onto the first coordinate axis, and then partition the data into m sets using hyperplanes

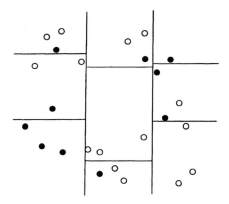

FIGURE 21.4. *Gessaman's partition with m = 3.*

perpendicular to that axis, in such a way that each set contains an equal number of points (except, possibly, the rightmost set, where fewer points may fall if n is not a multiple of m). We obtain m cylindrical sets. In the same fashion, cut each of these cylindrical sets, along the second axis, into m boxes such that each box contains the same number of data points. Continuing in the same way along the remaining coordinate axes, we obtain m^d rectangular cells, each of which (with the exception of those on the boundary) contains about k_n points. The classification rule g_n uses a majority vote among those Y_i's for which X_i lies within a given cell. Consistency of this classification rule can be established by an argument similar to that used for the k_n-spacing rule above. One needs to check that the conditions of Theorem 21.2 are satisfied. The only minor difference appears in the computation of Δ_n, which in this case is bounded from above by $2^{m^d} \binom{n+m}{n}^d$. The following theorem summarizes the result:

Theorem 21.4. *Assume that the marginal distributions of X in $\mathcal{R}^d$ are nonatomic. Then the partitioning classification rule based on Gessaman's rule is strongly consistent if $k_n \to \infty$ and $k_n/n \to 0$ as n tends to infinity.*

To consider distributions with possibly atomic marginals, the partitioning algorithm must be modified, since every atom has more than k_n points falling on it for large n. With a proper modification, a *strongly universally consistent* rule can be obtained. We leave the details to the reader (Problem 21.4).

REMARK. Consistency of Gessaman's classification scheme can also be derived from the results of Gordon and Olshen (1978) under the additional condition $k_n/\sqrt{n} \to \infty$. Results of Breiman, Friedman, Olshen, and Stone (1984) can be used to improve this condition to $k_n / \log n \to \infty$. Theorem 21.4 guarantees consistency under the weakest possible condition $k_n \to \infty$. $\square$

21.5 Partitioning Rules Based on Clustering

Clustering is one of the most widely used methods in statistical data analysis. Typical clustering schemes divide the data into a finite number of disjoint groups by minimizing some empirical error measure, such as the average squared distance from cluster centers (see Hartigan (1975)). In this section we outline the application of our results to classification rules based on k-means clustering of unlabeled observations.

As a first step, we divide $X_1, \ldots, X_n$ into k_n disjoint groups having cluster centers $a_1, \ldots, a_{k_n} \in \mathcal{R}^d$. The vectors $a_1, \ldots, a_{k_n}$ are chosen to minimize the *empirical squared Euclidean distance error*,

$$e_n(b_1, \ldots, b_{k_n}) = \frac{1}{n} \sum_{i=1}^{n} \min_{1 \leq j \leq k_n} \|X_i - b_j\|^2$$

over all the nearest-neighbor clustering rules having k_n representatives $b_1, \ldots, b_{k_n} \in \mathcal{R}^d$, where $\| \cdot \|$ denotes the usual Euclidean norm. Note that the choice of cluster centers depends only on the vectors X_i, not on their labels. For the behavior of $e_n(a_1, \ldots, a_{k_n})$, see Problem 29.4.

The vectors $a_1, \ldots, a_{k_n}$ give rise to a *Voronoi* partition $\mathcal{P}_n = \{A_1, \ldots, A_{k_n}\}$ in a natural way: for each $j \in \{1, \ldots, k_n\}$, let

$$A_j = \{x : \|x - a_j\| \leq \|x - a_i\| \text{ for all } i = 1, \ldots, k_n\}.$$

Ties are broken by assigning points on the boundaries to the vector that has the smallest index.

The classification rule g_n is defined in the usual way: $g_n(x)$ is a majority vote among those Y_j's such that X_j falls in $A_n(x)$. If the measure μ of X has a bounded support, Theorem 21.2 shows that the classification rule $\{g_n\}$ is strongly consistent if k_n grows with n at an appropriate rate. Note that this rule is just another of the *prototype nearest neighbor rules* that we discussed in Chapter 19.

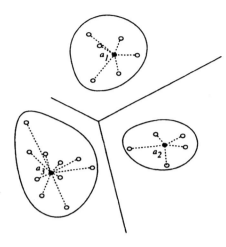

FIGURE 21.5. *Example of partitioning based on clustering with $k_n = 3$. The criterion we minimize is the sum of the squares of the distances of the X_i's to the a_j's.*

Theorem 21.5. (LUGOSI AND NOBEL (1996)). *Assume that there is a bounded set $A \subset \mathcal{R}^d$ such that $\mathbf{P}\{X \in A\} = 1$. Let $\{k_n\}$ be a sequence of integers for which*

$$k_n \to \infty \quad and \quad \frac{k_n^2 \log n}{n} \to 0 \quad as \; n \to \infty.$$

Let $g_n(\cdot, D_n)$ be the histogram classification rule based on the Voronoi partition of k_n cluster centers minimizing the empirical squared Euclidean distance error. Then

$$L(g_n) \to L^*$$

with probability one as n tends to infinity. If $d = 1$ or 2, then the second condition on k_n can be relaxed to

$$\frac{k_n \log n}{n} \to 0.$$

PROOF. Again, we check the conditions of Theorem 21.2. Let $\mathcal{F}_n$ consist of all Voronoi partitions of k_n points in $\mathcal{R}^d$. As each partition consists of k_n cells, we may use Lemma 21.1 to bound $\Delta_n(\mathcal{F}_n^{(M)})$. Clearly, boundaries between cells are subsets of hyperplanes. Since there are at most $k_n(k_n - 1)/2$ boundaries between the k_n Voronoi cells, each cell of a partition in $\mathcal{F}_n$ is a polytope with at most $k_n(k_n - 1)/2 < k_n^2$ faces. By Theorem 13.9, n fixed points in $\mathcal{R}^d$, $d \geq 2$, can be split by hyperplanes in at most n^{d+1} different ways. It follows that for each

$M < \infty$, $\Delta_n(\mathcal{F}_n^{(M)}) \leq 2^{k_n} n^{(d+1)k_n^2}$, and consequently

$$\frac{1}{n} \log \Delta_n(\mathcal{F}_n^{(M)}) \leq \frac{k_n}{n} + \frac{(d+1)k_n^2 \log n}{n} \to 0$$

by the second condition on the sequence $\{k_n\}$. Thus condition (i) of Theorem 21.2 is satisfied.

It remains to establish condition (ii) of Theorem 21.2. This time we need the weaker condition mentioned in the remark after the theorem, that is, that for every $\gamma > 0$ and $\delta \in (0, 1)$

$$\inf_{T:\mu(T)\geq 1-\delta} \mu(\{x : \mathrm{diam}(A_n(x) \cap T) > \gamma\}) \to 0 \quad \text{with probability one.}$$

Clearly, we are done if we can prove that there is a sequence of subsets T_n of $\mathcal{R}^d$ (possibly depending on the data D_n) such that $\mu(T_n) \to 1$ with probability one, and

$$\mu(\{x : \mathrm{diam}(A_n(x) \cap T_n) > \gamma\}) = 0.$$

To this end, let $a_1, \ldots, a_{k_n}$ denote the optimal cluster centers corresponding to D_n, and define

$$T_n = \bigcup_{j=1}^{k_n} S_{a_j, \gamma/2} \cap A_j,$$

where $S_{x,r}$ is the ball of radius r around the vector x. Clearly, $x \in T_n$ implies that $\|x - a(x)\| < \gamma/2$, where $a(x)$ denotes the closest cluster center to x among $a_1, \ldots, a_{k_n}$. But since

$$A_n(x) \cap T_n = A_n(x) \cap S_{a(x),\gamma/2},$$

it follows that

$$\mathrm{diam}(A_n(x) \cap T_n) \leq \mathrm{diam}(S_{a(x),\gamma/2}) = \gamma,$$

and

$$\mu(\{x : \mathrm{diam}(A_n(x) \cap T_n) > \gamma\}) = 0.$$

It remains to show that $\mu(T_n) \to 1$ with probability one as $n \to \infty$. Using Markov's inequality, we may write

$$
\begin{aligned}
1 - \mu(T_n) &= \mathbf{P}\{X \notin T_n | D_n\} \\
&\leq \mathbf{P}\left\{ \min_{1 \leq j \leq k_n} \|X - a_j\|^2 > \left(\frac{\gamma}{2}\right)^2 \middle| D_n \right\} \\
&\leq \frac{\mathbf{E}\left\{ \min_{1 \leq j \leq k_n} \|X - a_j\|^2 \middle| D_n \right\}}{(\gamma/2)^2}.
\end{aligned}
$$

Using a large-deviation inequality for the empirical squared error of nearest-neighbor clustering schemes, it can be shown (see Problem 29.4) that if X has bounded support, then

$$\mathbf{E}\left\{ \min_{1 \leq j \leq k_n} \|X - a_j\|^2 \middle| D_n \right\} - \min_{b_1,\ldots,b_{k_n} \in \mathcal{R}^d} \mathbf{E}\left\{ \min_{1 \leq j \leq k_n} \|X - b_j\|^2 \right\} \to 0$$

with probability one if $k_n \log n/n \to 0$ as $n \to \infty$. Moreover, it is easy to see that

$$\min_{b_1,\dots,b_{k_n} \in \mathcal{R}^d} \mathbf{E} \left\{ \min_{1 \leq j \leq k_n} \|X - b_j\|^2 \right\} \to 0$$

as $k_n \to \infty$. It follows that $\mu(T_n) \to 1$, as desired.

If $d = 1$, then the cells of the Voronoi partition are intervals on the real line, and therefore, $\Delta_n(\mathcal{F}_n^{(M)}) \leq 2^{k_n} n^{k_n}$. Similarly, if $d = 2$, then the number of hyperplanes defining the Voronoi partition increases *linearly* with k_n. To see this, observe that if we connect centers of neighboring clusters by edges, then we obtain a *planar graph*. From Euler's theorem (see, e.g., Edelsbrunner (1987, p.242)), the number of edges in a planar graph is bounded by $3N - 6$, $N \geq 3$, where N is the number of vertices. Thus, in order to satisfy condition (i), it suffices that $k_n \log n/n \to 0$ in both cases. $\square$

REMARK. In the theorem above we assume that the cluster centers $a_1, \dots, a_k$ are empirically optimal in the sense that they are chosen to minimize the empirical squared Euclidean distance error

$$\frac{1}{n} \sum_{i=1}^{n} \min_{1 \leq j \leq k_n} \|X_i - a_j\|^2.$$

Practically speaking, it is hard to determine minimum. To get around this difficulty, several fast algorithms have been proposed that approximate the optimum (see Hartigan (1975) for a survey). Perhaps the most popular algorithm is the so-called *k-means* clustering method, also known in the theory of quantization as the *Lloyd-Max algorithm* (Lloyd (1982), Max (1960), Linde, Buzo, and Gray (1980)). The iterative method works as follows:

STEP 1.   Take k initial centers $a_1^{(0)}, \dots, a_k^{(0)} \in \mathcal{R}^d$, and set $i = 0$.

STEP 2.   Cluster the data points $X_1, \dots, X_n$ around the centers $a_1^{(i)}, \dots, a_k^{(i)}$ into k sets such that the m-th set $C_m^{(i)}$ contains the X_j's that are closer to $a_m^{(i)}$ than to any other center. Ties are broken in favor of smaller indices.

STEP 3.   Determine the new centers $a_1^{(i+1)}, \dots, a_k^{(i+1)}$ as the averages of the data points within the clusters:

$$a_m^{(i+1)} = \frac{\sum_{j:X_j \in C_m^{(i)}} X_j}{|C_m^{(i)}|}.$$

STEP 4.   Increase i by one, and repeat Steps 1 and 2 until there are no changes in the cluster centers.

It is easy to see that each step of the algorithm decreases the empirical squared Euclidean distance error. On the other hand, the empirical squared error can take finitely many different values during the execution of the algorithm. Therefore the algorithm halts in finite time. By inspecting the proof of Theorem 21.5, it is not hard to see that consistency can also be proved for partitions given by the suboptimal cluster centers obtained by the k-means method, provided that it is initialized appropriately (see Problem 21.6). □

21.6 Data-Based Scaling

We now choose the grid size h in a cubic histogram rule in a data-dependent manner and denote its value by H_n. Theorem 21.2 implies the following general result:

Theorem 21.6. *Let g_n be the cubic histogram classifier based on a partition into cubes of size H_n. If*

   (a) $\lim_{n \to \infty} H_n = 0$ *and*

   (b) $\lim_{n \to \infty} n H_n^d = \infty$ *with probability one,*

then the partitioning rule is strongly universally consistent.

To prove the theorem, we need the following auxiliary result:

Lemma 21.2. *Let $Z_1, Z_2, \ldots$ be a sequence of nonnegative random variables. If $\lim_{n \to \infty} Z_n = 0$ with probability one, then there exists a sequence $a_n \downarrow 0$ of positive numbers such that $\lim_{n \to \infty} I_{\{Z_n \geq a_n\}} = 0$ with probability one.*

PROOF. Define $V_n = \sup_{m \geq n} Z_m$. Then clearly, $V_n \downarrow 0$ with probability one. We can find a subsequence $n_1, n_2, \ldots$ of positive integers such that for each k,

$$P\{V_{n_k} \geq 1/k\} \leq 2^{-k}.$$

Then the Borel-Cantelli lemma implies that

$$\lim_{k \to \infty} I_{\{V_{n_k} \geq 1/k\}} = 0 \text{ with probability one.} \tag{21.2}$$

Define $a_n = 1/k$ for $n_k \leq n < n_{k+1}$. Then, for $n_k \leq n < n_{k+1}$,

$$I_{\{V_n \geq a_n\}} = I_{\{V_n \geq 1/k\}} \leq I_{\{V_{n_k} \geq 1/k\}}.$$

The fact that $V_n \geq Z_n$ and (21.2) imply the statement. □

PROOF OF THEOREM 21.6.   Let $\{a_n\}$ and $\{b_n\}$ be sequences of positive numbers with $a_n < b_n$. Then

$$\left\{ \lim_{n \to \infty} L(g_n) = L^* \right\} \subseteq \left\{ \lim_{n \to \infty} L(g_n) = L^*, H_n \in [a_n, b_n] \right\} \bigcup \{H_n \notin [a_n, b_n]\}.$$

It follows from Lemma 21.2 that there exist sequences of positive numbers $\{a_n\}$ and $\{b_n\}$ satisfying $a_n < b_n$, $b_n \to 0$, and $na_n^d \to \infty$ as $n \to \infty$ such that

$$\lim_{n \to \infty} I_{\{H_n \notin [a_n, b_n]\}} = 0 \quad \text{with probability one.}$$

Therefore we may assume that for each n, $\mathbf{P}\{H_n \in [a_n, b_n]\} = 1$. Since $b_n \to 0$, condition (ii) of Theorem 21.2 holds trivially, as all diameters of all cells are inferior to $b_n \sqrt{d}$. It remains to check condition (i). Clearly, for each $M < \infty$, each partition in $\mathcal{F}_n^{(M)}$ contains less than $c_0(1/a_n)^d$ cells, where the constant c_0 depends on M and d only. On the other hand, it is easy to see that n points can not be partitioned more than $c_1 n(1/a_n)^d$ different ways by cubic-grid partitions with cube size $h \geq a_n$ for some other constant c_1. Therefore, for each $M < \infty$,

$$\Delta_n(\mathcal{F}^{(M)}) \leq 2^{c_0(1/a_n)^d} \frac{c_1 n}{a_n^d},$$

and condition (i) is satisfied. $\square$

In many applications, different components of the feature vector X correspond to different physical measurements. For example, in a medical application, the first coordinate could represent blood pressure, the second cholesterol level, and the third the weight of the patient. In such cases there is no reason to use cubic histograms, because the resolution of the partition h_n along the coordinate axes depends on the apparent scaling of the measurements, which is rather arbitrary. Then one can use scale-independent partitions such as methods based on order statistics described earlier. Alternatively, one might use rectangular partitions instead of cubic ones, and let the data decide the scaling along the different coordinate axes. Again, Theorem 21.2 can be used to establish conditions of universal consistency of the classification rule corresponding to data-based rectangular partitions:

Theorem 21.7. *Consider a data-dependent histogram rule when the cells of the partition are all rectangles of the form*

$$[k_1 H_{n1}, (k_1 + 1)H_{n1}) \times \ldots \times [k_d H_{nd}, (k_d + 1)H_{nd}),$$

where $k_1, \ldots, k_d$ run through the set of integers, and the edge sizes of the rectangles $H_{n1}, \ldots, H_{nd}$ are determined from the data D_n. If as $n \to \infty$

$$H_{ni} \to 0 \text{ for each } 1 \leq i \leq d, \text{ and } nH_{n1} \cdots H_{nd} \to \infty \text{ with probability one,}$$

then the data-dependent rectangular partitioning rule is strongly universally consistent.

To prove this, just check the conditions of Theorem 21.2 (Problem 21.7). We may pick, for example, $H_{n1}, \ldots, H_{nd}$ to minimize the *resubstitution* estimate

$$\sum_{i=1}^{n} I_{\{g_n(X_i) \neq Y_i\}}$$

subject of course to certain conditions, so that $\sum_{i=1}^{d} H_{ni} \to 0$ with probability one, yet $n \prod_{i=1}^{d} H_{ni} \to \infty$ with probability one. See Problem 21.8.

21.7 Classification Trees

Consider a partition of the space obtained by a binary classification tree in which each node dichotomizes its set by a hyperplane (see Chapter 20 for more on classification trees). The construction of the tree is stopped according to some unspecified rule, and classification is by majority vote over the convex polytopes of the partition.

The following corollary of Theorem 21.2 generalizes (somewhat) the consistency results in the book by Breiman, Friedman, Olshen, and Stone (1984):

Theorem 21.8. (LUGOSI AND NOBEL (1996)). *Let g_n be a binary tree classifier based upon at most $m_n - 1$ hyperplane splits, where $m_n = o(n / \log n)$. If, in addition, condition (ii) of Theorem 21.2 is satisfied, then g_n is strongly consistent. In particular, the rule is strongly consistent if condition (ii) of Theorem 21.2 holds and every cell of the partition contains at least k_n points, where $k_n / \log n \to \infty$.*

PROOF. To check condition (i) of Theorem 21.2, recall Theorem 13.9 which implies that $n \geq 2$ points in a d-dimensional Euclidean space can be dichotomized by hyperplanes in at most n^{d+1} different ways. From this, we see that the number of different ways n points of $\mathcal{R}^d$ can be partitioned by the rule g_n can be bounded by

$$\left(n^{d+1}\right)^{m_n} = 2^{(d+1)m_n \log n},$$

as there are not more than m_n cells in the partition. Thus,

$$\Delta_n(\mathcal{F}_n) \leq 2^{m_n} 2^{(d+1)m_n \log n}.$$

By the assumption that $m_n \log n / n \to 0$ we have

$$\frac{1}{n} \log \Delta_n(\mathcal{F}_n) \leq \frac{m_n}{n} + \frac{m_n(d+1)\log n}{n} \to 0,$$

so that condition (i) of Theorem 21.2 is satisfied.

For the second part of the statement, observe that there are no more than n/k_n cells in any partition, and that the tree-structured nature of the partitions assures that g_n is based on at most n/k_n hyperplane splits. This completes the proof. □

Problems and Exercises

PROBLEM 21.1. Let $\mathcal{P}$ be a partition of $\mathcal{R}^d$. Prove that

$$\sum_{A \in \mathcal{P}} |\mu_n(A) - \mu(A)| = 2 \sup_{B \in \mathcal{B}(\mathcal{P})} |\mu_n(B) - \mu(B)|,$$

where the class of sets $\mathcal{B}(\mathcal{P})$ contains all sets obtained by unions of cells of $\mathcal{P}$. This is Scheffé's (1947) theorem for partitions. See also Problem 12.13.

PROBLEM 21.2. Show that condition (ii) of Theorem 21.2 may be replaced by the following: for every $\gamma > 0$ and $\delta \in (0, 1)$

$$\lim_{n \to \infty} \inf_{T \subset \mathcal{R}^d : \mu(T) \geq 1 - \delta} \mu(\{x : \text{diam}(A_n(x) \cap T) > \gamma\}) = 0 \quad \text{with probability one.}$$

PROBLEM 21.3. Let X be uniformly distributed on the unit square $[0, 1]^2$. Let $\eta(x) = 1$ if $x^{(1)} \leq 2/3$, and $\eta(x) = 0$ otherwise (see Figure 21.3). Consider the algorithm when first the $x^{(1)}$-coordinate is cut at the k-th smallest $x^{(1)}$-value among the training data. Next the rectangle with $n - k$ points is cut according to the $x^{(2)}$-axis, isolating another k points. This is repeated on a rotational basis for the two coordinate axes. Show that the error probability of the obtained partitioning classifier does not tend to $L^* = 0$. Can you determine the asymptotic error probability?

PROBLEM 21.4. Modify Gessaman's rule based on statistically equivalent blocks so that the rule is strongly universally consistent.

PROBLEM 21.5. Cut each axis independently into intervals containing exactly k of the (projected) data points. The i-th axis has intervals $A_{1,i}, A_{2,i}, \ldots$. Form a histogram rule that takes a majority vote over the product sets $A_{i_1,1} \times \cdots \times A_{i_d,d}$.

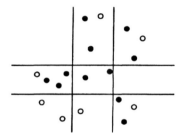

FIGURE 21.6. *A partition based upon the method obtained above with $k = 6$.*

This rule does *not* guarantee a minimal number of points in every cell. Nevertheless, if $k^d = o(n)$, $k \to \infty$, show that this decision rule is consistent, i.e., that $\mathbf{E}\{L_n\} \to L^*$ in probability.

PROBLEM 21.6. Show that each step of the k-means clustering algorithm decreases the empirical squared error. Conclude that Theorem 21.5 is also true if the clusters are given by the k-means algorithm. HINT: Observe that the only property of the clusters used in the proof of Theorem 21.5 is that

$$\mathbf{E}\left\{ \min_{1 \leq j \leq k_n} \|X - a_j\|^2 \,\Big|\, D_n \right\} \to 0 \quad \bullet$$

with probability one. This can be proven for clusters given by the k-means algorithm if it is appropriately initialized. To this end, use the techniques of Problem 29.4.

PROBLEM 21.7. Prove Theorem 21.7.

PROBLEM 21.8. Consider the H_{ni}'s in Theorem 21.7, the interval sizes for cubic histograms. Let the H_{ni}'s be found by minimizing

$$\sum_{i=1}^{n} I_{\{g_n(X_i) \neq Y_i\}}$$

subject to the condition that *each* marginal interval contain at least k data points (that is, at least k data points have that one coordinate in the interval). Under which condition on k is the rule consistent?

PROBLEM 21.9. Take a histogram rule with data-dependent sizes $H_{n1}, \ldots, H_{nd}$ as in Theorem 21.7, defined as follows:

(1) $H_{ni} = \left(\max_{1 \le j \le n} X_j^{(i)} - \min_{1 \le j \le n} X_j^{(i)} \right) / \sqrt{n}, \ 1 \le i \le d$ (where $X_j = (X_j^{(1)}, \ldots, X_j^{(d)})$;

(2) $H_{ni} = (W_{ni} - V_{ni})/n^{1/(2d)}, \ 1 \le i \le d$, where V_{ni} and W_{ni} are 25 and 75 percentiles of $X_1^{(i)}, \ldots, X_n^{(i)}$.

Assume for convenience that X has nonatomic marginals. Show that (1) leads sometimes to an inconsistent rule, even if $d = 1$. Show that (2) always yields a scale-invariant consistent rule.

22
Splitting the Data

22.1   The Holdout Estimate

Universal consistency gives us a partial satisfaction—without knowing the under-lying distribution, taking more samples is guaranteed to push us close to the Bayes rule in the long run. Unfortunately, we will never know just how close we are to the Bayes rule unless we are given more information about the unknown distribution (see Chapter 7). A more modest goal is to do as well as possible within a given class of rules. To fix the ideas, consider all nearest neighbor rules based upon metrics of the form

$$\|x\|^2 = \sum_{i=1}^{d} a_i x^{(i)^2},$$

where $a_i \geq 0$ for all i and $x = \left(x^{(1)}, \ldots, x^{(d)}\right)$. Here the a_i's are variable scale fac-tors. Let ϕ_n be a particular nearest neighbor rule for a given choice of $(a_1, \ldots, a_d)$, and let g_n be a data-based rule chosen from this class. The best we can hope for now is something like

$$\frac{L(g_n)}{\inf_{\phi_n} L(\phi_n)} \to 1 \quad \text{in probability}$$

for all distributions, where $L(g_n) = \mathbf{P}\{g_n(X) \neq Y | D_n\}$ is the conditional probabil-ity of error for g_n. This sort of optimality-within-a-class is definitely achievable. However, proving such optimality is generally not easy as g_n depends on the data. In this chapter we present one possible methodology for selecting provably good rules from restricted classes. This is achieved by splitting the data into a training sequence and a testing sequence. This idea was explored and analyzed in depth in Devroye (1988b) and is now formalized.

The data sequence D_n is split into a training sequence $D_m = (X_1, Y_1), \ldots,$ (X_m, Y_m) and a testing sequence $T_l = (X_{m+1}, Y_{m+1}), \ldots, (X_{m+l}, Y_{m+l})$, where $l + m = n$. The sequence D_m defines a class of classifiers C_m whose members are denoted by $\phi_m(\cdot) = \phi_m(\cdot, D_m)$. The testing sequence is used to select a classifier from C_m that minimizes the error count

$$\widehat{L}_{m,l}(\phi_m) = \frac{1}{l} \sum_{i=1}^{l} I_{\{\phi_m(X_{m+i}) \neq Y_{m+i}\}}.$$

This error estimate is called the *holdout estimate*, as the testing sequence is "held out" of the design of ϕ_m. Thus, the selected classifier $g_n \in C_m$ satisfies

$$\widehat{L}_{m,l}(g_n) \leq \widehat{L}_{m,l}(\phi_m)$$

for all $\phi_m \in C_m$. The subscript n in g_n may be a little confusing, since g_n is in C_m, a class of classifiers depending on the first m pairs D_m only. However, g_n depends on the *entire data* D_n, as the rest of the data is used for testing the classifiers in C_m. We are interested in the difference between the error probability

$$L(g_n) = \mathbf{P}\{g_n(X) \neq Y | D_n\},$$

and that of the best classifier in C_m, $\inf_{\phi_m \in C_m} L(\phi_m)$. Note that $L(\phi_m) = \mathbf{P}\{\phi_m(X) \neq Y | D_m\}$ denotes the error probability conditioned on D_m. The conditional probability

$$\mathbf{P}\left\{ L(g_n) - \inf_{\phi_m \in C_m} L(\phi_m) > \epsilon \,\middle|\, D_m \right\}$$

is small when most testing sequences T_l pick a rule g_n whose error probability is within ϵ of the best classifier in C_m. We have already addressed similar questions in Chapters 8 and 12. There we have seen (Lemma 8.2), that

$$L(g_n) - \inf_{\phi_m \in C_m} L(\phi_m) \leq 2 \sup_{\phi_m \in C_m} |\widehat{L}_{m,l}(\phi_m) - L(\phi_m)|.$$

If C_m contains finitely many rules, then the bound of Theorem 8.3 may be useful:

$$\mathbf{P}\left\{ \sup_{\phi_m \in C_m} |\widehat{L}_{m,l}(\phi_m) - L(\phi_m)| > \epsilon \,\middle|\, D_m \right\} \leq 2|C_m|e^{-2l\epsilon^2}.$$

If we take $m = l = n/2$, then Theorem 8.3 shows (see Problem 12.1) that on the average we are within $\sqrt{\log(2e|C_m|)/n}$ of the best possible error rate, whatever it is.

If C_m is of infinite cardinality, then we can use the Vapnik-Chervonenkis theory to get similar inequalities. For example, from Theorem 12.8 we get

$$\mathbf{P}\left\{ \sup_{\phi_m \in C_m} |\widehat{L}_{m,l}(\phi_m) - L(\phi_m)| > \epsilon \,\middle|\, D_m \right\} \leq 4e^8 S(C_m, l^2)e^{-2l\epsilon^2}, \tag{22.1}$$

and consequently,

$$\mathbf{P}\left\{ L(g_n) - \inf_{\phi_m \in C_m} L(\phi_m) > \epsilon \,\middle|\, D_m \right\} \le 4e^8 S(C_m, l^2)e^{-l\epsilon^2/2}, \qquad (22.2)$$

where $S(C_m, l)$ is the l-th shatter coefficient corresponding to the class of classifiers C_m (see Theorem 12.6 for the definition). Since C_m depends on the training data D_m, the shatter coefficients $S(C_m, l)$ may depend on D_m, too. However, usually it is possible to find upper bounds on the random variable $S(C_m, l)$ that depend on m and l only, but not on the actual values of the random variables $X_1, Y_1, \ldots, X_m, Y_m$. Both upper bounds above are *distribution-free*, and the problem now is purely combinatorial: count $|C_m|$ (this is usually trivial), or compute $S(C_m, l)$.

REMARK. With much more effort it is possible to obtain performance bounds of the holdout estimate in the form of bounds for

$$\mathbf{P}\{\widehat{L}_{m,l}(\phi_m) - L(\phi_n) > \epsilon\}$$

for some special rules where ϕ_m and ϕ_n are carefully defined. For example, Devroye and Wagner (1979a) give upper bounds when both ϕ_m and ϕ_n are k-nearest neighbor classifiers with the same k (but working with different sample size). □

REMARK. Minimizing the holdout estimate is not the only possibility. Other error estimates that do not split the data may be used in classifier selection as well. Such estimates are discussed in Chapters 23, 24, and 31. However, these estimates are usually tailored to work well for specific discrimination rules. The most general and robust method is certainly the data splitting described here. □

22.2 Consistency and Asymptotic Optimality

Typically, C_m becomes richer as m grows, and it is natural to ask whether the empirically best classifier in C_m is consistent.

Theorem 22.1. *Assume that from each C_m we can pick one ϕ_m such that the sequence of ϕ_m's is consistent for a certain class of distributions. Then the automatic rule g_n defined above is consistent for the same class of distributions (i.e., $EL(g_n) \to L^*$ as $n \to \infty$) if*

$$\lim_{n \to \infty} \frac{\log(E\{S(C_m, l)\})}{l} = 0.$$

PROOF. Decompose the difference between the actual error probability and the Bayes error as

$$L(g_n) - L^* = \left(L(g_n) - \inf_{\phi_m \in C_m} L(\phi_m) \right) + \left(\inf_{\phi_m \in C_m} L(\phi_m) - L^* \right).$$

The convergence of the first term on the right-hand side is a direct corollary of (22.2). The second term converges to zero by assumption. □

Theorem 22.1 shows that a consistent rule is picked if the sequence of C_m's contains a consistent rule, even if we do not know which functions from C_m lead to consistency. If we are just worried about consistency, Theorem 22.1 reassures us that nothing is lost as long as we take l much larger than $\log(\mathbf{E}\{S(C_m, l)\})$. Often, this reduces to a very weak condition on the size l of the testing set.

Let us now introduce the notion of asymptotic optimality. A sequence of rules g_n is said to be asymptotically optimal for a given distribution of (X, Y) when

$$\lim_{n \to \infty} \frac{\mathbf{E}\{L(g_n)\} - L^*}{\mathbf{E}\left\{\inf_{\phi_m \in C_m} L(\phi_m)\right\} - L^*} = 1.$$

Our definition is not entirely fair, because g_n uses n observations, whereas the family of rules in the denominator is restricted to using m observations. If g_n is not taken from the same C_m, then it is possible to have a ratio which is smaller than one. But if $g_n \in C_m$, then the ratio always is at least one. That is why the definition makes sense in our setup.

When our selected rule is asymptotically optimal, we have achieved something very strong: we have in effect picked a rule (or better, a sequence of rules) which has a probability of error converging at the optimal rate attainable within the sequence of C_m's. And we do not even have to know what the optimal rate of convergence is. This is especially important in nonparametric rules, where some researchers choose smoothing factors based on theoretical results about the optimal attainable rate of convergence for certain classes of distributions.

We are constantly faced with the problem of choosing between parametric and nonparametric discriminators. Parametric discriminators are based upon an underlying model in which a finite number of unknown parameters is estimated from the data. A case in point is the multivariate normal distribution, which leads to linear or quadratic discriminators. If the model is wrong, parametric methods can perform very poorly; when the model is right, their performance is difficult to beat. The method based on splitting the data chooses among the best discriminator depending upon which happens to be best for the given data. We can throw in C_m a variety of rules, including nearest neighbor rules, a few linear discriminators, a couple of tree classifiers and perhaps a kernel rule. The probability bounds above can be used when the complexity of C_m (measured by its shatter coefficients) does not get out of hand.

The notion of asymptotic optimality can be too strong in many cases. The reason for this is that in some rare lucky situations $\mathbf{E}\left\{\inf_{\phi_m \in C_m} L(\phi_m)\right\} - L^*$ may be very small. In these cases it is impossible to achieve asymptotic optimality. We can fix this problem by introducing the notion of ϵ_m-optimality, where ϵ_m is a positive sequence decreasing to 0 with m. A rule is said to be ϵ_m-optimal when

$$\lim_{n \to \infty} \frac{\mathbf{E}\{L(g_n)\} - L^*}{\mathbf{E}\left\{\inf_{\phi_m \in C_m} L(\phi_m)\right\} - L^*} = 1.$$

for all distributions of (X, Y) for which

$$\lim_{m \to \infty} \frac{\mathbf{E}\left\{\inf_{\phi_m \in C_m} L(\phi_m)\right\} - L^*}{\epsilon_m} = \infty.$$

In what follows we apply the idea of data splitting to scaled nearest neighbor rules and to rules closely related to the data-dependent partitioning classifiers studied in Chapter 21. Many more examples are presented in Chapters 25 and 26.

22.3   Nearest Neighbor Rules with Automatic Scaling

Let us work out the simple example introduced above. The $(a_1, \ldots, a_d)$-NN rule is the nearest neighbor rule for the metric

$$\|x\| = \sqrt{\sum_{i=1}^{d} a_i^2 x^{(i)2}},$$

where $x = \left(x^{(1)}, \ldots, x^{(d)}\right)$. The class C_m is the class of all $(a_1, \ldots, a_d)$-NN rules for $D_m = ((X_1, Y_1), \ldots, (X_m, Y_m))$. The testing sequence T_l is used to choose $(a_1, \ldots, a_d)$ so as to minimize the holdout error estimate. In order to have

$$L(g_n) - \inf_{\phi_m \in C_m} L(\phi_m) \to 0 \quad \text{in probability},$$

it suffices that

$$\lim_{n \to \infty} \frac{\log \mathbf{E}\left\{S(C_m, l)\right\}}{l} = 0.$$

This puts a lower bound on l. To get this lower bound, one must compute $S(C_m, l)$. Clearly, $S(C_m, l)$ is bounded by the number of ways of classifying $X_{m+1}, \ldots, X_{m+l}$ using rules picked from C_m, that is, the total number of different values for

$$\{(\phi_m(X_{m+1}), \ldots, \phi_m(X_{m+l})), \phi_m \in C_m\}.$$

We now show that regardless of D_m and $X_{m+1}, \ldots, X_{m+l}$, for $n \geq 4$ we have

$$S(C_m, l) \leq \left(l\binom{m}{2}\right)^d.$$

This sort of result is typical—the bound does not depend upon D_m or T_l. When plugged back into the condition of convergence, it yields the simple condition

$$\frac{\log m}{l} \to 0.$$

In fact, we may thus take l slightly larger than $\log m$. It would plainly be silly to take $l = n/2$, as we would thereby in fact throw away most of the data.

Set

$$\mathcal{A}_m = \{(\phi_m(X_{m+1}), \ldots, \phi_m(X_{m+l})), \phi_m \in \mathcal{C}_m\}.$$

To count the number of values in $\mathcal{A}_m$, note that all squared distances can be written as

$$\|X_i - X_j\|^2 = \sum_{k=1}^{d} a_k^2 \rho_{k,i,j},$$

where $\rho_{k,i,j}$ is a nonnegative number only depending upon X_i and X_j. Note that each squared distance is linear in $(a_1^2, \ldots, a_d^2)$. Now consider the space of $(a_1^2, \ldots, a_d^2)$. Observe that, in this space, $\phi_m(X_{m+1})$ is constant within each cell of the partition determined by the $\binom{m}{2}$ hyperplanes

$$\sum_{k=1}^{d} a_k^2 \rho_{k,i,m+1} = \sum_{k=1}^{d} a_k^2 \rho_{k,i',m+1}, \quad 1 \le i < i' \le m.$$

To see this, note that within each set in the partition, $\phi_m(X_{m+1})$ keeps the same nearest neighbor among $X_1, \ldots, X_m$. It is known that $k > 2$ hyperplanes in $\mathcal{R}^d$ create partitions of cardinality not exceeding k^d (see Problem 22.1). Now, overlay the l partitions obtained for $\phi_m(X_{m+1}), \ldots, \phi_m(X_{m+l})$ respectively. This yields at most

$$\left(l \binom{m}{2} \right)^d$$

sets, as the overlays are determined by $l\binom{m}{2}$ hyperplanes. But clearly, on each of these sets, $(\phi_m(X_{m+1}), \ldots, \phi_m(X_{m+l}))$ is constant. Therefore,

$$\mathcal{S}(\mathcal{C}_m, l) \le |\mathcal{A}_m| \le \left(l \binom{m}{2} \right)^d.$$

22.4   Classification Based on Clustering

Recall the classification rule based on clustering that was introduced in Chapter 21. The data points $X_1, \ldots, X_n$ are grouped into k clusters, where k is a predetermined integer, and a majority vote decides within the k clusters. If k is chosen such that $k \to \infty$ and $k^2 \log n / n \to 0$, then the rule is consistent. For a given finite n, however, these conditions give little guidance. Also, the choice of k could dramatically affect the performance of the rule, as there may be a mismatch between k and some unknown natural number of clusters. For example, one may construct distributions in which the optimal number of clusters does not increase with n.

Let us split the data and let the testing sequence decide the value of k. In the framework of this chapter, $\mathcal{C}_m$ contains the classifiers based on the first m pairs D_m of the data with all possible values of k. Clearly, $\mathcal{C}_m$ is a finite family with $|\mathcal{C}_m| = m$. In this case, by Problem 12.1, we have

$$\mathbf{E}\left\{ L(g_n) - \inf_{\phi_m \in \mathcal{C}_m} L(\phi_m) \,\middle|\, D_m \right\} \le \sqrt{\frac{2\log(2m)}{l}}.$$

The consistency result Theorem 21.5 implies that

$$\mathbf{E}\left\{\inf_{\phi_m \in C_m} L(\phi_m)\right\} \to L^*$$

for all distributions, if the X_i's have bounded support. Thus, we see that our strategy leads to a universally consistent rule whenever $(\log m)/l \to 0$. This is a very mild condition, since we can take l equal to a small fraction of n, without sacrificing consistency. If l is small compared to n, then m is close to n, so $\inf_{\phi_m \in C_m} L(\phi_m)$ is likely to be close to $\inf_{\phi_n \in C_n} L(\phi_n)$. Thus, we do not lose much by sacrificing a part of the data for testing purposes, but the gain can be tremendous, as we are guaranteed to be within $\sqrt{(\log m)/l}$ of the optimum in the class C_m.

That we cannot take $l = 1$ and hope to obtain consistency should be obvious. It should also be noted that for $l = m$, we are roughly within $\sqrt{\log(m)/m}$ of the best possible probability of error within the family. Also the empirical selection rule is $\sqrt{\log(m)/l}$-optimal.

22.5   Statistically Equivalent Blocks

Recall from Chapter 21 that classifiers based on statistically equivalent blocks typically partition the feature space $\mathcal{R}^d$ into rectangles such that each rectangle contains k points, where k is a certain positive integer, the parameter of the rule. This may be done in several different ways. One of many such rules—the rule introduced by Gessaman (1970)—is consistent if $k \to \infty$ and $k/n \to 0$ (Theorem 21.4). Again, we can let the data pick the value of k by minimizing the holdout estimate. Just as in the previous section, $|C_m| = m$, and every remark mentioned there about consistency and asymptotic optimality remains true for this case as well.

We can enlarge the family C_m by allowing partitions without restrictions on cardinalities of cells. This leads very quickly to oversized families of rules, and we have to impose reasonable restrictions. Consider cuts into at most k rectangles, where k is a number picked beforehand. Recall that for a fixed partition, the class assigned to every rectangle is decided upon by a majority vote among the training points. On the real line, choosing a partition into at most k sets is equivalent to choosing $k - 1$ cut positions from $m + l + 1 = n + 1$ spacings between all test and training points. Hence,

$$S(C_m, l) \le \binom{n + 1}{k - 1} \le (n + 1)^{k-1}.$$

For consistency, k has to grow as n grows. It is easily seen that

$$\mathbf{E}\left\{\inf_{\phi_m \in C_m} L(\phi_m)\right\} \to L^*$$

if $k \to \infty$ and $m \to \infty$. To achieve consistency of the selected rule, however, we also need

$$\frac{\log S(\mathcal{C}_m, l)}{l} \leq \frac{(k-1)\log(n+1)}{l} \to 0.$$

Now consider d-dimensional partitions defined by at most $k - 1$ consecutive orthogonal cuts, that is, the first cut divides $\mathcal{R}^d$ into two halfspaces along a hyperplane perpendicular to one of the coordinate axes. The second cut splits one of the halfspaces into two parts along another orthogonal hyperplane, and so forth. This procedure yields a partition of the space into k rectangles. We see that for the first cut, there are at most $1 + dn$ possible combinations to choose from. This yields the loose upper bound

$$S(\mathcal{C}_m, l) \leq (1 + dn)^{k-1}.$$

This bound is also valid for all grids defined by at most $k - 1$ cuts. The main difference here is that every cut defines two halfspaces of $\mathcal{R}^d$, and not two hyperrectangles of a cells, so that we usually end up with 2^k rectangles in the partition.

Assume that $\mathcal{C}_m$ contains all histograms with partitions into at most k (possibly infinite) rectangles. Then, considering that a rectangle in $\mathcal{R}^d$ requires choosing $2d$ spacings between all test and training points, two per coordinate axis,

$$S(\mathcal{C}_m, l) \leq (n + 1)^{2d(k-1)}.$$

See Feinholz (1979) for more work on such partitions.

22.6   Binary Tree Classifiers

We can analyze binary tree classifiers from the same viewpoint. Recall that such classifiers are represented by binary trees, where each internal node corresponds to a split of a cell by a hyperplane, and the terminal nodes represent the cells of the partition.

Assume that there are k cells (and therefore $k - 1$ splits leading to the partition). If every split is perpendicular to one of the axes, then, the situation is the same as in the previous section,

$$S(\mathcal{C}_m, l) \leq (1 + dn)^{k-1}.$$

For smaller families of rules whose cuts depend upon the training sequence only, the bound is pessimistic. Others have proposed generalizing orthogonal cuts by using general hyperplane cuts. Recall that there are at most

$$\sum_{j=0}^{d+1} \binom{n}{j} \leq n^{d+1} + 1$$

ways of dichotomizing n points in $\mathcal{R}^d$ by hyperplanes (see Theorem 13.9). Thus, if we allow up to $k - 1$ internal nodes (or hyperplane cuts),

$$S(C_m, l) \leq \left(1 + n^{d+1}\right)^{k-1}.$$

The number of internal nodes has to be restricted in order to obtain consistency from this bound. Refer to Chapter 20 for more details.

Problems and Exercises

PROBLEM 22.1. Prove that n hyperplanes partition $\mathcal{R}^d$ into at most $\sum_{i=0}^{d} \binom{n}{i}$ contiguous regions when $d \leq n$ (Schläffli (1950), Cover (1965)). HINT: Proceed by induction.

PROBLEM 22.2. Assume that g_n is selected so as to minimize the holdout error estimate $\widehat{L}_{l,m}(\phi_m)$ over C_m, the class of rules based upon the first m data points. Assume furthermore that we vary l over $[\log^2 n, n/2]$, and that we pick the best l (and $m = n - l$) by minimizing the holdout error estimate again. Show that if $S(C_m, l) = O(n^\gamma)$ for some $\gamma > 0$, then the obtained rule is strongly universally consistent.

PROBLEM 22.3. Let C_m be the class of $(a_1, \ldots, a_d)$-NN rules based upon D_m. Show that if $m/n \to 1$, then

$$\inf_{\phi_m \in C_m} L(\phi_m) - \inf_{\phi_n \in C_n} L(\phi_n) \to 0 \quad \text{in probability}$$

for all distributions of (X, Y) for which X has a density. Conclude that if $l/\log n \to \infty$, $m = n - l, l = o(n)$, then

$$L(g_n) - \inf_{\phi_n \in C_n} L(\phi_n) \to 0 \quad \text{in probability.}$$

PROBLEM 22.4. FINDING THE BEST SPLIT. This exercise is concerned with the automatic selection of m and $l = n - m$. If g_n is the selected rule minimizing the holdout estimate, then

$$\mathbf{E}\{L(g_n)|D_m\} - L^* \leq \sqrt{\frac{2\log\left(4S(C_m, l^2)\right) + 16}{l}} + \left(\inf_{\phi_m \in C_m} L(\phi_m) - L^*\right). \tag{22.3}$$

Since in most interesting cases $S(C_m, l)$ is bounded from above as a polynomial of m and l, the estimation error typically decreases as l increases. On the other hand, the approximation error $\inf_{\phi_m \in C_m} L(\phi_m) - L^*$ typically decreases as m increases, as the class C_m gets richer. Some kind of balance between the two terms is required to get optimum performance. We may use the empirical estimates $\widehat{L}_{m,l}(\phi_m)$ again to decide which value of m we wish to choose (Problem 22.2). However, as m gets large—and therefore l small—the class C_m will tend to overfit the data T_l, providing strongly optimistically biased estimates for $\inf_{\phi_m \in C_m} L(\phi_m)$. To prevent overfitting, we may apply the method of complexity regularization (Chapter 18). By (22.1), we may define the penalty term by

$$r(m, l) = \sqrt{\frac{2\log\left(4e^8\mathbf{E}\{S(C_m, l^2)\}\right) + \log n}{l}},$$

and minimize the penalized error estimate $\tilde{L}_{m,l}(\phi) = \widehat{L}_{m,l}(\phi) + r(m, l)$ over all $\phi \in \cup_{m=1}^{n} \mathcal{C}_m$. Denote the selected rule by ϕ_n^*. Prove that for every n and all distributions of (X, Y),

$$
\mathbf{E}\{L(\phi_n^*) - L^*\} \leq \min_{m,l} \left(3\sqrt{\frac{2 \log\left(4\mathbf{E}\{\mathcal{S}(\mathcal{C}_m, l^2)\}\right) + \log n + 16}{l}} \right.
$$
$$
\left. + \left(\mathbf{E}\left\{ \inf_{\phi_m \in \mathcal{C}_m} L(\phi_m) - L^* \right\} \right) \right) + \frac{2}{\sqrt{n}}.
$$

HINT: Proceed similarly to the proof of Theorem 18.2.

23
The Resubstitution Estimate

Estimating the error probability is of primordial importance for classifier selection. The method explored in the previous chapter attempts to solve this problem by using a testing sequence to obtain a reliable holdout estimate. The independence of testing and training sequences leads to a rather straightforward analysis. For a good performance, the testing sequence has to be sufficiently large (although we often get away with testing sequences as small as about $\log n$). When data are expensive, this constitutes a waste. Assume that we do not split the data and use the same sequence for testing and training. Often dangerous, this strategy nevertheless works if the class of rules from which we select is sufficiently restricted. The error estimate in this case is appropriately called the resubstitution estimate and it will be denoted by $L_n^{(R)}$. This chapter explores its virtues and pitfalls. A third error estimate, the deleted estimate, is discussed in the next chapter. Estimates based upon other paradigms are treated briefly in Chapter 31.

23.1 The Resubstitution Estimate

The *resubstitution estimate* $L_n^{(R)}$ counts the number of errors committed on the training sequence by the classification rule. Expressed formally,

$$L_n^{(R)} = \frac{1}{n} \sum_{i=1}^{n} I_{\{g_n(X_i) \neq Y_i\}}.$$

Sometimes $L_n^{(R)}$ is called the *apparent error rate*. It is usually strongly optimistically biased. Since the classifier g_n is tuned by D_n, it is intuitively clear that g_n may behave better on D_n than on independent data.

The best way to demonstrate this biasedness is to consider the 1-nearest neighbor rule. If X has a density, then the nearest neighbor of X_i among $X_1, \ldots, X_n$ is X_i itself with probability one. Therefore, $L_n^{(R)} \equiv 0$, regardless of the value of $L_n = L(g_n)$. In this case the resubstitution estimate is useless. For k-nearest neighbor rules with large k, $L_n^{(R)}$ is close to L_n. This was demonstrated by Devroye and Wagner (1979a), who obtained upper bounds on the performance of the resubstitution estimate for the k-nearest neighbor rule without posing any assumption on the distribution.

Also, if the classifier whose error probability is to be estimated is a member of a class of classifiers with finite Vapnik-Chervonenkis dimension (see the definitions in Chapter 12), then we can get good performance bounds for the resubstitution estimate. To see this, consider any generalized linear classification rule, that is, any rule that can be put into the following form:

$$g_n(x) = \begin{cases} 0 & \text{if } a_{0,n} + \sum_{i=1}^{d^*} a_{i,n} \psi_i(x) \le 0 \\ 1 & \text{otherwise,} \end{cases}$$

where the ψ_i's are fixed functions, and the coefficients $a_{i,n}$ depend on the data D_n in an arbitrary but measurable way. We have the following estimate for the performance of the resubstitution estimate $L_n^{(R)}$.

Theorem 23.1. (DEVROYE AND WAGNER (1976A)). *For all n and $\epsilon > 0$, the resubstitution estimate $L_n^{(R)}$ of the error probability L_n of a generalized linear rule satisfies*

$$\mathbf{P}\left\{|L_n^{(R)} - L_n| \ge \epsilon\right\} \le 8n^{d^*} e^{-n\epsilon^2/32}.$$

PROOF. Define the set $A_n \subset \mathcal{R}^d \times \{0, 1\}$ as the set of all pairs $(x, y) \in \mathcal{R}^d \times \{0, 1\}$, on which g_n errs:

$$A_n = \{(x, y) : g_n(x) \ne y\}.$$

Observe that

$$L_n = \mathbf{P}\{(X, Y) \in A_n | D_n\}$$

and

$$L_n^{(R)} = \frac{1}{n} \sum_{j=1}^{n} I_{\{(X_j, Y_j) \in A_n\}},$$

or denoting the measure of (X, Y) by ν, and the corresponding empirical measure by ν_n,

$$L_n = \nu(A_n),$$

and

$$L_n^{(R)} = \nu_n(A_n).$$

The set A_n depends on the data D_n, so that, for example, $\mathbf{E}\{v_n(A_n)\} \neq \mathbf{E}\{v(A_n)\}$. Fortunately, the powerful Vapnik-Chervonenkis theory comes to the rescue via the inequality

$$|L_n - L_n^{(R)}| \leq \sup_{C \in \mathcal{C}} |v_n(C) - v(C)|,$$

where $\mathcal{C}$ is the family of all sets of the form $\{(x, y) : \phi(x) \neq y\}$, where $\phi : \mathcal{R}^d \to \{0, 1\}$ is a generalized linear classifier based on the functions $\psi_1, \ldots, \psi_{d^*}$. By Theorems 12.6, 13.1, and 13.7 we have

$$\mathbf{P}\left\{\sup_{C \in \mathcal{C}} |v_n(C) - v(C)| \geq \epsilon\right\} \leq 8n^{d^*} e^{-n\epsilon^2/32}. \quad \square$$

Similar inequalities can be obtained for other classifiers. For example, for partitioning rules with fixed partitions we have the following:

Theorem 23.2. *Let g_n be a classifier whose value is constant over cells of a fixed partition of $\mathcal{R}^d$ into k cells. Then*

$$\mathbf{P}\left\{|L_n^{(R)} - L_n| \geq \epsilon\right\} \leq 8 \cdot 2^k e^{-n\epsilon^2/32}.$$

The proof is left as an exercise (Problem 23.1). From Theorems 23.1 and 23.2 we get bounds for the expected difference between the resubstitution estimate and the actual error probability L_n. For example, Theorem 23.1 implies (via Problem 12.1) that

$$\mathbf{E}\left\{|L_n^{(R)} - L_n|\right\} = O\left(\sqrt{\frac{d^* \log n}{n}}\right).$$

In some special cases the expected behavior of the resubstitution estimate can be analyzed in more detail. For example, McLachlan (1976) proved that if the conditional distributions of X given $Y = 0$ and $Y = 1$ are both normal with the same covariance matrices, and the rule is linear and based on the estimated parameters, then the bias of the estimate is of the order $1/n$:

$$\left|\mathbf{E}\left\{L_n^{(R)}\right\} - \mathbf{E}\{L_n\}\right| = O\left(\frac{1}{n}\right).$$

McLachlan also showed for this case that for large n the expected value of the resubstitution estimate is smaller than that of L_n, that is, the estimate is *optimistically biased*, as expected.

23.2 Histogram Rules

In this section, we explore the properties of the resubstitution estimate for histogram rules. Let $\mathcal{P} = \{A_1, A_2, \ldots\}$ be a fixed partition of $\mathcal{R}^d$, and let g_n be the corresponding histogram classifier (see Chapters 6 and 9). Introduce the notation

$$\nu_{0,n}(A) = \frac{1}{n} \sum_{j=1}^{n} I_{\{Y_j=0\}} I_{\{X_j \in A\}}, \quad \text{and} \quad \nu_{1,n}(A) = \frac{1}{n} \sum_{j=1}^{n} I_{\{Y_j=1\}} I_{\{X_j \in A\}}.$$

The analysis is simplified if we rewrite the estimate $L_n^{(R)}$ in the following form (see Problem 23.2):

$$L_n^{(R)} = \sum_{i} \min\{\nu_{0,n}(A_i), \nu_{1,n}(A_i)\}. \tag{23.1}$$

It is also interesting to observe that $L_n^{(R)}$ can be put in the following form:

$$L_n^{(R)} = \int \left(I_{\{\eta_{1,n}(x) \le \eta_{0,n}(x)\}} \eta_{1,n}(x) + I_{\{\eta_{1,n}(x) > \eta_{0,n}(x)\}} \eta_{0,n}(x) \right) \mu(dx),$$

where

$$\eta_{0,n}(x) = \frac{\frac{1}{n} \sum_{j=1}^{n} I_{\{Y_j=0, X_j \in A(x)\}}}{\mu(A(x))} \quad \text{and} \quad \eta_{1,n}(x) = \frac{\frac{1}{n} \sum_{j=1}^{n} I_{\{Y_j=1, X_j \in A(x)\}}}{\mu(A(x))}.$$

We can compare this form with the following expression of L_n:

$$L_n = \int \left(I_{\{\eta_{1,n}(x) \le \eta_{0,n}(x)\}} \eta(x) + I_{\{\eta_{1,n}(x) > \eta_{0,n}(x)\}} (1 - \eta(x)) \right) \mu(dx)$$

(see Problem 23.3). We begin the analysis of performance of the estimate by showing that its mean squared error is not larger than a constant times the number of cells in the partition over n.

Theorem 23.3. *For any distribution of (X, Y) and for all n, the estimate $L_n^{(R)}$ of the error probability of any histogram rule satisfies*

$$\mathbf{Var}\{L_n^{(R)}\} \le \frac{1}{n}.$$

Also, the estimate is optimistically biased, that is,

$$\mathbf{E}\{L_n^{(R)}\} \le \mathbf{E}\{L_n\}.$$

If, in addition, the histogram rule is based on a partition $\mathcal{P}$ of $\mathcal{R}^d$ into at most k cells, then

$$\mathbf{E}\left\{\left(L_n^{(R)} - L_n\right)^2\right\} \le \frac{6k}{n}.$$

PROOF. The first inequality is an immediate consequence of Theorem 9.3 and (23.1). Introduce the auxiliary quantity

$$R^* = \sum_{i} \min\{\nu_0(A_i), \nu_1(A_i)\},$$

where $v_0(A) = P\{Y = 0, X \in A\}$, and $v_1(A) = P\{Y = 1, X \in A\}$. We use the decomposition

$$E\left\{\left(L_n^{(R)} - L_n\right)^2\right\} \le 2\left(E\left\{\left(L_n^{(R)} - R^*\right)^2\right\} + E\left\{\left(R^* - L_n\right)^2\right\}\right). \qquad (23.2)$$

First we bound the second term on the right-hand side of (23.2). Observe that R^* is just the Bayes error corresponding to the pair of random variables $(T(X), Y)$, where the function T transforms X according to $T(x) = i$ if $x \in A_i$. Since the histogram classification rule $g_n(x)$ can be written as a function of $T(x)$, its error probability L_n cannot be smaller than R^*. Furthermore, by the first identity of Theorem 2.2, we have

$$0 \le L_n - R^*$$
$$= \sum_i I_{\{\text{sign}(v_{1,n}(A_i) - v_{0,n}(A_i)) \ne \text{sign}(v_1(A_i) - v_0(A_i))\}} |v_1(A_i) - v_0(A_i)|$$
$$\le \sum_i |v_1(A_i) - v_0(A_i) - \left(v_{1,n}(A_i) - v_{0,n}(A_i)\right)|.$$

If the partition has at most k cells, then by the Cauchy-Schwarz inequality,

$$E\left\{(L_n - R^*)^2\right\}$$
$$\le E\left\{\left(\sum_i |v_1(A_i) - v_0(A_i) - \left(v_{1,n}(A_i) - v_{0,n}(A_i)\right)|\right)^2\right\}$$
$$\le k \sum_i \text{Var}\left\{v_{1,n}(A_i) - v_{0,n}(A_i)\right\}$$
$$\le k \sum_i \frac{\mu(A_i)}{n} = \frac{k}{n}.$$

We bound the first term on the right-hand side of (23.2):

$$E\left\{\left(L_n^{(R)} - R^*\right)^2\right\} \le \text{Var}\{L_n^{(R)}\} + \left(R^* - E\left\{L_n^{(R)}\right\}\right)^2.$$

As we have seen earlier, $\text{Var}\{L_n^{(R)}\} \le 1/n$, so it suffices to bound $|R^* - E\{L_n^{(R)}\}|$. By Jensen's inequality,

$$E\{L_n^{(R)}\} = \sum_i E\left\{\min\{v_{0,n}(A_i), v_{1,n}(A_i)\}\right\}$$
$$\le \sum_i \min\left(E\{v_{0,n}(A_i)\}, E\{v_{1,n}(A_i)\}\right)$$
$$= \sum_i \min(v_0(A_i), v_1(A_i))$$
$$= R^*.$$

So we bound $R^* - \mathbf{E}\{L_n^{(R)}\}$ from above. By the inequality

$$|\min(a, b) - \min(c, d)| \le |a - c| + |b - d|,$$

we have

$$
\begin{aligned}
R^* - \mathbf{E}\{L_n^{(R)}\} &\le \sum_i \mathbf{E}\left\{|v_0(A_i) - v_{0,n}(A_i)| + |v_1(A_i) - v_{1,n}(A_i)|\right\} \\
&\le \sum_i \left(\sqrt{\mathbf{Var}\{v_{0,n}(A_i)\}} + \sqrt{\mathbf{Var}\{v_{1,n}(A_i)\}}\right)
\end{aligned}
$$

(by the Cauchy-Schwarz inequality)

$$
\begin{aligned}
&= \sum_i \left(\sqrt{\frac{v_0(A_i)(1 - v_0(A_i))}{n}} + \sqrt{\frac{v_1(A_i)(1 - v_1(A_i))}{n}}\right) \\
&\le \sum_i \left(\sqrt{\frac{2(v_0(A_i)(1 - v_0(A_i)) + v_1(A_i)(1 - v_1(A_i)))}{n}}\right) \\
&\le \sum_i \sqrt{\frac{2\mu(A_i)}{n}},
\end{aligned}
$$

where we used the elementary inequality $\sqrt{a} + \sqrt{b} \le \sqrt{2(a + b)}$. Therefore,

$$\left(R^* - \mathbf{E}\{L_n^{(R)}\}\right)^2 \le \left(\sum_i \sqrt{\frac{2\mu(A_i)}{n}}\right)^2.$$

To complete the proof of the third inequality, observe that if there are at most k cells, then

$$
\begin{aligned}
R^* - \mathbf{E}\{L_n^{(R)}\} &\le \sum_i \sqrt{\frac{2\mu(A_i)}{n}} \\
&\le k\sqrt{\frac{1}{k}\sum_i \frac{2\mu(A_i)}{n}} \quad \text{(by Jensen's inequality)} \\
&= \sqrt{\frac{2k}{n}}.
\end{aligned}
$$

Therefore, $(R^* - \mathbf{E}\{L_n^{(R)}\})^2 \le 2k/n$. Finally,

$$\mathbf{E}\{L_n\} - \mathbf{E}\{L_n^{(R)}\} = (\mathbf{E}\{L_n\} - R^*) + \left(R^* - \mathbf{E}\{L_n^{(R)}\}\right) \ge 0. \quad \square$$

We see that if the partition contains a small number of cells, then the resubstitution estimate performs very nicely. However, if the partition has a large number of cells, then the resubstitution estimate of L_n can be very misleading, as the next result indicates:

Theorem 23.4. (DEVROYE AND WAGNER (1976B)). *For every n there exists a partitioning rule and a distribution such that*

$$\mathbf{E}\left\{\left(L_n^{(R)} - L_n\right)^2\right\} \geq 1/4.$$

PROOF. Let $A_1, \ldots, A_{2n}$ be $2n$ cells of the partition such that $\mu(A_i) = 1/(2n)$, $i = 1, \ldots, 2n$. Assume further that $\eta(x) = 1$ for every $x \in \mathcal{R}^d$, that is, $Y = 1$ with probability one. Then clearly $L_n^{(R)} = 0$. On the other hand, if a cell does not contain any of the data points $X_1, \ldots, X_n$, then $g_n(x) = 0$ in that cell. But since the number of points is only half of the number of cells, at least half of the cells are empty. Therefore $L_n \geq 1/2$, and $|L_n^{(R)} - L_n| \geq 1/2$. This concludes the proof. $\square$

23.3 Data-Based Histograms and Rule Selection

Theorem 23.3 demonstrates the usefulness of $L_n^{(R)}$ for histogram rules with a fixed partition, provided that the number of cells in the partition is not too large. If we want to use $L_n^{(R)}$ to select a good classifier, the estimate should work *uniformly* well over the class from which we select a rule. In this section we explore such data-based histogram rules.

Let $\mathcal{F}$ be a class of partitions of $\mathcal{R}^d$. We will assume that each member of $\mathcal{F}$ partitions $\mathcal{R}^d$ into at most k cells. For each partition $\mathcal{P} \in \mathcal{F}$, define the corresponding histogram rule by

$$g_n^{(\mathcal{P})}(x) = \begin{cases} 0 & \text{if } \sum_{i=1}^n I_{\{Y_i=1, X_i \in A(x)\}} \leq \sum_{i=1}^n I_{\{Y_i=0, X_i \in A(x)\}} \\ 1 & \text{otherwise,} \end{cases}$$

where $A(x)$ is the cell of $\mathcal{P}$ that contains x. Denote the error probability of $g_n^{(\mathcal{P})}(x)$ by

$$L_n(\mathcal{P}) = \mathbf{P}\left\{g_n^{(\mathcal{P})}(X) \neq Y | D_n\right\}.$$

The corresponding error estimate is denoted by

$$L_n^{(R)}(\mathcal{P}) = \sum_{A \in \mathcal{P}} \min\{v_{0,n}(A), v_{1,n}(A)\}.$$

By analogy with Theorems 23.1 and 23.2, we can derive the following result, which gives a useful bound for the largest difference between the estimate and the error probability within the class of histogram classifiers defined by $\mathcal{F}$. The combinatorial coefficient $\Delta_n^*(\mathcal{F})$ defined in Chapter 21 appears as a coefficient in the upper bound. The computation of $\Delta_n^*(\mathcal{F})$ for several different classes of partitions is illustrated in Chapter 21.

Theorem 23.5. (FEINHOLZ (1979)). *Assume that each member of $\mathcal{F}$ partitions $\mathcal{R}^d$ into at most k cells. For every n and $\epsilon > 0$,*

$$\mathbf{P}\left\{\sup_{\mathcal{P}\in\mathcal{F}}|L_n^{(R)}(\mathcal{P}) - L_n(\mathcal{P})| > \epsilon\right\} \leq 8 \cdot 2^k \Delta_n^*(\mathcal{F})e^{-n\epsilon^2/32}.$$

PROOF. We can proceed as in the proof of Theorem 23.1. The shatter coefficient $\mathcal{S}(\mathcal{C}, n)$ corresponding to the class of histogram classifiers defined by partitions in $\mathcal{F}$ can clearly be bounded from above by the number of different ways in which n points can be partitioned by members of $\mathcal{F}$, times 2^k, as there are at most 2^k different ways to assign labels to cells of a partition of at most k cells. $\square$

The theorem has two interesting implications. The error estimate $L_n^{(R)}$ can also be used to estimate the performance of histogram rules based on data-dependent partitions (see Chapter 21). The argument of the proof of Theorem 23.3 is not valid for these rules. However, Theorem 23.5 provides performance guarantees for these rules in the following corollaries:

COROLLARY 23.1. *Let $g_n(x)$ be a histogram classifier based on a random partition $\mathcal{P}_n$ into at most k cells, which is determined by the data D_n. Assume that for any possible realization of the training data D_n, the partition $\mathcal{P}_n$ is a member of a class of partitions $\mathcal{F}$. If L_n is the error probability of g_n, then*

$$\mathbf{P}\left\{|L_n^{(R)} - L_n| > \epsilon\right\} \leq 8 \cdot 2^k \Delta_n^*(\mathcal{F})e^{-n\epsilon^2/32},$$

and

$$\mathbf{E}\left\{\left(L_n^{(R)} - L_n\right)^2\right\} \leq \frac{32k + 32\log(\Delta_n^*(\mathcal{F})) + 128}{n}.$$

PROOF. The first inequality follows from Theorem 23.5 by the obvious inequality

$$\mathbf{P}\left\{|L_n^{(R)} - L_n| > \epsilon\right\} \leq \mathbf{P}\left\{\sup_{\mathcal{P}\in\mathcal{F}}|L_n^{(R)}(\mathcal{P}) - L_n(\mathcal{P})| > \epsilon\right\}.$$

The second inequality follows from the first one via Problem 12.1. $\square$

Perhaps the most important application of Theorem 23.5 is in classifier selection. Let $\mathcal{C}_n$ be a class of (possibly data-dependent) histogram rules. We may use the error estimate $L_n^{(R)}$ to select a classifier that minimizes the estimated error probability. Denote the selected histogram rule by ϕ_n^*, that is,

$$L_n^{(R)}(\phi_n^*) \leq L_n^{(R)}(\phi_n) \quad \text{for all} \quad \phi_n \in \mathcal{C}_n.$$

Here $L_n^{(R)}(\phi_n)$ denotes the estimated error probability of the classification rule ϕ_n. The question is how well the selection method works, in other words, how close the error probability of the selected classifier $L(\phi_n^*)$ is to the error probability of the best rule in the class, $\inf_{\phi_n\in\mathcal{C}_n} L(\phi_n)$. It turns out that if the possible partitions are not too complex, then the method works very well:

COROLLARY 23.2. *Assume that for any realization of the data D_n, the possible partitions that define the histogram classifiers in C_n belong to a class of partitions $\mathcal{F}$, whose members partition $\mathcal{R}^d$ into at most k cells. Then*

$$\mathbf{P}\left\{L(\phi_n^*) - \inf_{\phi_n \in C_n} L(\phi_n) > \epsilon\right\} \leq 8 \cdot 2^k \Delta_n^*(\mathcal{F}) e^{-n\epsilon^2/128}.$$

PROOF. By Lemma 8.2,

$$L(\phi_n^*) - \inf_{\phi_n \in C_n} L(\phi_n) \leq 2 \sup_{\phi_n \in C_n} |L_n^{(R)}(\phi_n) - L(\phi_n)| \leq 2 \sup_{\mathcal{P} \in \mathcal{F}} |L_n^{(R)}(\mathcal{P}) - L(\mathcal{P})|.$$

Therefore, the statement follows from Theorem 23.5. $\square$

Problems and Exercises

PROBLEM 23.1. Prove Theorem 23.2. HINT: Proceed as in the proof of Theorem 23.1.

PROBLEM 23.2. Show that for histogram rules, the resubstitution estimate may be written as

$$L_n^{(R)} = \sum_i \min\{v_{0,n}(A_i), v_{1,n}(A_i)\}.$$

PROBLEM 23.3. Consider the resubstitution estimate $L_n^{(R)}$ of the error probability L_n of a histogram rule based on a fixed sequence of partitions $\mathcal{P}_n$. Show that if the regression function estimate

$$\eta_n(x) = \frac{\frac{1}{n}\sum_{j=1}^{n} I_{\{Y_j=1, X_j \in A(x)\}}}{\mu(A(x))}$$

is consistent, that is, it satisfies $\mathbf{E}\left\{\int |\eta(x) - \eta_n(x)|\mu(dx)\right\} \to 0$ as $n \to \infty$, then

$$\lim_{n\to\infty} \mathbf{E}\left\{|L_n^{(R)} - L_n|\right\} = 0,$$

and also $\mathbf{E}\left\{L_n^{(R)}\right\} \to L^*$.

PROBLEM 23.4. Let $(X_1', Y_1'), \ldots, (X_m', Y_m')$ be a sequence that depends in an arbitrary fashion on the data D_n, and let g_n be the nearest neighbor rule with $(X_1', Y_1'), \ldots, (X_m', Y_m')$, where m is fixed. Let $L_n^{(R)}$ denote the resubstitution estimate of $L_n = L(g_n)$. Show that for all $\epsilon > 0$ and all distributions,

$$\mathbf{P}\left\{|L_n - L_n^{(R)}| \geq \epsilon\right\} \leq 8n^{dm(m-1)} e^{-n\epsilon^2/32}.$$

PROBLEM 23.5. Find a rule for $(X, Y) \in \mathcal{R} \times \{0, 1\}$ such that for all nonatomic distributions with $L^* = 0$ we have $\mathbf{E}\{L_n\} \to 0$, yet $\mathbf{E}\left\{L_n^{(R)}\right\} \geq \mathbf{E}\{L_n\}$. (Thus, $L_n^{(R)}$ may be pessimistically biased even for a consistent rule.)

PROBLEM 23.6. For histogram rules on fixed partitions (partitions that do not change with n and are independent of the data), show that $\mathbf{E}\left\{L_n^{(R)}\right\}$ is monotonically nonincreasing.

PROBLEM 23.7. Assume that X has a density, and investigate the resubstitution estimate of the 3-NN rule. What is the limit of $\mathbf{E}\left\{L_n^{(R)}\right\}$ as $n \to \infty$?

24

Deleted Estimates of the Error Probability

what we call "leave-one-out" error estimation

The *deleted estimate* (also called *cross-validation, leave-one-out,* or *U-method*) attempts to avoid the bias present in the resubstitution estimate. Proposed and developed by Lunts and Brailovsky (1967), Lachenbruch (1967), Cover (1969), and Stone (1974), the method deletes the first pair (X_1, Y_1) from the training data and makes a decision g_{n-1} using the remaining $n - 1$ pairs. It tests for an an error on (X_1, Y_1), and repeats this procedure for all n pairs of the training data D_n. The estimate $L_n^{(D)}$ is the average the number of errors.

We formally denote the training set with (X_i, Y_i) deleted by

$$D_{n,i} = ((X_1, Y_1), \ldots, (X_{i-1}, Y_{i-1}), (X_{i+1}, Y_{i+1}), \ldots, (X_n, Y_n)).$$

Then we define

$$L_n^{(D)} = \frac{1}{n} \sum_{i=1}^{n} I_{\{g_{n-1}(X_i, D_{n,i}) \neq Y_i\}}.$$

Clearly, the deleted estimate is almost unbiased in the sense that

$$\mathbf{E} L_n^{(D)} = \mathbf{E} L_{n-1}.$$

Thus, $L_n^{(D)}$ should be viewed as an estimator of L_{n-1}, rather than of L_n. In most of the interesting cases L_n converges with probability one so that the difference between L_{n-1} and L_n becomes negligible for large n.

The designer has the luxury of being able to pick the most convenient g_{n-1}. In some cases the choice is very natural, in other cases it is not. For example, if g_n is the 1-nearest neighbor rule, then letting g_{n-1} be the 1-nearest neighbor rule based on $n - 1$ data pairs seems to be an obvious choice. We will see later that this indeed yields an extremely good estimator. But what should g_{n-1} be if g_n is, for example, a k-NN rule? Should it be the k-nearest neighbor classifier, the $k - 1$-nearest neighbor classifier, or maybe something else? Well, the choice is typically nontrivial and needs careful attention if the designer wants a distribution-free performance guarantee for the resulting estimate. Because of the variety of choices for g_{n-1}, we should not speak of *the* deleted estimate, but rather of *a* deleted estimate.

In this chapter we analyze the performance of deleted estimates for a few proto-type classifiers, such as the kernel, nearest neighbor, and histogram rules. In most cases studied here, deleted estimates have good distribution-free properties.

24.1   A General Lower Bound

We begin by exploring general limitations of error estimates for some important nonparametric rules. An error estimate $\widehat{L}_n$ is merely a function $\left(\mathcal{R}^d \times \{0, 1\}\right)^n \rightarrow [0, 1]$, which is applied to the data $D_n = ((X_1, Y_1), \ldots, (X_n, Y_n))$.

Theorem 24.1. *Let g_n be one of the following classifiers:*

(a)   *The kernel rule*

$$g_n(x) = \begin{cases} 0 & \text{if } \sum_{i=1}^{n} I_{\{Y_i=0\}} K\left(\frac{x-X_i}{h}\right) \geq \sum_{i=1}^{n} I_{\{Y_i=1\}} K\left(\frac{x-X_i}{h}\right) \\ 1 & \text{otherwise} \end{cases}$$

with a nonnegative kernel K of compact support and smoothing factor $h > 0$.

(b)   *The histogram rule*

$$g_n(x) = \begin{cases} 0 & \text{if } \sum_{i=1}^{n} I_{\{Y_i=1\}} I_{\{X_i \in A(x)\}} \leq \sum_{i=1}^{n} I_{\{Y_i=0\}} I_{\{X_i \in A(x)\}} \\ 1 & \text{otherwise} \end{cases}$$

based on a fixed partition $\mathcal{P} = \{A_1, A_2, \ldots\}$ containing at least n cells.

(c)   *The lazy histogram rule*

$$g_n(x) = Y_j, \quad x \in A_i,$$

where X_j is the minimum-index point among $X_1, \ldots, X_n$ for which $X_j \in A_i$, where $\mathcal{P} = \{A_1, A_2, \ldots\}$ is a fixed partition containing at least n cells.
Denote the probability of error for g_n by $L_n = \mathbf{P}\{g_n(X, D_n) \neq Y | D_n\}$. Then for every $n \geq 10$, and for every error estimator $\widehat{L}_n$, there exists a distribution of (X, Y) with $L^* = 0$ such that

$$\mathbf{E}\left\{\left|\widehat{L}_n(D_n) - L_n\right|\right\} \geq \frac{1}{\sqrt{32n}}.$$

The theorem says that for any estimate $\widehat{L}_n$, there exists a distribution with the property that $\mathbf{E}\left\{\left|\widehat{L}_n(D_n) - L_n\right|\right\} \geq 1/\sqrt{32n}$. For the rules g_n given in the theorem, *no* error estimate can possibly give better distribution-free performance guarantees. Error estimates are necessarily going to perform at least this poorly for some distributions.

PROOF OF THEOREM 24.1. Let $\widehat{L}_n$ be an arbitrary fixed error estimate. The proof relies on randomization ideas similar to those of the proofs of Theorems 7.1 and 7.2. We construct a family of distributions for (X, Y). For $b \in [0, 1)$ let $b = 0.b_1 b_2 b_3 \dots$ be its binary expansion. In all cases the distribution of X is uniform on n points $\{x_1, \dots, x_n\}$. For the histogram and lazy histogram rules choose $x_1, \dots, x_n$ such that they fall into different cells. For the kernel rule choose $x_1, \dots, x_n$ so that they are isolated from each other, that is, $K(\frac{x_i - x_j}{h}) = 0$ for all $i, j \in \{1, \dots, n\}, i \neq j$. To simplify the notation we will refer to these points by their indices, that is, we will write $X = i$ instead of $X = x_i$. For a fixed b define

$$Y = b_X.$$

We may create infinitely many samples (one for each $b \in [0, 1)$) drawn from the distribution of (X, Y) as follows. Let $X_1, \dots, X_n$ be i.i.d. and uniformly distributed on $\{1, \dots, n\}$. All the samples share the same $X_1, \dots, X_n$, but differ in their Y_i's. For given X_i, define $Y_i = b_{X_i}$. Write $Z_n = (X_1, \dots, X_n)$ and $N_i = \sum_{j=1}^{n} I_{\{X_j = i\}}$. Observe that D_n is a function of Z_n and b. It is clear that for all classifiers covered by our assumptions, for a fixed b,

$$L_n = \frac{1}{n} \sum_{i \in S} I_{\{b_i = 1\}},$$

where $S = \{i : 1 \leq i \leq n, N_i = 0\}$ is the set of empty bins. We randomize the distribution as follows. Let $B = 0.B_1 B_2 \dots$ be a uniform $[0, 1]$ random variable independent of (X, Y). Then clearly $B_1, B_2, \dots$ are independent uniform binary random variables. Note that

$$\sup_b \mathbf{E} \left\{ \left| \widehat{L}_n(D_n) - L_n \right| \right\} \quad \geq \quad \mathbf{E} \left\{ \left| \widehat{L}_n(D_n) - L_n \right| \right\}$$

$$(\text{where } b \text{ is replaced by } B)$$

$$= \quad \mathbf{E} \left\{ \mathbf{E} \left\{ \left| \widehat{L}_n(D_n) - L_n \right| \, \big| \, Z_n \right\} \right\}.$$

In what follows we bound the conditional expectation within the brackets. To make the dependence upon B explicit we write $\widehat{L}_n(Z_n, B) = \widehat{L}_n(D_n)$ and $L_n(B) = L_n$. Thus,

$$\mathbf{E} \left\{ \left| \widehat{L}_n(Z_n, B) - L_n(B) \right| \, \big| \, Z_n \right\}$$

$$= \quad \frac{1}{2} \mathbf{E} \left\{ \left| \widehat{L}_n(Z_n, B) - L_n(B) \right| + \left| \widehat{L}_n(Z_n, B^*) - L_n(B^*) \right| \, \big| \, Z_n \right\}$$

$$(\text{where } B_i^* = B_i \text{ for } i \text{ with } N_i > 0 \text{ and } B_i^* = 1 - B_i \text{ for } i \text{ with } N_i = 0)$$

$$\geq \quad \frac{1}{2} \mathbf{E} \left\{ \left| L_n(B) - L_n(B^*) \right| \, \big| \, Z_n \right\}$$

$$(\text{since } \widehat{L}_n(Z_n, B^*) = \widehat{L}_n(Z_n, B))$$

$$= \frac{1}{2}\mathbf{E}\left\{ \frac{1}{n}\left| \sum_{i \in S}\left(I_{\{B_i=1\}} - I_{\{B_i=0\}} \right) \right| \,\middle|\, Z_n \right\}$$

(recall the expression for L_n given above)

$$= \frac{1}{2n}\mathbf{E}\left\{ |2B(|S|, 1/2) - 1|\,|\,|S| \right\}$$

(where $B(|S|, 1/2)$ is a binomial $(|S|, 1/2)$ random variable)

$$\geq \frac{1}{2n}\sqrt{\frac{|S|}{2}} \quad \text{(by Khintchine's inequality, see Lemma A.5).}$$

In summary, we have

$$\sup_b \mathbf{E}\left\{ \left| \widehat{L}_n(D_n) - L_n \right| \right\} \geq \frac{1}{2n}\mathbf{E}\left\{ \sqrt{\frac{|S|}{2}} \right\}.$$

We have only to bound the right-hand side. We apply Lemma A.4 to the random variable $\sqrt{|S|} = \sqrt{\sum_{i=1}^n I_{\{N_i=0\}}}$. Clearly, $\mathbf{E}|S| = n(1 - 1/n)^n$, and

$$\mathbf{E}\left\{ |S|^2 \right\} = \mathbf{E}\left\{ \sum_{i=1}^n I_{\{N_i=0\}} + \sum_{i \neq j} I_{\{N_i=0, N_j=0\}} \right\}$$

$$= \mathbf{E}|S| + n(n - 1)\left(1 - \frac{2}{n} \right)^n,$$

so that

$$\mathbf{E}\left\{ \sqrt{|S|} \right\} \geq \frac{\left(n\left(1 - \frac{1}{n}\right)^n \right)^{3/2}}{\sqrt{n\left(1 - \frac{1}{n}\right)^n + n(n-1)\left(1 - \frac{2}{n}\right)^n}}$$

$$\geq \sqrt{n}\,\frac{\left(1 - \frac{1}{n}\right)^n}{\sqrt{\frac{1}{n} + \frac{(1-2/n)^n}{(1-1/n)^n}}}$$

$$\geq \sqrt{n}\,\frac{\left(1 - \frac{1}{n}\right)^n}{\sqrt{1 - \frac{1}{n} - \frac{1}{n^2} - \cdots}}$$

$$\geq \sqrt{n}\left(1 - \frac{1}{n}\right)^n\sqrt{e} \quad \left(\text{use } \left(1 - \frac{1}{n}\right)^n \leq \frac{1}{e} \right)$$

$$\geq \frac{1}{2}\sqrt{n} \quad \text{(for } n \geq 10\text{).}$$

The proof is now complete. $\square$

24.2 A General Upper Bound for Deleted Estimates

The following inequality is a general tool for obtaining distribution-free upper bounds for the difference between the deleted estimate and the true error probability L_n:

Theorem 24.2. (ROGERS AND WAGNER (1978); DEVROYE AND WAGNER (1976B)). *Assume that g_n is a symmetric classifier, that is, $g_n(x, D_n) = g_n(x, D'_n)$, where D'_n is obtained by permuting the pairs of D_n arbitrarily. Then*

$$\mathbf{E}\left\{\left(L_n^{(D)} - L_n\right)^2\right\} \le \frac{1}{n} + 6\mathbf{P}\left\{g_n(X, D_n) \ne g_{n-1}(X, D_{n-1})\right\}.$$

PROOF. First we express the three terms on the right-hand side of

$$\mathbf{E}\left\{\left(L_n^{(D)} - L_n\right)^2\right\} = \mathbf{E}\left\{L_n^{(D)2}\right\} - 2\mathbf{E}\left\{L_n^{(D)}L_n\right\} + \mathbf{E}\left\{L_n^2\right\}.$$

The first term can be bounded, by using symmetry of g_n, by

$$\mathbf{E}\left\{L_n^{(D)2}\right\}$$

$$= \mathbf{E}\left\{\left(\frac{1}{n}\sum_{i=1}^n I_{\{g_{n-1}(X_i, D_{n,i}) \ne Y_i\}}\right)^2\right\}$$

$$= \mathbf{E}\left\{\frac{1}{n^2}\sum_{i=1}^n I_{\{g_{n-1}(X_i, D_{n,i}) \ne Y_i\}}\right\}$$

$$\quad + \mathbf{E}\left\{\frac{1}{n^2}\sum_{i \ne j} I_{\{g_{n-1}(X_i, D_{n,i}) \ne Y_i\}} I_{\{g_{n-1}(X_j, D_{n,j}) \ne Y_j\}}\right\}$$

$$= \frac{\mathbf{E}L_{n-1}}{n} + \frac{n-1}{n}\mathbf{P}\{g_{n-1}(X_1, D_{n,1}) \ne Y_1, g_{n-1}(X_2, D_{n,2}) \ne Y_2\}$$

$$\le \frac{1}{n} + \mathbf{P}\{g_{n-1}(X_1, D_{n,1}) \ne Y_1, g_{n-1}(X_2, D_{n,2}) \ne Y_2\}.$$

The second term is written as

$$\mathbf{E}\left\{L_n^{(D)}L_n\right\}$$

$$= \mathbf{E}\left\{L_n\frac{1}{n}\sum_{i=1}^n I_{\{g_{n-1}(X_i, D_{n,i}) \ne Y_i\}}\right\}$$

$$= \frac{1}{n}\sum_{i=1}^n \mathbf{E}\left\{L_n I_{\{g_{n-1}(X_i, D_{n,i}) \ne Y_i\}}\right\}$$

$$= \frac{1}{n}\sum_{i=1}^n \mathbf{E}\left\{\mathbf{P}\{g_n(X, D_n) \ne Y | D_n\}\mathbf{P}\{g_{n-1}(X_i, D_{n,i}) \ne Y_i | D_n\}\right\}$$

$$= \frac{1}{n} \sum_{i=1}^{n} \mathbf{E} \left\{ \mathbf{P}\{g_n(X, D_n) \neq Y, g_{n-1}(X_i, D_{n,i}) \neq Y_i | D_n\} \right\}$$

$$= \mathbf{P}\{g_n(X, D_n) \neq Y, g_{n-1}(X_1, D_{n,1}) \neq Y_1\}.$$

For the third term, we introduce the pair (X', Y'), independent of X, Y, and D_n, having the same distribution as (X, Y). Then

$$\begin{aligned}
\mathbf{E}\{L_n^2\} &= \mathbf{E}\left\{\mathbf{P}\{g_n(X, D_n) \neq Y | D_n\}^2\right\} \\
&= \mathbf{E}\left\{\mathbf{P}\{g_n(X, D_n) \neq Y | D_n\}\mathbf{P}\{g_n(X', D_n) \neq Y' | D_n\}\right\} \\
&= \mathbf{E}\left\{\mathbf{P}\{g_n(X, D_n) \neq Y, g_n(X', D_n) \neq Y' | D_n\}\right\} \\
&= \mathbf{P}\{g_n(X, D_n) \neq Y, g_n(X', D_n) \neq Y'\},
\end{aligned}$$

where we used independence of (X', Y').

We introduce the notation

$$\begin{aligned}
D_n' &= (X_3, Y_3), \ldots, (X_n, Y_n), \\
A_{k,i,j} &= \left\{g_n(X_k; (X_i, Y_i), (X_j, Y_j), D_n') \neq Y_k\right\}, \\
B_{k,i} &= \left\{g_{n-1}(X_k; (X_i, Y_i), D_n') \neq Y_k\right\},
\end{aligned}$$

and we formally replace (X, Y) and (X', Y') by (X_α, Y_α) and (X_β, Y_β) so that we may work with the indices α and β. With this notation, we have shown thus far the following:

$$\mathbf{E}\left\{\left(L_n^{(D)} - L_n\right)^2\right\} \leq \frac{1}{n} + \mathbf{P}\{A_{\alpha,1,2}, A_{\beta,1,2}\} - \mathbf{P}\{A_{\alpha,1,2}, B_{1,2}\}$$
$$+ \mathbf{P}\{B_{1,2}, B_{2,1}\} - \mathbf{P}\{A_{\alpha,1,2}, B_{2,1}\}.$$

Note that

$$\begin{aligned}
&\mathbf{P}\{A_{\alpha,1,2}, A_{\beta,1,2}\} - \mathbf{P}\{A_{\alpha,1,2}, B_{1,2}\} \\
&= \mathbf{P}\{A_{\alpha,1,2}, A_{\beta,1,2}\} - \mathbf{P}\{A_{\alpha,\beta,2}, B_{\beta,2}\} \\
&\quad \text{(by symmetry)} \\
&= \mathbf{P}\{A_{\alpha,1,2}, A_{\beta,1,2}\} - \mathbf{P}\{A_{\alpha,\beta,2}, A_{\beta,1,2}\} \\
&\quad + \mathbf{P}\{A_{\alpha,\beta,2}, A_{\beta,1,2}\} - \mathbf{P}\{A_{\alpha,\beta,2}, B_{\beta,2}\} \\
&= I + II.
\end{aligned}$$

Also,

$$\begin{aligned}
&\mathbf{P}\{B_{1,2}, B_{2,1}\} - \mathbf{P}\{A_{\alpha,1,2}, B_{1,2}\} \\
&= \mathbf{P}\{B_{\alpha,\beta}, B_{\beta,\alpha}\} - \mathbf{P}\{A_{\alpha,\beta,2}, B_{\beta,2}\} \\
&\quad \text{(by symmetry)}
\end{aligned}$$

$$= \quad \mathbf{P}\{B_{\alpha,\beta}, B_{\beta,\alpha}\} - \mathbf{P}\{A_{\alpha,\beta,2}, B_{\beta,\alpha}\}$$
$$+ \mathbf{P}\{A_{\alpha,\beta,2}, B_{\beta,\alpha}\} - \mathbf{P}\{A_{\alpha,\beta,2}, B_{\beta,2}\}$$
$$= \quad III + IV.$$

Using the fact that for events $\{C_i\}$, $|\mathbf{P}\{C_i, C_j\} - \mathbf{P}\{C_i, C_k\}| \leq \mathbf{P}\{C_j \triangle C_k\}$, we bound

$$
\begin{aligned}
I &\leq \mathbf{P}\{A_{\alpha,1,2} \triangle A_{\alpha,\beta,2}\}, \\
II &\leq \mathbf{P}\{A_{\beta,1,2} \triangle B_{\beta,2}\} \stackrel{\text{def}}{=} V, \\
III &\leq \mathbf{P}\{B_{\alpha,\beta} \triangle A_{\alpha,\beta,2}\} = V, \\
IV &\leq \mathbf{P}\{B_{\beta,\alpha} \triangle B_{\beta,2}\}.
\end{aligned}
$$

The upper bounds for II and III are identical by symmetry. Also,

$$I \leq \mathbf{P}\{A_{\alpha,1,2} \triangle B_{\alpha,2}\} + \mathbf{P}\{B_{\alpha,2} \triangle A_{\alpha,\beta,2}\} = 2V$$

and

$$IV \leq \mathbf{P}\{B_{\beta,\alpha} \triangle A_{\beta,\alpha,2}\} + \mathbf{P}\{A_{\beta,\alpha,2} \triangle B_{\beta,2}\} = 2V,$$

for a grand total of $6V$. This concludes the proof. $\square$

24.3   Nearest Neighbor Rules

Theorem 24.2 can be used to obtain distribution-free upper bounds for specific rules. Here is the most important example.

Theorem 24.3. (ROGERS AND WAGNER (1978)). *Let g_n be the k-nearest neighbor rule with randomized tie-breaking. If $L_n^{(D)}$ is the deleted estimate with g_{n-1} chosen as the k-NN rule (with the same k and with the same randomizing random variables), then*

$$\mathbf{E}\left\{\left(L_n^{(D)} - L_n\right)^2\right\} \leq \frac{6k+1}{n}.$$

PROOF. Because of the randomized tie-breaking, the k-NN rule is symmetric, and Theorem 24.2 is applicable. We only have to show that

$$\mathbf{P}\{g_n(X, D_n) \neq g_{n-1}(X, D_{n-1})\} \leq \frac{k}{n}.$$

Clearly, $g_n(X, D_n) \neq g_{n-1}(X, D_{n-1})$ can happen only if X_n is among the k nearest neighbors of X. But the probability of this event is just k/n, since by symmetry, all points are equally likely to be among the k nearest neighbors. $\square$

REMARK. If g_n is the k-NN rule such that distance ties are broken by comparing indices, then g_n is not symmetric, and Theorem 24.2 is no longer applicable (unless, e.g., X has a density). Another nonsymmetric classifier is the lazy histogram rule. □

REMARK. Applying Theorem 24.3 to the 1-NN rule, the Cauchy-Schwarz inequality implies $\mathbf{E}\left|L_n^{(D)} - L_n\right| \leq \sqrt{7/n}$ for all distributions. □

REMARK. Clearly, the inequality of Theorem 24.3 holds for any rule that is some function of the k nearest points. For the k-nearest neighbor rule, with a more careful analysis, Devroye and Wagner (1979a) improved Theorem 24.3 to

$$\mathbf{E}\left\{\left(L_n^{(D)} - L_n\right)^2\right\} \leq \frac{1}{n} + \frac{24\sqrt{k}}{n\sqrt{2\pi}}$$

(see Problem 24.8). □

Probability inequalities for $|L_n^{(D)} - L_n|$ can also be obtained with further work. By Chebyshev's inequality we immediately get

$$\mathbf{P}\{|L_n^{(D)} - L_n| > \epsilon\} \leq \frac{\mathbf{E}\left\{\left(L_n^{(D)} - L_n\right)^2\right\}}{\epsilon^2},$$

so that the above bounds on the expected squared error can be used. Sharper distribution-free inequalities were obtained by Devroye and Wagner (1979a; 1979b) for several nonparametric rules. Here we present a result that follows immediately from what we have already seen:

Theorem 24.4. *Consider the k-nearest neighbor rule with randomized tie-breaking. If $L_n^{(D)}$ is the deleted estimate with g_{n-1} chosen as the k-NN rule with the same tie-breaking, then*

$$\mathbf{P}\{|L_n^{(D)} - \mathbf{E}L_{n-1}| > \epsilon\} \leq 2e^{-n\epsilon^2/(k^2\gamma_d^2)}.$$

PROOF. The result follows immediately from McDiarmid's inequality by the following argument: from Lemma 11.1, given n points in $\mathcal{R}^d$, a particular point can be among the k nearest neighbors of at most $k\gamma_d$ points. To see this, just set μ equal to the empirical measure of the n points in Lemma 11.1. Therefore, changing the value of one pair from the training data can change the value of the estimate by at most $2k\gamma_d$. Now, since $\mathbf{E}L_n^{(D)} = \mathbf{E}L_{n-1}$, Theorem 9.1 yields the result. □

Exponential upper bounds for the probability $\mathbf{P}\{|L_n^{(D)} - L_n| > \epsilon\}$ are typically much harder to obtain. We mention one result without proof.

Theorem 24.5. (DEVROYE AND WAGNER (1979A)). *For the k-nearest neighbor rule,*

$$P\{|L_n^{(D)} - L_n| > \epsilon\} \leq 2e^{-n\epsilon^2/18} + 6e^{-n\epsilon^3/(108k(\gamma_d+2))}.$$

One of the drawbacks of the deleted estimate is that it requires much more computation than the resubstitution estimate. If conditional on $Y = 0$ and $Y = 1$, X is gaussian, and the classification rule is the appropriate parametric rule, then the estimate can be computed quickly. See Lachenbruch and Mickey (1968), Fukunaga and Kessel (1971), and McLachlan (1992) for further references.

Another, and probably more serious, disadvantage of the deleted estimate is its large variance. This fact can be illustrated by the following example from Devroye and Wagner (1979b): let n be even, and let the distribution of (X, Y) be such that Y is independent of X with $P\{Y = 0\} = P\{Y = 1\} = 1/2$. Consider the k-nearest neighbor rule with $k = n - 1$. Then obviously, $L_n = 1/2$. Clearly, if the number of zeros and ones among the labels $Y_1, \ldots, Y_n$ are equal, then $L_n^{(D)} = 1$. Thus, for $0 < \epsilon < 1/2$,

$$P\{|L_n^{(D)} - L_n| > \epsilon\} \geq P\left\{\sum_{i=1}^{n} I_{\{Y_i=0\}} = \frac{n}{2}\right\} = \frac{1}{2^n}\binom{n}{n/2}.$$

By Stirling's formula (Lemma A.3), we have

$$P\{|L_n^{(D)} - L_n| > \epsilon\} \geq \frac{1}{\sqrt{2\pi n}}\frac{1}{e^{1/12}}.$$

Therefore, for this simple rule and certain distributions, the probability above can not decrease to zero faster than $1/\sqrt{n}$. Note that in the example above, $EL_n^{(D)} = EL_{n-1} = 1/2$, so the lower bound holds for $P\{|L_n^{(D)} - EL_n^{(D)}| > \epsilon\}$ as well. Also, in this example, we have

$$E\left\{(L_n^{(D)} - EL_n^{(D)})^2\right\} \geq \frac{1}{4}P\left\{\sum_{i=1}^{n} I_{\{Y_i=0\}} = \frac{n}{2}\right\} \geq \frac{1}{4\sqrt{2\pi n}}\frac{1}{e^{1/12}}.$$

In Chapter 31 we describe other estimates with much smaller variances.

24.4   Kernel Rules

Theorem 24.2 may also be used to obtain tight distribution-free upper bounds for the performance of the deleted estimate of the error probability of kernel rules. We have the following bound:

Theorem 24.6. *Assume that $K \geq 0$ is a regular kernel of bounded support, that is, it is a function satisfying*

$$\begin{aligned}
&\text{(i)} &&K(x) \geq \beta, \quad \|x\| \leq \rho, \\
&\text{(ii)} &&K(x) \leq B, \\
&\text{(iii)} &&K(x) = 0, \quad \|x\| > R,
\end{aligned}$$

for some positive finite constants β, ρ, B and R. Let the kernel rule be defined by

$$g_n(x) = \begin{cases} 0 & \text{if } \sum_{i=1}^n I_{\{Y_i=0\}} K\,(x - X_i) \geq \sum_{i=1}^n I_{\{Y_i=1\}} K\,(x - X_i) \\ 1 & \text{otherwise,} \end{cases}$$

and define g_{n-1} similarly. Then there exist constants $C_1(d)$ depending upon d only and $C_2(K)$ depending upon K only such that for all n,

$$\mathbf{E}\left\{ \left(L_n^{(D)} - L_n \right)^2 \right\} \leq \frac{1}{n} + \frac{C_1(d)C_2(K)}{\sqrt{n}}.$$

One may take $C_2(K) = 6(1 + R/\rho)^{d/2} \min(2, B/\beta)$.

REMARK. Since $C_2(K)$ is a scale-invariant factor, the theorem applies to the rule with $K(u)$ replaced by $K_h(u) = \frac{1}{h} K\left(\frac{u}{h}\right)$ for any smoothing factor. As it is, $C_2(K)$ is minimal and equal to 12 if we let K be the uniform kernel on the unit ball ($R = \rho$, $B = \beta$). The assumptions of the theorem require that g_{n-1} is defined with the same kernel and smoothing factor as g_n. $\square$

REMARK. The theorem applies to virtually any kernel of compact support that is of interest to the practitioners. Note, however, that the gaussian kernel is not covered by the result. The theorem generalizes an earlier result of Devroye and Wagner (1979b), in which a more restricted class of kernels was considered. They showed that if K is the uniform kernel then

$$\mathbf{E}\left\{ \left(L_n^{(D)} - L_n \right)^2 \right\} \leq \frac{1}{2n} + \frac{24}{\sqrt{n}}.$$

See Problem 24.4. $\square$

We need the following auxiliary inequality, which we quote without proof:

Lemma 24.1. (PETROV (1975), P.44). *Let $Z_1, \ldots, Z_n$ be real-valued i.i.d. random variables. For $\epsilon > 0$,*

$$\sup_x \mathbf{P}\left\{ \sum_{i=1}^n Z_i \in [x, x + \epsilon] \right\} \leq \frac{C\epsilon}{\sqrt{n}} \cdot \frac{1}{\sqrt{\sup_{0<\lambda\leq\epsilon} \lambda \cdot \mathbf{P}\{|Z_1| \geq \lambda/2\}}},$$

where C is a universal constant.

COROLLARY 24.1. *Let $Z_1, \ldots, Z_n$ be real-valued i.i.d. random variables. For $\epsilon \geq \lambda > 0$,*

$$\mathbf{P}\left\{ \left| \sum_{i=1}^n Z_i \right| < \frac{\epsilon}{2} \right\} \leq \frac{C}{\sqrt{n}} \frac{\epsilon}{\lambda} \frac{1}{\sqrt{\mathbf{P}\{|Z_1| \geq \lambda/2\}}},$$

where C is a universal constant.

PROOF OF THEOREM 24.6. We apply Theorem 24.2, by finding an upper bound for

$$\mathbf{P}\{g_n(X, D_n) \neq g_{n-1}(X, D_{n-1})\}\,.$$

For the kernel rule with kernel $K \geq 0$, in which h is absorbed,

$$\mathbf{P}\{g_n(X, D_n) \neq g_{n-1}(X, D_{n-1})\}$$

$$\leq \quad \mathbf{P}\left\{\left|\sum_{i=1}^{n-1}(2Y_i - 1)K(X - X_i)\right| \leq K(X - X_n), K(X - X_n) > 0\right\}.$$

Define $B' = \max(2\beta, B)$. We have

$$\mathbf{P}\left\{\left|\sum_{i=1}^{n-1}(2Y_i - 1)K(X - X_i)\right| \leq K(X - X_n), K(X - X_n) > 0\right\}$$

$$\leq \quad \mathbf{P}\left\{\left|\sum_{i=1}^{n-1}(2Y_i - 1)K(X - X_i)\right| \leq B', X_n \in S_{X,R}\right\}$$

(where $S_{X,R}$ is the ball of radius R centered at X)

$$= \quad \mathbf{E}\left\{I_{\{X_n \in S_{X,R}\}}\frac{2CB'}{2\beta\sqrt{n}\sqrt{\mathbf{P}\{|(2Y_1 - 1)K(X - X_1)| \geq \beta|X\}}}\right\}$$

(by Corollary 24.1, since $2\beta \leq B'$)

$$\leq \quad \mathbf{E}\left\{I_{\{X_n \in S_{X,R}\}}\frac{CB'}{\beta\sqrt{n\mu(S_{X,\rho})}}\right\}$$

(recall that $K(u) \geq \beta$ for $\|u\| \leq \rho$,

so that $\mathbf{P}\{|(2Y_1 - 1)K(x - X_1)| \geq \beta\} \geq \mathbf{P}\{X_1 \geq S_{x,\rho}\}$)

$$= \quad \frac{CB'}{\beta\sqrt{n}}\int\int_{S_{t,R}}\frac{\mu(dy)}{\sqrt{\mu(S_{x,\rho})}}\mu(dx)$$

$$\leq \quad \frac{Cc_dB'}{\beta\sqrt{n}}\left(1 + \frac{R}{\rho}\right)^{d/2}.$$

where we used Lemma 10.2. The constant c_d depends upon the dimension only.
□

24.5 Histogram Rules

In this section we discuss properties of the deleted estimate of the error probability of histogram rules. Let $\mathcal{P} = \{A_1, A_2, \ldots\}$ be a partition of $\mathcal{R}^d$, and let g_n be the corresponding histogram classifier (see Chapters 6 and 9). To get a performance bound for the deleted estimate, we can simply apply Theorem 24.2.

Theorem 24.7. *For the histogram rule g_n corresponding to any partition $\mathcal{P}$, and for all n,*

$$\mathbf{E}\left\{\left(L_n^{(D)} - L_n\right)^2\right\} \leq \frac{1}{n} + 6\sum_i \frac{\mu(A_i)^{3/2}}{\sqrt{\pi(n-1)}} + 6\sum_i \mu^2(A_i)e^{-n\mu(A_i)},$$

and in particular,

$$\mathbf{E}\left\{\left(L_n^{(D)} - L_n\right)^2\right\} \leq \frac{1 + 6/e}{n} + \frac{6}{\sqrt{\pi(n-1)}}.$$

PROOF. The first inequality follows from Theorem 24.2 if we can find an upper bound for $\mathbf{P}\{g_n(X) \neq g_{n-1}(X)\}$. We introduce the notation

$$\nu_{0,n}(A) = \frac{1}{n}\sum_{j=1}^n I_{\{Y_j=0\}}I_{\{X_j \in A\}}.$$

Clearly, $g_{n-1}(X)$ can differ from $g_n(X)$ only if both X_n and X fall in the same cell of the partition, and if the number of zeros in the cell is either equal, or less by one than the number of ones. Therefore, by independence, we have

$$\mathbf{P}\{g_n(X) \neq g_{n-1}(X)\}$$
$$= \sum_i \mathbf{P}\{g_n(X) \neq g_{n-1}(X) | X \in A_i, X_n \in A_i\}\mu(A_i)^2$$
$$\leq \sum_i \mathbf{P}\left\{\nu_{0,n-1}(A_i) = \left\lfloor\frac{\mu_{n-1}(A_i)}{2}\right\rfloor \Big| X \in A_i, X_n \in A_i\right\}\mu(A_i)^2.$$

The terms in the sum above may be bounded as follows:

$$\mathbf{P}\left\{\nu_{0,n-1}(A) = \left\lfloor\frac{\mu_{n-1}(A)}{2}\right\rfloor \Big| X \in A, X_n \in A\right\}$$
$$= \mathbf{P}\left\{\nu_{0,n-1}(A) = \left\lfloor\frac{\mu_{n-1}(A)}{2}\right\rfloor\right\} \quad \text{(by independence)}$$
$$\leq \mathbf{P}\{\mu_{n-1}(A) = 0\}$$
$$\quad + \mathbf{E}\left\{\mathbf{P}\left\{\nu_{0,n-1}(A) = \left\lfloor\frac{\mu_{n-1}(A)}{2}\right\rfloor \Big| X_1, \ldots, X_n\right\} I_{\{\mu_{n-1}(A)>0\}}\right\}$$
$$\leq (1 - \mu(A))^n + \mathbf{E}\left\{\frac{1}{\sqrt{2\pi(n-1)\mu_{n-1}(A)}} I_{\{\mu_{n-1}(A)>0\}}\right\}$$
$$\text{(by Lemma A.3)}$$
$$\leq (1 - \mu(A))^n + \sqrt{\mathbf{E}\left\{\frac{I_{\{\mu_{n-1}(A)>0\}}}{2\pi(n-1)\mu_{n-1}(A)}\right\}} \quad \text{(by Jensen's inequality)}$$
$$\leq e^{-n\mu(A)} + \sqrt{\frac{1}{\pi(n-1)\mu(A)}},$$

where in the last step we use Lemma A.2. This concludes the proof of the first inequality. The second one follows trivially by noting that $xe^{-x} \leq 1/e$ for all x.
□

REMARK. It is easy to see that the inequalities of Theorem 24.7 are tight, up to a constant factor, in the sense that for any partition $\mathcal{P}$, there exists a distribution such that

$$\mathbf{E}\left\{\left(L_n^{(D)} - L_n\right)^2\right\} \geq \frac{1}{4\sqrt{2\pi n}} \frac{1}{e^{1/12}}$$

(see Problem 24.5). □

REMARK. The second inequality in Theorem 23.3 points out an important difference between the behavior of the resubstitution and the deleted estimates for histogram rules. As mentioned above, for some distributions the variance of $L_n^{(D)}$ can be of the order $1/\sqrt{n}$. This should be contrasted with the much smaller variance of the resubstitution estimate. The small variance of $L_n^{(R)}$ comes often with a larger bias. Other types of error estimates with small variance are discussed in Chapter 31. □

REMARK. Theorem 24.3 shows that for any partition,

$$\sup_{(X,Y)} \mathbf{E}\left\{\left(L_n^{(D)} - L_n\right)^2\right\} = O\left(\frac{1}{\sqrt{n}}\right).$$

On the other hand, if $k = o(\sqrt{n})$, where k is the number of cells in the partition, then for the resubstitution estimate we have a better guaranteed distribution-free performance:

$$\sup_{(X,Y)} \mathbf{E}\left\{\left(L_n^{(R)} - L_n\right)^2\right\} = o\left(\frac{1}{\sqrt{n}}\right).$$

At first sight, the resubstitution estimate seems preferable to the deleted estimate. However, if the partition has a large number of cells, $L_n^{(R)}$ may be off the mark; see Theorem 23.4. □

Problems and Exercises

PROBLEM 24.1. Show the following variant of Theorem 24.2: For all symmetric classifiers

$$\begin{aligned}
\mathbf{E}\left\{\left(L_n^{(D)} - L_n\right)^2\right\} \leq \frac{1}{n} \quad &+ \quad 2\mathbf{P}\{g_n(X, D_n) \neq g_{n-1}(X, D_{n-1})\} \\
&+ \quad \mathbf{P}\{g_n(X, D_n) \neq g_n(X, D_n^*)\} \\
&+ \quad \mathbf{P}\{g_{n-1}(X, D_{n-1}) \neq g_{n-1}(X, D_{n-1}^*)\},
\end{aligned}$$

where D_n^* and D_{n-1}^* are just D_n and D_{n-1} with (X_1, Y_1) replaced by an independent copy (X_0, Y_0).

PROBLEM 24.2. Let g_n be the relabeling NN rule with the k-NN classifier as ancestral rule, as defined in Chapter 11. Provide an upper bound for the squared error $\mathbf{E}\left\{\left(L_n^{(D)} - L_n\right)^2\right\}$ of the deleted estimate.

PROBLEM 24.3. Let g_n be the rule obtained by choosing the best $k \leq k_0$ in the k-NN rule (k_0 is a constant) by minimizing the standard deleted estimate $L_n^{(D)}$ with respect to k. How would you estimate the probability of error for this rule? Give the best possible distribution-free performance guarantees you can find.

PROBLEM 24.4. Consider the kernel rule with the window kernel $K = S_{0.1}$. Show that

$$\mathbf{E}\left\{(L_n^{(D)} - L_n)^2\right\} \leq \frac{1 + 6/e}{n} + \frac{6}{\sqrt{\pi(n-1)}}.$$

HINT: Follow the line of the proof of Theorem 24.7.

PROBLEM 24.5. Show that for any partition $\mathcal{P}$, there exists a distribution such that for the deleted estimate of the error probability of the corresponding histogram rule,

$$\mathbf{E}\left\{\left(L_n^{(D)} - L_n\right)^2\right\} \geq \frac{1}{4\sqrt{2\pi n}} \frac{1}{e^{1/12}}.$$

HINT: Proceed as in the proof of the similar inequality for k-nearest neighbor rules.

PROBLEM 24.6. Consider the k-spacings method (see Chapter 21). We estimate the probability of error (L_n) by a modified deleted estimate $\widehat{L}_n$ as follows:

$$\widehat{L}_n = \frac{1}{n} \sum_{i=1}^n I_{\{Y_i \neq g_{n,i}(X_i, D_n)\}},$$

where $g_{n,i}$ is a histogram rule based upon the same k-spacings partition used for g_n—that is, the partition determined by k-spacings of the data points $X_1, \ldots, X_n$—but in which a majority vote is based upon the Y_j's in the same cell of the partition with Y_i deleted. Show that

$$\mathbf{E}\left\{(\widehat{L}_n - L_n)\right\} \leq \frac{6k+1}{n}.$$

HINT: Condition on the X_i's, and verify that the inequality of Theorem 24.2 remains valid.

PROBLEM 24.7. Consider a rule in which we rank the real-valued observations $X_1, \ldots, X_n$ from small to large, to obtain $X_{(1)}, \ldots, X_{(n)}$. Assume that X_1 has a density. Derive an inequality for the error $|L_n^{(D)} - L_n|$ for some deleted estimate $L_n^{(D)}$ (of your choice), when the rule is defined by a majority vote over the data-dependent partition

$$(-\infty, X_{(1)}], (X_{(1)}, X_{(2)}], (X_{(3)}, X_{(5)}], (X_{(6)}, X_{(9)}], \ldots$$

defined by 1, 2, 3, 4, ... points, respectively.

PROBLEM 24.8. Prove that for the k-nearest neighbor rule,

$$\mathbf{E}\left\{\left(L_n^{(D)} - L_n\right)^2\right\} \leq \frac{1}{n} + \frac{24\sqrt{k}}{n\sqrt{2\pi}}$$

(Devroye and Wagner (1979a)). HINT: Obtain a refined upper bound for

$$\mathbf{P}\left\{g_n(X, D_n) \neq g_{n-1}(X, D_{n-1})\right\},$$

using techniques not unlike those of the proof of Theorem 24.7.

PROBLEM 24.9. OPEN-ENDED PROBLEM. Investigate if Theorem 24.6 can be extended to kernels with unbounded support such as the gaussian kernel.

25
Automatic Kernel Rules

We saw in Chapter 10 that for a large class of kernels, if the smoothing parameter h converges to zero such that nh^d goes to infinity as $n \to \infty$, then the kernel classification rule is universally consistent. For a particular n, asymptotic results provide little guidance in the selection of h. On the other hand, selecting the wrong value of h may lead to catastrophic error rates—in fact, the crux of every nonparametric estimation problem is the choice of an appropriate smoothing factor. It tells us how far we generalize each data point X_i in the space. Purely atomic distributions require little smoothing ($h = 0$ will generally be fine), while distributions with densities require a lot of smoothing. As there are no simple tests for verifying whether the data are drawn from an absolutely continuous distribution—let alone a distribution with a Lipschitz density—it is important to let the data D_n determine h. A data-dependent smoothing factor is merely a mathematical function $H_n : \left(\mathcal{R}^d \times \{0, 1\} \right)^n \to [0, \infty)$. For brevity, we will simply write H_n to denote the random variable $H_n(D_n)$. This chapter develops results regarding such functions H_n.

This chapter is not a luxury but a necessity. Anybody developing software for pattern recognition must necessarily let the data do the talking—in fact, good universally applicable programs can have *only* data-dependent parameters.

Consider the family of kernel decision rules g_n and let the smoothing factor h play the role of parameter. The best parameter (H_{OPT}) is the one that minimizes L_n. Unfortunately, it is unknown, as is L_{OPT}, the corresponding minimal probability of error. The first goal of any data-dependent smoothing factor H_n should be to approach the performance of H_{OPT}. We are careful here to avoid saying that H_n should be close to H_{OPT}, as closeness of smoothing factors does not necessarily imply closeness of error probabilities and vice versa. Guarantees one might want in this respect are

$$E\{L_n - L_{\text{OPT}}\} \leq \alpha_n$$

for some suitable sequence $\alpha_n \to 0$, or better still,

$$E\{L_n - L^*\} \leq (1 + \beta_n)E\{L_{\text{OPT}} - L^*\},$$

for another sequence $\beta_n \to 0$. But before one even attempts to develop such data-dependent smoothing factors, one's first concern should be with consistency: is it true that with the given H_n, $L_n \to L^*$ in probability or with probability one? This question is dealt with in the next section. In subsequent sections, we give various examples of data-dependent smoothing factors.

25.1 Consistency

We start with consistency results that generalize Theorem 10.1. The first result assumes that the value of the smoothing parameter is picked from a discrete set.

Theorem 25.1. *Assume that the random variable H_n takes its values from the set of real numbers of the form $\frac{1}{(1+\delta_n)^k}$, where k is a nonnegative integer and $\delta_n > 0$. Let K be a regular kernel function. (Recall Definition 10.1.) Define the kernel classification rule corresponding to the random smoothing parameter H_n by*

$$g_n(x) = \begin{cases} 0 & \text{if } \sum_{i=1}^{n} I_{\{Y_i=0\}} K\left(\frac{x-X_i}{H_n}\right) \geq \sum_{i=1}^{n} I_{\{Y_i=1\}} K\left(\frac{x-X_i}{H_n}\right) \\ 1 & \text{otherwise.} \end{cases}$$

If

$$\frac{1}{\delta_n} = e^{o(n)}$$

and

$$H_n \to 0 \quad \text{and} \quad nH_n^d \to \infty \quad \text{with probability one as } n \to \infty,$$

then $L(g_n) \to L^*$ with probability one, that is, g_n is strongly universally consistent.

PROOF. The theorem is a straightforward extension of Theorem 10.1. Clearly, $L(g_n) \to L^*$ with probability one if and only if for every $\epsilon > 0$, $I_{\{L(g_n)-L^*>\epsilon\}} \to 0$ with probability one. Now, for any $\beta > 0$,

$$I_{\{L(g_n)-L^*>\epsilon\}} \leq I_{\{1/H_n>\beta,nH_n^d>\beta,L(g_n)-L^*>\epsilon\}} + I_{\{1/H_n\leq\beta\}} + I_{\{nH_n^d\leq\beta\}}.$$

We have to show that the random variables on the right-hand side converge to zero with probability one. The convergence of the second and third terms follows from the conditions on H_n. The convergence of the first term follows from Theorem 10.1, since it states that for any $\epsilon > 0$, there exist $\beta > 0$ and n_0 such that for the error probability $L_{n,k}$ of the kernel rule with smoothing parameter $h = \frac{1}{(1+\delta_n)^k}$,

$$P\{L_{n,k} - L^* > \epsilon\} \leq 4e^{-cn\epsilon^2}$$

for some constant c depending on the dimension only, provided that $n > n_0$, $h < 1/\beta$ and $nh^d > \beta$. Now clearly,

$$\mathbf{P}\left\{L_n - L^* > \epsilon, 1/H_n > \beta, nH_n^d > \beta\right\}$$

$$\leq \mathbf{P}\left\{\sup_{k:(1+\delta_n)^k > \beta, n/(1+\delta_n)^{kd} > \beta} L_{n,k} - L^* > \epsilon\right\}$$

$$\leq C_n \sup_{k:(1+\delta)^k > \beta, n/(1+\delta)^{kd} > \beta} \mathbf{P}\{L_{n,k} - L^* > \epsilon\},$$

by the union bound, where C_n is the number of possible values of H_n in the given range. As

$$\beta < (1 + \delta_n)^k < \left(\frac{n}{\beta}\right)^{1/d},$$

we note that

$$C_n \leq 2 + \frac{\frac{1}{d}\log\left(\frac{n}{\beta}\right) - \log\beta}{\log(1 + \delta_n)} = O\left(\log n/\delta_n^2\right) = e^{o(n)}$$

by the condition on the sequence $\{\delta_n\}$. Combining this with Theorem 10.1, for $n > n_0$, we get

$$\mathbf{P}\left\{L_n - L^* > \epsilon, 1/H_n > \beta, nH_n^d > \beta\right\} \leq 4C_n e^{-cn\epsilon^2},$$

which is summable in n. The Borel-Cantelli lemma implies that

$$I_{\{1/H_n > \beta, nH_n^d > \beta, L(g_n) - L^* > \epsilon\}} \to 0$$

with probability one, and the theorem is proved. $\square$

For weak consistency, it suffices to require convergence of H_n and nH_n^d in probability (Problem 25.1):

Theorem 25.2. *Assume that the random variable H_n takes its values from the set of real numbers of the form $\frac{1}{(1+\delta_n)^k}$, where k is a nonnegative integer and $\delta_n > 0$. Let K be a regular kernel. If*

$$\frac{1}{\delta_n} = e^{o(n)}$$

and

$$H_n \to 0 \quad and \quad nH_n^d \to \infty \quad in\ probability\ as\ n \to \infty,$$

then the kernel classification rule corresponding to the random smoothing parameter H_n is universally consistent, that is, $L(g_n) \to L^*$ in probability.

We are now prepared to prove a result similar to Theorem 25.1 without restricting the possible values of the random smoothing parameter H_n. For technical reasons, we need to assume some additional regularity conditions on the kernel function: K must be decreasing along rays starting from the origin, but it should not decrease too rapidly. Rapidly decreasing functions such as the Gaussian kernel, or functions of bounded support, such as the window kernel are excluded.

Theorem 25.3. *Let K be a regular kernel that is monotone decreasing along rays, that is, for any $x \in \mathcal{R}^d$ and $a > 1$, $K(ax) \leq K(x)$. Assume in addition that there exists a constant $c > 0$ such that for every sufficiently small $\delta > 0$, and $x \in \mathcal{R}^d$, $K((1+\delta)x) \geq (1-c\delta)K(x)$. Let $\{H_n\}$ be a sequence of random variables satisfying*

$$H_n \to 0 \quad and \quad nH_n^d \to \infty \quad with\ probability\ one,\ as\ n \to \infty.$$

Then the error probability $L(g_n)$ of the kernel classification rule with kernel K and smoothing parameter H_n converges to L^* with probability one, that is, the rule is strongly universally consistent.

REMARK. The technical condition on K is needed to ensure that small changes in h do not cause dramatic changes in $L(g_n)$. We expect some smooth behavior of $L(g_n)$ as a function of h. The conditions are rather restrictive, as the kernels must have infinite support and decrease slower than at a polynomial rate. An example satisfying the conditions is

$$K(x) = \begin{cases} 1 & \text{if } \|x\| \leq 1 \\ 1/\|x\|^r & \text{otherwise,} \end{cases}$$

where $r > 0$ (see Problem 25.2). The conditions on H_n are by no means necessary. We have already seen that consistency occurs for atomic distributions if $K(0) > 0$ and $H_n \equiv 0$, or for distributions with $L^* = 1/2$ when H_n takes any value. However, Theorem 25.3 provides us with a simple collection of sufficient conditions. $\square$

PROOF OF THEOREM 25.3.   First we discretize H_n. Define a sequence $\delta_n \to 0$ satisfying the condition in Theorem 25.1, and introduce the random variables $\underline{H}_n$ and $\overline{H}_n$ as follows: $\underline{H}_n = \frac{1}{(1+\delta_n)^{K_n}}$, where K_n is the smallest integer such that $H_n > \frac{1}{(1+\delta_n)^{K_n}}$, and let $\overline{H}_n = (1+\delta_n)\underline{H}_n$. Thus, $\underline{H}_n < H_n \leq \overline{H}_n$. Note that both $\underline{H}_n$ and $\overline{H}_n$ satisfy the conditions of Theorem 25.1. As usual, the consistency proof is based on Theorem 2.3. Here, however, we need a somewhat tricky choice of the denominator of the functions that approximate $\eta(x)$. Introduce

$$\widehat{\eta}_{n,H_n}(x) = \frac{\frac{1}{n} \sum_{i=1}^n I_{\{Y_i=1\}} K\left(\frac{x-X_i}{H_n}\right)}{\int K\left(\frac{x-z}{H_n}\right) \mu(dz)}.$$

Clearly, the value of the classification rule $g_n(x)$ equals one if and only if $\widehat{\eta}_{n,H_n}(x)$ is greater than the function defined similarly, with the $I_{\{Y_i=1\}}$'s replaced with $I_{\{Y_i=0\}}$. Then by Theorem 2.3 it suffices to show that

$$\int |\widehat{\eta}_{n,H_n}(x) - \eta(x)|\mu(dx) \to 0$$

with probability one. We use the following decomposition:

$$\int |\widehat{\eta}_{n,H_n}(x) - \eta(x)|\mu(dx)$$

$$\leq \int |\widehat{\eta}_{n,H_n}(x) - \widehat{\eta}_{n,\overline{H}_n}(x)|\mu(dx) + \int |\widehat{\eta}_{n,\overline{H}_n}(x) - \eta(x)|\mu(dx). \quad (25.1)$$

The second term on the right-hand side converges to zero with probability one, which can be seen by repeating the argument of the proof of Theorem 25.1, using the observation that in the proof of Theorem 10.1 we proved consistency via an exponential probability inequality for

$$\int |\eta_{n,h_n}(x) - \eta(x)|\mu(dx).$$

The first term may be bounded as the following simple chain of inequalities indicates:

$$\int |\widehat{\eta}_{n,H_n}(x) - \widehat{\eta}_{n,\overline{H}_n}(x)|\mu(dx)$$

$$= \int \left| \frac{\frac{1}{n}\sum_{i=1}^n Y_i K\left(\frac{x-X_i}{H_n}\right) - \frac{1}{n}\sum_{i=1}^n Y_i K\left(\frac{x-X_i}{\overline{H}_n}\right)}{\int K\left(\frac{x-z}{\overline{H}_n}\right)\mu(dz)} \right| \mu(dx)$$

$$\leq \int \frac{\frac{1}{n}\sum_{i=1}^n \left(K\left(\frac{x-X_i}{H_n}\right) - K\left(\frac{x-X_i}{\overline{H}_n}\right)\right)}{\int K\left(\frac{x-z}{\overline{H}_n}\right)\mu(dz)} \mu(dx)$$

(by the monotonicity of K)

$$\leq \int \frac{\frac{1}{n}\sum_{i=1}^n (1 - (1 - c\delta_n))K\left(\frac{x-X_i}{\overline{H}_n}\right)}{\int K\left(\frac{x-z}{\overline{H}_n}\right)\mu(dz)} \mu(dx)$$

(from $\overline{H}_n = (1 + \delta_n)\underline{H}_n$, and the condition on K, if n is large enough)

$$= c\delta_n \int \frac{\frac{1}{n}\sum_{i=1}^n K\left(\frac{x-X_i}{\overline{H}_n}\right)}{\int K\left(\frac{x-z}{\overline{H}_n}\right)\mu(dz)} \mu(dx).$$

Since $\overline{H}_n$ satisfies the conditions of Theorem 25.1, the integral on the right-hand side converges to one with probability one, just as we argued for the second term on the right-hand side of (25.1). But δ_n converges to zero. Therefore, the first term on the right-hand side of (25.1) tends to zero with probability one. □

REMARK. A quick inspection of the proof above shows that if $\{a_n\}$ and $\{b_n\}$ are deterministic sequences with the property that $a_n < b_n$, $b_n \to 0$, and $na_n^d \to \infty$, then for the kernel estimate with kernel as in Theorem 25.3, we have

$$\sup_{a_n \le h \le b_n} L_n(h) \to L^* \text{ with probability one}$$

for all distributions. One would never use the worst smoothing factor over the range $[a_n, b_n]$, but this corollary points out just how powerful Theorem 25.3 is. □

25.2 Data Splitting

Our first example of a data-dependent H_n is based upon the minimization of a suitable error estimate. You should have read Chapter 22 on data splitting if you want to understand the remainder of this section.

The data sequence $D_n = (X_1, Y_1), \ldots, (X_n, Y_n)$ is divided into two parts. The first part $D_m = (X_1, Y_1), \ldots, (X_m, Y_m)$ is used for training, while the remaining $l = n - m$ pairs constitute the testing sequence:

$$T_l = (X_{m+1}, Y_{m+1}), \ldots, (X_{m+l}, Y_{m+l}).$$

The training sequence D_m is used to design a class of classifiers C_m, which, in our case is the class of kernel rules based on D_m, with all possible values of $h > 0$, for fixed kernel K. Note that the value of the kernel rule $g_m(x)$ with smoothing parameter h is zero if and only if

$$f_m(x) = \sum_{i=1}^{m} (Y_i - 1/2) K \left(\frac{x - X_i}{h} \right) \le 0.$$

Classifiers in C_m are denoted by ϕ_m. A classifier is selected from C_m that minimizes the holdout estimate of the error probability:

$$\widehat{L}_{m,l}(\phi_m) = \frac{1}{l} \sum_{i=1}^{l} I_{\{\phi_m(X_{m+i}) \ne Y_{m+i}\}}.$$

The particular rule selected in this manner is called g_n. The question is how far the error probability $L(g_n)$ of the obtained rule is from that of the optimal rule in C_m.

FINITE COLLECTIONS. It is computationally attractive to restrict the possible values of h to a finite set of real numbers. For example, C_m could consist of all kernel rules with $h \in \{2^{-k_m}, 2^{-k_m+1}, \ldots, 1/2, 1, 2, \ldots, 2^{k_m}\}$, for some positive integer k_m. The advantage of this choice of C_m is that the best h in this class is within a factor of two of the best h among all possible real smoothing factors, unless the best h is smaller than 2^{-k_m-1} or larger than 2^{k_m+1}. Clearly, $|C_m| = 2k_m + 1$, and as pointed out in Chapter 22, for the selected rule g_n Hoeffding's inequality and the union bound imply that

$$\mathbf{P}\left\{ L(g_n) - \inf_{\phi_m \in C_m} L(\phi_m) > \epsilon \,\middle|\, D_m \right\} \le (4k_m + 2)e^{-l\epsilon^2/2}.$$

If $k_m = e^{o(l)}$, then the upper bound decreases exponentially in l, and, in fact,

$$\mathbf{E}\left\{L(g_n) - \inf_{\phi_m \in C_m} L(\phi_m)\right\} = O\left(\sqrt{\frac{\log(l)}{l}}\right).$$

By Theorem 10.1, C_m contains a subsequence of consistent rules if $m \to \infty$, and $k_m \to \infty$ as $n \to \infty$. To make sure that g_n is strongly universally consistent as well, we only need that $\lim_{n\to\infty} l = \infty$, and $k_m = e^{o(l)}$ (see Theorem 22.1). Under these conditions, the rule is $\sqrt{\log(k_m)/l}$-optimal (see Chapter 22).

The discussion above does little to help us with the selection of m, l, and k_m. Safe, but possibly suboptimal, choices might be $l = n/10, m = n - l, k_m = 2 \log_2 n$. Note that the argument above is valid for any regular kernel K.

INFINITE COLLECTIONS.   If we do not want to exclude any value of the smoothing parameter, and pick h from $[0, \infty)$, then C_m is of infinite cardinality. Here, we need something stronger, like the Vapnik-Chervonenkis theory. For example, from Chapter 22, we have

$$\mathbf{P}\left\{L(g_n) - \inf_{\phi_m \in C_m} L(\phi_m) > \epsilon \,\middle|\, D_m\right\} \le 4e^8 S(C_m, l^2)e^{-l\epsilon^2/2},$$

where $S(C_m, l)$ is the l-th shatter coefficient corresponding to the class of classifiers C_m. We now obtain upper bounds for $S(C_m, l)$ for different choices of K.

Define the function

$$f_m(x) = f_m(x, D_m) = \sum_{i=1}^{m}\left(Y_i - \frac{1}{2}\right) K\left(\frac{x - X_i}{h}\right).$$

Recall that for the kernel rule based on D_m,

$$g_m(x) = \begin{cases} 0 & \text{if } f_m(x, D_m) \le 0 \\ 1 & \text{otherwise.} \end{cases}$$

We introduce the *kernel complexity* κ_m:

$$\kappa_m = \sup_{x,(x_1,y_1),\dots,(x_m,y_m)} \{\text{Number of sign changes of}$$

$$f_m(x, (x_1, y_1), \dots, (x_m, y_m)) \text{ as } h \text{ varies from 0 to infinity}\}.$$

Suppose we have a kernel with kernel complexity κ_m. Then, as h varies from 0 to infinity, the binary l-vector

$$(\text{sign}(f_m(X_{m+1})), \dots, \text{sign}(f_m(X_{m+l})))$$

changes at most $l\kappa_m$ times. It can thus take at most $l\kappa_m + 1$ different values. Therefore,

$$S(C_m, l) \le l\kappa_m + 1.$$

We postpone the issue of computing kernel complexities until the next section. It suffices to note that if g_n is obtained by minimizing the holdout error estimate $\widehat{L}_{m,l}(\phi_m)$ by varying h, then

$$\mathbf{E}\left\{ L(g_n) - \inf_{\phi_m \in \mathcal{C}_m} L(\phi_m)\middle| D_m \right\} \tag{25.2}$$

$$\leq 16\sqrt{\frac{\log(8e\mathcal{S}(\mathcal{C}_m, l))}{2l}} \quad \text{(Corollary 12.1)}$$

$$\leq 16\sqrt{\frac{\log(8e(l\kappa_m + 1))}{2l}}. \tag{25.3}$$

Various probability bounds may also be derived from the results of Chapter 12. For example, we have

$$\mathbf{P}\left\{ L(g_n) - \inf_{\phi_m \in \mathcal{C}_m} L(\phi_m) > \epsilon\middle| D_m \right\} \leq 4e^8 \mathcal{S}(\mathcal{C}_m, l^2)e^{-l\epsilon^2/2}$$

$$\leq 4e^8(l^2\kappa_m + 1)e^{-l\epsilon^2/2}. \tag{25.4}$$

Theorem 25.4. *Assume that g_n minimizes the holdout estimate $\widehat{L}_{m,l}(\phi_m)$ over all kernel rules with fixed kernel K of kernel complexity κ_m, and over all (unrestricted) smoothing factors $h > 0$. Then g_n is strongly universally consistent if*

(i) $\lim_{n\to\infty} m = \infty$;

(ii) $\lim_{n \to \infty} \dfrac{\log \kappa_m}{l} = 0$;

(iii) $\lim_{n\to\infty} \dfrac{l}{\log n} = \infty$;

(iv) K is a regular kernel.

For weak universal consistency, (iii) may be replaced by (v): $\lim_{n\to\infty} l = \infty$.

PROOF. Note that $\mathcal{C}_m$ contains a strongly universally consistent subsequence— take $h = m^{-1/(2d)}$ for example, and apply Theorem 10.1, noting that $h \to 0$, yet $mh^d \to \infty$. Thus,

$$\lim_{n \to \infty} \inf_{\phi_m \in \mathcal{C}_m} L(\phi_m) = L^* \quad \text{with probability one.}$$

It suffices to apply Theorem 22.1 and to note that the bound in (25.4) is summable in n when $l/\log n \to \infty$ and $\log \kappa_m = o(l)$. For weak universal consistency, a simple application of (25.2) suffices to note that we only need $l \to \infty$ instead of $l/\log n \to \infty$. $\square$

Approximation errors decrease with m. For example, if class densities exist, we may combine the inequality of Problem 2.10 with bounds from Devroye and Györfi (1985) and Holmström and Klemelä (1992) to conclude that $\mathbf{E}\left\{ \inf_{\phi_m \in \mathcal{C}_m} L(\phi_m) \right\} - L^*$ is of the order of $m^{-2\alpha/(4+d)}$, with $\alpha \in [1, 2]$, under suitable conditions on K

and the densities. By (25.2), the estimation error is $O\left(\sqrt{\log(l\kappa_m)/l}\right)$, requiring instead large values for l. Clearly, some sort of balance is called for. Ignoring the logarithmic term for now, we see that l should be roughly $m^{4\alpha/(4+d)}$ if we are to balance errors of both kinds. Unfortunately, all of this is ad hoc and based upon unverifiable distributional conditions. Ideally, one should let the data select l and m. See Problem 25.3, and Problem 22.4 on optimal data splitting.

For some distributions, the estimation error is just too large to obtain asymptotic optimality as defined in Chapter 22. For example, the best bound on the estimation error is $O\left(\sqrt{\log n/n}\right)$, attained when $\kappa_m = 1, l = n$. If the distribution of X is atomic with finitely many atoms, then the expected approximation error is $O(1/\sqrt{m})$. Hence the error introduced by the selection process smothers the approximation error when m is linear in n. Similar conclusions may even be drawn when X has a density: consider the uniform distribution on $[0, 1] \cup [3, 4]$ with $\eta(x) = 1$ if $x \in [0, 1]$ and $\eta(x) = 0$ if $x \in [3, 4]$. For the kernel rule with $h = 1, K = I_{[-1,1]}$, the expected approximation error tends to $L^* = 0$ exponentially quickly in m, and this is always far better than the estimation error which at best is $O\left(\sqrt{\log n/n}\right)$.

25.3 Kernel Complexity

We now turn to the *kernel complexity* κ_m. The following lemmas are useful in our computations. If $l/\log n \to \infty$, we note that for strong consistency it suffices that $\kappa_m = O(m^\gamma)$ for some finite γ (just verify the proof again). This, as it turns out, is satisfied for nearly all practical kernels.

Lemma 25.1. *Let $0 \le b_1 < \cdots < b_m$ be fixed numbers, and let $a_i \in \mathcal{R}, 1 \le i \le m$, be fixed, with the restriction that $a_m \ne 0$. Then the function $f(x) = \sum_{i=1}^{m} a_i x^{b_i}$ has at most $m - 1$ nonzero positive roots.*

PROOF. Note first that f cannot be identically zero on any interval of nonzero length. Let $Z(g)$ denote the number of nonzero positive roots of a function g. We have

$$Z\left(\sum_{i=1}^{m} a_i x^{b_i}\right)$$

$$= Z\left(\sum_{i=1}^{m} a_i x^{c_i}\right) \quad \text{(where } c_i = b_i - b_1, \text{ all } i; \text{ thus, } c_1 = 0)$$

$$\leq \ Z\left(\sum_{i=2}^{m} a_i c_i x^{c_i-1}\right) + 1$$

(for a continuously differentiable g, we have $Z(g) \leq 1 + Z(g')$)

$$= \ Z\left(\sum_{i=2}^{m} a_i' x^{b_i'}\right) + 1$$

(where $a_i' = a_i c_i$, $b_i' = c_i - c_2$, all $i \geq 2$).

Note that the b_i' are increasing, $b_2' = 0$, and $a_i' \neq 0$. As $Z\left(a_m x^{b_m}\right) = 0$ for $a_m \neq 0$, we derive our claim by simple induction on m. $\square$

Lemma 25.2. *Let $a_1, \ldots, a_m$ be fixed real numbers, and let $b_1, \ldots, b_m$ be different nonnegative reals. Then if $\alpha \neq 0$, the function*

$$f(x) = \sum_{i=1}^{m} a_i e^{-b_i x^\alpha}, \quad x \geq 0,$$

is either identically zero, or takes the value 0 at most m times.

PROOF. Define $y = e^{-x^\alpha}$. If $\alpha \neq 0$, y ranges from 0 to 1. By Lemma 25.1, $g(y) = \sum_{i=1}^{m} a_i y^{b_i}$ takes the value 0 at at most $m - 1$ positive y-values, unless it is identically zero everywhere. This concludes the proof of the lemma. $\square$

A *star-shaped kernel* is one of the form $K(x) = I_{\{x \in A\}}$, where A is a star-shaped set of unit Lebesgue measure, that is, $x \notin A$ implies $cx \notin A$ for all $c \geq 1$. It is clear that $\kappa_m = m - 1$ by a simple thresholding argument. On the real line, the kernel $K(x) = \sum_{i=-\infty}^{\infty} a_i I_{\{x \in [2i, 2i+1]\}}$ for $a_i > 0$ oscillates infinitely often, and has $\kappa_m = \infty$ for all values of $m \geq 2$. We must therefore disallow such kernels. For the same reason, kernels such as $K(x) = (\sin x / x)^r$ on the real line are not good (see Problem 25.4).

If $K = \sum_{i=1}^{k} a_i I_{A_i}$ for some finite k, some numbers a_i and some star-shaped sets A_i, then $\kappa_m \leq k(m - 1)$.

Consider next kernels of the form

$$K(x) = \|x\|^{-r} I_{\{x \in A\}},$$

where A is star-shaped, and $r \geq 0$ is a constant (see Sebestyen (1962)). We see that

$$\begin{aligned}
f_m(x) &= \sum_{i=1}^{m} (Y_i - \frac{1}{2}) K\left(\frac{x - X_i}{h}\right) \\
&= h^r \sum_{i=1}^{m} (Y_i - \frac{1}{2}) \|x - X_i\|^{-r} I_{\{(x - X_i)/h \in A\}},
\end{aligned}$$

which changes sign at most as often as $f_m(x)/h^r$. From our earlier remarks, it is easy to see that $\kappa_m = m - 1$, as κ_m is the same as for the kernel $K = I_A$. If A is replaced by $\mathcal{R}^d$, then the kernel is not integrable, but clearly, $\kappa_m = 0$. Assume next that we have

$$K(x) = \begin{cases} 1 & \text{if } \|x\| \leq 1 \\ 1/\|x\|^r & \text{otherwise,} \end{cases}$$

where $r > 0$. For $r > d$, these kernels are integrable and thus regular. Note that

$$K\left(\frac{x}{h}\right) = I_{\{\|x\| \leq h\}} + I_{\{\|x\| > h\}} \frac{h^r}{\|x\|^r}.$$

As h increases, $f_m(x)$, which is of the form

$$\sum_{i:\|x - X_i\| \leq h} \left(Y_i - \frac{1}{2}\right) + h^r \sum_{i:\|x - X_i\| > h} \left(Y_i - \frac{1}{2}\right) \frac{1}{\|x - X_i\|^r},$$

transfers an X_i from one sum to the other at most m times. On an interval on which no such transfer occurs, f_m varies as $\alpha + \beta h^r$ and has at most one sign change. Therefore, κ_m cannot be more than $m + 1$ (one for each h-interval) plus m (one for each transfer), so that $\kappa_m \leq 2m + 1$. For more practice with such computations, we refer to the exercises. We now continue with a few important classes of kernels.

Consider next exponential kernels such as

$$K(x) = e^{-\|x\|^\alpha}$$

for some $\alpha > 0$, where $\|\cdot\|$ is any norm on $\mathcal{R}^d$. These kernels include the popular gaussian kernels. As the decision rule based on D_m is of the form

$$g_n(x) = \begin{cases} 1 & \text{if } \sum_{i=1}^m (2Y_i - 1)e^{-\frac{\|x - X_i\|^\alpha}{h^\alpha}} > 0 \\ 0 & \text{otherwise,} \end{cases}$$

a simple application of Lemma 25.2 shows that $\kappa_m \leq m$. The entire class of kernels behaves nicely.

Among compact support kernels, kernels of the form

$$K(x) = \left(\sum_{i=1}^k a_i \|x\|^{b_i}\right) I_{\{\|x\| \leq 1\}}$$

for real numbers a_i, and $b_i \geq 1$, are important. A particularly popular kernel in d-dimensional density estimation is Deheuvels' (1977) kernel

$$K(x) = \left(1 - \|x\|^2\right) I_{\{\|x\| \leq 1\}}.$$

If the kernel was $K(x) = \left(\sum_{i=1}^k a_i \|x\|^{b_i}\right)$, without the indicator function, then Lemma 25.1 would immediately yield the estimate $\kappa_m \leq k$, uniformly over all m. Such kernels would be particularly interesting. With the indicator function multiplied in, we have $\kappa_m \leq km$, simply because $f_m(x)$ at each h is based upon a subset of the X_j's, $1 \leq j \leq m$, with the subset growing monotonically with h. For each subset, the function $f_m(x)$ is a polynomial in $\|x\|$ with powers $b_1, \ldots, b_k$, and changes sign at most k times. Therefore, polynomial kernels of compact support also have small complexities. Observe that the "k" in the bound $\kappa_m \leq km$ refers to the number of terms in the polynomial and not the maximal power.

A large class of kernels of finite complexity may be obtained by applying the rich theory of total positivity. See Karlin (1968) for a thorough treatment. A real-valued function L of two real variables is said to be *totally positive* on $A \times B \subset \mathcal{R}^2$ if for all n, and all $s_1 < \cdots < s_n, s_i \in A, t_1 < \cdots < t_n, t_i \in B$, the determinant of the matrix with elements $L(s_i, t_j)$ is nonnegative. A key property of such functions is the following result, which we cite without proof:

Theorem 25.5. (SCHOENBERG (1950)). *Let L be a totally positive function on $A \times B \in \mathcal{R}^2$, and let $\alpha : B \to \mathcal{R}$ be a bounded function. Define the function*

$$\beta(s) = \int_B L(s, t)\alpha(t)\sigma(dt),$$

on A, where σ is a finite measure on B. Then $\beta(s)$ changes sign at most as many times as $\alpha(s)$ does. (The number of sign changes of a function β is defined as the supremum of sign changes of sequences of the form $\beta(s_1), \ldots, \beta(s_n)$, where n is arbitrary, and $s_1 < \cdots < s_n$.)

COROLLARY 25.1. *Assume that the kernel K is such that the function $L(s, t) = K(st)$ is totally positive for $s > 0$ and $t \in \mathcal{R}^d$. Then the kernel complexity of K satisfies $\kappa_m \leq m - 1$.*

PROOF. We apply Theorem 25.5. We are interested in the number of sign changes of the function

$$\beta(s) = \sum_{i=1}^{m} (2Y_i - 1)K((X_i - x)s)$$

on $s \in (0, \infty)$ (s plays the role of $1/h$). But $\beta(s)$ may be written as

$$\beta(s) = \int_B L(s, t)\alpha(t)\sigma(dt),$$

where $L(s, t) = K(st)$, the measure σ puts mass 1 on each $t = X_i - x, i = 1, \ldots, m$, and $\alpha(t)$ is defined at these points as

$$\alpha(t) = 2Y_i - 1 \quad \text{if} \quad t = X_i - x.$$

Other values of $\alpha(t)$ are irrelevant for the integral above. Clearly, $\alpha(t)$ can be defined such that it changes sign at most $m - 1$ times. Then Theorem 25.5 implies that $\beta(s)$ changes sign at most $m - 1$ times, as desired. $\square$

This corollary equips us with a whole army of kernels with small complexity. For examples, refer to the monograph of Karlin (1968).

25.4 Multiparameter Kernel Rules

Assume that in the kernel rules considered in C_m, we perform an optimization with respect to more than one parameter. Collect these parameters in θ, and write the discrimination function as

$$f_m(x) = \sum_{i=1}^{m} \left(Y_i - \frac{1}{2} \right) K_\theta(x - X_i).$$

EXAMPLES:

(i) Product kernels. Take

$$K_\theta(x) = \prod_{j=1}^{d} K \left(\frac{x}{h^{(j)}} \right),$$

where $\theta = (h^{(1)}, \ldots, h^{(d)})$ is a vector of smoothing factors—one per dimension—and K is a fixed one-dimensional kernel.

(ii) Kernels of variable form. Define

$$K_\theta(x) = e^{-\|x\|^\alpha / h^\alpha},$$

where $\alpha > 0$ is a shape parameter, and $h > 0$ is the standard smoothing parameter. Here $\theta = (\alpha, h)$ is two-dimensional.

(iii) Define

$$K_\theta(x) = \frac{1}{1 + x^T R R^T x} = \frac{1}{1 + \|Rx\|^2},$$

where x^T is the transpose of x, and R is an orthogonal transformation matrix, all of whose free components taken together are collected in θ. Kernels of this kind may be used to adjust automatically to a certain variance-covariance structure in the data.

We will not spend a lot of time on these cases. Clearly, one route is to properly generalize the definition of kernel complexity. In some cases, it is more convenient to directly find upper bounds for $S(C_m, l)$. In the product kernel case, with one-dimensional kernel $I_{[-1,1]}$, we claim for example that

$$S(C_m, l) \le (lm)^d + 1.$$

The corresponding rule takes a majority vote over centered rectangles with sides equal to $2h^{(1)}, 2h^{(2)}, \ldots, 2h^{(d)}$. To see why the inequality is true, consider the d-dimensional quadrant of lm points obtained by taking the absolute values of the vectors $X_j - X_i$, $m < j \le m + l = n$, $1 \le i \le m$, where the absolute value of a vector is a vector whose components are the absolute values of the components of the vector. To compute $S(C_m, l)$, it suffices to count how many different subsets can be obtained from these lm points by considering all possible rectangles with one vertex at the origin, and the diagonally opposite vertex in the quadrant. This is $1 + (lm)^d$. The strong universal consistency of the latter family C_m is insured when $m \to \infty$ and $l / \log n \to \infty$.

25.5 Kernels of Infinite Complexity

In this section we demonstrate that not every kernel function supports smoothing factor selection based on data splitting. These kernels have infinite kernel complexity, or even worse, infinite VC dimension. Some of the examples may appear rather artificial, but some "nice" kernels will surprise us by misbehaving.

We begin with a kernel K having the property that the class of sets

$$\mathcal{A} = \left\{ \left\{ x : K\left(\frac{x - X_1}{h}\right) > 0 \right\} : h > 0 \right\}$$

has VC dimension $V_{\mathcal{A}} = \infty$ for any fixed value of X_1 (i.e., using the notation of the previous sections, $V_{C_1} = \infty$). Hence, with one sample point, all hope is lost to use the Vapnik-Chervonenkis inequality in any meaningful way. Unfortunately, the kernel K takes alternately positive and negative values. In the second part of this section, a kernel is constructed that is unimodal and symmetric and has $V_{C_m} = \infty$ for $m = 4$ when $D_4 = ((X_1, Y_1), (X_2, Y_2), (X_3, Y_3), (X_4, Y_4))$ takes certain values. Finally, in the last part, we construct a positive kernel with the property that for any m and any nondegenerate nonatomic distribution, $\lim_{m \to \infty} \mathbf{P}\left\{ V_{C_m} = \infty \right\} = 1$.

We return to $\mathcal{A}$. Our function is picked as follows:

$$K(x) = \alpha(x)g(i), \quad 2^i \le x < 2^{i+1}, \quad i \ge 1,$$

where $g(i) \in \{-1, 1\}$ for all i, $\alpha(x) > 0$ for all x, $\alpha(x) \downarrow 0$ as $x \uparrow \infty$, $x > 0$, and $\alpha(x) \downarrow 0$ as $x \downarrow -\infty$, $x < 0$ (the monotonicity is not essential and may be dropped but the resulting class of kernels will be even less interesting). We enumerate all binary strings in lexicographical order, replace all 0's by -1's, and map the bit sequence to $g(1), g(2), \ldots$. Hence, the bit sequence $0, 1, 00, 01, 10, 11, 000, 001, 010, \ldots$ becomes

$$-1, 1, -1, -1, -1, 1, 1, -1, 1, 1, -1, -1, -1, -1, -1, 1, -1, 1, -1, \ldots.$$

Call this sequence S. For every n, we can find a set $\{x_1, \ldots, x_n\}$ that can be shattered by sets from $\mathcal{A}$. Take $\{x_1, \ldots, x_n\} = X_1 + \{2^1, \ldots, 2^n\}$. A subset of this may be characterized by a string s_n from $\{-1, 1\}^n$, 1's denoting membership, and -1's denoting absence. We find the first occurrence of s_n in S and let the starting point be $g(k)$. Take $h = 2^{1-k}$. Observe that

$$\left(K\left(\frac{x_1 - X_1}{h}\right), \ldots, K\left(\frac{x_n - X_1}{h}\right) \right)$$
$$= \left(K\left(2^k\right), \ldots, K\left(2^{k+n-1}\right) \right)$$
$$= \left(\alpha(2^k)g(k), \alpha(2^{k+1})g(k+1), \ldots, \alpha(2^{k+n-1})g(k+n-1) \right),$$

which agrees in sign with s_n as desired. Hence, the VC dimension is infinite.

The next kernel is symmetric, unimodal, and piecewise quadratic. The intervals into which $[0, \infty)$ is divided are denoted by $A_0, A_1, A_2, \ldots$, from left to right, where

$$A_i = \begin{cases} \left[\frac{3}{4}2^i, \frac{3}{2}2^i\right) & i \geq 1 \\ \left[0, \frac{3}{2}\right) & i = 0. \end{cases}$$

On each A_i, K is of the form $ax^2 + bx + c$. Observe that $K'' = 2a$ takes the sign of a. Also, any finite symmetric difference of order 2 has the sign of a, as

$$K(x + \delta) + K(x - \delta) - 2K(x) = 2a\delta^2, \quad x - \delta, x, x + \delta \in A_i.$$

We take four points and construct the class

$$\mathcal{A}_4 = \left\{ \left\{ x : \sum_{i=1}^{4} K\left(\frac{x - X_i}{h}\right)(2Y_i - 1) > 0 \right\} : h > 0 \right\},$$

where $X_1 = -\delta$, $X_2 = \delta$, $X_3 = X_4 = 0$, $Y_1 = Y_2 = 0$, $Y_3 = Y_4 = 1$ are fixed for now. On A_i, we let the "a" coefficient have the same sign as $g(i)$, known from the previous example. All three quadratic coefficients are picked so that $K \geq 0$ and K is unimodal. For each n, we show that the set $\{2^1, \ldots, 2^n\}$ can be shattered by intersecting with sets from $\mathcal{A}_4$. A subset is again identified by a $\{-1, 1\}^n$-valued string s_n, and its first match in S is $g(k), \ldots, g(k + n - 1)$. We take $h = 2^{1-k}$. Note that for $1 \leq i \leq n$,

$$\text{sign}\left(K\left(\frac{2^i - \delta}{h}\right) + K\left(\frac{2^i + \delta}{h}\right) - 2K\left(\frac{2^i}{h}\right)\right)$$

$$= \text{sign}\left(K\left(2^{k+i-1} - \delta 2^{k-1}\right) + K\left(2^{k+i-1} + \delta 2^{k-1}\right) - 2K\left(2^{k+i-1}\right)\right)$$

$$= \text{sign}\left(K''\left(2^{k+i-1}\right)\right)$$

(if $\delta \leq 1/4$, by the finite-difference property of quadratics)

$$= g(k + i - 1),$$

as desired. Hence, any subset of $\{2^1, \ldots, 2^n\}$ can be picked out.

The previous example works whenever $\delta \leq 1/4$. It takes just a little thought to see that if $(X_1, Y_1), \ldots, (X_4, Y_4)$ are i.i.d. and drawn from the distribution of (X, Y), then

$$\mathbf{P}\left\{V_{\mathcal{A}_4} = \infty\right\} \geq \mathbf{P}\{X_1 = -X_2, X_3 = X_4 = 0, Y_1 = Y_2 = 0, Y_3 = Y_4 = 1\},$$

and this is positive if given $Y = 0$, X has atoms at δ and $-\delta$ for some $\delta > 0$; and given $Y = 1$, X has an atom at 0. However, with some work, we may even remove these restrictions (see Problem 25.12).

We also draw the reader's attention to an 8-point example in which K is symmetric, unimodal, and convex on $[0, \infty)$, yet $V_{\mathcal{C}_8} = \infty$. (See Problem 25.11.)

Next we turn to our general m-point example. Let the class of rules be given by

$$g_{m,h}(x) = \begin{cases} 1 & \text{if } \sum_{i=1}^{m}(2Y_i - 1)K\left(\frac{x - X_i}{h}\right) > 0 \\ 0 & \text{otherwise,} \end{cases}$$

$h > 0$, where the data $(X_1, Y_1), \ldots, (X_m, Y_m)$ are fixed for now. We exhibit a nonnegative kernel such that if the X_i's are different and not all the Y_i's are the same, $V_{C_m} = \infty$. This situation occurs with probability tending to one whenever X is nonatomic and $\mathbf{P}\{Y = 1\} \in (0, 1)$. It is stressed here that the *same* kernel K is used regardless of the data.

The kernel is of the form $K(x) = K_0(x)I_A(x)$, where $K_0(x) = e^{-x^2}$, and $A \subset \mathcal{R}$ will be specially picked. Order $X_1, \ldots, X_m$ into $X_{(1)} < \cdots < X_{(m)}$ and find the first pair $X_{(i)}, X_{(i+1)}$ with opposite values for $Y_{(i)}$. Without loss of generality, assume that $X_{(i)} = -1$, $X_{(i+1)} = 1$, $2Y_{(i)} - 1 = -1$, and $2Y_{(i+1)} - 1 = 1$ ($Y_{(i)}$ is the Y-value that corresponds to $X_{(i)}$). Let the smallest value of $|X_{(j)}|$, $j \neq i$, $j \neq i + 1$ be denoted by $\delta > 1$. The contribution of all such $X_{(j)}$'s for $x \in [-\delta/3, \delta/3]$ is not more than $(m - 2)K_0((1 + 2\delta/3)/h) \leq mK_0((1 + 2\delta/3)/h)$. The contribution of either $X_{(i)}$ or $X_{(i+1)}$ is at least $K_0((1 + \delta/3)/h)$. For fixed δ and m (after all, they are given), we first find h^* such that $h \leq h^*$ implies

$$K_0 \left(\frac{1 + \delta/3}{h} \right) > mK_0 \left(\frac{1 + 2\delta/3}{h} \right).$$

For $h \leq h^*$, the rule, for $x \in [-\delta/3, \delta/3]$, is equivalent to a 2-point rule based on

$$g_{m,h}(x) = \begin{cases} 1 & \text{if } -K \left(\frac{x+1}{h} \right) + K \left(\frac{x-1}{h} \right) > 0 \\ 0 & \text{otherwise.} \end{cases}$$

We define the set A by

$$A = \bigcup_{k=1}^{\infty} \bigcup_{l=0}^{2^k-1} 3^{2^k+l} A_{k,l},$$

where the sets $A_{k,l}$ are specified later. (For $c \neq 0$ and $B \subset \mathcal{R}$, $cB = \{x \in \mathcal{R} : x/c \in B\}$.) Here k represents the length of a bit string, and l cycles through all 2^k bit strings of length k. For a particular such bit string, $b_1, \ldots, b_k$, represented by l, we define $A_{k,l}$ as follows. First of all, $A_{k,l} \subseteq [1/2, 3/2]$, so that all sets $3^{2^k+l} A_{k,l}$ are nonoverlapping. $A_{k,l}$ consists of the sets

$$\left(\bigcup_{i:b_i=0} \left[1 - \frac{1}{(i+1)2^k}, 1 - \frac{1}{(i+2)2^k} \right) \right)$$

$$\cup \left(\bigcup_{i:b_i=1} \left(1 + \frac{1}{(i+2)2^k}, 1 + \frac{1}{(i+1)2^k} \right] \right).$$

$A_{k,l}$ is completed by symmetry.

We now exhibit for each integer k a set of that size that can be shattered. These sets will be positioned in $[-\delta/3, 0]$ at coordinate values $-1/(2 \cdot 2^\rho), -1/(3 \cdot 2^\rho), \ldots, -1/((k + 1) \cdot 2^\rho)$, where ρ is a suitable large integer such that $1/(2^{\rho+1}) < \delta/3$. Assume that we wish to extract the set indexed by the bit vector $(b_1, \ldots, b_k)$ ($b_i = 1$ means that $-1/(i + 1)2^\rho$ must be extracted). To do so, we are only allowed to vary h. First, we find a pair (k', l), where k' is at least k and $1/(2^{k'+1}) < \delta/3$.

$3^{-2^{k'}} \leq h^*$ is needed too. Also, l is such that it matches $b_1, \ldots, b_k$ in its first $k \leq k'$ bits. Take $h = 3^{-2^{k'}-l}$ and $\rho = k'$ and observe that

$$
g_{n,h}\left(-\frac{1}{(i+1)2^{k'}}\right)
$$

$$
= \begin{cases} 1 & \text{if } K\left(\frac{-\frac{1}{(i+1)2^{k'}}-1}{3^{2^{k'}+l}}\right) > K\left(\frac{-\frac{1}{(i+1)2^{k'}}+1}{3^{2^{k'}+l}}\right) \\ 0 & \text{otherwise} \end{cases}
$$

$$
= \begin{cases} 1 & \text{if } 1 + \frac{1}{(i+1)2^{k'}} \in A_{k,l}, \; 1 - \frac{1}{(i+1)2^{k'}} \notin A_{k,l} \\ 0 & \text{if } 1 + \frac{1}{(i+1)2^{k'}} \notin A_{k,l}, \; 1 - \frac{1}{(i+1)2^{k'}} \in A_{k,l} \\ \text{unimportant} & \text{otherwise} \end{cases}
$$

$$
= \begin{cases} 1 & \text{if } b_i = 1 \\ 0 & \text{if } b_i = 0, \end{cases}
$$

$1 \leq i \leq k$. Thus, we pick the desired set. This construction may be repeated for all values of l of course. We have shown the following:

Theorem 25.6. *If X is nonatomic, $\mathbf{P}\{Y = 1\} \in (0, 1)$, and if V_{C_m} denotes the vc dimension of the class of kernel rules based on m i.i.d. data drawn from the distribution of (X, Y), and with the kernel specified above, then*

$$
\lim_{m \to \infty} \mathbf{P}\{V_{C_m} = \infty\} = 1.
$$

25.6   On Minimizing the Apparent Error Rate

In this section, we look more closely at kernel rules that are picked by minimizing the resubstitution estimate $L_n^{(R)}$ over kernel rules with smoothing factor $h > 0$. We make two remarks in this respect:

    (i) The procedure is generally inconsistent if X is nonatomic.

    (ii) The method is consistent if X is purely atomic.

To see (i), take $K = I_{\{S_{0,1}\}}$. We note that $L_n^{(R)} = 0$ if $h < \min_{i \neq j} \|X_i - X_j\|$ as $g_n(X_i) = Y_i$ for such h. In fact, if H_n is the minimizing h, it may take any value on

$$
\left[0, \min_{i,j:Y_i=1,Y_j=0} \|X_i - X_j\|\right].
$$

If X is independent of Y, $\mathbf{P}\{Y = 1\} = 2/3$, and X has a density, then,

$$
\lim_{c \to \infty} \liminf_{n \to \infty} \mathbf{P}\left\{\min_{i,j:Y_i=1,Y_j=0} \|X_i - X_j\|^d \leq \frac{c}{n^2}\right\} = 1
$$

(Problem 25.13). Thus, $nH^d \to 0$ in probability, and therefore, in this case, $\mathbf{E}\{L(g_n)\} \to 2/3$ (since ties are broken by favoring the "0" class). Hence the inconsistency.

Consider next X purely atomic. Fix $(X_1, Y_1), \ldots, (X_n, Y_n)$, the data; and consider the class C_n of kernel rules in which $h > 0$ is a free parameter. If a typical rule is $g_{n,h}$, let A_n be the class of sets $\{x : g_{n,h}(x) = 1\}$, $h > 0$. If g_n is the rule that minimizes $L_n^{(R)}(g_{n,h})$, we have

$$L(g_n) - \inf_{g_{n,h} \in C_n} L(g_{n,h})$$

$$\leq \; 2 \sup_{g_{n,h} \in C_n} \left| L_n^{(R)}(g_{n,h}) - L(g_{n,h}) \right| \quad \text{(Theorem 8.4)}$$

$$\leq \; 2 \sup_{A \in A_n} |\mu_n(A) - \mu(A)| + 2 \sup_{A \in A_n^c} |\mu_n(A) - \mu(A)|$$

$$(A_n^c \text{ denotes the collection of sets } \{x : g_{n,h}(x) = 0\})$$

$$\leq \; 4 \sup_{B \in B} |\mu_n(B) - \mu(B)|,$$

where B is the collection of all Borel sets. However, the latter quantity tends to zero with probability one—as denoting the set of the atoms of X by T, we have

$$\sup_{B \in B} |\mu_n(B) - \mu(B)|$$

$$= \; \frac{1}{2} \sum_{x \in T} |\mu_n(\{x\}) - \mu(\{x\})|$$

$$\leq \; \frac{1}{2} \sum_{x \in A} |\mu_n(\{x\}) - \mu(\{x\})| + \mu_n(A^c) + \mu(A^c)$$

(where A is an arbitrary finite subset of T)

$$\leq \; 2\mu(A^c) + \frac{1}{2} \sum_{x \in A} |\mu_n(\{x\}) - \mu(\{x\})| + |\mu_n(A^c) - \mu(A^c)|.$$

The first term is small by choice of A and the last two terms are small by applying Hoeffding's inequality to each of the $|A| + 1$ terms. For yet a different proof, the reader is referred to Problems 25.16 and 25.17.

Note next that

$$L(g_n) - L^* \leq \left(L(g_n) - \inf_{g_{n,h} \in C_n} L(g_{n,h}) \right) + \left(\inf_{g_{n,h} \in C_n} L(g_{n,h}) - L^* \right).$$

But if $K(0) > 0$ and $K(x) \to 0$ as $\|x\| \to \infty$ along any ray, then we have $\inf_{g_{n,h} \in C_n} L(g_{n,h}) \leq L(g_n')$, where g_n' is the fundamental rule discussed in Chapter 27 (in which $h = 0$). Theorem 27.1 shows that $L(g_n') \to L^*$ with probability one, and therefore, $L(g_n) \to 0$ with probability one, proving

Theorem 25.7. *Take any kernel K with $K(0) > 0$ and $K(x) \to 0$ as $\|x\| \to \infty$ along any ray, and let g_n be selected by minimizing $L_n^{(R)}$ over all $h \geq 0$. Then $L(g_n) \to L^*$ almost surely whenever the distribution of X is purely atomic.*

REMARK. The above theorem is all the more surprising since we may take just about any kernel, such as $K(x) = e^{-\|x\|^2}$, $K(x) = \sin(\|x\|)/\|x\|$, or $K(x) = (1 + \|x\|)^{-1/3}$. Also, if X puts its mass on a dense subset of $\mathcal{R}^d$, the data will look and feel like data from an absolutely continuous distribution (well, very roughly speaking); yet there is a dramatic difference with rules that minimize $L_n^{(R)}$ when the X_i's are indeed drawn from a distribution with a density! $\square$

25.7   Minimizing the Deleted Estimate

We have seen in Chapter 24 that the deleted estimate of the error probability of a kernel rule is generally very reliable. This suggests using the estimate as a basis of selecting a smoothing factor for the kernel rule. Let $L_n^{(D)}(g_{n,h})$ be the deleted estimate of $L(g_{n,h})$, obtained by using in $g_{n-1,h}$ the same kernel K and smoothing factor h. Define the set of h's for which

$$L_n^{(D)}(g_{n,h}) = \inf_{h' \in [0,\infty)} L_n^{(D)}(g_{n,h'})$$

by A. Set

$$H_n = \inf\{h : h \in A\}.$$

Two fundamental questions regarding H_n must be asked:

(a) If we use H_n in the kernel estimate, is the rule universally consistent? Note in this respect that Theorem 25.3 cannot be used because it is not true that $H_n \to 0$ in probability in all cases.

(b) If H_n is used as smoothing factor, how does $E\{L(g_n)\} - L^*$ compare to $E\left\{\inf_h L(g_{n,h})\right\} - L^*$?

To our knowledge, both questions have been unanswered so far. We believe that this way of selecting h is very effective. Below, generalizing a result by Tutz (1986) (Theorem 27.6), we show that the kernel rule obtained by minimizing the deleted estimate is consistent when the distribution of X is atomic with finitely many atoms.

REMARK. If $\eta \equiv 1$ everywhere, so that $Y_i \equiv 1$ for all i, and if K has support equal to the unit ball in $\mathcal{R}^d$, and $K > 0$ on this ball, then $L_n^{(D)}(g_{n,h})$ is minimal and zero if

$$h \geq \max_{1 \leq j \leq n} \|X_j - X_j^{NN}\|,$$

where X_j^{NN} denotes the nearest neighbor of X_j among the data points $X_1, \ldots, X_{j-1}, X_{j+1}, \ldots, X_n$. But this shows that

$$H_n = \max_{1 \leq j \leq n} \|X_j - X_j^{NN}\|.$$

We leave it as an exercise to show that for any nonatomic distribution of X, $n H_n^d \to \infty$ with probability one. However, it is also true that for some distributions, $H_n \to \infty$ with probability one (Problem 25.19). Nevertheless, for $\eta \equiv 1$, the rule is consistent!

If $\eta \equiv 0$ everywhere, then $Y_i = 0$ for all i. Under the same condition on K as above, with tie breaking in favor of class "0" (as in the entire book), it is clear that $H_n \equiv 0$. Interestingly, here too the rule is consistent despite the strange value of H_n. $\square$

We state the following theorem for window kernels, though it can be extended to include kernels taking finitely many different values (see Problem 25.18):

Theorem 25.8. *Let g_n be the kernel rule with kernel $K = I_A$ for some bounded set $A \subset \mathcal{R}^d$ containing the origin, and with smoothing factor chosen by minimizing the deleted estimate as described above. Then $\mathbf{E}\{L(g_n)\} \to L^*$ whenever X has a discrete distribution with finitely many atoms.*

PROOF. Let $U_1, \ldots, U_n$ be independent, uniform [0, 1] random variables, independent of the data D_n, and consider the augmented sample

$$D'_n = ((X'_1, Y_1), \ldots, (X'_n, Y_n)),$$

where $X'_i = (X_i, U_i)$. For each $h \geq 0$, introduce the rule

$$g'_{n,h}(x')$$
$$= \begin{cases} 1 & \text{if } \sum_{i=1}^n K\left(\frac{x-X_i}{h}\right)(2Y_i - 1) - \sum_{i=1}^n K(0)(2Y_i - 1)I_{\{u=U_i\}} > 0 \\ 0 & \text{otherwise,} \end{cases}$$

where $x' = (x, u) \in \mathcal{R}^d \times [0, 1]$. Observe that minimizing the resubstitution estimate $L_n^{(R)}(g'_{n,h})$ over these rules yields the same h as minimizing $L_n^{(D)}(g_{n,h})$ over the original kernel rules. Furthermore, $\mathbf{E}\{L(g_n)\} = \mathbf{E}\{L(g'_n)\}$ if g'_n is obtained by minimizing $L_n^{(R)}(g'_{n,h})$. It follows from the universal consistency of kernel rules (Theorem 10.1) that

$$\lim_{n \to \infty} \inf_h L(g'_{n,h}) \to L^* \quad \text{with probability one.}$$

Thus, it suffices to investigate $L(g'_n) - \inf_h L(g'_{n,h})$. For a given D'_n, define the subsets $A'_h \subset \mathcal{R}^d \times [0, 1]$ by

$$A'_h = \left\{x' : g'_{n,h}(x') = 1\right\}, \quad h \in [0, \infty).$$

Arguing as in the previous section, we see that

$$L(g'_n) - \inf_h L(g'_{n,h}) \leq 4 \sup_h |\nu_n(A'_h) - \nu(A'_h)|,$$

where v is the measure of X' on $\mathcal{R}^d \times [0, 1]$, and v_n is the empirical measure determined by D'_n. Observe that each A'_h can be written as

$$A'_h = A_h \times \{[0, 1] - \{U_1, \ldots, U_n\}\} \cup B'_h,$$

where

$$A_h = \left\{ x \in \mathcal{R}^d : \sum_{i=1}^{n} K\left(\frac{x - X_i}{h}\right)(2Y_i - 1) > 0 \right\},$$

and

$$B'_h = \left\{ (x, u) \in \mathcal{R}^d \times \{U_1, \ldots, U_n\} : \right.$$
$$\left. \sum_{i=1}^{n} K\left(\frac{x - X_i}{h}\right)(2Y_i - 1) - K(0)\sum_{i=1}^{n}(2Y_i - 1)I_{\{u=U_i\}} > 0 \right\}.$$

Clearly, as

$$v_n(A'_h) = v_n(B'_h) \quad \text{and} \quad v(A'_h) = \mu(A_h),$$

it suffices to prove that $\sup_h |v_n(B'_h) - \mu(A_h)| \to 0$ in probability, as $n \to \infty$. Then

$$\sup_h |v_n(B'_h) - \mu(A_h)|$$
$$\leq \sup_h |v_n(B'_h) - v(B'_h)| + \sup_h |v(B'_h) - \mu(A_h)|$$
$$\leq \sup_{B' \subset \mathcal{R}^d \times \{U_1, \ldots, U_n\}} |v_n(B') - v(B')| + \sup_h |v(B'_h) - \mu(A_h)|.$$

The first term on the right-hand side tends to zero in probability, which may be seen by Problem 25.17 and the independence of the U_i's of D_n. To bound the second term, observe that for each h,

$$|v(B'_h) - \mu(A_h)| \leq \mu\left(\left\{ x : \left| \sum_{i=1}^{n} K\left(\frac{x - X_i}{h}\right)(2Y_i - 1) \right| \leq K(0) \right\}\right)$$
$$= \sum_{j=1}^{N} P\{X = x_j\} I_{\left\{ \left| \sum_{i=1}^{n} K\left(\frac{x_j - X_i}{h}\right)(2Y_i - 1) \right| \leq K(0) \right\}},$$

where $x_1, \ldots, x_N$ are the atoms of X. Thus, the proof is finished if we show that for each x_j,

$$E\left\{ \sup_h I_{\left\{ \left| \sum_{i=1}^{n} K\left(\frac{x_j - X_i}{h}\right)(2Y_i - 1) \right| \leq K(0) \right\}} \right\} \to 0.$$

Now, we exploit the special form $K = I_A$ of the kernel. Observe that

$$\sup_h I_{\left\{\left|\sum_{i=1}^n K\left(\frac{x_j - x_i}{h}\right)(2Y_i - 1)\right| \leq K(0)\right\}}$$

$$= \sup_h I_{\left\{\left|\sum_{i=1}^n I_A\left(\frac{x_j - x_i}{h}\right)(2Y_i - 1)\right| \leq 1\right\}}$$

$$\leq \sum_{k=1}^N I_{\{|V_1 + \cdots + V_k| \leq 1\}},$$

where

$$V_k = \sum_{i:X_i = x_{(k)}} (2Y_i - 1)$$

and $x_{(1)}, x_{(2)}, \ldots, x_{(N)}$ is an ordering of the atoms of X such that $x_{(1)} = x_j$ and $(x_j - x_{(k)})/h \in A$ implies $(x_j - x_{(k-1)})/h \in A$ for $1 < k \leq N$. By properties of the binomial distribution, as $n \to \infty$,

$$\mathbf{E}\left\{\sum_{k=1}^N I_{\{|V_1 + \cdots + V_k| \leq 1\}}\right\} = O\left(\frac{1}{\sqrt{n}}\right).$$

This completes the proof. $\square$

HISTORICAL REMARK. In an early paper by Habbema, Hermans, and van den Broek (1974), the deleted estimate is used to select an appropriate subspace for the kernel rule. The kernel rules in turn have smoothing factors that are selected by maximum likelihood. $\square$

25.8   Sieve Methods

Sieve methods pick a best estimate or rule from a limited class of rules. For example, our sieve $\mathcal{C}_k$ might consist of rules of the form

$$\phi(x) = \begin{cases} 1 & \text{if } \sum_{i=1}^k a_i K\left(\frac{x - x_i}{h_i}\right) > 0 \\ 0 & \text{otherwise,} \end{cases}$$

where the a_i's are real numbers, the x_i's are points from $\mathcal{R}^d$, and the h_i's are positive numbers. There are formally $(d + 2)k$ free scalar parameters. If we pick a rule ϕ_n^* that minimizes the empirical error on $(X_1, Y_1), \ldots, (X_n, Y_n)$, we are governed by the theorems of Chapters 12 and 18, and we will need to find the vc dimension of $\mathcal{C}_k$. For this, conditions on K will be needed. We will return to this question in Chapter 30 on neural networks, as they are closely related to the sieves described here.

25.9   Squared Error Minimization

Many researchers have considered the problem of selecting h in order to minimize the L_2 error (integrated squared error) of the kernel estimate

$$\widehat{\eta}_{n,h} = \frac{\frac{1}{n} \sum_{i=1}^{n} Y_i K \left(\frac{X_i - x}{h}\right)}{\frac{1}{n} \sum_{i=1}^{n} K \left(\frac{X_i - x}{h}\right)}$$

of the regression function $\eta(x) = \mathbf{P}\{Y = 1 | X = x\}$,

$$\int \left(\widehat{\eta}_{n,h}(x) - \eta(x)\right)^2 \mu(dx).$$

For example, Härdle and Marron (1985) proposed and studied a cross-validation method for choosing the optimal h for the kernel regression estimate. They obtain asymptotic optimality for the integrated squared error. Although their method gives us a choice for h if we consider $\eta(x) = \mathbf{P}\{Y = 1 | X = x\}$ as the regression function, it is not clear that the h thus obtained is optimal for the probability of error. In fact, as the following theorem illustrates, for some distributions, the smoothing parameter that minimizes the L_2 error yields a rather poor error probability compared to that corresponding to the optimal h.

Theorem 25.9. *Let $d = 1$. Consider the kernel classification rule with the window kernel $K = I_{[-1,1]}$, and smoothing parameter $h > 0$. Denote its error probability by $L_n(h)$. Let h^* be the smoothing parameter that minimizes the mean integrated squared error*

$$\mathbf{E} \left\{ \int \left(\widehat{\eta}_{n,h}(x) - \eta(x)\right)^2 \mu(dx) \right\}.$$

Then for some distributions

$$\lim_{n \to \infty} \frac{\mathbf{E}\{L_n(h^*)\}}{\inf_h \mathbf{E}\{L_n(h)\}} = \infty,$$

and the convergence is exponentially fast.

We leave the details of the proof to the reader (Problem 25.20). Only a rough sketch is given here. Consider X uniform on $[0, 1] \cup [3, 4]$, and define

$$\eta(x) = \begin{cases} 1 - x/2 & \text{if } x \in [0, 1] \\ 2 - x/2 & \text{otherwise.} \end{cases}$$

The optimal value of h (i.e., the value minimizing the error probability) is one. It is constant, independent of n. This shows that we should not a priori exclude any values of h, as is commonly done in studies on regression and density estimation. The minimal error probability can be bounded from above (using Hoeffding's inequality) by $e^{-c_1 n}$ for some constant $c_1 > 0$. On the other hand, straightforward calculations show that the smoothing factor h^* that minimizes the mean integrated squared error goes to zero as $n \to \infty$ as $O\left(n^{-1/4}\right)$. The corresponding error probability is larger than $e^{-c_2 h^* n}$ for some constant c_2. The order-of-magnitude difference between the exponents explains the exponential speed of convergence to infinity.

Problems and Exercises

PROBLEM 25.1. Prove Theorem 25.2.

PROBLEM 25.2. Show that for any $r > 0$, the kernel

$$K(x) = \begin{cases} 1 & \text{if } \|x\| \leq 1 \\ 1/\|x\|^r & \text{otherwise,} \end{cases}$$

satisfies the conditions of Theorem 25.3.

PROBLEM 25.3. Assume that K is a regular kernel with kernel complexity $\kappa_m \leq m^\gamma$ for some constant $\gamma > 0$. Let g_n be selected so as to minimize the holdout error estimate $\widehat{L}_{l,m}(\phi_m)$ over $\mathcal{C}_m$, the class of kernel rules based upon the first m data points, with smoothing factor $h > 0$. Assume furthermore that we vary l over $[\log^2 n, n/2]$, and that we pick the best l (and $m = n - l$) by minimizing the holdout error estimate again. Show that the obtained rule is strongly universally consistent.

PROBLEM 25.4. Prove that the kernel complexity κ_m of the de la Vallée-Poussin kernel $K(x) = (\sin x/x)^2$, $x \in \mathcal{R}$, is infinite when $m \geq 2$.

PROBLEM 25.5. Show that $\kappa_m = \infty$ for $m \geq 2$ when $K(x) = \cos^2(x)$, $x \in \mathcal{R}$.

PROBLEM 25.6. Compute an upper bound for the kernel complexity κ_m for the following kernels, where $x = (x^{(1)}, \ldots, x^{(d)}) \in \mathcal{R}^d$:

$$A. \qquad K(x) = \prod_{i=1}^{d} \left(1 - x^{(i)^2}\right) I_{\{|x^{(i)}| \leq 1\}}.$$

$$B. \qquad K(x) = \frac{1}{\left(1 + \|x\|^2\right)^\alpha}, \quad \alpha > 0.$$

$$C. \qquad K(x) = \prod_{i=1}^{d} \frac{1}{1 + x^{(i)^2}}.$$

$$D. \qquad K(x) = \prod_{i=1}^{d} \cos(x^{(i)}) I_{\{|x^{(i)}| \leq \pi/2\}}.$$

PROBLEM 25.7. Can you construct a kernel on $\mathcal{R}$ with the property that its complexity satisfies $2^m \leq \kappa_m < \infty$ for all m? Prove your claim.

PROBLEM 25.8. Show that for kernel classes $\mathcal{C}_m$ with kernel rules having a fixed training sequence D_m but variable $h > 0$, we have $V_{\mathcal{C}_m} \leq \log_2 \kappa_m$.

PROBLEM 25.9. Calculate upper bounds for $S(C_m, l)$ when C_m is the class of kernel rules based on fixed training data but with variable parameter θ in the following cases (for definitions, see Section 25.4):

  (1) K_θ is a product kernel of $\mathcal{R}^d$, where the unidimensional kernel K has kernel complexity κ_m.
  (2) $K_\theta = \frac{1}{1+\|x\|^\alpha/h^\alpha}$, where $\theta = (h, \alpha)$, $h > 0$, $\alpha > 0$ are two parameters.
  (3) $K_\theta(x) = I_{\{x \in A\}}$, where A is any ellipsoid of $\mathcal{R}^d$ centered at the origin.

PROBLEM 25.10. Prove or disprove: if D_m is fixed and C_m is the class of all kernel rules based on D_m with $K = I_A$, A being any convex set of $\mathcal{R}^d$ containing the origin, is it possible that $V_{C_m} = \infty$, or is $V_{C_m} < \infty$ for all possible configurations of D_m?

PROBLEM 25.11. Let

$$A_8 = \left\{ \left\{ x : \sum_{i=1}^{8} K\left(\frac{x - X_i}{h}\right)(2Y_i - 1) > 0 \right\}, h > 0 \right\}$$

with $X_1 = -3\delta$, $X_2 = X_3 = X_4 = -\delta$, $X_5 = X_6 = X_7 = \delta$, $X_8 = 3\delta$, $Y_1 = 1$, $Y_2 = Y_3 = Y_4 = -1$, $Y_5 = Y_6 = Y_7 = 1$, $Y_8 = -1$, and let K be piecewise cubic. Extending the quadratic example in the text, show that the vc dimension of A_8 is infinite for some K in this class that is symmetric, unimodal, positive, and convex on $[0, \infty)$.

PROBLEM 25.12. Draw $(X_1, Y_1), \ldots, (X_4, Y_4)$ from the distribution of (X, Y) on $\mathcal{R}^d \times \{0, 1\}$, where X has a density and $\eta(x) \in (0, 1)$ at all x. Find a symmetric unimodal K such that

$$A_4 = \left\{ \left\{ x : \sum_{i=1}^{4} K\left(\frac{x - X_i}{h}\right)(2Y_i - 1) > 0 \right\}, h > 0 \right\}$$

has vc dimension satisfying $P\{V_{A_4} = \infty\} > 0$. Can you find such a kernel K with a bounded support?

PROBLEM 25.13. Let X have a density f on $\mathcal{R}^d$ and let $X_1, \ldots, X_n$ be i.i.d., drawn from f. Show that

$$\lim_{c \to \infty} \liminf_{n \to \infty} P\left\{ \min_{i \neq j} \|X_i - X_j\|^d \leq \frac{c}{n^2} \right\} = 1.$$

Apply this result in the following situation: define $H_n = \min_{i,j:Y_i=1,Y_j=0} \|X_i - X_j\|$, where $(X_1, Y_1), \ldots, (X_n, Y_n)$ are i.i.d. $\mathcal{R}^d \times \{0, 1\}$-valued random variables distributed as (X, Y), with X absolutely continuous, Y independent of X, and $P\{Y = 1\} \in (0, 1)$. Show that

$$\lim_{c \to \infty} \liminf_{n \to \infty} P\left\{ H_n^d \leq \frac{c}{n^2} \right\} = 1.$$

Conclude that $nH^d \to 0$ in probability. If you have a kernel rule with kernel $S_{0,1}$ on $\mathcal{R}^d$, and if the smoothing factor H_n is random but satisfies $nH_n^d \to 0$ in probability, then

$$\lim_{n \to \infty} E\{L_n\} = P\{Y = 1\}$$

whenever X has a density. Show this.

PROBLEM 25.14. Consider the variable kernel rule based upon the variable kernel density estimate of Breiman, Meisel, and Purcell (1977) and studied by Krzyzak (1983)

$$g_n(x) = \begin{cases} 1 & \text{if } \sum_{i:Y_i=1} K_{H_i}(x - X_i) > \sum_{i:Y_i=0} K_{H_i}(x - X_i) \\ 0 & \text{otherwise} . \end{cases}$$

Here K is a positive-valued kernel, $K_u = (1/u^d)K(x/u)$ for $u > 0$, and H_i is the distance between X_i and the k-th nearest neighbor of X_i among X_j, $j \neq i$, $1 \leq j \leq n$. Investigate the consistency of this rule when $k/n \to 0$ and $k \to \infty$ and X_1 has a density.

PROBLEM 25.15. CONTINUATION. Fix $k = 1$, and let K be the normal density in $\mathcal{R}^d$. If X_1 has a density, what can you say about the asymptotic probability of error of the variable kernel rule? Is the inequality of Cover and Hart still valid? Repeat the exercise for the uniform kernel on the unit ball of $\mathcal{R}^d$.

PROBLEM 25.16. If $X_1, \ldots, X_n$ are discrete i.i.d. random variables and N denotes the number of different values taken by $X_1, \ldots, X_n$, then

$$\lim_{n \to \infty} \frac{\mathbf{E}\{N\}}{n} = 0,$$

and $N/n \to 0$ with probability one. HINT: For the weak convergence, assume without loss of generality that the probabilities are monotone. For the strong convergence, use McDiarmid's inequality.

PROBLEM 25.17. Let $\mathcal{B}$ be the class of all Borel subsets of $\mathcal{R}^d$. Using the previous exercise, show that for any *discrete* distribution,

$$\sup_{A \in \mathcal{B}} |\mu_n(A) - \mu(A)| \to 0 \quad \text{with probability one.}$$

HINT: Recall the necessary and sufficient condition $\mathbf{E}\{\log N_A(X_1, \ldots, X_n)\}/n \to 0$ from Chapter 12.

PROBLEM 25.18. Prove Theorem 25.8 allowing kernel functions taking finitely many different values.

PROBLEM 25.19. Let $X_1, \ldots, X_n$ be an i.i.d. sample drawn from the distribution of X. Let X_j^{NN} denote the nearest neighbor of X_j among $X_1, \ldots, X_{j-1}, X_{j+1}, \ldots, X_n$. Define

$$B_n = \max_j \left\| X_j - X_j^{NN} \right\|.$$

(1) Show that for *all* nonatomic distributions of X, $nB_n^d \to \infty$ with probability one.
(2) Is it true that for every X with a density, there exists a constant $c > 0$ such that with probability one, $nB_n^d \geq c \log n$ for all n large enough?
(3) Exhibit distributions on the real line for which $B_n \to \infty$ with probability one.

HINT: Look at the difference between the first and second order statistics.

PROBLEM 25.20. Prove Theorem 25.9. HINT: Use the example given in the text. Get an upper bound for the error probability corresponding to $h = 1$ by Hoeffding's inequality. The mean integrated squared error can be computed for every h in a straightforward way by observing that

$$\mathbf{E}\left\{ \int \left(\widehat{\eta}_{n,h}(x) - \eta(x)\right)^2 \mu(dx) \right\} = \int \mathbf{Var}\{\widehat{\eta}_{n,h}(x)\} + \left(\mathbf{E}\{\widehat{\eta}_{n,h}(x)\} - \eta(x)\right)^2 \mu(dx).$$

Split the integral between 0 and 1 in three parts, 0 to h, h to $1 - h$, and $1 - h$ to 1. Setting the derivative of the obtained expression with respect to h equal to zero leads to a third-order equation in h, whose roots are $O(n^{-1/4})$. To get a lower bound for the corresponding error probability, use the crude bound

$$\mathbf{E}\{L_n(h)\} \geq \frac{1}{2} \int_{1-h}^{1} \mathbf{P}\{g_n(X) \neq Y | X = x\} dx.$$

Now, estimate the tail of a binomial distribution from below; and use Stirling's formula to show that, modulo polynomial factors, the error probability is larger than $2^{-n^{3/4}}$.

26
Automatic Nearest Neighbor Rules

The error probability of the k-nearest neighbor rule converges to the Bayes risk for all distributions when $k \to \infty$, and $k/n \to 0$ as $n \to \infty$. The convergence result is extended here to include data-dependent choices of k. We also look at the data-based selection of a metric and of weights in weighted nearest neighbor rules.

To keep the notation consistent with that of earlier chapters, random (data-based) values of k are denoted by K_n. In most instances, K_n is merely a function of D_n, the data sequence $(X_1, Y_1), \ldots, (X_n, Y_n)$. The reader should not confuse K_n with the kernel K in other chapters.

26.1 Consistency

We start with a general theorem assessing strong consistency of the k-nearest neighbor rule with data-dependent choices of k. For the sake of simplicity, we assume the existence of the density of X. The general case can be taken care of by introducing an appropriate tie-breaking method as in Chapter 11.

Theorem 26.1. *Let $K_1, K_2, \ldots$ be integer valued random variables, and let g_n be the K_n-nearest neighbor rule. If X has a density, and*

$$K_n \to \infty \quad and \quad K_n/n \to 0 \quad with \; probability \; one \; as \quad n \to \infty,$$

then $L(g_n) \to L^*$ with probability one.

PROOF. $L(g_n) \to L^*$ with probability one if and only if for every $\epsilon > 0$, $I_{\{L(g_n)-L^*>\epsilon\}} \to 0$ with probability one. Clearly, for any $\beta > 0$,

$$I_{\{L(g_n)-L^*>\epsilon\}} \leq I_{\{1/K_n+K_n/n<2\beta,\,L(g_n)-L^*>\epsilon\}} + I_{\{1/K_n+K_n/n\geq2\beta\}}.$$

We are done if both random variables on the right-hand side converge to zero with probability one. The convergence of the second term for all $\beta > 0$ follows trivially from the conditions of the theorem. The convergence of the first term follows from the remark following Theorem 11.1, which states that for any $\epsilon > 0$, there exist $\beta > 0$ and n_0 such that for the error probability $L_{n,k}$ of the k-nearest neighbor rule,

$$\mathbf{P}\{L_{n,k} - L^* > \epsilon\} \leq 4e^{-cn\epsilon^2}$$

for some constant c depending on the dimension only, provided that $n > n_0$, $k > 1/\beta$ and $k/n < \beta$. Now clearly,

$$\mathbf{P}\{L(g_n) - L^* > \epsilon,\, 1/K_n + K_n/n < 2\beta\} \leq \mathbf{P}\{\sup_{1/\beta \leq k \leq n\beta} L_{n,k} - L^* > \epsilon\}$$

$$\leq n \sup_{1/\beta \leq k \leq n\beta} \mathbf{P}\{L_{n,k} - L^* > \epsilon\},$$

by the union bound. Combining this with Theorem 11.1, we get

$$\mathbf{P}\{L(g_n) - L^* > \epsilon,\, 1/K_n + K_n/n < 2\beta\} \leq 4ne^{-cn\epsilon^2},\quad n \geq n_0.$$

The Borel-Cantelli lemma implies that

$$I_{\{1/K_n+K_n/n<2\beta,\,L(g_n)-L^*>\epsilon\}} \to 0$$

with probability one, and the theorem is proved. $\square$

Sometimes we only know that $K_n \to \infty$ and $K_n/n \to 0$ in probability. In such cases weak consistency is guaranteed. The proof is left as an exercise (Problem 26.1).

Theorem 26.2. *Let $K_1, K_2, \ldots$ be integer valued random variables, and let g_n be the K_n-nearest neighbor rule. If X has a density, and*

$$K_n \to \infty \quad and \quad K_n/n \to 0 \quad in\ probability\ as \quad n \to \infty,$$

then $\lim_{n\to\infty} \mathbf{E}L(g_n) = L^*$, that is, g_n is weakly consistent.

26.2 Data Splitting

Consistency by itself may be obtained by choosing $k = \lfloor\sqrt{n}\rfloor$, but few—if any—users will want to blindly use such recipes. Instead, a healthy dose of feedback from the data is preferable. If we proceed as in Chapter 22, we may split the data sequence $D_n = (X_1, Y_1), \ldots, (X_n, Y_n)$ into a training sequence $D_m = (X_1, Y_1), \ldots, (X_m, Y_m)$, and a testing sequence $T_l = (X_{m+1}, Y_{m+1}), \ldots, (X_n, Y_n)$, where $m + l = n$. The training sequence D_m is used to design a class of classifiers C_m. The testing sequence is used to select a classifier from C_m that minimizes the holdout estimate of the error probability,

$$\widehat{L}_{m,l}(\phi_m) = \frac{1}{l} \sum_{i=1}^{l} I_{\{\phi_m(X_{m+i}) \neq Y_{m+i}\}}.$$

If C_m contains all k-nearest neighbor rules with $1 \leq k \leq m$, then $|C_m| = m$. Therefore, we have

$$\mathbf{P}\left\{ L(g_n) - \inf_{\phi_m \in C_m} L(\phi_m) > \epsilon \,\middle|\, D_m \right\} \leq 2m e^{-l\epsilon^2/2},$$

where g_n is the selected k-nearest neighbor rule. By combining Theorems 11.1 and 22.1, we immediately deduce that g_n is universally consistent if

$$\lim_{n \to \infty} m = \infty, \qquad \lim_{n \to \infty} \frac{l}{\log m} = \infty.$$

It is strongly universally consistent if $\lim_{n \to \infty} l/\log n = \infty$ also. Note too that

$$\mathbf{E}\left\{ \left| L(g_n) - \inf_{\phi_m \in C_m} L(\phi_m) \right| \right\} \leq 2\sqrt{\frac{\log(2m) + 1}{2l}}$$

(by Problem 12.1), so that it is indeed important to pick l much larger than $\log m$.

26.3 Data Splitting for Weighted NN Rules

Royall (1966) introduced the weighted NN rule in which the i-th nearest neighbor receives weight w_i, where $w_1 \geq w_2 \geq \ldots \geq w_k \geq 0$ and the w_i's sum to one. We assume that $w_{k+1} = \cdots = w_n = 0$ if there are n data points. Besides the natural appeal of attaching more weight to nearer neighbors, there is also a practical by-product: if the w_i's are all of the form $1/z_i$, where z_i is a prime integer, then no two subsums of w_i's are equal, and therefore voting ties are avoided altogether.

Consider now data splitting in which C_m consists of all weighted k-NN rules as described above—clearly, $k \leq m$ now. As $|C_m| = \infty$, we compute the shatter coefficients $S(C_m, l)$. We claim that

$$S(C_m, l) \leq \begin{cases} \sum_{i=1}^{k} \binom{l}{i} \leq l^k & \text{if } l \geq k \\ 2^l & \text{if } l < k. \end{cases} \tag{26.1}$$

This result is true even if we do not insist that $w_1 \geq w_2 \geq \ldots \geq w_k \geq 0$.

PROOF OF (26.1).   Each X_j in the testing sequence is classified based upon the sign of $\sum_{i=1}^{k} a_{ij} w_i$, where $a_{ij} \in \{-1, 1\}$ depends upon the class of the i-th nearest neighbor of X_j among $X_1, \ldots, X_m$ (and does not depend upon the w_i's). Consider the l-vector of signs of $\sum_{i=1}^{k} a_{ij} w_i$, $m < j \leq n$. In the computation of $\mathcal{S}(\mathcal{C}_m, l)$, we consider the a_{ij}'s as fixed numbers, and vary the w_i's subject to the condition laid out above. Here is the crucial step in the argument: the collection of all vectors $(w_1, \ldots, w_k)$ for which X_j is assigned to class 1 is a linear halfspace of $\mathcal{R}^k$. Therefore, $\mathcal{S}(\mathcal{C}_m, l)$ is bounded from above by the number of cells in the partition of $\mathcal{R}^k$ defined by l linear halfspaces. This is bounded by $\sum_{i=1}^{k} \binom{l}{i}$ (see Problem 22.1) if $k \leq l$. $\square$

Let g_n be the rule in $|\mathcal{C}_m|$ that minimizes the empirical error committed on the test sequence $(X_{m+1}, Y_{m+1}), \ldots, (X_n, Y_n)$. Then by (22.1), if $l = n - m \geq k$, we have

$$\mathbf{P}\left\{ L(g_n) - \inf_{\phi_m \in \mathcal{C}_m} L(\phi_m) > \epsilon \,\middle|\, D_m \right\} \leq 4 e^8 l^{2k} e^{-l\epsilon^2/2}.$$

If $k \to \infty$ (which implies $m \to \infty$), we have universal consistency when $k \log(l)/l \to 0$. The estimation error is of the order of $\sqrt{k \log l / l}$—in the terminology of Chapter 22, the rule is $\sqrt{k \log l / l}$-optimal. This error must be weighed against the unknown approximation error. Let us present a quick heuristic argument. On the real line, there is compelling evidence to suggest that when X has a smooth density, $k = cm^{4/5}$ is nearly optimal. With this choice, if both l and m grow linearly in n, the estimation error is of the order of $\sqrt{\log n}/n^{1/10}$. This is painfully large—to reduce this error by a factor of two, sample sizes must rise by a factor of about 1000. The reason for this disappointing result is that $\mathcal{C}_m$ is just too rich for the values of k that interest us. Automatic selection may lead to rules that overfit the data.

If we restrict $\mathcal{C}_m$ by making m very small, the following rough argument may be used to glean information about the size of m and $l = n - m$. We will take $k = m \ll n$. For smooth regression function η, the estimation error may be anywhere between $m^{-2/5}$ and $m^{-4/5}$ on the real line. As the estimation error is of the order of $\sqrt{m \log n / n}$, equating the errors leads to the rough recipe that $m \approx (n/\log n)^{5/9}$, and $m \approx (n/\log n)^{5/13}$, respectively. Both errors are then about $(n/\log n)^{-2/9}$ and $(n/\log n)^{-4/9}$, respectively. This is better than with the previous example with m linear in n. Unfortunately, it is difficult to test whether the conditions on η that guarantee certain errors are satisfied. The above procedure is thus doomed to remain heuristic.

26.4   Reference Data and Data Splitting

Split the data into D_m and T_l as is done in the previous section. Let $\mathcal{C}_m$ contain all 1-NN rules that are based upon the data $(x_1, y_1), \ldots, (x_k, y_k)$, where $k \leq m$ is to be picked, $\{x_1, \ldots, x_k\} \subset \{X_1, \ldots X_m\}$, and $\{y_1, \ldots, y_k\} \in \{0, 1\}^k$. Note that because the y_i's are free parameters, $\{(x_1, y_1), \ldots, (x_k, y_k)\}$ is not necessarily a subset of $\{(X_1, Y_1), \ldots, (X_m, Y_m)\}$—this allows us to flip certain y-values at some data points. Trivially, $|C_m| = \binom{m}{k}2^k$. Hence,

$$\mathbf{P}\left\{L(g_n) - \inf_{\phi_m \in C_m} L(\phi_m) > \epsilon \,\bigg|\, D_m\right\} \le 2^{k+1}\binom{m}{k}e^{-l\epsilon^2/2},$$

where $l = n - m$, and $g_n \in C_m$ minimizes the empirical error on the test sequence T_l. The best rule in C_m is universally consistent when $k \to \infty$ (see Theorem 19.4). Therefore, g_n is universally consistent when the above bound converges to zero. Sufficient conditions are

(i) $\lim_{n \to \infty} l = \infty$;

(ii) $\lim_{n \to \infty} \dfrac{k \log m}{l} = 0.$

As the estimation error is $O\left(\sqrt{k \log m/l}\right)$, it is important to make l large, while keeping k small.

The selected sequence $(x_1, y_1), \ldots, (x_k, y_k)$ may be called reference data, as it captures the information in the larger data set. If k is sufficiently small, the computation of $g_n(x)$ is extremely fast. The idea of selecting reference data or throwing out useless or "bad" data points has been proposed and studied by many researchers under names such as condensed NN rules, edited NN rules, and selective NN rules. See Hart (1968), Gates (1972), Wilson (1972), Wagner (1973), Ullmann (1974), Ritter et al. (1975), Tomek (1976b), and Devijver and Kittler (1980). See also Section 19.1.

26.5   Variable Metric NN Rules

The data may also be used to select a suitable metric for use with the k-NN rule. The metric adapts itself for certain scale information gleaned from the data. For example, we may compute the distance between x_1 and x_2 by the formula

$$\begin{aligned}
\|A^T(x_1 - x_2)\| &= \left((x_1 - x_2)^T A A^T (x_1 - x_2)\right)^{1/2} \\
&= \left((x_1 - x_2)^T \Sigma (x_1 - x_2)\right)^{1/2},
\end{aligned}$$

where $(x_1 - x_2)$ is a column vector, $(\cdot)^T$ denotes its transpose, A is a $d \times d$ transformation matrix, and $\Sigma = AA^T$ is a positive definite matrix. The elements of A or Σ may be determined from the data according to some heuristic formulas. We refer to Fukunaga and Hostetler (1973), Short and Fukunaga (1981), Fukunaga and Flick (1984), and Myles and Hand (1990) for more information.

For example, the object of principal component analysis is to find a transformation matrix A such that the components of the vector $A^T X$ have unit variance and are uncorrelated. These methods are typically based on estimating the eigenvalues of the covariance matrix of X. For such situations, we prove the following general consistency result:

Theorem 26.3. *Let the random metric ρ_n be of the form*

$$\rho_n(x, y) = \|A_n^T(x - y)\|,$$

where the matrix A_n is a function of $X_1, \ldots, X_n$. Assume that distance ties occur with zero probability, and there are two sequences of nonnegative random variables $\{m_n\}$ and $\{M_n\}$ such that for any n and $x, y \in \mathcal{R}^d$,

$$m_n \|x - y\| \le \rho_n(x, y) \le M_n \|x - y\|$$

and

$$\mathbf{P} \left\{ \liminf_{n \to \infty} \frac{m_n}{M_n} = 0 \right\} = 0.$$

If

$$\lim_{n \to \infty} k_n = \infty \quad and \quad \lim_{n \to \infty} \frac{k_n}{n} = 0,$$

then the k_n-nearest neighbor rule based on the metric ρ_n is consistent.

PROOF. We verify the three conditions of Theorem 6.3. In this case, $W_{ni}(X) = 1/k_n$ if X_i is one of the k_n nearest neighbors of X (according to ρ_n), and zero otherwise. Condition (iii) holds trivially. Just as in the proof of consistency of the ordinary k_n-nearest neighbor rule, for condition (i) we need the property that the number of data points that can be among the k_n nearest neighbors of a particular point is at most $k_n \gamma_d$, where the constant γ_d depends on the dimension only. This is a deterministic property, and it can be proven exactly the same way as for the standard nearest neighbor rule. The only condition of Theorem 6.3 whose justification needs extra work is condition (ii): we need to show that for any $a > 0$,

$$\lim_{n \to \infty} \mathbf{E} \left\{ \sum_{i=1}^{n} W_{ni}(X) I_{\{\|X - X_i\| > a\}} \right\} = 0.$$

Denote the k-th nearest neighbor of X according to ρ_n by $\tilde{X}_{(k)}$, and the k-th nearest neighbor of X according to the Euclidean metric by $X_{(k)}$. Then

$$\mathbf{E} \left\{ \sum_{i=1}^{n} W_{ni}(X) I_{\{\|X - X_i\| > a\}} \right\}$$

$$= \frac{1}{k_n} \mathbf{E} \left\{ \sum_{i=1}^{n} I_{\{\rho_n(X, X_i) \le \rho_n(X, \tilde{X}_{(k_n)}), \|X - X_i\| > a\}} \right\}$$

$$\le \frac{1}{k_n} \mathbf{E} \left\{ \sum_{i=1}^{n} I_{\{\rho_n(X, X_i) \le \rho_n(X, \tilde{X}_{(k_n)}), \ \rho_n(X, X_i) > m_n a\}} \right\}$$

$$\text{(since } m_n \|x - y\| \le \rho_n(x, y))$$

$$= \frac{1}{k_n} \mathbf{E} \left\{ \sum_{i=1}^{n} I_{\{m_n a \le \rho_n(X, X_i) \le \rho_n(X, \tilde{X}_{(k_n)})\}} \right\}$$

$$= \frac{1}{k_n} \mathbf{E} \left\{ \sum_{j=1}^{k_n} I_{\{m_n a \le \rho_n(X, \tilde{X}_{(j)})\}} \right\}$$

$$\le \mathbf{P} \left\{ m_n a \le \rho_n(X, \tilde{X}_{(k_n)}) \right\}$$

$$= \mathbf{P} \left\{ \sum_{i=1}^{n} I_{\{\rho_n(X, X_i) \le m_n a\}} < k_n \right\}$$

$$\le \mathbf{P} \left\{ \sum_{i=1}^{n} I_{\{M_n \|X - X_i\| \le m_n a\}} < k_n \right\}$$

(since $M_n \|x - y\| \ge \rho_n(x, y)$)

$$= \mathbf{P} \left\{ \|X - X_{(k_n)}\| > a \frac{m_n}{M_n} \right\}.$$

But we know from the consistency proof of the ordinary k_n-nearest neighbor rule in Chapter 11 that for each $a > 0$,

$$\lim_{n \to \infty} \mathbf{P} \left\{ \|X - X_{(k_n)}\| > a \right\} = 0.$$

It follows from the condition on m_n/M_n that the probability above converges to zero as well. $\square$

The conditions of the theorem hold if, for example, the elements of the matrix A_n converge to the elements of an invertible matrix A in probability. In that case, we may take m_n as the smallest, and M_n as the largest eigenvalues of $A_n^T A_n$. Then m_n/M_n converges to the ratio of the smallest and largest eigenvalues of $A^T A$, a positive number.

If we pick the elements of A by minimizing the empirical error of a test sequence T_l over C_m, where C_m contains all k-NN rules based upon a training sequence D_m (thus, the elements of A are the free parameters), the value of $S(C_m, l)$ is too large to be useful—see Problem 26.3. Furthermore, such minimization is not computationally feasible.

26.6   Selection of k Based on the Deleted Estimate

If you wish to use all the available data in the training sequence, without splitting, then empirical selection based on minimization of other estimates of the error probability may be your solution. Unfortunately, performance guarantees for the selected rule are rarely available. If the class of rules is finite, as when k is selected from $\{1, \ldots, n\}$, there are useful inequalities. We will show you how this works.

Let C_n be the class of all k-nearest neighbor rules based on a fixed training sequence D_n, but with k variable. Clearly, $|C_n| = n$. Assume that the deleted estimate is used to pick a classifier g_n from C_n:

$$L_n^{(D)}(g_n) \leq L_n^{(D)}(\phi_n) \quad \text{for all} \ \phi_n \in C_n.$$

We can derive performance bounds for g_n from Theorem 24.5. Since the result gives poor bounds for large values of k, the range of k's has to be restricted—see the discussion following Theorem 24.5. Let k_0 denote the value of the largest k allowed, that is, C_n now contains all k-nearest neighbor rules with k ranging from 1 to k_0.

Theorem 26.4. *Let g_n be the classifier minimizing the deleted estimate of the error probability over C_n, the class of k-nearest neighbor rules with $1 \leq k \leq k_0$. Then*

$$\mathbf{P}\left\{ L(g_n) - \inf_{\phi_n \in C_n} L(\phi_n) > \epsilon \right\} \leq k_0 e^{-cn\epsilon^3/k_0},$$

where c is a constant depending on the dimension only. If $k_0 \to \infty$ and $k_0 \log n/n \to 0$, then g_n is strongly universally consistent.

PROOF. From Theorem 8.4 we recall

$$L(g_n) - \inf_{\phi_n \in C_n} L(\phi_n) \leq 2 \sup_{\phi_n \in C_n} \left| L(\phi_n) - L_n^{(D)}(\phi_n) \right|.$$

The inequality now follows from Theorem 24.5 via the union bound. Universal consistency follows from the previous inequality, and the fact that the k_0-NN rule is strongly universally consistent (see Theorem 11.1). $\square$

Problems and Exercises

PROBLEM 26.1. Let $K_1, K_2, \ldots$ be integer valued random variables, and let g_n be the K_n-nearest neighbor rule. Show that if X has a density, and $K_n \to \infty$ and $K_n/n \to 0$ in probability as $n \to \infty$, then $\mathbf{E}\{L(g_n)\} \to L^*$.

PROBLEM 26.2. Let C be the class of all 1-NN rules based upon pairs $(x_1, y_1), \ldots, (x_k, y_k)$, where k is a fixed parameter (possibly varying with n), and the (x_i, y_i)'s are variable pairs from $\mathcal{R}^d \times \{0, 1\}$. Let g_n be the rule that minimizes the empirical error over C, or, equivalently, let g_n be the rule that minimizes the resubstitution estimate $L_n^{(R)}$ over C.
   (1)   Compute a suitable upper bound for $S(C, n)$.
   (2)   Compute a good upper bound for V_C as a function of k and d.
   (3)   If $k \to \infty$ with n, show that the sequence of classes C contains a strongly universally consistent subsequence of rules (you may assume for convenience that X has a density to avoid distance ties).
   (4)   Under what condition on k can you guarantee strong universal consistency of g_n?

(5)   Give an upper bound for

$$\mathbf{E}\left\{ L(g_n) - \inf_{\phi \in C} L(\phi) \right\}.$$

PROBLEM 26.3. Let C_m contain all k-NN rules based upon data pairs $(X_1, Y_1), \ldots, (X_m, Y_m)$. The metric used in computing the neighbors is derived from the norm

$$\|x\|_\Sigma = \sum_{i=1}^d \sum_{j=1}^d \sigma_{ij} x^{(i)} x^{(j)}, \quad x = (x^{(1)}, \ldots, x^{(d)}),$$

where $\{\sigma_{ij}\}$ forms a positive definite matrix Σ. The elements of Σ are the free parameters in C_m. Compute upper and lower bounds for $S(C_m, l)$ as a function of m, l, k, and d.

PROBLEM 26.4. Let g_n be the rule obtained by minimizing $L_n^{(D)}$ over all k-NN rules with $1 \le k \le n$. Prove or disprove: g_n is strongly universally consistent. Note that in view of Theorem 26.4, it suffices to consider $\epsilon n / \log n \le k \le n - 1$ for all $\epsilon > 0$.

27
Hypercubes and Discrete Spaces

In many situations, the pair (X, Y) is purely binary, taking values in $\{0, 1\}^d \times \{0, 1\}$. Examples include boolean settings (each component of X represents "on" or "off"), representations of continuous variables through quantization (continuous variables are always represented by bit strings in computers), and ordinal data (a component of X is 1 if and only if a certain item is present). The components of X are denoted by $X^{(1)}, \ldots, X^{(d)}$. In this chapter, we review pattern recognition briefly in this setup.

Without any particular structure in the distribution of (X, Y) or the function $\eta(x) = \mathbf{P}\{Y = 1 | X = x\}, x \in \{0, 1\}^d$, the pattern recognition problem might as well be cast in the space of the first 2^d positive integers: $(X, Y) \in \{1, \ldots, 2^d\} \times \{0, 1\}$. This is dealt with in the first section. However, things become more interesting under certain structural assumptions, such as the assumption that the components of X be independent. This is dealt with in the third section. General discrimination rules on hypercubes are treated in the rest of the chapter.

27.1 Multinomial Discrimination

At first sight, discrimination on a finite set $\{1, \ldots, k\}$—called multinomial discrimination—may seem utterly trivial. Let us call the following rule the *fundamental rule*, as it captures what most of us would do in the absence of any additional information.

$$g_n^*(x) = \begin{cases} 1 & \text{if } \sum_{i=1}^n I_{\{X_i=x, Y_i=1\}} > \sum_{i=1}^n I_{\{X_i=x, Y_i=0\}}, \\ 0 & \text{otherwise.} \end{cases}$$

The fundamental rule coincides with the standard kernel and histogram rules if the smoothing factor or bin width are taken small enough. If $p_i = \mathbf{P}\{X = i\}$ and η is as usual, then it takes just a second to see that

$$L_n \geq \sum_{x:\sum_{i=1}^{n} I_{\{X_i = x\}} = 0} \eta(x) p_x$$

and

$$\mathbf{E}L_n \geq \sum_x \eta(x) p_x (1 - p_x)^n,$$

where $L_n = L(g_n^*)$. Assume $\eta(x) = 1$ at all x. Then $L^* = 0$ and

$$\mathbf{E}L_n \geq \sum_x p_x (1 - p_x)^n.$$

If $p_x = 1/k$ for all x, we have $\mathbf{E}L_n \geq (1 - 1/k)^n \geq 1/2$ if $k \geq 2n$. This simple calculation shows that we cannot say anything useful about fundamental rules unless $k \leq 2n$ *at the very least*. On the positive side, the following universal bound is useful.

Theorem 27.1. *For the fundamental rule, we have $L_n \to L^*$ with probability one as $n \to \infty$, and, in fact, for all distributions,*

$$\mathbf{E}L_n \leq L^* + \sqrt{\frac{k}{2(n+1)}} + \frac{k}{en}$$

and

$$\mathbf{E}L_n \leq L^* + 1.075\sqrt{\frac{k}{n}}.$$

PROOF. The first statement follows trivially from the strong universal consistency of histogram rules (see Theorem 9.4). It is the universal inequality that is of interest here. If $B(n, p)$ denotes a binomial random variable with parameters n and p, then we have

$$\mathbf{E}L_n = \sum_{x=1}^{k} p_x \Big(\eta(x) + (1 - 2\eta(x)) \mathbf{P}\{B(N(x), \eta(x)) > N(x)/2\} \Big)$$

(here $N(x)$ is a binomial $(n, p(x))$ random variable)

$$= \sum_{x=1}^{k} p_x \Big(1 - \eta(x) + (2\eta(x) - 1) \mathbf{P}\{B(N(x), \eta(x)) \leq N(x)/2\} \Big).$$

From this, if $\xi(x) = \min(\eta(x), 1 - \eta(x))$,

$$EL_n$$

$$\leq \sum_{x=1}^{k} p_x \left(\xi(x) + (1 - 2\xi(x))\mathbf{P}\{B(N(x), \xi(x)) \geq N(x)/2\} \right)$$

$$\leq L^* + \sum_{x=1}^{k} p_x (1 - 2\xi(x))\mathbf{E} \left\{ \frac{N(x)\xi(x)(1 - \xi(x))}{N(x)\xi(x)(1 - 2\xi(x)) + \left(\frac{1}{2} - \xi(x)\right)^2 N(x)^2} \right\}$$

(by the Chebyshev-Cantelli inequality—Theorem A.17)

$$\leq L^* + \sum_{x=1}^{k} p_x (1 - 2\xi(x))\mathbf{E} \left\{ \frac{1}{1 + (1 - 2\xi(x))^2 N(x)} \right\}$$

(since $\xi(x)(1 - \xi(x)) \leq 1/4$)

$$\leq L^* + \sum_{x=1}^{k} p_x \mathbf{E} \left\{ \frac{1}{2\sqrt{N(x)}} I_{\{N(x)>0\}} + (1 - 2\xi(x))I_{\{N(x)=0\}} \right\}$$

(since the function $u/(1 + Nu^2) \leq 1/(2\sqrt{N})$ for $0 \leq u \leq 1$)

$$\leq L^* + \sum_{x=1}^{k} p_x (1 - p_x)^n + \frac{1}{2} \sum_{x=1}^{k} p_x \sqrt{\mathbf{E} \left\{ \frac{1}{N(x)} I_{\{N(x)=0\}} \right\}}$$

(by Jensen's inequality)

$$\leq L^* + \left(1 - \frac{1}{k}\right)^n + \frac{1}{2} \sum_{x=1}^{k} p_x \sqrt{\frac{2}{(n + 1)p_x}}$$

(by Lemma A.2 and the fact that the worst distribution

has $p_x = 1/k$ for all x)

$$\leq L^* + e^{-n/k} + \frac{1}{\sqrt{2(n + 1)}} \sum_{x=1}^{k} \sqrt{p_x}$$

$$\leq L^* + \frac{k}{en} + \sqrt{\frac{k}{2(n + 1)}} \quad \text{(since } e^{-u} \leq 1/(eu) \text{ for } u \geq 0,$$

and by the Cauchy-Schwarz inequality).

This concludes the proof, as $\sqrt{1/2} + 1/e \leq 1.075$. $\square$

For several key properties of the fundamental rule, the reader is referred to Glick (1973) and to Problem 27.1. Other references include Krzanowski (1987), Goldstein (1977), and Goldstein and Dillon (1978). Note also the following extension of Theorem 27.1, which shows that the fundamental rule can handle all discrete distributions.

Theorem 27.2. *If X is purely atomic (with possibly infinitely many atoms), then $L_n \to L^*$ with probability one for the fundamental rule.*

PROOF. Number the atoms $1, 2, \ldots$. Define $X' = \min(X, k)$ and replace (X_i, Y_i) by (X_i', Y_i), where $X_i' = \min(X_i, k)$. Apply the fundamental rule to the new problem and note that by Theorem 27.1, if k is fixed, $L_n \to L^*$ with probability one for the new rule (and new distribution). However, the difference in L_n's and in L^*'s between the truncated and nontruncated versions cannot be more than $\mathbf{P}\{X \geq k\}$, which may be made as small as desired by choice of k. $\square$

27.2 Quantization

Consider a fixed partition of space $\mathcal{R}^d$ into k sets $\{A_1, \ldots, A_k\}$, and let g_n be the standard partitioning rule based upon majority votes in the A_i's (ties are broken by favoring the response "0," as elsewhere in the book). We consider two rules:

(1) The rule g_n considered above; the data are $(X_1, Y_1), \ldots, (X_n, Y_n)$, with $(X, Y) \in \mathcal{R}^d \times \{0, 1\}$. The probability of error is denoted by L_n, and the Bayes probability of error is $L^* = \mathbf{E}\{\min(\eta(X), 1 - \eta(X))\}$ with $\eta(x) = \mathbf{P}\{Y = 1 | X = x\}$, $x \in \mathcal{R}^d$.

(2) The fundamental rule g_n' operating on the quantized data $(X_1', Y_1), \ldots, (X_n', Y_n)$, with $(X', Y) \in \{1, \ldots, k\} \times \{0, 1\}$, $X_i' = j$ if $X \in A_j$. The Bayes probability of error is $L'^* = \mathbf{E}\{\min(\eta'(X'), 1 - \eta'(X'))\}$ with $\eta'(x') = \mathbf{P}\{Y = 1 | X' = x'\}$, $x' \in \{1, \ldots, k\}$. The probability of error is denoted by L_n'.

Clearly, g_n' is nothing but the fundamental rule for the quantized data. As $g_n(x) = g_n'(x')$, where $x' = j$ if $x \in A_j$, we see that

$$
\begin{aligned}
L_n &= \mathbf{P}\{Y \neq g_n(X) | X_1, Y_1, \ldots, X_n, Y_n\} \\
&= \mathbf{P}\{Y \neq g_n'(X') | X_1', Y_1, \ldots, X_n', Y_n\} \\
&= L_n'.
\end{aligned}
$$

However, the Bayes error probabilities are different in the two situations. We claim, however, the following:

Theorem 27.3. *For the standard partitioning rule,*

$$
\mathbf{E}\{L_n\} \leq L^* + 1.075\sqrt{\frac{k}{n}} + \delta,
$$

where $\delta = \mathbf{E}\left\{\left|\eta(X) - \eta'(X')\right|\right\}$. Furthermore,

$$
L^* \leq L'^* \leq L^* + \delta.
$$

PROOF. Clearly, $L^* \leq L'^*$ (see Problem 2.1). Also,

$$
\begin{aligned}
L'^* &= \mathbf{E}\{\min(\eta'(X'), 1 - \eta'(X'))\} \\
&\leq \mathbf{E}\{\min(\eta(X) + |\eta(X) - \eta'(X')|, 1 - \eta(X) + |\eta(X) - \eta'(X')|)\} \\
&\leq \delta + \mathbf{E}\{\min(\eta(X), 1 - \eta(X))\} \\
&= \delta + L^*.
\end{aligned}
$$

Furthermore,

$$
\mathbf{E}\{L_n\} \leq L'^* + 1.075\sqrt{\frac{k}{n}} \leq L^* + \delta + 1.075\sqrt{\frac{k}{n}}
$$

by Theorem 27.1. $\square$

As an immediate corollary, we show how to use the last bound to get useful distribution-free performance bounds.

Theorem 27.4. *Let $\mathcal{F}$ be a class of distributions of (X, Y) for which $X \in [0, 1]^d$ with probability one, and for some constants $c > 0$, $1 \geq \alpha > 0$,*

$$
|\eta(x) - \eta(z)| \leq c\|x - z\|^\alpha, \quad x, z \in \mathcal{R}^d.
$$

Then, if we consider all cubic histogram rules g_n (see Chapter 6 for a definition), we have

$$
\inf_{\text{cubic histogram rule } g_n} \mathbf{E}\{L(g_n) - L^*\} \leq \frac{a}{\sqrt{n}} + \frac{b}{n^{\frac{\alpha}{d+2\alpha}}},
$$

where a and b are constants depending upon c, α, and d only (see the proof for explicit expressions).

Theorem 27.4 establishes the existence of rules that perform uniformly at rate $O\left(n^{-\alpha/(d+2\alpha)}\right)$ over $\mathcal{F}$. Results like this have an impact on the number of data points required to guarantee a given performance for any $(X, Y) \in \mathcal{F}$.

PROOF. Consider a cubic grid with cells of volume h^d. As the number of cells covering $[0, 1]^d$ does not exceed $(1/h + 2)^d$, we apply Theorem 27.3 to obtain

$$
\mathbf{E}\{L(g_n)\} \leq L^* + 1.075\left(2 + \frac{1}{h}\right)^{d/2} + \delta,
$$

where

$$
\delta \leq \sup_x |\eta(x) - \eta'(x')| \leq c \sup_{A_i} \sup_{x, z \in A_i} \|z - x\|^\alpha \leq c\left(h\sqrt{d}\right)^\alpha.
$$

The right-hand side is approximately maximal when

$$
h = \left(\frac{1.075d}{2\alpha\left(c\sqrt{d}\right)^\alpha \sqrt{n}}\right)^{\frac{2}{d+2\alpha}} \overset{\text{def}}{=} \frac{c'}{n^{\frac{1}{d+2\alpha}}}.
$$

Resubstitution yields the result. $\square$

27.3   Independent Components

Let $X = (X^{(1)}, \ldots, X^{(d)})$ have components that are conditionally independent, given $\{Y = 1\}$, and also, given $\{Y = 0\}$. Introduce the notation

$$
\begin{aligned}
p(i) &= \mathbf{P}\{X^{(i)} = 1 | Y = 1\}, \\
q(i) &= \mathbf{P}\{X^{(i)} = 1 | Y = 0\}, \\
p &= \mathbf{P}\{Y = 1\}.
\end{aligned}
$$

With $x = (x^{(1)}, \ldots, x^{(d)}) \in \{0, 1\}^d$, we see that

$$
\eta(x) = \frac{\mathbf{P}\{Y = 1, X = x\}}{\mathbf{P}\{X = x\}} = \frac{p\mathbf{P}\{X = x | Y = 1\}}{p\mathbf{P}\{X = x | Y = 1\} + (1 - p)\mathbf{P}\{X = x | Y = 0\}},
$$

and

$$
\begin{aligned}
\mathbf{P}\{X = x | Y = 1\} &= \prod_{i=1}^{d} p(i)^{x^{(i)}}(1 - p(i))^{1 - x^{(i)}}, \\
\mathbf{P}\{X = x | Y = 0\} &= \prod_{i=1}^{d} q(i)^{x^{(i)}}(1 - q(i))^{1 - x^{(i)}}.
\end{aligned}
$$

Simple consideration shows that the Bayes rule is given by

$$
g^*(x) = \begin{cases} 1 & \text{if } p \prod_{i=1}^{d} p(i)^{x^{(i)}}(1 - p(i))^{1 - x^{(i)}} \\ & \quad > (1 - p) \prod_{i=1}^{d} q(i)^{x^{(i)}}(1 - q(i))^{1 - x^{(i)}} \\ 0 & \text{otherwise.} \end{cases}
$$

Taking logarithms, it is easy to see that this is equivalent to the following rule:

$$
g^*(x) = \begin{cases} 1 & \text{if } \alpha_0 + \sum_{i=1}^{d} \alpha_i x^{(i)} > 0 \\ 0 & \text{otherwise,} \end{cases}
$$

where

$$
\begin{aligned}
\alpha_0 &= \log\left(\frac{p}{1 - p}\right) + \sum_{i=1}^{d} \log\left(\frac{1 - p(i)}{1 - q(i)}\right), \\
\alpha_i &= \log\left(\frac{p(i)}{q(i)} \cdot \frac{1 - q(i)}{1 - p(i)}\right) \quad i = 1, \ldots, d.
\end{aligned}
$$

In other words, the Bayes classifier is linear! This beautiful fact was noted by Minsky (1961), Winder (1963), and Chow (1965).

Having identified the Bayes rule as a linear discrimination rule, we may apply the full force of the Vapnik-Chervonenkis theory. Let g_n be the rule that minimizes the empirical error over the class of all linear discrimination rules. As the class of linear halfspaces of $\mathcal{R}^d$ has VC dimension $d + 1$ (see Corollary 13.1), we recall from Theorem 13.11 that for g_n,

$$\mathbf{P}\{L(g_n) - L^* > \epsilon\} \leq 8n^{d+1}e^{-n\epsilon^2/128},$$

$$\mathbf{E}\{L(g_n) - L^*\} \leq 16\sqrt{\frac{(d+1)\log n + 4}{2n}} \qquad \text{(by Corollary 12.1)},$$

$$\mathbf{P}\{L(g_n) - L^* > \epsilon\} \leq 16\left(\sqrt{n}\epsilon\right)^{4096(d+1)}e^{-n\epsilon^2/2}, \quad n\epsilon^2 \geq 64,$$

and

$$\mathbf{E}\{L(g_n) - L^*\} \leq \frac{c}{\sqrt{n}},$$

where

$$c = 16 + \sqrt{10^{13}(d+1)\log(10^{12}(d+1))} \quad \text{(Problem 12.10)}.$$

These bounds are useful (i.e., they tend to zero with n) if $d = o(n/\log n)$. In contrast, without the independence assumption, we have pointed out that no non-trivial guarantee can be given about $\mathbf{E}\{L(g_n)\}$ unless $2^d < n$. The independence assumption has led us out of the high-dimensional quagmire.

One may wish to attempt to estimate the $p(i)$'s and $q(i)$'s by $\widehat{p}(i)$ and $\widehat{q}(i)$ and to use these in the *plug-in rule* $g_n(x)$ that decides 1 if

$$\widehat{p}\prod_{i=1}^{d} \widehat{p}(i)^{x^{(i)}}(1 - \widehat{p}(i))^{1-x^{(i)}} > (1 - \widehat{p})\prod_{i=1}^{d} \widehat{q}(i)^{x^{(i)}}(1 - \widehat{q}(i))^{1-x^{(i)}},$$

where $\widehat{p}$ is the standard sample-based estimate of p. The maximum-likelihood estimate of $p(i)$ is given by

$$\widehat{p}(i) = \frac{\sum_{j=1}^{n} I_{\{X_j^{(i)}=1, Y_j=1\}}}{\sum_{j=1}^{n} I_{\{Y_j=1\}}},$$

while

$$\widehat{q}(i) = \frac{\sum_{j=1}^{n} I_{\{X_j^{(i)}=1, Y_j=0\}}}{\sum_{j=1}^{n} I_{\{Y_j=0\}}},$$

with 0/0 equal to 0 (see Problem 27.6). Note that the plug-in rule too is linear, with

$$g_n(x) = \begin{cases} 1 & \text{if } \alpha_0 + \sum_{i=1}^{d} \alpha_i x^{(i)} > 0 \\ 0 & \text{otherwise,} \end{cases}$$

where

$$\alpha_0 = \log\left(\frac{\widehat{p}}{1 - \widehat{p}}\right) + \sum_{i=1}^{d} \log\left(\frac{N_{01}(i)}{N_{01}(i) + N_{11}(i)} \cdot \frac{N_{00}(i) + N_{10}(i)}{N_{00}(i)}\right),$$

$$\alpha_i = \log\left(\frac{N_{11}(i)}{N_{01}(i)} \cdot \frac{N_{00}(i)}{N_{10}(i)}\right) \quad i = 1, \ldots, d,$$

and

$$N_{00}(i) = \sum_{j=1}^{n} I_{\{X_j^{(i)}=0, Y_j=0\}}, \qquad N_{01}(i) = \sum_{j=1}^{n} I_{\{X_j^{(i)}=0, Y_j=1\}},$$

$$N_{10}(i) = \sum_{j=1}^{n} I_{\{X_j^{(i)}=1, Y_j=0\}}, \qquad N_{11}(i) = \sum_{j=1}^{n} I_{\{X_j^{(i)}=1, Y_j=1\}},$$

$$\widehat{p} = \sum_{i=1}^{d} (N_{01}(i) + N_{11}(i)).$$

For all this, we refer to Warner, Toronto, Veasey, and Stephenson (1961) (see also McLachlan (1992)).

Use the inequality

$$\mathbf{E}\{L(g_n) - L^*\} \leq 2\mathbf{E}\{|\eta_n(X) - \eta(X)|\},$$

where η is as given in the text, and η_n is as η but with p, $p(i)$, $q(i)$ replaced by $\widehat{p}$, $\widehat{p}(i)$, $\widehat{q}(i)$, to establish consistency:

Theorem 27.5. *For the plug-in rule, $L(g_n) \to L^*$ with probability one and $\mathbf{E}\{L(g_n)\} - L^* \to 0$, whenever the components are independent.*

We refer to the Problem section for an evaluation of an upper bound for $\mathbf{E}\{L(g_n) - L^*\}$ (see Problem 27.7).

Linear discrimination on the hypercube has of course limited value. The world is full of examples that are not linearly separable. For example, on $\{0, 1\}^2$, if $Y = X^{(1)}(1 - X^{(2)}) + X^{(2)}(1 - X^{(1)})$ (so that Y implements the boolean "xor" or "exclusive or" function), the problem is not linearly separable if all four possible values of X have positive probability. However, the exclusive or function may be dealt with very nicely if one considers quadratic discriminants dealt with in a later section (27.5) on series methods.

27.4   Boolean Classifiers

By a *boolean classification problem*, we mean a pattern recognition problem on the hypercube for which $L^* = 0$. This setup relates to the fact that if we consider $Y = 1$ as a circuit failure and $Y = 0$ as an operable circuit, then Y is a deterministic function of the $X^{(i)}$'s, which may be considered as gates or switches. In that case, Y may be written as a boolean function of $X^{(1)}, \ldots, X^{(d)}$. We may limit boolean classifiers in various ways by partially specifying this function. For example, following Natarajan (1991), we first consider all *monomials*, that is, all functions $g : \{0, 1\}^d \to \{0, 1\}$ of the form

$$x^{(i_1)} x^{(i_2)} \cdots x^{(i_k)}$$

for some $k \leq d$ and some indices $1 \leq i_1 < \cdots < i_k \leq d$ (note that algebraic multiplication corresponds to a boolean "and"). In such situations, one might attempt to minimize the empirical error. As we know that g^* is also a monomial, it is clear that the minimal empirical error is zero. One such minimizing monomial is given by

$$g_n(x^{(1)}, \ldots, x^{(d)}) = x^{(i_1)} \cdots x^{(i_k)},$$

where

$$\min_{1 \leq i \leq n} X_i^{(i_1)} = \cdots = \min_{1 \leq i \leq n} X_i^{(i_k)} = 1, \quad \text{and} \quad \min_{1 \leq i \leq n} X_i^{(j)} = 0, \quad j \notin \{i_1, \ldots, i_k\}.$$

Thus, g_n picks those components for which every data point has a "1". Clearly, the empirical error is zero. The number of possible functions is 2^d. Therefore, by Theorem 12.1,

$$\mathbf{P}\{L(g_n) > \epsilon\} \leq 2^d e^{-n\epsilon},$$

and

$$\mathbf{E}\{L(g_n)\} \leq \frac{d+1}{n}.$$

Here again, we have avoided the curse of dimensionality. For good performance, it suffices that n be a bit larger than d, regardless of the distribution of the data!

Assume that we limit the complexity of a boolean classifier g by requiring that g must be written as an expression having at most k operations "not," "and," or "or," with the $x^{(i)}$'s as inputs. To avoid problems with precedence rules, we assume that any number of parentheses is allowed in the expression. One may visualize each expression as an expression tree, that is, a tree in which internal nodes represent operations and leaves represent operands (inputs). The number of such binary trees with k internal nodes (and thus $k+1$ leaves) is given by the Catalan number

$$\frac{1}{k+1}\binom{2k}{k}$$

(see, e.g., Kemp (1984)). Furthermore, we may associate any of the $x^{(i)}$'s with the leaves (possibly preceded by "not") and "and" or "or" with each binary internal node, thus obtaining a total of not more than

$$2^k(2d)^{k+1}\frac{1}{k+1}\binom{2k}{k}$$

possible boolean functions of this kind. As $k \to \infty$, this bound is not more than

$$(4d)^{k+1}\frac{4^k}{\sqrt{\pi k^3}} \leq (16d)^k$$

for k large enough. Again, for k large enough,

$$\mathbf{P}\{L(g_n) > \epsilon\} \leq (16d)^k e^{-n\epsilon},$$

and

$$E\{L(g_n)\} \le \frac{k \log(16d) + 1}{n}.$$

Note that k is much more important than d in determining the sample size. For historic reasons, we mention that if g is any boolean expression consisting of at most k "not," "and," or "or" operations, then the number of such functions was shown by Pippenger (1977) not to exceed

$$\left(\frac{16(d + k)^2}{k} \right)^k.$$

Pearl (1979) used Pippenger's estimate to obtain performance bounds such as the ones given above.

27.5   Series Methods for the Hypercube

It is interesting to note that we may write any function η on the hypercube as a linear combination of Rademacher-Walsh polynomials

$$\psi_i(x) = \begin{cases} 1 & i = 0 \\ 2x^{(1)} - 1 & i = 1 \\ \vdots & \vdots \\ 2x^{(d)} - 1 & i = d \\ (2x^{(1)} - 1)(2x^{(2)} - 1) & i = d + 1 \\ \vdots & \vdots \\ (2x^{(d-1)} - 1)(2x^{(d)} - 1) & i = d + 1 + \binom{d}{2} \\ (2x^{(1)} - 1)(2x^{(2)} - 1)(2x^{(3)} - 1) & i = d + 2 + \binom{d}{2} \\ \vdots & \vdots \\ (2x^{(1)} - 1) \cdots (2x^{(d)} - 1) & i = 2^d - 1. \end{cases}$$

We verify easily that

$$\sum_x \psi_i(x)\psi_j(x) = \begin{cases} 2^d & \text{if } i = j \\ 0 & \text{if } i \ne j, \end{cases}$$

so that the ψ_i's form an orthogonal system. Therefore, we may write

$$\mu(\{x\})\eta(x) = \sum_{i=0}^{2^d - 1} a_i \psi_i(x),$$

where

$$a_i = \frac{1}{2^d} \sum_x \psi_i(x)\eta(x)\mu(\{x\}) = E \left\{ \frac{\psi_i(X)}{2^d} I_{\{Y=1\}} \right\},$$

and $\mu(\{x\}) = \mathbf{P}\{X = x\}$. Also,

$$\mu(\{x\})(1 - \eta(x)) = \sum_{i=0}^{2^d-1} b_i \psi_i(x),$$

with

$$b_i = \frac{1}{2^d} \sum_x \psi_i(x)(1 - \eta(x))\mu(\{x\}) = \mathbf{E}\left\{\frac{\psi_i(X)}{2^d} I_{\{Y=0\}}\right\}.$$

Sample-based estimates of a_i and b_i are

$$\widehat{a}_i = \frac{1}{n}\sum_{j=1}^n \frac{\psi_i(X_j)}{2^d} I_{\{Y_j=1\}},$$

$$\widehat{b}_i = \frac{1}{n}\sum_{j=1}^n \frac{\psi_i(X_j)}{2^d} I_{\{Y_j=0\}}.$$

The Bayes rule is given by

$$g^*(x) = \begin{cases} 1 & \text{if } \sum_{i=0}^{2^d-1}(a_i - b_i)\psi_i(x) > 0 \\ 0 & \text{otherwise.} \end{cases}$$

Replacing $a_i - b_i$ formally by $\widehat{a}_i - \widehat{b}_i$ yields the plug-in rule. Observe that this is just a discrete version of the Fourier series rules discussed in Chapter 17. This rule requires the estimation of 2^d differences $a_i - b_i$. Therefore, we might as well have used the fundamental rule.

When our hand is forced by the dimension, we may wish to consider only rules in the class $\mathcal{C}$ given by

$$g(x) = \begin{cases} 1 & \text{if } \sum_{i=0}^{\binom{d}{0}+\binom{d}{1}+\cdots+\binom{d}{k}}(a_i' - b_i')\psi_i(x) > 0 \\ 0 & \text{otherwise,} \end{cases}$$

where $k \le d$ is a positive integer, and $a_0', b_0', a_1', b_1', \ldots$ are arbitrary constants. We have seen that the VC dimension of this class is not more than $\binom{d}{0} + \binom{d}{1} + \cdots + \binom{d}{k} \le d^k + 1$ (Theorem 13.9). Within this class, estimation errors of the order of $O\left(\sqrt{d^k \log n/n}\right)$ are thus possible if we minimize the empirical error (Theorem 13.12). This, in effect, forces us to take $k \ll \log_d n$. For larger k, pattern recognition is all but impossible.

As an interesting side note, observe that for a given parameter k, each member of $\mathcal{C}$ is a k-th degree polynomial in x.

REMARK. PERFORMANCE OF THE PLUG-IN RULE. Define the plug-in rule by

$$g_n(x) = \begin{cases} 1 & \text{if } \sum_{i=0}^{\binom{d}{0}+\binom{d}{1}+\cdots+\binom{d}{k}}(\widehat{a}_i - \widehat{b}_i)\psi_i(x) > 0 \\ 0 & \text{otherwise.} \end{cases}$$

As $n \to \infty$, by the law of large numbers, g_n approaches g_∞:

$$g_\infty(x) = \begin{cases} 1 & \text{if } \sum_{i=0}^{\binom{d}{0}+\binom{d}{1}+\cdots+\binom{d}{k}} (a_i - b_i)\psi_i(x) > 0 \\ 0 & \text{otherwise,} \end{cases}$$

where a_i, b_i are as defined above. Interestingly, g_∞ may be much worse than the best rule in $\mathcal{C}$! Consider the simple example for $d = 2$, $k = 1$:

Table of $\eta(x)$:

	$x^{(1)} = 0$	$x^{(1)} = 1$
$x^{(2)} = 1$	1/8	6/8
$x^{(2)} = 0$	5/8	1/8

Table of $\mu(\{x\}) = \mathbf{P}\{X = x\}$:

	$x^{(1)} = 0$	$x^{(1)} = 1$
$x^{(2)} = 1$	p	$\frac{5-10p}{9}$
$x^{(2)} = 0$	$\frac{4-8p}{9}$	p

A simple calculation shows that for any choice $p \in [0, 1/2]$, either $g_\infty \equiv 1$ or $g_\infty \equiv 0$. We have $g_\infty \equiv 1$ if $p < 10/47$, and $g_\infty \equiv 0$ otherwise. However, in obvious notation, $L(g \equiv 1) = (41p + 11)/36$, $L(g \equiv 0) = (50 - 82p)/72$. The best constant rule is $g \equiv 1$ when $p < 7/41$ and $g \equiv 0$ otherwise. For $7/41 < p < 10/47$, the plug-in rule does not even pick the best constant rule, let alone the best rule in $\mathcal{C}$ with $k = 1$, which it was intended to pick.

This example highlights the danger of parametric rules or plug-in rules when applied to incorrect or incomplete models. $\square$

REMARK. HISTORICAL NOTES. The Rademacher-Walsh expansion occurs frequently in switching theory, and was given in Duda and Hart (1973). The Bahadur-Lazarsfeld expansion (Bahadur (1961)) is similar in nature. Ito (1969) presents error bounds for discrimination based upon a k-term truncation of the series. Ott and Kronmal (1976) provide further statistical properties. The rules described here with k defining the number of interactions are also obtained if we model $\mathbf{P}\{X = x|Y = 1\}$ and $\mathbf{P}\{X = x|Y = 1\}$ by functions of the form

$$\exp\left(\sum_{i=0}^{\binom{d}{0}+\binom{d}{1}+\cdots+\binom{d}{k}} a_i \psi_i(x) \right) \quad \text{and} \quad \exp\left(\sum_{i=0}^{\binom{d}{0}+\binom{d}{1}+\cdots+\binom{d}{k}} b_i \psi_i(x) \right).$$

The latter model is called the log-linear model (see McLachlan (1992, section 7.3)). $\square$

27.6   Maximum Likelihood

The maximum likelihood method (see Chapter 15) should not be used for picking the best rule from a class $\mathcal{C}$ that is not guaranteed to include the Bayes rule. Perhaps a simple example will suffice to make the point. Consider the following hypercube setting with $d = 2$:

$\eta(x)$:

	$x^{(1)} = 0$	$x^{(1)} = 1$
$x^{(2)} = 1$	0	1
$x^{(2)} = 0$	1	0

$\mu(\{x\})$:

	$x^{(1)} = 0$	$x^{(1)} = 1$
$x^{(2)} = 1$	p	q
$x^{(2)} = 0$	r	s

In this example, $L^* = 0$. Apply maximum likelihood to a class $\mathcal{F}$ with two members $\{\eta_A, \eta_B\}$, where

$$\eta_A(x) = \begin{cases} 4/5 & \text{if } x^{(1)} = 0 \\ 1/5 & \text{if } x^{(1)} = 1, \end{cases}$$

$$\eta_B(x) = \begin{cases} 0 & \text{if } x^{(1)} = 0 \\ 1 & \text{if } x^{(1)} = 1. \end{cases}$$

Then maximum likelihood won't even pick the best member from $\mathcal{F}$! To verify this, with $g_A(x) = I_{\{\eta_A(x) > 1/2\}}$, $g_B(x) = I_{\{\eta_B(x) > 1/2\}}$, we see that $L(g_A) = p + q$, $L(g_B) = r + s$. However, if η_{ML} is given by

$$\eta_{\text{ML}} = \arg\max_{\eta' \in \{\eta_A, \eta_B\}} \left(\prod_{i:Y_i=1} \eta'(X_i) \cdot \prod_{i:Y_i=0} (1 - \eta'(X_i)) \right),$$

and if we write N_{ij} for the number of data pairs (X_k, Y_k) having $X_k = i$, $Y_k = j$, then $\eta_{\text{ML}} = \eta_A$ if

$$\left(\frac{4}{5}\right)^{N_{01}+N_{10}} \left(\frac{1}{5}\right)^{N_{11}+N_{00}} > \begin{cases} 0 & \text{if } N_{01} + N_{10} > 0 \\ 1 & \text{if } N_{01} + N_{10} = 0 \end{cases}$$

and $\eta_{\text{ML}} = \eta_B$ otherwise. Equivalently, $\eta_{\text{ML}} = \eta_A$ if and only if $N_{01} + N_{10} > 0$. Apply the strong law of large numbers to note that $N_{01}/n \to r$, $N_{10}/n \to s$, $N_{00}/n \to p$, and $N_{11}/n \to q$ with probability one, as $n \to \infty$. Thus,

$$\lim_{n \to \infty} \mathbf{P}\{\eta_{\text{ML}} = \eta_A\} = \begin{cases} 1 & \text{if } r + s > 0 \\ 0 & \text{otherwise.} \end{cases}$$

Take $r + s = \epsilon$ very small, $p + q = 1 - \epsilon$. Then, for the maximum likelihood rule,

$$\mathbf{E}L_n \to L(g_A) = 1 - \epsilon > \epsilon = \min(L(g_A), L(g_B)).$$

However, when $\mathcal{F}$ contains the Bayes rule, maximum likelihood is consistent (see Theorem 15.1).

27.7   Kernel Methods

Sometimes, d is so large with respect to n that the atoms in the hypercube are sparsely populated. Some amount of smoothing may help under some circumstances. Consider a kernel K, and define the kernel rule

$$g_n(x) = \begin{cases} 1 & \text{if } \frac{1}{n} \sum_{i=1}^{n} (2Y_i - 1) K \left(\frac{\|x - X_i\|}{h} \right) > 0 \\ 0 & \text{otherwise,} \end{cases}$$

where $\|x - z\|$ is just the Hamming distance (i.e., the number of disagreements between components of x and z). With $K(u) = e^{-u^2}$, the rule above reduces to a rule given in Aitchison and Aitken (1976). In their paper, different h's are considered for the two classes, but we won't consider that distinction here. Observe that at $h = 0$, we obtain the fundamental rule. As $h \to \infty$, we obtain a (degenerate) majority rule over the entire sample. The weight given to an observation X_i decreases exponentially in $\|x - X_i\|^2$. For consistency, we merely need $h \to 0$ (the condition $nh^d \to \infty$ of Theorem 10.1 is no longer needed). And in fact, we even have consistency with $h \equiv 0$ as this yields the fundamental rule.

The data-based choice of h has been the object of several papers, including Hall (1981) and Hall and Wand (1988). In the latter paper, a mean squared error criterion is minimized. We only mention the work of Tutz (1986; 1988; 1989), who picks h so as to minimize the deleted estimate $L_n^{(D)}$.

Theorem 27.6. (TUTZ (1986)). *Let H_n be the smoothing factor in the Aitchison-Aitken rule that minimizes $L_n^{(D)}$. Then the rule is weakly consistent for all distributions of (X, Y) on the hypercube.*

PROOF. See Theorem 25.8. □

Problems and Exercises

PROBLEM 27.1. THE FUNDAMENTAL RULE. Let g_n^* be the fundamental rule on a finite set $\{1, \ldots, k\}$, and define $L_n = L(g_n^*)$. Let g^* be the Bayes rule (with error probability L^*), and let

$$\Delta = \inf_{x: \eta(x) \neq 1/2} \left(\frac{1}{2} - \min(\eta(x), 1 - \eta(x)) \right).$$

Let $L_n^{(R)}$ be the resubstitution error estimate (or apparent error rate). Show the following:
  (1) $E\left\{ L_n^{(R)} \right\} \leq EL_n$ (the apparent error rate is always optimistic; Hills (1966)).
  (2) $E\left\{ L_n^{(R)} \right\} \leq L^*$ (the apparent error rate is in fact very optimistic; Glick (1973)).
  (3) $EL_n \leq L^* + e^{-2n\Delta^2}$ (for a similar inequality related to Hellinger distances, see Glick (1973)). This is an exponential but distribution-dependent error rate.

PROBLEM 27.2. DISCRETE LIPSCHITZ CLASSES. Consider the class of regression functions $\eta \in [0, 1]$ with $|\eta(x) - \eta(z)| \leq c\rho(x, z)^\alpha$, where $x, z \in \{0, 1\}^d$, $\rho(\cdot, \cdot)$ denotes the Hamming distance, and $c > 0$ and $\alpha > 0$ are constants (note that α is not bounded from above). The purpose is to design a discrimination rule for which uniformly over all distributions of (X, Y) on $\{0, 1\}^d \times \{0, 1\}$ with such $\eta(x) = P\{Y = 1 | X = x\}$, we have

$$E\{L_n - L^*\} \leq \frac{\Psi(c, \alpha, d)}{\sqrt{n}},$$

where the function $\Psi(c, \alpha, d)$ is as small as possible. Note: for $c = 1$, the class contains all regression functions on the hypercube, and thus $\Psi(1, \alpha, d) = 1.075 \cdot 2^{d/2}$ (Theorem 27.1). How small should c be to make Ψ polynomial in d?

PROBLEM 27.3. With a cubic histogram partition of $[0, 1]^d$ into k^d cells (of volume $1/k^d$ each), we have, for the Lipschitz (c, α) class $\mathcal{F}$ of Theorem 27.4,

$$\sup_{(X,Y)\in\mathcal{F}} \delta = c \left(\frac{\sqrt{d}}{k}\right)^\alpha.$$

This grows as $d^{\alpha/2}$. Can you define a partition into k^d cells for which $\sup_{(X,Y)\in\mathcal{F}} \delta$ is smaller? HINT: Consult Conway and Sloane (1993).

PROBLEM 27.4. Consider the following randomized histogram rule: $X_1, \ldots, X_k$ partition $[0, 1]^d$ into polyhedra based on the nearest neighbor rule. Within each cell, we employ a majority rule based upon $X_{k+1}, \ldots, X_n$. If X is uniform on $[0, 1]^d$ and η is Lipschitz (c, α) (as in Theorem 27.4), then can you derive an upper bound for $E\{L_n - L^*\}$ as a function of k, n, c, α, and d? How does your bound compare with the cubic histogram rule that uses the same number (k) of cells?

PROBLEM 27.5. Let $\mathcal{F}$ be the class of all Lipschitz (c, α) functions $\eta' \in \mathcal{R}^d \to [0, 1]$. Let $(X, Y) \in \mathcal{F}$ denote the fact that (X, Y) has regression function $\eta(x) = P\{Y = 1 | X = x\}$ in $\mathcal{F}$. Then, for any cubic histogram rule, show that

$$\sup_{(X,Y)\in\mathcal{F}} E\{L_n - L^*\} \geq \frac{1}{2} - L^*.$$

Thus, the compactness condition on the space is essential for the distribution-free error bound given in Theorem 27.4.

PROBLEM 27.6. INDEPENDENT MODEL. Show that in the independent model, the maximum likelihood estimate $\widehat{p}(i)$ of $p(i)$ is given by

$$\frac{\sum_{j=1}^n I_{\{X_j^{(i)}=1, Y_j=1\}}}{\sum_{j=1}^n I_{\{Y_j=1\}}}.$$

PROBLEM 27.7. INDEPENDENT MODEL. For the plug-in rule in the independent model, is it true that $E\{L_n - L^*\} = O(1/\sqrt{n})$ uniformly over all pairs (X, Y) on $\{0, 1\}^d \times \{0, 1\}$? If so, find a constant c depending upon d only, such that $E\{L_n - L^*\} \leq c/\sqrt{n}$. If not, provide a counterexample.

PROBLEM 27.8. Consider a hypercube problem in which $X = (X^{(1)}, \ldots, X^{(d)})$ and each $X^{(i)} \in \{-1, 0, 1\}$ (a ternary generalization). Assume that the $X^{(i)}$'s are independent but not necessarily identically distributed. Show that there exists a quadratic Bayes rule, i.e., $g^*(x)$ is 1 on the set

$$\alpha_0 + \sum_{i=1}^{d} \alpha_i x^{(i)} + \sum_{i,j=1}^{d} \alpha_{ij} x^{(i)} x^{(j)} > 0,$$

where α_0, $\{\alpha_i\}$, and $\{\alpha_{ij}\}$ are some weights (Kazmierczak and Steinbuch (1963)).

PROBLEM 27.9. Let $\mathcal{A}$ be the class of all sets on the hypercube $\{0, 1\}^d$ of the form $x^{(i_1)} \cdots x^{(i_k)} = 1$, where $(x^{(1)}, \ldots, x^{(d)}) \in \{0, 1\}^d$, $1 \le i_1 < \cdots < i_k \le d$. (Thus, $\mathcal{A}$ is the class of all sets carved out by the monomials.) Show that the VC dimension of $\mathcal{C}$ is d. HINT: The set $\{(0, 1, 1, \ldots, 1), (1, 0, 1, \ldots, 1), \ldots, (1, 1, 1, \ldots, 0)\}$ is shattered by $\mathcal{A}$. No set of size $d + 1$ can be shattered by $\mathcal{A}$ by the pigeonhole principle.

PROBLEM 27.10. Show that the Catalan number

$$\frac{1}{n+1} \binom{2n}{n} \sim \frac{4^n}{\sqrt{\pi n^3}}.$$

(See, e.g., Kemp (1984).)

PROBLEM 27.11. Provide an argument to show that the number of boolean functions with at most k operations "and" or "or" and d operands of the form $x^{(i)}$ or $1 - x^{(i)}$, $x^{(i)} \in \{0, 1\}$, is not more than

$$2(2d)^{k+1} \frac{1}{k+1} \binom{2k}{k}$$

(this is 2^k times less than the bound given in the text).

PROBLEM 27.12. Provide upper and lower bounds on the VC dimension of the class of sets $\mathcal{A}$ on the hypercube $\{0, 1\}^d$ that can be described by a boolean expression with the $x^{(i)}$'s or $1 - x^{(i)}$'s as operands and with at most k operations "and" or "or."

PROBLEM 27.13. LINEAR DISCRIMINATION ON THE HYPERCUBE. Let g_n be the rule that minimizes the empirical error $\widehat{L}_n(\phi)$ over all linear rules ϕ, when the data are drawn from any distribution on $\{0, 1\}^d \times \{0, 1\}$. Let L_n be its probability of error, and let L be the minimum error probability over all linear rules. Show that for $\epsilon > 0$,

$$\mathbf{P}\{L_n - L > \epsilon\} \le 4 \binom{2^d}{d} e^{-n\epsilon^2/2}.$$

Deduce that

$$\mathbf{E}\{L_n - L\} \le 2 \frac{\sqrt{1 + \log 4\binom{2^d}{d}}}{\sqrt{2n}} \le 2 \frac{d+1}{\sqrt{2n}}.$$

Compare this result with the general Vapnik-Chervonenkis bound for linear rules (Theorem 13.11) and deduce when the bound given above is better. HINT: Count the number of possible linear rules.

PROBLEM 27.14. On the hypercube $\{0, 1\}^d$, show that the kernel rule of Aitchison and Aitken (1976) is strongly consistent when $\lim_{n \to \infty} h = 0$.

PROBLEM 27.15. Pick h in the kernel estimate by minimizing the resubstitution estimate $L_n^{(R)}$, and call it $H_n^{(R)}$. For $L_n^{(D)}$, we call it $H_n^{(D)}$. Assume that the kernel function is of the form $K(\|\cdot\|/h)$ with $K \ge 0$, $K(u) \downarrow 0$ as $u \uparrow \infty$. Let $\widehat{L}_n$ be the error estimate for the kernel rule with one of these two choices. Is it possible to find a constant c, depending upon d and K only, such that

$$\mathbf{E}\left\{\widehat{L}_n - \inf_{h \geq 0} L_n(h)\right\} \leq \frac{c}{\sqrt{n}} \ ?$$

If so, give a proof. If not, provide a counterexample. Note: if the answer is positive, a minor corollary of this result is Tutz's theorem. However, an explicit constant c may aid in determining appropriate sample sizes. It may also be minimized with respect to K.

PROBLEM 27.16. (Simon (1991).) Construct a partition of the hypercube $\{0, 1\}^d$ in the following manner, based upon a binary classification tree with perpendicular splits: Every node at level i splits the subset according to $x^{(i)} = 0$ or $x^{(i)} = 1$, so that there are at most d levels of nodes. (In practi... the most important component should be $x^{(1)}$.) For example, all possible partitions of $\{0, 1\}^2$ obtainable with 2 cuts are

Assign to each internal node (each region) a class. Define the Horton-Strahler number ξ of a tree as follows: if a tree has one node, then $\xi = 0$. If the root of the tree has left and right subtrees with Horton-Strahler numbers ξ_1 and ξ_2, then set

$$\xi = \max(\xi_1, \xi_2) + I_{\{\xi_1 = \xi_2\}}.$$

Let C be the class of classifiers g described above with Horton-Strahler number $\leq \xi$.
   (1)   Let $S = \{x \in \{0, 1\}^d : \|x\| \leq \xi\}$, where $\|\cdot\|$ denotes Hamming distance from the all-zero vector. Show that S is shattered by the class of sets $\{g = 1 : g \in C\}$.
   (2)   Show that $|S| = \sum_{i=0}^{\xi} \binom{d}{i}$.
   (3)   Conclude that the VC dimension of C is at least $\sum_{i=0}^{\xi} \binom{d}{i}$. (Simon has shown that the VC dimension of C is exactly this, but that proof is more involved.)
   (4)   Assuming $L^* = 0$, obtain an upper bound for $\mathbf{E}\{L_n\}$ as a function of ξ and d, where L_n is the probability of error for the rule picked by minimizing the empirical error over C.
   (5)   Interpret ξ as the height of the largest complete binary tree that can be embedded in the classification tree.

28

Epsilon Entropy and Totally Bounded Sets

28.1 Definitions

This chapter deals with discrimination rules that are picked from a certain class of classifiers by minimizing the empirical probability of error over a finite set of carefully selected rules. We begin with a class $\mathcal{F}$ of regression functions (i.e., a posteriori probability functions) $\eta : \mathcal{R}^d \to [0, 1]$ from which η_n will be picked by the data. The massiveness of $\mathcal{F}$ can be measured in many ways—the route followed here is suggested in the work of Kolmogorov and Tikhomirov (1961). We will depart from their work only in details. We suggest comparing the results here with those from Chapters 12 and 15.

Let $\mathcal{F}_\epsilon = \{\eta^{(1)}, \ldots, \eta^{(N)}\}$ be a finite collection of functions $\mathcal{R}^d \to [0, 1]$ such that

$$\mathcal{F} \subset \bigcup_{\eta' \in \mathcal{F}_\epsilon} S_{\eta', \epsilon},$$

where $S_{\eta', \epsilon}$ is the ball of all functions $\xi : \mathcal{R}^d \to [0, 1]$ with

$$\|\xi - \eta'\|_\infty = \sup_x |\xi(x) - \eta'(x)| < \epsilon.$$

In other words, for each $\eta' \in \mathcal{F}$, there exists an $\eta^{(i)} \in \mathcal{F}_\epsilon$ with $\sup_x |\eta'(x) - \eta^{(i)}(x)| < \epsilon$. The fewer $\eta^{(i)}$'s needed to cover $\mathcal{F}$, the smaller $\mathcal{F}$ is, in a certain sense. $\mathcal{F}_\epsilon$ is called an ϵ-cover of $\mathcal{F}$. The minimal value of $|\mathcal{F}_\epsilon|$ over all ϵ-covers is called the ϵ-covering number ($\mathcal{N}_\epsilon$). Following Kolmogorov and Tikhomirov (1961),

$$\log_2 \mathcal{N}_\epsilon$$

is called the ϵ-*entropy* of $\mathcal{F}$. We will also call it the metric entropy. A collection $\mathcal{F}$ is *totally bounded* if $\mathcal{N}_\epsilon < \infty$ for all $\epsilon > 0$. It is with such classes that we are concerned in this chapter. The next section gives a few examples. In the following section, we define the *skeleton estimate* based upon picking the empirically best member from $\mathcal{F}_\epsilon$.

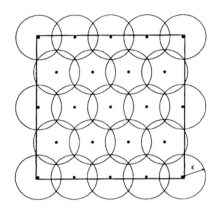

FIGURE 28.1. *An ϵ-cover of the unit square.*

28.2   Examples: Totally Bounded Classes

The simple scalar parametric class $\mathcal{F} = \left\{ e^{-\theta|x|}, x \in \mathcal{R}; \theta > 0 \right\}$ is not totally bounded (Problem 28.1). This is due simply to the presence of an unrestricted scale factor. It would still fail to be totally bounded if we restricted θ to $[1, \infty)$ or $[0, 1]$. However, if we force $\theta \in [0, 1]$ and change the class $\mathcal{F}$ to have functions

$$\eta'(x) = \begin{cases} e^{-\theta|x|} & \text{if } |x| \leq 1 \\ 0 & \text{otherwise,} \end{cases}$$

then the class is totally bounded. While it is usually difficult to compute $\mathcal{N}_\epsilon$ exactly, it is often simple to obtain matching upper and lower bounds. Here is a simple argument. Take $\theta_i = 2\epsilon i$, $0 \leq i \leq \lfloor 1/(2\epsilon) \rfloor$ and define $\theta^* = 1$. Each of these θ-values defines a function η'. Collect these in $\mathcal{F}_\epsilon$. Note that with $\eta' \in \mathcal{F}$, with parameter θ, if $\widehat{\theta}$ is the nearest value among $\{\theta_i, 0 \leq i \quad \text{\tiny\textbullet} \; 1/(2\epsilon) \rfloor\} \cup \{\theta^*\}$, then $|\widehat{\theta} - \theta| \leq \epsilon$. But then

$$\sup_{|x| \leq 1} \left| e^{-\theta|x|} - e^{-\widehat{\theta}|x|} \right| \leq |\theta - \widehat{\theta}| \leq \epsilon.$$

Hence $\mathcal{F}_\epsilon$ is an ϵ-cover of $\mathcal{F}$ of cardinality $\lfloor 1/(2\epsilon) \rfloor + 2$. We conclude that $\mathcal{F}$ is totally bounded, and that

$$\mathcal{N}_\epsilon \leq 2 + \lfloor 1/(2\epsilon) \rfloor.$$

(See Problem 28.2 for a d-dimensional generalization.)

For a lower bound, we use once again an idea from Kolmogorov and Tikhomirov (1961). Let $\mathcal{G}_\epsilon = \left\{\eta^{(1)}, \ldots, \eta^{(M)}\right\} \subset \mathcal{F}$ be a subset with the property that for every $i \neq j$, $\sup_x \left|\eta^{(i)}(x) - \eta^{(j)}(x)\right| \geq \epsilon$. The set $\mathcal{G}_\epsilon$ is thus ϵ-*separated*. The maximal cardinality ϵ-separated set is called the ϵ-*packing number* (or ϵ-*separation number*) $\mathcal{M}_\epsilon$. It is easy to see that

$$\mathcal{M}_{2\epsilon} \leq \mathcal{N}_\epsilon \leq \mathcal{M}_\epsilon$$

(Problem 28.3). With this in hand, we see that $\mathcal{G}_\epsilon$ may be constructed as follows for our example: Begin with $\theta_0 = 0$. Then define θ_1 by $e^{-\theta_1} = 1 - \epsilon$, $e^{-\theta_2} = 1 - 2\epsilon$, etcetera, until $\theta_i > 1$. It is clear that this way an ϵ-separated set $\mathcal{G}_\epsilon$ may be constructed with $|\mathcal{G}_\epsilon| = \lfloor (1 - 1/e)/\epsilon \rfloor + 1$. Thus,

$$\mathcal{N}_\epsilon \geq \mathcal{M}_{2\epsilon} \geq \left\lfloor \frac{1 - 1/e}{2\epsilon} \right\rfloor + 1.$$

The ϵ-entropy of $\mathcal{F}$ grows as $\log_2(1/\epsilon)$ as $\epsilon \downarrow 0$.

Consider next a larger class, not of a parametric nature: let $\mathcal{F}$ be a class of functions η on $[0, \Delta]$ satisfying the Lipschitz condition

$$|\eta(x) - \eta(x')| \leq c|x - x'|$$

and taking values on $[0, 1]$. Kolmogorov and Tikhomirov (1961) showed that if

$$\epsilon \leq \min\left(\frac{1}{4}, \frac{1}{16\Delta c}\right),$$

then

$$\frac{\Delta c}{\epsilon} + \log_2 \frac{1}{\epsilon} - 3 \ \leq \ \log_2 \mathcal{M}_{2\epsilon}$$
$$\leq \ \log_2 \mathcal{N}_\epsilon$$
$$\leq \ \frac{\Delta c}{\epsilon} + \log_2 \frac{1}{\epsilon} + 3$$

(see Problem 28.5). Observe that the metric entropy is exponentially larger than for the parametric class considered above. This has a major impact on sample sizes needed for similar performances (see the next sections).

Another class of functions of interest is that containing functions $\eta : [0, 1] \to [0, 1]$ that are s-times differentiable and for which the s-th derivative $\eta^{(s)}$ satisfies a Hölder condition of order α,

$$\left|\eta^{(s)}(x) - \eta^{(s)}(x')\right| \leq c|x - x'|^\alpha, \quad x, x' \in [0, 1],$$

where c is a constant. In that case, $\log_2 \mathcal{M}_\epsilon$ and $\log_2 \mathcal{N}_\epsilon$ are both $\Theta\left(\epsilon^{-1/(s+\alpha)}\right)$ as $\epsilon \downarrow 0$, where $a_n = \Theta(b_n)$ means that $a_n = O(b_n)$ and $b_n = O(a_n)$. This result, also due to Kolmogorov and Tikhomirov (1961), establishes a continuum of rates of increase of the ϵ-entropy. In $\mathcal{R}^d$, with functions $\eta : [0, 1]^d \to [0, 1]$, if the Hölder condition holds for all derivatives of order s, then $\log_2 \mathcal{N}_\epsilon = \Theta\left(\epsilon^{-d/(s+\alpha)}\right)$.

Here we have an interesting interpretation of dimension: doubling the dimension roughly offsets the effect of doubling the number of derivatives (or the degree of smoothness) of the η's. Working with Lipschitz functions on $\mathcal{R}^1$ is roughly equivalent to working with functions on $\mathcal{R}^{25}$ for which all 24-th order derivatives are Lipschitz! As there are 25^{24} such derivatives, we note immediately how much we must pay for certain performances in high dimensions.

Let $\mathcal{F}$ be the class of all entire analytic functions $\eta : [0, 2\pi] \to [0, 1]$ whose periodic continuation satisfies

$$|\eta(z)| \leq ce^{\sigma|\Im(z)|}$$

for some constants c, σ (z is a complex variable, $\Im(z)$ is its imaginary part). For this class, we know that

$$\log_2 \mathcal{N}_\epsilon \sim (4\lfloor \sigma \rfloor + 2) \log \frac{1}{\epsilon} \quad \text{as } \epsilon \downarrow 0$$

(Kolmogorov and Tikhomirov (1961)). The class appears to be as small as our parametric class. See also Vitushkin (1961).

28.3   Skeleton Estimates

The members of $\mathcal{F}_\epsilon$ form a representative skeleton of $\mathcal{F}$. We assume that $\mathcal{F}_\epsilon \subset \mathcal{F}$ (this condition was not imposed in the definition of an ϵ-cover). For each $\eta' \in \mathcal{F}$, we define its discrimination rule as

$$g(x) = \begin{cases} 1 & \text{if } \eta'(x) > 1/2 \\ 0 & \text{otherwise.} \end{cases}$$

Thus, we will take the liberty of referring to η' as a rule. For each such η', we define the probability of error as usual:

$$L(\eta') = \mathbf{P}\{g(X) \neq Y\}.$$

The empirical probability of error of η' is denoted by

$$\widehat{L}_n(\eta') = \frac{1}{n} \sum_{i=1}^{n} I_{\{g(X_i) \neq Y_i\}}.$$

We define the *skeleton estimate* η_n by

$$\eta_n = \arg \min_{\eta' \in \mathcal{F}_\epsilon} \widehat{L}_n(\eta').$$

One of the best rules in $\mathcal{F}$ is denoted by η^*:

$$L(\eta^*) \leq L(\eta'), \quad \eta' \in \mathcal{F}.$$

The first objective, as in standard empirical risk minimization, is to ensure that $L(\eta_n)$ is close to $L(\eta^*)$. If the true a posteriori probability function η is in $\mathcal{F}$ (recall that $\eta(x) = \mathbf{P}\{Y = 1 | X = x\}$), then it is clear that $L^* = L(\eta^*)$. It will be seen from the theorem below that under this condition, the skeleton estimate has nice consistency and rate-of-convergence properties. The result is distribution-free in the sense that no structure on the distribution of X is assumed. Problems 6.9 and 28.9 show that convergence of $L(\eta_n)$ to $L(\eta^*)$ for all η's—that is, not only for those in $\mathcal{F}$—holds if X has a density. In any case, because the skeleton estimate is selected from a finite deterministic set (that may be constructed before data are collected!), probability bounding is trivial: for all $\epsilon > 0, \delta > 0$, we have

$$
\mathbf{P}\left\{L(\eta_n) - \inf_{\eta' \in \mathcal{F}_\epsilon} L(\eta') > \delta\right\}
$$

$$
\leq \quad |\mathcal{F}_\epsilon| \sup_{\eta' \in \mathcal{F}_\epsilon} \mathbf{P}\left\{|\widehat{L}_n(\eta') - L(\eta')| > \delta/2\right\} \quad \text{(by Lemma 8.2)}
$$

$$
\leq \quad 2|\mathcal{F}_\epsilon| e^{-n\delta^2/2} \quad \text{(Hoeffding's inequality).}
$$

Theorem 28.1. *Let $\mathcal{F}$ be a totally bounded class of functions $\eta' : \mathcal{R}^d \to [0, 1]$. There is a sequence $\{\epsilon_n > 0\}$ and a sequence of skeletons $\mathcal{F}_{\epsilon_n} \subset \mathcal{F}$ such that if η_n is the skeleton estimate drawn from $\mathcal{F}_{\epsilon_n}$, then*

$$
L(\eta_n) \to L^* \quad \text{with probability one,}
$$

whenever the true regression function $\eta(x) = \mathbf{P}\{Y = 1 | X = x\}$ is in $\mathcal{F}$.

It suffices to take $\mathcal{F}_\epsilon$ as an ϵ-cover of $\mathcal{F}$ (note that $|\mathcal{F}_\epsilon|$ need not equal the ϵ-covering number $\mathcal{N}_\epsilon$), and to define ϵ_n as the smallest positive number for which

$$
2n \geq \frac{\log |\mathcal{F}_{\epsilon_n}|}{\epsilon_n^2}.
$$

Finally, with ϵ_n picked in this manner,

$$
\mathbf{E}\left\{L(\eta_n) - L^*\right\} \leq (2 + \sqrt{8})\epsilon_n + \sqrt{\frac{\pi}{n}}.
$$

PROOF. We note that $\inf_{\eta' \in \mathcal{F}_\epsilon} L(\eta') \leq L^* + 2\epsilon$, because if $\eta' \in S_{\eta,\epsilon}$, then

$$
\mathbf{E}\{|\eta'(X) - \eta(X)|\} \leq \sup_x |\eta'(x) - \eta(x)| \leq \epsilon
$$

and thus, by Theorem 2.2, $L(\eta') - L^* \leq 2\epsilon$. Then for any $\delta \geq 2\epsilon_n$,

$$
\mathbf{P}\{L(\eta_n) - L^* > \delta\} \quad \leq \quad \mathbf{P}\left\{L(\eta_n) - \inf_{\eta' \in \mathcal{F}_{\epsilon_n}} L(\eta') > \delta - 2\epsilon_n\right\}
$$

$$
\leq \quad 2|\mathcal{F}_{\epsilon_n}| e^{-n(\delta - 2\epsilon_n)^2/2} \quad \text{(see above)}
$$

$$
\leq \quad 2e^{2n\epsilon_n^2 - n(\delta - 2\epsilon_n)^2/2},
$$

which is summable in n, as $\epsilon_n = o(1)$. This shows the first part of the theorem. For the second part, we have

$$
\begin{aligned}
\mathbf{E}\{L(\eta_n)\} &- L^* \\
&\leq\ 2\epsilon_n + \mathbf{E}\{L(\eta_n)\} - \inf_{\eta' \in \mathcal{F}_{\epsilon_n}} L(\eta') \\
&\leq\ 2\epsilon_n + \int_0^\infty \min\left(2e^{2n\epsilon_n^2 - nt^2/2},\, 1\right) dt \\
&\leq\ (2 + \sqrt{8})\epsilon_n + 2\int_{\sqrt{8}\epsilon_n}^\infty e^{2n\epsilon_n^2 - nt^2/2} dt \\
&\leq\ (2 + \sqrt{8})\epsilon_n + \int_0^\infty e^{-nt^2/4} dt \\
&\qquad (\text{since } \epsilon_n^2 - t^2/4 \leq -t^2/8 \text{ for } t \geq \sqrt{8}\epsilon_n) \\
&=\ (2 + \sqrt{8})\epsilon_n + \sqrt{\frac{\pi}{n}}.
\end{aligned}
$$

The proof is completed. $\square$

Observe that the estimate η_n is picked from a deterministic class. This, of course, requires quite a bit of preparation and knowledge on behalf of the user. Knowledge of the ϵ-entropy (or at least an upper bound) is absolutely essential. Furthermore, one must be able to construct $\mathcal{F}_\epsilon$. This is certainly not computationally simple. Skeleton estimates should therefore be mainly of theoretical importance. They may be used, for example, to establish the existence of estimates with a guaranteed error bound as given in Theorem 28.1. A similar idea in nonparametric density estimation was proposed and worked out in Devroye (1987).

REMARK. In the first step of the proof, we used the inequality

$$
\mathbf{E}\{|\eta'(X) - \eta(X)|\} \leq \sup_x |\eta'(x) - \eta(x)| < \epsilon.
$$

It is clear from this that what we really need is not an ϵ-cover of $\mathcal{F}$ with respect to the supremum norm, but rather, with respect to the $L_1(\mu)$ norm. In other words, the skeleton estimate works equally well if the skeleton is an ϵ-cover of $\mathcal{F}$ with respect to the $L_1(\mu)$ norm, that is, it is a list of finitely many candidates with the property that for each $\eta' \in \mathcal{F}$ there exists an $\eta^{(i)}$ in the list such that $\mathbf{E}\{|\eta'(X) - \eta^{(i)}(X)|\} < \epsilon$. It follows from the inequality above that the smallest such covering has fewer elements than that of any ϵ-cover of $\mathcal{F}$ with respect to the supremum norm. Therefore, estimates based on such skeletons perform better. In fact, the difference may be essential. As an example, consider the class $\mathcal{F}$ of all regression functions on $[0, 1]$ that are monotone increasing. For $\epsilon < 1/2$, this class does not have a finite ϵ-cover with respect to the supremum norm. However, for any μ it is possible to find an ϵ-cover of $\mathcal{F}$, with respect to $L_1(\mu)$, with not more than $4^{\lceil 1/\epsilon \rceil}$ elements (Problem 28.6). Unfortunately, an ϵ-cover with respect to $L_1(\mu)$ depends on μ, the distribution of X. Since μ is not known in advance, we cannot construct this better skeleton. However, in some cases, we may have some a priori information about μ. For example, if we know that μ is a member of a known class of distributions, then we may be able to construct a skeleton that is an ϵ-cover for all measures in the class. In the above example, if we know that μ has a density bounded by a known number, then there is a finite skeleton with this property (see Problem 28.7). We note here that the basic idea behind the empirical covering method of Buescher and Kumar (1996a), described in Problem 12.14, is to find a good skeleton based on a fraction of the data. Investigating this question further, we notice that even covering in $L_1(\mu)$ is more than what we really need. From the proof of Theorem 28.1, we see that all we need is a skeleton $\mathcal{F}_\epsilon$ such that $\inf_{\eta' \in \mathcal{F}_\epsilon} L(\eta') \leq L^* + \epsilon$ for all $\eta \in \mathcal{F}$. Staying with the example of the class of monotonically increasing η's, we see that we may take in $\mathcal{F}_\epsilon$ the functions $\eta^{(i)}(x) = I_{\{x \geq q_i\}}$, where q_i is the i-th ϵ-quantile of μ, that is, q_i is the smallest number z such that $P\{X \leq z\} \geq i/\epsilon$. This collection of functions forms a skeleton in the required sense with about $(1/\epsilon)$ elements, instead of the $4^{\lceil 1/\epsilon \rceil}$ obtained by covering in $L_1(\mu)$, a significant improvement. Problem 28.8 illustrates another application of this idea. For more work on this we refer to Vapnik (1982), Benedek and Itai (1988), Kulkarni (1991), Dudley, Kulkarni, Richardson, and Zeitouni (1994), and Buescher and Kumar (1996a). □

28.4 Rate of Convergence

In this section, we take a closer look at the distribution-free upper bound

$$\mathbf{E}\left\{L(\eta_n) - L^*\right\} \leq (2 + \sqrt{8})\epsilon_n + \sqrt{\frac{\pi}{n}}.$$

For typical parametric classes (such as the one discussed in a Section 28.2), we have

$$\mathcal{N}_\epsilon = \Theta\left(\frac{1}{\epsilon}\right).$$

If we take $|\mathcal{F}_\epsilon|$ close enough to $\mathcal{N}_\epsilon$, then ϵ_n is the solution of

$$\frac{\log \mathcal{N}_\epsilon}{\epsilon^2} \cong n,$$

or $\epsilon_n = \Theta(\log n/\sqrt{n})$, and we achieve a guaranteed rate of $O(\log n/\sqrt{n})$. The same is true for the example of the class of analytic functions discussed earlier.

The situation is different for massive classes such as the Lipschitz functions on $[0, 1]^d$. Recalling that $\log \mathcal{N}_\epsilon = \Theta(1/\epsilon^d)$ as $\epsilon \downarrow 0$, we note that $\epsilon_n = \Theta(n^{-1/(2+d)})$. For this class, we have

$$\mathbf{E}\left\{L(\eta_n) - L^*\right\} = O\left(n^{-\frac{1}{2+d}}\right).$$

Here, once again, we encounter the phenomenon called the "curse of dimensionality." In order to achieve the performance $\mathbf{E}\{L(\eta_n) - L^*\} \leq \epsilon$, we need a sample of size $n \geq (1/\epsilon)^{2+d}$, exponentially large in d. Note that the class of classifiers defined by this class of functions has infinite VC dimension. The skeleton estimates thus provide a vehicle for dealing with very large classes.

Finally, if we take all η on $[0, 1]$ for which s derivatives exist, and $\eta^{(s)}$ is Lipschitz with a given constant c, similar considerations show that the rate of convergence is $O\left(n^{-(s+1)/(2s+3)}\right)$, which ranges from $O\left(n^{-1/3}\right)$ (at $s = 0$) to $O\left(n^{-1/2}\right)$ (as $s \to \infty$). As the class becomes smaller, we can guarantee better rates of convergence. Of course, this requires more a priori knowledge about the true η.

We also note that if $\log \mathcal{N}_\epsilon / \epsilon^2 = 2n$ in Theorem 28.1, then the bound is

$$\sqrt{\frac{\pi}{n}} + (2 + \sqrt{8})\epsilon_n = \sqrt{\frac{\pi}{n}} + (2 + \sqrt{8})\sqrt{\frac{\log \mathcal{N}_{\epsilon_n}}{2n}}.$$

The error grows only sub-logarithmically in the size of the skeleton set. It grows as the *square root of the ϵ-entropy*. Roughly speaking (and ignoring the dependence of ϵ_n upon n), we may say that for the same performance guarantees, doubling the ϵ-entropy implies that we should double the sample size (to keep $\log \mathcal{N}_\epsilon / n$ constant). When referring to ϵ-entropy, it is important to keep this sample size interpretation in mind.

Problems and Exercises

PROBLEM 28.1. Show that the class of functions $e^{-\theta|x|}$ on the real line, with parameter $\theta > 0$, is not totally bounded.

PROBLEM 28.2. Compute a good upper bound for $\mathcal{N}_\epsilon$ as a function of d and ϵ for the class $\mathcal{F}$ of all functions on $\mathcal{R}^d$ given by

$$\eta'(x) = \begin{cases} e^{-\theta\|x\|} & \text{if } \|x\| \leq 1 \\ 0 & \text{otherwise,} \end{cases}$$

where $\theta \in [0, 1]$ is a parameter. Repeat this question if $\theta_1, \ldots, \theta_d$ are in $[0, 1]$ and

$$\eta'(x) = \begin{cases} e^{-\sqrt{\sum_{i=1}^d \theta_i x_i^2}} & \text{if } \max_{1 \leq i \leq d} |x_i| \leq 1 \\ 0 & \text{otherwise.} \end{cases}$$

HINT: Both answers are polynomial in $1/\epsilon$ as $\epsilon \downarrow 0$.

PROBLEM 28.3. Show that $\mathcal{M}_{2\epsilon} \leq \mathcal{N}_\epsilon \leq \mathcal{M}_\epsilon$ for any totally bounded set $\mathcal{F}$ (Kolmogorov and Tikhomirov (1961)).

PROBLEM 28.4. Find a class $\mathcal{F}$ of functions $\eta' : \mathcal{R}^d \to [0, 1]$ such that
(a) for every $\epsilon > 0$, it has a finite ϵ-cover;
(b) the VC dimension of $\mathcal{A} = \{\{x : \eta'(x) > 1/2\}; \eta' \in \mathcal{F}\}$ is infinite.

PROBLEM 28.5. Show that if $\mathcal{F} = \{\eta' : [0, \Delta] \to [0, 1], \eta'$ is Lipschitz with constant $c\}$, then for ϵ small enough, $\log_2 \mathcal{N}_\epsilon \leq A/\epsilon$, where A is a constant depending upon Δ and c. (This is not as precise as the statement in the text obtained by Kolmogorov and Tikhomirov, but it will give you excellent practice.)

PROBLEM 28.6. Obtain an estimate for the cardinality of the smallest ϵ-cover with respect to the $L_1(\mu)$ norm for the class of η's on $[0, 1]$ that are increasing. In particular, show that for any μ it is possible to find an ϵ-cover with $4^{\lceil 1/\epsilon \rceil}$ elements. Can you do something similar for η's on $[0, 1]^d$ that are increasing in each coordinate?

PROBLEM 28.7. Consider the class of η's on $[0, 1]$ that are increasing. Show that for every $\epsilon > 0$, there is a finite list $\eta^{(1)}, \ldots \eta^{(N)}$ such that for all η in the class,

$$\inf_{i \leq N} \mathbf{E}\{|\eta(X) - \eta^{(i)}(X)|\} < \epsilon,$$

whenever X has a density bounded by $B < \infty$. Estimate the smallest such N.

PROBLEM 28.8. Assume that X has a bounded density on $[0, 1]^2$, and that η is monotonically increasing in both coordinates. (This is a reasonable assumption in many applications.) Then the set $\{x : g^*(x) = 0\}$ is a monotone layer. Consider the following classification rule: take a $k \times k$ grid in $[0, 1]^2$, and minimize the empirical error over all classifiers ϕ such that $\{x : \phi(x) = 0\}$ is a monotone layer, and it is a union of cells in the $k \times k$ grid. What is the optimal choice of k? Obtain an upper bound for $L(g_n) - L^*$. Compare your result with that obtained for empirical error minimization over the class of all monotone layers in Section 13.4. HINT: Count the number of different classifiers in the class. Use Hoeffding's inequality and the union-of-events bound for the estimation error. Bound the approximation error using the bounded density assumption.

PROBLEM 28.9. Apply Problem 6.9 to extend the consistency result in Theorem 28.1 as follows. Let $\mathcal{F}$ be a totally bounded class of functions $\eta' : \mathcal{R}^d \to [0, 1]$ such that $\lambda(\{x : \eta'(x) = 1/2\}) = 0$ for each $\eta' \in \mathcal{F}$ (λ is the Lebesgue measure on $\mathcal{R}^d$). Show that there is a sequence $\{\epsilon_n > 0\}$ and a sequence of skeletons $\mathcal{F}_{\epsilon_n}$ such that if η_n is the skeleton estimate drawn from $\mathcal{F}_{\epsilon_n}$, then $L(\eta_n) \to \inf_{\eta' \in \mathcal{F}} L(\eta')$ with probability one, whenever X has a density. In particular, $L(\eta_n) \to L^*$ with probability one if the Bayes rule takes the form $g^*(x) = I_{\{\eta'(x) > 1/2\}}$ for some $\eta' \in \mathcal{F}$. Note: the true regression function η is not required to be in $\mathcal{F}$.

29

Uniform Laws of Large Numbers

29.1 Minimizing the Empirical Squared Error

In Chapter 28 the data D_n were used to select a function η_n from a class $\mathcal{F}$ of candidate regression functions $\eta' : \mathcal{R}^d \to [0, 1]$. The corresponding classification rule g_n is $I_{\{\eta_n > 1/2\}}$. Selecting η_n was done in two steps: a skeleton—an ϵ-covering—of $\mathcal{F}$ was formed, and the empirical error count was minimized over the skeleton. This method is computationally cumbersome. It is tempting to use some other empirical quantity to select a classifier. Perhaps the most popular among these measures is the empirical squared error:

$$e_n(\eta') = \frac{1}{n} \sum_{i=1}^{n} (\eta'(X_i) - Y_i)^2.$$

Assume now that the function η_n is selected by minimizing the empirical squared error over $\mathcal{F}$, that is,

$$e_n(\eta_n) \le e_n(\eta'), \quad \eta' \in \mathcal{F}.$$

As always, we are interested in the error probability,

$$L(\eta_n) = \mathbf{P}\left\{ I_{\{\eta_n(x) > 1/2\}} \ne Y | D_n \right\},$$

of the resulting classifier. If the true regression function $\eta(x) = \mathbf{P}\{Y = 1 | X = x\}$ is not in the class $\mathcal{F}$, then it is easy to see that empirical squared error minimization may fail miserably (see Problem 29.1). However, if $\eta \in \mathcal{F}$, then for every $\eta' \in \mathcal{F}$

we have

$$L(\eta') - L^* \leq 2\sqrt{\mathbf{E}\left\{(\eta'(X) - \eta(X))^2\right\}} \quad \text{(by Corollary 6.2)}$$

$$= 2\sqrt{\mathbf{E}\left\{(\eta'(X) - Y)^2\right\} - \mathbf{E}\left\{(\eta(X) - Y)^2\right\}}$$

$$= 2\sqrt{\mathbf{E}\left\{(\eta'(X) - Y)^2\right\} - \inf_{\tilde\eta \in \mathcal{F}} \mathbf{E}\left\{(\tilde\eta(X) - Y)^2\right\}},$$

where the two equalities follow from the fact that $\eta(X) = \mathbf{E}\{Y|X\}$. Thus, we have

$$L(\eta_n) - L^*$$

$$\leq 2\sqrt{\mathbf{E}\left\{(\eta_n(X) - Y)^2|D_n\right\} - \inf_{\eta' \in \mathcal{F}} \mathbf{E}\left\{(\eta'(X) - Y)^2\right\}}$$

$$\leq 2\sqrt{2 \sup_{\eta' \in \mathcal{F}} \left|\frac{1}{n}\sum_{i=1}^{n}(\eta'(X_i) - Y_i)^2 - \mathbf{E}\left\{(\eta'(X) - Y)^2\right\}\right|}, \qquad (29.1)$$

by an argument as in the proof of Lemma 8.2. Thus, the method is consistent if the supremum above converges to zero. If we define $Z_i = (X_i, Y_i)$ and $f(Z_i) = (\eta'(X_i) - Y_i)^2$, then we see that we need only to bound

$$\sup_{f \in \mathcal{F}} \left|\frac{1}{n}\sum_{i=1}^{n} f(Z_i) - \mathbf{E}\{f(Z_1)\}\right|,$$

where $\mathcal{F}$ is a class of bounded functions. In the next four sections we develop upper bounds for such uniform deviations of averages from their expectations. Then we apply these techniques to establish consistency of generalized linear classifiers obtained by minimization of the empirical squared error.

29.2 Uniform Deviations of Averages from Expectations

Let $\mathcal{F}$ be a class of real-valued functions defined on $\mathcal{R}^d$, and let $Z_1, \ldots, Z_n$ be i.i.d. $\mathcal{R}^d$-valued random variables. We assume that for each $f \in \mathcal{F}$, $0 \leq f(x) \leq M$ for all $x \in \mathcal{R}^d$ and some $M < \infty$. By Hoeffding's inequality,

$$\mathbf{P}\left\{\left|\frac{1}{n}\sum_{i=1}^{n} f(Z_i) - \mathbf{E}\{f(Z_1)\}\right| > \epsilon\right\} \leq 2e^{-2n\epsilon^2/M^2}$$

for any $f \in \mathcal{F}$. However, it is much less trivial to obtain information about the probabilities

$$\mathbf{P}\left\{\sup_{f \in \mathcal{F}}\left|\frac{1}{n}\sum_{i=1}^{n} f(Z_i) - \mathbf{E}\{f(Z_1)\}\right| > \epsilon\right\}.$$

Vapnik and Chervonenkis (1981) were the first to obtain bounds for the probability above. For example, the following simple observation makes Theorems 12.5, 12.8, and 12.10 easy to apply in the new situation:

Lemma 29.1.

$$\sup_{f \in \mathcal{F}} \left| \frac{1}{n} \sum_{i=1}^{n} f(Z_i) - \mathbf{E}\{f(Z)\} \right| \leq M \sup_{f \in \mathcal{F}, t > 0} \left| \frac{1}{n} \sum_{i=1}^{n} I_{\{f(Z_i) > t\}} - \mathbf{P}\{f(Z) > t\} \right|.$$

PROOF. Exploiting the identity $\int_0^\infty \mathbf{P}\{X > t\} dt = \mathbf{E} X$ for nonnegative random variables, we have

$$\sup_{f \in \mathcal{F}} \left| \frac{1}{n} \sum_{i=1}^{n} f(Z_i) - \mathbf{E}\{f(Z)\} \right|$$

$$= \sup_{f \in \mathcal{F}} \left| \int_0^\infty \left(\frac{1}{n} \sum_{i=1}^{n} I_{\{f(Z_i) > t\}} - \mathbf{P}\{f(Z) > t\} \right) dt \right|$$

$$\leq M \sup_{f \in \mathcal{F}, t > 0} \left| \frac{1}{n} \sum_{i=1}^{n} I_{\{f(Z_i) > t\}} - \mathbf{P}\{f(Z) > t\} \right|. \quad \square$$

For example, from Theorem 12.5 and Lemma 29.1 we get

COROLLARY 29.1. *Define the collection of sets*

$$\widehat{\mathcal{F}} = \left\{ A_{f,t} : f \in \mathcal{F}, t \in [0, M] \right\},$$

where for every $f \in \mathcal{F}$ *and* $t \in [0, M]$ *the set* $A_{f,t} \in \mathcal{R}^d$ *is defined as*

$$A_{f,t} = \{z : f(z) > t\}.$$

Then

$$\mathbf{P} \left\{ \sup_{f \in \mathcal{F}} \left| \frac{1}{n} \sum_{i=1}^{n} f(Z_i) - \mathbf{E}\{f(Z_1)\} \right| > \epsilon \right\} \leq 8s(\widehat{\mathcal{F}}, n) e^{-n\epsilon^2/(32M^2)}.$$

EXAMPLE. Consider the empirical squared error minimization problem sketched in the previous section. Let $\mathcal{F}$ be the class of monotone increasing functions η' : $\mathcal{R} \to [0, 1]$, and let η_n be the function selected by minimizing the empirical squared error. By (29.1), if $\eta(x) = \mathbf{P}\{Y = 1 | X = x\}$ is also monotone increasing, then

$$\mathbf{P}\{L(\eta_n) - L^* > \epsilon\}$$

$$\leq \mathbf{P} \left\{ \sup_{\eta' \in \mathcal{F}} \left| \frac{1}{n} \sum_{i=1}^{n} (\eta'(X_i) - Y_i)^2 - \mathbf{E} \left\{ (\eta'(X) - Y)^2 \right\} \right| > \frac{\epsilon^2}{8} \right\}.$$

If $\widehat{\mathcal{F}}$ contains all subsets of $\mathcal{R} \times \{0, 1\}$ of the form

$$A_{\eta', t} = \left\{ (x, y) : (\eta'(x) - y)^2 > t \right\}, \quad \eta' \in \mathcal{F}, t \in [0, 1],$$

then it is easy to see that its n-th shatter coefficient satisfies $s(\widehat{\mathcal{F}}, n) \leq (n/2 + 1)^2$. Thus, Corollary 29.1 can be applied, and the empirical squared error minimization is consistent. $\square$

In many cases, Corollary 29.1 does not provide the best possible bound. To state a similar, but sometimes more useful result, we introduce l_1-*covering numbers*. The notion is very similar to that of covering numbers discussed in Chapter 28. The main difference is that here the balls are defined in terms of an l_1-distance, rather than the supremum norm.

DEFINITION 29.1. *Let A be a bounded subset of $\mathcal{R}^d$. For every $\epsilon > 0$, the l_1-covering number, denoted by $\mathcal{N}(\epsilon, A)$, is defined as the cardinality of the smallest finite set in $\mathcal{R}^d$ such that for every $z \in A$ there is a point $t \in \mathcal{R}^d$ in the finite set such that $(1/d)\|z - t\|_1 < \epsilon$. ($\|x\|_1 = \sum_{i=1}^{d} |x^{(i)}|$ denotes the l_1-norm of the vector $x = (x^{(1)}, \dots, x^{(d)})$ in $\mathcal{R}^d$.) In other words, $\mathcal{N}(\epsilon, A)$ is the smallest number of l_1-balls of radius ϵd, whose union contains A. $\log \mathcal{N}(\epsilon, A)$ is often called the metric entropy of A.*

We will mainly be interested in covering numbers of special sets. Let $z_1^n = (z_1, \dots, z_n)$ be n fixed points in $\mathcal{R}^d$, and define the following set:

$$\mathcal{F}(z_1^n) = \{(f(z_1), \dots, f(z_n)); f \in \mathcal{F}\} \subset \mathcal{R}^n.$$

The l_1-covering number of $\mathcal{F}(z_1^n)$ is $\mathcal{N}(\epsilon, \mathcal{F}(z_1^n))$.

If $Z_1^n = (Z_1, \dots, Z_n)$ is a sequence of i.i.d. random variables, then $\mathcal{N}(\epsilon, \mathcal{F}(Z_1^n))$ is a random variable, whose expected value plays a central role in our problem:

Theorem 29.1. (POLLARD (1984)). *For any n and $\epsilon > 0$,*

$$\mathbf{P}\left\{ \sup_{f \in \mathcal{F}} \left| \frac{1}{n} \sum_{i=1}^{n} f(Z_i) - \mathbf{E}\{f(Z_1)\} \right| > \epsilon \right\} \leq 8\mathbf{E}\left\{ \mathcal{N}(\epsilon/8, \mathcal{F}(Z_1^n)) \right\} e^{-n\epsilon^2/(128M^2)}.$$

The proof of the theorem is given in Section 29.4.

REMARK. Theorem 29.1 is a generalization of the basic Vapnik-Chervonenkis inequality. To see this, define l_∞-covering numbers based on the maximum norm (Vapnik and Chervonenkis (1981)): $\mathcal{N}_\infty(\epsilon, A)$ is the cardinality of the smallest finite set in $\mathcal{R}^d$ such that for every $z \in A$ there is a point $t \in \mathcal{R}^d$ in the set such that $\max_{1 \leq i \leq d} |z^{(i)} - t^{(i)}| < \epsilon$. If the functions in $\mathcal{F}$ are indicators of sets from a class $\mathcal{A}$ of subsets of $\mathcal{R}^d$, then it is easy to see that for every $\epsilon \in (0, 1/2)$,

$$\mathcal{N}_\infty(\epsilon, \mathcal{F}(z_1^n)) = N_{\mathcal{A}}(z_1, \dots, z_n),$$

where $N_A(z_1, \ldots, z_n)$ is the combinatorial quantity that was used in Definition 12.1 of shatter coefficients. Since

$$\mathcal{N}(\epsilon, \mathcal{F}(z_1^n)) \leq \mathcal{N}_\infty(\epsilon, \mathcal{F}(z_1^n)),$$

Theorem 29.1 remains true with l_∞-covering numbers, therefore, it is a generalization of Theorem 12.5. To see this, notice that if $\mathcal{F}$ contains indicators of sets of the class $\mathcal{A}$, then

$$\sup_{f \in \mathcal{F}} \left| \frac{1}{n} \sum_{i=1}^n f(Z_i) - \mathbf{E}\{f(Z_1)\} \right| = \sup_{A \in \mathcal{A}} \left| \frac{1}{n} \sum_{i=1}^n I_{\{Z_i \in A\}} - \mathbf{P}\{Z_1 \in A\} \right|. \quad \square$$

For inequalities sharper and more general than Theorem 29.1 we refer to Vapnik (1982), Pollard (1984; 1986), Haussler (1992), and Anthony and Shawe-Taylor (1990).

29.3 Empirical Squared Error Minimization

We return to the minimization of the empirical squared error. Let $\mathcal{F}$ be a class of functions $\eta' : \mathcal{R}^d \to [0, 1]$, containing the true a posteriori function η. The empirical squared error

$$\frac{1}{n} \sum_{i=1}^n (\eta'(X_i) - Y_i)^2$$

is minimized over $\eta' \in \mathcal{F}$, to obtain the estimate η_n. The next result shows that empirical squared error minimization is consistent under general conditions. Observe that these are the same conditions that we assumed in Theorem 28.1 to prove consistency of skeleton estimates.

Theorem 29.2. *Assume that $\mathcal{F}$ is a totally bounded class of functions. (For the definition see Chapter 28.) If $\eta \in \mathcal{F}$, then the classification rule obtained by minimizing the empirical squared error over $\mathcal{F}$ is strongly consistent, that is,*

$$\lim_{n \to \infty} L(\eta_n) = L^* \quad \text{with probability one.}$$

PROOF. Recall that by (29.1),

$$\mathbf{P}\{L(\eta_n) - L^* > \epsilon\}$$
$$\leq \mathbf{P}\left\{ \sup_{\eta' \in \mathcal{F}} \left| \frac{1}{n} \sum_{i=1}^n (\eta'(X_i) - Y_i)^2 - \mathbf{E}\left\{(\eta'(X) - Y)^2\right\} \right| > \frac{\epsilon^2}{8} \right\}.$$

We apply Theorem 29.1 to show that for every $\epsilon > 0$, the probability on the right-hand side converges to zero exponentially as $n \to \infty$. To this end, we need to find a suitable upper bound on $\mathbf{E}\{\mathcal{N}(\epsilon, \mathcal{J}(Z_1^n))\}$, where $\mathcal{J}$ is the class of functions

$f'(x, y) = (\eta'(x) - y)^2$ from $\mathcal{R}^d \times \{0, 1\}$ to $[0, 1]$, where $\eta' \in \mathcal{F}$, and $Z_i = (X_i, Y_i)$. Observe that for any $f', \bar{f} \in \mathcal{J}$,

$$
\begin{aligned}
\frac{1}{n} \sum_{i=1}^{n} |f'(Z_i) - \bar{f}(Z_i)| &= \frac{1}{n} \sum_{i=1}^{n} |(\eta'(X_i) - Y_i)^2 - (\bar{\eta}(X_i) - Y_i)^2| \\
&\leq \frac{2}{n} \sum_{i=1}^{n} |\eta'(X_i) - \bar{\eta}(X_i)| \\
&\leq 2 \sup_x |\eta'(x) - \bar{\eta}(x)|.
\end{aligned}
$$

This inequality implies that for every $\epsilon > 0$,

$$
\mathbf{E}\{\mathcal{N}(\epsilon, \mathcal{J}(Z_1^n))\} \leq \mathcal{N}_{\epsilon/2},
$$

where $\mathcal{N}_\epsilon$ is the covering number of $\mathcal{F}$ defined in Chapter 28. By the assumption of total boundedness, for every $\epsilon > 0, \mathcal{N}_{\epsilon/2} < \infty$. Since $\mathcal{N}_{\epsilon/2}$ does not depend on n, the theorem is proved. $\square$

REMARK. The nonasymptotic exponential nature of the inequality in Theorem 29.1 makes it possible to obtain upper bounds for the rate of convergence of $L(\eta_n)$ to L^* in terms of the covering numbers $\mathcal{N}_\epsilon$ of the class $\mathcal{F}$. However, since we started our analysis by the loose inequality $L(\eta') - L^* \leq 2\sqrt{\mathbf{E}\{(\eta'(X) - \eta(X))^2\}}$, the resulting rates are likely to be suboptimal (see Theorem 6.5). Also, the inequality of Theorem 29.1 may be loose in this case. In a somewhat different setup, Barron (1991) developed a proof method based on Bernstein's inequality that is useful for obtaining tighter upper bounds for $L(\eta_n) - L^*$ in certain cases. $\square$

29.4   Proof of Theorem 29.1

The main tricks in the proof resemble those of Theorem 12.5. We can show that for $n\epsilon^2 \geq 2M^2$,

$$
\mathbf{P}\left\{\sup_{f \in \mathcal{F}} \left| \frac{1}{n} \sum_{i=1}^{n} f(Z_i) - \mathbf{E}\{f(Z_1)\} \right| > \epsilon \right\} \leq 4\mathbf{P}\left\{\sup_{f \in \mathcal{F}} \left| \frac{1}{n} \sum_{i=1}^{n} \sigma_i f(Z_i) \right| > \frac{\epsilon}{4} \right\},
$$

where $\sigma_1, \ldots, \sigma_n$ are i.i.d. $\{-1, 1\}$-valued random variables, independent of the Z_i's, with $\mathbf{P}\{\sigma_i = 1\} = \mathbf{P}\{\sigma_i = -1\} = 1/2$. The only minor difference with Theorem 12.5 appears when Chebyshev's inequality is applied. We use the fact that by boundedness, $\mathbf{Var}(f(Z_1)) \leq M^2/4$ for every $f \in \mathcal{F}$.

Now, take a minimal $\epsilon/8$-covering of $\mathcal{F}(Z_1^n)$, that is, $M = \mathcal{N}(\epsilon/8, \mathcal{F}(Z_1^n))$ functions $g_1, \ldots, g_M$ such that for every $f \in \mathcal{F}$ there is a $g^* \in \{g_1, \ldots, g_M\}$ with

$$
\frac{1}{n} \sum_{i=1}^{n} |f(Z_i) - g^*(Z_i)| \leq \frac{\epsilon}{8}.
$$

For any function f, we have

$$\left| \frac{1}{n} \sum_{i=1}^{n} \sigma_i f(Z_i) \right| \leq \frac{1}{n} \sum_{i=1}^{n} |\sigma_i f(Z_i)| = \frac{1}{n} \sum_{i=1}^{n} |f(Z_i)|,$$

and thus

$$\left| \frac{1}{n} \sum_{i=1}^{n} \sigma_i f(Z_i) \right| \leq \left| \frac{1}{n} \sum_{i=1}^{n} \sigma_i g^*(Z_i) \right| + \left| \frac{1}{n} \sum_{i=1}^{n} \sigma_i (f(Z_i) - g^*(Z_i)) \right|$$

$$\leq \left| \frac{1}{n} \sum_{i=1}^{n} \sigma_i g^*(Z_i) \right| + \frac{1}{n} \sum_{i=1}^{n} |f(Z_i) - g^*(Z_i)|.$$

As $(1/n) \sum_{i=1}^{n} |g^*(Z_i) - f(Z_i)| \leq \epsilon/8$,

$$\mathbf{P} \left\{ \sup_{f \in \mathcal{F}} \left| \frac{1}{n} \sum_{i=1}^{n} \sigma_i f(Z_i) \right| > \frac{\epsilon}{4} \middle| Z_1, \ldots, Z_n \right\}$$

$$\leq \mathbf{P} \left\{ \sup_{f \in \mathcal{F}} \left| \frac{1}{n} \sum_{i=1}^{n} \sigma_i g^*(Z_i) \right| + \frac{1}{n} \sum_{i=1}^{n} |f(Z_i) - g^*(Z_i)| > \frac{\epsilon}{4} \middle| Z_1, \ldots, Z_n \right\}$$

$$\leq \mathbf{P} \left\{ \max_{g_j} \left| \frac{1}{n} \sum_{i=1}^{n} \sigma_i g_j(Z_i) \right| > \frac{\epsilon}{8} \middle| Z_1, \ldots, Z_n \right\}.$$

Now that we have been able to convert the "sup" into a "max," we can use the union bound:

$$\mathbf{P} \left\{ \max_{g_j} \left| \frac{1}{n} \sum_{i=1}^{n} \sigma_i g_j(Z_i) \right| > \frac{\epsilon}{8} \middle| Z_1, \ldots, Z_n \right\}$$

$$\leq \mathcal{N} \left(\frac{\epsilon}{8}, \mathcal{F}(Z_1^n) \right) \max_{g_j} \mathbf{P} \left\{ \left| \frac{1}{n} \sum_{i=1}^{n} \sigma_i g_j(Z_i) \right| > \frac{\epsilon}{8} \middle| Z_1, \ldots, Z_n \right\}.$$

We need only find a uniform bound for the probability following the "max." This, however, is easy, since after conditioning on $Z_1, \ldots, Z_n$, $\sum_{i=1}^{n} \sigma_i g_j(Z_i)$ is the sum of independent bounded random variables whose expected value is zero. Therefore, Hoeffding's inequality gives

$$\mathbf{P} \left\{ \left| \frac{1}{n} \sum_{i=1}^{n} \sigma_i g_j(Z_i) \right| > \frac{\epsilon}{8} \middle| Z_1, \ldots, Z_n \right\} \leq 2e^{-n\epsilon^2/(128M^2)}.$$

In summary,

$$\mathbf{P} \left\{ \sup_{f \in \mathcal{F}} \left| \frac{1}{n} \sum_{i=1}^{n} \sigma_i f(Z_i) \right| > \frac{\epsilon}{4} \right\}$$

$$\leq \mathbf{E} \left\{ \mathbf{P} \left\{ \max_{g_j} \left| \frac{1}{n} \sum_{i=1}^{n} \sigma_i g_j(Z_i) \right| > \frac{\epsilon}{8} \middle| Z_1, \ldots, Z_n \right\} \right\}$$

$$\leq \; 2\mathbf{E}\left\{\mathcal{N}\left(\frac{\epsilon}{8},\mathcal{F}(Z_1^n)\right)\right\}e^{-n\epsilon^2/(128M^2)}.$$

The theorem is proved. □

Properties of $\mathcal{N}(\epsilon,\mathcal{F}(Z_1^n))$ will be studied in the next section.

29.5   Covering Numbers and Shatter Coefficients

In this section we study covering numbers, and relate them to shatter coefficients of certain classes of sets. As in Chapter 28, we introduce l_1-*packing numbers*. Let $\mathcal{F}$ be a class of functions on $\mathcal{R}^d$, taking their values in $[0, M]$. Let μ be an arbitrary probability measure on $\mathcal{R}^d$. Let $g_1, \ldots, g_m$ be a finite collection of functions from $\mathcal{F}$ with the property that for any two of them

$$\int_{\mathcal{R}^d}|g_i(x)-g_j(x)|\mu(dx)\geq\epsilon.$$

The largest m for which such a collection exists is called the packing number of $\mathcal{F}$ (relative to μ), and is denoted by $\mathcal{M}(\epsilon,\mathcal{F})$. If μ places probability $1/n$ on each of $z_1, \ldots, z_n$, then by definition $\mathcal{M}(\epsilon,\mathcal{F}) = \mathcal{M}(\epsilon,\mathcal{F}(z_1^n))$, and it is easy to see (Problem 28.3) that

$$\mathcal{M}(2\epsilon,\mathcal{F}(z_1^n))\leq\mathcal{N}(\epsilon,\mathcal{F}(z_1^n))\leq\mathcal{M}(\epsilon,\mathcal{F}(z_1^n)).$$

An important feature of a class of functions $\mathcal{F}$ is the VC dimension $V_{\mathcal{F}^+}$ of

$$\mathcal{F}^+=\{\{(x,t):t\leq f(x)\};f\in\mathcal{F}\}.$$

This is clarified by the following theorem, which is a slight refinement of a result by Pollard (1984), which is based on Dudley's (1978) work. It connects the packing number of $\mathcal{F}$ with the shatter coefficients of $\mathcal{F}^+$. See also Haussler (1992) for a somewhat different argument.

Theorem 29.3. *Let $\mathcal{F}$ be a class of $[0, M]$-valued functions on $\mathcal{R}^d$. For every $\epsilon > 0$ and probability measure μ,*

$$\mathcal{M}(\epsilon,\mathcal{F})\leq s\left(\mathcal{F}^+,k\right),$$

where $k=\left\lceil\dfrac{M}{\epsilon}\log\dfrac{e\epsilon\mathcal{M}^2(\epsilon,\mathcal{F})}{2M}\right\rceil$.

PROOF. Let $\{g_1, g_2, \ldots, g_m\}$ be an arbitrary ϵ-packing of $\mathcal{F}$ of size $m \leq \mathcal{M}(\epsilon,\mathcal{F})$. The proof is in the spirit of the probabilistic method of combinatorics (see, e.g., Spencer (1987)). To prove the inequality, we create k random points on $\mathcal{R}^d \times [0, M]$ in the following way, where k is a positive integer specified later. We generate k independent random variables $S_1, \ldots, S_k$ on $\mathcal{R}^d$ with common distribution

μ, and independently of this, we generate another k independent random variables $T_1, \ldots, T_k$, uniformly distributed on $[0, M]$. This yields k random pairs $(S_1, T_1), \ldots, (S_k, T_k)$. For any two functions g_i and g_j in an ϵ-packing, the probability that the sets $G_i = \{(x, t) : t \leq g_i(x)\}$ and $G_j = \{(x, t) : t \leq g_j(x)\}$ pick the same points from our random set of k points is bounded as follows:

$$\mathbf{P}\{G_i \text{ and } G_j \text{ pick the same points}\}$$

$$= \prod_{l=1}^{k} \left(1 - \mathbf{P}\{(S_l, T_l) \in G_i \triangle G_j\}\right)$$

$$= \left(1 - \mathbf{E}\left\{\mathbf{P}\left\{(S_1, T_1) \in G_i \triangle G_j\right\} \big| S_1\right\}\right)^k$$

$$= \left(1 - \frac{1}{M}\mathbf{E}\{|g_i(S_1) - g_j(S_1)|\}\right)^k$$

$$\leq (1 - \epsilon/M)^k$$

$$\leq e^{-k\epsilon/M},$$

where we used the definition of the functions $g_1, \ldots, g_m$. Observe that the expected number of pairs (g_i, g_j) of these functions, such that the corresponding sets $G_i = \{(x, t) : t \leq g_i(x)\}$ and $G_j = \{(x, t) : t \leq g_j(x)\}$ pick the same points, is bounded by

$$\mathbf{E}\left\{\left|\left\{(g_i, g_j); G_i \text{ and } G_j \text{ pick the same points}\right\}\right|\right\}$$

$$= \binom{m}{2}\mathbf{P}\{G_i \text{ and } G_j \text{ pick the same points}\} \leq \binom{m}{2}e^{-k\epsilon/M}.$$

Since for k randomly chosen points the average number of pairs that pick the same points is bounded by $\binom{m}{2}e^{-k\epsilon/M}$, there exist k points in $\mathcal{R}^d \times [0, M]$, such that the number of pairs (g_i, g_j) that pick the same points is actually bounded by $\binom{m}{2}e^{-k\epsilon/M}$. For each such pair we can add one more point in $\mathcal{R}^d \times [0, M]$ such that the point is contained in $G_i \triangle G_j$. Thus, we have obtained a set of no more than $k + \binom{m}{2}e^{-k\epsilon/M}$ points such that the sets $G_1, \ldots, G_m$ pick different subsets of it. Since k was arbitrary, we can choose it to minimize this expression. This yields $\lfloor \frac{M}{\epsilon} \log\left(e\epsilon\binom{m}{2}/M\right)\rfloor$ points, so the shatter coefficient of $\mathcal{F}^+$ corresponding to this number must be greater than m, which proves the statement. $\square$

The meaning of Theorem 29.3 is best seen from the following simple corollary:

COROLLARY 29.2. *Let $\mathcal{F}$ be a class of $[0, M]$-valued functions on $\mathcal{R}^d$. For every $\epsilon > 0$ and probability measure μ,*

$$\mathcal{M}(\epsilon, \mathcal{F}) \leq \left(\frac{4eM}{\epsilon} \log \frac{2eM}{\epsilon}\right)^{V_{\mathcal{F}^+}}.$$

PROOF. Recall that Theorem 13.2 implies

$$s(\mathcal{F}^+, k) \leq \left(\frac{ek}{V_{\mathcal{F}^+}}\right)^{V_{\mathcal{F}^+}}.$$

The inequality follows from Theorem 29.3 by straightforward calculation. The details are left as an exercise (Problem 29.2). □

Recently Haussler (1991) was able to get rid of the "log" factor in the above upper bound. He proved that if $\epsilon = k/n$ for an integer k, then

$$\mathcal{M}(\epsilon, \mathcal{F}) \leq e(d+1)\left(\frac{2e}{\epsilon}\right)^{V_{\mathcal{F}^+}}.$$

The quantity $V_{\mathcal{F}^+}$ is sometimes called the *pseudo dimension* of $\mathcal{F}$ (see Problem 29.3). It follows immediately from Theorem 13.9 that if $\mathcal{F}$ is a linear space of functions of dimension r, then its pseudo dimension is at most $r + 1$. A few more properties are worth mentioning:

Theorem 29.4. (WENOCUR AND DUDLEY, (1981)). *Let $g : \mathcal{R}^d \to \mathcal{R}$ be an arbitrary function, and consider the class of functions $\mathcal{G} = \{g + f; f \in \mathcal{F}\}$. Then*

$$V_{\mathcal{G}^+} = V_{\mathcal{F}^+}.$$

PROOF. If the points $(s_1, t_1), \ldots, (s_k, t_k) \in \mathcal{R}^d \times \mathcal{R}$ are shattered by $\mathcal{F}^+$, then the points $(s_1, t_1 + g(s_1)), \ldots, (s_k, t_k + g(s_k))$ are shattered by $\mathcal{G}^+$. This proves

$$V_{\mathcal{G}^+} \geq V_{\mathcal{F}^+}.$$

The proof of the other inequality is similar. □

Theorem 29.5. (NOLAN AND POLLARD (1987); DUDLEY, (1987)). *Let $g: [0, M] \to \mathcal{R}$ be a fixed nondecreasing function, and define the class $\mathcal{G} = \{g \circ f; f \in \mathcal{F}\}$. Then*

$$V_{\mathcal{G}^+} \leq V_{\mathcal{F}^+}.$$

PROOF. Assume that $n \leq V_{\mathcal{G}^+}$, and let the functions $f_1, \ldots, f_{2^n} \in \mathcal{F}$ be such that the binary vector

$$\left(I_{\{g(f_j(s_1)) \geq t_1\}}, \ldots, I_{\{g(f_j(s_n)) \geq t_n\}}\right)$$

takes all 2^n values if $j = 1, \ldots, 2^n$. For all $1 \leq i \leq n$ define the numbers

$$u_i = \min_{1 \leq j \leq 2^n} \left\{f_j(s_i) : g(f_j(s_i)) \geq t_i\right\}$$

and

$$l_i = \max_{1 \leq j \leq 2^n} \left\{f_j(s_i) : g(f_j(s_i)) < t_i\right\}.$$

By the monotonicity of g, $u_i > l_i$. Then the binary vector

$$\left(I_{\{f_j(s_1) \geq \frac{u_1 + l_1}{2}\}}, \ldots, I_{\{f_j(s_n) \geq \frac{u_n + l_n}{2}\}} \right)$$

takes the same value as

$$\left(I_{\{g(f_j(s_1)) \geq t_1\}}, \ldots, I_{\{g(f_j(s_n)) \geq t_n\}} \right)$$

for every $j \leq 2^n$. Therefore, the pairs

$$\left(s_1, \frac{u_1 + l_1}{2} \right), \ldots, \left(s_n, \frac{u_n + l_n}{2} \right)$$

are shattered by $\mathcal{F}^+$, which proves the theorem. $\square$

Next we present a few results about covering numbers of classes of functions whose members are sums or products of functions from other classes. Similar results can be found in Nobel (1992), Nolan and Pollard (1987), and Pollard (1990).

Theorem 29.6. *Let $\mathcal{F}_1, \ldots, \mathcal{F}_k$ be classes of real functions on $\mathcal{R}^d$. For n arbitrary, fixed points $z_1^n = (z_1, \ldots, z_n)$ in $\mathcal{R}^d$, define the sets $\mathcal{F}_1(z_1^n), \ldots, \mathcal{F}_k(z_1^n)$ in $\mathcal{R}^n$ by*

$$\mathcal{F}_j(z_1^n) = \left\{ (f_j(z_1), \ldots, f_j(z_n)); f_j \in \mathcal{F}_j \right\},$$

$j = 1, \ldots, k$. *Also, introduce*

$$\mathcal{F}(z_1^n) = \{(f(z_1), \ldots, f(z_n)); f \in \mathcal{F}\}$$

for the class of functions

$$\mathcal{F} = \left\{ f_1 + \ldots + f_k; f_j \in \mathcal{F}_j, \quad j = 1, \ldots, k \right\}.$$

Then for every $\epsilon > 0$ and z_1^n

$$\mathcal{N}(\epsilon, \mathcal{F}(z_1^n)) \leq \prod_{j=1}^{k} \mathcal{N}(\epsilon/k, \mathcal{F}_j(z_1^n)).$$

PROOF. Let $S_1, \ldots, S_k \subset \mathcal{R}^n$ be minimal ϵ/k-coverings of $\mathcal{F}_1(z_1^n), \ldots, \mathcal{F}_k(z_1^n)$, respectively. This implies that for any $f_j \in \mathcal{F}_j$ there is a vector $s_j = (s_j^{(1)}, \ldots, s_j^{(n)}) \in S_j$ such that

$$\frac{1}{n} \sum_{i=1}^{n} \left| f_j(z_i) - s_j^{(i)} \right| < \epsilon/k,$$

for every $j = 1, \ldots, k$. Moreover, $|S_j| = \mathcal{N}(\epsilon/k, \mathcal{F}_j(z_1^n))$. We show that

$$S = \left\{ s_1 + \ldots + s_k; s_j \in S_j, j = 1, \ldots, k \right\}$$

is an ϵ-covering of $\mathcal{F}(z_1^n)$. This follows immediately from the triangle inequality, since for any $f_1, \ldots, f_k$ there is $s_1, \ldots, s_k$ such that

$$\frac{1}{n} \sum_{i=1}^{n} \left| f_1(z_i) + \ldots + f_k(z_i) - (s_1^{(i)} + \ldots + s_k^{(i)}) \right|$$

$$\leq \frac{1}{n} \sum_{i=1}^{n} \left| f_1(z_i) - s_1^{(i)} \right| + \ldots + \frac{1}{n} \sum_{i=1}^{n} \left| f_k(z_i) - s_k^{(i)} \right| < k\frac{\epsilon}{k}. \quad \square$$

Theorem 29.7. (POLLARD (1990)). *Let $\mathcal{F}$ and $\mathcal{G}$ be classes of real functions on $\mathcal{R}^d$, bounded by M_1 and M_2, respectively. (That is, e.g., $|f(x)| \leq M_1$ for every $x \in \mathcal{R}^d$ and $f \in \mathcal{F}$.) For arbitrary fixed points $z_1^n = (z_1, \ldots, z_n)$ in $\mathcal{R}^d$ define the sets $\mathcal{F}(z_1^n)$ and $\mathcal{G}(z_1^n)$ in $\mathcal{R}^n$ as in Theorem 29.6. Introduce*

$$\mathcal{J}(z_1^n) = \{(h(z_1), \ldots, h(z_n)); h \in \mathcal{J}\}$$

for the class of functions

$$\mathcal{J} = \{fg; f \in \mathcal{F}, g \in \mathcal{G}\}.$$

Then for every $\epsilon > 0$ and z_1^n

$$\mathcal{N}(\epsilon, \mathcal{J}(z_1^n)) \leq \mathcal{N}\left(\frac{\epsilon}{2M_2}, \mathcal{F}(z_1^n)\right) \cdot \mathcal{N}\left(\frac{\epsilon}{2M_1}, \mathcal{G}(z_1^n)\right).$$

PROOF. Let $S \subset [-M_1, M_1]^n$ be an $\epsilon/(2M_2)$-covering of $\mathcal{F}(z_1^n)$, that is, for any $f \in \mathcal{F}$ there is a vector $s = (s^{(1)}, \ldots, s^{(n)}) \in S$ such that

$$\frac{1}{n} \sum_{i=1}^{n} \left| f(z_i) - s^{(i)} \right| < \frac{\epsilon}{2M_2}.$$

It is easy to see that S can be chosen such that $|S| = \mathcal{N}(\epsilon/(2M_2), \mathcal{F}(z_1^n))$. Similarly, let $T \subset [-M_2, M_2]$ be an $\epsilon/(2M_1)$-covering of $\mathcal{G}(z_1^n)$ with $|T| = \mathcal{N}(\epsilon/(2M_1), \mathcal{G}(z_1^n))$ such that for any $g \in \mathcal{G}$ there is a $t = (t^{(1)}, \ldots, t^{(n)}) \in T$ with

$$\frac{1}{n} \sum_{i=1}^{n} \left| g(z_i) - t^{(i)} \right| < \frac{\epsilon}{2M_1}.$$

We show that the set

$$U = \{st; s \in S, t \in T\}$$

is an ϵ-covering of $\mathcal{J}(z_1^n)$. Let $f \in \mathcal{F}$ and $g \in \mathcal{G}$ be arbitrary and $s \in S$ and $t \in T$ the corresponding vectors such that

$$\frac{1}{n} \sum_{i=1}^{n} \left| f(z_i) - s^{(i)} \right| < \frac{\epsilon}{2M_2} \quad \text{and} \quad \frac{1}{n} \sum_{i=1}^{n} \left| g(z_i) - t^{(i)} \right| < \frac{\epsilon}{2M_1}.$$

Then

$$\frac{1}{n} \sum_{i=1}^{n} |f(z_i)g(z_i) - s^{(i)}t^{(i)}|$$

$$= \frac{1}{n} \sum_{i=1}^{n} |f(z_i)(t^{(i)} + g(z_i) - t^{(i)}) - s^{(i)}t^{(i)}|$$

$$= \frac{1}{n} \sum_{i=1}^{n} |t^{(i)}(f(z_i) - s^{(i)}) + f(z_i)(g(z_i) - t^{(i)})|$$

$$\leq \frac{M_2}{n} \sum_{i=1}^{n} |f(z_i) - s^{(i)}| + \frac{M_1}{n} \sum_{i=1}^{n} |g(z_i) - t^{(i)}| < \epsilon. \quad \square$$

29.6   Generalized Linear Classification

In this section we use the uniform laws of large numbers discussed in this chapter to prove that squared error minimization over an appropriately chosen class of generalized linear classifiers yields a universally consistent rule. Consider the class $C^{(k_n)}$ of generalized linear classifiers, whose members are functions $\phi : \mathcal{R}^d \to \{0, 1\}$ of the form

$$\phi(x) = \begin{cases} 0 & \text{if } \sum_{j=1}^{k_n} a_j \psi_j(x) \leq 0 \\ 1 & \text{otherwise,} \end{cases}$$

where $\psi_1, \ldots, \psi_{k_n}$ are fixed basis functions, and the coefficients $a_1, \ldots, a_{k_n}$ are arbitrary real numbers. The training sequence D_n is used to determine the coefficients a_i. In Chapter 17 we studied the behavior of the classifier whose coefficients are picked to minimize the empirical error probability

$$\widehat{L}_n = \frac{1}{n} \sum_{i=1}^{n} I_{\{\phi(X_i) \neq Y_i\}}.$$

Instead of minimizing the empirical error probability $\widehat{L}_n(\phi)$, several authors suggested minimizing the empirical squared error

$$\frac{1}{n} \sum_{i=1}^{n} \left(\sum_{j=1}^{k_n} a_j \psi_j(X_i) - (2Y_i - 1) \right)^2$$

(see, e.g., Duda and Hart (1973), Vapnik (1982)). This is rather dangerous. For example, for $k = 1$ and $d = 1$ it is easy to find a distribution such that the error probability of the linear classifier that minimizes the empirical squared error converges to $1 - \epsilon$, while the error probability of the best linear classifier is ϵ, where ϵ is an arbitrarily small positive number (Theorem 4.7). Clearly, similar examples can be found for any fixed k. This demonstrates powerfully the danger of minimizing squared error instead of error count. Minimizing the latter yields a classifier whose average error probability is always within $O\left(\sqrt{\log n/n}\right)$ of the optimum in the class, for fixed k. We note here that in some special cases minimization of the two types of error are equivalent (see Problem 29.5). Interestingly though, if $k_n \to \infty$ as n increases, we can obtain universal consistency by minimizing the empirical squared error.

Theorem 29.8. *Let $\psi_1, \psi_2, \ldots$ be a sequence of bounded functions with $|\psi_j(x)| \leq 1$ such that the set of all finite linear combinations of the ψ_j's*

$$\bigcup_{k=1}^{\infty}\left\{\sum_{j=1}^{k} a_j\psi_j(x); a_1, a_2, \ldots \in \mathcal{R}\right\}$$

is dense in $L_2(\mu)$ on all balls of the form $\{x : \|x\| \leq M\}$ for any probability measure μ. Let the coefficients $a_1^*, \ldots, a_{k_n}^*$ minimize the empirical squared error

$$\frac{1}{n}\sum_{i=1}^{n}\left(\sum_{j=1}^{k_n} a_j\psi_j(X_i) - (2Y_i - 1)\right)^2$$

under the constraint $\sum_{j=1}^{k_n}|a_j| \leq b_n, b_n \geq 1$. Define the generalized linear classifier g_n by

$$g_n(x) = \begin{cases} 0 & \text{if } \sum_{j=1}^{k_n} a_j^*\psi_j(x) \leq 0 \\ 1 & \text{otherwise.} \end{cases}$$

If k_n and b_n satisfy

$$k_n \to \infty, \quad b_n \to \infty \quad \text{and} \quad \frac{k_n b_n^4 \log(b_n)}{n} \to 0,$$

then $\mathbf{E}\{L(g_n)\} \to L^*$ for all distributions of (X, Y), that is, the rule g_n is universally consistent. If we assume additionally that $b_n^4 \log n = o(n)$, then g_n is strongly universally consistent.

PROOF. Let $\delta > 0$ be arbitrary. Then there exists a constant M such that $\mathbf{P}\{\|X\| > M\} < \delta$. Thus,

$$L(g_n) - L^* \leq \delta + \mathbf{P}\{g_n(X) \neq Y, \|X\| \leq M | D_n\} - \mathbf{P}\{g^*(X) \neq Y, \|X\| \leq M\}.$$

It suffices to show that $\mathbf{P}\{g_n(X) \neq Y, \|X\| \leq M | D_n\} - \mathbf{P}\{g^*(X) \neq Y, \|X\| \leq M\} \to 0$ in the required sense for every $M > 0$. Introduce the notation $f_n^*(x) = \sum_{j=1}^{k_n} a_j^*\psi_j(x)$. By Corollary 6.2, we see that

$$\mathbf{P}\{g_n(X) \neq Y, \|X\| \leq M | D_n\} - \mathbf{P}\{g^*(X) \neq Y, \|X\| \leq M\}$$

$$\leq \sqrt{\int_{\|x\| \leq M} \left(f_n^*(x) - (2\eta(x) - 1)\right)^2 \mu(dx)}.$$

We prove that the right-hand side converges to zero in probability. Observe that since $\mathbf{E}\{2Y - 1|X = x\} = 2\eta(x) - 1$, for any function $h(x)$,

$$(h(x) - (2\eta(x) - 1))^2$$
$$= \mathbf{E}\{(h(X) - (2Y - 1))^2|X = x\} - \mathbf{E}\{((2Y - 1) - (2\eta(X) - 1))^2|X = x\}$$

(see Chapter 2), therefore, denoting the class of functions over which we minimize by

$$\mathcal{F}_n = \left\{ \sum_{j=1}^{k_n} a_j \psi_j; \sum_{j=1}^{k_n} |a_j| \leq b_n \right\},$$

we have

$$\int_{\|x\| \leq M} \left(f_n^*(x) - (2\eta(x) - 1) \right)^2 \mu(dx)$$

$$= \mathbf{E}\left\{ \left(f_n^*(X) - (2Y - 1) \right)^2 I_{\{\|X\| \leq M\}} \Big| D_n \right\}$$
$$- \mathbf{E}\left\{ ((2Y - 1) - (2\eta(X) - 1))^2 I_{\{\|X\| \leq M\}} \right\}$$

$$= \left(\mathbf{E}\left\{ \left(f_n^*(X) - (2Y - 1) \right)^2 I_{\{\|X\| \leq M\}} \Big| D_n \right\} \right.$$
$$\left. - \inf_{f \in \mathcal{F}_n} \mathbf{E}\left\{ (f(X) - (2Y - 1))^2 I_{\{\|X\| \leq M\}} \right\} \right)$$
$$+ \inf_{f \in \mathcal{F}_n} \mathbf{E}\left\{ (f(X) - (2Y - 1))^2 I_{\{\|X\| \leq M\}} \right\}$$
$$- \mathbf{E}\left\{ ((2Y - 1) - (2\eta(X) - 1))^2 I_{\{\|X\| \leq M\}} \right\} .$$

The last two terms may be combined to yield

$$\inf_{f \in \mathcal{F}_n} \int_{\|x\| \leq M} (f(x) - (2\eta(x) - 1))^2 \mu(dx),$$

which converges to zero by the denseness assumption. To prove that the first term converges to zero in probability, observe that we may assume without loss of generality that $\mathbf{P}\{\|X\| > M\} = 0$. As in the proof of Lemma 8.2, it is easy to show that

$$\mathbf{E}\left\{ \left(f_n^*(X) - (2Y - 1) \right)^2 I_{\{\|X\| \leq M\}} \Big| D_n \right\}$$
$$- \inf_{f \in \mathcal{F}_n} \mathbf{E}\left\{ (f(X) - (2Y - 1))^2 I_{\{\|X\| \leq M\}} \right\}$$

$$= \mathbf{E}\left\{ \left(f_n^*(X) - (2Y - 1) \right)^2 \Big| D_n \right\} - \inf_{f \in \mathcal{F}_n} \mathbf{E}\left\{ (f(X) - (2Y - 1))^2 \right\}$$

$$\leq 2 \sup_{f \in \mathcal{F}_n} \left| \frac{1}{n} \sum_{i=1}^{n} (f(X_i) - (2Y_i - 1))^2 - \mathbf{E}\left\{ (f(X) - (2Y - 1))^2 \right\} \right|$$

$$= 2 \sup_{h \in \mathcal{J}} \left| \frac{1}{n} \sum_{i=1}^{n} h(X_i, Y_i) - \mathbf{E}\{h(X, Y)\} \right|,$$

where the class of functions $\mathcal{J}$ is defined by

$$\mathcal{J} = \left\{ h(x, y) = (f(x) - (2y - 1))^2 ; f \in \mathcal{F}_n \right\}.$$

Observe that since $|2y - 1| = 1$ and $|\psi_j(x)| \leq 1$, we have

$$0 \leq h(x, y) \leq \left(\sum_{j=1}^{k_n} |a_j| + 1 \right)^2 \leq 2(b_n^2 + 1) \leq 4b_n^2.$$

Therefore, Theorem 29.1 asserts that

$$\mathbf{P} \left\{ \mathbf{E} \left\{ \left(f_n^*(X) - (2Y - 1) \right)^2 \Big| D_n \right\} \right.$$

$$\left. - \inf_{f \in \mathcal{F}_n} \mathbf{E} \left\{ (f(X) - (2Y - 1))^2 \right\} > \epsilon \right\}$$

$$\leq \mathbf{P} \left\{ \sup_{h \in \mathcal{J}} \left| \frac{1}{n} \sum_{i=1}^{n} h(X_i, Y_i) - \mathbf{E}\{h(X, Y)\} \right| > \epsilon/2 \right\}$$

$$\leq 8\mathbf{E} \left\{ \mathcal{N} \left(\frac{\epsilon}{16}, \mathcal{J}(Z_1^n) \right) \right\} e^{-n\epsilon^2/(512(4b_n^2)^2)},$$

where $Z_1^n = (X_1, Y_1), \ldots, (X_n, Y_n)$. Next, for fixed z_1^n, we estimate the covering number $\mathcal{N}\left(\epsilon/16, \mathcal{J}(z_1^n)\right)$. For arbitrary $f_1, f_2 \in \mathcal{F}_n$, consider the functions $h_1(x, y) = (f_1(x) - (2y - 1))^2$ and $h_2(x, y) = (f_2(x) - (2y - 1))^2$. Then for any probability measure v on $\mathcal{R}^d \times \{0, 1\}$,

$$\int |h_1(x, y) - h_2(x, y)| v(d(x, y))$$

$$= \int \left| (f_1(x) - (2y - 1))^2 - (f_2(x) - (2y - 1))^2 \right| v(d(x, y))$$

$$\leq \int 2|f_1(x) - f_2(x)|(b_n + 1) v(d(x, y))$$

$$\leq 4b_n \int |f_1(x) - f_2(x)| \mu(dx),$$

where μ is the marginal measure for v on $\mathcal{R}^d$. Thus, for any $z_1^n = (x_1, y_1), \ldots, (x_n, y_n)$ and ϵ,

$$\mathcal{N}(\epsilon, \mathcal{J}(z_1^n)) \leq \mathcal{N} \left(\frac{\epsilon}{4b_n}, \mathcal{F}_n(x_1^n) \right).$$

Therefore, it suffices to estimate the covering number corresponding to $\mathcal{F}_n$. Since $\mathcal{F}_n$ is a subset of a linear space of functions, we have $V_{\mathcal{F}_n^+} \leq k_n + 1$ (Theorem 13.9). By Corollary 29.2,

$$\mathcal{N} \left(\frac{\epsilon}{4b_n}, \mathcal{F}_n(x_1^n) \right) \leq \left(\frac{8eb_n}{\epsilon/(4b_n)} \log \frac{4eb_n}{\epsilon/(4b_n)} \right)^{k_n+1} \leq \left(\frac{32eb_n^2}{\epsilon} \right)^{2(k_n+1)}.$$

Summarizing, we have

$$\mathbf{P}\left\{ \mathbf{E}\left\{ \left. \left(f_n^*(X) - (2Y - 1) \right)^2 \right| D_n \right\} \right.$$

$$\left. - \inf_{f \in \mathcal{F}_n} \mathbf{E}\left\{ (f(X) - (2Y - 1))^2 \right\} > \epsilon \right\}$$

$$\leq \ 8 \left(\frac{2^{10} e b_n^2}{\epsilon} \right)^{2(k_n+1)} e^{-n\epsilon^2/(2^{13} b_n^4)},$$

which goes to zero if $k_n b_n^4 \log(b_n)/n \to 0$. The proof of the theorem is completed.

It is easy to see that if we assume additionally that $b_n^4 \log n/n \to 0$, then strong universal consistency follows by applying the Borel-Cantelli lemma to the last probability. $\Box$

REMARK. Minimization of the squared error is attractive because there are efficient algorithms to find the minimizing coefficients, while minimizing the number of errors committed on the training sequence is computationally more difficult. If the dimension k of the generalized linear classifier is fixed, then *stochastic approximation* asymptotically provides the minimizing coefficients. For more information about this we refer to Robbins and Monro (1951), Kiefer and Wolfowitz (1952), Dvoretzky (1956), Fabian (1971), Tsypkin (1971), Nevelson and Khasminskii (1973), Kushner (1984), Ruppert (1991), and Ljung, Pflug, and Walk (1992). For example, Györfi (1984) proved that if $(U_1, V_1), (U_2, V_2), \ldots$ form a stationary and ergodic sequence, in which each pair is distributed as the bounded random variable pair $(U, V) \in \mathcal{R}^k \times \mathcal{R}$, and the vector of coefficients $a = (a_1, \ldots, a_k)$ minimizes

$$R(a) = \mathbf{E}\left\{ \left(a^T U - V \right)^2 \right\},$$

and $a^{(0)} \in \mathcal{R}^k$ is arbitrary, then the sequence of coefficient vectors defined by

$$a^{(n+1)} = a^{(n)} - \frac{1}{n+1} \left(a^{(n)^T} U_{n+1} - V_{n+1} \right) U_{n+1}$$

satisfies

$$\lim_{n \to \infty} R(a^{(n)}) = R(a) \ \text{a.s.} \ \Box$$

Problems and Exercises

PROBLEM 29.1. Find a class $\mathcal{F}$ containing two functions $\eta_1, \eta_2 : \mathcal{R} \to [0, 1]$ and a distribution of (X, Y) such that $\min(L(\eta_1), L(\eta_2)) = L^*$, but as $n \to \infty$, the probability

$$\mathbf{P}\{L(\eta_n) = \max(L(\eta_1), L(\eta_2))\}$$

converges to one, where η_n is selected from $\mathcal{F}$ by minimizing the empirical squared error.

PROBLEM 29.2. Prove Corollary 29.2.

PROBLEM 29.3. Let $\mathcal{F}$ be a class of functions on $\mathcal{R}^d$, taking their values in $[0, M]$. Haussler (1992) defines the pseudo dimension of $\mathcal{F}$ as the largest integer n for which there exist n points in $\mathcal{R}^d$, $z_1, \ldots, z_n$, and a vector $v = \left(v^{(1)}, \ldots, v^{(n)}\right) \in \mathcal{R}^n$ such that the binary n-vector

$$\left(I_{\{v^{(1)}+f(z_1)>0\}}, \ldots, I_{\{v^{(n)}+f(z_n)>0\}}\right)$$

takes all 2^n possible values as f ranges through $\mathcal{F}$. Prove that the pseudo dimension of $\mathcal{F}$ equals the quantity $V_{\mathcal{F}^+}$ defined in the text.

PROBLEM 29.4. CONSISTENCY OF CLUSTERING. Let $X, X_1, \ldots, X_n$ be i.i.d. random variables in $\mathcal{R}^d$, and assume that there is a number $0 < M < \infty$ such that $P\{X \in [-M, M]^d\} = 1$. Take the *empirically optimal clustering* of $X_1, \ldots, X_n$, that is, find the points $a_1, \ldots, a_k$ that minimize the *empirical squared error*:

$$e_n(a_1, \ldots, a_k) = \frac{1}{n} \sum_{i=1}^{n} \min_{1 \le j \le k} \|a_j - X_i\|^2.$$

The error of the clustering is defined by the mean squared error

$$e(a_1, \ldots, a_k) = \mathbf{E}\left\{ \min_{1 \le j \le k} \|a_j - X\|^2 \,\Big|\, X_1, \ldots, X_n \right\}.$$

Prove that if $a_1, \ldots, a_k$ denote the empirically optimal cluster centers, then

$$e(a_1, \ldots, a_k) - \inf_{b_1, \ldots, b_k \in \mathcal{R}^d} e(b_1, \ldots, b_k) \le 2 \sup_{b_1, \ldots, b_k \in \mathcal{R}^d} |e_n(b_1, \ldots, b_k) - e(b_1, \ldots, b_k)|,$$

and that for every $\epsilon > 0$

$$\mathbf{P}\left\{ \sup_{b_1, \ldots, b_k \in \mathcal{R}^d} |e_n(b_1, \ldots, b_k) - e(b_1, \ldots, b_k)| > \epsilon \right\} \le 4e^8 n^{2k(d+1)} e^{-n\epsilon^2/(32M^4)}.$$

Conclude that the error of the empirically optimal clustering converges to that of the truly optimal one as $n \to \infty$. (Pollard (1981; 1982), Linder, Lugosi, and Zeger (1994)). HINT: For the first part proceed as in the proof of Lemma 8.2. For the second part use the technique shown in Corollary 29.1. To compute the VC dimension, exploit Corollary 13.2.

PROBLEM 29.5. Let $\psi_1, \ldots, \psi_k$ be indicator functions of cells of a k-way partition of $\mathcal{R}^d$. Consider generalized linear classifiers based on these functions. Show that the classifier obtained by minimizing the number of errors made on the training sequence is the same as for the classifier obtained by minimizing the empirical squared error. Point out that this is just the histogram classifier based on the partition defined by the ψ_i's (Csibi (1975)).

30

Neural Networks

30.1 / Multilayer Perceptrons

The linear discriminant or perceptron (see Chapter 4) makes a decision

$$\phi(x) = \begin{cases} 0 & \text{if } \psi(x) \le 1/2 \\ 1 & \text{otherwise,} \end{cases}$$

based upon a linear combination $\psi(x)$ of the inputs,

$$\psi(x) = c_0 + \sum_{i=1}^{d} c_i x^{(i)} = c_0 + c^T x, \tag{30.1}$$

where the c_i's are weights, $x = (x^{(1)}, \ldots, x^{(d)})^T$, and $c = (c_1, \ldots, c_d)^T$. This is called a neural network without hidden layers (see Figure 4.1).

In a (feed-forward) neural network with one hidden layer, one takes

$$\psi(x) = c_0 + \sum_{i=1}^{k} c_i \sigma(\psi_i(x)), \tag{30.2}$$

where the c_i's are as before, and each ψ_i is of the form given in (30.1): $\psi_i(x) = b_i + \sum_{j=1}^{d} a_{ij} x^{(j)}$ for some constants b_i and a_{ij}. The function σ is called a *sigmoid*. S 状的

We define sigmoids to be nondecreasing functions with $\sigma(x) \to -1$ as $x \downarrow -\infty$ and $\sigma(x) \to 1$ as $x \uparrow \infty$. Examples include:

(1) the *threshold* sigmoid

$$\sigma(x) = \begin{cases} -1 & \text{if } x \le 0 \\ 1 & \text{if } x > 0; \end{cases}$$

(2) the *standard*, or *logistic*, sigmoid

$$\sigma(x) = \frac{1 - e^{-x}}{1 + e^{-x}};$$

(3) the *arctan* sigmoid

$$\sigma(x) = \frac{2}{\pi} \arctan(x);$$

(4) the *gaussian* sigmoid

$$\sigma(x) = 2 \int_{-\infty}^{x} \frac{1}{\sqrt{2\pi}} e^{-u^2/2} du - 1.$$

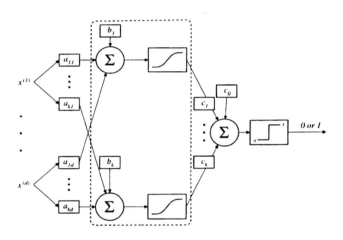

FIGURE 30.1. *A neural network with one hidden layer. The hidden neurons are those within the frame.*

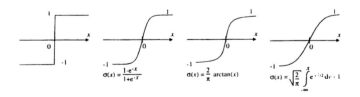

FIGURE 30.2. *The threshold, standard, arctan, and gaussian sigmoids.*

For early discussion of multilayer perceptrons, see Rosenblatt (1962), Barron (1975), Nilsson (1965), and Minsky and Papert (1969). Surveys may be found in Barron and Barron (1988), Ripley (1993; 1994), Hertz, Krogh, and Palmer (1991), and Weiss and Kulikowski (1991).

In the perceptron with one hidden layer, we say that there are k hidden neurons—the output of the i-th hidden neuron is $u_i = \sigma(\psi_i(x))$. Thus, (30.2) may be rewritten as

$$\psi(x) = c_0 + \sum_{i=1}^{k} c_i u_i,$$

which is similar in form to (30.1). We may continue this process and create multi-layer feed-forward neural networks. For example, a two-hidden-layer perceptron uses

$$\psi(x) = c_0 + \sum_{i=1}^{l} c_i z_i,$$

where

$$z_i = \sigma\left(d_{i0} + \sum_{j=1}^{k} d_{ij} u_j\right), \quad 1 \le i \le l,$$

and

$$u_j = \sigma\left(b_j + \sum_{i=1}^{d} a_{ji} x^{(i)}\right), \quad 1 \le j \le k,$$

and the d_{ij}'s, b_j's, and a_{ji}'s are constants. The first hidden layer has k hidden neurons, while the second hidden layer has l hidden neurons.

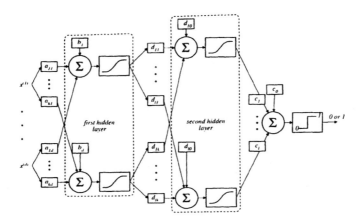

FIGURE 30.3. *A feed-forward neural network with two hidden layers.*

The step from perceptron to a one-hidden-layer neural network is nontrivial. We know that linear discriminants cannot possibly lead to universally consistent rules. Fortunately, one-hidden-layer neural networks yield universally consistent discriminants provided that we allow k, the number of hidden neurons, to grow unboundedly with n. The interest in neural networks is undoubtedly due to the possibility of implementing them directly via processors and circuits. As the hardware is fixed beforehand, one does not have the luxury to let k become a function of n, and thus, the claimed universal consistency is a moot point. We will deal with both fixed architectures and variable-sized neural networks. Because of the universal consistency of one-hidden-layer neural networks, there is little theoretical gain in considering neural networks with more than one hidden layer. There may, however, be an information-theoretic gain as the number of hidden neurons needed to achieve the same performance may be substantially reduced. In fact, we will make a case for two hidden layers, and show that after two hidden layers, little is gained for classification.

For theoretical analysis, the neural networks are rooted in a classical theorem by Kolmogorov (1957) and Lorentz (1976) which states that every continuous function f on $[0, 1]^d$ can be written as

$$
f(x) = \sum_{i=1}^{2d+1} F_i \left(\sum_{j=1}^{d} G_{ij}(x^{(j)}) \right),
$$

where the G_{ij}'s and the F_i's are continuous functions whose form depends on f. We will see that neural networks approximate any measurable function with arbitrary precision, despite the fact that the form of the sigmoids is fixed beforehand.

As an example, consider $d = 2$. The function $x^{(1)}x^{(2)}$ is rewritten as

$$
\frac{1}{4} \left(\left(x^{(1)} + x^{(2)}\right)^2 - \left(x^{(1)} - x^{(2)}\right)^2 \right),
$$

which is in the desired form. However, it is much less obvious how one would rewrite more general continuous functions. In fact, in neural networks, we approximate the G_{ij}'s and F_i's by functions of the form $\sigma(b + a^T x)$ and allow the number of tunable coefficients to be high enough such that any continuous function may be represented—though no longer rewritten exactly in the form of Kolmogorov and Lorentz. We discuss other examples of approximations based upon such representations in a later section.

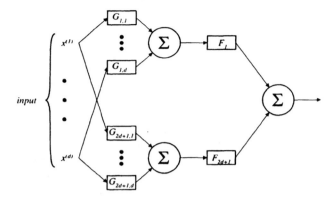

FIGURE 30.4. *The general Kolmogorov-Lorentz representation of a continuous function.*

30.2 Arrangements

A finite set $\mathcal{A}$ of hyperplanes in $\mathcal{R}^d$ partitions the space into connected convex polyhedral pieces of various dimensions. Such a partition $\mathcal{P} = \mathcal{P}(\mathcal{A})$ is called an *arrangement*. An arrangement is called *simple* if any d hyperplanes of $\mathcal{A}$ have a unique point in common and if $d + 1$ hyperplanes have no point in common.

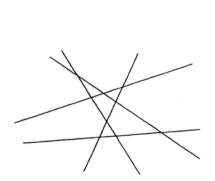

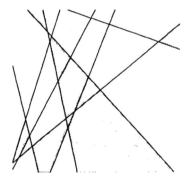

FIGURE 30.5. *An arrangement of five lines in the plane.*

FIGURE 30.6. *An arrangement classifier.*

A simple arrangement creates polyhedral cells. Interestingly, the number of these cells is independent of the actual configuration of the hyperplanes. In particular, the number of cells is exactly 2^k if $d \geq k$, and

$$\sum_{i=0}^{d} \binom{k}{i} \quad \text{if } d < k,$$

where $|\mathcal{A}| = k$. For a proof of this, see Problem 22.1, or Lemma 1.2 of Edelsbrunner (1987). For general arrangements, this is merely an upper bound.

We may of course use arrangements for designing classifiers. We let $g_{\mathcal{A}}$ be the natural classifier obtained by taking majority votes over all Y_i's for which X_i is in the same cell of the arrangement $\mathcal{P} = \mathcal{P}(\mathcal{A})$ as x.

All classifiers discussed in this section possess the property that they are invariant under linear transformations and universally consistent (in some cases, we assume that X has a density, but that is only done to avoid messy technicalities).

If we fix k and find that $\mathcal{A}$ with $|\mathcal{A}| = k$ for which the empirical error

$$L_n^{(R)} = \frac{1}{n} \sum_{i=1}^{n} I_{\{g_{\mathcal{A}}(X_i) \neq Y_i\}}$$

is minimal, we obtain—perhaps at great computational expense—the empirical risk optimized classifier. There is a general theorem for such classifiers—see, for example, Corollary 23.2—the conditions of which are as follows:

(1) It must be possible to select a given sequence of $\mathcal{A}$'s for which $L_n(g_{\mathcal{A}})$ (the conditional probability of error with $g_{\mathcal{A}}$) tends to L^* in probability. But if $k \to \infty$, we may align the hyperplanes with the axes, and create a cubic histogram, for which, by Theorem 6.2, we have consistency if the grid expands to ∞ and the cell sizes in the grid shrink to 0. Thus, as $k \to \infty$, this condition holds trivially.

(2) The collection $\mathcal{G} = \{g_{\mathcal{A}}\}$ is not too rich, in the sense that $n/\log \mathcal{S}(\mathcal{G}, n) \to \infty$, where $\mathcal{S}(\mathcal{G}, n)$ denotes the shatter coefficient of $\mathcal{G}$, that is, the maximal number of ways $(x_1, y_1), \ldots, (x_n, y_n)$ can be split by sets of the form

$$\left(\bigcup_{A \in \mathcal{P}(\mathcal{A})} A \times \{0\} \right) \cup \left(\bigcup_{A \in \mathcal{P}(\mathcal{A})} A \times \{1\} \right).$$

If $|\mathcal{A}| = 1$, we know that $\mathcal{S}(\mathcal{G}, n) \leq 2(n^d + 1)$ (see Chapter 13). For $|\mathcal{A}| = k$, a trivial upper bound is

$$\left(2(n^d + 1) \right)^k.$$

The consistency condition is fulfilled if $k = o(n/\log n)$.

We have

Theorem 30.1. *The empirical-risk-optimized arrangement classifier based upon arrangements with $|\mathcal{A}| \leq k$ has $\mathbf{E}\{L_n\} \to L^*$ for all distributions if $k \to \infty$ and $k = o(n/\log n)$.*

Arrangements can also be made from the data at hand in a simpler way. Fix k points $X_1, \ldots, X_k$ in general position and look at all possible $\binom{k}{d}$ hyperplanes you can form with these points. These form your collection $\mathcal{A}$, which defines your arrangement. No optimization of any kind is performed. We take the natural classifier obtained by a majority vote within the cells of the partition over $(X_{k+1}, Y_{k+1}), \ldots, (X_n, Y_n)$.

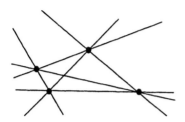

FIGURE 30.7. *Arrangement determined by* $k = 4$ *data points on the plane.*

Here we cannot apply the powerful consistency theorem mentioned above. Also, the arrangement is no longer simple. Nevertheless, the partition of space depends on the X_i's only, and thus Theorem 6.1 (together with Lemma 20.1) is useful. The rule thus obtained is consistent if $\text{diam}(A(X)) \to 0$ in probability and the number of cells is $o(n)$, where $A(X)$ is the cell to which X belongs in the arrangement. As the number of cells is certainly not more than

$$\sum_{i=0}^{d} \binom{k'}{i},$$

where $k' = \binom{k}{d}$, we see that the number of cells divided by n tends to zero if

$$k^{d^2}/n \to 0.$$

This puts a severe restriction on the growth of k. However, it is easy to prove the following:

Lemma 30.1. *If $k \to \infty$, then* $\text{diam}(A(X)) \to 0$ *in probability whenever X has a density.*

PROOF. As noted in Chapter 20 (see Problem 20.6), the set of all x for which for all $\epsilon > 0$, we have $\mu(x + \epsilon Q_i) > 0$ for all quadrants $Q_1, \ldots, Q_{2^d}$ having one vertex at $(0, 0, \ldots, 0)$ and sides of length one, has μ-measure one. For such x, if at least one of the X_i's ($i \leq k$) falls in each of the 2^d quadrants $x + \epsilon Q_i$, then $\text{diam}(A(x)) \leq 2d\epsilon$ (see Figure 30.8).

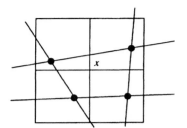

FIGURE 30.8. *The diameter of the cell containing* x *is less than* 4ϵ *if there is a data point in each of the four quadrants of size* ϵ *around* x.

Therefore, for arbitrary $\epsilon > 0$,

$$\mathbf{P}\{\operatorname{diam}(A(x)) > 2d\epsilon\} \leq 2^d \left(1 - \min_{1 \leq i \leq 2^d} \mu(x + \epsilon Q_i)\right)^k \to 0.$$

Thus, by the Lebesgue dominated convergence theorem,

$$\mathbf{P}\{\operatorname{diam}(A(X)) > 2d\epsilon\} \to 0. \quad \square$$

Theorem 30.2. *The arrangement classifier defined above is consistent whenever* X *has a density and*

$$k \to \infty, \quad k^{d^2} = o(n).$$

The theorem points out that empirical error minimization over a finite set of arrangements can also be consistent. Such a set may be formed as the collection of arrangements consisting of hyperplanes through d points of $X_1, \ldots, X_k$. As nothing new is added here to the discussion, we refer the reader to Problem 30.1.

So how do we deal with arrangements in a computer? Clearly, to reach a cell, we find for each hyperplane $A \in \mathcal{A}$ the side to which x belongs. If $f(x) = a^T x + a_0$, then $f(x) > 0$ in one halfplane, $f(x) = 0$ on the hyperplane, and $f(x) < 0$ in the other halfplane. If $\mathcal{A} = \{A_1, \ldots, A_k\}$, the vector $(I_{\{H_1(x)>0\}}, \ldots, I_{\{H_k(x)>0\}})$ describes the cell to which x belongs, where $H_i(x)$ is a linear function that is positive if x is on one side of the hyperplane A_i, negative if x is on the other side of A_i, and 0 if $x \in A_i$. A decision is thus reached in time $O(kd)$. More importantly, the whole process is easily parallelizable and can be pictured as a battery of perceptrons. It is easy to see that the classifier depicted in Figure 30.9 is identical to the arrangement classifier. In neural network terminology, the first hidden layer of neurons corresponds to just k perceptrons (and has $k(d+1)$ weights or parameters, if you wish). The first layer outputs a k-vector of bits that pinpoints the precise location of x in the cells of the arrangement. The second layer only assigns a class (decision) to each cell of the arrangement by firing up one neuron. It has 2^k neurons (for class assignments), but of course, in natural classifiers, these neurons do not require training or learning—the majority vote takes care of that.

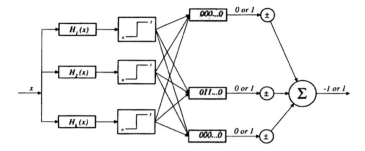

FIGURE 30.9. *Arrangement classifier realized by a two-hidden-layer neural network. Each of the 2^k cells in the second hidden layer performs an "and" operation: the output of node "101" is 1 if its three inputs are 1,0, and 1, respectively. Otherwise its output is 0. Thus, one and only one of the 2^k outputs is 1.*

If a more classical second layer is needed—without boolean operations—let $b = (b_1, \ldots, b_k)$ be the k-vector of bits seen at the output of the first layer. Assign a perceptron in the second layer to each region of the arrangement and define the output $z \in \{-1, 1\}$ to be $\sigma \left(\sum_{i=1}^{k} c_j b_j - k + 1/2 \right)$, where $c_j \in \{-1, 1\}$ are weights. For each region of the arrangement, we have a description in terms of $c = (c_1, \ldots, c_k)$. The argument of the sigmoid function is $1/2$ if $2b_j - 1 = c_j$ for all j and is negative otherwise. Hence $z = 1$ if and only if $2b - 1 = c$. Assume we now take a decision based upon the sign of

$$\sum_l w_l z_l + w_0,$$

where the w_l's are weights and the z_l's are the outputs of the second hidden layer. Assume that we wish to assign class 1 to s regions in the arrangement and class 0 to t other regions. For a class 1 region l, set $w_l = 1$, and for a class 0 region, set $w_l = -1$. Define $w_0 = 1 + s - t$. Then, if $z_j = 1$, $z_i = -1$, $i \neq j$,

$$\sum_l w_l z_l + w_0 = w_j + w_0 - \sum_{i \neq j} w_i = \begin{cases} 1 & \text{if } w_j = 1 \\ -1 & \text{if } w_j = -1. \end{cases}$$

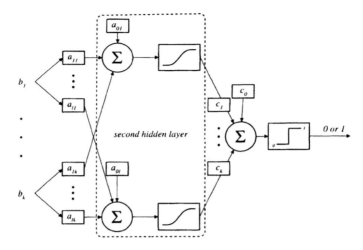

FIGURE 30.10. *The second hidden layer of a two-hidden-layer neural network with threshold sigmoids in the first layer. For each k-vector of bits $b = (b_1, \ldots, b_k)$ at the output of the first layer, we may find a decision $g(b) \in \{0, 1\}$. Now return to the two-hidden-layer network of Figure 30.9 and assign the values $g(b)$ to the neurons in the second hidden layer to obtain an equivalent network.*

Thus, every arrangement classifier corresponds to a neural network with two hidden layers, and threshold units. The correspondence is also reciprocal. Assume someone shows a two-hidden-layer neural network with the first hidden layer as above—thus, outputs consist of a vector of k bits—and with the second hidden layer consisting once again a battery of perceptrons (see Figure 30.10). Whatever happens in the second hidden layer, the decision is just a function of the configuration of k input bits. The output of the first hidden layer is constant over each region of the arrangement defined by the hyperplanes given by the input weights of the units of the first layer. Thus, the neural network classifier with threshold units in the first hidden layer is equivalent to an arrangement classifier with the same number of hyperplanes as units in the first hidden layer. The equivalence with tree classifiers is described in Problem 30.2.

Of course, equivalence is only valid up to a certain point. If the number of neurons in the second layer is small, then neural networks are more restricted. This could be an advantage in training. However, the majority vote in an arrangement classifier avoids training of the second layer's neurons altogether, and offers at the same time an easier interpretation of the classifier. Conditions on consistency of general two-hidden-layer neural networks will be given in Section 30.4.

30.3 Approximation by Neural Networks

Consider first the class $C^{(k)}$ of classifiers (30.2) that contains all neural network classifiers with the threshold sigmoid and k hidden nodes in two hidden layers. The training data D_n are used to select a classifier from $C^{(k)}$. For good performance of the selected rule, it is necessary that the best rule in $C^{(k)}$ has probability of error close to L^*, that is, that

$$\inf_{\phi \in C^{(k)}} L(\phi) - L^*$$

is small. We call this quantity the *approximation error*. Naturally, for fixed k, the approximation error is positive for most distributions. However, for large k, it is expected to be small. The question is whether the last statement is true for all distributions of (X, Y). We showed in the section on arrangements that the class of two-hidden-layer neural network classifiers with m nodes in the first layer and 2^m nodes in the second layer contains all arrangement classifiers with m hyperplanes. Therefore, for $k = m + 2^m$, the class of all arrangement classifiers with m hyperplanes is a subclass of $C^{(k)}$. From this, we easily obtain the following approximation result:

Theorem 30.3. *If $C^{(k)}$ is the class of all neural network classifiers with the threshold sigmoid and k neurons in two hidden layers, then*

$$\lim_{k \to \infty} \inf_{\phi \in C^{(k)}} L(\phi) - L^* = 0$$

for all distributions of (X, Y).

It is more surprising that the same property holds if $C^{(k)}$ is the class of one-hidden-layer neural networks with k hidden neurons, and an arbitrary sigmoid. More precisely, $C^{(k)}$ is the class of classifiers

$$\phi(x) = \begin{cases} 0 & \text{if } \psi(x) \leq 1/2 \\ 1 & \text{otherwise,} \end{cases}$$

where ψ is as in (30.2).

By Theorem 2.2, we have

$$L(\phi) - L^* \leq 2\mathbf{E}\{|\psi(X) - \eta(X)|\},$$

where $\eta(x) = \mathbf{P}\{Y = 1 | X = x\}$. Thus, $\inf_{\phi \in C^{(k)}} L(\phi) - L^* \to 0$ as $k \to \infty$ if

$$\mathbf{E}\{|\psi_k(x) - \eta(x)|\} \to 0$$

for some sequence $\{\psi_k\}$ with $\phi_k \in C^{(k)}$ for $\phi_k(x) = I_{\{\psi_k(x) > 1/2\}}$. For universal consistency, we need only assure that the family of ψ's can approximate any η in the $L_1(\mu)$ sense. In other words, the approximation error $\inf_{\phi \in C^{(k)}} L(\phi) - L^*$ converges to zero if the class of functions ψ is dense in $L_1(\mu)$ for every μ. Another sufficient condition for this—but of course much too severe—is that the class

$\mathcal{F}$ of functions ψ becomes dense in the L_∞ (supremum-) norm in the space of continuous functions $C[a, b]^d$ on $[a, b]^d$, where $[a, b]^d$ denotes the hyperrectangle of $\mathcal{R}^d$ defined by its opposite vertices a and b, for any a and b.

Lemma 30.2. *Assume that a sequence of classes of functions $\mathcal{F}_k$ becomes dense in the L_∞ norm in the space of continuous functions $C[a, b]^d$ (where $[a, b]^d$ is the hyperrectangle of $\mathcal{R}^d$ defined by a, b). In other words, assume that for every $a, b \in \mathcal{R}^d$, and every bounded function g,*

$$\lim_{k \to \infty} \inf_{f \in \mathcal{F}_k} \sup_{x \in [a,b]^d} |f(x) - g(x)| = 0.$$

Then for any distribution of (X, Y),

$$\lim_{k \to \infty} \inf_{\phi \in \mathcal{C}^{(k)}} L(\phi) - L^* = 0,$$

where $\mathcal{C}^{(k)}$ is the class of classifiers $\phi(x) = I_{\{\psi(x) > 1/2\}}$ for $\psi \in \mathcal{F}_k$.

PROOF. For fixed $\epsilon > 0$, find a, b such that $\mu([a, b]^d) \geq 1 - \epsilon/3$, where μ is the probability measure of X. Choose a continuous function $\widehat{\eta}$ vanishing off $[a, b]^d$ such that

$$\mathbf{E}\{|\eta(X) - \widehat{\eta}(X)|\} \leq \frac{\epsilon}{6}.$$

Find k and $f \in \mathcal{F}_k$ such that

$$\sup_{x \in [a,b]^d} |f(x) - \widehat{\eta}(x)| \leq \frac{\epsilon}{6}.$$

For $\phi(x) = I_{\{f(x) > 1/2\}}$, we have, by Theorem 2.2,

$$
\begin{aligned}
L(\phi) - L^* &\leq 2\mathbf{E}\left\{|f(X) - \eta(X)|I_{\{X \in [a,b]^d\}}\right\} + \frac{\epsilon}{3} \\
&\leq 2\mathbf{E}\left\{|f(X) - \widehat{\eta}(X)|I_{\{X \in [a,b]^d\}}\right\} + 2\mathbf{E}\{|\widehat{\eta}(X) - \eta(X)|\} + \frac{\epsilon}{3} \\
&\leq 2\sup_{x \in [a,b]^d} |f(x) - \widehat{\eta}(x)| + +2\mathbf{E}\{|\widehat{\eta}(X) - \eta(X)|\} + \frac{\epsilon}{3} \\
&\leq \epsilon. \quad \square
\end{aligned}
$$

This text is basically about all such good families, such as families that are obtainable by summing kernel functions, and histogram families. The first results for approximation with neural networks with one hidden layer appeared in 1989, when Cybenko (1989), Hornik, Stinchcombe, and White (1989), and Funahashi (1989) proved independently that feedforward neural networks with one hidden layer are dense with respect to the supremum norm on bounded sets in the set of continuous functions. In other words, by taking k large enough, every continuous function on $\mathcal{R}^d$ can be approximated arbitrarily closely, uniformly over any bounded set by functions realized by neural networks with one hidden layer. For a survey of various denseness results we refer to Barron (1989) and Hornik (1993). The proof given here uses ideas of Chen, Chen, and Liu (1990). It uses the denseness of the class of trigonometric polynomials in the L_∞ sense for $C[0, 1]^d$ (this is a special case of the Stone-Weierstrass theorem; see Theorem A.9 in the Appendix), that is, by functions of the form

$$\sum_i \left(\alpha_i \cos(2\pi a_i^T x) + \beta_i \sin(2\pi b_i^T x) \right),$$

where a_i, b_i are integer-valued vectors of $\mathcal{R}^d$.

Theorem 30.4. *For every continuous function $f : [a, b]^d \to \mathcal{R}$ and for every $\epsilon > 0$, there exists a neural network with one hidden layer and function $\psi(x)$ as in (30.2) such that*

$$\sup_{x \in [a,b]^d} |f(x) - \psi(x)| < \epsilon.$$

PROOF. We prove the theorem for the threshold sigmoid

$$\sigma(x) = \begin{cases} -1 & \text{if } x \leq 0 \\ 1 & \text{if } x > 0. \end{cases}$$

The extension to general nondecreasing sigmoids is left as an exercise (Problem 30.3). Fix $\epsilon > 0$. We take the Fourier series approximation of $f(x)$. By the Stone-Weierstrass theorem (Theorem A.9), there exists a large positive integer M, nonzero real coefficients $\alpha_1, \ldots, \alpha_M, \beta_1, \ldots, \beta_M$, and integers $m_{i,j}$ for $i = 1, \ldots, M$, $j = 1, \ldots, d$, such that

$$\sup_{x \in [a,b]^d} \left| \sum_{i=1}^M \left(\alpha_i \cos\left(\frac{\pi}{a} m_i^T x\right) + \beta_i \sin\left(\frac{\pi}{a} m_i^T x\right) \right) - f(x) \right| < \frac{\epsilon}{2},$$

where $m_i = (m_{i,1}, \ldots, m_{i,d})$, $i = 1, \ldots, M$. It is clear that every continuous function on the real line that is zero outside some bounded interval can be arbitrarily closely approximated uniformly on the interval by one-dimensional neural networks, that is, by functions of the form

$$\sum_{i=1}^k c_i \sigma(a_i x + b_i) + c_0.$$

Just observe that the indicator function of an interval $[b, c]$ may be written as $\sigma(x - b) + \sigma(-x + c)$. This implies that bounded functions such as sin and cos can be approximated arbitrarily closely by neural networks. In particular, there exist neural networks $u_i(x), v_i(x)$ with $i = 1, \ldots, M$, (i.e., mappings from $\mathcal{R}^d$ to $\mathcal{R}$) such that

$$\sup_{x \in [a,b]^d} \left| u_i(x) - \cos\left(\frac{\pi}{a} m_i^T x\right) \right| < \frac{\epsilon}{4M |\alpha_i|}$$

and

$$\sup_{x \in [a,b]^d} \left| v_i(x) - \sin\left(\frac{\pi}{a} m_i^T x\right) \right| < \frac{\epsilon}{4M|\beta_i|}.$$

Therefore, applying the triangle inequality we get

$$\sup_{x \in [a,b]^d} \left| \sum_{i=1}^{M} \left(\alpha_i \cos\left(\frac{\pi}{a} m_i^T x\right) + \beta_i \sin\left(\frac{\pi}{a} m_i^T x\right) \right) - \sum_{i=1}^{M} (\alpha_i u_i(x) + \beta_i v_i(x)) \right| < \frac{\epsilon}{2}.$$

Since the u_i's and v_i's are neural networks, their linear combination

$$\psi(x) = \sum_{i=1}^{M} (\alpha_i u_i(x) + \beta_i v_i(x))$$

is a neural network too and, in fact,

$$\sup_{x \in [a,b]^d} |f(x) - \psi(x)|$$

$$\leq \sup_{x \in [a,b]^d} \left| f(x) - \sum_{i=1}^{M} \left(\alpha_i \cos\left(\frac{\pi}{a} m_i^T x\right) + \beta_i \sin\left(\frac{\pi}{a} m_i^T x\right) \right) \right|$$

$$+ \sup_{x \in [a,b]^d} \left| \sum_{i=1}^{M} \left(\alpha_i \cos\left(\frac{\pi}{a} m_i^T x\right) + \beta_i \sin\left(\frac{\pi}{a} m_i^T x\right) \right) - \psi(x) \right|$$

$$< \frac{2\epsilon}{2} = \epsilon. \quad \square$$

The convergence may be arbitrarily slow for some f. By restricting the class of functions, it is possible to obtain upper bounds for the rate of convergence. For an example, see Barron (1993). The following corollary of Theorem 30.4 is obtained via Lemma 30.2:

COROLLARY 30.1. *Let $C^{(k)}$ contain all neural network classifiers defined by networks of one hidden layer with k hidden nodes, and an arbitrary sigmoid σ. Then for any distribution of (X, Y),*

$$\lim_{k \to \infty} \inf_{\phi \in C^{(k)}} L(\phi) - L^* = 0.$$

The above convergence also holds if the range of the parameters a_{ij}, b_i, c_i is restricted to an interval $[-\beta_k, \beta_k]$, where $\lim_{k \to \infty} \beta_k = \infty$.

REMARK. It is also true that the class of one-hidden-layer neural networks with k hidden neurons becomes dense in $L_1(\mu)$ for every probability measure μ on $\mathcal{R}^d$ as $k \to \infty$ (see Problem 30.4). Then Theorem 2.2 may be used directly to prove Corollary 30.1. $\square$

In practice, the network architecture (i.e., k in our case) is given to the designer, who can only adjust the parameters a_{ij}, b_i, and c_i, depending on the data D_n. In this respect, the above results are only of theoretical interest. It is more interesting to find out how far the error probability of the chosen rule is from $\inf_{\phi \in C_k} L(\phi)$. We discuss this problem in the next few sections.

30.4   VC Dimension   *Thm 30.7*

Assume now that the data $D_n = ((X_1, Y_1), \ldots, (X_n, Y_n))$ are used to tune the parameters of the network. To choose a classifier from $C^{(k)}$, we focus on the difference between the probability of error of the selected rule and that of the best classifier in $C^{(k)}$. Recall from Chapters 12 and 14 that the VC dimension $V_{C^{(k)}}$ of the class $C^{(k)}$ determines the performance of some learning algorithms. Theorem 14.5 tells us that no method of picking a classifier from $C^{(k)}$ can guarantee better than $\Omega\left(\sqrt{V_{C^{(k)}}/n}\right)$ performance uniformly for all distributions. Thus, for meaningful distribution-free performance guarantees, the sample size n has to be significantly larger than the VC dimension. On the other hand, by Corollary 12.1, there exists a way of choosing the parameters of the network—namely, by minimization of the empirical error probability—such that the obtained classifier ϕ_n^* satisfies

$$\mathbf{E}\left\{L(\phi_n^*)\right\} - \inf_{\phi \in C^{(k)}} L(\phi) \leq 16\sqrt{\frac{V_{C^{(k)}} \log n + 4}{2n}}$$

for all distributions. On the other hand, if $V_{C^{(k)}} = \infty$, then for any n and any rule, some bad distributions exist that induce very large error probabilities (see Theorem 14.3).

We start with a universal lower bound on the VC dimension of networks with one hidden layer.

Theorem 30.5. (BAUM (1988)). *Let σ be an arbitrary sigmoid and consider the class $C^{(k)}$ of neural net classifiers with k nodes in one hidden layer. Then*

$$V_{C^{(k)}} \geq 2 \left\lfloor \frac{k}{2} \right\rfloor d.$$

PROOF. We prove the statement for the threshold sigmoid, and leave the extension as an exercise (Problem 30.7). We need to show that there is a set of $n = 2\lfloor k/2 \rfloor d$ points in $\mathcal{R}^d$ that can be shattered by sets of the form $\{x : \psi(x) > 1/2\}$, where ψ is a one-layer neural network of k hidden nodes. Clearly, it suffices to prove this for even k. In fact, we prove more: if k is even, any set of $n = kd$ points in general position can be shattered (points are in general position if no $d + 1$ points fall on the same $d - 1$-dimensional hyperplane). Let $\{x_1, \ldots, x_n\}$ be a set of $n = kd$ such points. For each subset of this set, we construct a neural network ψ with k hidden nodes such that $\psi(x_i) > 1/2$ if and only if x_i is a member of this subset. We may assume without loss of generality that the cardinality of the subset to be picked out is at most $n/2$, since otherwise we can use $1/2 - \psi(x)$, where ψ picks out the complement of the subset. Partition the subset into at most $n/(2d) = k/2$ groups, each containing at most d points. For each such group, there exists a hyperplane $a^T x + b = 0$ that contains these points, but no other point from $\{x_1, \ldots, x_n\}$. Moreover, there exists a small positive number h such that $a^T x_i + b \in [-h, h]$ if and only if x_i is among this group of at most d points. Therefore, the simple network

$$\sigma(a^T x + b + h) + \sigma(-a^T x - b + h)$$

is larger than 0 on x_i for exactly these x_i's. Denote the vectors a, and parameters b, h obtained for the $k/2$ groups by $a_1, \ldots, a_{k/2}, b_1, \ldots, b_{k/2}$, and $h_1, \ldots, h_{k/2}$. Let $h = \min_{j \leq k/2} h_j$. It is easy to see that

$$\psi(x) = \sum_{j=1}^{k/2} \left(\sigma(a_j^T x + b_j + h) + \sigma(-a_j^T x - b_j + h) \right) + \frac{1}{2}$$

is larger than $1/2$ for exactly the desired x_i's. This network has k hidden nodes. $\square$

Theorem 30.5 implies that there is no hope for good performance guarantees unless the sample size is much larger than kd. Recall Chapter 14, where we showed that $n \gg V_{C^{(k)}}$ is necessary for a guaranteed small error probability, regardless of the method of tuning the parameters. Bartlett (1993) improved Theorem 30.5 in several ways. For example, he proved that

$$V_{C^{(k)}} \geq d \min\left(k, \frac{2^d}{d^2/2 + d + 1} \right) + 1.$$

Bartlett also obtained similar lower bounds for not fully connected networks—see Problem 30.9—and for two-hidden-layer networks.

Next we show that for the threshold sigmoid, the bound of Theorem 30.5 is tight up to a logarithmic factor, that is, the vc dimension is at most of the order of $kd \log k$.

Theorem 30.6. (BAUM AND HAUSSLER (1989)). *Let σ be the threshold sigmoid and let $C^{(k)}$ be the class of neural net classifiers with k nodes in the hidden layer. Then the shatter coefficients satisfy*

$$S(C^{(k)}, n) \leq \left(\sum_{i=0}^{d+1} \binom{n}{i} \right)^k \left(\sum_{i=0}^{k+1} \binom{n}{i} \right)$$

$$\leq \left(\frac{ne}{d+1} \right)^{k(d+1)} \left(\frac{ne}{k+1} \right)^{k+1} \leq (ne)^{kd+2k+1},$$

which implies that for all $k, d \geq 1$,

$$V_{C^{(k)}} \leq (2kd + 4k + 2) \log_2(e(kd + 2k + 1)).$$

PROOF. Fix n points $x_1, \ldots, x_n \in \mathcal{R}^d$. We bound the number of different values of the vector $(\phi(x_1), \ldots, \phi(x_n))$ as ϕ ranges through $C^{(k)}$. A node j in the hidden layer realizes a dichotomy of the n points by a hyperplane split. By Theorems 13.9 and 13.3, this can be done at most $\sum_{i=0}^{d+1} \binom{n}{i} \leq (ne/(d+1))^{d+1}$ different ways. The different splittings obtained at the k nodes determine the k-dimensional input of the output node. Different choices of the parameters $c_0, c_1, \ldots, c_k$ of the output node determine different k-dimensional linear splits of the n input vectors. This cannot be done in more than $\sum_{i=0}^{k+1} \binom{n}{i} \leq (ne/(k+1))^{k+1}$ different ways for a fixed setting of the a_{ij} and b_i parameters. This altogether yields at most

$$\left(\sum_{i=0}^{d+1} \binom{n}{i} \right)^k \left(\sum_{i=0}^{k+1} \binom{n}{i} \right)$$

different dichotomies of the n points $x_1, \ldots, x_n$, as desired. The bound on the VC dimension follows from the fact that $V_{C^{(k)}} \leq n$ if $S(C^{(k)}, n) \geq 2^n$. $\square$

For threshold sigmoids, the gap between the lower and upper bounds above is logarithmic in kd. Notice that the VC dimension is about the number of weights (or tunable parameters) $w = kd + 2k + 1$ of the network. Surprisingly, Maass (1994) proved that for networks with at least two hidden layers, the upper bound has the right order of magnitude, that is, the VC dimension is $\Omega(w \log w)$. A simple application of Theorems 30.4 and 30.6 provides the next consistency result that was pointed out in Faragó and Lugosi (1993):

Theorem 30.7. *Let σ be the threshold sigmoid. Let g_n be a classifier from $C^{(k)}$ that minimizes the empirical error*

$$\widehat{L}_n(\phi) = \frac{1}{n} \sum_{i=1}^{n} I_{\{\phi(X_i) \neq Y_i\}}$$

over $\phi \in C^{(k)}$. If $k \to \infty$ such that $k \log n / n \to 0$ as $n \to \infty$, then g_n is strongly universally consistent, that is,

$$\lim_{n \to \infty} L(g_n) = L^*$$

with probability one for all distributions of (X, Y).

PROOF. By the usual decomposition into approximation and estimation errors,

$$L(g_n) - L^* = \left(L(g_n) - \inf_{\phi \in C^{(k)}} L(\phi) \right) + \left(\inf_{\phi \in C^{(k)}} L(\phi) - L^* \right).$$

The second term on the right-hand side tends to zero by Corollary 30.1. For the estimation error, by Theorems 12.6 and 30.6,

$$P \left\{ L(g_n) - \inf_{\phi \in C^{(k)}} L(\phi) > \epsilon \right\} \leq 8 S(C^{(k)}, n) e^{-n\epsilon^2/128}$$

$$\leq\ \ 8(ne)^{(kd+2k+1)}e^{-n\epsilon^2/128},$$

which is summable if $k = o(n/\log n)$. $\square$

The theorem assures us that if σ is the threshold sigmoid, then a sequence of properly sized networks may be trained to asymptotically achieve the optimum probability of error, regardless of what the distribution of (X, Y) is. For example, $k \approx \sqrt{n}$ will do the job. However, this is clearly not the optimal choice in the majority of cases. Since Theorem 30.6 provides suitable upper bounds on the VC dimension of each class $C^{(k)}$, one may use complexity regularization as described in Chapter 18 to find a near-optimum size network.

Unfortunately, the situation is much less clear for more general, continuous sigmoids. The VC dimension then depends on the specific sigmoid. It is not hard to see that the VC dimension of $C^{(k)}$ with an arbitrary nondecreasing sigmoid is always larger than or equal to that with the threshold sigmoid (Problem 30.8). Typically, the VC dimension of a class of such networks is significantly larger than that for the threshold sigmoid. In fact, it can even be infinite! Macintyre and Sontag (1993) demonstrated the existence of continuous, infinitely many times differentiable monotone increasing sigmoids such that the VC dimension of $C^{(k)}$ is infinite if $k \geq 2$. Their sigmoids have little squiggles, creating the large variability. It is even more surprising that infinite VC dimension may occur for even smoother sigmoids, whose second derivative is negative for $x > 0$ and positive for $x < 0$. In Chapter 25 (see Problem 25.11) we basically proved the following result. The details are left to the reader (Problem 30.13).

Theorem 30.8. *There exists a sigmoid σ that is monotone increasing, continuous, concave on $(0, \infty)$, and convex on $(-\infty, 0)$, such that $V_{C^{(k)}} = \infty$ for each $k \geq 8$.*

We recall once again that infinite VC dimension implies that there is no hope of obtaining nontrivial distribution-free upper bounds on

$$L(g_n) - \inf_{\phi \in C^{(k)}} L(\phi),$$

no matter how the training sequence D_n is used to select the parameters of the neural network. However, as we will see later, it is still possible to obtain universal consistency. Finiteness of the VC dimension has been proved for many types of sigmoids. Maass (1993) and Goldberg and Jerrum (1993) obtain upper bounds for piecewise polynomial sigmoids. The results of Goldberg and Jerrum (1993) apply for general classes parametrized by real numbers, e.g., for classes of neural networks with the sigmoid

$$\sigma(x) = \begin{cases} 1 - 1/(2x + 2) & \text{if } x \geq 0 \\ 1/(2 - 2x) & \text{if } x < 0. \end{cases}$$

Macintyre and Sontag (1993) prove $V_{C^{(k)}} < \infty$ for a large class of sigmoids, which includes the standard, arctan, and gaussian sigmoids. While finiteness is useful, the lack of an explicit tight upper bound on $V_{C^{(k)}}$ prevents us from getting meaningful upper bounds on the performance of g_n, and also from applying the structural risk minimization of Chapter 18. For the standard sigmoid, and for networks with k hidden nodes and w tunable weights Karpinski and Macintyre (1994) recently reported the upper bound

$$V \le \frac{kw(kw - 1)}{2} + w(1 + 2k) + w(1 + 3k)\log(3w + 6kw + 3).$$

See also Shawe-Taylor (1994).

Unfortunately, the consistency result of Theorem 30.7 is only of theoretical interest, as there is no efficient algorithm to find a classifier that minimizes the empirical error probability. Relatively little effort has been made to solve this important problem. Faragó and Lugosi (1993) exhibit an algorithm that finds the empirically optimal network. However, their method takes time exponential in kd, which is intractable even for the smallest toy problems. Much more effort has been invested in the tuning of networks by minimizing the empirical squared error, or the empirical L_1 error. These problems are also computationally demanding, but numerous suboptimal hill-climbing algorithms have been used with some success. Most famous among these is the back propagation algorithm of Rumelhart, Hinton, and Williams (1986). Nearly all known algorithms that run in reasonable time may get stuck at local optima, which results in classifiers whose probability of error is hard to predict. In the next section we study the error probability of neural net classifiers obtained by minimizing empirical L_p errors.

We end this section with a very simple kind of one-layer network. The *committee machine* (see, e.g., Nilsson (1965) and Schmidt (1994)) is a special case of a one-hidden-layer neural network of the form (30.2) with $c_0 = 0, c_1 = \cdots = c_k = 1$, and the threshold sigmoid.

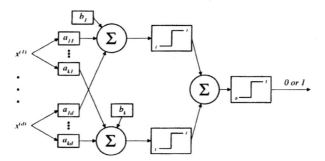

FIGURE 30.11. *The committee machine has fixed weights at the output of the hidden layer.*

Committee machines thus use a majority vote over the outcomes of the hidden neurons. It is interesting that the lower bound of Theorem 30.5 remains valid when $C^{(k)}$ is the class of all committee machines with k neurons in the hidden layer. It is less obvious, however, that the class of committee machines is large enough for the asymptotic property

$$\lim_{k \to \infty} \inf_{\phi \in C^{(k)}} L(\phi) = L^*$$

for all distributions of (X, Y) (see Problem 30.6).

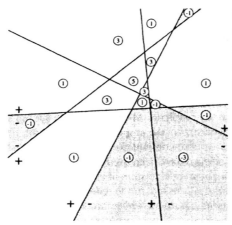

FIGURE 30.12. *A partition of the plane determined by a committee machine with 5 hidden neurons. The total vote is shown in each region. The region in which we decide "0" is shaded.*

30.5 L_1 Error Minimization

In the previous section, we obtained consistency for the standard threshold sigmoid networks by empirical risk minimization. We could not apply the same methodology for general sigmoids simply because the VC dimension for general sigmoidal networks is not bounded. It is bounded for certain classes of sigmoids, and for those, empirical risk minimization yields universally consistent classifiers. Even if the VC dimension is infinite, we may get consistency, but this must then be proved by other methods, such as methods based upon metric entropy and covering numbers (see Chapters 28 and 29, as well as the survey by Haussler (1992)). One could also train the classifier by minimizing another empirical criterion, which is exactly what we will do in this section. We will be rewarded with a general consistency theorem for *all* sigmoids.

For $1 \leq p < \infty$, the empirical L_p error of a neural network ψ is defined by

$$J_n^{(p)}(\psi) = \left(\frac{1}{n} \sum_{i=1}^{n} |\psi(X_i) - Y_i|^p \right)^{1/p}.$$

The most interesting cases are $p = 1$ and $p = 2$. For $p = 2$ this is just the empirical squared error, while $p = 1$ yields the empirical absolute error. Often it makes sense to attempt to choose the parameters of the network ψ such that $J_n^{(p)}(\psi)$ is minimized. In situations where one is not only interested in the number of errors, but also how *robust* the decision is, such error measures may be meaningful. In other words, these error measures penalize even good decisions if ψ is close to the threshold value 0. Minimizing $J_n^{(p)}$ is like finding a good regression function estimate. Our concern is primarily with the error probability. In Chapter 4 we already highlighted the dangers of squared error minimization and L_p errors in general. Here we will concentrate on the consistency properties. We minimize the empirical error over a class of functions, which should not be too large to avoid overfitting. However, the class should be large enough to contain a good approximation of the target function. Thus, we let the class of candidate functions grow with the sample size n, as in Grenander's "method of sieves" (Grenander (1981)). Its consistency and rates of convergence have been widely studied primarily for least squares regression function estimation and nonparametric maximum likelihood density estimation—see Geman and Hwang (1982), Gallant (1987), and Wong and Shen (1992).

REMARK. REGRESSION FUNCTION ESTIMATION. In the regression function estimation setup, White (1990) proved consistency of neural network estimates based on squared error minimization. Barron (1991; 1994) used a complexity-regularized modification of these error measures to obtain the fastest possible rate of convergence for nonparametric neural network estimates. Haussler (1992) provides a general framework for empirical error minimization, and provides useful tools for handling neural networks. Various consistency properties of nonparametric neural network estimates have been proved by White (1991), Mielniczuk and Tyrcha (1993), and Lugosi and Zeger (1995). □

We only consider the $p = 1$ case, as the generalization to other values of p is straightforward. Define the L_1 error of a function $\psi : \mathcal{R}^d \to \mathcal{R}$ by

$$J(\psi) = \mathbf{E}\{|\psi(X) - Y|\}.$$

We pointed out in Problem 2.12 that one of the functions minimizing $J(\psi)$ is the Bayes rule g^* whose error is denoted by

$$J^* = \inf_\psi J(\psi) = J(g^*).$$

Then clearly, $J^* = L^*$. We have also seen that if we define a decision by

$$g(x) = \begin{cases} 0 & \text{if } \psi(x) \leq 1/2 \\ 1 & \text{otherwise,} \end{cases}$$

then its error probability $L(g) = \mathbf{P}\{g(X) \neq Y\}$ satisfies the inequality

$$L(g) - L^* \leq J(\psi) - J^*.$$

Our approach is to select a neural network from a suitably chosen class of networks by minimizing the empirical error

$$J_n(\psi) = \frac{1}{n} \sum_{i=1}^{n} |\psi(X_i) - Y_i|.$$

Denoting this function by ψ_n, according to the inequality above, the classifier

$$g_n(x) = \begin{cases} 0 & \text{if } \psi_n(x) \leq 1/2 \\ 1 & \text{otherwise,} \end{cases}$$

is consistent if the L_1 error

$$J(\psi_n) = \mathbf{E}\{|\psi_n(X) - Y|| D_n\}$$

converges to J^* in probability. Convergence with probability one provides strong consistency. For universal convergence, the class over which the minimization is performed has to be defined carefully. The following theorem shows that this may be achieved by neural networks with k nodes, in which the range of the output weights $c_0, c_1, \ldots, c_k$ is restricted.

Theorem 30.9. (LUGOSI AND ZEGER (1995)). *Let σ be an arbitrary sigmoid. Define the class $\mathcal{F}_n$ of neural networks by*

$$\mathcal{F}_n = \left\{ \sum_{i=1}^{k_n} c_i \sigma(a_i^T x + b_i) + c_0 : a_i \in \mathcal{R}^d, b_i \in \mathcal{R}, \sum_{i=0}^{k_n} |c_i| \leq \beta_n \right\},$$

and let ψ_n be a function that minimizes the empirical L_1 error

$$J_n(\psi) = \frac{1}{n} \sum_{i=1}^{n} |\psi(X_i) - Y_i|$$

over $\psi \in \mathcal{F}_n$. If k_n and β_n satisfy

$$\lim_{n \to \infty} k_n = \infty, \quad \lim_{n \to \infty} \beta_n = \infty, \quad \text{and} \quad \lim_{n \to \infty} \frac{k_n \beta_n^2 \log(k_n \beta_n)}{n} = 0,$$

then the classification rule

$$g_n(x) = \begin{cases} 0 & \text{if } \psi_n(x) \leq 1/2 \\ 1 & \text{otherwise} \end{cases}$$

is universally consistent.

REMARK. Strong universal consistency may also be shown by imposing slightly more restrictive conditions on k_n and β_n (see Problem 30.16). $\square$

PROOF. By the argument preceding the theorem, it suffices to prove that $J(\psi_n) - J^* \to 0$ in probability. Write

$$J(\psi_n) - J^* = \left(J(\psi_n) - \inf_{\psi \in \mathcal{F}_n} J(\psi) \right) + \left(\inf_{\psi \in \mathcal{F}_n} J(\psi) - J^* \right).$$

To handle the approximation error—the second term on the right-hand side—let $\psi' \in \mathcal{F}_n$ be a function such that

$$\mathbf{E}\{|\psi'(X) - g^*(X)|\} \leq \mathbf{E}\{|\psi(X) - g^*(X)|\}$$

for each $\psi \in \mathcal{F}_n$. The existence of such a function may be seen by noting that $\mathbf{E}\{|\psi(X) - g^*(X)|\}$ is a continuous function of the parameters a_i, b_i, c_i of the neural network ψ. Clearly,

$$
\begin{aligned}
\inf_{\psi \in \mathcal{F}_n} J(\psi) - J^* &\leq J(\psi') - J^* \\
&= \mathbf{E}\{|\psi'(X) - Y|\} - \mathbf{E}\{|g^*(X) - Y|\} \\
&\leq \mathbf{E}\{|\psi'(X) - g^*(X)|\},
\end{aligned}
$$

which converges to zero as $n \to \infty$, by Problem 30.4. We start the analysis of the estimation error by noting that as in Lemma 8.2, we have

$$
\begin{aligned}
J(\psi_n) - \inf_{\psi \in \mathcal{F}_n} J(\psi) &\leq 2 \sup_{\psi \in \mathcal{F}_n} |J(\psi) - J_n(\psi)| \\
&= 2 \sup_{\psi \in \mathcal{F}_n} \left| \mathbf{E}\{|\psi(X) - Y|\} - \frac{1}{n} \sum_{i=1}^n |\psi(X_i) - Y_i| \right|.
\end{aligned}
$$

Define the class $\mathcal{M}_n$ of functions on $\mathcal{R}^d \times \{0, 1\}$ by

$$\mathcal{M}_n$$
$$= \left\{ m(x, y) = \left| \sum_{i=1}^{k_n} c_i \sigma(a_i^T x + b_i) + c_0 - y \right| : a_i \in \mathcal{R}^d, b_i \in \mathcal{R}, \sum_{i=0}^{k_n} |c_i| \leq \beta_n \right\}.$$

Then the previous bound becomes

$$2 \sup_{m \in \mathcal{M}_n} \left| \mathbf{E}\{m(X, Y)\} - \frac{1}{n} \sum_{i=1}^n m(X_i, Y_i) \right|.$$

Such quantities may be handled by the uniform law of large numbers of Theorem 29.1, which applies to classes of uniformly bounded functions. Indeed, for each $m \in \mathcal{M}_n$

$$
\begin{aligned}
m(x, y) &= \left| \sum_{i=1}^{k_n} c_i \sigma(a_i^T x + b_i) + c_0 - y \right| \\
&\leq 2 \max \left(\sum_{i=1}^{k_n} c_i \sigma(a_i^T x + b_i) + c_0, 1 \right) \\
&\leq 2 \max \left(\sum_{i=1}^{k_n} |c_i|, 1 \right) \\
&\leq 2 \beta_n,
\end{aligned}
$$

if n is so large that $\beta_n \geq 1$. For such n's, Theorem 29.1 states that

$$\mathbf{P}\left\{\sup_{m\in\mathcal{M}_n}\left|\mathbf{E}\{m(X,Y)\} - \frac{1}{n}\sum_{i=1}^{n}m(X_i,Y_i)\right| > \epsilon\right\}$$

$$\leq 8\mathbf{E}\left\{\mathcal{N}(\epsilon/8, \mathcal{M}_n(D_n))\right\}e^{-n\epsilon^2/(512\beta_n^2)},$$

where $\mathcal{N}(\epsilon, \mathcal{M}_n(D_n))$ denotes the l_1-covering number of the random set

$$\mathcal{M}_n(D_n) = \{(m(X_1,Y_1), \ldots, m(X_n,Y_n)) : m \in \mathcal{M}_n\} \subset \mathcal{R}^n,$$

defined in Chapter 29. All we need now is to estimate these covering numbers. Observe that for $m_1, m_2 \in \mathcal{M}_n$ with $m_1(x,y) = |\psi_1(x) - y|$ and $m_2(x,y) = |\psi_2(x) - y|$, for any probability measure ν on $\mathcal{R}^d \times \{0,1\}$,

$$\int |m_1(x,y) - m_2(x,y)|\nu(d(x,y)) \leq \int |\psi_1(x) - \psi_2(x)|\mu(dx),$$

where μ is the marginal of ν on $\mathcal{R}^d$. Therefore, it follows that $\mathcal{N}(\epsilon, \mathcal{M}_n(D_n)) \leq \mathcal{N}(\epsilon, \mathcal{F}_n(X_1^n))$, where $X_1^n = (X_1, \ldots, X_n)$. It means that an upper bound on the covering number of the class of neural networks $\mathcal{F}_n$ is also an upper bound on the quantity that interests us. This bounding may be done by applying lemmas from Chapters 13 and 30. Define the following three collections of functions:

$$\begin{aligned}
\mathcal{G}_1 &= \left\{a^T x + b; \ a \in \mathcal{R}^d, b \in \mathcal{R}\right\}, \\
\mathcal{G}_2 &= \left\{\sigma(a^T x + b); \ a \in \mathcal{R}^d, b \in \mathcal{R}\right\}, \\
\mathcal{G}_3 &= \left\{c\sigma(a^T x + b); \ a \in \mathcal{R}^d, b \in \mathcal{R}, c \in [-\beta_n, \beta_n]\right\}.
\end{aligned}$$

By Theorem 13.9, the VC dimension of the class of sets

$$\mathcal{G}_1^+ = \{(x,t) : t \leq \psi(x), \psi \in \mathcal{G}_1\}$$

is $V_{\mathcal{G}_1^+} \leq d + 2$. This implies by Lemma 29.5 that $V_{\mathcal{G}_2^+} \leq d + 2$, so by Corollary 29.2, for any $x_1^n = (x_1, \ldots, x_n)$,

$$\mathcal{N}(\epsilon, \mathcal{G}_2(x_1^n)) \leq 2\left(\frac{4e}{\epsilon}\right)^{2(d+2)},$$

where $\mathcal{G}_2(x_1^n) = \{z \in \mathcal{R}^n : z = (g(x_1), \ldots, g(x_n)), g \in \mathcal{G}_2\}$. Now, using similar notations, Theorem 29.7 allows us to estimate covering numbers of $\mathcal{G}_3(x_1^n)$:

$$\mathcal{N}(\epsilon, \mathcal{G}_3(x_1^n)) \leq \frac{4}{\epsilon}\mathcal{N}(\epsilon/(2\beta_n), \mathcal{G}_2(x_1^n)) \leq \left(\frac{8e\beta_n}{\epsilon}\right)^{2d+5}$$

if $\beta_n > 2/e$. Finally, we can apply Lemma 29.6 to obtain

$$\begin{aligned}
\mathcal{N}(\epsilon, \mathcal{F}_n(x_1^n)) &\leq \frac{2\beta_n(k_n+1)}{\epsilon}\mathcal{N}(\epsilon/(k_n+1), \mathcal{G}_3(x_1^n))^{k_n} \\
&\leq \left(\frac{8e(k_n+1)\beta_n}{\epsilon}\right)^{k_n(2d+5)+1}
\end{aligned}$$

Thus, substituting this bound into the probability inequality above, we get for n large enough,

$$\mathbf{P}\left\{\sup_{\psi \in \mathcal{F}_n} \left| \mathbf{E}|\psi(X) - Y| - \frac{1}{n}\sum_{j=1}^{n} |\psi(X_j) - Y_j| \right| > \epsilon \right\}$$

$$\leq 8\left(\frac{64e(k_n + 1)\beta_n}{\epsilon}\right)^{k_n(2d+5)+1} e^{-n\epsilon^2/(512\beta_n^2)},$$

which tends to zero if

$$\frac{k_n\beta_n^2 \log(k_n\beta_n)}{n} \to 0,$$

concluding the proof. $\square$

There are yet other ways to obtain consistency for general sigmoidal networks. We may restrict the network by discretizing the values of the coefficients in some way—thus creating a sieve with a number of members that is easy to enumerate—and applying complexity regularization (Chapter 18). This is the method followed by Barron (1988; 1991).

30.6   The Adaline and Padaline

Widrow (1959) and Widrow and Hoff (1960) introduced the Adaline, and Specht (1967; 1990) studied polynomial discriminant functions such as the Padaline. Looked at formally, the discriminant function ψ used in the decision $g(x) = I_{\{\psi(x)>0\}}$ is of a polynomial nature, with ψ consisting of sums of monomials like

$$a\left(x^{(1)}\right)^{i_1} \cdots \left(x^{(d)}\right)^{i_d},$$

where $i_1, \ldots, i_d \geq 0$ are integers, and a is a coefficient. The order of a monomial is $i_1 + \cdots + i_d$. Usually, all terms up to, and including those of order r are included. Widrow's Adaline (1959) has $r = 1$. The total number of monomials of order r or less does not exceed $(r + 1)^d$. The motivation for developing these discriminants is that only up to $(r + 1)^d$ coefficients need to be trained and stored. In applications in which data continuously arrive, the coefficients may be updated on-line and the data can be discarded. This property is, of course, shared with standard neural networks. In most cases, order r polynomial discriminants are not translation invariant. Minimizing a given criterion on-line is a phenomenal task, so Specht noted that training is not necessary if the a's are chosen so as to give decisions that are close to those of the kernel method with normal kernels.

For example, if $K(u) = e^{-u^2/2}$, the kernel method picks a smoothing factor h

(such that $h \to 0$ and $nh^d \to \infty$; see Chapter 10) and uses

$$
\psi(x) = \frac{1}{n} \sum_{i=1}^{n} (2Y_i - 1) K \left(\frac{\|x - X_i\|}{h} \right)
$$

$$
= \frac{1}{n} \sum_{i=1}^{n} (2Y_i - 1) e^{-x^T x/(2h^2)} e^{x^T X_i/h^2} e^{-X_i^T X_i/(2h^2)} .
$$

The *same* decision is obtained if we use

$$
\psi(x) = \frac{1}{n} \sum_{i=1}^{n} (2Y_i - 1) e^{x^T X_i/h^2 - X_i^T X_i/(2h^2)} .
$$

Now, approximate this by using Taylor's series expansion and truncating to the order r terms. For example, the coefficient of $\left(x^{(1)}\right)^{i_1} \cdots \left(x^{(d)}\right)^{i_d}$ in the expansion of $\psi(x)$ would be, if $\sum_{j=1}^{d} i_j = i$,

$$
\frac{1}{i_1! \cdots i_d!} \cdot \frac{1}{h^{2i}} \cdot \sum_{j=1}^{n} (2Y_j - 1) \left(X_j^{(1)}\right)^{i_1} \cdots \left(X_j^{(d)}\right)^{i_d} e^{-X_j^T X_j/(2h^2)} .
$$

These sums are easy to update on-line, and decisions are based on the sign of the order r truncation ψ_r of ψ. The classifier is called the *Padaline*. Specht notes that overfitting in ψ_r does not occur due to the fact that overfitting does not occur for the kernel method based on ψ. His method interpolates between the latter method, generalized linear discrimination, and generalizations of the perceptron.

For fixed r, the Padaline defined above is not universally consistent (for the same reason linear discriminants are not universally consistent), but if r is allowed to grow with n, the decision based on the sign of ψ_r becomes universally consistent (Problem 30.14). Recall, however, that Padaline was not designed with a variable r in mind.

30.7   Polynomial Networks

Besides Adaline and Padaline, there are several ways of constructing polynomial many-layered networks in which basic units are of the form $\left(x^{(1)}\right)^{i_1} \cdots \left(x^{(k)}\right)^{i_k}$ for inputs $x^{(1)}, \dots, x^{(k)}$ to that level. Pioneers in this respect are Gabor (1961), Ivakhnenko (1968; 1971) who invented the GMDH method—the group method of data handling—and Barron (1975). See also Ivakhnenko, Konovalenko, Tulupchuk, and Tymchenko (1968) and Ivakhnenko, Petrache, and Krasyts'kyy (1968). These networks can be visualized in the following way:

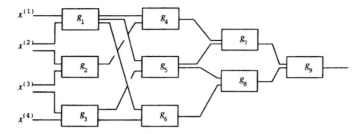

FIGURE 30.13. *Simple polynomial network: Each g_i represents a simple polynomial function of its input. In Barron's work (Barron (1975), Barron and Barron (1988)), the g_i's are sometimes 2-input elements of the form $g_i(x, y) = a_i + b_i x + c_i y + d_i x y$.*

If the g_i's are Barron's quadratic elements, then the network shown in Figure 30.13 represents a particular polynomial of order 8 in which the largest degree of any $x^{(i)}$ is at most 4. The number of unknown coefficients is 36 in the example shown in the figure, while in a full-fledged order-8 polynomial network, it would be much larger.

For training these networks, many strategies have been proposed by Barron and his associates. Ivakhnenko (1971), for example, trains one layer at a time and lets only the best neurons in each layer survive for use as input in the next layer.

It is easy to see that polynomial networks, even with only two inputs per node, and with degree in each cell restricted to two, but with an unrestricted number of layers, can implement any polynomial in d variables. As the polynomials are dense in the L_∞ sense on $C[a, b]^d$ for all $a, b \in \mathcal{R}^d$, we note by Lemma 30.2 that such networks include a sequence of classifiers approaching the Bayes error for any distribution. Consider several classes of polynomials:

$$\mathcal{G}_1 = \{\psi = a_1\psi_1 + \cdots + a_k\psi_k\},$$

where $\psi_1, \ldots, \psi_k$ are fixed monomials, but the a_i's are free coefficients.

$$\mathcal{G}_2 = \{\psi \in \mathcal{G}_1 : \psi_1, \ldots, \psi_k \text{ are monomials of order } \leq r\}.$$

(The order of a monomial $\left(x^{(1)}\right)^{i_1} \cdots \left(x^{(d)}\right)^{i_d}$ is $i_1 + \cdots + i_d$.)

$$\mathcal{G}_3 = \{\psi \in \mathcal{G}_1 : \psi_1, \ldots, \psi_k \text{ are monomials of order } \leq r, \text{ but } k \text{ is not fixed}\}.$$

As $k \to \infty$, $\mathcal{G}_1$ does not generally become dense in $C[a, b]^d$ in the L_∞ sense unless we add $\psi_{k+1}, \psi_{k+2}, \ldots$ in a special way. However, $\mathcal{G}_3$ becomes dense as $r \to \infty$ by Theorem A.9 and $\mathcal{G}_2$ becomes dense as both $k \to \infty$ and $r \to \infty$ (as $\mathcal{G}_2$ contains a subclass of $\mathcal{G}_3$ for a smaller r depending upon k obtained by including in $\psi_1, \ldots, \psi_k$ all monomials in increasing order).

The VC dimension of the class of classifiers based on $\mathcal{G}_1$ is not more than k (see Theorem 13.9). The VC dimension of classifiers based upon $\mathcal{G}_2$ does not exceed those of $\mathcal{G}_3$, which in turn is nothing but the number of possible monomials of order $\leq r$. A simple counting argument shows that this is bounded by $(r + 1)^d$. See also Anthony and Holden (1993).

These simple bounds may be used to study the consistency of polynomial networks.

Let us take a *fixed structure network* in which all nodes are fixed—they have at most k inputs with $k \geq 2$ fixed and represent polynomials of order $\leq r$ with $r \geq 2$ fixed. For example, with $r = 2$, each cell with input $z_1, \ldots, z_k$ computes $\sum_i a_i \psi_i(z_1, \ldots, z_k)$, where the a_i's are coefficients and the ψ_i's are fixed monomials of order r or less, and all such monomials are included. Assume that the layout is fixed and is such that it can realize all polynomials of order $\leq s$ on the input $x^{(1)}, \ldots, x^{(d)}$. One way of doing this is to realize all polynomials of order $\leq r$ by taking all $\binom{d}{k}$ possible input combinations, and to repeat at the second level with $\binom{\binom{d}{k}}{k}$ cells of neurons, and so forth for a total of s/r layers of cells. This construction is obviously redundant but it will do for now. Then note that the VC dimension is not more than $(s + 1)^d$, as noted above. If we choose the best coefficients in the cells by empirical risk minimization, then the method is consistent:

Theorem 30.10. *In the fixed-structure polynomial network described above, if L_n is the probability of error of the empirical risk minimizer, then $\mathbf{E}\{L_n\} \to L^*$ if $s \to \infty$ and $s = o(n^{1/d})$.*

PROOF. Apply Lemma 30.2 and Theorem 12.6. $\square$

Assume a fixed-structure network as above such that all polynomials of order $\leq s$ are realized plus some other ones, while the number of layers of cells is not more than l. Then the VC dimension is not more than $(rl + 1)^d$ because the maximal order is not more than lr. Hence, we have consistency under the same conditions as above, that is, $s \to \infty$, and $l = o(n^{1/d})$. Similar considerations can now be used in a variety of situations.

30.8   Kolmogorov-Lorentz Networks and Additive Models

Answering one of Hilbert's famous questions, Kolmogorov (1957) and Lorentz (1976) (see also Sprecher (1965) and Hecht-Nielsen (1987)) obtained the following interesting representation of any continuous function on $[0, 1]^d$.

Theorem 30.11. (KOLMOGOROV (1957); LORENTZ (1976)). *Let f be continuous on $[0, 1]^d$. Then f can be rewritten as follows: let $\delta > 0$ be an arbitrary constant, and choose $0 < \epsilon \leq \delta$ rational. Then*

$$f = \sum_{k=1}^{2d+1} g(z_k),$$

where $g : \mathcal{R} \rightarrow \mathcal{R}$ is a continuous function (depending upon f and ϵ), and each z_k is rewritten as

$$z_k = \sum_{j=1}^{d} \lambda^k \psi \left(x^{(j)} + \epsilon k \right) + k.$$

Here λ is real and ψ is monotonic and increasing in its argument. Also, both λ and ψ are universal (independent of f), and ψ is Lipschitz: $|\psi(x) - \psi(y)| \leq c|x - y|$ for some $c > 0$.

The Kolmogorov-Lorentz theorem states that f may be represented by a very simple network that we will call the Kolmogorov-Lorentz network. What is amazing is that the first layer is fixed and known beforehand. Only the mapping g depends on f. This representation immediately opens up new revenues of pursuit— we need not mix the input variables. Simple additive functions of the input variables suffice to represent all continuous functions.

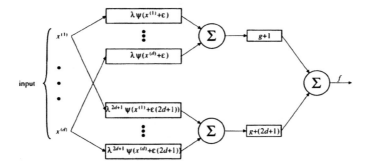

FIGURE 30.14. *The Kolmogorov-Lorentz network of Theorem* 30.11.

To explain what is happening here, we look at the interleaving of bits to make one-dimensional numbers out of d-dimensional vectors. For the sake of simplicity, let $f : [0, 1]^2 \rightarrow \mathcal{R}$. Let

$$x = 0.x_1 x_2 x_3 \ldots \quad \text{and} \quad y = 0.y_1 y_2 y_3 \ldots$$

be the binary expansions of x and y, and consider a representation for the function $f(x, y)$. The bit-interleaved number $z \in [0, 1]$ has binary expansion

$$z = 0.x_1 y_1 x_2 y_2 x_3 y_3 \ldots$$

and may thus be written as $z = \phi(x) + (1/2)\phi(y)$, where

$$\phi(x) = 0.x_1 0 x_2 0 x_3 0 \ldots$$

and thus

$$\frac{1}{2}\phi(y) = 0.0y_10y_20y_3\ldots.$$

We can also retrieve x and y from z by noting that

$$x = \psi_1(z), \quad \psi_1(z) = 0.z_1z_3z_5z_7\ldots,$$

and

$$y = \psi_2(z), \quad \psi_2(z) = 0.z_2z_4z_6z_8\ldots.$$

Therefore,

$$
\begin{aligned}
f(x, y) &= f(\psi_1(z), \psi_2(z)) \\
&\overset{\text{def}}{=} g(z) \quad \text{(a one-dimensional function of } z) \\
&= g\left(\phi(x) + \frac{1}{2}\phi(y)\right).
\end{aligned}
$$

The function ϕ is strictly monotone increasing. Unfortunately, ϕ, ψ_1, and ψ_2 are not continuous. Kolmogorov's theorem for this special case is as follows:

Theorem 30.12. (KOLMOGOROV (1957)). *There exist five monotone functions* $\phi_1, \ldots, \phi_5 : [0, 1] \rightarrow \mathcal{R}$ *satisfying* $|\phi_i(x_1) - \phi_i(x_2)| \leq |x_1 - x_2|$, *with the following property: for every* $f \in C[0, 1]^2$ *(the continuous functions on* $[0, 1]^2$*), there exists a continuous function* g *such that for all* $(x_1, x_2) \in [0, 1]^2$,

$$f(x_1, x_2) = \sum_{i=1}^{5} g\left(\phi_i(x_1) + \frac{1}{2}\phi_i(x_2)\right).$$

The difference with pure bit-interleaving is that now the ϕ_i's are continuous and g is continuous whenever f is. Also, just as in our simle example, Kolmogorov gives an explicit construction for $\phi_1, \ldots, \phi_5$.

Kolmogorov's theorem may be used to show the denseness of certain classes of functions that may be described by networks. There is one pitfall however: any such result must involve at least one neuron or cell that has a general function in it, and we are back at square one, because a general function, even on only one input, may be arbitrarily complicated and wild.

Additive models include, for example, models such as

$$\alpha + \sum_{i=1}^{d} \psi_i(x^{(i)}),$$

where the ψ_i's are unspecified univariate functions (Friedman and Silverman (1989), Hastie and Tibshirani (1990)). These are not powerful enough to approximate all functions. A generalized additive model is

$$\sigma\left(\alpha + \sum_{i=1}^{d} \psi_i(x^{(i)})\right)$$

(Hastie and Tibshirani (1990)), where σ is now a given or unspecified function. From Kolmogorov's theorem, we know that the model

$$\sum_{k=1}^{2d+1} \psi \left(\alpha_k + \sum_{i=1}^{d} \psi_{i,k}(x^{(i)}) \right)$$

with $\psi_{i,k}$, ψ unspecified functions, includes all continuous functions on compact sets and is thus ideally suited for constructing networks. In fact, we may take $\alpha_k = k$ and take all $\psi_{i,k}$'s as specified in Kolmogorov's theorem. This leaves only ψ as the unknown.

Now, consider the following: any continuous univariate function f can be approximated on bounded sets to within ϵ by simple combinations of threshold sigmoids σ of the form

$$\sum_{i=1}^{k} a_i \sigma(x - c_i),$$

where a_i, c_i, k are variable. This leads to a two-hidden-layer neural network representation related to that of Kůrková (1992), where only the last layer has unknown coefficients for a total of $2k$.

Theorem 30.13. *Consider a network classifier of the form described above in which $\sigma(\cdot)$ is the threshold sigmoid, and the a_i's and c_i's are found by empirical error minimization. Then $\mathbf{E}\{L_n\} \to L^*$ for all distributions of (X, Y), if $k \to \infty$ and $k \log n / n \to 0$.*

PROOF. We will only outline the proof. First observe that we may approximate all functions on $C[a, b]^d$ by selecting k large enough. By Lemma 30.2 and Theorem 12.6, it suffices to show that the VC dimension of our class of classifiers is $o(n/\log n)$. Considering $\alpha_k + \sum_{i=1}^{d} \psi_{i,k}(x^{(i)})$ as new input elements, called y_k, $1 \leq k \leq 2d + 1$, we note that the VC dimension is not more than that of the classifiers based on

$$\sum_{k=1}^{2d+1} \sum_{i=1}^{k} a_i \sigma(y_j - c_i),$$

which in turn is not more than that of the classifiers given by

$$\sum_{i=1}^{k(2d+1)} b_l \sigma(z_l - d_l),$$

where $\{b_l\}$, $\{d_l\}$ are parameters, and $\{z_l\}$ is an input sequence. By Theorem 13.9, the VC dimension is not more than $k(2d + 1)$. This concludes the proof. $\square$

For more results along these lines, we refer to Kůrková (1992).

30.9   Projection Pursuit

In projection pursuit (Friedman and Tukey (1974), Friedman and Stuetzle (1981), Friedman, Stuetzle, and Schroeder (1984), Huber (1985), Hall (1989), Flick, Jones, Priest, and Herman (1990)), one considers functions of the form

$$\psi(x) = \sum_{j=1}^{k} \psi_j(b_j + a_j^T x), \tag{30.3}$$

where $b_j \in \mathcal{R}$, $a_j \in \mathcal{R}^d$ are constants and $\psi_1, \ldots, \psi_k$ are fixed functions. This is related to, but not a special case of, one-hidden-layer neural networks. Based upon the Kolmogorov-Lorentz representation theorem, we may also consider

$$\psi(x) = \sum_{j=1}^{d} \psi_j(x^{(j)}), \tag{30.4}$$

for fixed functions ψ_j (Friedman and Silverman (1989), Hastie and Tibshirani (1990)). In (30.4) and (30.3), the ψ_j's may be approximated in turn by spline functions or other nonparametric constructs. This approach is covered in the literature on generalized additive models (Stone (1985), Hastie and Tibshirani (1990)).

The class of functions $e^{a^T x}$, $a \in \mathcal{R}^d$, satisfies the conditions of the Stone-Weierstrass theorem (Theorem A.9) and is therefore dense in the L_∞ norm on $C[a, b]^d$ for any $a, b \in \mathcal{R}^d$ (see, e.g., Diaconis and Shahshahani (1984)).

As a corollary, we note that the same denseness result applies to the family

$$\sum_{i=1}^{k} \psi_i(a_i^T x), \tag{30.5}$$

where $k \geq 1$ is arbitrary and $\psi_1, \psi_2, \ldots$ are general functions. The latter result is at the basis of projection pursuit methods for approximating functions, where one tries to find vectors a_i and functions ψ_i that approximate a given function very well.

REMARK. In some cases, approximations by functions as in (30.5) may be exact. For example,

$$x^{(1)}x^{(2)} = \frac{1}{4}(x^{(1)} + x^{(2)})^2 - \frac{1}{4}(x^{(1)} - x^{(2)})^2$$

and

$$\max(x^{(1)}, x^{(2)}) = \frac{1}{2}|x^{(1)} + x^{(2)}| + \frac{1}{2}|x^{(1)} - x^{(2)}|. \quad \square$$

Theorem 30.14. (DIACONIS AND SHAHSHAHANI (1984)). *Let m be a positive integer. There are $\binom{m+d-1}{m}$ distinct vectors $a_j \in \mathcal{R}^d$ such that any homogeneous polynomial f of order m can be written as*

$$f(x) = \sum_{j=1}^{\binom{m+d-1}{m}} c_j (a_j^T x)^m$$

for some real numbers c_j.

Every polynomial of order m over $\mathcal{R}^d$ is a homogeneous polynomial of order m over $\mathcal{R}^{d+1}$ by replacing the constant 1 by a component $x^{(d+1)}$ raised to an appropriate power. Thus, any polynomial of order m over $\mathcal{R}^d$ may be decomposed exactly by

$$f(x) = \sum_{j=1}^{\binom{m+d}{m}} c_j \left(a_j^T x + b_j\right)^m$$

for some real numbers b_j, c_j and vectors $a_j \in \mathcal{R}^d$.

Polynomials may thus be represented exactly in the form $\sum_{i=1}^k \phi_i(a_i^T x)$ with $k = \binom{m+d}{m}$. As the polynomials are dense in $C[a, b]^d$, we have yet another proof that $\left\{ \sum_{i=1}^k \phi_i(a_i^T x) \right\}$ is dense in $C[a, b]^d$. See Logan and Shepp (1975) or Logan (1975) for other proofs.

The previous discussion suggests at least two families from which to select a discriminant function. As usual, we let $g(x) = I_{\{\psi(x) > 0\}}$ for a discriminant function ψ. Here ψ could be picked from

$$\mathcal{F}_m = \left\{ \sum_{j=1}^m c_j(a_j^T x + b_j)^m; b_j, c_j \in \mathcal{R}, a_j \in \mathcal{R}^d \right\}$$

or

$$\mathcal{F}_m' = \left\{ \sum_{j=1}^m c_j e^{a_j^T x}; c_j \in \mathcal{R}, a_j \in \mathcal{R}^d \right\},$$

where m is sufficiently large. If we draw ψ by minimizing the empirical error (admittedly at a tremendous computational cost), then convergence may result if m is not too large. We need to know the VC dimension of the classes of classifiers corresponding to $\mathcal{F}_m$ and $\mathcal{F}_m'$. Note that $\mathcal{F}_m$ coincides with all polynomials of order m and each such polynomial is the sum of at most $(m+1)^d$ monomials. If we invoke Lemma 30.2 and Theorem 12.6, then we get

Theorem 30.15. *Empirical risk minimization to determine $\{a_j, b_j, c_j\}$ in $\mathcal{F}_m$ leads to a universally consistent classifier provided that $m \to \infty$ and $m = o(n^{1/d} / \log n)$.*

Projection pursuit is very powerful and not at all confined to our limited discussion above. In particular, there are many other ways of constructing good consistent classifiers that do not require extensive computations such as empirical error minimization.

30.10   Radial Basis Function Networks

We may perform discrimination based upon networks with functions of the form

$$\psi(x) = \sum_{i=1}^{k} a_i K\left(\frac{x - x_i}{h_i}\right)$$

(and decision $g(x) = I_{\{\psi(x)>0\}}$), where k is an integer, $a_1, \ldots, a_k\, h_1, \ldots, h_k$ are constants, $x_1, \ldots, x_k \in \mathcal{R}^d$, and K is a kernel function (such as $K(u) = e^{-\|u\|^2}$ or $K(u) = 1/(1 + \|u\|^2)$). In this form, ψ covers several well-known methodologies:

(1) The *kernel rule* (Chapter 10): Take $k = n, a_i = 2Y_i - 1, h_i = h, x_i = X_i$. With this choice, for a large class of kernels, we are guaranteed convergence if $h \to 0$ and $nh^d \to \infty$. This approach is attractive as no difficult optimization problem needs to be solved.

(2) The *potential function method*. In Bashkirov, Braverman, and Muchnik (1964), the parameters are $k = n, h_i = h, x_i = X_i$. The weights a_i are picked to minimize the empirical error on the data, and h is held fixed. The original kernel suggested there is $K(u) = 1/(1 + \|u\|^2)$.

(3) *Linear discrimination.* For $k = 2, K(u) = e^{-\|u\|^2}, h_i \equiv h, a_1 = 1, a_2 = -1$, the set $\{x : \psi(x) > 0\}$ is a linear halfspace. This, of course, is not universally consistent. Observe that the separating hyperplane is the collection of all points x at equal distance from x_1 and x_2. By varying x_1 and x_2, all hyperplanes may be obtained.

(4) *Radial basis function (RBF) neural networks* (e.g., Powell (1987), Broomhead and Lowe (1988), Moody and Darken (1989), Poggio and Girosi (1990), Xu, Krzyżak, and Oja (1993), Xu, Krzyżak, and Yuille (1994), and Krzyżak, Linder, and Lugosi (1993)). An even more general function ψ is usually employed here:

$$\psi(x) = \sum_{i=1}^{k} c_i K\left((x - x_i)^T A_i(x - x_i)\right) + c_0,$$

where the A_i's are tunable $d \times d$ matrices.

(5) *Sieve methods.* Grenander (1981) and Geman and Hwang (1982) advocate the use of maximum likelihood methods to find suitable values for the tunable parameters in ψ (for k, K fixed beforehand) subject to certain compactness constraints on these parameters to control the abundance of choices one may have. If we were to use empirical error minimization, we would find, if $k \geq n$, that all data points can be correctly classified (take the h_i's small enough, set $a_i = 2Y_i - 1, x_i = X_i, k = n$), causing overfitting. Hence, k must be smaller than n if parameters are picked in this manner. Practical ways of choosing the parameters are discussed by Kraaijveld and Duin (1991),

and Chou and Chen (1992). In both (4) and (5), the x_i's may be thought of as representative prototypes, the a_i's as weights, and the h_i's as the radii of influence. As a rule, k is much smaller than n as $x_1, \ldots, x_k$ summarizes the information present at the data.

To design a consistent RBF neural network classifier, we may proceed as in (1). We may also take $k \to \infty$ but $k = o(n)$. Just let $(x_1, \ldots, x_k) \equiv (X_1, \ldots, X_k)$, $(a_1, \ldots, a_k \equiv 2Y_1 - 1, \ldots, 2Y_k - 1)$, and choose A_i or h_i to minimize a given error criterion based upon $X_{k+1}, \ldots, X_n$, such as

$$\hat{L}_n(g) = \frac{1}{n-k} \sum_{i=k+1}^{n} I_{\{g(X_i) \neq Y_i\}},$$

where $g(x) = I_{\{\psi(x) > 0\}}$, and ψ is as in (1). This is nothing but data splitting (Chapter 22). Convergence conditions are described in Theorem 22.1.

A more ambitious person might try empirical risk minimization to find the best $x_1, \ldots, x_k, a_1, \ldots, a_k, A_1, \ldots, A_k$ ($d \times d$ matrices as in (4)) based upon

$$\frac{1}{n} \sum_{i=1}^{n} I_{\{g(X_i) \neq Y_i\}}.$$

If $k \to \infty$, the class of rules contains a consistent subsequence, and therefore, it suffices only to show that the VC dimension is $o(n/\log n)$. This is a difficult task and some kernels yield infinite VC dimension, even if $d = 1$ and k is very small (see Chapter 25). However, there is a simple argument if $K = I_R$ for a simple set R. Let

$$\mathcal{A} = \left\{ \{x : x = a + Ay, y \in R\} : a \in \mathcal{R}^d, A \text{ a } d \times d \text{ matrix} \right\}.$$

If R is a sphere, then $\mathcal{A}$ is the class of all ellipsoids. The number of ways of shattering a set $\{x_1, \ldots, x_n\}$ by intersecting with members from $\mathcal{A}$ is not more than

$$n^{d(d+1)/2+1}$$

(see Theorem 13.9, Problem 13.10). The number of ways of shattering a set $\{x_1, \ldots, x_n\}$ by intersecting with sets of the form

$$\left\{ a_1 I_{R_1} + \cdots + a_k I_{R_k} > 0 : R_1, \ldots, R_k \in \mathcal{A}, a_1, \ldots, a_k \in \mathcal{R} \right\}$$

is not more than the product of all ways of shattering by intersections with R_1, with R_2, and so forth, that is,

$$n^{k(d(d+1)/2+1)}.$$

The logarithm of the shatter coefficient is $o(n)$ if $k = o(n/\log n)$. Thus, by Corollary 12.1, we have

Theorem 30.16. (KRZYŻAK, LINDER, AND LUGOSI (1993)). *If we take $k \to \infty$, $k = o(n/\log n)$ in the RBF classifier (4) in which $K = I_R$, R being the unit ball of*

$\mathcal{R}^d$, and in which all the parameters are chosen by empirical risk minimization, then $\mathbf{E}\{L_n\} \to L^*$ for all distributions of (X, Y). Furthermore, if $\mathcal{G}_k$ is the class of all RBF classifiers with k prototypes,

$$\mathbf{E}\{L_n\} - \inf_{g \in \mathcal{G}_k} L(g) \le 16\sqrt{\frac{k(d(d+1)/2 + 1)\log n + \log(8e)}{2n}}.$$

The theorem remains valid (with modified constants in the error estimate) when R is a hyperrectangle or polytope with a bounded number of faces. However, for more general K, the VC dimension is more difficult to evaluate.

For general kernels, consistent RBF classifiers can be obtained by empirical L_1 or L_2 error minimization (Problem 30.32). However, no efficient practical algorithms are known to compute the minima.

Finally, as suggested by Chou and Chen (1992) and Kraaijveld and Duin (1991), it is a good idea to place $x_1, \ldots, x_k$ by k-means clustering or another clustering method and to build an RBF classifier with those values or by optimization started at the given cluster centers.

Problems and Exercises

PROBLEM 30.1. Let k, l be integers, with $d \le k < n$, and $l \le \binom{k}{d}$. Assume X has a density. Let $\mathcal{A}$ be a collection of hyperplanes drawn from the $\binom{k}{d}$ possible hyperplanes through d points of $\{X_1, \ldots, X_k\}$, and let $g_{\mathcal{A}}$ be the corresponding natural classifier based upon the arrangement $\mathcal{P}(\mathcal{A})$. Take l such collections $\mathcal{A}$ at random and with replacement, and pick the best $\mathcal{A}$ by minimizing $\widehat{L}_n(g_{\mathcal{A}})$, where

$$\widehat{L}_n(g_{\mathcal{A}}) = \frac{1}{n-k} \sum_{i=k+1}^{n} I_{\{Y_i \ne g_{\mathcal{A}}(X_i)\}}.$$

Show that the selected classifier is consistent if $l \to \infty$, $l^d = o(n)$, $n/(l \log k) \to \infty$. (Note: this is applicable with $k = \lfloor n/2 \rfloor$, that is, half the sample is used to define $\mathcal{A}$, and the other half is used to pick a classifier empirically.)

PROBLEM 30.2. We are given a tree classifier with k internal linear splits and $k + 1$ leaf regions (a BSP tree). Show how to combine the neurons in a two-hidden-layer perceptron with k and $k + 1$ hidden neurons in the two hidden layers so as to obtain a decision that is identical to the tree-based classifier (Brent (1991), Sethi (1990; 1991)). For more on the equivalence of decision trees and neural networks, see Meisel (1990), Koutsougeras and Papachristou (1989), or Golea and Marchand (1990). HINT: Mimic the argument for arrangements in text.

PROBLEM 30.3. Extend the proof of Theorem 30.4 so that it includes any nondecreasing sigmoid with $\lim_{x \to -\infty} \sigma(x) = -1$ and $\lim_{x \to \infty} \sigma(x) = 1$. HINT: If t is large, $\sigma(tx)$ approximates the threshold sigmoid.

PROBLEM 30.4. This exercise states denseness in $L_1(\mu)$ for any probability measure μ. Show that for every probability measure μ on $\mathcal{R}^d$, every measurable function $f : \mathcal{R}^d \to \mathcal{R}$

with $\int |f(x)|\mu(dx) < \infty$, and every $\epsilon > 0$, there exists a neural network with one hidden layer and function $\psi(x)$ as in (30.2) such that

$$\int |f(x) - \psi(x)|\mu(dx) < \epsilon$$

(Hornik (1991)). HINT: Proceed as in Theorem 30.4, considering the following:

(1) Approximate f in $L_1(\mu)$ by a continuous function $g(x)$ that is zero outside some bounded set $B \subset \mathcal{R}^d$. Since $g(x)$ is bounded, its maximum $\beta = \max_{x \in \mathcal{R}^d} |g(x)|$ is finite.

(2) Now, choose a to be a positive number large enough so that both $B \subset [-a, a]^d$ and $\mu\left([-a, a]^d\right)$ is large.

(3) Extend the restriction of $g(x)$ to $[-a, a]^d$ periodically by tiling over all of $\mathcal{R}^d$. The obtained function, $\widehat{g}(x)$ is a good approximation of $g(x)$ in $L_1(\mu)$.

(4) Take the Fourier series approximation of $\widehat{g}(x)$, and use the Stone-Weierstrass theorem as in Theorem 30.4.

(5) Observe that every continuous function on the real line that is zero outside some bounded interval can be arbitrarily closely approximated uniformly *over the whole real line* by one-dimensional neural networks. Thus, bounded functions such as the sine and cosine functions can be approximated arbitrarily closely by neural networks in $L_1(\nu)$ for any probability measure ν on $\mathcal{R}$.

(6) Apply the triangle inequality to finish the proof.

PROBLEM 30.5. Generalize the previous exercise for denseness in $L_p(\mu)$. More precisely, let $1 \leq p < \infty$. Show that for every probability measure μ on $\mathcal{R}^d$, every measurable function $f : \mathcal{R}^d \to \mathcal{R}$ with $\int |f(x)|^p \mu(dx) < \infty$, and every $\epsilon > 0$, there exists a neural network with one hidden layer $h(x)$ such that

$$\left(\int |f(x) - h(x)|^p \mu(dx)\right)^{1/p} < \epsilon$$

(Hornik (1991)).

PROBLEM 30.6. COMMITTEE MACHINES. Let $C^{(k)}$ be the class of all committee machines. Prove that for all distributions of (X, Y),

$$\lim_{k \to \infty} \inf_{\phi \in C^{(k)}} L(\phi) = L^*.$$

HINT: For a one-hidden-layer neural network with coefficients c_i in (30.2), approximate the c_i's by discretization (truncation to a grid of values), and note that $c_i \psi_i(x)$ may thus be approximated in a committee machine by a sufficient number of identical copies of $\psi_i(x)$. This only forces the number of neurons to be a bit larger.

PROBLEM 30.7. Prove Theorem 30.5 for arbitrary sigmoids. HINT: Approximate the threshold sigmoid by $\sigma(tx)$ for a sufficiently large t.

PROBLEM 30.8. Let σ be a nondecreasing sigmoid with $\sigma(x) \to -1$ if $x \to -\infty$ and $\sigma(x) \to 1$ if $x \to \infty$. Denote by $C_\sigma^{(k)}$ the class of corresponding neural network classifiers with k hidden layers. Show that $V_{C_\sigma^{(k)}} \geq V_{C^{(k)}}$, where $C^{(k)}$ is the class corresponding to the threshold sigmoid.

PROBLEM 30.9. This exercise generalizes Theorem 30.5 for not fully connected networks with one hidden layer. Consider the class $\mathcal{C}^{(k)}$ of one-hidden-layer neural networks with the threshold sigmoid such that each of the k nodes in the hidden layer are connected to $d_1, d_2, \ldots, d_k$ inputs, where $1 \leq d_i \leq d$. More precisely, $\mathcal{C}^{(k)}$ contains all classifiers based on functions of the form

$$\psi(x) = c_0 + \sum_{i=1}^{k} c_i \sigma(\psi_i(x)), \quad \text{where} \quad \psi_i(x) = \sum_{j=1}^{d_i} a_j x^{(m_{i,j})},$$

where for each i, $(m_{i,1}, \ldots, m_{i,d_i})$ is a vector of distinct positive integers not exceeding d. Show that

$$V_{\mathcal{C}^{(k)}} \geq \sum_{i=1}^{k} d_i + 1$$

if the a_j's, c_i's and $m_{i,j}$'s are the tunable parameters (Bartlett (1993)).

PROBLEM 30.10. Let σ be a sigmoid that takes m different values. Find upper bounds on the VC dimension of the class $\mathcal{C}_\sigma^{(k)}$.

PROBLEM 30.11. Consider a one-hidden-layer neural network ψ. If $(X_1, Y_1), \ldots, (X_n, Y_n)$ are fixed and all X_i's are different, show that with n hidden neurons, we are always able to tune the weights such that $Y_i = \psi(X_i)$ for all i. (This remains true if $Y \in \mathcal{R}$ instead of $Y \in \{0, 1\}$.) The property above describes a situation of overfitting that occurs when the neural network becomes too "rich"—recall also that the VC dimension, which is at least d times the number of hidden neurons, must remain smaller than n for any meaningful training.

PROBLEM 30.12. THE BERNSTEIN PERCEPTRON. Consider the following perceptron for one-dimensional data:

$$\phi(x) = \begin{cases} 1 & \text{if } \sum_{i=1}^{k} a_i x^i (1 - x)^{k-i} > 1/2 \\ 0 & \text{otherwise.} \end{cases}$$

Let us call this the Bernstein perceptron since it involves Bernstein polynomials. If n data points are collected, how would you choose k (as a function of n) and how would you adjust the weights (the a_i's) to make sure that the Bernstein perceptron is consistent for all distributions of (X, Y) with $\mathbf{P}\{X \in [0, 1]\} = 1$? Can you make the Bernstein perceptron consistent for all distributions of (X, Y) on $\mathcal{R} \times \{0, 1\}$?

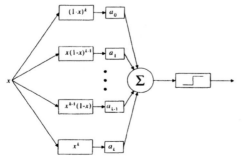

FIGURE 30.15. The Bernstein perceptron.

PROBLEM 30.13. Use the ideas of Section 25.5 and Problem 25.11 to prove Theorem 30.8.

PROBLEM 30.14. Consider Specht's Padaline with $r = r_n \uparrow \infty$. Let $h = h_n \downarrow 0$, and $nh^d \to \infty$. Show that for any distribution of (X, Y), $E\{L_n\} \to L^*$.

PROBLEM 30.15. BOUNDED FIRST LAYERS. Consider a feed-forward neural network with any number of layers, with only one restriction, that is, the first layer has at most $k < d$ outputs $z_1, \ldots, z_k$, where

$$z_j = \sigma \left(b_j + \sum_{i=1}^{d} a_{ji} x^{(i)} \right),$$

$x = \left(x^{(1)}, \ldots, x^{(d)} \right)$ is the input, and σ is an arbitrary function $\mathcal{R} \to \mathcal{R}$ (not just a sigmoid). The integer k remains fixed. Let A denote the $k \times (d + 1)$ matrix of weights a_{ji}, b_j.

  (1) If $L^*(A)$ is the Bayes error for a recognition problem based upon (Z, Y), with $Z = (Z_1, \ldots, Z_k)$ and

$$Z_j = \sigma \left(b_j + \sum_{i=1}^{d} a_{ji} X^{(i)} \right), \qquad X = \left(X^{(1)}, \ldots, X^{(d)} \right),$$

then show that for some distribution of (X, Y), $\inf_A L^*(A) > L^*$, where L^* is the Bayes probability of error for (X, Y).
  (2) If $k \geq d$ however, show that for any strictly monotonically increasing sigmoid σ, $\inf_A L^*(A) = L^*$.
  (3) Use (1) to conclude that any neural network based upon a first layer with $k < d$ outputs is not consistent for some distribution, regardless of how many layers it has (note however, that the inputs of each layer are restricted to be the outputs of the previous layer).

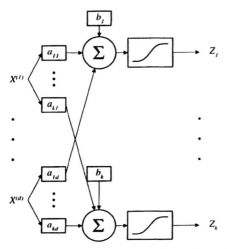

FIGURE 30.16. *The first layer is restricted to have k outputs. It has $k(d + 1)$ tunable parameters.*

PROBLEM 30.16. Find conditions on k_n and β_n that guarantee strong universal consistency in Theorem 30.9.

PROBLEM 30.17. BARRON NETWORKS. Call a *Barron network* a network of any number of layers (as in Figure 30.13) with 2 inputs per cell and cells that perform the operation $\alpha + \beta x + \gamma y + \delta xy$ on inputs $x, y \in \mathcal{R}$, with trainable weights $\alpha, \beta, \gamma, \delta$. If ψ is the output of a network with d inputs and k cells (arranged in any way), compute an upper bound for the VC dimension of the classifier $g(x) = I_{\{\psi(x)>0\}}$, as a function of d and k. Note that the structure of the network (i.e., the positions of the cells and the connections) is variable.

PROBLEM 30.18. CONTINUED. Restrict the Barron network to l layers and k cells per layer (for kl cells total), and repeat the previous exercise.

PROBLEM 30.19. CONTINUED. Find conditions on k and l in the previous exercises that would guarantee the universal consistency of the Barron network, if we train to minimize the empirical error.

PROBLEM 30.20. Consider the family of functions $\mathcal{F}_{k,l}$ of the form

$$\sum_{i=1}^{k} \sum_{j=1}^{l} w_{ij} (\psi_i(x))^j, \quad \psi_i(x) = \sum_{i'=1}^{d} \sum_{j'=1}^{l} w'_{ii'j'} \left(x^{(i')}\right)^{j'}, \quad 1 \le i \le k,$$

where all the w_{ij}'s and $w'_{ii'j'}$'s are tunable parameters. Show that for every $f \in C[0,1]^d$ and $\epsilon > 0$, there exist k, l large enough so that for some $g \in \mathcal{F}_{k,l}$, $\sup_{x \in [0,1]^d} |f(x) - g(x)| \le \epsilon$.

PROBLEM 30.21. CONTINUED. Obtain an upper bound on the VC dimension of the above two-hidden-layer network. HINT: The VC dimension is usually about equal to the number of degrees of freedom (which is $(1+d)lk$ here).

PROBLEM 30.22. CONTINUED. If g_n is the rule obtained by empirical error minimization over $\mathcal{F}_{k,l}$, then show that $L(g_n) \to L^*$ in probability if $k \to \infty$, $l \to \infty$, and $kl = o(n/\log n)$.

PROBLEM 30.23. How many different monomials of order r in $x^{(1)}, \ldots, x^{(d)}$ are there? How does this grow with r when d is held fixed?

PROBLEM 30.24. Show that ψ_1, ψ_2, and ϕ in the bit-interleaving example in the section on Kolmogorov-Lorentz representations are not continuous. Which are the points of discontinuity?

PROBLEM 30.25. Let

$$\mathcal{F}'_m = \left\{ \sum_{j=1}^{m} c_j e^{a_j^T x} ; c_j \in \mathcal{R}, a_j \in \mathcal{R}^d \right\}.$$

Pick $\{a_j, c_j\}$ by empirical risk minimization for the classifier $g(x) = I_{\{\psi(x)>0\}}$, $\psi \in \mathcal{F}'_m$. Show that $\mathbf{E}\{L_n\} \to L^*$ for all distributions of (X, Y) when $m \to \infty$ and $m = o(n^{1/d}/\log n)$.

PROBLEM 30.26. Write $\left(x^{(1)}x^{(2)}\right)^2$ as $\sum_{i=1}^{4} \left(a_{i1}x^{(1)} + a_{i2}x^{(2)}\right)^2$ and identify the coefficients $\{a_{i1}, a_{i2}\}$. Show that there is an entire subspace of solutions (Diaconis and Shahshahani (1984)).

PROBLEM 30.27. Show that $e^{x^{(1)}x^{(2)}}$ and $\sin(x^{(1)}x^{(2)})$ cannot be written in the form $\sum_{i=1}^{k} \psi_i(a_i^T x)$ for any finite k, where $x = (x^{(1)}, x^{(2)})$ and $a_i \in \mathcal{R}^2$, $1 \le i \le k$ (Diaconis and Shahshahani (1984)). Thus, the projection pursuit representation of functions can only at best approximate all continuous functions on bounded sets.

PROBLEM 30.28. Let $\mathcal{F}$ be the class of classifiers of the form $g = I_{\{\psi>0\}}$, where $\psi(x) = a_1 f_1(x^{(1)}) + a_2 f_2(x^{(2)})$, for arbitrary functions f_1, f_2, and coefficients $a_1, a_2 \in \mathcal{R}$. Show that for some distribution of (X, Y) on $\mathcal{R}^2 \times \{0, 1\}$, $\inf_{g \in \mathcal{F}} L(g) > L^*$, so that there is no hope of meaningful distribution-free classification based on additive functions only.

PROBLEM 30.29. CONTINUED. Repeat the previous exercise for functions of the form $\psi(x) = a_1 f_1(x^{(2)}, x^{(3)}) + a_2 f_2(x^{(1)}, x^{(3)}) + a_3 f_3(x^{(1)}, x^{(2)})$ and distributions of (X, Y) on $\mathcal{R}^3 \times \{0, 1\}$. Thus, additive functions of pairs do not suffice either.

PROBLEM 30.30. Let $K : \mathcal{R} \rightarrow \mathcal{R}$ be a nonnegative bounded kernel with $\int K(x)dx < \infty$. Show that for any $\epsilon > 0$, any measurable function $f : \mathcal{R}^d \rightarrow \mathcal{R}$ and probability measure μ such that $\int |f(x)|\mu(dx) < \infty$, there exists a function ψ of the form $\psi(x) = \sum_{i=1}^k c_i K \left((x - b_i)^T A_i(x - b_i) \right) + c_0$ such that $\int |f(x) - \psi(x)\mu(dx) < \epsilon$ (see Poggio and Girosi (1990), Park and Sandberg (1991; 1993), Darken, Donahue, Gurvits, and Sontag (1993), Krzyżak, Linder, and Lugosi (1993) for such denseness results). HINT: Relate the problem to a similar result for kernel estimates.

PROBLEM 30.31. Let $C^{(k)}$ be the class of classifiers defined by the functions

$$\sum_{i=1}^k c_i K \left((x - b_i)^T A_i(x - b_i) \right) + c_0.$$

Find upper bounds on its VC dimension when K is an indicator of an interval containing the origin.

PROBLEM 30.32. Consider the class of radial basis function networks

$$\mathcal{F}_n = \left\{ \sum_{i=1}^{k_n} c_i K \left((x - b_i)^T A_i(x - b_i) \right) + c_0 : \sum_{i=0}^{k_n} |c_i| \le \beta_n \right\},$$

where K is nonnegative, unimodal, bounded, and continuous. Let ψ_n be a function that minimizes $J_n(\psi) = n^{-1} \sum_{i=1}^n |\psi(X_i) - Y_i|$ over $\psi \in \mathcal{F}_n$, and define g_n as the corresponding classifier. Prove that if $k_n \rightarrow \infty$, $\beta_n \rightarrow \infty$, and $k_n \beta_n^2 \log(k_n \beta_n) = o(n)$ as $n \rightarrow \infty$, then g_n is universally consistent (Krzyżak, Linder, and Lugosi (1993)). HINT: Proceed as in the proof of Theorem 30.9. Use Problem 30.30 to handle the approximation error. Bounding the covering numbers needs a little additional work.

31
Other Error Estimates

In this chapter we discuss some alternative error estimates that have been introduced to improve on the performance of the standard estimates—holdout, resubstitution, and deleted—we have encountered so far. The first group of these estimates—smoothed and posterior probability estimates—are used for their small variance. However, we will give examples that show that classifier selection based on the minimization of these estimates may fail even in the simplest situations. Among other alternatives, we deal briefly with the rich class of bootstrap estimates.

31.1   Smoothing the Error Count

The resubstitution, deleted, and holdout estimates of the error probability (see Chapters 22 and 23) are all based on counting the number of errors committed by the classifier to be tested. This is the reason for the relatively large variance inherently present in these estimates. This intuition is based on the following. Most classification rules can be written into the form

$$g_n(x) = \begin{cases} 0 & \text{if } \eta_n(x, D_n) \geq 1/2 \\ 1 & \text{otherwise,} \end{cases}$$

where $\eta_n(x, D_n)$ is either an estimate of the a posteriori probability $\eta(x)$, as in the case of histogram, kernel, or nearest neighbor rules; or something else, as for generalized linear, or neural network classifiers. In any case, if $\eta_n(x, D_n)$ is close to $1/2$, then we feel that the decision is less robust compared to when the value of $\eta_n(x, D_n)$ is far from $1/2$. In other words, intuitively, inverting the value of the decision $g_n(x)$ at a point x, where $\eta_n(x, D_n)$ is close to $1/2$ makes less difference in the error probability, than if $|\eta_n(x, D_n) - 1/2|$ is large. The error estimators based on counting the number of errors do not take the value of $\eta_n(x)$ into account: they "penalize" errors with the same amount, no matter what the value of η_n is. For example, in the case of the resubstitution estimate

$$L_n^{(R)} = \frac{1}{n} \sum_{i=1}^{n} I_{\{g_n(X_i) \neq Y_i\}},$$

each error contributes with $1/n$ to the overall count. Now, if $\eta_n(X_i, D_n)$ is close to $1/2$, a small perturbation of X_i can flip the decision $g_n(X_i)$, and therefore change the value of the estimate by $1/n$, although the error probability of the rule g_n probably does not change by this much. This phenomenon is what causes the relatively large variance of error counting estimators.

Glick (1978) proposed a modification of the counting error estimates. The general form of his "smoothed" estimate is

$$L_n^{(S)} = \frac{1}{n} \sum_{i=1}^{n} \left(I_{\{Y_i=0\}} r(\eta_n(X_i, D_n)) + I_{\{Y_i=1\}}(1 - r(\eta_n(X_i, D_n))) \right),$$

where r is a monotone increasing function satisfying $r(1/2 - u) = 1 - r(1/2 + u)$. Possible choices of the smoothing function $r(u)$ are $r(u) = u$, or $r(u) = 1/(1 + e^{1/2-cu})$, where the parameter $c > 0$ may be adjusted to improve the behavior of the estimate (see also Glick (1978), Knoke (1986), or Tutz (1985)). Both of these estimates give less penalty to errors close to the decision boundary, that is, to errors where η_n is close to $1/2$. Note that taking $r(u) = I_{\{u \geq 1/2\}}$ corresponds to the resubstitution estimate. We will see in Theorem 31.2 below that if r is smooth, then $L_n^{(S)}$ indeed has a very small variance in many situations.

Just like the resubstitution estimate, the estimate $L_n^{(S)}$ may be strongly optimistically biased. Just consider the 1-nearest neighbor rule, when $L_n^{(S)} = 0$ with probability one, whenever X has a density. To combat this defect, one may define the deleted version of the smoothed estimate,

$$L_n^{(SD)} = \frac{1}{n} \sum_{i=1}^{n} \left(I_{\{Y_i=0\}} r(\eta_{n-1}(X_i, D_{n,i})) + I_{\{Y_i=1\}}(1 - r(\eta_{n-1}(X_i, D_{n,i}))) \right),$$

where $D_{n,i}$ is the training sequence with the i-th pair (X_i, Y_i) deleted. The first thing we notice is that this estimate is still biased, even asymptotically. To illustrate this point, consider $r(u) = u$. In this case,

$$
\begin{aligned}
\mathbf{E}&\left\{L_n^{(SD)}\right\} \\
&= \mathbf{E}\left\{ I_{\{Y=0\}} \eta_{n-1}(X, D_{n-1}) + I_{\{Y=1\}}(1 - \eta_{n-1}(X, D_{n-1})) \right\} \\
&= \mathbf{E}\{(1 - \eta(X))\eta_{n-1}(X, D_{n-1}) + \eta(X)(1 - \eta_{n-1}(X, D_{n-1}))\}.
\end{aligned}
$$

If the estimate $\eta_{n-1}(x, D_{n-1})$ was perfect, that is, equal to $\eta(x)$ for every x, then the expected value above would be $2\mathbf{E}\{\eta(X)(1 - \eta(X))\}$, which is the asymptotic error probability L_{NN} of the 1-NN rule. In fact,

$$
\begin{aligned}
&\left| \mathbf{E}\left\{L_n^{(SD)}\right\} - L_{NN} \right| \\
&= \left| \mathbf{E}\left\{ I_{\{Y=0\}} \eta_{n-1}(X, D_{n-1}) + I_{\{Y=1\}}(1 - \eta_{n-1}(X, D_{n-1})) \right. \right. \\
&\qquad\qquad \left. \left. - \left(I_{\{Y=0\}} \eta(X) + I_{\{Y=1\}}(1 - \eta(X)) \right) \right\} \right| \\
&\le 2 \left| \mathbf{E}\left\{ \eta_{n-1}(X, D_{n-1}) - \eta(X) \right\} \right|.
\end{aligned}
$$

This means that when the estimate $\eta_n(x)$ of the a posteriori probability of $\eta(x)$ is consistent in $L_1(\mu)$, then $L_n^{(SD)}$ converges to L_{NN}, and not to L^*!

Biasedness of an error estimate is not necessarily a bad property. In most applications, all we care about is how the classification rule selected by minimizing the error estimate works. Unfortunately, in this respect, smoothed estimates perform poorly, even compared to other strongly biased error estimates such as the empirical squared error (see Problem 31.4). The next example illustrates our point.

Theorem 31.1. *Let the distribution of X be concentrated on two values such that $P\{X = a\} = P\{X = b\} = 1/2$, and let $\eta(a) = 3/8$ and $\eta(b) = 5/8$. Assume that the smoothed error estimate*

$$
L_n^{(S)}(\eta') = \frac{1}{n} \sum_{i=1}^{n} \left(I_{\{Y_i=0\}} \eta'(X_i) + I_{\{Y_i=1\}}(1 - \eta'(X_i)) \right)
$$

is minimized over $\eta' \in \mathcal{F}$, to select a classifier from $\mathcal{F}$, where the class $\mathcal{F}$ contains two functions, the true a posteriori probability function η, and $\tilde{\eta}$, where

$$
\tilde{\eta}(a) = 0, \quad \tilde{\eta}(b) = \frac{3}{8}.
$$

Then the probability that $\tilde{\eta}$ is selected converges to one as $n \to \infty$.

PROOF. Straightforward calculation shows that

$$
\mathbf{E}\left\{L_n^{(S)}(\eta)\right\} = \frac{15}{32} \quad \text{and} \quad \mathbf{E}\left\{L_n^{(S)}(\tilde{\eta})\right\} = \frac{6}{32}.
$$

The statement follows from the law of large numbers. $\square$

REMARK. The theorem shows that even if the true a posteriori probability function is contained in a finite class of candidates, the smoothed estimate with $r(u) = u$ is unable to select a good discrimination rule. The result may be extended to general smooth r's. As Theorems 15.1 and 29.2 show, empirical squared error minimization or maximum likelihood never fail in this situation. $\square$

Finally we demonstrate that if r is smooth, then the variance of $L_n^{(S)}$ is indeed small. Our analysis is based on the work of Lugosi and Pawlak (1994). The bounds for the variance of $L_n^{(S)}$ remain valid for $L_n^{(SD)}$. Consider classification rules of the form

$$g_n(x) = \begin{cases} 0 & \text{if } \eta_n(x, D_n) \geq 1/2 \\ 1 & \text{otherwise,} \end{cases}$$

where $\eta_n(x, D_n)$ is an estimate of the a posteriori probability $\eta(x)$. Examples include the histogram rule, where

$$\eta_n(x, D_n) = \frac{\sum_{i=1}^n Y_i I_{\{X_i \in A_n(x)\}}}{\sum_{i=1}^n I_{\{X_i \in A_n(x)\}}}$$

(see Chapters 6 and 9), the k-nearest neighbor rule, where

$$\eta_n(x, D_n) = \frac{1}{k} \sum_{i=1}^k Y_{(i)}(x)$$

(Chapters 5 and 11), or the kernel rule, where

$$\eta_n(x, D_n) = \frac{\sum_{i=1}^n Y_i K_h(x - X_i)}{n\mathbf{E}\{K_h(x - X)\}}$$

(Chapter 10). In the sequel, we concentrate on the performance of the smoothed estimate of the error probability of these nonparametric rules. The next theorem shows that for these rules, the variance of the smoothed error estimate is $O(1/n)$, no matter what the distribution is. This is a significant improvement over the variance of the deleted estimate, which, as pointed out in Chapter 23, can be larger than $1/\sqrt{2\pi n}$.

Theorem 31.2. *Assume that the smoothing function $r(u)$ satisfies $0 \leq r(u) \leq 1$ for $u \in [0, 1]$, and is uniformly Lipschitz continuous, that is,*

$$|r(u) - r(v)| \leq c|u - v|$$

for all u, v, and for some constant c. Then the smoothed estimate $L_n^{(S)}$ of the histogram, k-nearest neighbor, and moving window rules (with kernel $K = I_{S_{0,1}}$) satisfies

$$\mathbf{P}\left\{\left|L_n^{(S)} - \mathbf{E}\left\{L_n^{(S)}\right\}\right| \geq \epsilon\right\} \leq 2e^{-2n\epsilon^2/C},$$

and

$$\mathbf{Var}\{L_n^{(S)}\} \leq \frac{C}{4n},$$

where C is a constant depending on the rule only. In the case of the histogram rule the value of C is $C = (1 + 4c)^2$, for the k-nearest neighbor rule $C = (1 + 2c\gamma_d)^2$, and for the moving window rule $C = (1 + 2c\beta_d)^2$. Here c is the constant in the Lipschitz condition, γ_d is the minimal number of cones centered at the origin of angle $\pi/6$ that cover $\mathcal{R}^d$, and β_d is the minimal number of balls of radius $1/2$ that cover the unit ball in $\mathcal{R}^d$.

REMARK. Notice that the inequalities of Theorem 31.2 are valid for *all* n, ϵ, and $h > 0$ for the histogram and moving window rules, and k for the nearest neighbor rules. Interestingly, the constant C does not change with the dimension in the histogram case, but grows exponentially with d for the k-nearest neighbor and moving window rules. □

PROOF. The probability inequalities follow from appropriate applications of Mc-Diarmid's inequality. The upper bound on the variance follows similarly from Theorem 9.3. We consider each of the three rules in turn.

THE HISTOGRAM RULE. Let $(x_1, y_1), \ldots, (x_n, y_n)$ be a fixed training sequence. If we can show that by replacing the value of a pair (x_i, y_i) in the training sequence by some (x_i', y_i') the value of the estimate $L_n^{(S)}$ can change by at most $(1 + 2c)/n$, then the inequality follows by applying Theorem 9.2 with $c_i = 5/n$.

The i-th term of the sum in $L_n^{(S)}$ can change by one, causing $1/n$ change in the average. Obviously, all the other terms in the sum that can change are the ones corresponding to the x_j's that are in the same set of the partition as either x_i or x_i'. Denoting the number of points in the same set with x_i and x_i' by k and k' respectively, it is easy to see that the estimate of the a posteriori probabilities in these points can change by at most $2/k$ and $2/k'$, respectively. It means that the overall change in the value of the sum can not exceed $(1+k\frac{2c}{k}+k'\frac{2c}{k'})/n = (1+4c)/n$.

THE K-NEAREST NEIGHBOR RULE. To avoid difficulties caused by breaking distance ties, assume that X has a density. Then recall that the application of Lemma 11.1 for the empirical distribution implies that no X_j can be one of the k nearest neighbors of more than $k\gamma_d$ points from D_n. Thus, changing the value of one pair in the training sequence can change at most $1 + 2k\gamma_d$ terms in the expression of $L_n^{(D)}$, one of them by at most 1, and all the others by at most c/k. Theorem 9.2 yields the result.

THE MOVING WINDOW RULE. Again, we only have to check the condition of Theorem 9.2 with $c_i = (1 + 2c\beta_d)/n$. Fix a training sequence $(x_1, y_1), \ldots, (x_n, y_n)$ and replace the pair (x_i, y_i) by (x_i', y_i'). Then the i-th term in the sum of the expression of $L_n^{(S)}$ can change by at most one. Clearly, the j-th term, for which $x_j \notin S_{x_i,h}$ and $x_j \notin S_{x_i',h}$, keeps its value. It is easy to see that all the other terms can change by at most $c \cdot \max\{1/k_j, 1/k_j'\}$, where k_j and k_j' are the numbers of points x_k, $k \neq i, j$, from the training sequence that fall in $S_{x_i,h}$ and $S_{x_i',h}$, respectively. Thus, the overall change in the sum does not exceed

$$1 + \sum_{x_j \in S_{x_i,h}} \frac{c}{k_j} + \sum_{x_j \in S_{x_i',h}} \frac{c}{k_j'}.$$

It suffices to show by symmetry that $\sum_{x_j \in S_{x_i,h}} 1/k_j \leq \beta_d$. Let $n_j = |\{x_k, \ k \neq i, j : x_k \in S_{x_i,h} \cap S_{x_j,h}\}|$. Then clearly,

$$\sum_{x_j \in S_{x_i,h}} \frac{1}{k_j} \leq \sum_{x_j \in S_{x_i,h}} \frac{1}{n_j}.$$

To bound the right-hand side from above, cover $S_{x_i,h}$ by β_d balls $S_1, \ldots, S_{\beta_d}$ of radius $h/2$. Denote the number of points falling in them by l_m $(m = 1, \ldots, \beta_d)$:

$$l_m = |\{x_k, \ k \neq i : x_k \in S_{x_i,h} \cap S_m\}|.$$

Then

$$\sum_{x_j \in S_{x_i,h}} \frac{1}{n_j} \leq \sum_{x_j \in S_{x_i,h}} \sum_{m:x_j \in S_m} \frac{1}{l_m} = \beta_d,$$

and the theorem is proved. $\square$

31.2   Posterior Probability Estimates

The error estimates discussed in this section improve on the biasedness of smoothed estimates, while preserving their small variance. Still, these estimates are of questionable utility in classifier selection. Considering the formula

$$L(g_n) = \mathbf{E}\left\{ I_{\{\eta_n(X,D_n)\leq 1/2\}}\eta(X) + I_{\{\eta_n(X,D_n)>1/2\}}(1 - \eta(X))\Big| D_n\right\}$$

for the error probability of a classification rule $g_n(x) = I_{\{\eta_n(x,D_n)>1/2\}}$, it is plausible to introduce the estimate

$$L_n^{(P)} = \frac{1}{n}\sum_{i=1}^{n}\left(I_{\{\eta_n(X_i,D_n)\leq 1/2\}}\eta_n(X_i, D_n)\right.$$
$$\left. + I_{\{\eta_n(X_i,D_n)>1/2\}}(1 - \eta_n(X_i, D_n))\right),$$

that is, the expected value is estimated by a sample average, and instead of the (unknown) a posteriori probability $\eta(x)$, its estimate $\eta_n(x, D_n)$ is plugged into the formula of L_n. The estimate $L_n^{(P)}$ is usually called the *posterior probability* error estimate. In the case of nonparametric rules such as histogram, kernel, and k-NN rules it is natural to use the corresponding nonparametric estimates of the a posteriori probabilities for plugging in the expression of the error probability. This, and similar estimates of L_n have been introduced and studied by Fukunaga and Kessel (1973), Hora and Wilcox (1982), Fitzmaurice and Hand (1987), Ganesalingam and McLachlan (1980), Kittler and Devijver (1981), Matloff and Pruitt (1984), Moore, Whitsitt, and Landgrebe (1976), Pawlak (1988), Schwemer and Dunn (1980), and Lugosi and Pawlak (1994). It is interesting to notice the similarity between the estimates $L_n^{(S)}$ and $L_n^{(P)}$, although they were developed from different scenarios.

To reduce the bias, we can use the leave-one-out, (or deleted) version of the estimate,

$$L_n^{(PD)} = \frac{1}{n}\sum_{i=1}^{n}\left(I_{\{\eta_{n-1}(X_i,D_{n,i})\leq 1/2\}}\eta_{n-1}(X_i, D_{n,i})\right)$$
$$+ I_{\{\eta_{n-1}(X_i,D_{n,i})>1/2\}}(1 - \eta_{n-1}(X_i, D_{n,i}))).$$

The deleted version $L_n^{(PD)}$, has a much better bias than $L_n^{(SD)}$. We have the bound

$$
\begin{aligned}
&\left|\mathbf{E}\left\{L_n^{(PD)}\right\} - \mathbf{E}L_{n-1}\right| \\
&= \left|\mathbf{E}\left\{I_{\{g_n(X)=0\}}\eta_{n-1}(X, D_{n-1}) + I_{\{g_n(X)=1\}}(1 - \eta_{n-1}(X, D_{n-1}))\right.\right. \\
&\qquad \left.\left. - \left(I_{\{g_n(X)=0\}}\eta(X) + I_{\{g_n(X)=1\}}(1 - \eta(X))\right)\right\}\right| \\
&\leq 2\left|\mathbf{E}\left\{\eta_{n-1}(X, D_{n-1}) - \eta(X)\right\}\right|.
\end{aligned}
$$

This means that if the estimate $\eta_n(x)$ of the a posteriori probability of $\eta(x)$ is consistent in $L_1(\mu)$, then $\mathbf{E}L_n^{(PD)}$ converges to L^*. This is the case for all distributions for the histogram and moving window rules if $h \to 0$ and $nh^d \to \infty$, and the k-nearest neighbor rule if $k \to \infty$ and $k/n \to 0$, as it is seen in Chapters 9, 10, and 11. For specific cases it is possible to obtain sharper bounds on the bias of $L_n^{(PD)}$. For the histogram rule, Lugosi and Pawlak (1994) carried out such analysis. They showed for example that the estimate $L_n^{(PD)}$ is optimistically biased (see Problem 31.2).

Posterior probability estimates of L_n share the good stability properties of smoothed estimates (Problem 31.1).

Finally, let us select a function $\eta' : \mathcal{R}^d \to [0, 1]$—and a corresponding rule $g(x) = I_{\{\eta'(x)>1/2\}}$—from a class $\mathcal{F}$ based on the minimization of the posterior probability error estimate

$$
L_n^{(P)}(\eta') = \frac{1}{n}\sum_{i=1}^{n}\left(I_{\{\eta'(X_i)\leq 1/2\}}\eta'(X_i) + I_{\{\eta'(X_i)>1/2\}}(1 - \eta'(X_i))\right).
$$

Observe that $L_n^{(P)}(\eta') = 0$ when $\eta'(x) \in \{0, 1\}$ for all x, that is, rule selection based on this estimate just does not make sense. The reason is that $L_n^{(P)}$ ignores the Y_i's of the data sequence!

Fukunaga and Kessel (1973) argued that efficient posterior probability estimators can be obtained if additional unclassified observations are available. Very often in practice, in addition to the training sequence D_n, further feature vectors $X_{n+1}, \ldots, X_{n+l}$ are given without their labels $Y_{n+1}, \ldots, Y_{n+l}$, where the X_{n+i} are i.i.d., independent from X and D_n, and they have the same distribution as that of X. This situation is typical in medical applications, when large sets of medical records are available, but it is usually very expensive to get their correct diagnosis. These unclassified samples can be efficiently used for testing the performance of a classifier designed from D_n by using the estimate

$$
\begin{aligned}
L_{n,l}^{(U)} &= \frac{1}{l}\sum_{i=1}^{l}\left(I_{\{\eta_n(X_{n+i}, D_n)\leq 1/2\}}\eta_n(X_{n+i}, D_n)\right. \\
&\qquad \left. + I_{\{\eta_n(X_{n+i}, D_n)>1/2\}}(1 - \eta_n(X_{n+i}, D_n))\right).
\end{aligned}
$$

Again, using $L_{n,l}^{(U)}$ for rule selection is meaningless.

31.3   Rotation Estimate

This method, suggested by Toussaint and Donaldson (1970), is a combination of the holdout and deleted estimates. It is sometimes called the Π-estimate. Let $s < n$ be a positive integer (typically much smaller than n), and assume, for the sake of simplicity, that $q = n/s$ is an integer. The rotation method forms the holdout estimate, by holding the first s pairs of the training data out, then the second s pairs, and so forth. The estimate is defined by averaging the q numbers obtained this way. To formalize, denote by $D_{n,j}^{(s)}$ the training data, with the j-th s-block held out $(j = 1, \ldots, q)$:

$$D_{n,j}^{(s)} = ((X_1, Y_1), \ldots, (X_{s(j-1)}, Y_{s(j-1)}), (X_{sj+1}, Y_{sj+1}), \ldots, (X_n, Y_n)).$$

The estimate is defined by

$$L_{n,s}^{(\Pi)} = \frac{1}{q} \sum_{j=1}^{q} \frac{1}{s} \sum_{i=s(j-1)+1}^{sj} I_{\left\{g_{n-s}(X_i, D_{n,j}^{(s)}) \neq Y_i\right\}}.$$

$s = 1$ yields the deleted estimate. If $s > 1$, then the estimate is usually more biased than the deleted estimate, as

$$\mathbf{E}L_{n,s}^{(\Pi)} = \mathbf{E}L_{n-s},$$

but usually exhibits smaller variance.

31.4   Bootstrap

Bootstrap methods for estimating the misclassification error became popular following the revolutionary work of Efron (1979; 1983). All bootstrap estimates introduce artificial randomization. The *bootstrap sample*

$$D_m^{(b)} = \left((X_1^{(b)}, Y_1^{(b)}), \ldots, (X_m^{(b)}, Y_m^{(b)})\right)$$

is a sequence of random variable pairs drawn randomly *with replacement* from the set $\{(X_1, Y_1), \ldots, (X_n, Y_n)\}$. In other words, conditionally on the training sample $D_n = ((X_1, Y_1), \ldots, (X_n, Y_n))$, the pairs $(X_i^{(b)}, Y_i^{(b)})$ are drawn independently from ν_n, the empirical distribution based on D_n in $\mathcal{R}^d \times \{0, 1\}$.

One of the standard bootstrap estimates aims to compensate the (usually optimistic) bias

$$B(g_n) = \mathbf{E}\{L(g_n)\} - L_n^{(R)}(g_n)$$

of the resubstitution estimate $L_n^{(R)}$. To estimate $B(g_n)$, a bootstrap sample of size $m = n$ may be used:

$$\widehat{B}_n^{(b)}(g_n) = \sum_{i=1}^{n} \left(\frac{1}{n} - \frac{1}{n} \sum_{j=1}^{n} I_{\left\{X_j^{(b)} \neq X_i\right\}}\right) I_{\left\{g_n(X_i, D_n^{(b)}) \neq Y_i\right\}}.$$

Often, bootstrap sampling is repeated several times to average out effects of the additional randomization. In our case,

$$\bar{B}_n(g_n) = \frac{1}{B} \sum_{b=1}^{B} \widehat{B}_n^{(b)}(g_n)$$

yields the estimator

$$L_n^{(B)}(g_n) = L_n^{(R)}(g_n) - \bar{B}_n(g_n).$$

Another instance of a bootstrap sample of size n, the so-called E0 estimator, uses resubstitution on the training pairs not appearing in the bootstrap sample. The estimator is defined by

$$\frac{\sum_{j=1}^{n} I_{\left\{X_j \notin \{X_1^{(b)}, \dots, X_n^{(b)}\}\right\}} I_{\left\{g_n(X_i, D_n^{(b)}) \neq Y_i\right\}}}{\sum_{j=1}^{n} I_{\left\{X_j \notin \{X_1^{(b)}, \dots, X_n^{(b)}\}\right\}}}.$$

Here, too, averages may be taken after generating the bootstrap sample several times.

Many other versions of bootstrap estimates have been reported, such as the "0.632 estimate," "double bootstrap," and "randomized bootstrap" (see Hand (1986), Jain, Dubes, and Chen (1987), and McLachlan (1992) for surveys and additional references). Clearly, none of these estimates provides a universal remedy, but for several specific classification rules, bootstrap estimates have been experimentally found to outperform the deleted and resubstitution estimates. However, one point has to be made clear: we always lose information with the additional randomization. We summarize this in the following simple general result:

Theorem 31.3. *Let $X_1, \dots, X_n$ be drawn from an unknown distribution μ, and let $\alpha(\mu)$ be a functional to be estimated. Let $r(\cdot)$ be a convex risk function (such as $r(u) = u^2$ or $r(u) = |u|$). Let $X_1^{(b)}, \dots X_m^{(b)}$ be a bootstrap sample drawn from $X_1, \dots, X_n$. Then*

$$\inf_{\psi} \mathbf{E}\left\{ r(\psi(X_1^{(b)}, \dots X_m^{(b)}) - \alpha(\mu)) \right\} \geq \inf_{\psi} \mathbf{E}\left\{ r(\psi(X_1, \dots, X_n) - \alpha(\mu)) \right\}.$$

The theorem states that no matter how large m is, the class of estimators that are functions of the original sample is always at least as good as the class of all estimators that are based upon bootstrap samples. In our case, $\alpha(\mu)$ plays the role of the expected error probability $\mathbf{E}L_n = \mathbf{P}\{g_n(X) \neq Y\}$. If we take $r(u) = u^2$, then it follows from the theorem that there is no estimator $\widehat{L}_m$ based on the bootstrap sample $D_m^{(b)}$ whose squared error $\mathbf{E}\left\{(\mathbf{E}L_n - \widehat{L}_m)^2\right\}$ is smaller than that of some nonbootstrap estimate. In the proof of the theorem we construct such a non-bootstrap estimator. It is clear, however, that, in general, the latter estimator is too complex to have any practical value. The randomization of bootstrap methods may provide a useful tool to overcome the computational difficulties in finding good estimators.

PROOF. Let ψ be any mapping taking m arguments. Then

$$\mathbf{E}\left\{r\left(\psi(X_1^{(b)}, \ldots X_m^{(b)}) - \alpha(\mu)\right) | X_1, \ldots, X_n\right\}$$

$$= \frac{1}{n^m} \sum_{(i_1,\ldots,i_m)\subset\{1,\ldots,n\}^m} r(\psi(X_{i_1}, \ldots, X_{i_m}) - \alpha(\mu))$$

$$\geq r\left(\frac{1}{n^m} \sum_{(i_1,\ldots,i_m)\subset\{1,\ldots,n\}^m} (\psi(X_{i_1}, \ldots, X_{i_m}) - \alpha(\mu))\right)$$

(by Jensen's inequality and the convexity of r)

$$= r\left(\frac{1}{n^m} \sum_{(i_1,\ldots,i_m)\subset\{1,\ldots,n\}^m} \psi(X_{i_1}, \ldots, X_{i_m}) - \alpha(\mu)\right)$$

$$= r\left(\psi^*(X_1, \ldots, X_n) - \alpha(\mu)\right),$$

where

$$\psi^*(x_1, \ldots, x_n) = \frac{1}{n^m} \sum_{(i_1,\ldots,i_m)\subset\{1,\ldots,n\}^m} \psi(x_{i_1}, \ldots, x_{i_m}).$$

Now, after taking expectations with respect to $X_1, \ldots, X_n$, we see that for every ψ we start out with, there is a ψ^* that is at least as good. $\square$

If $m = O(n)$, then the bootstrap has an additional problem related to the coupon collector problem. Let N be the number of different pairs in the bootstrap sample $(X_1^{(b)}, Y_1^{(b)}), \ldots, (X_m^{(b)}, Y_m^{(b)})$. Then, if $m \sim cn$ for some constant c, then $N/n \to 1 - e^{-c}$ with probability one. To see this, note that

$$\mathbf{E}\{n - N\} = n\left(1 - \frac{1}{n}\right)^{cn} \sim ne^{-c},$$

so $\mathbf{E}\{N/n\} \to 1 - e^{-c}$. Furthermore, if one of the m drawings is varied, N changes by at most one. Hence, by McDiarmid's inequality, for $\epsilon > 0$,

$$\mathbf{P}\{|N - \mathbf{E}\{N\}| > n\epsilon\} \leq 2e^{-2n\epsilon^2},$$

from which we conclude that $N/n \to 1 - e^{-c}$ with probability one. As $n\left(1 - e^{-c}\right)$ of the original data pairs do not appear in the bootstrap sample, a considerable loss of information takes place that will be reflected in the performance. This phenomenon is well-known, and motivated several modifications of the simplest bootstrap estimate. For more information, see the surveys by Hand (1986), Jain, Dubes, and Chen (1987), and McLachlan (1992).

Problems and Exercises

PROBLEM 31.1. Show that the posterior probability estimate $L_n^{(P)}$ of the histogram, k-nearest neighbor, and moving window rules satisfies

$$\mathbf{P}\left\{\left|L_n^{(P)} - \mathbf{E}\left\{L_n^{(P)}\right\}\right| \ge \epsilon\right\} \le 2e^{-2n\epsilon^2/C},$$

and

$$\mathbf{Var}\{L_n^{(P)}\} \le \frac{C}{4n},$$

where C is a constant, depending on the rule. In the case of the histogram rule the value of C is $C = 25$, for the k-nearest neighbor rule $C = (1 + 2\gamma_d)^2$, and for the moving window rule $C = (1 + 2\beta_d)^2$. Also show that the deleted version $L_n^{(PD)}$ of the estimate satisfies the same inequalities (Lugosi and Pawlak (1994)). HINT: Proceed as in the proof of Theorem 31.2.

PROBLEM 31.2. Show that the deleted posterior probability estimate of the error probability of a histogram rule is always optimistically biased, that is, for all n, and all distributions, $\mathbf{E}\left\{L_n^{(PD)}\right\} \le \mathbf{E}L_{n-1}$.

PROBLEM 31.3. Show that for any classification rule and any estimate $0 \le \eta_n(x, D_n) \le 1$ of the a posteriori probabilities, for all distributions of (X, Y) for all l, n, and $\epsilon > 0$

$$\mathbf{P}\left\{\left|L_{n,l}^{(U)} - \mathbf{E}\left\{L_{n,l}^{(U)}\right\}\right| \ge \epsilon \,\middle|\, D_n\right\} \le 2e^{-2l\epsilon^2},$$

and $\mathbf{Var}\left\{L_{n,l}^{(U)} | D_n\right\} \le 1/l$. Further, show that for all l, $\mathbf{E}\left\{L_{n,l}^{(U)}\right\} = \mathbf{E}\left\{L_{n+1}^{(PD)}\right\}$.

PROBLEM 31.4. EMPIRICAL SQUARED ERROR. Consider the deleted empirical squared error

$$e_n = \frac{1}{n}\sum_{i=1}^{n}\left(Y_i - \eta_{n-1}(X_i, D_{n,i})\right)^2.$$

Show that

$$\mathbf{E}\{e_n\} = \frac{L_{NN}}{2} + \mathbf{E}\left\{(\eta(X) - \eta_{n-1}(X, D_{n-1}))^2\right\},$$

where L_{NN} is the asymptotic error probability on the 1-nearest neighbor rule. Show that if η_{n-1} is the histogram, kernel, or k-NN estimate, then $\mathbf{Var}\{e_n\} \le c/n$ for some constant depending on the dimension only. We see that e_n is an asymptotically optimistically biased estimate of $L(g_n)$ when η_{n-1} is an $L_2(\mu)$-consistent estimate of η. Still, this estimate is useful in classifier selection (see Theorem 29.2).

32
Feature Extraction

32.1 Dimensionality Reduction

So far, we have not addressed the question of how the components of the feature vector X are obtained. In general, these components are based on d measurements of the object to be classified. How many measurements should be made? What should these measurements be? We study these questions in this chapter. General recipes are hard to give as the answers depend on the specific problem. However, there are some rules of thumb that should be followed. One such rule is that noisy measurements, that is, components that are independent of Y, should be avoided. Also, adding a component that is a function of other components is useless. A necessary and sufficient condition for measurements providing additional information is given in Problem 32.1.

Our goal, of course, is to make the error probability $L(g_n)$ as small as possible. This depends on many things, such as the joint distribution of the selected components and the label Y, the sample size, and the classification rule g_n. To make things a bit simpler, we first investigate the Bayes errors corresponding to the selected components. This approach makes sense, since the Bayes error is the theoretical limit of the performance of any classifier. As Problem 2.1 indicates, collecting more measurements cannot increase the Bayes error. On the other hand, having too many components is not desirable. Just recall the curse of dimensionality that we often faced: to get good error rates, the number of training samples should be exponentially large in the number of components. Also, computational and storage limitations may prohibit us from working with many components.

We may formulate the feature selection problem as follows: Let $X^{(1)}, \ldots, X^{(d)}$

be random variables representing d measurements. For a set $A \subseteq \{1, \ldots, d\}$ of indices, let X_A denote the $|A|$-dimensional random vector, whose components are the $X^{(i)}$'s with $i \in A$ (in the order of increasing indices). Define

$$L^*(A) = \inf_{g:\mathcal{R}^{|A|} \to \{0,1\}} P\{g(X_A) \neq Y\}$$

as the Bayes error corresponding to the pair (X_A, Y).

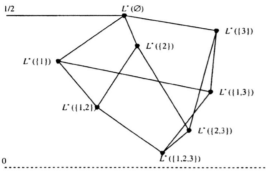

FIGURE 32.1. *A possible arrangement of $L^*(A)$'s for $d = 3$.*

Obviously, $L^*(A) \leq L^*(B)$ whenever $B \subset A$, and $L^*(\emptyset) = \min(P\{Y = 0\}, P\{Y = 1\})$. The problem is to find an efficient way of selecting an index set A with $|A| = k$, whose corresponding Bayes error is the smallest. Here $k < d$ is a fixed integer. Exhaustive search through the $\binom{d}{k}$ possibilities is often undesirable because of the imposed computational burden. Many attempts have been made to find fast algorithms to obtain the best subset of features. See Fu, Min, and Li (1970), Kanal (1974), Ben-Bassat (1982), and Devijver and Kittler (1982) for surveys. It is easy to see that the best k individual features—that is, components corresponding to the k smallest values of $L^*(\{i\})$—do not necessarily constitute the best k-dimensional vector: just consider a case in which $X^{(1)} \equiv X^{(2)} \equiv \cdots \equiv X^{(k)}$. Cover and Van Campenhout (1977) showed that any ordering of the 2^d subsets of $\{1, \ldots, d\}$ consistent with the obvious requirement $L^*(A) \leq L^*(B)$ if $B \subseteq A$ is possible. More precisely, they proved the following surprising result:

Theorem 32.1. (COVER AND VAN CAMPENHOUT (1977)). *Let $A_1, A_2, \ldots, A_{2^d}$ be an ordering of the 2^d subsets of $\{1, \ldots, d\}$, satisfying the consistency property $i < j$ if $A_i \subset A_j$. (Therefore, $A_1 = \emptyset$ and $A_{2^d} = \{1, \ldots, d\}$.) Then there exists a distribution of the random variables $(X, Y) = (X^{(1)}, \ldots, X^{(d)}, Y)$ such that*

$$L^*(A_1) > L^*(A_2) > \cdots > L^*(A_{2^d}).$$

The theorem shows that every feature selection algorithm that finds the best k-element subset has to search exhaustively through all k-element subsets for some distributions. Any other method is doomed to failure for some distribution. Many suboptimal, heuristic algorithms have been introduced trying to avoid the computational demand of exhaustive search (see, e.g., Sebestyen (1962), Meisel (1972),

Chang (1973), Vilmansen (1973), and Devijver and Kittler (1982)). Narendra and Fukunaga (1977) introduced an efficient branch-and-bound method that finds the optimal set of features. Their method avoids searching through all subsets in many cases by making use of the monotonicity of the Bayes error with respect to the partial ordering of the subsets. The key of our proof of the theorem is the following simple lemma:

Lemma 32.1. *Let* $A_1, \ldots, A_{2^d}$ *be as in Theorem 32.1. Let* $1 < i < 2^d$. *Assume that the distribution of* (X, Y) *on* $\mathcal{R}^d \times \{0, 1\}$ *is such that the distribution of* X *is concentrated on a finite set,* $L^*(A_{2^d}) = L^*(\{1, \ldots, 2^d\}) = 0$, *and* $L^*(A_j) < 1/2$ *for each* $i < j \leq 2^d$. *Then there exists another finite distribution such that* $L^*(A_j)$ *remains unchanged for each* $j > i$, *and*

$$L^*(A_j) < L^*(A_i) < \frac{1}{2} \quad for\ each\ j > i.$$

PROOF. We denote the original distribution of X by μ and the a posteriori probability function by η. We may assume without loss of generality that every atom of the distribution of X is in $[0, M)^d$ for some $M > 0$. Since $L^*(\{1, \ldots, d\}) = 0$, the value of $\eta(x)$ is either zero or one at each atom. We construct the new distribution by duplicating each atom in a special way. We describe the new distribution by defining a measure μ' on $\mathcal{R}^d$ and an a posteriori function $\eta' : \mathcal{R}^d \to \{0, 1\}$.

Define the vector $v_{A_i} \in \mathcal{R}^d$ such that its m-th component equals M if $m \notin A_i$, and zero if $m \in A_i$. The new measure μ' has twice as many atoms as μ. For each atom $x \in \mathcal{R}^d$ of μ, the new measure μ' has two atoms, namely, $x_1 = x$ and $x_2 = x + v_{A_i}$. The new distribution is specified by

$$\mu'(x_1) = q\mu(x), \quad \mu'(x_2) = (1 - q)\mu(x), \quad \eta'(x_1) = \eta(x), \quad and \quad \eta'(x_2) = 1 - \eta(x),$$

where $q \in (0, 1/2)$ is specified later. It remains for us to verify that this distribution satisfies the requirements of the lemma for some q. First observe that the values $L^*(A_j)$ remain unchanged for all $j > i$. This follows from the fact that there is at least one component in A_j along which the new set of atoms is strictly separated from the old one, leaving the corresponding contribution to the Bayes error unchanged. On the other hand, as we vary q from zero to $1/2$, the new value of $L^*(A_i)$ grows continuously from the old value of $L^*(A_i)$ to $1/2$. Therefore, since by assumption $\max_{j > i} L^*(A_j) < 1/2$, there exists a value of q such that the new $L^*(A_i)$ satisfies $\max_{j > i} L^*(A_j) < L^*(A_i) < 1/2$ as desired. □

PROOF OF THEOREM 32.1.   We construct the desired distribution in $2^d - 2$ steps, applying Lemma 32.1 in each step. The procedure for $d = 3$ is illustrated in Figure 32.2.

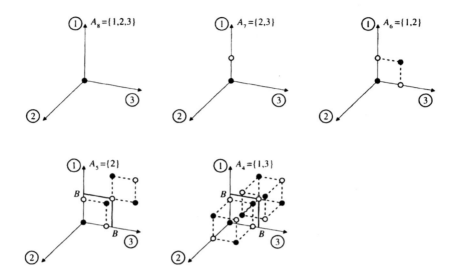

FIGURE 32.2. *Construction of a prespecified ordering. In this three-dimensional example, the first four steps of the procedure are shown, when the desired ordering is* $L^*(\{1, 2, 3\}) \leq L^*(\{2, 3\}) \leq L^*(\{1, 2\}) \leq L^*(\{2\}) \leq L^*(\{1, 3\}) \leq \cdots$. *Black circles represent atoms with* $\eta = 1$ *and white circles are those with* $\eta = 0$.

We start with a monoatomic distribution concentrated at the origin, with $\eta(0) = 1$. Then clearly, $L^*(A_{2^d}) = 0$. By Lemma 32.1, we construct a distribution such that $L^*(A_{2^d}) = 0$ and $0 = L^*(A_{2^d}) < L^*(A_{2^d-1}) < 1/2$. By applying the lemma again, we can construct a distribution with $0 = L^*(A_{2^d}) < L^*(A_{2^d-1}) < L^*(A_{2^d-2}) < 1/2$. After i steps, we have a distribution satisfying the last i inequalities of the desired ordering. The construction is finished after $2^d - 2$ steps. $\square$

REMARK. The original example of Cover and Van Campenhout (1977) uses the multidimensional gaussian distribution. Van Campenhout (1980) developed the idea further by showing that not only all possible orderings, but all possible *values* of the $L^*(A_i)$'s can be achieved by some distributions. The distribution constructed in the above proof is discrete. It has 2^{2^d-2} atoms. $\square$

One may suspect that feature extraction is much easier if given Y, the components $X^{(1)}, \ldots, X^{(d)}$ are conditionally independent. However, three and four-dimensional examples given by Elashoff, Elashoff, and Goldman (1967), Toussaint (1971), and Cover (1974) show that even the individually best two independent components are not the best pair of components. We do not know if Theorem 32.1

generalizes to the case when the components are conditionally independent. In the next example, the pair of components consisting of the two worst single features is the best pair, and vice versa.

Theorem 32.2. (TOUSSAINT (1971)). *There exist binary-valued random variables* $X_1, X_2, X_3, Y \in \{0, 1\}$ *such that* X_1, X_2, *and* X_3 *are conditionally independent (given Y), and*

$$L^*(\{1\}) < L^*(\{2\}) < L^*(\{3\}),$$

but

$$L^*(\{1, 2\}) > L^*(\{1, 3\}) > L^*(\{2, 3\}).$$

PROOF. Let $\mathbf{P}\{Y = 1\} = 1/2$. Then the joint distribution of X_1, X_2, X_3, Y is specified by the conditional probabilities $\mathbf{P}\{X_i = 1 | Y = 0\}$ and $\mathbf{P}\{X_i = 1 | Y = 1\}$, $i = 1, 2, 3$. Straightforward calculation shows that the values

$$\mathbf{P}\{X_1 = 1 | Y = 0\} = 0.1, \quad \mathbf{P}\{X_1 = 1 | Y = 1\} = 0.9,$$

$$\mathbf{P}\{X_2 = 1 | Y = 0\} = 0.05, \quad \mathbf{P}\{X_2 = 1 | Y = 1\} = 0.8,$$

$$\mathbf{P}\{X_3 = 1 | Y = 0\} = 0.01, \quad \mathbf{P}\{X_3 = 1 | Y = 1\} = 0.71$$

satisfy the stated inequalities. □

As our ultimate goal is to minimize the error probability, finding the feature set minimizing the Bayes error is not the best we can do. For example, if we know that we will use the 3-nearest neighbor rule, then it makes more sense to select the set of features that minimizes the asymptotic error probability $L_{3\text{NN}}$ of the 3-nearest neighbor rule. Recall from Chapter 5 that

$$L_{3\text{NN}} = \mathbf{E}\left\{\eta(X)(1 - \eta(X))\big(1 + 4\eta(X)(1 - \eta(X))\big)\right\}.$$

The situation here is even messier than for the Bayes error. As the next example shows, it is not even true that $A \subset B$ implies $L_{3\text{NN}}(A) \geq L_{3\text{NN}}(B)$, where $A, B \subseteq \{1, \ldots, d\}$ are two subsets of components, and $L_{3\text{NN}}(A)$ denotes the asymptotic error probability of the 3-nearest neighbor rule for (X_A, Y). In other words, adding components may increase $L_{3\text{NN}}$! This can never happen to the Bayes error—and in fact, to any f-error (Theorem 3.3). The anomaly is due to the fact that the function $x(1 - x)(1 + 4x(1 - x))$ is *convex* near zero and one.

EXAMPLE. Let the joint distribution of $X = (X_1, X_2)$ be uniform on $[0, 2]^2$. The joint distribution of (X, Y) is defined by the a posteriori probabilities given by

$$\eta(x) = \begin{cases} 0.1 & \text{if } x \in [0, 1/2) \times [0, 1/2) \\ 0 & \text{if } x \in [1/2, 1] \times [0, 1/2) \\ 1 & \text{if } x \in [0, 1/2) \times [1/2, 1] \\ 0.9 & \text{if } x \in [1/2 \times 1] \times [1/2, 1] \end{cases}$$

(Figure 32.3). Straightforward calculations show that $L_{3NN}(\{1, 2\}) = 0.0612$, while $L_{3NN}(\{2\}) = 0.056525$, a smaller value! $\square$

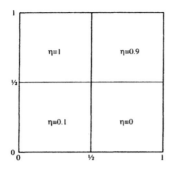

$L_{3NN}(\{1,2\})= 0.0612$
$L_{3NN}(\{2\}) = 0.056525$

FIGURE 32.3. *An example when an additional feature increases the error probability of the 3-NN rule.*

Of course, the real measure of the goodness of the selected feature set is the error probability $L(g_n)$ of the classifier designed by using training data D_n. If the classification rule g_n is not known at the stage of feature selection, then the best one can do is to estimate the Bayes errors $L^*(A)$ for each set A of features, and select a feature set by minimizing the estimate. Unfortunately, as Theorem 8.5 shows, no method of estimating the Bayes errors can guarantee good performance. If we know what classifier will be used after feature selection, then the best strategy is to select a set of measurements based on comparing estimates of the error probabilities. We do not pursue this question further.

For special cases, we do not need to mount a big search for the best features. Here is a simple example: given $Y = i$, let $X = (X^{(1)}, \ldots, X^{(d)})$ have d independent components, where given $Y = i$, $X^{(j)}$ is normal (m_{ji}, σ_j^2). It is easy to verify (Problem 32.2) that if $\mathbf{P}\{Y = 1\} = 1/2$, then

$$L^* = \mathbf{P}\left\{N > \frac{r}{2}\right\},$$

where N is a standard normal random variable, and

$$r^2 = \sum_{j=1}^{d} \left(\frac{m_{j1} - m_{j0}}{\sigma_j}\right)^2$$

is the square of the Mahalanobis distance (see also Duda and Hart (1973, pp. 66–67)). For this case, the quality of the j-th feature is measured by

$$\left(\frac{m_{j1} - m_{j0}}{\sigma_j}\right)^2.$$

We may as well rank these values, and given that we need only $d' < d$ features, we are best off taking the d' features with the highest quality index.

It is possible to come up with analytic solutions in other special cases as well. For example, Raudys (1976) and Raudys and Pikelis (1980) investigated the dependence of $\mathbf{E}\{L(g_n)\}$ on the dimension of the feature space in the case of certain linear classifiers and normal distributions. They point out that for a fixed n, by increasing the number of features, the expected error probability $\mathbf{E}\{L(g_n)\}$ first decreases, and then, after reaching an optimum, grows again.

32.2 Transformations with Small Distortion

One may view the problem of feature extraction in general as the problem of finding a transformation (i.e., a function) $T : \mathcal{R}^d \rightarrow \mathcal{R}^d$ so that the Bayes error L_T^* corresponding to the pair $(T(X), Y)$ is close to the Bayes error L^* corresponding to the pair (X, Y). One typical example of such transformations is fine quantization (i.e., discretization) of X, when T maps the observed values into a set of finitely, or countably infinitely many values. Reduction of dimensionality of the observations can be put in this framework as well. In the following result we show that small distortion of the observation cannot cause large increase in the Bayes error probability.

Theorem 32.3. (FARAGÓ AND GYÖRFI (1975)). *Assume that for a sequence of transformations* T_n, $n = 1, 2, \ldots$

$$\|T_n(X) - X\| \rightarrow 0$$

in probability, where $\| \cdot \|$ *denotes the Euclidean norm in* $\mathcal{R}^d$. *Then, if* L^* *is the Bayes error for* (X, Y) *and* $L_{T_n}^*$ *is the Bayes error for* $(T_n(X), Y)$,

$$L_{T_n}^* \rightarrow L^*.$$

PROOF. For arbitrarily small $\epsilon > 0$ we can choose a uniformly continuous function $0 \leq \bar{\eta}(x) \leq 1$ such that

$$2\mathbf{E}\{|\eta(X) - \bar{\eta}(X)|\} < \epsilon.$$

For any transformation T, consider now the decision problem of $(T(X), \tilde{Y})$, where the random variable $\tilde{Y}$ satisfies $\mathbf{P}\{\tilde{Y} = 1 | X = x\} = \bar{\eta}(x)$ for every $x \in \mathcal{R}^d$. Denote the corresponding Bayes error by $\tilde{L}_T^*$, and the Bayes error corresponding to the pair $(X, \tilde{Y})$ by $\tilde{L}^*$. Obviously

$$0 \leq L_{T_n}^* - L^* \leq |L_{T_n}^* - \tilde{L}_{T_n}^*| + |\tilde{L}_{T_n}^* - \tilde{L}^*| + |\tilde{L}^* - L^*|. \qquad (32.1)$$

To bound the first and third terms on the right-hand side, observe that for any transformation T,

$$\mathbf{P}\{Y = 1 | T(X)\} = \mathbf{E}\{\mathbf{P}\{Y = 1 | X\} | T(X)\} = \mathbf{E}\{\eta(X) | T(X)\}.$$

Therefore, the Bayes error corresponding to the pair $(T(X), Y)$ equals

$$L_T^* = \mathbf{E}\{\min(\mathbf{E}\{\eta(X)|T(X)\}, 1 - \mathbf{E}\{\eta(X)|T(X)\})\}.$$

Thus,

$$
\begin{aligned}
|L_T^* - \tilde{L}_T^*| &= |\mathbf{E}\{\min(\mathbf{E}\{\eta(X)|T(X)\}, 1 - \mathbf{E}\{\eta(X)|T(X)\})\} \\
&\quad - \mathbf{E}\{\min(\mathbf{E}\{\tilde{\eta}(X)|T(X)\}, 1 - \mathbf{E}\{\tilde{\eta}(X)|T(X)\})\}| \\
&\leq \mathbf{E}\{|\mathbf{E}\{\tilde{\eta}(X)|T(X)\} - \mathbf{E}\{\eta(X)|T(X)\}|\} \\
&\leq \mathbf{E}\{|\tilde{\eta}(X) - \eta(X)|\} \\
&\quad \text{(by Jensen's inequality)} \\
&< \epsilon,
\end{aligned}
$$

so the first and third terms of (32.1) are less than ϵ. For the second term, define the decision function

$$g_n(x) = \begin{cases} 0 & \text{if } \tilde{\eta}(x) \leq 1/2 \\ 1 & \text{otherwise,} \end{cases}$$

which has error probability $\tilde{L}(g_n) = \mathbf{P}\{g_n(T_n(X)) \neq \tilde{Y}\}$. Then $\tilde{L}(g_n) \geq \tilde{L}_{T_n}^*$, and we have

$$0 \leq \tilde{L}_{T_n}^* - \tilde{L}^* \leq \tilde{L}(g_n) - \tilde{L}^* \leq 2\mathbf{E}\{|\tilde{\eta}(T_n(X)) - \tilde{\eta}(X)|\},$$

by Theorem 2.2. All we have to show now is that the limit supremum of the above quantity does not exceed ϵ. Let $\delta(\epsilon)$ be the inverse modulus of continuity of the a posteriori probability $\tilde{\eta}$, that is,

$$\delta(\epsilon) = \sup\{\|x - y\| : 2|\tilde{\eta}(x) - \tilde{\eta}(y)| < \epsilon\}.$$

For every $\epsilon > 0$, we have $\delta(\epsilon) > 0$ by the uniform continuity of $\tilde{\eta}$. Now, we have

$$
\begin{aligned}
&2\mathbf{E}\{|\tilde{\eta}(T_n(X)) - \tilde{\eta}(X)|\} \\
&\leq 2\mathbf{E}\left\{I_{\{\|X-T_n(X)\|>\delta(\epsilon)\}}\right\} + 2\mathbf{E}\left\{I_{\{\|X-T_n(X)\|\leq\delta(\epsilon)\}}|\tilde{\eta}(T_n(X)) - \tilde{\eta}(X)|\right\}.
\end{aligned}
$$

Clearly, the first term on the right-hand side converges to zero by assumption, while the second term does not exceed ϵ by the definition of $\delta(\epsilon)$. □

REMARK. It is clear from the proof of Theorem 32.3 that everything remains true if the observation X takes its values from a separable metric space with metric ρ, and the condition of the theorem is modified to $\rho(T(X), X) \to 0$ in probability. This generalization has significance in curve recognition, when X is a stochastic process. Then Theorem 32.3 asserts that one does not lose much information by using usual discretization methods, such as, for example, Karhunen-Loève series expansion (see Problems 32.3 to 32.5). □

32.3   Admissible and Sufficient Transformations

Sometimes the cost of guessing zero while the true value of Y is one is different from the cost of guessing one, while $Y = 0$. These situations may be handled as follows. Define the costs

$$C(m, l), \quad m, l = 0, 1.$$

Here $C(Y, g(X))$ is the cost of deciding on $g(X)$ when the true label is Y. The risk of a decision function g is defined as the expected value of the cost:

$$R_g = E\{C(Y, g(X))\}.$$

Note that if

$$C(m, l) = \begin{cases} 1 & \text{if } m \neq l \\ 0 & \text{otherwise,} \end{cases}$$

then the risk is just the probability of error. Introduce the notation

$$Q_m(x) = \eta(x)C(1, m) + (1 - \eta(x))C(0, m), \quad m = 0, 1.$$

Then we have the following extension of Theorem 2.1:

Theorem 32.4. *Define*

$$\widehat{g}(x) = \begin{cases} 1 & \text{if } Q_1(x) > Q_0(x) \\ 0 & \text{otherwise.} \end{cases}$$

Then for all decision functions g we have

$$R_{\widehat{g}} \leq R_g.$$

$R_{\widehat{g}}$ is called the Bayes risk. The proof is left as an exercise (see Problem 32.7). Which transformations preserve all the necessary information in the sense that the Bayes error probability corresponding to the pair $(T(X), Y)$ equals that of (X, Y)? Clearly, every invertible mapping T has this property. However, the practically interesting transformations are the ones that provide some compression of the data. The most efficient of such transformations is the Bayes decision g^* itself: g^* is specifically designed to minimize the error probability. If the goal is to minimize the Bayes risk with respect to some other cost function than the error probability, then g^* generally fails to preserve the Bayes risk. It is natural to ask what transformations preserve the Bayes risk for *all possible* cost functions. This question has a practical significance, when collecting data and construction of the decision are separated in space or in time. In such cases the data should be transmitted via a communication channel (or should be stored). In both cases there is a need for an efficient data compression rule. In this problem formulation, when getting the data, one may not know the final cost function. Therefore a desirable data compression (transformation) does not increase the Bayes risk for any cost function $C(\cdot, \cdot)$. Here T is a measurable function mapping from $\mathcal{R}^d$ to $\mathcal{R}^{d'}$ for some positive integer d'.

DEFINITION 32.1. *Let $R^*_{C,T}$ denote the Bayes risk for the cost function C and transformed observation $T(X)$. A transformation T is called admissible if for any cost function C,*

$$R^*_{C,T} = R^*_C,$$

where R^*_C denotes the Bayes risk for the original observation.

Obviously each invertible transformation T is admissible. A nontrivial example of an admissible transformation is

$$T^*(X) = \eta(X),$$

since according to Theorem 32.4, the Bayes decision for any cost function can be constructed by the a posteriori probability $\eta(x)$ and by the cost function. Surprisingly, this is basically the only such transformation in the following sense:

Theorem 32.5. *A transformation T is admissible if and only if there is a mapping G such that*

$$G(T(X)) = \eta(X) \quad \text{with probability one.}$$

PROOF. The converse is easy since for such G

$$R^*_C \leq R^*_{C,T(X)} \leq R^*_{C,G(T(X))} = R^*_C.$$

Assume now that T is admissible but such function G does not exist. Then there is a set $A \subset \mathcal{R}^d$ such that $\mu(A) > 0$, $T(x)$ is constant on A, while all values of $\eta(x)$ are different on A, that is, if $x, y \in A$, then $x \neq y$ implies $\eta(x) \neq \eta(y)$. Then there are real numbers $0 < c < 1$ and $\epsilon > 0$, and sets $B, D \subset A$ such that $\mu(B), \mu(D) > 0$, and

$$1 - \eta(x) \quad > \quad c + \epsilon \ \text{ if } \ x \in B,$$
$$1 - \eta(x) \quad < \quad c - \epsilon \ \text{ if } \ x \in D.$$

Now, choose a cost function with the following values:

$$c_{0,0} = c_{1,1} = 0, \qquad c_{0,1} = 1, \quad \text{and} \quad c_{1,0} = \frac{1 - c}{c}.$$

Then,

$$Q_0(x) \quad = \quad \eta(x),$$
$$Q_1(x) \quad = \quad c_{1,0}(1 - \eta(x)),$$

and the Bayes decision on $B \cup D$ is given by

$$g^*(x) = \begin{cases} 0 & \text{if } c < 1 - \eta(x) \\ 1 & \text{otherwise.} \end{cases}$$

Now, let $g(T(x))$ be an arbitrary decision. Without loss of generality we can assume that $g(T(x)) = 0$ if $x \in A$. Then the difference between the risk of $g(T(x))$ and the Bayes risk is

$$
\begin{aligned}
R_{g(T(X))} - R_{g^*(X)} &= \int_{\mathcal{R}^d} Q_{g(T(x))}(x)\mu(dx) - \int_{\mathcal{R}^d} Q_{g^*(x)}(x)\mu(dx) \\
&= \int_{\mathcal{R}^d} (Q_{g(T(x))}(x) - Q_{g^*(x)}(x))\mu(dx) \\
&\geq \int_D (Q_0(x) - Q_1(x))\mu(dx) \\
&= \int_D (\eta(x) - c_{1,0}(1 - \eta(x))\mu(dx) \\
&= \int_D \left(1 - \frac{1 - \eta(x)}{c}\right)\mu(dx) \geq \frac{\epsilon}{c}\mu(D) > 0. \quad \square
\end{aligned}
$$

We can give another characterization of admissible transformations by virtue of a well-known concept of mathematical statistics:

DEFINITION 32.2. $T(X)$ is called a *sufficient statistic* if the random variables X, $T(X)$, and Y form a Markov chain in this order. That is, for any set A, $P\{Y \in A|T(X), X\} = P\{Y \in A|T(X)\}$.

Theorem 32.6. *A transformation T is admissible if and only if $T(X)$ is a sufficient statistic.*

PROOF. Assume that T is admissible. Then according to Theorem 32.5 there is a mapping G such that

$$G(T(X)) = \eta(X) \quad \text{with probability one.}$$

Then

$$P\{Y = 1|T(X), X\} = P\{Y = 1|X\} = \eta(X) = G(T(X)),$$

and

$$
\begin{aligned}
P\{Y = 1|T(X)\} &= \mathbf{E}\{P\{Y = 1|T(X), X\}|T(X)\} \\
&= \mathbf{E}\{G(T(X))|T(X)\} \\
&= G(T(X)),
\end{aligned}
$$

thus, $P\{Y = 1|T(X), X\} = P\{Y = 1|T(X)\}$, therefore $T(X)$ is sufficient. On the other hand, if $T(X)$ is sufficient, then

$$P\{Y = 1|X\} = P\{Y = 1|T(X), X\} = P\{Y = 1|T(X)\},$$

so for the choice

$$G(T(X)) = \mathbf{P}\{Y = 1 | T(X)\}$$

we have the desired function $G(\cdot)$, and therefore T is admissible. □

Theorem 32.6 states that we may replace X by any sufficient statistic $T(X)$ without altering the Bayes error. The problem with this, in practice, is that we do not know the sufficient statistics because we do not know the distribution. If the distribution of (X, Y) is known to some extent, then Theorem 32.6 may be useful.

EXAMPLE. Assume that it is known that $\eta(x) = e^{-c\|x\|}$ for some unknown $c > 0$. Then $\|X\|$ is a sufficient statistic. Thus, for discrimination, we may replace the d-dimensional vector X by the 1-dimensional random variable $\|X\|$ without loss of discrimination power. □

EXAMPLE. If $\eta\left(x^{(1)}, x^{(2)}, x^{(3)}\right) = \eta_0\left(x^{(1)}x^{(2)}, x^{(2)}x^{(3)}\right)$ for some function η_0, then $\left(X^{(1)}X^{(2)}, X^{(2)}X^{(3)}\right)$ is a 2-dimensional sufficient statistic. For discrimination, there is no need to deal with $X^{(1)}$, $X^{(2)}$, $X^{(3)}$. It suffices to extract the features $X^{(1)}X^{(2)}$ and $X^{(2)}X^{(3)}$. □

EXAMPLE. If given $Y = i$, X is normal with unknown mean m_i and diagonal covariance matrix $\sigma^2 I$ (for unknown σ), then $\eta(x)$ is a function of $\|x - m_1\|^2 - \|x - m_0\|^2$ only for unknown m_0, m_1. Here we have no obvious sufficient statistic. However, if m_0 and m_1 are both known, then a quick inspection shows that $x^T(m_1 - m_0)$ is a 1-dimensional sufficient statistic. Again, it suffices to look for the simplest possible argument for η. If $m_0 = m_1 = 0$ but the covariance matrices are $\sigma_0^2 I$ and $\sigma_1^2 I$ given that $Y = 0$ or $Y = 1$, then $\|X\|^2$ is a sufficient statistic. □

In summary, the results of this section are useful for picking out features when some theoretical information is available regarding the distribution of (X, Y).

Problems and Exercises

PROBLEM 32.1. Consider the pair $(X, Y) \in \mathcal{R}^d \times \{0, 1\}$ of random variables, and let $X^{(d+1)} \in \mathcal{R}$ be an additional component. Define the augmented random vector $X' = (X, X^{(d+1)})$. Denote the Bayes errors corresponding to the pairs (X, Y) and (X', Y) by L_X^* and $L_{X'}^*$ respectively. Clearly, $L_X^* \geq L_{X'}^*$. Prove that equality holds if and only if

$$\mathbf{P}\left\{I_{\{\eta'(X') > 1/2\}} \neq I_{\{\eta(X) > 1/2\}}\right\} = 0,$$

where the a posteriori probability functions $\eta : \mathcal{R}^d \to \{0, 1\}$ and $\eta' : \mathcal{R}^{d+1} \to \{0, 1\}$ are defined as $\eta(x) = \mathbf{P}\{Y = 1 | X = x\}$ and $\eta'(x') = \mathbf{P}\{Y = 1 | X' = x'\}$. HINT: Consult with the proof of Theorem 32.5.

PROBLEM 32.2. Let $\mathbf{P}\{Y = 1\} = 1/2$, and given $Y = i, i = 0, 1$, let $X = \left(X^{(1)}, \ldots, X^{(d)}\right)$ have d independent components, where $X^{(j)}$ is normal $\left(m_{ji}, \sigma_j^2\right)$. Prove that $L^* = \mathbf{P}\left\{N > r/2\right\}$, where N is a standard normal random variable, and r^2 is the square of the Mahalanobis distance: $r^2 = \sum_{j=1}^d \left((m_{j1} - m_{j0})/\sigma_j\right)^2$ (Duda and Hart (1973), pp. 66–67).

PROBLEM 32.3. SAMPLING OF A STOCHASTIC PROCESS. Let $X(t)$, $t \in [0, 1]$, be a stochastic process (i.e., a collection of real-valued random variables indexed by t), and let Y be a binary random variable. For integer $N > 0$, define $X_N^{(i)} = X(i/N)$, $i = 1, \ldots, N$. Find sufficient conditions on the function $m(t) = \mathbf{E}\{X(t)\}$ and on the covariance function $K(t, s) = \mathbf{E}\{(X(t) - \mathbf{E}\{X(t)\})(X(s) - \mathbf{E}\{X(s)\})\}$ such that

$$\lim_{N \to \infty} L^*(X_N^{(\cdot)}) = L^*(X(\cdot)),$$

where $L^*(X_N^{(\cdot)})$ is the Bayes error corresponding to $(X_N^{(1)}, \ldots, X_N^{(N)})$, and $L^*(X(\cdot))$ is the infimum of the error probabilities of decision functions that map measurable functions into $\{0, 1\}$. HINT: Introduce the stochastic process $X_N(t)$ as the linear interpolation of $X_n^{(1)}, \ldots, X^{(N)}$. Find conditions under which

$$\lim_{N \to \infty} \mathbf{E}\left\{\int_0^1 (X_N(t) - X(t))^2 dt\right\} = 0,$$

and use Theorem 32.3, and the remark following it.

PROBLEM 32.4. EXPANSION OF A STOCHASTIC PROCESS. Let $X(t)$, $m(t)$, and $K(t, s)$ be as in the previous problem. Let $\psi_1, \psi_2, \ldots$ be a complete orthonormal system of functions on $[0, 1]$. Define

$$X^{(i)} = \int_0^1 X(t)\psi_i(t)dt.$$

Find conditions under which

$$\lim_{N \to \infty} L^*(X^{(1)}, \ldots, X^{(N)}) = L^*(X(\cdot)).$$

PROBLEM 32.5. Extend Theorem 32.3 such that the transformations $T_n(X, D_n)$ are allowed to depend on the training data. This extension has significance, because feature extraction algorithms use the training data.

PROBLEM 32.6. For discrete X prove that $T(X)$ is sufficient iff $Y, T(X), X$ form a Markov chain (in this order).

PROBLEM 32.7. Prove Theorem 32.4.

PROBLEM 32.8. Recall the definition of F-errors from Chapter 3. Let F be a strictly concave function. Show that $d_F(X, Y) = d_F(T(X), Y)$ if and only if the transformation T is admissible. Conclude that $L_{NN}(X, Y) = L_{NN}(T(X), Y)$ Construct a T and a distribution of (X, Y) such that $L^*(X, Y) = L^*(T(X), Y)$, but T is not admissible.

PROBLEM 32.9. Find sufficient statistics of minimal dimension for the following discrimination problems:
  (1) It is known that for two given sets A and B with $A \cap B = \emptyset$, if $Y = 1$, we have $X \in A$ and if $Y = 0$, then $X \in B$, or vice versa.
  (2) Given Y, X is a vector of independent gamma random variables with common unknown shape parameter a and common unknown scale parameter b (i.e., the marginal density of each component of X is of the form $x^{a-1}e^{-x/b}/(\Gamma(a)b^a)$, $x > 0$). The parameters a and b depend upon Y.

PROBLEM 32.10. Let X have support on the surface of a ball of $\mathcal{R}^d$ centered at the origin of unknown radius. Find a sufficient statistic for discrimination of dimension smaller than d.

PROBLEM 32.11. Assume that the distribution of $X = \left(X^{(1)}, X^{(2)}, X^{(3)}, X^{(4)}\right)$ is such that $X^{(2)}X^{(4)} = 1$ and $X^{(1)} + X^{(2)} + X^{(3)} = 0$ with probability one. Find a simple sufficient statistic.

Appendix

In this appendix we summarize some basic definitions and results from the theory of probability. Most proofs are omitted as they may be found in standard textbooks on probability, such as Ash (1972), Shiryayev (1984), Chow and Teicher (1978), Durrett (1991), Grimmett and Stirzaker (1992), and Zygmund (1959). We also give a list of useful inequalities that are used in the text.

A.1 Basics of Measure Theory

DEFINITION A.1. *Let S be a set, and let $\mathcal{F}$ be a family of subsets of S. $\mathcal{F}$ is called a σ-algebra if*
  (i) $\emptyset \in \mathcal{F}$,
  (ii) $A \in \mathcal{F}$ *implies* $A^c \in \mathcal{F}$,
  (iii) $A_1, A_2, \ldots \in \mathcal{F}$ *implies* $\cup_{i=1}^{\infty} A_i \in \mathcal{F}$.

A σ-algebra is closed under complement and union of countably infinitely many sets. Conditions (i) and (ii) imply that $S \in \mathcal{F}$. Moreover, (ii) and (iii) imply that a σ-algebra is closed under countably infinite intersections.

DEFINITION A.2. *Let S be a set, and let $\mathcal{F}$ be a σ-algebra of subsets of S. Then $(S, \mathcal{F})$ is called a measurable space. The elements of $\mathcal{F}$ are called measurable sets.*

DEFINITION A.3. *If $S = \mathcal{R}^d$ and $\mathcal{B}$ is the smallest σ-algebra containing all rectangles, then $\mathcal{B}$ is called the Borel σ-algebra. The elements of $\mathcal{B}$ are called Borel sets.*

DEFINITION A.4. *Let $(S, \mathcal{F})$ be a measurable space and let $f : S \to \mathcal{R}$ be a function. f is called measurable if for all $B \in \mathcal{B}$*

$$f^{-1}(B) = \{s : f(s) \in B\} \in \mathcal{F},$$

that is, the inverse image of any Borel set B is in $\mathcal{F}$.

Obviously, if A is a measurable set, then the indicator variable I_A is a measurable function. Moreover, finite linear combinations of indicators of measurable sets (called simple functions) are also measurable functions. It can be shown that the set of measurable functions is closed under addition, subtraction, multiplication, and division. Moreover, the supremum and infimum of a sequence of measurable functions, as well as its pointwise limit supremum and limit infimum are measurable.

DEFINITION A.5. *Let $(S, \mathcal{F})$ be a measurable space and let $v : \mathcal{F} \to [0, \infty)$ be a function. v is a measure on $\mathcal{F}$ if*
  (i) $v(\emptyset) = 0,$
  (ii) *v is σ-additive, that is, $A_1, A_2, \ldots \in \mathcal{F}$, and $A_i \cap A_j = 0, i \neq j$ imply that $v(\cup_{i=1}^{\infty} A_i) = \sum_{i=1}^{\infty} v(A_i)$.*

In other words, a measure is a nonnegative, σ-additive set function.

DEFINITION A.6. *v is a finite measure if $v(S) < \infty$. v is a σ-finite measure if there are countably many measurable sets $A_1, A_2, \ldots$ such that $\cup_{i=1}^{\infty} A_i = S$ and $v(A_i) < \infty, i = 1, 2, \ldots$.*

DEFINITION A.7. *The triple $(S, \mathcal{F}, v)$ is a measure space if $(S, \mathcal{F})$ is a measurable space and v is a measure on $\mathcal{F}$.*

DEFINITION A.8. *The Lebesgue measure λ on $\mathcal{R}^d$ is a measure on the Borel σ-algebra of $\mathcal{R}^d$ such that the λ measure of each rectangle equals to its volume.*

A.2   The Lebesgue Integral

DEFINITION A.9. *Let $(S, \mathcal{F}, v)$ be a measure space and $f = \sum_{i=1}^{n} x_i I_{A_i}$ a simple function such that the measurable sets $A_1, \ldots, A_n$ are disjoint. Then the (Lebesgue) integral of f with respect to v is defined by*

$$\int f dv = \int_S f(s)v(ds) = \sum_{i=1}^{n} x_i v(A_i).$$

If f is a nonnegative-valued measurable function, then introduce a sequence of simple functions as follows:

$$f_n(s) = \begin{cases} (k-1)/n & \text{if} \quad (k-1)/n \leq f(s) < k/n, \quad k = 1, 2, \ldots n2^n \\ 2^n & \text{if} \quad f(s) \geq 2^n. \end{cases}$$

Then the f_n's are simple functions, and $f_n(s) \rightarrow f(s)$ in a monotone non-decreasing fashion. Therefore, the sequence of integrals $\int f_n d\nu$ is monotone non-decreasing, with a limit. The integral f is then defined by

$$\int f d\nu = \int_S f(s)\nu(ds) = \lim_{n \to \infty} \int f_n d\nu.$$

If f is an arbitrary measurable function, then decompose it as a difference of its positive and negative parts,

$$f(s) = f(s)^+ - f(s)^- = f(s)^+ - (-f(s))^+,$$

where f^+ and f^- are both nonnegative functions. Define the integral of f by

$$\int f d\nu = \int f^+ d\nu - \int f^- d\nu,$$

if at least one term on the right-hand side is finite. Then we say that the integral exists. If the integral is finite then f is integrable.

DEFINITION A.10. *If $\int f d\nu$ exists and A is a measurable function, then $\int_A f d\nu$ is defined by*

$$\int_A f d\nu = \int f I_A d\nu.$$

DEFINITION A.11. *We say that $f_n \rightarrow f$ (mod ν) if*

$$\nu\left(\left\{s : \lim_{n \to \infty} f_n(s) \neq f(s)\right\}\right) = 0.$$

Theorem A.1. (BEPPO-LEVY THEOREM). *If $f_n(s) \rightarrow f(s)$ in a monotone increasing way for some nonnegative integrable function f, then*

$$\int \lim_{n \to \infty} f_n d\nu = \lim_{n \to \infty} \int f_n d\nu.$$

Theorem A.2. (LEBESGUE'S DOMINATED CONVERGENCE THEOREM). *Assume that $f_n \rightarrow f$ (mod ν) and $|f_n(s)| \leq g(s)$ for $s \in S$, $n = 1, 2, \ldots$, where $\int g d\nu < \infty$. Then*

$$\int \lim_{n \to \infty} f_n d\nu = \lim_{n \to \infty} \int f_n d\nu.$$

Theorem A.3. (FATOU'S LEMMA). *Let $f_1, f_2, \ldots$ be measurable functions.*

(i) If there exists a measurable function g with $\int g d\nu > -\infty$ such that for every n, $f_n(s) \geq g(s)$, then

$$\liminf_{n \to \infty} \int f_n d\nu \geq \int \liminf_{n \to \infty} f_n d\nu.$$

(ii) *If there is a a measurable function g with $\int g d\nu < \infty$, such that $f_n(s) \leq g(s)$ for every n, then*

$$\limsup_{n \to \infty} \int f_n d\nu \leq \int \limsup_{n \to \infty} f_n d\nu.$$

DEFINITION A.12. *Let ν_1 and ν_2 be measures on a measurable space $(S, \mathcal{F})$. We say that ν_1 is absolutely continuous with respect to ν_2 if and only if $\nu_2(A) = 0$ implies $\nu_1(A) = 0$ ($A \in \mathcal{F}$). We denote this relation by $\nu_1 \ll \nu_2$.*

Theorem A.4. (RADON-NIKODYM THEOREM). *Let ν_1 and ν_2 be measures on the measurable space $(S, \mathcal{F})$ such that $\nu_1 \ll \nu_2$ and ν_2 is σ-finite. Then there exists a measurable function f such that for all $A \in \mathcal{F}$*

$$\nu_1(A) = \int_A f d\nu_2.$$

f is unique (mod ν_2). If ν_1 is finite, then f has a finite integral.

DEFINITION A.13. *f is called the density, or Radon-Nikodym derivative of ν_1 with respect to ν_2. We use the notation $f = d\nu_1/d\nu_2$.*

DEFINITION A.14. *Let ν_1 and ν_2 be measures on a measurable space $(S, \mathcal{F})$. If there exists a set $A \in \mathcal{F}$ such that $\nu_1(A^c) = 0$ and $\nu_2(A) = 0$, then ν_1 is singular with respect to ν_2 (and vice versa).*

Theorem A.5. (LEBESGUE DECOMPOSITION THEOREM). *If μ is a σ-finite measure on a measurable space $(S, \mathcal{F})$, then there exist two unique measures ν_1, ν_2 such that $\mu = \nu_1 + \nu_2$, where $\nu_1 \ll \mu$ and ν_2 is singular with respect to μ.*

DEFINITION A.15. *Let $(S, \mathcal{F}, \nu)$ be a measure space and let f be a measurable function. Then f induces a measure μ on the Borel σ-algebra as follows:*

$$\mu(B) = \nu(f^{-1}(B)), \quad B \in \mathcal{B}.$$

Theorem A.6. *Let ν be a measure on the Borel σ-algebra $\mathcal{B}$ of $\mathcal{R}$, and let f and g be measurable functions. Then for all $B \in \mathcal{B}$,*

$$\int_B g(x)\mu(dx) = \int_{f^{-1}(B)} g(f(s))\nu(ds),$$

where μ is induced by f.

DEFINITION A.16. *Let ν_1 and ν_2 be measures on the measurable spaces $(S_1, \mathcal{F}_1)$ and $(S_2, \mathcal{F}_2)$, respectively. Let $(S, \mathcal{F})$ be a measurable space such that $S = S_1 \times S_2$, and $F_1 \times F_2 \in \mathcal{F}$ whenever $F_1 \in \mathcal{F}_1$ and $F_2 \in \mathcal{F}_2$. ν is called the product measure of ν_1 and ν_2 on $\mathcal{F}$ if for $F_1 \in \mathcal{F}_1$ and $F_2 \in \mathcal{F}_2$, $\nu(F_1 \times F_2) = \nu_1(F_1)\nu_2(F_2)$. The product of more than two measures can be defined similarly.*

Theorem A.7. (FUBINI'S THEOREM). *Let h be a measurable function on the product space $(S, \mathcal{F})$. Then*

$$\int_S h(u, v)\nu(d(u, v)) \;=\; \int_{S_1} \left(\int_{S_2} h(u, v)\nu_2(dv) \right) \nu_1(du)$$

$$=\; \int_{S_2} \left(\int_{S_1} h(u, v)\nu_1(du) \right) \nu_2(dv),$$

assuming that one of the three integrals is finite.

A.3   Denseness Results

Lemma A.1. (COVER AND HART (1967)). *Let μ be a probability measure on $\mathcal{R}^d$, and define its support set by*

$$A = support(\mu) = \left\{ x : \text{ for all } r > 0, \ \mu(S_{x,r}) > 0 \right\}.$$

Then $\mu(A) = 1$.

PROOF. By the definition of A,

$$A^c = \left\{ x : \mu(S_{x,r_x}) = 0 \text{ for some } r_x > 0 \right\}.$$

Let Q denote the set of vectors in $\mathcal{R}^d$ with rational components (or any countable dense set). Then for each $x \in A^c$, there is a $y_x \in Q$ with $\|x - y_x\| \leq r_x/3$. This implies $S_{y_x,r_x/2} \subset S_{x,r_x}$. Therefore, $\mu(S_{y_x,r_x/2}) = 0$, and

$$A^c \subset \bigcup_{x \in A^c} S_{y_x,r_x/2}.$$

The right-hand side is a union of countably many sets of zero measure, and therefore $\mu(A^c) = 1$. □

DEFINITION A.17. *Let $(S, \mathcal{F}, \nu)$ be a measure space. For a fixed number $p \geq 1$, $L_p(\nu)$ denotes the set of all measurable functions satisfying $\int |f|^p d\nu < \infty$.*

Theorem A.8. *For every probability measure ν on $\mathcal{R}^d$, the set of continuous functions with bounded support is dense in $L_p(\nu)$. In other words, for every $\epsilon > 0$ and $f \in L_p$ there is a continuous function with bounded support $g \in L_p$ such that*

$$\int |f - g|^p d\nu < \epsilon.$$

The following theorem is a rich source of denseness results:

Theorem A.9. (STONE-WEIERSTRASS THEOREM). *Let $\mathcal{F}$ be a family of real-valued continuous functions on a closed bounded subset B of $\mathcal{R}^d$. Assume that $\mathcal{F}$ is an algebra, that is, for any $f_1, f_2 \in \mathcal{F}$ and $a, b \in \mathcal{R}$, we have $af_1 + bf_2 \in \mathcal{F}$, and $f_1 f_2 \in \mathcal{F}$. Assume furthermore that if $x \neq y$ then there is an $f \in \mathcal{F}$ such that $f(x) \neq f(y)$, and that for each $x \in B$ there exists an $f \in \mathcal{F}$ with $f(x) \neq 0$. Then for every $\epsilon > 0$ and continuous function $g : B \to \mathcal{R}$, there exists an $f \in \mathcal{F}$ such that*

$$\sup_{x \in B} |g(x) - f(x)| < \epsilon.$$

The following two theorems concern differentiation of integrals. Good general references are Wheeden and Zygmund (1977) and de Guzmán (1975):

Theorem A.10. (THE LEBESGUE DENSITY THEOREM). *Let f be a density on $\mathcal{R}^d$. Let $\{Q_k(x)\}$ be a sequence of closed cubes centered at x and contracting to x. Then*

$$\lim_{k \to \infty} \frac{\int_{Q_k(x)} |f(x) - f(y)| dy}{\lambda(Q_k(x))} = 0$$

at almost all x, where λ denotes the Lebesgue measure. Note that this implies

$$\lim_{k \to \infty} \frac{\int_{Q_k(x)} f(y) dy}{\lambda(Q_k(x))} = f(x)$$

at almost all x.

COROLLARY A.1. *Let $\mathcal{A}$ be a collection of subsets of $S_{0,1}$ with the property that for all $A \in \mathcal{A}$, $\lambda(A) \geq c\lambda(S_{0,1})$ for some fixed $c > 0$. Then for almost all x, if $x + rA = \{y : (y - x)/r \in A\}$,*

$$\lim_{r \to 0} \sup_{A \in \mathcal{A}} \left| \frac{\int_{x+rA} f(y) dy}{\lambda(x + rA)} - f(x) \right| = 0.$$

The Lebesgue density theorem also holds if $\{Q_k(x)\}$ is replaced by a sequence of contracting balls centered at x, or indeed by any sequence of sets that satisfy

$$x + br_k S \subseteq Q_k(x) \subseteq x + ar_k S,$$

where S is the unit ball of $\mathcal{R}^d$, $r_k \downarrow 0$, and $0 < b \leq a < \infty$ are fixed constants. This follows from the Lebesgue density theorem. It does not hold in general when $\{Q_k(x)\}$ is a sequence of hyperrectangles containing x and contracting to x. For that, an additional restriction is needed:

Theorem A.11. (THE JESSEN-MARCINKIEWICZ-ZYGMUND THEOREM). *Let f be a density on $\mathcal{R}^d$ with $\int f \log^{d-1}(1 + f) dx < \infty$. Let $\{Q_k(x)\}$ be a sequence of hyperrectangles containing x and for which $\operatorname{diam}(Q_k(x)) \to 0$. Then*

$$\lim_{k \to \infty} \frac{\int_{Q_k(x)} |f(x) - f(y)| dy}{\lambda(Q_k(x))} = 0$$

at almost all x.

COROLLARY A.2. *If f is a density and $\{Q_k(x)\}$ is as in Theorem A.11, then*

$$\liminf_{k \to \infty} \frac{1}{\lambda(Q_k(x))} \int_{Q_k(x)} f(y)dy \geq f(x)$$

at almost all x.

To see this, take $g = \min(f, M)$ for large fixed M. As $\int g \log^{d-1}(1 + g) < \infty$, by the Jessen-Marcinkiewicz-Zygmund theorem,

$$\liminf_{k \to \infty} \frac{1}{\lambda(Q_k(x))} \int_{Q_k(x)} g(y)dy \geq g(x)$$

at almost all x. Conclude by letting $M \to \infty$ along the integers.

A.4   Probability

DEFINITION A.18. *A measure space $(\Omega, \mathcal{F}, \mathbf{P})$ is called a probability space if $\mathbf{P}\{\Omega\} = 1$. Ω is the sample space or sure event, the measurable sets are called events, and the measurable functions are called random variables. If $X_1, \ldots, X_n$ are random variables then $X = (X_1, \ldots, X_n)$ is a vector-valued random variable.*

DEFINITION A.19. *Let X be a random variable, then X induces the measure μ on the Borel σ-algebra of $\mathcal{R}$ by*

$$\mu(B) = \mathbf{P}\{\{\omega : X(\omega) \in B\}\} = \mathbf{P}\{X \in B\}, \quad B \in \mathcal{B}.$$

The probability measure μ is called the distribution of the random variable X.

DEFINITION A.20. *Let X be a random variable. The expectation of X is the integral of x with respect to the distribution μ of X:*

$$\mathbf{E}\{X\} = \int_{\mathcal{R}} x\mu(dx)$$

if it exists.

DEFINITION A.21. *Let X be a random variable. The variance of X is*

$$\mathbf{Var}\{X\} = \mathbf{E}\left\{(X - \mathbf{E}\{X\})^2\right\}$$

if $\mathbf{E}\{X\}$ is finite, and ∞ if $\mathbf{E}\{X\}$ is not finite or does not exist.

DEFINITION A.22. *Let $X_1, \ldots, X_n$ be random variables. They induce the measure $\mu^{(n)}$ on the Borel σ-algebra of $\mathcal{R}^n$ with the property*

$$\mu^{(n)}(B_1 \times \ldots \times B_n) = \mathbf{P}\{X_1 \in B_1, \ldots, X_n \in B_n\}, \quad B_1, \ldots, B_n \in \mathcal{B}.$$

$\mu^{(n)}$ is called the joint distribution of the random variables $X_1, \ldots, X_n$. Let μ_i be the distribution of X_i $(i = 1, \ldots, n)$. The random variables $X_1, \ldots, X_n$ are independent if their joint distribution $\mu^{(n)}$ is the product measure of $\mu_1, \ldots, \mu_n$. The events $A_1, \ldots, A_n \in \mathcal{F}$ are independent if the random variables $I_{A_1}, \ldots, I_{A_n}$ are independent.

Fubini's theorem implies the following:

Theorem A.12. *If the random variables $X_1, \ldots, X_n$ are independent and have finite expectations then*

$$\mathbf{E}\{X_1 X_2 \ldots X_n\} = \mathbf{E}\{X_1\}\mathbf{E}\{X_2\} \cdots \mathbf{E}\{X_n\}.$$

A.5 Inequalities

Theorem A.13. (CAUCHY-SCHWARZ INEQUALITY). *If the random variables X and Y have finite second moments ($\mathbf{E}\{X^2\} < \infty$ and $\mathbf{E}\{Y^2\} < \infty$), then*

$$|\mathbf{E}\{XY\}| \leq \sqrt{\mathbf{E}\{X^2\}\mathbf{E}\{Y^2\}}.$$

Theorem A.14. (HÖLDER'S INEQUALITY). *Let $p, q \in (1, \infty)$ such that $(1/p) + (1/q) = 1$. Let X and Y be random variables such that $(\mathbf{E}\{|X^p|\})^{1/p} < \infty$ and $(\mathbf{E}\{|Y^q|\})^{1/q} < \infty$. Then*

$$\mathbf{E}\{|XY|\} \leq \left(\mathbf{E}\{|X^p|\}\right)^{1/p} \left(\mathbf{E}\{|Y^q|\}\right)^{1/q}.$$

Theorem A.15. (MARKOV'S INEQUALITY). *Let X be a nonnegative-valued random variable. Then for each $t > 0$,*

$$\mathbf{P}\{X \geq t\} \leq \frac{\mathbf{E}\{X\}}{t}.$$

Theorem A.16. (CHEBYSHEV'S INEQUALITY). *Let X be a random variable. Then for each $t > 0$,*

$$\mathbf{P}\{|X - \mathbf{E}\{X\}| \geq t\} \leq \frac{\text{Var}\{X\}}{t^2}.$$

Theorem A.17. (CHEBYSHEV-CANTELLI INEQUALITY). *Let $t \geq 0$. Then*

$$\mathbf{P}\{X - \mathbf{E}X > t\} \leq \frac{\text{Var}\{X\}}{\text{Var}\{X\} + t^2}.$$

PROOF. We may assume without loss of generality that $\mathbf{E}X = 0$. Then for all t

$$t = \mathbf{E}\{t - X\} \leq \mathbf{E}\{(t - X)I_{\{X \leq t\}}\}.$$

Thus for $t \geq 0$ from the Cauchy-Schwarz inequality,

$$
\begin{aligned}
t^2 &\leq \mathbf{E}\{(t - X)^2\}\mathbf{E}\{I^2_{\{X \leq t\}}\} \\
&= \mathbf{E}\{(t - X)^2\}\mathbf{P}\{X \leq t\} \\
&= (\mathbf{Var}\{X\} + t^2)\mathbf{P}\{X \leq t\},
\end{aligned}
$$

that is,

$$
\mathbf{P}\{X \leq t\} \geq \frac{t^2}{\mathbf{Var}\{X\} + t^2},
$$

and the claim follows. $\square$

Theorem A.18. (JENSEN'S INEQUALITY). *If f is a real-valued convex function on a finite or infinite interval of $\mathcal{R}$, and X is a random variable with finite expectation, taking its values in this interval, then*

$$
f(\mathbf{E}\{X\}) \leq \mathbf{E}\{f(X)\}.
$$

Theorem A.19. (ASSOCIATION INEQUALITIES). *Let X be a real-valued random variable and let $f(x)$ and $g(x)$ be monotone nondecreasing real-valued functions. Then*

$$
\mathbf{E}\{f(X)g(X)\} \geq \mathbf{E}\{f(X)\}\mathbf{E}\{g(X)\},
$$

provided that all expectations exist and are finite. If f is monotone increasing and g is monotone decreasing, then

$$
\mathbf{E}\{f(X)g(X)\} \leq \mathbf{E}\{f(X)\}\mathbf{E}\{g(X)\}.
$$

PROOF. We prove the first inequality. The second follows by symmetry. Let X have distribution μ. Then we write

$$
\begin{aligned}
&\mathbf{E}\{f(X)g(X)\} - \mathbf{E}\{f(X)\}\mathbf{E}\{g(X)\} \\
&= \int f(x)g(x)\,\mu(dx) - \int f(y)\,\mu(dy) \int g(x)\,\mu(dx) \\
&= \int \left(\int [f(x) - f(y)]g(x)\,\mu(dx) \right) \mu(dy) \\
&= \int \left(\int h(x, y)g(x)\,\mu(dx) \right) \mu(dy),
\end{aligned}
$$

where $h(x, y) = f(x) - f(y)$. By Fubini's theorem the last integral equals

$$
\begin{aligned}
&\int_{\mathcal{R}^2} h(x, y)g(x)\,\mu^2(dxdy) \\
&= \int_{x>y} h(x, y)g(x)\,\mu^2(dxdy) + \int_{x<y} h(x, y)g(x)\,\mu^2(dxdy),
\end{aligned}
$$

since $h(x, x) = 0$ for all x. Here $\mu^2(dxdy) = \mu(dx) \cdot \mu(dy)$. The second integral on the right-hand side is just

$$\int_x \left(\int_{y>x} h(x, y)g(x)\mu(dy) \right) \mu(dx) = \int_y \left(\int_{x>y} h(y, x)g(y)\mu(dx) \right) \mu(dy).$$

Thus, we have

$$\mathbf{E}\{f(X)g(X)\} - \mathbf{E}\{f(X)\}\mathbf{E}\{g(X)\}$$

$$= \int_{\mathcal{R}^2} h(x, y)g(x)\mu^2(dxdy)$$

$$= \int_y \left(\int_{x>y} [h(x, y)g(x) + h(y, x)g(y)]\mu(dx) \right) \mu(dy)$$

$$= \int_y \left(\int_{x>y} h(x, y)[g(x) - g(y)]\mu(dx) \right) \mu(dy)$$

$$\geq 0,$$

since $h(y, x) = -h(x, y)$, and by the fact that $h(x, y) \geq 0$ and $g(x) - g(y) \geq 0$ if $x > y$. $\square$

A.6   Convergence of Random Variables

DEFINITION A.23. *Let $\{X_n\}$, $n = 1, 2, \ldots$, be a sequence of random variables. We say that*

$$\lim_{n \to \infty} X_n = X \quad \text{in probability}$$

if for each $\epsilon > 0$

$$\lim_{n \to \infty} \mathbf{P}\{|X_n - X| \geq \epsilon\} = 0.$$

We say that

$$\lim_{n \to \infty} X_n = X \quad \text{with probability one (or almost surely),}$$

if $X_n \to X \pmod{\mathbf{P}}$, that is,

$$\mathbf{P}\left\{ \omega : \lim_{n \to \infty} X_n(\omega) = X(\omega) \right\} = 1.$$

For a fixed number $p \geq 1$ we say that

$$\lim_{n \to \infty} X_n = X \quad \text{in } L_p,$$

if

$$\lim_{n \to \infty} \mathbf{E}\left\{ |X_n - X|^p \right\} = 0.$$

Theorem A.20. *Convergence in L_p implies convergence in probability.*

Theorem A.21. $\lim_{n \to \infty} X_n = X$ *with probability one if and only if*

$$\lim_{n \to \infty} \sup_{n \leq m} |X_m - X| = 0$$

in probability. Thus, convergence with probability one implies convergence in probability.

Theorem A.22. (BOREL-CANTELLI LEMMA). *Let A_n, $n = 1, 2, \ldots$, be a sequence of events. Introduce the notation*

$$[A_n \ \text{i.o.}] = \limsup_{n \to \infty} A_n = \cap_{n=1}^{\infty} \cup_{m=n}^{\infty} A_m.$$

("i.o." stands for "infinitely often.") If

$$\sum_{n=1}^{\infty} P\{A_n\} < \infty$$

then

$$P\{[A_n \ \text{i.o.}]\} = 0.$$

By Theorems A.21 and A.22, we have

Theorem A.23. *If for each $\epsilon > 0$*

$$\sum_{n=1}^{\infty} P\{|X_n - X| \geq \epsilon\} < \infty,$$

then $\lim_{n \to \infty} X_n = X$ with probability one.

A.7 Conditional Expectation

If Y is a random variable with finite expectation and A is an event with positive probability, then the conditional expectation of Y given A is defined by

$$E\{Y|A\} = \frac{E\{Y I_A\}}{P\{A\}}.$$

The conditional probability of an event B given A is

$$P\{B|A\} = E\{I_B|A\} = \frac{P\{A \cap B\}}{P\{A\}}.$$

DEFINITION A.24. *Let Y be a random variable with finite expectation and X be a d-dimensional vector-valued random variable. Let $\mathcal{F}_X$ be the σ-algebra generated by X:*

$$\mathcal{F}_X = \{X^{-1}(B); B \in \mathcal{B}^n\}.$$

The conditional expectation $E\{Y|X\}$ of Y given X is a random variable with the property that for all $A \in \mathcal{F}_X$

$$\int_A Y d\mathbf{P} = \int_A E\{Y|X\} d\mathbf{P}.$$

The existence and uniqueness (with probability one) of $E\{Y|X\}$ is a consequence of the Radon-Nikodym theorem if we apply it to the measures

$$v^+(A) = \int_A Y^+ d\mathbf{P}, \qquad v^-(A) = \int_A Y^- d\mathbf{P}, \quad A \in \mathcal{F}_X,$$

such that

$$E\{Y^+|X\} = \frac{dv^+}{d\mathbf{P}}, \qquad E\{Y^-|X\} = \frac{dv^-}{d\mathbf{P}},$$

and

$$E\{Y|X\} = E\{Y^+|X\} - E\{Y^-|X\}.$$

DEFINITION A.25. *Let C be an event and X be a d-dimensional vector-valued random variable. Then the conditional probability of C given X is $P\{C|X\} = E\{I_C|X\}$.*

Theorem A.24. *Let Y be a random variable with finite expectation. Let C be an event, and let X and Z be vector-valued random variables. Then*

(i)   *There is a measurable function g on $\mathcal{R}^d$ such that $E\{Y|X\} = g(X)$ with probability one.*

(ii) $E\{Y\} = E\{E\{Y|X\}\}, \quad P\{C\} = E\{P\{C|X\}\}.$

(iii) $E\{Y|X\} = E\{E\{Y|X, Z\}|X\}, \quad P\{C|X\} = E\{P\{C|X, Y\}|X\}.$

(iv)   *If Y is a function of X then $E\{Y|X\} = Y$.*

(v)   *If (Y, X) and Z are independent, then $E\{Y|X, Z\} = E\{Y|X\}$.*

(vi)   *If $Y = f(X, Z)$ for a measurable function f, and X and Z are independent, then $E\{Y|X\} = g(X)$, where $g(x) = E\{f(x, Z)\}$.*

A.8   The Binomial Distribution

An integer-valued random variable X is said to be binomially distributed with parameters n and p if

$$P\{X = k\} = \binom{n}{k} p^k (1 - p)^{n-k}, \quad k = 0, 1, \ldots, n.$$

If $A_1, \ldots, A_n$ are independent events with $P\{A_i\} = p$, then $X = \sum_{i=1}^n I_{A_i}$ is binomial (n, p). I_{A_i} is called a *Bernoulli* random variable with parameter p.

Lemma A.2. *Let the random variable $B(n, p)$ be binomially distributed with parameters n and p. Then*

(i)

$$E\left\{\frac{1}{1 + B(n, p)}\right\} \leq \frac{1}{(n + 1)p}$$

and

(ii)

$$E\left\{\frac{1}{B(n, p)} I_{\{B(n,p)>0\}}\right\} \leq \frac{2}{(n + 1)p}.$$

PROOF. (i) follows from the following simple calculation:

$$\begin{aligned}
E\left\{\frac{1}{1 + B(n, p)}\right\} &= \sum_{k=0}^{n} \frac{1}{k + 1}\binom{n}{k} p^k (1 - p)^{n-k} \\
&= \frac{1}{(n + 1)p} \sum_{k=0}^{n} \binom{n + 1}{k + 1} p^{k+1} (1 - p)^{n-k} \\
&\leq \frac{1}{(n + 1)p} \sum_{k=0}^{n+1} \binom{n + 1}{k} p^k (1 - p)^{n-k+1} = \frac{1}{(n + 1)p}.
\end{aligned}$$

For (ii) we have

$$E\left\{\frac{1}{B(n, p)} I_{\{B(n,p)>0\}}\right\} \leq E\left\{\frac{2}{1 + B(n, p)}\right\} \leq \frac{2}{(n + 1)p}$$

by (i). $\square$

Lemma A.3. *Let B be a binomial random variable with parameters n and p. Then for every $0 \leq p \leq 1$,*

$$P\left\{B = \left\lfloor\frac{n}{2}\right\rfloor\right\} \leq \sqrt{\frac{1}{2n\pi}},$$

and for $p = 1/2$

$$P\left\{B = \left\lfloor\frac{n}{2}\right\rfloor\right\} \geq \sqrt{\frac{1}{2n\pi}} \frac{1}{e^{1/12}}.$$

PROOF. The lemma follows from Stirling's formula:

$$\sqrt{2\pi n}\left(\frac{n}{e}\right)^n e^{1/(12n)} \leq n! \leq \sqrt{2\pi n}\left(\frac{n}{e}\right)^n e^{1/(12n+1)}$$

(see, e.g., Feller (1968)). $\square$

Lemma A.4. (DEVROYE AND GYÖRFI (1985), P. 194). *For any random variable* X *with finite fourth moment,*

$$E\{|X|\} \geq \frac{\left(E\{X^2\}\right)^{3/2}}{\left(E\{X^4\}\right)^{1/2}}.$$

PROOF. Fix $a > 0$. The function $1/x + ax^2$ is minimal on $(0, \infty)$ when $x^3 = 1/(2a)$. Thus,

$$\frac{x + ax^4}{x^2} \geq (2a)^{1/3} + \frac{a}{(2a)^{2/3}} = \frac{3}{2}(2a)^{1/3}.$$

Replace x by $|X|$ and take expectations:

$$E\{|X|\} \geq \frac{3}{2}(2a)^{1/3}E\{X^2\} - aE\{X^4\}.$$

The lower bound, considered as a function of a, is maximized if we take $a = \frac{1}{2}\left(E\{X^2\}/E\{X^4\}\right)^{3/2}$. Resubstitution yields the given inequality. $\square$

Lemma A.5. *Let* B *be a binomial* $(n, 1/2)$ *random variable. Then*

$$E\left\{\left|B - \frac{n}{2}\right|\right\} \geq \sqrt{\frac{n}{8}}.$$

PROOF. This bound is a special case of Khintchine's inequality (see Szarek (1976), Haagerup (1978), and also Devroye and Györfi (1985), p. 139). Rather than proving the given inequality, we will show how to apply the previous lemma to get (without further work) the inequality

$$E\left\{\left|B - \frac{n}{2}\right|\right\} \geq \sqrt{\frac{n}{12}}.$$

Indeed, $E\left\{(B - n/2)^2\right\} = n/4$ and $E\left\{(B - n/2)^4\right\} = 3n^2/16 - n/8 \leq 3n^2/16$. Thus,

$$E\left\{\left|B - \frac{n}{2}\right|\right\} \geq \frac{(n/4)^{3/2}}{(3n^2/16)^{1/2}} = \sqrt{\frac{n}{12}}. \quad \square$$

Lemma A.6. (SLUD (1977)). *Let* B *be a binomial* (n, p) *random variable with* $p \leq 1/2$. *Then for* $n(1 - p) \geq k \geq np$,

$$P\{B \geq k\} \geq P\left\{N \geq \frac{k - np}{\sqrt{np(1 - p)}}\right\},$$

where N *is normal* $(0, 1)$.

A.9   The Hypergeometric Distribution

Let N, b, and n be positive integers with $N > n$ and $N > b$. A random variable X taking values on the integers $0, 1, \ldots, b$ is *hypergeometric* with parameters N, b and n, if

$$P\{X = k\} = \frac{\binom{b}{k}\binom{N-b}{n-k}}{\binom{N}{n}}, \quad k = 1, \ldots, b.$$

X models the number of blue balls in a sample of n balls drawn without replacement from an urn containing b blue and $N - b$ red balls.

Theorem A.25. (HOEFFDING (1963)). *Let the set A consist of N numbers $a_1, \ldots, a_N$. Let $Z_1, \ldots, Z_n$ denote a random sample taken without replacement from A, where $n \leq N$. Denote*

$$m = \frac{1}{N} \sum_{i=1}^{N} a_i \quad and \quad c = \max_{i, j \leq N} |a_i - a_j|.$$

Then for any $\epsilon > 0$ we have

$$P\left\{ \left| \frac{1}{n} \sum_{i=1}^{n} Z_i - m \right| \geq \epsilon \right\} \leq 2e^{-2n\epsilon^2/c^2}.$$

Specifically, if X is hypergeometrically distributed with parameters N, b, and n, then

$$P\left\{ \left| X - \frac{bn}{N} \right| \geq n\epsilon \right\} \leq 2e^{-2n\epsilon^2}.$$

For more inequalities of this type, see Hoeffding (1963) and Serfling (1974).

A.10   The Multinomial Distribution

A vector $(N_1, \ldots, N_k)$ of integer-valued random variables is *multinomially distributed* with parameters $(n, p_1, \ldots, p_k)$ if

$$P\{N_1 = i_1, \ldots, N_k = i_k\} = \begin{cases} \frac{n!}{i_1! \cdots i_k!} p_1^{i_1} \cdots p_k^{i_k} & \text{if } \sum_{j=1}^{k} i_j = k, \ i_j \geq 0 \\ 0 & \text{otherwise.} \end{cases}$$

Lemma A.7. *The moment-generating function of a multinomial $(k, p_1, \ldots, p_k)$ vector is*

$$E\left\{ e^{t_1 N_1 + \cdots + t_k N_k} \right\} = \left(p_1 e^{t_1} + \cdots + p_k e^{t_k} \right)^n.$$

A.11   The Exponential and Gamma Distributions

A nonnegative random variable has *exponential* distribution with parameter $\lambda > 0$ if it has a density

$$f(x) = \lambda e^{-\lambda x}, \quad x \geq 0.$$

A nonnegative-valued random variable has the *gamma distribution* with parameters $a, b \geq 0$ if it has density

$$f(x) = \frac{x^{a-1}e^{-x/b}}{\Gamma(a)b^a}, \quad x \geq 0, \quad \text{where} \quad \Gamma(a) = \int_0^\infty x^{a-1}e^{-x}dx.$$

The sum of n i.i.d. exponential(λ) random variables has gamma distribution with parameters n and $1/\lambda$.

A.12   The Multivariate Normal Distribution

A d-dimensional random variable $X = \left(X^{(1)}, \ldots, X^{(d)}\right)$ has the *multivariate normal* distribution if it has a density

$$f(x) = \frac{1}{\sqrt{(2\pi)^d \det(\Sigma)}} e^{-\frac{1}{2}(x-m)^T \Sigma^{-1}(x-m)},$$

where $m \in \mathcal{R}^d$, Σ is a positive definite symmetric $d \times d$ matrix with entries σ_{ij}, and $\det(\Sigma)$ denotes the determinant of Σ. Then $\mathbf{E}X = m$, and for all $i, j = 1, \ldots, d$,

$$\mathbf{E}\left\{(X^{(i)} - \mathbf{E}X^{(i)})(X^{(j)} - \mathbf{E}X^{(j)})\right\} = \sigma_{ij}.$$

Σ is called the *covariance matrix* of X.

Notation

- I_A indicator of an event A.
- $I_B(x) = I_{\{x \in B\}}$ indicator function of a set B.
- $|A|$ cardinality of a finite set A.
- A^c complement of a set A.
- $A \triangle B$ symmetric difference of sets A, B.
- $f \circ g$ composition of functions f, g.
- log   natural logarithm (base e).
- $\lfloor x \rfloor$ integer part of the real number x.
- $\lceil x \rceil$ upper integer part of the real number x.
- $X \stackrel{\mathcal{L}}{=} Z$ if X and Z have the same distribution.
- $x^{(1)}, \ldots, x^{(d)}$ components of the d-dimensional column vector x.
- $\|x\| = \sqrt{\sum_{i=1}^{d} \left(x^{(i)}\right)^2}$ L_2-norm of $x \in \mathcal{R}^d$.
- $X \in \mathcal{R}^d$ observation, vector-valued random variable.
- $Y \in \{0, 1\}$ label, binary random variable.
- $D_n = ((X_1, Y_1), \ldots, (X_n, Y_n))$ training data, sequence of i.i.d. pairs that are independent of (X, Y), and have the same distribution as that of (X, Y).
- $\eta(x) = \mathbf{P}\{Y = 1 | X = x\}, 1 - \eta(x) = \mathbf{P}\{Y = 0 | X = x\}$ a posteriori probabilities.
- $p = \mathbf{P}\{Y = 1\}, 1 - p = \mathbf{P}\{Y = 0\}$ class probabilities.
- $g^* : \mathcal{R}^d \to \{0, 1\}$ Bayes decision function.
- $\phi : \mathcal{R}^d \to \{0, 1\}, g_n : \mathcal{R}^d \times \left\{\mathcal{R}^d \times \{0, 1\}\right\}^n \to \{0, 1\}$ classification functions. The short notation $g_n(x) = g_n(x, D_n)$ is also used.
- $L^* = \mathbf{P}\{g^*(X) \neq Y\}$ Bayes risk, the error probability of the Bayes decision.

- $L_n = L(g_n) = \mathbf{P}\{g_n(X, D_n) \neq Y | D_n\}$ error probability of a classification function g_n.
- $\hat{L}_n(\phi) = \frac{1}{n} \sum_{i=1}^{n} I_{\{\phi(X_i) \neq Y_i\}}$ empirical error probability of a classifier ϕ.
- $\mu(A) = \mathbf{P}\{X \in A\}$ probability measure of X.
- $\mu_n(A) = \frac{1}{n} \sum_{i=1}^{n} I_{\{X_i \in A\}}$ empirical measure corresponding to $X_1, \ldots, X_n$.
- λ Lebesgue measure on $\mathcal{R}^d$.
- $f(x)$ density of X, Radon-Nikodym derivative of μ with respect to λ (if it exists).
- $f_0(x)$, $f_1(x)$ conditional densities of X given $Y = 0$ and $Y = 1$, respectively (if they exist).
- $\mathcal{P}$ partition of $\mathcal{R}^d$.
- $X_{(k)}(x)$, $X_{(k)}$ k-th nearest neighbor of x among $X_1, \ldots, X_n$.
- $K : \mathcal{R}^d \to \mathcal{R}$ kernel function.
- $h, h_n > 0$ smoothing factor for a kernel rule.
- $K_h(x) = (1/h)K(x/h)$ scaled kernel function.
- $T_m = ((X_{n+1}, Y_{n+1}), ..., (X_{n+m}, Y_{n+m}))$ testing data, sequence of i.i.d. pairs that are independent of (X, Y) and D_n, and have the same distribution as that of (X, Y).
- $\mathcal{A}$ class of sets.
- $\mathcal{C}, \mathcal{C}_n$ classes of classification functions.
- $s(\mathcal{A}, n)$ n-th shatter coefficient of the class of sets $\mathcal{A}$.
- $V_{\mathcal{A}}$ Vapnik-Chervonenkis dimension of the class of sets $\mathcal{A}$.
- $\mathcal{S}(\mathcal{C}, n)$ n-th shatter coefficient of the class of classifiers $\mathcal{C}$.
- $V_{\mathcal{C}}$ Vapnik-Chervonenkis dimension of the class of classifiers $\mathcal{C}$.
- $S_{x,r} = \{y \in \mathcal{R}^d : \|y - x\| \leq r\}$ closed Euclidean ball in $\mathcal{R}^d$ centered at $x \in \mathcal{R}^d$, with radius $r > 0$.

References

Abou-Jaoude, S. (1976a). Conditions nécessaires et suffisantes de convergence L_1 en probabilité de l'histogramme pour une densité. *Annales de l'Institut Henri Poincaré*, 12:213–231.

Abou-Jaoude, S. (1976b). La convergence L_1 et L_∞ de l'estimateur de la partition aléatoire pour une densité. *Annales de l'Institut Henri Poincaré*, 12:299–317.

Abou-Jaoude, S. (1976c). Sur une condition nécessaire et suffisante de L_1-convergence presque complète de l'estimateur de la partition fixe pour une densité. *Comptes Rendus de l'Académie des Sciences de Paris*, 283:1107–1110.

Aitchison, J. and Aitken, C. (1976). Multivariate binary discrimination by the kernel method. *Biometrika*, 63:413–420.

Aizerman, M., Braverman, E., and Rozonoer, L. (1964a). The method of potential functions for the problem of restoring the characteristic of a function converter from randomly observed points. *Automation and Remote Control*, 25:1546–1556.

Aizerman, M., Braverman, E., and Rozonoer, L. (1964b). The probability problem of pattern recognition learning and the method of potential functions. *Automation and Remote Control*, 25:1307–1323.

Aizerman, M., Braverman, E., and Rozonoer, L. (1964c). Theoretical foundations of the potential function method in pattern recognition learning. *Automation and Remote Control*, 25:917–936.

Aizerman, M., Braverman, E., and Rozonoer, L. (1970). Extrapolative problems in automatic control and the method of potential functions. *American Mathematical Society Translations*, 87:281–303.

Akaike, H. (1954). An approximation to the density function. *Annals of the Institute of Statistical Mathematics*, 6:127–132.

Akaike, H. (1974). A new look at the statistical model identification. *IEEE Transactions on Automatic Control*, 19:716–723.

Alexander, K. (1984). Probability inequalities for empirical processes and a law of the iterated logarithm. *Annals of Probability*, 4:1041–1067.

Anderson, A. and Fu, K. (1979). Design and development of a linear binary tree classifier for leukocytes. Technical Report TR-EE-79-31, Purdue University, Lafayette, IN.

Anderson, J. (1982). Logistic discrimination. In *Handbook of Statistics*, Krishnaiah, P. and Kanal, L., editors, volume 2, pages 169–191. North-Holland, Amsterdam.

Anderson, M. and Benning, R. (1970). A distribution-free discrimination procedure based on clustering. *IEEE Transactions on Information Theory*, 16:541–548.

Anderson, T. (1958). *An Introduction to Multivariate Statistical Analysis*. John Wiley, New York.

Anderson, T. (1966). Some nonparametric multivariate procedures based on statistically equivalent blocks. In *Multivariate Analysis*, Krishnaiah, P., editor, pages 5–27. Academic Press, New York.

Angluin, D. and Valiant, L. (1979). Fast probabilistic algorithms for Hamiltonian circuits and matchings. *Journal of Computing System Science*, 18:155–193.

Anthony, M. and Holden, S. (1993). On the power of polynomial discriminators and radial basis function networks. In *Proceedings of the Sixth Annual ACM Conference on Computational Learning Theory*, pages 158–164. Association for Computing Machinery, New York.

Anthony, M. and Shawe-Taylor, J. (1990). A result of Vapnik with applications. Technical Report CSD-TR-628, University of London, Surrey.

Argentiero, P., Chin, R., and Beaudet, P. (1982). An automated approach to the design of decision tree classifiers. *IEEE Transactions on Pattern Analysis and Machine Intelligence*, 4:51–57.

Arkadjew, A. and Braverman, E. (1966). *Zeichenerkennung und Maschinelles Lernen*. Oldenburg Verlag, München, Wien.

Ash, R. (1972). *Real Analysis and Probability*. Academic Press, New York.

Assouad, P. (1983a). Densité et dimension. *Annales de l'Institut Fourier*, 33:233–282.

Assouad, P. (1983b). Deux remarques sur l'estimation. *Comptes Rendus de l'Académie des Sciences de Paris*, 296:1021–1024.

Azuma, K. (1967). Weighted sums of certain dependent random variables. *Tohoku Mathematical Journal*, 68:357–367.

Bahadur, R. (1961). A representation of the joint distribution of responses to n dichotomous items. In *Studies in Item Analysis and Prediction*, Solomon, H., editor, pages 158–168. Stanford University Press, Stanford, CA.

Bailey, T. and Jain, A. (1978). A note on distance-weighted k-nearest neighbor rules. *IEEE Transactions on Systems, Man, and Cybernetics*, 8:311–313.

Barron, A. (1985). Logically smooth density estimation. Technical Report TR 56, Department of Statistics, Stanford University, Stanford, CA.

Barron, A. (1989). Statistical properties of artificial neural networks. In *Proceedings of the 28th Conference on Decision and Control*, pages 280–285. Tampa, FL.

Barron, A. (1991). Complexity regularization with application to artificial neural networks. In *Nonparametric Functional Estimation and Related Topics*, Roussas, G., editor, pages 561–576. NATO ASI Series, Kluwer Academic Publishers, Dordrecht.

Barron, A. (1993). Universal approximation bounds for superpositions of a sigmoidal function. *IEEE Transactions on Information Theory*, 39:930–944.

Barron, A. (1994). Approximation and estimation bounds for artificial neural networks. *Machine Learning*, 14:115–133.

Barron, A. and Barron, R. (1988). Statistical learning networks: a unifying view. In *Proceedings of the 20th Symposium on the Interface: Computing Science and Statistics*, Wegman, E., Gantz, D., and Miller, J., editors, pages 192–203. AMS, Alexandria, VA.

Barron, A. and Cover, T. (1991). Minimum complexity density estimation. *IEEE Transactions on Information Theory*, 37:1034–1054.

Barron, A., Györfi, L., and van der Meulen, E. (1992). Distribution estimation consistent in total variation and in two types of information divergence. *IEEE Transactions on Information Theory*, 38:1437–1454.

Barron, R. (1975). Learning networks improve computer-aided prediction and control. *Computer Design*, 75:65–70.

Bartlett, P. (1993). Lower bounds on the Vapnik-Chervonenkis dimension of multi-layer threshold networks. In *Proceedings of the Sixth annual ACM Conference on Computational Learning Theory*, pages 144–150. Association for Computing Machinery, New York.

Bashkirov, O., Braverman, E., and Muchnik, I. (1964). Potential function algorithms for pattern recognition learning machines. *Automation and Remote Control*, 25:692–695.

Baum, E. (1988). On the capabilities of multilayer perceptrons. *Journal of Complexity*, 4:193–215.

Baum, E. and Haussler, D. (1989). What size net gives valid generalization? *Neural Computation*, 1:151–160.

Beakley, G. and Tuteur, F. (1972). Distribution-free pattern verification using statistically equivalent blocks. *IEEE Transactions on Computers*, 21:1337–1347.

Beck, J. (1979). The exponential rate of convergence of error for k_n-NN nonparametric regression and decision. *Problems of Control and Information Theory*, 8:303–311.

Becker, P. (1968). *Recognition of Patterns*. Polyteknisk Forlag, Copenhagen.

Ben-Bassat, M. (1982). Use of distance measures, information measures and error bounds in feature evaluation. In *Handbook of Statistics*, Krishnaiah, P. and Kanal, L., editors, volume 2, pages 773–792. North-Holland, Amsterdam.

Benedek, G. and Itai, A. (1988). Learnability by fixed distributions. In *Computational Learning Theory: Proceedings of the 1988 Workshop*, pages 80–90. Morgan Kaufman, San Mateo, CA.

Benedek, G. and Itai, A. (1994). Nonuniform learnability. *Journal of Computer and Systems Sciences*, 48:311–323.

Bennett, G. (1962). Probability inequalities for the sum of independent random variables. *Journal of the American Statistical Association*, 57:33–45.

Beran, R. (1977). Minimum Hellinger distance estimates for parametric models. *Annals of Statistics*, 5:445–463.

Beran, R. (1988). Comments on "a new theoretical and algorithmical basis for estimation, identification and control" by P. Kovanic. *Automatica*, 24:283–287.

Bernstein, S. (1946). *The Theory of Probabilities*. Gastehizdat Publishing House, Moscow.

Bhattacharya, P. and Mack, Y. (1987). Weak convergence of k–NN density and regression estimators with varying k and applications. *Annals of Statistics*, 15:976–994.

Bhattacharyya, A. (1946). On a measure of divergence between two multinomial populations. *Sankhya, Series A*, 7:401–406.

Bickel, P. and Breiman, L. (1983). Sums of functions of nearest neighbor distances, moment bounds, limit theorems and a goodness of fit test. *Annals of Probability*, 11:185–214.

Birgé, L. (1983). Approximation dans les espaces métriques et théorie de l'estimation. *Zeitschrift für Wahrscheinlichkeitstheorie und verwandte Gebiete*, 65:181–237.

Birgé, L. (1986). On estimating a density using Hellinger distance and some other strange facts. *Probability Theory and Related Fields*, 71:271–291.

Blumer, A., Ehrenfeucht, A., Haussler, D., and Warmuth, M. (1989). Learnability and the Vapnik-Chervonenkis dimension. *Journal of the ACM*, 36:929–965.

Braverman, E. (1965). The method of potential functions. *Automation and Remote Control*, 26:2130–2138.

Braverman, E. and Pyatniskii, E. (1966). Estimation of the rate of convergence of algorithms based on the potential function method. *Automation and Remote Control*, 27:80–100.

Breiman, L., Friedman, J., Olshen, R., and Stone, C. (1984). *Classification and Regression Trees*. Wadsworth International, Belmont, CA.

Breiman, L., Meisel, W., and Purcell, E. (1977). Variable kernel estimates of multivariate densities. *Technometrics*, 19:135–144.

Brent, R. (1991). Fast training algorithms for multilayer neural nets. *IEEE Transactions on Neural Networks*, 2:346–354.

Broder, A. (1990). Strategies for efficient incremental nearest neighbor search. *Pattern Recognition*, 23:171–178.

Broomhead, D. and Lowe, D. (1988). Multivariable functional interpolation and adaptive networks. *Complex Systems*, 2:321–323.

Buescher, K. and Kumar, P. (1996a). Learning by canonical smooth estimation, Part I: Simultaneous estimation. *IEEE Transactions on Automatic Control*, 41:545–556.

Buescher, K. and Kumar, P. (1996b). Learning by canonical smooth estimation, Part II: Learning and choice of model complexity. *IEEE Transactions on Automatic Control*, 41:557–569.

Burbea, J. (1984). The convexity with respect to gaussian distributions of divergences of order α. *Utilitas Mathematica*, 26:171–192.

Burbea, J. and Rao, C. (1982). On the convexity of some divergence measures based on entropy functions. *IEEE Transactions on Information Theory*, 28:48–495.

Burshtein, D., Della Pietra, V., Kanevsky, D., and Nádas, A. (1992). Minimum impurity partitions. *Annals of Statistics*, 20:1637–1646.

Cacoullos, T. (1965). Estimation of a multivariate density. *Annals of the Institute of Statistical Mathematics*, 18:179–190.

Carnal, H. (1970). Die konvexe Hülle von n rotationssymmetrisch verteilten Punkten. *Zeitschrift für Wahrscheinlichkeitstheorie und verwandte Gebiete*, 15:168–176.

Casey, R. and Nagy, G. (1984). Decision tree design using a probabilistic model. *IEEE Transactions on Information Theory*, 30:93–99.

Cencov, N. (1962). Evaluation of an unknown distribution density from observations. *Soviet Math. Doklady*, 3:1559–1562.

Chang, C. (1974). Finding prototypes for nearest neighbor classifiers. *IEEE Transactions on Computers*, 26:1179–1184.

Chang, C. (1973). Dynamic programming as applied to feature selection in pattern recognition systems. *IEEE Transactions on Systems, Man, and Cybernetics*, 3:166–171.

Chen, T., Chen, H., and Liu, R. (1990). A constructive proof and an extension of Cybenko's approximation theorem. In *Proceedings of the 22nd Symposium of the Interface: Computing Science and Statistics*, pages 163–168. American Statistical Association, Alexandria, VA.

Chen, X. and Zhao, L. (1987). Almost sure L_1-norm convergence for data-based histogram density estimates. *Journal of Multivariate Analysis*, 21:179–188.

Chen, Z. and Fu, K. (1973). Nonparametric Bayes risk estimation for pattern classification. In *Proceedings of the IEEE Conference on Systems, Man, and Cybernetics*. Boston, MA.

Chernoff, H. (1952). A measure of asymptotic efficiency of tests of a hypothesis based on the sum of observations. *Annals of Mathematical Statistics*, 23:493–507.

Chernoff, H. (1971). A bound on the classification error for discriminating between populations with specified means and variances. In *Studi di probabilità, statistica e ricerca operativa in onore di Giuseppe Pompilj*, pages 203–211. Oderisi, Gubbio.

Chou, P. (1991). Optimal partitioning for classification and regression trees. *IEEE Transactions on Pattern Analysis and Machine Intelligence*, 13:340–354.

Chou, W. and Chen, Y. (1992). A new fast algorithm for effective training of neural classifiers. *Pattern Recognition*, 25:423–429.

Chow, C. (1965). Statistical independence and threshold functions. *IEEE Transactions on Computers*, E-14:66–68.

Chow, C. (1970). On optimum recognition error and rejection tradeoff. *IEEE Transactions on Information Theory*, 16:41–46.

Chow, Y. and Teicher, H. (1978). *Probability Theory, Independence, Interchangeability, Martingales*. Springer-Verlag, New York.

Ciampi, A. (1991). Generalized regression trees. *Computational Statistics and Data Analysis*, 12:57–78.

Collomb, G. (1979). Estimation de la regression par la méthode des k points les plus proches: propriétés de convergence ponctuelle. *Comptes Rendus de l'Académie des Sciences de Paris*, 289:245–247.

Collomb, G. (1980). *Estimation de la regression par la méthode des k points les plus proches avec noyau*. Lecture Notes in Mathematics #821, Springer-Verlag, Berlin. 159–175.

Collomb, G. (1981). Estimation non parametrique de la regression: revue bibliographique. *International Statistical Review*, 49:75–93.

Conway, J. and Sloane, N. (1993). *Sphere-Packings, Lattices and Groups*. Springer-Verlag, Berlin.

Coomans, D. and Broeckaert, I. (1986). *Potential Pattern Recognition in Chemical and Medical Decision Making*. Research Studies Press, Letchworth, Hertfordshire, England.

Cormen, T., Leiserson, C., and Rivest, R. (1990). *Introduction to Algorithms*. MIT Press, Boston, MA.

Cover, T. (1965). Geometrical and statistical properties of systems of linear inequalities with applications in pattern recognition. *IEEE Transactions on Electronic Computers*, 14:326–334.

Cover, T. (1968a). Estimation by the nearest neighbor rule. *IEEE Transactions on Information Theory*, 14:50–55.

Cover, T. (1968b). Rates of convergence for nearest neighbor procedures. In *Proceedings of the Hawaii International Conference on Systems Sciences*, pages 413–415. Honolulu.

Cover, T. (1969). Learning in pattern recognition. In *Methodologies of Pattern Recognition*, Watanabe, S., editor, pages 111–132. Academic Press, New York.

Cover, T. (1974). The best two independent measurements are not the two best. *IEEE Transactions on Systems, Man, and Cybernetics*, 4:116–117.

Cover, T. and Hart, P. (1967). Nearest neighbor pattern classification. *IEEE Transactions on Information Theory*, 13:21–27.

Cover, T. and Thomas, J. (1991). *Elements of Information Theory*. John Wiley, New York.

Cover, T. and Van Campenhout, J. (1977). On the possible orderings in the measurement selection problem. *IEEE Transactions on Systems, Man, and Cybernetics*, 7:657–661.

Cover, T. and Wagner, T. (1975). Topics in statistical pattern recognition. *Communication and Cybernetics*, 10:15–46.

Cramér, H. and Wold, H. (1936). Some theorems on distribution functions. *Journal of the London Mathematical Society*, 11:290–294.

Csibi, S. (1971). Simple and compound processes in iterative machine learning. Technical Report, CISM Summer Course, Udine, Italy.

Csibi, S. (1975). Using indicators as a base for estimating optimal decision functions. In *Colloquia Mathematica Societatis János Bolyai: Topics in Information Theory*, pages 143–153. Keszthely, Hungary.

Csiszár, I. (1967). Information-type measures of difference of probability distributions and indirect observations. *Studia Scientiarium Mathematicarum Hungarica*, 2:299–318.

Csiszár, I. (1973). Generalized entropy and quantization problems. In *Transactions of the Sixth Prague Conference on Information Theory, Statistical Decision Functions, Random Processes*, pages 159–174. Academia, Prague.

Csiszár, I. and Körner, J. (1981). *Information Theory: Coding Theorems for Discrete Memoryless Systems*. Academic Press, New York.

Cybenko, G. (1989). Approximations by superpositions of sigmoidal functions. *Math. Control, Signals, Systems*, 2:303–314.

Darken, C., Donahue, M., Gurvits, L., and Sontag, E. (1993). Rate of approximation results motivated by robust neural network learning. In *Proceedings of the Sixth ACM Workshop on Computational Learning Theory*, pages 303–309. Association for Computing Machinery, New York.

Das Gupta, S. (1964). Nonparametric classification rules. *Sankhya Series A*, 26:25–30.

Das Gupta, S. and Lin, H. (1980). Nearest neighbor rules of statistical classification based on ranks. *Sankhya Series A*, 42:419–430.

Dasarathy, B. (1991). *Nearest Neighbor Pattern Classification Techniques*. IEEE Computer Society Press, Los Alamitos, CA.

Day, N. and Kerridge, D. (1967). A general maximum likelihood discriminant. *Biometrics*, 23:313–324.

de Guzman, M. (1975). *Differentiation of Integrals in R^n*. Lecture Notes in Mathematics #481, Springer-Verlag, Berlin.

Deheuvels, P. (1977). Estimation nonparamétrique de la densité par histogrammes generalisés. *Publications de l'Institut de Statistique de l'Université de Paris*, 22:1–23.

Devijver, P. (1978). A note on ties in voting with the k-NN rule. *Pattern Recognition*, 10:297–298.

Devijver, P. (1979). New error bounds with the nearest neighbor rule. *IEEE Transactions on Information Theory*, 25:749–753.

Devijver, P. (1980). An overview of asymptotic properties of nearest neighbor rules. In *Pattern Recognition in Practice*, Gelsema, E. and Kanal, L., editors, pages 343–350. Elsevier Science Publishers, Amsterdam.

Devijver, P. and Kittler, J. (1980). On the edited nearest neighbor rule. In *Proceedings of the Fifth International Conference on Pattern Recognition*, pages 72–80. Pattern Recognition Society, Los Alamitos, CA.

Devijver, P. and Kittler, J. (1982). *Pattern Recognition: A Statistical Approach*. Prentice-Hall, Englewood Cliffs, NJ.

Devroye, L. (1978). A universal k-nearest neighbor procedure in discrimination. In *Proceedings of the 1978 IEEE Computer Society Conference on Pattern Recognition and Image Processing*, pages 142–147. IEEE Computer Society, Long Beach, CA.

Devroye, L. (1981a). On the almost everywhere convergence of nonparametric regression function estimates. *Annals of Statistics*, 9:1310–1309.

Devroye, L. (1981b). On the asymptotic probability of error in nonparametric discrimination. *Annals of Statistics*, 9:1320–1327.

Devroye, L. (1981c). On the inequality of Cover and Hart in nearest neighbor discrimination. *IEEE Transactions on Pattern Analysis and Machine Intelligence*, 3:75–78.

Devroye, L. (1982a). Bounds for the uniform deviation of empirical measures. *Journal of Multivariate Analysis*, 12:72–79.

Devroye, L. (1982b). Necessary and sufficient conditions for the almost everywhere convergence of nearest neighbor regression function estimates. *Zeitschrift für Wahrscheinlichkeitstheorie und verwandte Gebiete*, 61:467–481.

Devroye, L. (1983). The equivalence of weak, strong and complete convergence in L_1 for kernel density estimates. *Annals of Statistics*, 11:896–904.

Devroye, L. (1987). *A Course in Density Estimation*. Birkhäuser, Boston, MA.

Devroye, L. (1988a). Applications of the theory of records in the study of random trees. *Acta Informatica*, 26:123–130.

Devroye, L. (1988b). Automatic pattern recognition: A study of the probability of error. *IEEE Transactions on Pattern Analysis and Machine Intelligence*, 10:530–543.

Devroye, L. (1988c). The expected size of some graphs in computational geometry. *Computers and Mathematics with Applications*, 15:53–64.

Devroye, L. (1988d). The kernel estimate is relatively stable. *Probability Theory and Related Fields*, 77:521–536.

Devroye, L. (1991a). Exponential inequalities in nonparametric estimation. In *Nonparametric Functional Estimation and Related Topics*, Roussas, G., editor, pages 31–44. NATO ASI Series, Kluwer Academic Publishers, Dordrecht.

Devroye, L. (1991b). On the oscillation of the expected number of points on a random convex hull. *Statistics and Probability Letters*, 11:281–286.

Devroye, L. and Györfi, L. (1983). Distribution-free exponential bound on the L_1 error of partitioning estimates of a regression function. In *Proceedings of the Fourth Pannonian Symposium on Mathematical Statistics*, Konecny, F., Mogyoródi, J., and Wertz, W., editors, pages 67–76. Akadémiai Kiadó, Budapest, Hungary.

Devroye, L. and Györfi, L. (1985). *Nonparametric Density Estimation: The L_1 View*. John Wiley, New York.

Devroye, L. and Györfi, L. (1992). No empirical probability measure can converge in the total variation sense for all distributions. *Annals of Statistics*, 18:1496–1499.

Devroye, L., Györfi, L., Krzyżak, A., and Lugosi, G. (1994). On the strong universal consistency of nearest neighbor regression function estimates. *Annals of Statistics*, 22:1371–1385.

Devroye, L. and Krzyżak, A. (1989). An equivalence theorem for L_1 convergence of the kernel regression estimate. *Journal of Statistical Planning and Inference*, 23:71–82.

Devroye, L. and Laforest, L. (1990). An analysis of random d-dimensional quadtrees. *SIAM Journal on Computing*, 19:821–832.

Devroye, L. and Lugosi, G. (1995). Lower bounds in pattern recognition and learning. *Pattern Recognition*, 28:1011–1018.

Devroye, L. and Machell, F. (1985). Data structures in kernel density estimation. *IEEE Transactions on Pattern Analysis and Machine Intelligence*, 7:360–366.

Devroye, L. and Wagner, T. (1976a). A distribution-free performance bound in error estimation. *IEEE Transactions Information Theory*, 22:586–587.

Devroye, L. and Wagner, T. (1976b). Nonparametric discrimination and density estimation. Technical Report 183, Electronics Research Center, University of Texas.

Devroye, L. and Wagner, T. (1979a). Distribution-free inequalities for the deleted and hold-out error estimates. *IEEE Transactions on Information Theory*, 25:202–207.

Devroye, L. and Wagner, T. (1979b). Distribution-free performance bounds for potential function rules. *IEEE Transactions on Information Theory*, 25:601–604.

Devroye, L. and Wagner, T. (1979c). Distribution-free performance bounds with the resubstitution error estimate. *IEEE Transactions on Information Theory*, 25:208–210.

Devroye, L. and Wagner, T. (1980a). Distribution-free consistency results in nonparametric discrimination and regression function estimation. *Annals of Statistics*, 8:231–239.

Devroye, L. and Wagner, T. (1980b). On the L_1 convergence of kernel estimators of regression functions with applications in discrimination. *Zeitschrift für Wahrscheinlichkeitstheorie und verwandte Gebiete*, 51:15–21.

Devroye, L. and Wagner, T. (1982). Nearest neighbor methods in discrimination. In *Handbook of Statistics*, Krishnaiah, P. and Kanal, L., editors, volume 2, pages 193–197. North Holland, Amsterdam.

Devroye, L. and Wise, G. (1980). Consistency of a recursive nearest neighbor regression function estimate. *Journal of Multivariate Analysis*, 10:539–550.

Diaconis, P. and Shahshahani, M. (1984). On nonlinear functions of linear combinations. *SIAM Journal on Scientific and Statistical Computing*, 5:175–191.

Do-Tu, H. and Installé, M. (1975). On adaptive solution of piecewise linear approximation problem—application to modeling and identification. In *Milwaukee Symposium on Automatic Computation and Control*, pages 1–6. Milwaukee, WI.

Duda, R. and Hart, P. (1973). *Pattern Classification and Scene Analysis*. John Wiley, New York.

Dudani, S. (1976). The distance-weighted k-nearest-neighbor rule. *IEEE Transactions on Systems, Man, and Cybernetics*, 6:325–327.

Dudley, R. (1978). Central limit theorems for empirical measures. *Annals of Probability*, 6:899–929.

Dudley, R. (1979). Balls in R^k do not cut all subsets of $k+2$ points. *Advances in Mathematics*, 31 (3):306–308.

Dudley, R. (1984). Empirical processes. In *Ecole de Probabilité de St. Flour 1982*. Lecture Notes in Mathematics #1097, Springer-Verlag, New York.

Dudley, R. (1987). Universal Donsker classes and metric entropy. *Annals of Probability*, 15:1306–1326.

Dudley, R., Kulkarni, S., Richardson, T., and Zeitouni, O. (1994). A metric entropy bound is not sufficient for learnability. *IEEE Transactions on Information Theory*, 40:883–885.

Durrett, R. (1991). *Probability: Theory and Examples*. Wadsworth and Brooks/Cole, Pacific Grove, CA.

Dvoretzky, A. (1956). On stochastic approximation. In *Proceedings of the First Berkeley Symposium on Mathematical Statistics and Probability*, Neyman, J., editor, pages 39–55. University of California Press, Berkeley, Los Angeles.

Dvoretzky, A., Kiefer, J., and Wolfowitz, J. (1956). Asymptotic minimax character of a sample distribution function and of the classical multinomial estimator. *Annals of Mathematical Statistics*, 33:642–669.

Edelsbrunner, H. (1987). *Algorithms for Computational Geometry*. Springer-Verlag, Berlin.

Efron, B. (1979). Bootstrap methods: another look at the jackknife. *Annals of Statistics*, 7:1–26.

Efron, B. (1983). Estimating the error rate of a prediction rule: improvement on cross validation. *Journal of the American Statistical Association*, 78:316–331.

Efron, B. and Stein, C. (1981). The jackknife estimate of variance. *Annals of Statistics*, 9:586–596.

Ehrenfeucht, A., Haussler, D., Kearns, M., and Valiant, L. (1989). A general lower bound on the number of examples needed for learning. *Information and Computation*, 82:247–261.

Elashoff, J., Elashoff, R., and Goldmann, G. (1967). On the choice of variables in classification problems with dichotomous variables. *Biometrika*, 54:668–670.

Fabian, V. (1971). Stochastic approximation. In *Optimizing Methods in Statistics*, Rustagi, J., editor, pages 439–470. Academic Press, New York, London.

Fano, R. (1952). *Class Notes for Transmission of Information*. Course 6.574. MIT, Cambridge, MA.

Faragó, A., Linder, T., and Lugosi, G. (1993). Fast nearest neighbor search in dissimilarity spaces. *IEEE Transactions on Pattern Analysis and Machine Intelligence*, 15:957–962.

Faragó, A. and Lugosi, G. (1993). Strong universal consistency of neural network classifiers. *IEEE Transactions on Information Theory*, 39:1146–1151.

Faragó, T. and Györfi, L. (1975). On the continuity of the error distortion function for multiple hypothesis decisions. *IEEE Transactions on Information Theory*, 21:458–460.

Feinholz, L. (1979). Estimation of the performance of partitioning algorithms in pattern classification. Master's Thesis, Department of Mathematics, McGill University, Montreal.

Feller, W. (1968). *An Introduction to Probability Theory and its Applications, Vol.1*. John Wiley, New York.

Finkel, R. and Bentley, J. (1974). Quad trees: A data structure for retrieval on composite keys. *Acta Informatica*, 4:1–9.

Fisher, R. (1936). The case of multiple measurements in taxonomic problems. *Annals of Eugenics*, 7, part II:179–188.

Fitzmaurice, G. and Hand, D. (1987). A comparison of two average conditional error rate estimators. *Pattern Recognition Letters*, 6:221–224.

Fix, E. and Hodges, J. (1951). Discriminatory analysis. Nonparametric discrimination: Consistency properties. Technical Report 4, Project Number 21-49-004, USAF School of Aviation Medicine, Randolph Field, TX.

Fix, E. and Hodges, J. (1952). Discriminatory analysis: small sample performance. Technical Report 21-49-004, USAF School of Aviation Medicine, Randolph Field, TX.

Fix, E. and Hodges, J. (1991a). Discriminatory analysis, nonparametric discrimination, consistency properties. In *Nearest Neighbor Pattern Classification Techniques*, Dasarathy, B., editor, pages 32–39. IEEE Computer Society Press, Los Alamitos, CA.

Fix, E. and Hodges, J. (1991b). Discriminatory analysis: small sample performance. In *Nearest Neighbor Pattern Classification Techniques*, Dasarathy, B., editor, pages 40–56. IEEE Computer Society Press, Los Alamitos, CA.

Flick, T., Jones, L., Priest, R., and Herman, C. (1990). Pattern classification using protection pursuit. *Pattern Recognition*, 23:1367–1376.

Forney, G. (1968). Exponential error bounds for erasure, list, and decision feedback schemes. *IEEE Transactions on Information Theory*, 14:41–46.

Friedman, J. (1977). A recursive partitioning decision rule for nonparametric classification. *IEEE Transactions on Computers*, 26:404–408.

Friedman, J., Baskett, F., and Shustek, L. (1975). An algorithm for finding nearest neighbor. *IEEE Transactions on Computers*, 24:1000–1006.

Friedman, J., Bentley, J., and Finkel, R. (1977). An algorithm for finding best matches in logarithmic expected time. *ACM Transactions on Mathematical Software*, 3:209–226.

Friedman, J. and Silverman, B. (1989). Flexible parsimonious smoothing and additive modeling. *Technometrics*, 31:3–39.

Friedman, J. and Stuetzle, W. (1981). Projection pursuit regression. *Journal of the American Statistical Association*, 76:817–823.

Friedman, J., Stuetzle, W., and Schroeder, A. (1984). Projection pursuit density estimation. *Journal of the American Statistical Association*, 79:599–608.

Friedman, J. and Tukey, J. (1974). A projection pursuit algorithm for exploratory data analysis. *IEEE Transactions on Computers*, 23:881–889.

Fritz, J. and Györfi, L. (1976). On the minimization of classification error probability in statistical pattern recognition. *Problems of Control and Information Theory*, 5:371–382.

Fu, K., Min, P., and Li, T. (1970). Feature selection in pattern recognition. *IEEE Transactions on Systems Science and Cybernetics*, 6:33–39.

Fuchs, H., Abram, G., and Grant, E. (1983). Near real-time shaded display of rigid objects. *Computer Graphics*, 17:65–72.

Fuchs, H., Kedem, Z., and Naylor, B. (1980). On visible surface generation by a priori tree structures. In *Proceedings SIGGRAPH '80*, pages 124–133. Published as Computer Graphics, volume 14.

Fukunaga, K. and Flick, T. (1984). An optimal global nearest neighbor metric. *IEEE Transactions on Pattern Analysis and Machine Intelligence*, 6:314–318.

Fukunaga, K. and Hostetler, L. (1973). Optimization of k-nearest neighbor density estimates. *IEEE Transactions on Information Theory*, 19:320–326.

Fukunaga, K. and Hummels, D. (1987). Bayes error estimation using Parzen and k-NN procedures. *IEEE Transactions on Pattern Analysis and Machine Intelligence*, 9:634–643.

Fukunaga, K. and Kessel, D. (1971). Estimation of classification error. *IEEE Transactions on Computers*, 20:1521–1527.

Fukunaga, K. and Kessel, D. (1973). Nonparametric Bayes error estimation using unclassified samples. *IEEE Transactions Information Theory*, 19:434–440.

Fukunaga, K. and Mantock, J. (1984). Nonparametric data reduction. *IEEE Transactions on Pattern Analysis and Machine Intelligence*, 6:115–118.

Fukunaga, K. and Narendra, P. (1975). A branch and bound algorithm for computing k-nearest neighbors. *IEEE Transactions on Computers*, 24:750–753.

Funahashi, K. (1989). On the approximate realization of continuous mappings by neural networks. *Neural Networks*, 2:183–192.

Gabor, D. (1961). A universal nonlinear filter, predictor, and simulator which optimizes itself. *Proceedings of the Institute of Electrical Engineers*, 108B:422–438.

Gabriel, K. and Sokal, R. (1969). A new statistical approach to geographic variation analysis. *Systematic Zoology*, 18:259–278.

Gaenssler, P. (1983). *Empirical processes*. Lecture Notes–Monograph Series, Institute of Mathematical Statistics, Hayward, CA.

Gaenssler, P. and Stute, W. (1979). Empirical processes: A survey of results for independent identically distributed random variables. *Annals of Probability*, 7:193–243.

Gallant, A. (1987). *Nonlinear Statistical Models*. John Wiley, New York.

Ganesalingam, S. and McLachlan, G. (1980). Error rate estimation on the basis of posterior probabilities. *Pattern Recognition*, 12:405–413.

Garnett, J. and Yau, S. (1977). Nonparametric estimation of the Bayes error of feature extractors using ordered nearest neighbour sets. *IEEE Transactions on Computers*, 26:46–54.

Gates, G. (1972). The reduced nearest neighbor rule. *IEEE Transactions on Information Theory*, 18:431–433.

Gelfand, S. and Delp, E. (1991). On tree structured classifiers. In *Artificial Neural Networks and Statistical Pattern Recognition, Old and New Connections*, Sethi, I. and Jain, A., editors, pages 71–88. Elsevier Science Publishers, Amsterdam.

Gelfand, S., Ravishankar, C., and Delp, E. (1989). An iterative growing and pruning algorithm for classification tree design. In *Proceedings of the 1989 IEEE International Conference on Systems, Man, and Cybernetics*, pages 818–823. IEEE Press, Piscataway, NJ.

Gelfand, S., Ravishankar, C., and Delp, E. (1991). An iterative growing and pruning algorithm for classification tree design. *IEEE Transactions on Pattern Analysis and Machine Intelligence*, 13:163–174.

Geman, S. and Hwang, C. (1982). Nonparametric maximum likelihood estimation by the method of sieves. *Annals of Statistics*, 10:401–414.

Gessaman, M. (1970). A consistent nonparametric multivariate density estimator based on statistically equivalent blocks. *Annals of Mathematical Statistics*, 41:1344–1346.

Gessaman, M. and Gessaman, P. (1972). A comparison of some multivariate discrimination procedures. *Journal of the American Statistical Association*, 67:468–472.

Geva, S. and Sitte, J. (1991). Adaptive nearest neighbor pattern classification. *IEEE Transactions on Neural Networks*, 2:318–322.

Giné, E. and Zinn, J. (1984). Some limit theorems for empirical processes. *Annals of Probability*, 12:929–989.

Glick, N. (1973). Sample-based multinomial classification. *Biometrics*, 29:241–256.

Glick, N. (1976). Sample-based classification procedures related to empiric distributions. *IEEE Transactions on Information Theory*, 22:454–461.

Glick, N. (1978). Additive estimators for probabilities of correct classification. *Pattern Recognition*, 10:211–222.

Goldberg, P. and Jerrum, M. (1993). Bounding the Vapnik-Chervonenkis dimension of concept classes parametrized by real numbers. In *Proceedings of the Sixth ACM Workshop on Computational Learning Theory*, pages 361–369. Association for Computing Machinery, New York.

Goldstein, M. (1977). A two-group classification procedure for multivariate dichotomous responses. *Multivariate Behavorial Research*, 12:335–346.

Goldstein, M. and Dillon, W. (1978). *Discrete Discriminant Analysis*. John Wiley, New York.

Golea, M. and Marchand, M. (1990). A growth algorithm for neural network decision trees. *Europhysics Letters*, 12:205–210.

Goodman, R. and Smyth, P. (1988). Decision tree design from a communication theory viewpoint. *IEEE Transactions on Information Theory*, 34:979–994.

Gordon, L. and Olshen, R. (1984). Almost surely consistent nonparametric regression from recursive partitioning schemes. *Journal of Multivariate Analysis*, 15:147–163.

Gordon, L. and Olshen, R. (1978). Asymptotically efficient solutions to the classification problem. *Annals of Statistics*, 6:515–533.

Gordon, L. and Olshen, R. (1980). Consistent nonparametric regression from recursive partitioning schemes. *Journal of Multivariate Analysis*, 10:611–627.

Gowda, K. and Krishna, G. (1979). The condensed nearest neighbor rule using the concept of mutual nearest neighborhood. *IEEE Transactions on Information Theory*, 25:488–490.

Greblicki, W. (1974). Asymptotically optimal probabilistic algorithms for pattern recognition and identification. Technical Report, Monografie No. 3, Prace Naukowe Instytutu Cybernetyki Technicznej Politechniki Wroclawsjiej No. 18, Wroclaw, Poland.

Greblicki, W. (1978a). Asymptotically optimal pattern recognition procedures with density estimates. *IEEE Transactions on Information Theory*, 24:250–251.

Greblicki, W. (1978b). Pattern recognition procedures with nonparametric density estimates. *IEEE Transactions on Systems, Man and Cybernetics*, 8:809–812.

Greblicki, W. (1981). Asymptotic efficiency of classifying procedures using the Hermite series estimate of multivariate probability densities. *IEEE Transactions on Information Theory*, 27:364–366.

Greblicki, W., Krzyżak, A., and Pawlak, M. (1984). Distribution-free pointwise consistency of kernel regression estimate. *Annals of Statistics*, 12:1570–1575.

Greblicki, W. and Pawlak, M. (1981). Classification using the Fourier series estimate of multivariate density functions. *IEEE Transactions on Systems, Man and Cybernetics*, 11:726–730.

Greblicki, W. and Pawlak, M. (1982). A classification procedure using the multiple Fourier series. *Information Sciences*, 26:115–126.

Greblicki, W. and Pawlak, M. (1983). Almost sure convergence of classification procedures using Hermite series density estimates. *Pattern Recognition Letters*, 2:13–17.

Greblicki, W. and Pawlak, M. (1985). Pointwise consistency of the Hermite series density estimate. *Statistics and Probability Letters*, 3:65–69.

Greblicki, W. and Pawlak, M. (1987). Necessary and sufficient conditions for Bayes risk consistency of a recursive kernel classification rule. *IEEE Transactions on Information Theory*, 33:408–412.

Grenander, U. (1981). *Abstract Inference*. John Wiley, New York.

Grimmett, G. and Stirzaker, D. (1992). *Probability and Random Processes*. Oxford University Press, Oxford.

Guo, H. and Gelfand, S. (1992). Classification trees with neural network feature extraction. *IEEE Transactions on Neural Networks*, 3:923–933.

Györfi, L. (1975). An upper bound of error probabilities for multihypothesis testing and its application in adaptive pattern recognition. *Problems of Control and Information Theory*, 5:449–457.

Györfi, L. (1978). On the rate of convergence of nearest neighbor rules. *IEEE Transactions on Information Theory*, 29:509–512.

Györfi, L. (1981). Recent results on nonparametric regression estimate and multiple classification. *Problems of Control and Information Theory*, 10:43–52.

Györfi, L. (1984). Adaptive linear procedures under general conditions. *IEEE Transactions on Information Theory*, 30:262–267.

Györfi, L. and Györfi, Z. (1975). On the nonparametric estimate of a posteriori probabilities of simple statistical hypotheses. In *Colloquia Mathematica Societatis János Bolyai: Topics in Information Theory*, pages 299–308. Keszthely, Hungary.

Györfi, L. and Györfi, Z. (1978). An upper bound on the asymptotic error probability of the *k*-nearest neighbor rule for multiple classes. *IEEE Transactions on Information Theory*, 24:512–514.

Györfi, L., Györfi, Z., and Vajda, I. (1978). Bayesian decision with rejection. *Problems of Control and Information Theory*, 8:445–452.

Györfi, L. and Vajda, I. (1980). Upper bound on the error probability of detection in nongaussian noise. *Problems of Control and Information Theory*, 9:215–224.

Haagerup, U. (1978). Les meilleures constantes de l'inégalite de Khintchine. *Comptes Rendus de l'Académie des Sciences de Paris A*, 286:259–262.

Habbema, J., Hermans, J., and van den Broek, K. (1974). A stepwise discriminant analysis program using density estimation. In *COMPSTAT 1974*, Bruckmann, G., editor, pages 101–110. Physica Verlag, Wien.

Hagerup, T. and Rüb, C. (1990). A guided tour of Chernoff bounds. *Information Processing Letters*, 33:305–308.

Hall, P. (1981). On nonparametric multivariate binary discrimination. *Biometrika*, 68:287–294.

Hall, P. (1989). On projection pursuit regression. *Annals of Statistics*, 17:573–588.

Hall, P. and Wand, M. (1988). On nonparametric discrimination using density differences. *Biometrika*, 75:541–547.

Hand, D. (1981). *Discrimination and Classification*. John Wiley, Chichester, U.K.

Hand, D. (1986). Recent advances in error rate estimation. *Pattern Recognition Letters*, 4:335–346.

Härdle, W. and Marron, J. (1985). Optimal bandwidth selection in nonparametric regression function estimation. *Annals of Statistics*, 13:1465–1481.

Hart, P. (1968). The condensed nearest neighbor rule. *IEEE Transactions on Information Theory*, 14:515–516.

Hartigan, J. (1975). *Clustering Algorithms*. John Wiley, New York.

Hartmann, C., Varshney, P., Mehrotra, K., and Gerberich, C. (1982). Application of information theory to the construction of efficient decision trees. *IEEE Transactions on Information Theory*, 28:565–577.

Hashlamoun, W., Varshney, P., and Samarasooriya, V. (1994). A tight upper bound on the Bayesian probability of error. *IEEE Transactions on Pattern Analysis and Machine Intelligence*, 16:220–224.

Hastie, T. and Tibshirani, R. (1990). *Generalized Additive Models*. Chapman and Hall, London, U.K.

Haussler, D. (1991). Sphere packing numbers for subsets of the boolean n-cube with bounded Vapnik-Chervonenkis dimension. Technical Report, Computer Research Laboratory, University of California, Santa Cruz.

Haussler, D. (1992). Decision theoretic generalizations of the PAC model for neural net and other learning applications. *Information and Computation*, 100:78–150.

Haussler, D., Littlestone, N., and Warmuth, M. (1988). Predicting {0, 1} functions from randomly drawn points. In *Proceedings of the 29th IEEE Symposium on the Foundations of Computer Science*, pages 100–109. IEEE Computer Society Press, Los Alamitos, CA.

Hecht-Nielsen, R. (1987). Kolmogorov's mapping network existence theorem. In *IEEE First International Conference on Neural Networks*, volume 3, pages 11–13. IEEE, Piscataway, NJ.

Hellman, M. (1970). The nearest neighbor classification rule with a reject option. *IEEE Transactions on Systems, Man and Cybernetics*, 2:179–185.

Henrichon, E. and Fu, K. (1969). A nonparametric partitioning procedure for pattern classification. *IEEE Transactions on Computers*, 18:614–624.

Hertz, J., Krogh, A., and Palmer, R. (1991). *Introduction to the Theory of Neural Computation*. Addison-Wesley, Redwood City, CA.

Hills, M. (1966). Allocation rules and their error rates. *Journal of the Royal Statistical Society*, B28:1–31.

Hjort, N. (1986a). Contribution to the discussion of a paper by P. Diaconis and Freeman. *Annals of Statistics*, 14:49–55.

Hjort, N. (1986b). Notes on the theory of statistical symbol recognition. Technical Report 778, Norwegian Computing Centre, Oslo.

Hoeffding, W. (1963). Probability inequalities for sums of bounded random variables. *Journal of the American Statistical Association*, 58:13–30.

Holmström, L. and Klemelä, J. (1992). Asymptotic bounds for the expected l_1 error of a multivariate kernel density estimator. *Journal of Multivariate Analysis*, 40:245–255.

Hora, S. and Wilcox, J. (1982). Estimation of error rates in several-population discriminant analysis. *Journal of Marketing Research*, 19:57–61.

Horibe, Y. (1970). On zero error probability of binary decisions. *IEEE Transactions on Information Theory*, 16:347–348.

Horne, B. and Hush, D. (1990). On the optimality of the sigmoid perceptron. In *International Joint Conference on Neural Networks*, volume 1, pages 269–272. Lawrence Erlbaum Associates, Hillsdale, NJ.

Hornik, K. (1991). Approximation capabilities of multilayer feedforward networks. *Neural Networks*, 4:251–257.

Hornik, K. (1993). Some new results on neural network approximation. *Neural Networks*, 6:1069–1072.

Hornik, K., Stinchcombe, M., and White, H. (1989). Multi-layer feedforward networks are universal approximators. *Neural Networks*, 2:359–366.

Huber, P. (1985). Projection pursuit. *Annals of Statistics*, 13:435–525.

Hudimoto, H. (1957). A note on the probability of the correct classification when the distributions are not specified. *Annals of the Institute of Statistical Mathematics*, 9:31–36.

Ito, T. (1969). Note on a class of statistical recognition functions. *IEEE Transactions on Computers*, C-18:76–79.

Ito, T. (1972). Approximate error bounds in pattern recognition. In *Machine Intelligence*, Meltzer, B. and Mitchie, D., editors, pages 369–376. Edinburgh University Press, Edinburgh.

Ivakhnenko, A. (1968). The group method of data handling—a rival of the method of stochastic approximation. *Soviet Automatic Control*, 1:43–55.

Ivakhnenko, A. (1971). Polynomial theory of complex systems. *IEEE Transactions on Systems, Man, and Cybernetics*, 1:364–378.

Ivakhnenko, A., Konovalenko, V., Tulupchuk, Y., and Tymchenko, I. (1968). The group method of data handling in pattern recognition and decision problems. *Soviet Automatic Control*, 1:31–41.

Ivakhnenko, A., Petrache, G., and Krasyts'kyy, M. (1968). A GMDH algorithm with random selection of pairs. *Soviet Automatic Control*, 5:23–30.

Jain, A., Dubes, R., and Chen, C. (1987). Bootstrap techniques for error estimation. *IEEE Transactions on Pattern Analysis and Machine Intelligence*, 9:628–633.

Jeffreys, H. (1948). *Theory of Probability*. Clarendon Press, Oxford.

Johnson, D. and Preparata, F. (1978). The densest hemisphere problem. *Theoretical Computer Science*, 6:93–107.

Kůrková, V. (1992). Kolmogorov's theorem and multilayer neural networks. *Neural Networks*, 5:501–506.

Kailath, T. (1967). The divergence and Bhattacharyya distance measures in signal detection. *IEEE Transactions on Communication Technology*, 15:52–60.

Kanal, L. (1974). Patterns in pattern recognition. *IEEE Transactions on Information Theory*, 20:697–722.

Kaplan, M. (1985). The uses of spatial coherence in ray tracing. *SIGGRAPH '85 Course Notes*, 11:22–26.

Karlin, A. (1968). *Total Positivity, Volume 1*. Stanford University Press, Stanford, CA.

Karp, R. (1988). *Probabilistic Analysis of Algorithms*. Class Notes, University of California, Berkeley.

Karpinski, M. and Macintyre, A. (1994). Quadratic bounds for VC dimension of sigmoidal neural networks. Submitted to the ACM Symposium on Theory of Computing.

Kazmierczak, H. and Steinbuch, K. (1963). Adaptive systems in pattern recognition. *IEEE Transactions on Electronic Computers*, 12:822–835.

Kemp, R. (1984). *Fundamentals of the Average Case Analysis of Particular Algorithms*. B.G. Teubner, Stuttgart.

Kemperman, J. (1969). On the optimum rate of transmitting information. In *Probability and Information Theory*, pages 126–169. Springer Lecture Notes in Mathematics, Springer-Verlag, Berlin.

Kiefer, J. and Wolfowitz, J. (1952). Stochastic estimation of the maximum of a regression function. *Annals of Mathematical Statistics*, 23:462–466.

Kim, B. and Park, S. (1986). A fast k-nearest neighbor finding algorithm based on the ordered partition. *IEEE Transactions on Pattern Analysis and Machine Intelligence*, 8:761–766.

Kittler, J. and Devijver, P. (1981). An efficient estimator of pattern recognition system error probability. *Pattern Recognition*, 13:245–249.

Knoke, J. (1986). The robust estimation of classification error rates. *Computers and Mathematics with Applications*, 12A:253–260.

Kohonen, T. (1988). *Self-Organization and Associative Memory*. Springer-Verlag, Berlin.

Kohonen, T. (1990). Statistical pattern recognition revisited. In *Advanced Neural Computers*, Eckmiller, R., editor, pages 137–144. North-Holland, Amsterdam.

Kolmogorov, A. (1957). On the representation of continuous functions of many variables by superposition of continuous functions of one variable and addition. *Doklady Akademii Nauk USSR*, 114:953–956.

Kolmogorov, A. and Tikhomirov, V. (1961). ϵ-entropy and ϵ-capacity of sets in function spaces. *Translations of the American Mathematical Society*, 17:277–364.

Koutsougeras, C. and Papachristou, C. (1989). Training of a neural network for pattern classification based on an entropy measure. In *Proceedings of IEEE International Conference on Neural Networks*, volume 1, pages 247–254. IEEE San Diego Section, San Diego, CA.

Kraaijveld, M. and Duin, R. (1991). Generalization capabilities of minimal kernel-based methods. In *International Joint Conference on Neural Networks*, volume 1, pages 843–848. Piscataway, NJ.

Kronmal, R. and Tarter, M. (1968). The estimation of probability densities and cumulatives by Fourier series methods. *Journal of the American Statistical Association*, 63:925–952.

Krzanowski, W. (1987). A comparison between two distance-based discriminant principles. *Journal of Classification*, 4:73–84.

Krzyżak, A. (1983). Classification procedures using multivariate variable kernel density estimate. *Pattern Recognition Letters*, 1:293–298.

Krzyżak, A. (1986). The rates of convergence of kernel regression estimates and classification rules. *IEEE Transactions on Information Theory*, 32:668–679.

Krzyżak, A. (1991). On exponential bounds on the bayes risk of the kernel classification rule. *IEEE Transactions on Information Theory*, 37:490–499.

Krzyżak, A., Linder, T., and Lugosi, G. (1993). Nonparametric estimation and classification using radial basis function nets and empirical risk minimization. *IEEE Transactions on Neural Networks*. To appear.

Krzyżak, A. and Pawlak, M. (1983). Universal consistency results for Wolverton-Wagner regression function estimate with application in discrimination. *Problems of Control and Information Theory*, 12:33–42.

Krzyżak, A. and Pawlak, M. (1984a). Almost everywhere convergence of recursive kernel regression function estimates. *IEEE Transactions on Information Theory*, 31:91–93.

Krzyżak, A. and Pawlak, M. (1984b). Distribution-free consistency of a nonparametric kernel regression estimate and classification. *IEEE Transactions on Information Theory*, 30:78–81.

Kulkarni, S. (1991). *Problems of computational and information complexity in machine vision and learning*. PhD Thesis, Department of Electrical Engineering and Computer Science, MIT, Cambridge, MA.

Kullback, S. (1967). A lower bound for discrimination information in terms of variation. *IEEE Transactions on Information Theory*, 13:126–127.

Kullback, S. and Leibler, A. (1951). On information and sufficiency. *Annals of Mathematical Statistics*, 22:79–86.

Kushner, H. (1984). *Approximation and Weak Convergence Methods for Random Processes with Applications to Stochastic Systems Theory*. MIT Press, Cambridge, MA.

Lachenbruch, P. (1967). An almost unbiased method of obtaining confidence intervals for the probability of misclassification in discriminant analysis. *Biometrics*, 23:639–645.

Lachenbruch, P. and Mickey, M. (1968). Estimation of error rates in discriminant analysis. *Technometrics*, 10:1–11.

LeCam, L. (1970). On the assumptions used to prove asymptotic normality of maximum likelihood estimates. *Annals of Mathematical Statistics*, 41:802–828.

LeCam, L. (1973). Convergence of estimates under dimensionality restrictions. *Annals of Statistics*, 1:38–53.

Li, X. and Dubes, R. (1986). Tree classifier design with a permutation statistic. *Pattern Recognition*, 19:229–235.

Lin, Y. and Fu, K. (1983). Automatic classification of cervical cells using a binary tree classifier. *Pattern Recognition*, 16:69–80.

Linde, Y., Buzo, A., and Gray, R. (1980). An algorithm for vector quantizer design. *IEEE Transactions on Communications*, 28:84–95.

Linder, T., Lugosi, G., and Zeger, K. (1994). Rates of convergence in the source coding theorem, empirical quantizer design, and universal lossy source coding. *IEEE Transactions on Information Theory*, 40:1728–1740.

Lissack, T. and Fu, K. (1976). Error estimation in pattern recognition via l^α distance between posterior density functions. *IEEE Transactions on Information Theory*, 22:34–45.

Ljung, L., Pflug, G., and Walk, H. (1992). *Stochastic Approximation and Optimization of Random Systems*. Birkhäuser, Basel, Boston, Berlin.

Lloyd, S. (1982). Least squares quantization in PCM. *IEEE Transactions on Information Theory*, 28:129–137.

Loftsgaarden, D. and Quesenberry, C. (1965). A nonparametric estimate of a multivariate density function. *Annals of Mathematical Statistics*, 36:1049–1051.

Logan, B. (1975). The uncertainty principle in reconstructing functions from projections. *Duke Mathematical Journal*, 42:661–706.

Logan, B. and Shepp, L. (1975). Optimal reconstruction of a function from its projections. *Duke Mathematical Journal*, 42:645–660.

Loh, W. and Vanichsetakul, N. (1988). Tree-structured classification via generalized discriminant analysis. *Journal of the American Statistical Association*, 83:715–728.

Loizou, G. and Maybank, S. (1987). The nearest neighbor and the Bayes error rates. *IEEE Transactions on Pattern Analysis and Machine Intelligence*, 9:254–262.

Lorentz, G. (1976). The thirteenth problem of Hilbert. In *Proceedings of Symposia in Pure Mathematics*, volume 28, pages 419–430. Providence, RI.

Lugosi, G. (1992). Learning with an unreliable teacher. *Pattern Recognition*, 25:79–87.

Lugosi, G. and Nobel, A. (1996). Consistency of data-driven histogram methods for density estimation and classification. *Annals of Statistics*, 24:687–706.

Lugosi, G. and Pawlak, M. (1994). On the posterior-probability estimate of the error rate of nonparametric classification rules. *IEEE Transactions on Information Theory*, 40:475–481.

Lugosi, G. and Zeger, K. (1995). Nonparametric estimation via empirical risk minimization. *IEEE Transactions on Information Theory*, 41:677–678.

Lugosi, G. and Zeger, K. (1996). Concept learning using complexity regularization. *IEEE Transactions on Information Theory*, 42:48–54.

Lukács, E. and Laha, R. (1964). *Applications of Characteristic Functions in Probability Theory*. Griffin, London.

Lunts, A. and Brailovsky, V. (1967). Evaluation of attributes obtained in statistical decision rules. *Engineering Cybernetics*, 3:98–109.

Maass, W. (1993). Bounds for the computational power and learning complexity of analog neural nets. In *Proceedings of the 25th Annual ACM Symposium on the Theory of Computing*, pages 335–344. Association of Computing Machinery, New York.

Maass, W. (1994). Neural nets with superlinear vc-dimension. *Neural Computation*, 6:875–882.

Macintyre, A. and Sontag, E. (1993). Finiteness results for sigmoidal "neural" networks. In *Proceedings of the 25th Annual ACM Symposium on the Theory of Computing*, pages 325–334. Association of Computing Machinery, New York.

Mack, Y. (1981). Local properties of k–nearest neighbor regression estimates. *SIAM Journal on Algebraic and Discrete Methods*, 2:311–323.

Mahalanobis, P. (1936). On the generalized distance in statistics. *Proceedings of the National Institute of Sciences of India*, 2:49–55.

Mahalanobis, P. (1961). A method of fractile graphical analysis. *Sankhya Series A*, 23:41–64.

Marron, J. (1983). Optimal rates of convergence to Bayes risk in nonparametric discrimination. *Annals of Statistics*, 11:1142–1155.

Massart, P. (1983). *Vitesse de convergence dans le theoréme de la limite centrale pour le processus empirique*. PhD Thesis, Université Paris-Sud, Orsay, France.

Massart, P. (1990). The tight constant in the Dvoretzky-Kiefer-Wolfowitz inequality. *Annals of Probability*, 18:1269–1283.

Mathai, A. and Rathie, P. (1975). *Basic Concepts in Information Theory and Statistics.* Wiley Eastern Ltd., New Delhi.

Matloff, N. and Pruitt, R. (1984). The asymptotic distribution of an estimator of the bayes error rate. *Pattern Recognition Letters,* 2:271–274.

Matula, D. and Sokal, R. (1980). Properties of Gabriel graphs relevant to geographic variation research and the clustering of points in the plane. *Geographical Analysis,* 12:205–222.

Matushita, K. (1956). Decision rule, based on distance, for the classification problem. *Annals of the Institute of Statistical Mathematics,* 8:67–77.

Matushita, K. (1973). Discrimination and the affinity of distributions. In *Discriminant Analysis and Applications,* Cacoullos, T., editor, pages 213–223. Academic Press, New York.

Max, J. (1960). Quantizing for minimum distortion. *IEEE Transactions on Information Theory,* 6:7–12.

McDiarmid, C. (1989). On the method of bounded differences. In *Surveys in Combinatorics 1989,* pages 148–188. Cambridge University Press, Cambridge.

McLachlan, G. (1976). The bias of the apparent error rate in discriminant analysis. *Biometrika,* 63:239–244.

McLachlan, G. (1992). *Discriminant Analysis and Statistical Pattern Recognition.* John Wiley, New York.

Meisel, W. (1969). Potential functions in mathematical pattern recognition. *IEEE Transactions on Computers,* 18:911–918.

Meisel, W. (1972). *Computer Oriented Approaches to Pattern Recognition.* Academic Press, New York.

Meisel, W. (1990). Parsimony in neural networks. In *International Conference on Neural Networks,* pages 443–446. Lawrence Erlbaum Associates, Hillsdale, NJ.

Meisel, W. and Michalopoulos, D. (1973). A partitioning algorithm with application in pattern classification and the optimization of decision tree. *IEEE Transactions on Computers,* 22:93–103.

Michel-Briand, C. and Milhaud, X. (1994). Asymptotic behavior of the AID method. Technical Report, Université Montpellier 2, Montpellier.

Mielniczuk, J. and Tyrcha, J. (1993). Consistency of multilayer perceptron regression estimators. *Neural Networks,* To appear.

Minsky, M. (1961). Steps towards artificial intelligence. In *Proceedings of the IRE,* volume 49, pages 8–30.

Minsky, M. and Papert, S. (1969). *Perceptrons: An Introduction to Computational Geometry.* MIT Press, Cambridge, MA.

Mitchell, A. and Krzanowski, W. (1985). The Mahalanobis distance and elliptic distributions. *Biometrika,* 72:464–467.

Mizoguchi, R., Kizawa, M., and Shimura, M. (1977). Piecewise linear discriminant functions in pattern recognition. *Systems-Computers-Controls,* 8:114–121.

Moody, J. and Darken, J. (1989). Fast learning in networks of locally-tuned processing units. *Neural Computation,* 1:281–294.

Moore, D., Whitsitt, S., and Landgrebe, D. (1976). Variance comparisons for unbiased estimators of probability of correct classification. *IEEE Transactions on Information Theory,* 22:102–105.

Morgan, J. and Sonquist, J. (1963). Problems in the analysis of survey data, and a proposal. *Journal of the American Statistical Association,* 58:415–434.

Mui, J. and Fu, K. (1980). Automated classification of nucleated blood cells using a binary tree classifier. *IEEE Transactions on Pattern Analysis and Machine Intelligence*, 2:429–443.

Myles, J. and Hand, D. (1990). The multi-class metric problem in nearest neighbour discrimination rules. *Pattern Recognition*, 23:1291–1297.

Nadaraya, E. (1964). On estimating regression. *Theory of Probability and its Applications*, 9:141–142.

Nadaraya, E. (1970). Remarks on nonparametric estimates for density functions and regression curves. *Theory of Probability and its Applications*, 15:134–137.

Narendra, P. and Fukunaga, K. (1977). A branch and bound algorithm for feature subset selection. *IEEE Transactions on Computers*, 26:917–922.

Natarajan, B. (1991). *Machine Learning: A Theoretical Approach*. Morgan Kaufmann, San Mateo, CA.

Nevelson, M. and Khasminskii, R. (1973). *Stochastic Approximation and Recursive Estimation*. Translations of Mathematical Monographs, Vol. 47. American Mathematical Society, Providence, RI.

Niemann, H. and Goppert, R. (1988). An efficient branch-and-bound nearest neighbour classifier. *Pattern Recognition Letters*, 7:67–72.

Nilsson, N. (1965). *Learning Machines: Foundations of Trainable Pattern Classifying Systems*. McGraw-Hill, New York.

Nishizeki, T. and Chiba, N. (1988). *Planar Graphs: Theory and Algorithms*. North Holland, Amsterdam.

Nobel, A. (1994). Histogram regression estimates using data-dependent partitions. Technical Report, Beckman Institute, University of Illinois, Urbana-Champaign.

Nobel, A. (1992). *On uniform laws of averages*. PhD Thesis, Department of Statistics, Stanford University, Stanford, CA.

Nolan, D. and Pollard, D. (1987). U-processes: Rates of convergence. *Annals of Statistics*, 15:780–799.

Okamoʾo, M. (1958). Some inequalities relating to the partial sum of binomial probabilities. *Annals of the Institute of Statistical Mathematics*, 10:29–35.

Olshen, R. (1977). Comments on a paper by C.J. Stone. *Annals of Statistics*, 5:632–633.

Ott, J. and Kronmal, R. (1976). Some classification procedures for multivariate binary data using orthogonal functions. *Journal of the American Statistical Association*, 71:391–399.

Papadimitriou, C. and Bentley, J. (1980). A worst-case analysis of nearest neighbor searching by projection. In *Automata, Languages and Programming 1980*, pages 470–482. Lecture Notes in Computer Science #85, Springer-Verlag, Berlin.

Park, J. and Sandberg, I. (1991). Universal approximation using radial-basis-function networks. *Neural Computation*, 3:246–257.

Park, J. and Sandberg, I. (1993). Approximation and radial-basis-function networks. *Neural Computation*, 5:305–316.

Park, Y. and Sklansky, J. (1990). Automated design of linear tree classifiers. *Pattern Recognition*, 23:1393–1412.

Parthasarathy, K. and Bhattacharya, P. (1961). Some limit theorems in regression theory. *Sankhya Series A*, 23:91–102.

Parzen, E. (1962). On the estimation of a probability density function and the mode. *Annals of Mathematical Statistics*, 33:1065–1076.

Patrick, E. (1966). Distribution-free minimum conditional risk learning systems. Technical Report TR-EE-66-18, Purdue University, Lafayette, IN.

Patrick, E. and Fischer, F. (1970). A generalized k-nearest neighbor rule. *Information and Control*, 16:128–152.

Patrick, E. and Fisher, F. (1967). Introduction to the performance of distribution-free conditional risk learning systems. Technical Report TR-EE-67-12, Purdue University, Lafayette, IN.

Pawlak, M. (1988). On the asymptotic properties of smoothed estimators of the classification error rate. *Pattern Recognition*, 21:515–524.

Payne, H. and Meisel, W. (1977). An algorithm for constructing optimal binary decision trees. *IEEE Transactions on Computers*, 26:905–916.

Pearl, J. (1979). Capacity and error estimates for boolean classifiers with limited complexity. *IEEE Transactions on Pattern Analysis and Machine Intelligence*, 1:350–355.

Penrod, C. and Wagner, T. (1977). Another look at the edited nearest neighbor rule. *IEEE Transactions on Systems, Man and Cybernetics*, 7:92–94.

Peterson, D. (1970). Some convergence properties of a nearest neighbor decision rule. *IEEE Transactions on Information Theory*, 16:26–31.

Petrov, V. (1975). *Sums of Independent Random Variables*. Springer-Verlag, Berlin.

Pippenger, N. (1977). Information theory and the complexity of boolean functions. *Mathematical Systems Theory*, 10:124–162.

Poggio, T. and Girosi, F. (1990). A theory of networks for approximation and learning. *Proceedings of the IEEE*, 78:1481–1497.

Pollard, D. (1981). Strong consistency of k-means clustering. *Annals of Statistics*, 9:135–140.

Pollard, D. (1982). Quantization and the method of k-means. *IEEE Transactions on Information Theory*, 28:199–205.

Pollard, D. (1984). *Convergence of Stochastic Processes*. Springer-Verlag, New York.

Pollard, D. (1986). Rates of uniform almost sure convergence for empirical processes indexed by unbounded classes of functions. Manuscript.

Pollard, D. (1990). *Empirical Processes: Theory and Applications*. NSF-CBMS Regional Conference Series in Probability and Statistics, Institute of Mathematical Statistics, Hayward, CA.

Powell, M. (1987). Radial basis functions for multivariable interpolation: a review. *Algorithms for Approximation*. Clarendon Press, Oxford.

Preparata, F. and Shamos, M. (1985). *Computational Geometry—An Introduction*. Springer-Verlag, New York.

Psaltis, D., Snapp, R., and Venkatesh, S. (1994). On the finite sample performance of the nearest neighbor classifier. *IEEE Transactions on Information Theory*, 40:820–837.

Qing-Yun, S. and Fu, K. (1983). A method for the design of binary tree classifiers. *Pattern Recognition*, 16:593–603.

Quesenberry, C. and Gessaman, M. (1968). Nonparametric discrimination using tolerance regions. *Annals of Mathematical Statistics*, 39:664–673.

Quinlan, J. (1993). *C4.5: Programs for Machine Learning*. Morgan Kaufmann, San Mateo.

Rabiner, L., Levinson, S., Rosenberg, A., and Wilson, J. (1979). Speaker-independent recognition of isolated words using clustering techniques. *IEEE Transactions on Acoustics, Speech, and Signal Processing*, 27:339–349.

Rao, R. (1962). Relations between weak and uniform convergence of measures with applications. *Annals of Mathematical Statistics*, 33:659–680.

Raudys, S. (1972). On the amount of a priori information in designing the classification algorithm. *Technical Cybernetics*, 4:168–174.

Raudys, S. (1976). On dimensionality, learning sample size and complexity of classification algorithms. In *Proceedings of the 3rd International Conference on Pattern Recognition*, pages 166–169. IEEE Computer Society, Long Beach, CA.

Raudys, S. and Pikelis, V. (1980). On dimensionality, sample size, classification error, and complexity of classification algorithm in pattern recognition. *IEEE Transactions on Pattern Analysis and Machine Intelligence*, 2:242–252.

Raudys, S. and Pikelis, V. (1982). Collective selection of the best version of a pattern recognition system. *Pattern Recognition Letters*, 1:7–13.

Rejtő, L. and Révész, P. (1973). Density estimation and pattern classification. *Problems of Control and Information Theory*, 2:67–80.

Rényi, A. (1961). On measures of entropy and information. In *Proceedings of the Fourth Berkeley Symposium*, pages 547–561. University of California Press, Berkeley.

Révész, P. (1973). Robbins-Monroe procedures in a Hilbert space and its application in the theory of learning processes. *Studia Scientiarium Mathematicarum Hungarica*, 8:391–398.

Ripley, B. (1993). Statistical aspects of neural networks. In *Networks and Chaos—Statistical and Probabilistic Aspects*, Barndorff-Nielsen, O., Jensen, J., and Kendall, W., editors, pages 40–123. Chapman and Hall, London, U.K.

Ripley, B. (1994). Neural networks and related methods for classification. *Journal of the Royal Statistical Society*, 56:409–456.

Rissanen, J. (1983). A universal prior for integers and estimation by minimum description length. *Annals of Statistics*, 11:416–431.

Ritter, G., Woodruff, H., Lowry, S., and Isenhour, T. (1975). An algorithm for a selective nearest neighbor decision rule. *IEEE Transactions on Information Theory*, 21:665–669.

Robbins, H. and Monro, S. (1951). A stochastic approximation method. *Annals of Mathematical Statistics*, 22:400–407.

Rogers, W. and Wagner, T. (1978). A finite sample distribution-free performance bound for local discrimination rules. *Annals of Statistics*, 6:506–514.

Rosenblatt, F. (1962). *Principles of Neurodynamics: Perceptrons and the Theory of Brain Mechanisms*. Spartan Books, Washington, DC.

Rosenblatt, M. (1956). Remarks on some nonparametric estimates of a density function. *Annals of Mathematical Statistics*, 27:832–837.

Rounds, E. (1980). A combined nonparametric approach to feature selection and binary decision tree design. *Pattern Recognition*, 12:313–317.

Royall, R. (1966). *A Class of Nonparametric Estimators of a Smooth Regression Function*. PhD Thesis, Stanford University, Stanford, CA.

Rumelhart, D., Hinton, G., and Williams, R. (1986). Learning internal representations by error propagation. In *Parallel Distributed Processing Vol. I*, Rumelhart, D., J.L.McCelland, and the PDP Research Group, editors. MIT Press, Cambridge, MA. Reprinted in: J.A. Anderson and E. Rosenfeld, *Neurocomputing—Foundations of Research*, MIT Press, Cambridge, MA., pp. 673–695, 1988.

Ruppert, D. (1991). Stochastic approximation. In *Handbook of Sequential Analysis*, Ghosh, B. and Sen, P., editors, pages 503–529. Marcel Dekker, New York.

Samet, H. (1984). The quadtree and related hierarchical data structures. *Computing Surveys*, 16:187–260.

Samet, H. (1990a). *Applications of Spatial Data Structures*. Addison-Wesley, Reading, MA.

Samet, H. (1990b). *The Design and Analysis of Spatial Data Structures*. Addison-Wesley, Reading, MA.

Sansone, G. (1969). *Orthogonal Functions*. Interscience, New York.

Sauer, N. (1972). On the density of families of sets. *Journal of Combinatorial Theory Series A*, 13:145–147.

Scheffé, H. (1947). A useful convergence theorem for probability distributions. *Annals of Mathematical Statistics*, 18:434–458.

Schläffli, L. (1950). *Gesammelte Mathematische Abhandlungen*. Birkhäuser-Verlag, Basel.

Schmidt, W. (1994). *Neural Pattern Classifying Systems*. PhD Thesis, Technical University, Delft, The Netherlands.

Schoenberg, I. (1950). On Pólya frequency functions, II: variation-diminishing integral operators of the convolution type. *Acta Scientiarium Mathematicarum Szeged*, 12:97–106.

Schwartz, S. (1967). Estimation of probability density by an orthogonal series. *Annals of Mathematical Statistics*, 38:1261–1265.

Schwemer, G. and Dunn, O. (1980). Posterior probability estimators in classification simulations. *Communications in Statistics—Simulation*, B9:133–140.

Sebestyen, G. (1962). *Decision-Making Processes in Pattern Recognition*. Macmillan, New York.

Serfling, R. (1974). Probability inequalities for the sum in sampling without replacement. *Annals of Statistics*, 2:39–48.

Sethi, I. (1981). A fast algorithm for recognizing nearest neighbors. *IEEE Transactions on Systems, Man and Cybernetics*, 11:245–248.

Sethi, I. (1990). Entropy nets: from decision trees to neural nets. *Proceedings of the IEEE*, 78:1605–1613.

Sethi, I. (1991). Decision tree performance enhancement using an artificial neural network interpretation. In *Artificial Neural Networks and Statistical Pattern Recognition, Old and New Connections*, Sethi, I. and Jain, A., editors, pages 71–88. Elsevier Science Publishers, Amsterdam.

Sethi, I. and Chatterjee, B. (1977). Efficient decision tree design for discrete variable pattern recognition problems. *Pattern Recognition*, 9:197–206.

Sethi, I. and Sarvarayudu, G. (1982). Hierarchical classifier design using mutual information. *IEEE Transactions on Pattern Analysis and Machine Intelligence*, 4:441–445.

Shannon, C. (1948). A mathematical theory of communication. *Bell Systems Technical Journal*, 27:379–423.

Shawe-Taylor, J. (1994). Sample sizes for sigmoidal neural networks. Technical Report, Department of Computer Science, Royal Holloway, University of London, Egham, England.

Shawe-Taylor, J., Anthony, M., and Biggs, N.L. (1993). Bounding sample size with the Vapnik-Chervonenkis dimension. *Discrete Applied Mathematics*, 42:65–73.

Shiryayev, A. (1984). *Probability*. Springer-Verlag, New York.

Short, R. and Fukunaga, K. (1981). The optimal distance measure for nearest neighbor classification. *IEEE Transactions on Information Theory*, 27:622–627.

Simon, H. (1991). The Vapnik-Chervonenkis dimension of decision trees with bounded rank. *Information Processing Letters*, 39:137–141.

Simon, H. (1993). General lower bounds on the number of examples needed for learning probabilistic concepts. In *Proceedings of the Sixth Annual ACM Conference on Computational Learning Theory*, pages 402–412. Association for Computing Machinery, New York.

Sklansky, J. and Michelotti (1980). Locally trained piecewise linear classifiers. *IEEE Transactions on Pattern Analysis and Machine Intelligence*, 2:101–111.

Sklansky, J. and Wassel, G. (1979). *Pattern Classifiers and Trainable Machines*. Springer-Verlag, New York.

Slud, E. (1977). Distribution inequalities for the binomial law. *Annals of Probability*, 5:404–412.

Specht, D. (1967). Generation of polynomial discriminant functions for pattern classification. *IEEE Transactions on Electronic Computers*, 15:308–319.

Specht, D. (1971). Series estimation of a probability density function. *Technometrics*, 13:409–424.

Specht, D. (1990). Probabilistic neural networks and the polynomial Adaline as complementary techniques for classification. *IEEE Transactions on Neural Networks*, 1:111–121.

Spencer, J. (1987). *Ten Lectures on the Probabilistic Method*. SIAM, Philadelphia, PA.

Sprecher, D. (1965). On the structure of continuous functions of several variables. *Transactions of the American Mathematical Society*, 115:340–355.

Steele, J. (1975). *Combinatorial entropy and uniform limit laws*. PhD Thesis, Stanford University, Stanford, CA.

Steele, J. (1986). An Efron-Stein inequality for nonsymmetric statistics. *Annals of Statistics*, 14:753–758.

Stengle, G. and Yukich, J. (1989). Some new Vapnik-Chervonenkis classes. *Annals of Statistics*, 17:1441–1446.

Stoffel, J. (1974). A classifier design technique for discrete variable pattern recognition problems. *IEEE Transactions on Computers*, 23:428–441.

Stoller, D. (1954). Univariate two-population distribution-free discrimination. *Journal of the American Statistical Association*, 49:770–777.

Stone, C. (1977). Consistent nonparametric regression. *Annals of Statistics*, 5:595–645.

Stone, C. (1982). Optimal global rates of convergence for nonparametric regression. *Annals of Statistics*, 10:1040–1053.

Stone, C. (1985). Additive regression and other nonparametric models. *Annals of Statistics*, 13:689–705.

Stone, M. (1974). Cross-validatory choice and assessment of statistical predictions. *Journal of the Royal Statistical Society*, 36:111–147.

Stute, W. (1984). Asymptotic normality of nearest neighbor regression function estimates. *Annals of Statistics*, 12:917–926.

Sung, K. and Shirley, P. (1992). Ray tracing with the BSP tree. In *Graphics Gems III*, Kirk, D., editor, pages 271–274. Academic Press, Boston, MA.

Swonger, C. (1972). Sample set condensation for a condensed nearest neighbor decision rule for pattern recognition. In *Frontiers of Pattern Recognition*, Watanabe, S., editor, pages 511–519. Academic Press, New York.

Szarek, S. (1976). On the best constants in the Khintchine inequality. *Studia Mathematica*, 63:197–208.

Szegő, G. (1959). *Orthogonal Polynomials*, volume 32. American Mathematical Society, Providence, RI.

Talagrand, M. (1987). The Glivenko-Cantelli problem. *Annals of Probability*, 15:837–870.

Talagrand, M. (1994). Sharper bounds for Gaussian and empirical processes. *Annals of Probability*, 22:28–76.

Talmon, J. (1986). A multiclass nonparametric partitioning algorithm. In *Pattern Recognition in Practice II*, Gelsema, E. and Kanal, L., editors. Elsevier Science Publishers, Amsterdam.

Taneja, I. (1983). On characterization of J-divergence and its generalizations. *Journal of Combinatorics, Information and System Sciences*, 8:206–212.

Taneja, I. (1987). Statistical aspects of divergence measures. *Journal of Statistical Planning and Inference*, 16:137–145.

Tarter, M. and Kronmal, R. (1970). On multivariate density estimates based on orthogonal expansions. *Annals of Mathematical Statistics*, 41:718–722.

Tomek, I. (1976a). A generalization of the k-NN rule. *IEEE Transactions on Systems, Man and Cybernetics*, 6:121–126.

Tomek, I. (1976b). Two modifications of CNN. *IEEE Transactions on Systems, Man and Cybernetics*, 6:769–772.

Toussaint, G. (1971). Note on optimal selection of independent binary-valued features for pattern recognition. *IEEE Transactions on Information Theory*, 17:618.

Toussaint, G. (1974a). Bibliography on estimation of misclassification. *IEEE Transactions on Information Theory*, 20:472–479.

Toussaint, G. (1974b). On the divergence between two distributions and the probability of misclassification of several decision rules. In *Proceedings of the Second International Joint Conference on Pattern Recognition*, pages 27–34. Copenhagen.

Toussaint, G. and Donaldson, R. (1970). Algorithms for recognizing contour-traced hand-printed characters. *IEEE Transactions on Computers*, 19:541–546.

Tsypkin, Y. (1971). *Adaptation and Learning in Automatic Systems*. Academic Press, New York.

Tutz, G. (1985). Smoothed additive estimators for non-error rates in multiple discriminant analysis. *Pattern Recognition*, 18:151–159.

Tutz, G. (1986). An alternative choice of smoothing for kernel-based density estimates in discrete discriminant analysis. *Biometrika*, 73:405–411.

Tutz, G. (1988). Smoothing for discrete kernels in discrimination. *Biometrics Journal*, 6:729–739.

Tutz, G. (1989). On cross-validation for discrete kernel estimates in discrimination. *Communications in Statistics—Theory and Methods*, 18:4145–4162.

Ullmann, J. (1974). Automatic selection of reference data for use in a nearest-neighbor method of pattern classification. *IEEE Transactions on Information Theory*, 20:541–543.

Vajda, I. (1968). The estimation of minimal error probability for testing finite or countable number of hypotheses. *Problemy Peredaci Informacii*, 4:6–14.

Vajda, I. (1989). *Theory of Statistical Inference and Information*. Kluwer Academic Publishers, Dordrecht.

Valiant, L. (1984). A theory of the learnable. *Communications of the ACM*, 27:1134–1142.

Van Campenhout, J. (1980). The arbitrary relation between probability of error and measurement subset. *Journal of the American Statistical Association*, 75:104–109.

Van Ryzin, J. (1966). Bayes risk consistency of classification procedures using density estimation. *Sankhya Series A*, 28:161–170.

Vapnik, V. (1982). *Estimation of Dependencies Based on Empirical Data*. Springer-Verlag, New York.

Vapnik, V. and Chervonenkis, A. (1971). On the uniform convergence of relative frequencies of events to their probabilities. *Theory of Probability and its Applications*, 16:264–280.

Vapnik, V. and Chervonenkis, A. (1974a). Ordered risk minimization. I. *Automation and Remote Control*, 35:1226–1235.

Vapnik, V. and Chervonenkis, A. (1974b). Ordered risk minimization. II. *Automation and Remote Control*, 35:1403–1412.

Vapnik, V. and Chervonenkis, A. (1974c). *Theory of Pattern Recognition*. Nauka, Moscow. (in Russian); German translation: *Theorie der Zeichenerkennung*, Akademie Verlag, Berlin, 1979.

Vapnik, V. and Chervonenkis, A. (1981). Necessary and sufficient conditions for the uniform convergence of means to their expectations. *Theory of Probability and its Applications*, 26:821–832.

Vidal, E. (1986). An algorithm for finding nearest neighbors in (approximately) constant average time. *Pattern Recognition Letters*, 4:145–157.

Vilmansen, T. (1973). Feature evaluation with measures of probabilistic dependence. *IEEE Transactions on Computers*, 22:381–388.

Vitushkin, A. (1961). The absolute ϵ-entropy of metric spaces. *Translations of the American Mathematical Society*, 17:365–367.

Wagner, T. (1971). Convergence of the nearest neighbor rule. *IEEE Transactions on Information Theory*, 17:566–571.

Wagner, T. (1973). Convergence of the edited nearest neighbor. *IEEE Transactions on Information Theory*, 19:696–699.

Wang, Q. and Suen, C. (1984). Analysis and design of decision tree based on entropy reduction and its application to large character set recognition. *IEEE Transactions on Pattern Analysis and Machine Intelligence*, 6:406–417.

Warner, H., Toronto, A., Veasey, L., and Stephenson, R. (1961). A mathematical approach to medical diagnosis. *Journal of the American Medical Association*, 177:177–183.

Wassel, G. and Sklansky, J. (1972). Training a one-dimensional classifier to minimize the probability of error. *IEEE Transactions on Systems, Man, and Cybernetics*, 2:533–541.

Watson, G. (1964). Smooth regression analysis. *Sankhya Series A*, 26:359–372.

Weiss, S. and Kulikowski, C. (1991). *Computer Systems that Learn*. Morgan Kaufmann, San Mateo, CA.

Wenocur, R. and Dudley, R. (1981). Some special Vapnik-Chervonenkis classes. *Discrete Mathematics*, 33:313–318.

Wheeden, R. and Zygmund, A. (1977). *Measure and Integral*. Marcel Dekker, New York.

White, H. (1990). Connectionist nonparametric regression: multilayer feedforward networks can learn arbitrary mappings. *Neural Networks*, 3:535–549.

White, H. (1991). Nonparametric estimation of conditional quantiles using neural networks. In *Proceedings of the 23rd Symposium of the Interface: Computing Science and Statistics*, pages 190–199. American Statistical Association, Alexandria, VA.

Widrow, B. (1959). Adaptive sampled-data systems—a statistical theory of adaptation. In *IRE WESCON Convention Record*, volume part 4, pages 74–85.

Widrow, B. and Hoff, M. (1960). Adaptive switching circuits. In *IRE WESCON Convention Record*, volume part 4, pages 96–104. Reprinted in J.A. Anderson and E. Rosenfeld, *Neurocomputing: Foundations of Research*, MIT Press, Cambridge, MA., 1988.

Wilson, D. (1972). Asymptotic properties of nearest neighbor rules using edited data. *IEEE Transactions on Systems, Man and Cybernetics*, 2:408–421.

Winder, R. (1963). Threshold logic in artificial intelligence. In *Artificial Intelligence*, pages 107–128. IEEE Special Publication S–142.

Wolverton, C. and Wagner, T. (1969a). Asymptotically optimal discriminant functions for pattern classification. *IEEE Transactions on Systems, Science and Cybernetics*, 15:258–265.

Wolverton, C. and Wagner, T. (1969b). Recursive estimates of probability densities. *IEEE Transactions on Systems, Science and Cybernetics*, 5:307.

Wong, W. and Shen, X. (1992). Probability inequalities for likelihood ratios and convergence rates of sieve MLE's. Technical Report 346, Department of Statistics, University of Chicago, Chicago, IL.

Xu, L., Krzyżak, A., and Oja, E. (1993). Rival penalized competitive learning for clustering analysis, RBF net and curve detection. *IEEE Transactions on Neural Networks*, 4:636–649.

Xu, L., Krzyżak, A., and Yuille, A. (1994). On radial basis function nets and kernel regression: Approximation ability, convergence rate and receptive field size. *Neural Networks*. To appear.

Yatracos, Y. (1985). Rates of convergence of minimum distance estimators and kolmogorov's entropy. *Annals of Statistics*, 13:768–774.

Yau, S. and Lin, T. (1968). On the upper bound of the probability of error of a linear pattern classifier for probabilistic pattern classes. *Proceedings of the IEEE*, 56:321–322.

You, K. and Fu, K. (1976). An approach to the design of a linear binary tree classifier. In *Proceedings of the Symposium of Machine Processing of Remotely Sensed Data*, pages 3A–10. Purdue University, Lafayette, IN.

Yukich, J. (1985). Laws of large numbers for classes of functions. *Journal of Multivariate Analysis*, 17:245–260.

Yunck, T. (1976). A technique to identify nearest neighbors. *IEEE Transactions on Systems, Man, and Cybernetics*, 6:678–683.

Zhao, L. (1987). Exponential bounds of mean error for the nearest neighbor estimates of regression functions. *Journal of Multivariate Analysis*, 21:168–178.

Zhao, L. (1989). Exponential bounds of mean error for the kernel estimates of regression functions. *Journal of Multivariate Analysis*, 29:260–273.

Zhao, L., Krishnaiah, P., and Chen, X. (1990). Almost sure L_r-norm convergence for data-based histogram estimates. *Theory of Probability and its Applications*, 35:396–403.

Zygmund, A. (1959). *Trigonometric Series I*. University Press, Cambridge.

Author Index

Subject Index

Applications of Mathematics

(continued from page ii)

Printed in the United States
99980LV00001B/1/A

9 780387 946184